Optical Fiber Amplifiers: Materials, Devices, and Applications

Optical Fiber Amplifiers:
Materials, Devices, and Applications

Shoichi Sudo
Editor

Artech House, Inc.
Boston • London

Library of Congress Cataloging-in-Publication Data
Sudo, Shoichi
    Optical fiber amplifiers: materials, deivces, and applications / Shoichi Sudo, editor.
      p.   cm. (Artech House optoelectronics library)
    Includes bibliographical references and index.
    ISBN 0-89006-809-7   (alk. paper)
    1. Optical communications—Equipment and supplies. 2. Optical amplifiers. 3. Fiber
optics.   I. Sudo, S. (Shoichi)
  TK5103.59.O674   1997
  621.382'75—dc21   97-23661
                                            CIP

British Library Cataloguing in Publication Data
Optical fiber amplifiers: materials, devices, and applications
1. Optical communications 2. Optical amplifiers
I. Sudo, S. (Shoichi)
621.3'8275

ISBN 0-89006-809-7

Cover design by Deborah Dutton and Joseph Sherman Design

© 1997 ARTECH HOUSE, INC.
685 Canton Street
Norwood, MA 02062

International Standard Book Number: 0-89006-809-7
Library of Congress Catalog Card Number: 97-23661

10 9 8 7 6 5 4 3 2 1

MANAGER.
You two who often stood by me
in times of hardship and of gloom,
what do you think our enterprise
should bring to German lands and people?
I want the crowd to be well satisfied,
for, as you know, it lives and lets us live.
The boards are nailed, the stage is set,
and all the world looks for a lavish feast.
There they sit, with eyebrows raised,
and calmly wait to be astounded.
I have my ways to keep people well disposed,
but never was I in a fix like this.
It's true, they're not accustomed to the best,
yet they have read an awful lot of things.
How shall we plot a new and fresh approach
and make things pleasant and significant?
I'll grant, it pleases me to watch the crowds,
as they stream and hustle to our tent
and with mighty and repeated labors
press onward through the narrow gate of grace;
while the sun still shines—it's scarcely four o'clock
they fight and scramble for the ticket window
and as if in famine begging at the baker's door,
they almost break their necks to gain admission.
The poet alone can work this miracle
on such a diverse group. My friend, the time is now!
—Johann Wolfgang Goethe, *Faust*

Contents

Preface

A worldwide revolution has taken place in the way information is handled. This has been brought about by the advent of optical devices such as semiconductor lasers and low-loss optical fibers. The speed with which the merger of these two technologies has led to the optical information age has been more rapid than almost any other technological advance in history. In large part this has been due to the increasing realization that the ability to communicate and handle information is directly related to human activities and well-being. Today, the "Information Age" is clearly upon us. Optical fiber communication has had a dramatic impact on this ability to transmit information. In addition, impelled by the progress in fiber communications, activity has been stimulated in almost every sector of optoelectronic science and technology: condensed matter physics, materials science, quantum optics, optical storage, optical integrated circuits, optical sensing, medical optics, optical computing, and optical interconnection, to name but a few. In fact, conferences on either lasers or optical fiber now typically attract several thousand attendees.

At the current stage in the development of optical fiber communications, the optical fiber amplifier is having a most exciting effect on the topology of optical communication systems and will allow information technology to be accessed by a greater portion of the population. It is commonly recognized that no single subject in the field has received more attention than erbium-doped fiber amplifiers. However, the basic mechanism of the optical fiber amplifier is incredibly simple: it amplifies an optical signal in a fiber by using the stimulated emission of optically excited rare-earth ions in the fiber core. The operating principle of fiber amplifiers is the same of that of lasers, except that amplifiers do not need a cavity whereas lasers need one for oscillation.

Interestingly, however, optical fiber amplifiers received little attention for a long time, despite the fact that they were first demonstrated in 1962 just after the development of the laser in 1960. In fact, Dieter Ross (*Lasers: Light Amplifiers and Oscillators*, New York: Academic Press, 1969) stated in 1969 that:

> Amplification by stimulated emission is used nowadays in laser-oscillators for the production of coherent light. In comparison, the use of laser materials for light amplification is of less practical importance and has been investigated experimentally in detail only for helium-neon and ruby.

In 1972, just after the successful fabrication of high-silica fibers with a 20-dB loss and the room temperature operation of laser diodes, which had a great impact on the development of optical communication using glass fiber, a representative opinion on optical amplifiers was given in a paper by R. Kompfner ("Optics at Bell Laboratories-Optical Communications," *Appl. Optics*, Vol. 11, 1972, p. 2421):

> So far, laser amplifiers have not found many uses nor have they been proposed for any practical communication systems. This is undoubtedly due to the fact that most lasers have little gain; moreover, the more gain they have the more susceptible they are to instability because of even minor reflections in the transmission path. Until laser amplifiers are invented that incorporate unilateral gain or, even better, have gain in the forward direction and loss in the backward direction, we do not expect them to play an important part in communications systems.

Of course, this opinion originated from the poor characteristics of optical amplifiers in those days. It is also a reflection of the research trend at that time, which focused mainly on the realization of low-loss fibers and highly reliable laser diodes.

Many years later, in 1987, the history of the development of optical communication systems and fiber amplifiers changed dramatically when a group from Southampton University reported on high-gain fiber amplifiers operating in the 1.5-μm wavelength region, which they had achieved using erbium-doped silica-based fibers. This development of high-gain fiber amplifiers was a kind of rediscovery. However, this rediscovery was made possible because of the well-established fiber technology and mature laser diode technology, which were the results of sustained research efforts throughout the world over the preceding two decades. This development of practical fiber amplifiers conveyed an important message: a well-established technology can be used in a straightforward manner to generate another new and versatile technology with which to open a wide range of applications. The fiber amplifiers are exactly this kind of new and versatile technology, and they promise to revolutionize the field of optical communications. Needless to say, the revolutionary progress made with fiber amplifiers derives from their excellent features such as high gain, high power, low noise, broadband, and compactness.

Several books have been devoted to fiber amplifier technology, but never with quite the scope, objectives, and content described here. The main objective of this book is to provide readers with informative and practical knowledge, covering fields ranging from basic phenomena to the practical technologies of fiber amplifiers, together with the scientific and historical background to fiber amplifiers and related technologies.

This book is divided into five parts. The first part describes the developmental history of fiber amplifiers and the science and technologies related to them. These include optical fiber, optical communications, quantum electronics, optics, and physics, viewed in a broad historical context. In particular, a pattern of alternating growth in science and technology will be revealed, which has had an increasing effect on development and progress. The fiber amplifier is a typical example of a product whose origin is related to the alternating growth of science and technology. In Chapter 2, we outline fiber amplifier technology and include a description of the requirements for fiber amplifiers, a brief description of the behavior of rare-earth ions in amplification, key issues in fiber amplification, and in particular the fabrication and materials technology of fiber amplifiers. Chapters 3 to 5 are all devoted to a detailed description of fiber amplifier technology, in which we describe (1) the energy states of rare-earth ions in condensed materials and the transitions for optical amplification, (2) the various fabrication technologies used for rare-earth-doped fibers and the microscopic material structure of rare-earth-doped glasses, and (3) the device technology and amplification characteristics of fiber amplifiers together with their applications.

The contents of this book are intended to be neither too fundamental nor too fragmentary. In brief, this work is the result of lengthy and careful thought and is designed to be both informative and of practical value to readers. We cover fields ranging from basic phenomena and mechanisms to practical structures and methods for fabricating fiber amplifiers together with the scientific and technological background to fiber amplifiers and related technologies. In this sense, this book is not designed to be a standard textbook on fiber amplifiers because it actually includes too much information to fill this role. We hope this will make it more attractive to those working in fields related to fiber amplifiers, such as researchers, engineers, teachers, and even students, than any previously published book. Therefore, we sincerely trust that this book will enable undergraduate and graduate students, junior and senior scientists, and engineers and teachers to gain both basic and practical knowledge, to study fiber amplifiers, to fabricate novel devices, and to develop advanced application systems. Unfortunately we have no crystal ball with which to see the future, but we will be very happy if this book is able to provide a contribution to readers' work and study.

I would like to express my thanks to Mark Walsh of Artech House, Inc., for his critical comments and his encouragement with the manuscript development.

Shoichi Sudo
September 1997

CHAPTER 1

Introduction

S. Sudo

It seems that the human mind has first to construct forms independently before we can find them in things. Kepler's marvelous achievement is a particularly fine example of the truth that knowledge cannot spring from experience alone, but only from the comparison of the inventions of the mind with observed fact [1].

1.1 BRIEF HISTORY OF OPTICS AND QUANTUM ELECTRONICS

The history of man's relationship with light is, in some senses, both very old and quite new. In particular, man has only been able to utilize coherent light for a comparatively short time, that is, only since the development of lasers (light amplification by stimulated emission of radiation) in 1960 [2,3]. This is, as you know, because coherent light (radiation) has never been observed in nature.

Studies on light date back almost 400 years. This work includes that undertaken by Galileo Galilei, Johannes Kepler, and Snel van Royen (Snellius); followed by the studies of Rene Descartes, Sir Isaac Newton [4], and Christian Huygens in the 1600s; and also that of Johann Wolfgang Goethe [5] in the 1700s. In 1874, on the basis of his study of electric and magnetic fields, James Clerk Maxwell discovered that light is an infinite chain of vibrating electric fields turning into magnetic fields, that is, an electromagnetic wave that travels at the speed of light [6]. The essence of this discovery is simple, easy to picture, and ranks in importance alongside Newton's discovery of the universal law of gravitation. As a result of Maxwell's path-breaking work, light was recognized as an electromagnetic wave created by the vibration of electric and magnetic fields rapidly turning into each other. However, for a long time there had been an essential difference between electromagnetic waves and the light that was actually generated. For example, a radio (electromag-

1

netic) wave generated by a vacuum tube or a high-frequency generator has a constant frequency and a constant intensity and is able to travel great distances. In contrast, the light generated by an incandescent or a fluorescent lamp does not have a constant frequency and a stable amplitude. Furthermore, a radio wave has a much longer wavelength and much lower frequency than a light wave. Consequently, both waves cover a different area with little overlap in terms of their wavelengths and natural properties as waves; both waves and the light wave are essentially electromagnetic.

In 1917, Albert Einstein introduced the phenomenon of stimulated emission to explain Planck's black body radiation [7], which was described in his renowned 1916 paper [8]. He showed in this paper that the Planck radiation formula results in thermal equilibrium from the interactions between spontaneous emission, stimulated emission, and stimulated absorption. Nowadays, this concept of stimulated emission is recognized as a fundamental concept in the field of quantum electronics and lasers. However, 40 years passed after the publication of Einstein's theory before the first laser appeared. There are two main reasons for this long interval between the publication of the fundamental concept and the development of lasers. One is that quantum electronics research had to reach maturity before quantum electronics and lasers could develop. Research on quantum mechanics related to materials began in the early 1920s and provided a more accurate description of the behavior of atoms and molecules. The other reason was the difficulty in realizing laser operation because it required an excellent (low-loss) laser cavity and the formation of a population inversion in the light wavelength region.

It is generally recognized that quantum electronics was born in the period between the end of 1954 and early 1955, with the successful operation of an ammonia maser (microwave amplification by stimulated emission of radiation) [9,10]. Prior to this achievement, several outstanding pieces of work had been undertaken: one was the formation of a population inversion by E. M. Purcell and H. V. Pound [11], and another was the idea of the maser proposed by C. H. Townes [12]. In 1958, A. L. Schawlow and C. H. Townes published the theory of the "Infrared and Optical Maser" [13], that is, the theory of the laser. They tried to realize an optical maser by using potassium vapor pumped by a potassium lamp, but the minimum power needed for oscillation was impractically high. However, in June 1960, the first successful laser operation, which was announced by T. Maiman using a solid-state ruby crystal laser [2], was a three- rather than four-energy-level system. The story of the birth of the laser has been described in detail by J. L. Brombery in *Physics Today* [14]. The first laser operated with four-level ions was successfully demonstrated in December 1960 using uranium ions (U^{3+}) in a CaF_2 crystal [15]. Interestingly, however, Maiman did not use the term "optical maser" for his ruby laser, at least before 1961 [16,17]. It seems that the word "laser" became part of the nomenclature in 1963 [18,19].

It is worthwhile to say something about "quantum electronics" [20]. Electronics can be defined as the branch of technology related to the interaction of electrons (or ions) with electromagnetic fields (or radiation) or the control of electric-charge flow. This has led to the key features of electronics such as the functions of amplification and quick nonlinear response. The motion of the electrons in early electronic devices such as vacuum tubes can be adequately described by classical mechanics. However, the advent of semiconductor materials and devices and the maser required a precise description using quantum mechanics. Specifically, classical mechanics cannot completely describe the behavior of electrons moving through the crystal lattice of a semiconductor material and the physics of bound electrons. Therefore, a new branch of electronics known as quantum electronics came into being. Quantum electronics can be defined as an extension of electronics to phenomena whose nature can no longer be described by classical physics. It concerns itself particularly with the interaction of electromagnetic radiation with bound electrons. The need for this approach in electronics arose with the invention of the maser because a maser utilizes bound electrons interacting with a radiation field, so quantum mechanics must be used to describe the response of such atomic systems to radiation. It is, therefore, generally said that "quantum electronics" was born at the time the ammonia maser was invented. In addition, with the invention of the laser, which operates at frequencies in the optical region of the spectrum, classical electrodynamics no longer suffices for describing the radiation field. So, the expression of "quantum electronics" started to be commonly used after the invention of lasers in 1960. Although this term meant different things to different people, today it is commonly used in connection with devices and systems that rely principally on interactions between light and matter such as lasers and nonlinear optical devices used for optical amplification and wave mixing.

Three major developments achieved in the last 30 years, that is, the invention of the laser, the fabrication of low-loss optical fibers, and the introduction of other optical devices, have led to the introduction of many terms to describe new disciplines such as electro-optics, optoelectronics, quantum optics, lightwave technology, and photonics, as well as quantum electronics [21]. Basically, optics is an old and venerable subject involving the generation, propagation, and detection of light. Although there is a lack of complete agreement about the precise usage and meaning of these new terms, electro-optics is generally used for optical devices in which electrical effects play a role, such as lasers, electro-optic modulators, and switches. Optoelectronics typically refers to devices and systems that cover the ultraviolet to infrared wavelength region, just as electronic devices and systems operate at optical wavelengths. Studies of the quantum and coherence properties of light fall in the realm of quantum optics. The term *lightwave technology* has been used to describe devices and systems that are mainly used in optical communications and optical signal processing. Finally, the word *photonics,* which was coined by analogy with the word *electronics,* involves the control of photons (in free space or in matter), just as electronics involves the control of an electric-charge flow (in a vacuum or in matter).

1.2 BRIEF HISTORY OF PHYSICS AND ALTERNATING GROWTH OF SCIENCE AND TECHNOLOGY

Next, we offer a brief history of physics and the interaction between science, especially physics, and technology. Looking back throughout history, there have been many interesting interactions between science and technology. The first half of the present century was one of the most fruitful periods in the development of science, in particular, physics. It saw the theory of relativity, quantum theory, and modern ideas of the structure of atoms, molecules, and solids come into existence. In contrast, the last half of the present century has been a period of rapid technological growth, in particular, in the field of electronics and quantum electronics, which have been developed based on quantum mechanics. The resulting industrial products—televisions, computers, and lasers—have all come into being during this period.

There have been several other exciting times in the history of physics. One came in the 17th century with the development of classical mechanics by Galileo, Kepler, and Newton, which was initiated in the 16th century, when the book entitled *On the Revolution of the Heavenly Spheres* by Copernicus appeared [22]. That period of classical physics was completed by Isaac Newton in the late 17th century [23], when he summarized and systematized the previous great work undertaken by Johannes Kepler and Galileo Galilei [24]. The development of classical mechanics made possible the technological innovation in the 18th century known as the Industrial Revolution. Still another was the development of our ideas of electromagnetism, starting with Oersted, Faraday, and others and culminating in the theoretical work of Maxwell, which showed that electromagnetic waves and light were identical [6] (a discovery of the middle of the 19th century, as mentioned earlier). All these great advances in science were essentially complete by 1900. However, two big clouds hung over physics in the late 19th century that were related to the limitations of classical mechanics. One concerned the propagation of electromagnetic waves and light, and the other was related to a phenomenon called "black body radiation." Neither phenomenon could be explained by conventional theory. The stage was set for another and different development.

In the early 20th century, two great pieces of work attempted to solve the difficult problems that had developed by the late 19th century. These were the "relativity theorem" [25–27] and "quantum mechanics" [28–41]. The leading names attached to these discoveries are Planck [28], Einstein [25,26], Rutherford [29], Bohr [30–32], Heisenberg [34,35], and Schrödinger [36–40]. The early 20th century was one of the most exciting periods of scientific development in human history. Based on these great discoveries, a variety of important advances were made in many fields, including physics, chemistry, and biology. The development of lasers in 1960 was one of the finest of these achievements. In fact, lasers opened the new scientific and technological field of "quantum electronics," which nowadays offers marvelous new products for home and industrial use. Quantum electronics has become a highly successful field in the last half-century.

Quantum mechanics itself—which gave us lasers—has, however, confronted "the unanswered question" or the "speakable and unspeakable question" relating to the most fundamental concept and framework of quantum mechanics [42–48]. This began with the publications of papers in 1935 [49–55]. The most severe question or criterion for physical reality proposed by the three physicists—A. Einstein, B. Podolsky, and N. Rosen—which was later called the *Einstein–Podolsky–Rosen* (EPR) paradox, is as follows [49]:

> First, whatever the meaning assigned to the term complete, the following requirement for the complete theory seems to be a necessary one: Every element of the physical reality must have a counterpart in the physical theory. We shall call this the condition of completeness.
>
> Second, we shall be satisfied with the following criterion, which we regard as reasonable. If, without in any way disturbing a system, we can predict with certainty (i.e., with probability equal to unity) the value of a physical reality corresponding to this physical quantity.

As many readers know, Niels Bohr responded to that question almost immediately [56,57], but his answer was insufficient to unravel the paradox. Even Erwin Schrödinger remarked in reference to this discussion that: "It is rather discomforting that the quantum theory should allow a system to be steered or piloted into one or the other type of state at the experimenter's mercy in spite of his having no access to it" [58]. Immediately, many physicists realized that this peculiar situation raises a critical question. Prior to the EPR paradox, there had been other types of paradox such as Schrödinger's cat and particle dualism, which was already well known at the birth of modern quantum theory. However, the EPR paradox was more problematic than the others.

In short, this question posed by Einstein, Podolsky, and Rosen originated from the conceptual difference between quantum mechanics and relativity theory [59]. The concept of the relativity theory developed by Einstein is based on the principle of "local causes." The principle of local causes states that what happens in one area does not depend upon variables subject to the control of an experimenter in a distant spacelike separated area. This principle of "local causes could almost be called common sense. The results of an experiment undertaken in an area separated from us by space and distance should not depend on what we decide to do or not to do right here. Einstein explained his standpoint in his autobiography [60]:

> . . . on one supposition we should, in my opinion, absolutely hold fast: the real factual situation of the system S2 [the particle in area B] is independent of what is done with the system S1 [the particle in area A], which is spatially separated from the former.

However, according to quantum theory, changing the measuring device in area A changes the wave function, which describes the particles in area B. In addition, if we suppose that areas A and B are very far apart, it will take a certain amount of time for a light signal to travel from one to the other. Consequently, quantum mechanics insists that information must be communicated at superluminal (faster than light) speeds, that is, the signal must travel faster than the speed of light, contrary to the accepted ideas of physicists. Of course, Einstein did accept the mathematical equation of quantum mechanics. However, he felt that quantum mechanics was an incomplete manifestation of an underlying theory (the unified field theory) where an objectively real description is possible. Furthermore, Einstein stated that [60]:

> One can escape from this conclusion [that quantum theory is incomplete] only by either assuming that the measurement of S1 "telepathically" changes the real situation of S2 or by denying an independent real situation as such to things which are spatially separated from each other. Both alternatives appear to me entirely unacceptable.

Of course, few physicists believe in telepathy or superluminal communication. However, some physicists do believe either that, at a deep and fundamental level, there are no such things as "independent real situations" of things that have interacted in the past but are spatially separated from each other or that changing the measuring device in area A does change "the real factual situation" in area B. This led to Bell's theorem about 30 years after the appearance of the EPR paradox.

Bell's theorem is a mathematical proof regarding the statistical prediction of quantum theory [61]. J. S. Bell discovered that no matter what the settings of the polarizers, the number of clicks in area A correlate too closely to the number of clicks in area B to be explained by chance. If they are correct, however, then the principle of local causes (which states that what happens in one area does not depend upon variables subject to the control of an experimenter in a distant spacelike area) is an illusion. In other words, Bell's theorem indicates that the principle of local causes, no matter how reasonable it sounds, is mathematically incompatible with the statistical predictions of quantum theory [62]. In 1969, J. F. Clauser et al. proposed a realizable experiment that can prove Bell's theorem [62]. S. J. Freedman and J. F. Clauser performed this experiment and found that the statistical predictions upon which Bell based his theorem are correct [63]. The experimental results closest to the theoretical prediction were obtained by A. Aspect [64]. Bell's theorem not only suggests that the world is quite different from the way it seems, it insists upon it. Henry Stapp wrote [65,66]:

> Bell's theorem is the most profound discovery of science. The theorem of this paper supports this view of Nature by showing that superluminal transfer of information is necessary. . . . Indeed, the reasonable philosophical position of Bohr seems to lead to the rejection of the other possibilities, and hence, by inference, to the conclusion that superluminal transfer of information *is necessary*.

There have been several other important papers published to explain Bell's theorem that should be referenced [67–70].

Consequently, 95 years after Planck presented his quantum hypothesis, physicists have been forced to consider the possibility, among others, that the superluminal transfer of information between spacelike separated events may be an integral aspect of our physical reality or at least forced to consider nonseparability and nonlocality in quantum theory. In this picture, what happens here is intimately and immediately connected to what happens elsewhere in the universe. This means that we have reached the idea in which the "separate parts" of the universe are not separate parts or what David Bohm and others described as the "undivided universe," "unbroken wholeness," or "unseparable" universe [71–77].

Anyway, the EPR paradox has become far more problematic in recent times, as shown in the centenary edition of the *Physical* [78] and other publications [42–47,79–82].

Another possible way of solving the difficulty regarding the framework of quantum mechanics is called the "superstring theory," which is an attempt to marry relativity and quantum mechanics [83–94]. Of course, the main goals of the superstring theory are to build a unified field theory as well as to solve the above problem. Michio Kaku, however, wrote [82]:

> By the way, the superstring theory provides perhaps the most comprehensive way of looking at Schrödinger's cat.

At present, we are not sure whether the superstring theory can provide a comprehensive and satisfactory answer to the quest for a framework for quantum mechanics with as much success as the construction of the unified field theory or whether other approaches such as the "many-world interpretation" [95,96], "quantum trajectories" [97,98], and "quantum cosmology" [48,99] can provide such answers. Many theories about the interpretation of quantum theory or the description of nature have been developed; however, a detailed description of those theories is beyond the scope and purpose of this book. Of course, even today the "Copenhagen interpretation" of quantum theory is a major one. It has two pictures—the particle picture and the wave picture—as two complementary descriptions of the same reality, and each of these descriptions can only be partially true. Therefore, there must be limitations to the use of both the particle and the wave concepts, otherwise it would be impossible to avoid contradictions [41,100–106].

However, at this point it is worthwhile to relate other great physicists' opinions on this difficult problem of quantum theory. Heisenberg wrote in 1961 [107]:

> The strangest experience of those years was that the paradoxes of quantum theory did not disappear during this process of classification; on the contrary, they became even more marked and more exciting.

P. A. M. Dirac's opinion is as follows [108]:

> There are great difficulties . . . in connection with the present quantum
> mechanics. It is the best that one can do up till now. But, one should
> not suppose that it will survive indefinitely into the future. And I think
> that it is quite likely that at some future time we may get an improved
> quantum mechanics in which there will be a return to determinism and
> which will, therefore, justify the Einstein point of view.

Dirac presented this future view of quantum mechanics in a 1975 conference in Australia [108].

Nevertheless, most working physicists who have enjoyed 50 years of spectacular success with quantum mechanics simply coexist with its strange philosophical implications in a pragmatic approach or simply ignore such difficult problems. As a result, the successful consequences of quantum mechanics are all around us. The development of lasers, as previously described, is a typical example of such successful products. Similarly, it would have been impossible to develop many now familiar objects, such as televisions, computers, and radios, without quantum mechanics. Nor could such devices as electron microscopes have been realized to unlock the DNA molecule. Without question, the success of quantum mechanics has also altered the foundations of contemporary medicine, industry, and commerce. Quantum mechanics is undoubtedly the most successful theory yet devised by the human mind.

However, the irony is that quantum mechanics, which seems so definitive and clear-cut in its practical applications, is actually based on uncertainties, probabilities, and philosophically bizarre ideas. I think that this ironic situation has motivated a fascinating, interesting, and alternating growth interaction between science and technology. When we take a look at the history of science and technology, Newtonian mechanics constructed in the 17th century made a marvelous contribution to the growth of technology in the 18th and 19th centuries. At the same time the limitations and unsolved problems in Newtonian mechanics were developed in parallel with this technological growth. We may now be experiencing such a period of alternating interaction between quantum mechanics and the technologies which it has sired. We are making many attractive, technological products based on the scientific results provided by quantum mechanics, whereas quantum mechanics is currently confronting a difficulty as regards its fundamental framework. This means that today we could be nearing the birth of a new science [109–116]. The superstring theory might be one path in this direction. Finally, it is useful to heed the words of Joseph Campbell, which well express the alternate-growing concept of history of the interaction between science and technology [117]:

> One thing that comes out in myths is that at the bottom of the abyss
> comes the voice of salvation. The black moment is the moment when
> the real message of transformation is going to come. At the darkest
> moment comes the light.

The optical fiber amplifier described in this book is one of the most attractive and fantastic quantum electronics products to appear in technology, in particular, for optical communications. We do hope that the optical fiber amplifier can also make a significant contribution to the scientific work of today and tomorrow.

1.3 BRIEF HISTORY OF OPTICAL FIBER AMPLIFIERS

The basic concept of the optical fiber amplifier is incredibly simple: it amplifies an optical signal in a fiber by using the stimulated emission of optically excited rare-earth ions in the fiber core. The operation principle of fiber amplifiers is the same of that of lasers, except that amplifiers do not need a cavity whereas lasers need one for oscillation. Nevertheless, as with the aforementioned developmental history of lasers, it took about 30 years from the first attempt to construct optical fiber amplifiers in the 1960s before efficient and low-noise fiber amplifiers were developed in 1987 [118–128]. The fact that it took such a long time to complete the seemingly small step from lasers to amplifiers is a kind of mystery. However, advances in science and technology often involve such nonuniform steps. Sometimes a long interval is needed to achieve a small advance, whereas at other times great progress can be made very quickly. Some people feel that a period of little progress is like time spent asleep. Interestingly, however, such sleeping time is as important for progress in science and technology as it is for the human brain, because as history has shown there are many things going on beneath the surface. Technological advances are usually the result of incremental innovations that move existing knowledge forward in relatively small steps and so major breakthroughs occur much less frequently.

Looking back briefly at the history of amplifiers operating at optical frequencies, we find that the basic ideas that led to the unidirectional traveling-wave optical amplifier were presented in 1962 by J. E. Geusic and H. E. D. Scovil [129] just after the invention of the laser in 1960 [2] and just before the first experiments on fiber amplifiers by C. J. Koester and E. Snitzer in 1963 [118] and 1964 [119]. The ideas and experiment presented by Geusic and Scovil mark the first construction of a stable high-gain amplifier with a nonreciprocal configuration. It consisted of a small ruby rod, a xenon flash lamp, and an optical isolator to enable the realization of a nonreciprocal optical configuration. A stable high-gain amplifier with a nonreciprocal configuration had been accomplished in the form of the microwave traveling-wave maser [130], but no amplifier had been realized at optical frequencies at that time. Figure 1.1 shows the optical amplifier presented in 1962 by Geusic and Scovil: (a) is a schematic diagram of the unidirectional traveling-wave optical amplifier configuration, and (b) shows the structure of the ruby optical amplifier in detail [129]. The ruby rod was 3-in long, 0.25 of an inch in diameter, c-axis oriented, and composed of 0.065% by weight of Cr_2O_3 in Al_2O_3, and a xenon flash lamp was installed at the focus of a 3-in-long elliptical cylinder. They constructed two ruby

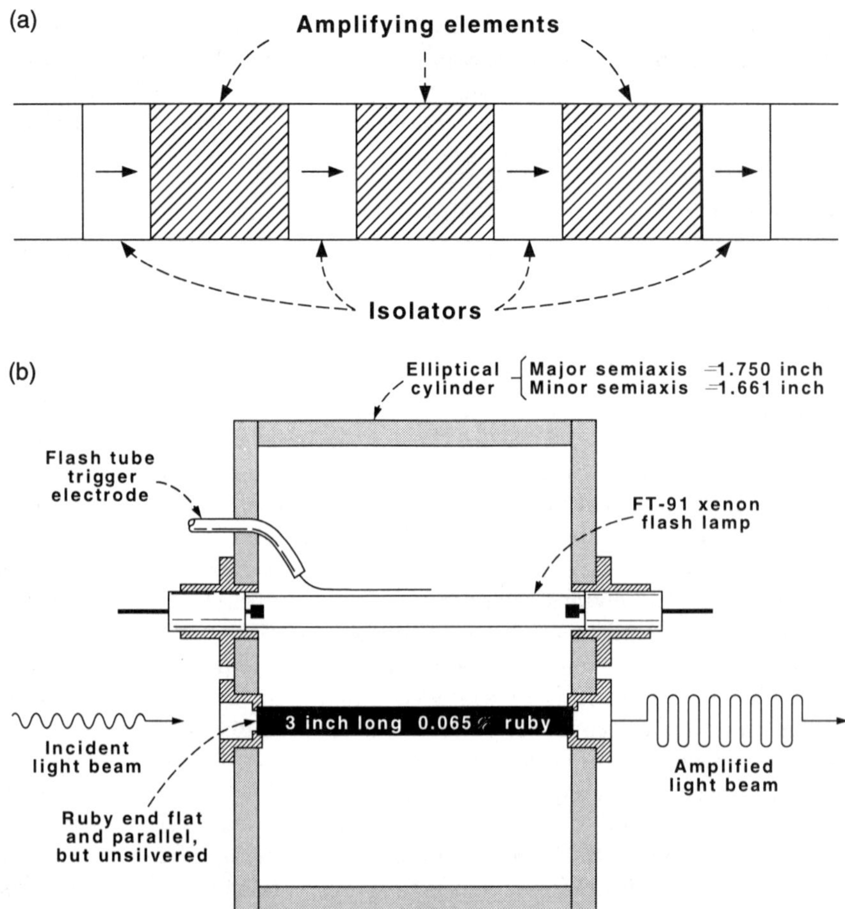

Figure 1.1 Optical amplifier presented in 1962 by Geusic and Scovil: (a) a schematic diagram of the unidirectional traveling-wave optical amplifier configuration and (b) the detailed structure of the ruby optical amplifier [129].

rod amplifiers, which had net gains of 10 and 6 dB, respectively [129]. To test the feasibility of realizing high-gain characteristics, a two-stage amplifier consisting of two ruby amplifiers separated by an isolator was constructed. This isolator with a 3-dB insertion loss was also prepared by Geusic and Scovil. It consisted of a Faraday material, that is, a PbO glass rod whose end faces were attached to dichroic polarizers, and a magnetic coil [129]. This two-stage amplifier exhibited a net gain of 12.2 dB [129]. The signal source was a ruby laser, and the detector was a photo-multiplier.

Additionally, it is quite interesting to note that they also attempted image amplification using their ruby amplifier. Figure 1.2 shows a schematic diagram of

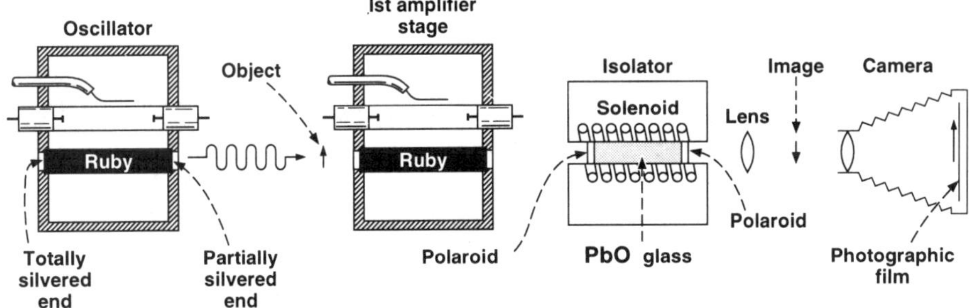

Figure 1.2 Schematic diagram of the image amplification experiment carried out by Geusic and Scovil [129].

the image amplification experiment they carried out [129]. The total physical length of this setup was about 20 in. Since the diameter of the ruby rods used in the amplifiers was 0.25 in, the ruby amplifier was a multimode amplifier capable of handling approximately 10^8 spatial modes. They tried to send image information through this amplifier supported by the 10^8 spatial modes and eventually succeeded in amplifying the image information with a 5-dB gain and obtained a greater contrast in the amplified image than in the unamplified image [129].

It can be said that these ideas and experiments were a very early attempt to construct a fiber amplifier, but the real history of fiber amplifiers began with the experiments performed in 1963 [118] and 1964 [119] by C. J. Koester and E. Snitzer because these experiments were the first to use a doped fiber with a core-cladding structure capable of confining a propagating light in the core. However, although the fiber they used initially was multimode, it was not an unclad rod. This core-cladding waveguide structure features the characteristics of fiber amplifiers and fiber lasers, and the gain in these devices increases in accordance with the optical densities of the pump and signal. Typical parameters of the neodymium-doped glass fiber they prepared initially were 5 to 15 μm for the core diameter, 26 to 45 μm for the first cladding and 0.75 to 3.0 mm for the overall diameter [118,119,122,125].

Figure 1.3 shows the configuration of the first-generation fiber amplifier developed by C. J. Koester, B. Ross, G. C. Holst, and E. Snitzer in the 1960s [118,119,122,125]. In this experiment [125], a signal from an 1.06-μm InAs$_{0.17}$P$_{0.83}$ laser diode was amplified in neodymium (Nd^{3+})-doped fiber. Neodymium-ion-doped fiber was excited with a linear flash lamp. The parameters of the doped fiber used in this experiment were a core diameter of 15 μm ($n_1 = 1.5638$), a first cladding diameter of 45 μm ($n_2 = 1.5630$), and an overall diameter of 3 mm [125]. The fiber was 45-cm long with its ends beveled at 20 degrees in order to suppress feedback and self-oscillation. At the beginning of their experiment, the fiber assembly was wound hot into the form of a helix (as shown in the inset of Figure 1.3)

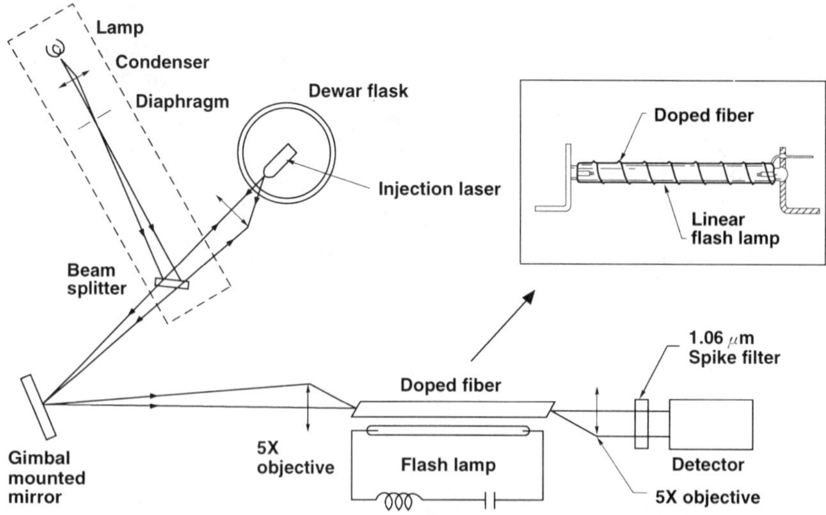

Figure 1.3 Configuration of the first-generation fiber amplifier developed by C. J. Koester, B. Ross, G. C. Holst, and E. Snitzer in the 1960s [118,119,122,125].

to improve the pumping efficiency with a xenon-arc flash lamp [118]. However, in a more advanced experiment they used a straight configuration [125]. The maximum signal gain achieved in their experiment was 5.7×10^4 or 47 dB for a 1.06-μm pulsed signal with a duration of several hundred microseconds [125].

The key features of this first-generation fiber amplifier can be summarized as follows: The active ion (Nd^{3+}) was efficient, but the pumping scheme (flash lamp) was not because its fluorescence spectrum was too wide for the narrow absorption spectrum of neodymium ions. In addition, the host fiber (multicomponent glass fiber) in this experiment was not a low-loss medium and did not have a high Δn. Consequently, efficient amplification was not achieved in this first-generation fiber amplifier.

This work was followed by the demonstration of the first laser diode-pumped neodymium-doped silica fiber laser by Stone and Burrus in the mid-1970s [131]. However, these pioneering research efforts on optical amplifiers and lasers using rare-earth-doped fibers received little attention at the time. One fact worth mentioning here is that an end pumping scheme using a semiconductor laser diode (GaAs diode) was proposed and investigated at this time for a Nd:YAG laser rod [132,133] and for neodymium-doped silica fiber [134], but this basic idea originated from an earlier attempt in 1962 reported by C. J. Koester and D. A. LaMarre [135].

Another approach for achieving fiber amplifiers and lasers is to use single-crystal fibers. In 1975, C. A. Burrus and J. Stone prepared Nd:YAG single-crystal fibers using a CO_2 laser beam as a heat source and succeeded in constructing an

LD-pumped miniature fiber laser [136–139]. In 1982, M. Fejer, R. L. Byer, R. Feigelson, and W. Kway developed a stable fabrication method for fabricating single-crystal fibers and prepared such fibers with various materials such as Nd:YAG, LiNbO$_3$, and Al$_2$O$_3$ with a view to various applications such as passive, active, and nonlinear devices [140–144]. Collaborating with them, M. J. F. Digonnet, C. J. Gaeta, and H. J. Shaw attempted to construct Nd:YAG single-crystal fiber lasers operating at lower pump power levels [145,146]. In addition to such experimental demonstrations, they carried out a theoretical analysis of optical fiber amplifiers and lasers in 1985 [147] and, eventually, in 1990, established a fundamental theory for the gain in three- and four-level fiber amplifiers and lasers [148].

The real breakthrough in fiber amplifier technology came in 1987 when a group from Southampton University reported high-gain fiber amplifiers operating in the 1.5-μm wavelength region, which they achieved using erbium-doped silica-based fibers [127,128]. This can be called a second-generation fiber amplifier. They had made great improvements, especially in relation to the pumping scheme and host fiber preparation, and achieved a gain of over 20 dB in the 1.5-μm region. Figure 1.4 shows the configuration of the second-generation fiber amplifier they developed [127,128].

The key features of this amplifier are as follows. First, *erbium* (Er^{3+}) ions were used as the active ions for efficient amplification at 1.5 μm. Second, an Ar^{+}-pumped DCM dye laser operating at 0.65 μm replaced the flash lamp as the pumping source, enabling the narrow absorption band (^{4}F$_{9/2}$ level) of the erbium ions to be pumped very efficiently. Third, the most important feature enabling this breakthrough was the development of low-loss, high-NA, and rare-earth-doped fiber fabricated by an extension of the well-established *modified chemical vapor deposition* (MCVD) technique. The MCVD method, which is an excellent technique for preparing low-loss

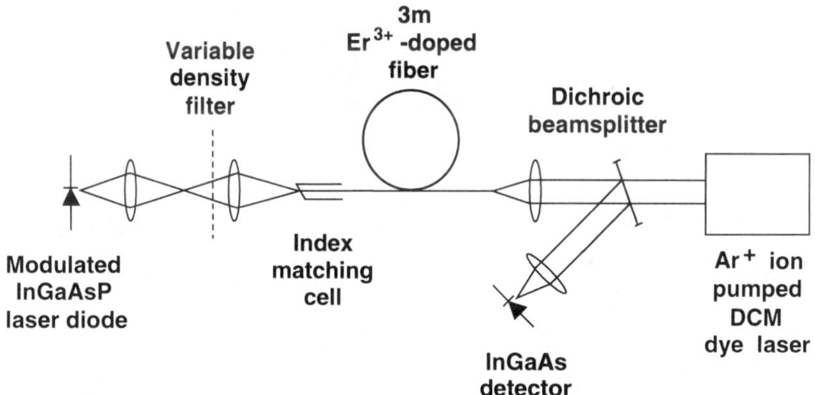

Figure 1.4 Configuration of the second-generation fiber amplifier developed by a group from Southampton University in 1987 [127,128].

high-silica fibers for optical fiber transmission, was developed in 1974 by J. B. MacChesney et al. [149,150]; and rare-earth-doped fibers were prepared using an extended MCVD method in 1985 [151,152]. This is a fascinating method for preparing low-loss neodymium-doped fibers.

Figure 1.5 shows a schematic diagram of neodymium-ion-doped fiber fabrication by the extended MCVD method developed in 1985 by S. B. Poole, D. N. Payne, and M. E. Fermann [151]. Unlike the standard MCVD method [149], this setup has an additional dopant carrier chamber in which to evaporate $NdCl_3$ and a stationary second burner to incorporate Nd_2O_3 in the core glass since the vapor pressure of $NdCl_3$ is so low at room temperature that $NdCl_3$ must be heated to several hundred degrees centigrade in order to transport it to the glass deposition region in the silica tube. This setup is an elegant extension of the transmission fiber fabrication method used for making doped fiber. With this method, they succeeded in fabricating neodymium-doped fibers with a very low background loss. Details of this fabrication technique will be described in Chapter 5.

In short, the excellent characteristics of erbium-doped fibers prepared by the extended MCVD method, that is, low-loss and high-NA, enabled efficient amplification to be achieved even when using three-energy-level transition ions such as erbium. In fact, until their success, it was widely believed to be almost impossible to realize efficient amplification using three-level transition ions because the three-level transition is not easily pumped by a flash lamp. Simon Poole recently stated that approximately two years before they tried amplification at 1.5 μm using erbium, they prepared neodymium-doped fiber and demonstrated a low-threshold cw fiber laser in 1985. He continued that this was because erbium ions have the three-level transition for 1.5-μm amplification, which led them to believe that a very high pump power would be needed to form the population inversion necessary for achieving amplification at 1.5 μm, and that it was very difficult to obtain amplifica-

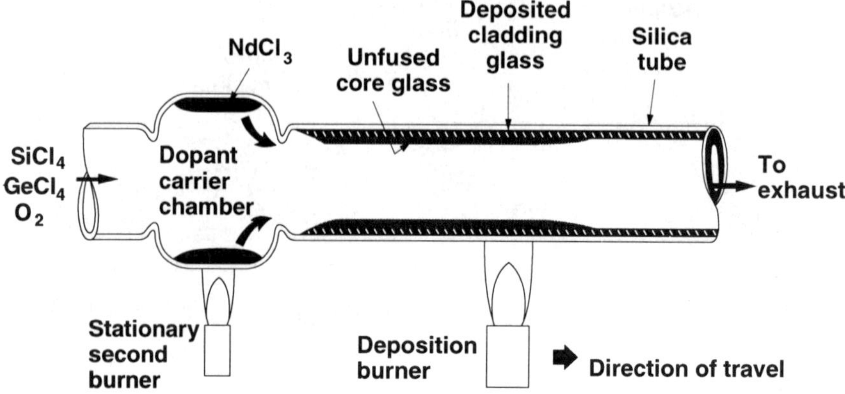

Figure 1.5 Schematic diagram of neodymium-ion-doped fiber fabrication by the MCVD method [151].

tion gain using erbium. He also stated that it was commonly held in the field of solid-state laser technology at that time that erbium was very difficult to use to achieve amplification gain. As it now turns out, however, such problems can be overcome by using a low-loss erbium-doped fiber and an efficient pumping scheme with a laser.

Another important milestone in the history of fiber amplifier technology was the realization of laser diodes pumping at 1.48 μm [153–155] or 0.98 μm [156–159]. This gave us a compact fiber amplifier with high gain for practical use, which can be called the third-generation fiber amplifier. Figure 1.6 shows a schematic diagram of this third-generation fiber amplifier with a laser diode pumping configuration. In Figure 1.6, the pump light from a fiber-pigtailed laser diode is launched into a rare-earth-doped fiber through a fiber coupler, together with a signal light. Consequently, all components used in this fiber amplifier are fiber-type devices, yielding a compact and robust setup. This was another breakthrough in the development of practical fiber amplifiers for use in optical fiber communication systems.

The recent development of practical fiber amplifiers is a kind of rediscovery. However, this rediscovery was made possible because of the well-established fiber technology and mature laser diode technology that have been developed as the result of sustained research efforts throughout the world over the last two decades. This development of practical fiber amplifiers conveyed an important message: that well-established technology can be used in a straightforward manner to generate another new and versatile technology with which to open a wide range of applications. The fiber amplifiers are such a new and versatile technology and they promise to revolutionize the field of optical communications. This is probably the most significant development since low-loss optical fibers and semiconductor laser diodes.

In the next section, we will review the developmental history of optical fiber communications and explain why an optical fiber amplifier is now needed. In

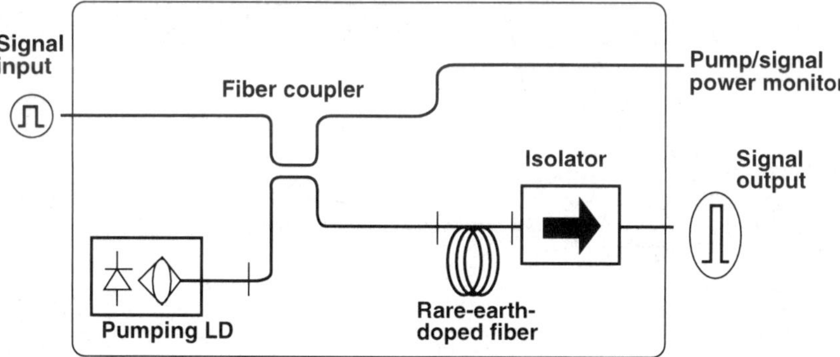

Figure 1.6 Schematic diagram of the third-generation fiber amplifier with a laser diode pumping configuration.

addition, we will describe the alternating growth of transmission fiber technology and active fiber technology.

1.4 BRIEF HISTORY OF OPTICAL COMMUNICATIONS, TRANSMISSION MEDIA, AND OPTICAL FIBERS

1.4.1 Early History of Optical Communications

Today, the Information Age is being ushered in by such devices and technology as television, computers, and lasers, which are all the successful consequences of quantum mechanics. Among the above, optical fiber communication technology stands out as an indispensable signal transmission tool for this Information Age.

However, optical communication—that is, the transmission of information by means of light—is not a new idea. More than a century ago in 1881, Alexander Graham Bell sought to send speech over a visible light beam, shortly after his invention of the "Telephone" in 1876. Actually, his "Photophone" was capable of transmitting speech information over distances of several hundred meters. Figure 1.7 shows the configuration of the "Photophone" proposed and demonstrated by Bell in 1881 [160]. This setup consists of a transmitter (light modulator activated by voice vibration) with a plane mirror of flexible silvered mica, a lens to collimate the light beam, and a photoconductive selenium receiver, which converts the light signal into an electrical signal [160,161]. Using this setup and sunlight, Bell succeeded in sending a voice signal over a distance of 213m and hearing distinctly from the illuminating receiver the words, "Mr. Bell, if you hear what I say, come to the window and wave your hat" [160]. This is the first message

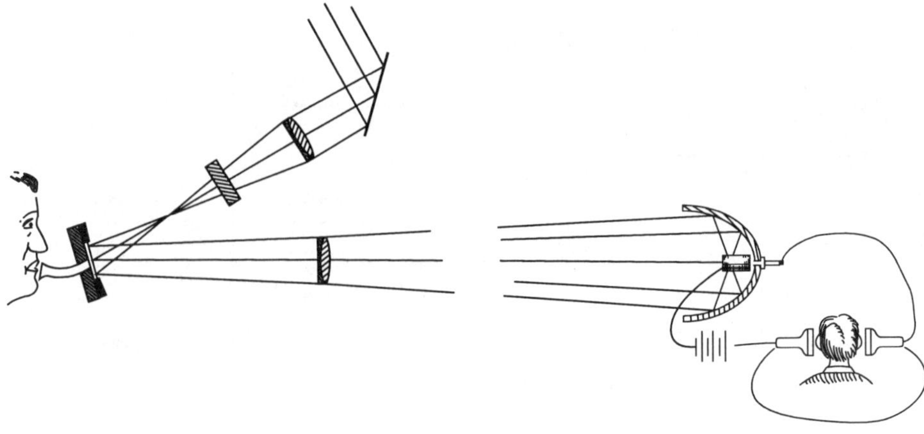

Figure 1.7 Configuration of the "Photophone" proposed and demonstrated in 1881 by A. G. Bell [160].

transmitted by means of light, and this demonstration is clear evidence of Bell's fertile inventive genius in the development of new technology.

Alexander Graham Bell's schemes for modulating a light beam and detecting signals were crude, and his system proved impractical. Nevertheless, the basic elements employed in all forms of optical telecommunications were there: (1) a visible or near-infrared light source that is modulated by the information-bearing signal, (2) a transmission medium, and (3) a detector that recovers the modulation as a baseband signal practically identical to that used as the input to the system.

Another early idea for optical communications was proposed by Hiroshi Seki in 1936 [162]. His concept is illustrated in Figure 1.8. His idea features the use of a light guide formed of a quartz rod or metal tube to eliminate external light and to prevent the transmitted light from scattering due to small particles in addition to the use of a light source, a lens, and a detector.

However, the fundamental advantage of using optical frequencies for telecommunications was not realized until the first laser was developed by Maiman in 1960 [2,3], which was the first generation of coherent light. Of course, the whole electromagnetic spectrum has been available, in principle, as a medium for communications since the epochal discoveries made in the middle of the 19th century by James Clerk Maxwell and Michael Faraday, which was followed by the discovery of long wavelength radiation by Heinrich Hertz in 1887 and its subsequent use in radio by Guglielmo Marconi in 1895. Since then, the history of the transmission of communications signals has largely been a record of advances to progressively shorter wavelengths, eventually to microwaves, and beyond to optical frequency.

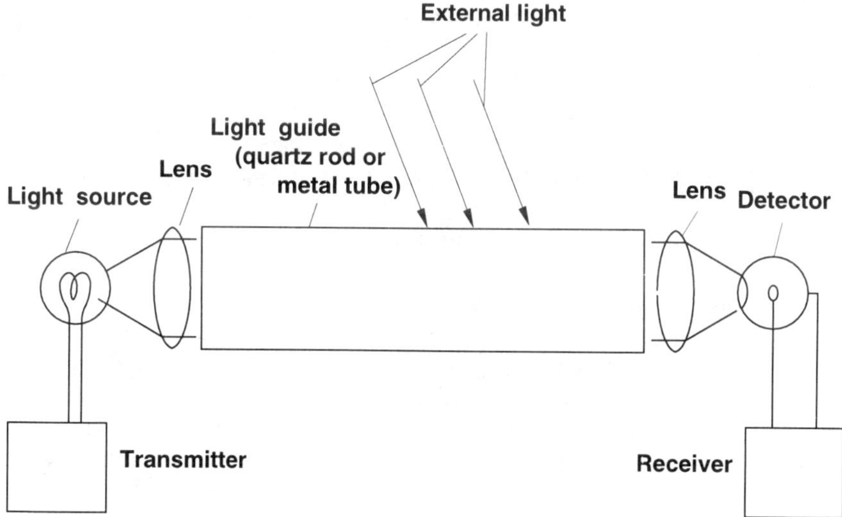

Figure 1.8 The optical communication system proposed in 1936 by H. Seki [162].

However, there was still an essential difference between electromagnetic (radio) waves and light, despite the fact that radio waves and light waves are both essentially electromagnetic waves. As explained in Section 1.1, this difference was that, at that time, radio waves were coherent and light waves were incoherent. Therefore, even when ideas for using light for the transmission of information were considered, they were so difficult to realize that little attention was paid to optical communication.

1.4.2 Various Transmission Media 1960–1969

The invention of the laser in 1960 generated considerable interest in the practical feasibility of the preceding ideas and gave them immediate importance, providing the impetus for a range of studies on transmission media, principles, methods, and components for possible systems. For example, an intense research effort was initiated by solid-state physicists and chemists on previously unknown or unexplored optical properties of solids and on the discovery and development of new lasers, light-emitting diodes, and nonlinear optical materials. This research effort provided the basis for new optoelectronic technology and brought lightwave communications within reach. In fact, we can find impressive evidence of this intense work on the development of new lasers and amplifiers in the book *Lasers: Light Amplifiers and Oscillators* published in 1969 by D. Ross, which provides over 4,300 references to related papers and books [163].

Nevertheless the question of an appropriate transmission medium remained unsettled even at the end of the 1960s. Figure 1.9 shows the candidates for light transmission media explored in the 1960s [164–180]. At this time they were categorized mainly in terms of two possible methods: one was transmission via media protected from the atmosphere (i.e., the waveguide method), and the other was transmission in and through the atmosphere (i.e., free beam propagation) [164,166,167]. The transmission media candidates shown in Figure 1.9(b–h)— namely, hollow, unclad fiber, clad fiber, iris, mirror, lens, and gas lens—all fall into the former category, that is, the waveguide method, with only free beam propagation, shown in Figure 1.9(a), falling into the latter.

At this juncture, it is useful to explain a little about light transmission in and through the atmosphere as the first step in an overview of the history of transmission media. Basically, optical communication through the Earth's atmosphere seems an attractive idea because there are windows in the atmosphere in the visible wavelength range and around 2 μm, 3.5 μm, and between 8 and 12 μm. However, investigations revealed that even with a clear atmosphere the pattern of light produced on a screen by a narrow beam of light coming from a distant coherent source was of a very unsteady, fluctuating kind. This is because of inhomogeneities in the refractive index of the atmosphere, which are in turn due to spatial and temporal fluctuations in the air temperature. This atmospheric turbulence resulted in a noiselike low-frequency modulation in the coherent light [169] and the random spreading and

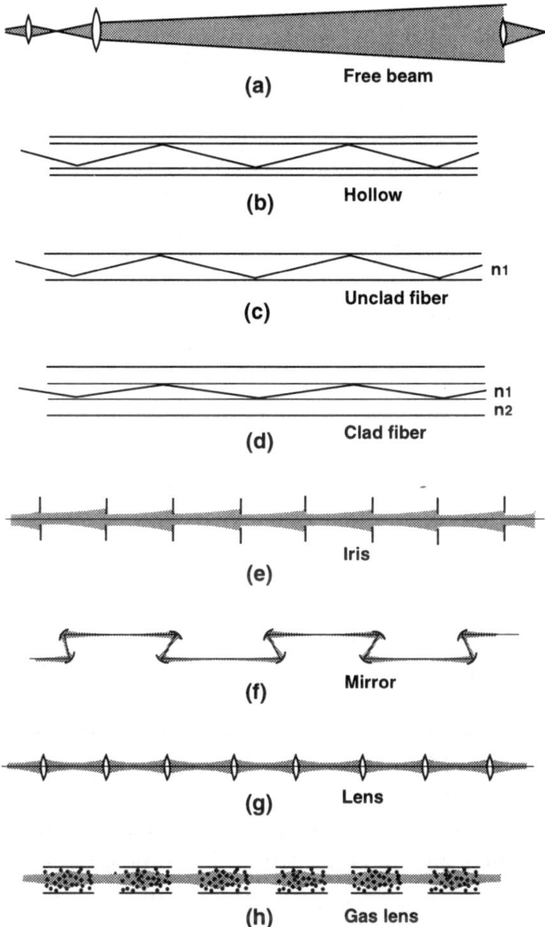

Figure 1.9 Candidates for light transmission media studied in the 1960s [164–180].

deflection of a long narrow beam that sometimes caused the beam to miss the receiving aperture altogether [170]. Other obstacles to the use of light beams for communication in the atmosphere were various kinds of water precipitation such as fog, rain, or snow. Eventually, the investigation concluded that typical amounts of fog and snow resulted in poor transmissions over paths of longer than half a kilometer at a wavelength of 3.5 μm and very poor transmission at all other wavelengths, and that special-purpose links for distances of 100m to 300m or "fair weather" communication applications seemed to be the only cases for which atmospheric optical transmission was feasible [171–173].

Second, hollow waveguides [as shown in Figure 1.9(b)] consisting of a metallic or glass wall have been investigated since the late 1950s [174–177]. For example, a 100-m glass tube, 2 cm in diameter, was prepared that had a reflective coating of aluminum inside it. However, the measured optical loss was 57 dB/km at 0.63 μm [177], which was much larger than the theoretical estimation [175]. This effectively ended discussions on the feasibility of hollow waveguides.

Third, research on unclad fiber as a light guide began in the 1930s [178]. [See Figure 1.9(c).] This owed its origin to the scientific demonstration given by John Tyndall at the Royal Society in England in 1854 [179]. In this experiment, Tyndall used an illuminated vessel of water and demonstrated that, when a stream of water was allowed to flow through a hole in the side of the vessel, light was conducted along the curved path of the stream. Tyndall described: "The lecture concluded with an experimental illustration of the total reflection of light at the common surface of two media of different refractive indexes" [179]. In 1930, Lamm of Germany used a crude assembly of quartz fibers to demonstrate the basic image and light transmission properties of fibers. But, he did not pursue his experiments actively, presumably because a very slight bend in unclad fibers would double the optical loss. In the 1960s, the possibility of using unclad fiber was discussed anew from a practical viewpoint. However, it was concluded that this type of waveguide is impractical.

Fourth, glass-clad glass fiber [as shown in Figure 1.9(d)] was first successfully prepared in 1958 by B. I. Hirschowitz et al. [180] and in 1959 by N. S. Kapany [181]. However, until the mid-1960s, few people thought that this glass-clad glass fiber could be used in long-distance optical communications. This was probably for two reasons [182,183]. The first is that, until the mid-1960s, the optical losses of glass fibers were far too high for them to seem a practical possibility. The second lies in the very high dispersion property that the glass fibers prepared at that time exhibited. This caused considerable broadening of the transmitted light pulse at the receiving end, which meant that it would be impossible to realize a practical communication channel even if the losses were reduced. This pulse broadening property is more apparent in the larger core and higher index difference fibers, especially in unclad fiber that has the largest index difference between the core glass and the air cladding. For communication purposes, which require low dispersion as well as low loss, a structure where both core and cladding are made of glass with a small index difference is preferable to an unclad structure or a structure with a high index difference. This is because such a core-cladding structure results in less multimode dispersion, and therefore less pulse spread at the receiving end, in addition to exhibiting lower optical losses. However, at least until the publication of a paper by K. C. Kao and G. A. Hockham in the mid-1960s [184], people had been unaware of the advantage of this core-cladding structure. Furthermore, it was not until the publication of a paper by D. Gloge in 1971 that the detailed characteristics of optical fibers with this structure were discussed in terms of pulse transmission [185]. Gloge named this structure the "weakly guiding fiber," and this concept was essential in achieving a small dispersion and a low loss for practical optical communication purposes. In 1970, F. P. Kapany, D. B. Keck, and R. D. Maurer achieved high-silica glass optical fibers with a low loss of 20 dB/km [186,187]. At the time, this value was believed to be sufficient for practical communication use and was close to the value predicted by Kao and Hockham in 1966 [184]. The history of this work will be described in the latter part of this section. However, it should be noted here that in the early 1960s detailed investigations of the propagating modes in optical fibers, that is, cylindrical dielectric waveguides, were carried out by E. Snitzer [188,189]. This provided researchers in fiber optics then and still today with a clear vision of the propagating modes in fibers.

Fifth, the iris guide for millimeter waves was discussed theoretically in 1959 by G. Goubau and J. R. Christian [183,190]. As shown in Figure 1.9(e), an iris guide consists of many optical irises (apertures) placed at regular intervals. Since light energy at the beam edges is repeatedly removed by the irises, the equilibrium state approaches with the fundamental (TEM_{00}) Gaussian mode which has the least loss resulting from the irises. In fact, when the iris diameter is 3.4 cm, the wavelength of light is 1 μm, and the iris spacing is 5 cm, the theoretical transmission loss is as low as 1 dB/km [164]. The validity of this theoretical estimation has been proved by J. W. Mink using a millimeter-wave simulation at a wavelength of 8.6 mm [191]. However, it was also shown by Mink's experiment that this system requires very exact alignment as regards positioning the irises when the wavelength is short. In addition, the direction of the light beam cannot be changed without the use of a mirror or prism, which would introduce additional attenuation. These problems are practically insurmountable. Consequently, until the end of the 1960s, the iris guide was thought to be useless for practical applications, although theoretically interesting.

Sixth, in the iris guide previously described, the propagating beam diverges and its phasefront changes as a result of diffraction. Therefore, schemes for the periodic correction of the phasefront, which diffracts gradually during propagation, were proposed in 1961 by G. Goubau and F. Schwering [192] and G. D. Boyd and J. P. Gordon [193]. This method involves the introduction of a mirror or lens to correct the phasefront curve periodically. Based on the theoretical calculation, the spacing of the phase-transforming elements is limited to about 100m by the topography; thus, it has been estimated that a typical target value for the loss coefficient achievable by these waveguides is 0.2 dB/km [164]. Diagrams of these schemes are shown in Figure 1.9(f,g). In 1964–1968, several mirror and lens guide experiments were conducted [194–198]. For example, in 1965 De Lange designed the first experiment using a mirror guide where a laser beam was shuttled back and forth 400 times between two concave mirrors spaced 100m apart through an aluminum pipe of 15-cm diameter in the laboratory. The 400 round trips covered a distance of 80 km. He found that alignment with respect to the beam reflectors was critical but not impractically so. An experiment using a lens guide was carried out in 1967 by D. Gloge. He used a 0.84-km-long straight iron pipe, installed outdoors at an average depth of 150 cm, with the intention of simulating actual future field conditions for a lens guide [196,197]. The pipe was filled with air and had a diameter of 10 cm. Six biconvex confocal glass lenses were located every 140m in the pipe, and a laser beam was transmitted in one direction for a number of weeks [197]. The beam was not noticeably affected by air turbulence. Gloge and others reported several important results [164,166,196,197]. One was that the transmission loss was about 0.4 dB/km; another was that the beam had a Gaussian profile when it was injected and was still approximately Gaussian after covering a distance of 120 km and passing 900 lenses. However, the most important finding was that a gas-filled underground installation has three major instabilities: (1) slowly varying thermal gradients in the ground that deflect the beam several centimeters, (2) ground drift and ground vibration, and (3) mechanical instabilities in the lens or mirror arrangements. So these results were at once promising and disappointing. In order to overcome these problems, various ray-stabilizing schemes were proposed and investigated [166]. However, in the late 1960s it became widely recognized that such schemes would not be suitable for practical use because the complicated mirror or lens guide systems had to be well-aligned and that alignment maintained over years.

Seventh, in addition to the aforementioned difficulties, another problem with the lens guide is how to bend the optical path of the beam. For example, for a bending curvature radius of 1 km, a lens spacing of at least 1m is required in order to minimize excess loss. An excess loss of 1% per lens, with one lens per meter, corresponds to 40 dB/km, which is far beyond practical acceptability. Further, if we reduce the bending curvature, the excess loss becomes much larger. Then, we recall that most of the excess loss associated with lenses is due to surface imperfections that cause a fraction of the light to be scattered out of the beam. Therefore, in order to reduce the excess loss caused by these imperfections, the idea arose of making a lens that has no surface, namely, the "gas lens" developed by D. W. Berreman in 1964 [198,199] and shown in Figure 1.9(h). The idea of the gas lens is simple: it is based on a distribution of gases that exhibits a higher density and refractive index at its axis than at the periphery. It can be implemented in a number of ways. One is to allow cool gas to flow axially through a heated cylinder and thus obtain a parabolic refractive-index distribution along the axis direction [200,201]. Another is to use appropriately disposed nozzles to mix two gases with the same temperature but different refractive indexes [202]. However, several problems appeared as the experiments progressed. For example, it turned out that, after the gas had flowed a certain distance, the temperature gradient and the resultant guiding structure based on the refractive-index gradient both disappeared. Besides, in most cases another problem arose before the temperature gradient had completely disappeared, and this was due to the axially asymmetric temperature distribution caused by the effect of gravity [203]. To overcome such difficulties, the gas must be removed before it flows beyond a limited range within which the effect of gravity is not noticeable. Thermal gas lenses were improved structurally, thus allowing beams to be guided stably [204]. In addition, the experimental demonstration of a curved array of such gas-lens units revealed that a practical curved waveguide could be obtained using the gas-lens guide [204]. Consequently, in the late 1960s, some researchers investigated the possibility of an optical transmission system consisting of lens waveguides for straight sections and gas-lens waveguides for curved sections.

1.4.3 Optical Fibers From 1966

From the beginning of the 1960s, glass fiber became an appealing choice as a laser beam transmission medium. It was thought that glass fiber was a promising candidate because it was assumed that the continuously guiding structure of glass fiber might provide a more-flexible, less-sensitive, and consequently, less-expensive transmission medium than other types of guiding structures [164]. But optical losses were still far too high to make glass fibers appear to be a practical possibility.

In parallel with the research on atmospheric transmission and optical beam guides previously described, research on glass fiber was carried out as an alternative laser waveguiding medium in many countries. For example, in 1964, J. Nishizawa and S. Sasaki developed optical glass fiber with a parabolic refractive-index profile, like that of a gas lens [205]. However, the specific origin of the use of glass fiber as a laser transmission medium as we know it today is found in the work undertaken at Standard Telecommunications Laboratories in England. It is widely recognized that the modern fiber optics era started in 1966 with the publication of the Kao and Hockham paper, which suggested that optical fibers based on silica glass could provide transmission media for lightwave communications if impurity effects could be overcome [206]. Although the most exciting fibers had losses of greater than 1,000 dB/km, Kao and Hockham speculated that silica-based fiber losses as low as 20 dB/km should be achievable if metal impurities could be removed to the 10^{-6} ppm (parts per million) level and if the fiber drawing process could be improved. They further suggested that such fibers could provide useful transmission media for lightwave communications. In addition, shortly after this landmark publication, they made a more precise measurement of a commercial silica glass rod and obtained an estimated optical loss of only 5 dB/km within an accuracy of 0.5 dB/km [207,208]. No materials expert is on record as predicting that losses as low as 20 dB/km or 5 dB/km would be achieved, but several organizations in several countries began research to explore this potentiality. Both multicomponent glasses and high-silica glass systems were explored. Multicomponent glasses offered the attraction of lower working temperatures, thereby lessening the chances of contamination, but purification of the starting materials and homogeneity of the glass proved to be major technical obstacles. Nevertheless, impressive progress was made, especially in the successful preparation of a graded-index profile fiber with optical losses as low as 100 dB/km in 1969 [209]. On the other hand, Corning Glass Works emphasized research on the development of a method for preparing high-silica glass fibers. At this time Bell Laboratories and other organizations started to pay more attention to this glass system after the difficulties of working with multicomponent glasses became more apparent.

The breakthrough in work on glass fibers came in 1970, when F. P. Kapron, D. B. Keck, and R. D. Maurer of Corning Glass Works announced the achievement of low losses of under 20 dB/km (at λ =0.63 μm) in single-mode fibers hundreds of meters long [210]. They used the so-called soot process for high-silica glass fabrication [211], which originated in the work undertaken in the 1930s by J. F. Hyde and M. E. Nordberg [212,213]. The fabrication process is as follows [211]. First, a thin layer of TiO_2-doped SiO_2 glass particles (soot) is deposited on the inside wall of a thick-walled fused silica tube by aiming the soot stream from an oxy-hydrogen burner down through the tube hole. Then this soot layer is sintered into a bubble-free and transparent SiO_2-TiO_2 glass film for the core. Finally, the fiber is drawn from the core/cladding composite structure obtained by collapsing the tube. In addition, the fibers prepared with TiO_2-doped SiO_2 core glass required heat treatment at 800° to 1,000°C in an oxygen gas atmosphere in order to reduce the absorption loss due to T^{3+} ions [211]. Based on this fabrication process, this method could be called the *inside soot deposition* (ISD) method. The most important feature of this method is that it adopts a vapor-phase reaction to synthesize SiO_2-TiO_2 glass particles and to deposit them inside the silica tube. This minimizes both impurity absorption losses and core/cladding interface scattering losses [211]. The development of glass fiber with a 20-dB/km loss raised the possibility of constructing a practical optical communication system because 20 dB/km was regarded at that time as the threshold of usefulness for the communications application of optical fibers and other waveguide media. Furthermore, it was widely recognized that the maximum permissible loss when designing an optical transmission system of more than one kilometer without amplification was around 20 dB/km [214].

Since then, progress in all areas of optical fiber communications has been both rapid and extensive. After 1970, major research efforts on the development of high-silica fibers shifted toward how to prepare a core-cladding structure with high-silica glasses, which is relatively more difficult than with multicomponent glass fibers. These approaches could be categorized into four types from the materials point of view and three types in terms of the fiber preform preparation method. With regard to glass materials, five main systems were investigated. These were the pure-silica (SiO_2) glass system [215,216], the SiO_2-B_2O_3 glass system [217–219], the SiO_2-TiO_2 glass system [210,211,220], the SiO_2-P_2O_5 glass system [221], and the SiO_2-Al_2O_3 glass system [222–224]. Three main fiber preform preparation methods were investigated: ISD [210,211], *chemical vapor deposition* (CVD) [218,219], and direct deposition using a plasma torch [224–227], a oxy-hydrogen flame [222], and a CO_2 laser [223] as melting heat sources. However, these high-silica glass systems and fiber preform preparation methods posed a variety of difficulties regarding: (1) controlling the refractive index of synthesized glasses for core/cladding structure formation; (2) crystallizing glass materials, especially at higher dopant concentrations; and (3) greatly reducing transition metal and OH ion contamination.

In short, the results of various investigations regarding both materials and methods revealed that there were two keys to preparing low-loss optical fibers based on high-silica glass systems. One was to find a way to achieve precise control of the refractive index of high-silica glass systems in terms of material composition and preparation method in order to enable the formation of the desired core/cladding structure in the prepared fiber. The other was to develop a technique for drastically reducing transition metal and OH ion contamination in synthesized high-silica glasses.

After several years of sustained worldwide efforts to achieve low-loss glass fibers, another great advance was made in 1974 when the MCVD method was developed by a group at Bell Laboratories [228–230]. With this method, they succeeded in fabricating a 1.1-dB/km (at $\lambda = 1.06$ μm) fiber. This MCVD method gave a powerful impetus to research programs on fiber fabrication techniques throughout the world because the MCVD method was the first fiber fabrication method that completely solved the technical problems of low-loss optical fiber fabrication using high-silica glass systems, in which sense the MCVD method can be called a refined and well-established method.

Figure 1.10 is a schematic diagram of the MCVD method [230,231]. As shown, raw halide material vapors carried by oxygen gas are introduced into a substrate tube (pure silica-glass tube), which is externally heated to about 1,400°C by a traversing oxy-hydrogen burner. As the burner traverses, the halide vapor materials oxidize to form fine glass particles (soot). The soot is deposited on the inside wall of the silica tube and immediately sintered into transparent glass with doped silica glasses. After multiglass layers have been formed by passing the flame repeatedly along the tube on a glass-working lathe, the tube is then thermally collapsed to

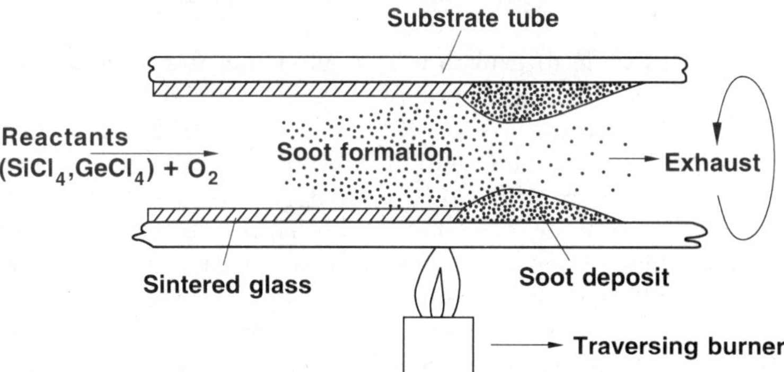

Figure 1.10 Schematic diagram of the MCVD method [230,231]. The soot is deposited on the inside wall of the silica tube and immediately sintered into transparent glass with doped silica glasses.

provide a preform rod. This silica tube becomes the outside cladding part, and the deposited glass layers become the core part of the fiber preform. This preform is subsequently drawn into a fiber using a high-temperature furnace. This process gave the MCVD method various advantages. First, by adopting the soot process the MCVD method made it possible to dope GeO_2 glass into SiO_2 glass. This enables the refractive-index of high-silica glasses to be controlled precisely without crystallization up to a high GeO_2 doping concentration. GeO_2 was also found to be the best dopant for SiO_2 in that there was no crystallization, no extra absorption loss, and no great change in the melting temperature or thermal expansion coefficient. It was also found to be the best in terms of refractive index [232,233]. Second, by using a closed chemical reaction setup for the CVD process based on semiconductor material deposition and by replacing SiH_4 and GeH_4 with $SiCl_4$ and $GeCl_4$ as the raw materials, the contamination by transition metals and OH^- ions was dramatically reduced to a very low level. For example, the OH^- ion contamination level was reduced from 100 ppm [218,219] to 10 ppm [228,229]. Third, by replacing the electrical heater used in the CVD method with the traversing burner, the glass deposition rate was greatly improved; hence a thick cladding layer was easily prepared, and it was possible to reduce both the scattering and absorption losses [233].

In 1976 a group at NTT Laboratories used the MCVD method to fabricate fibers with a low OH content and a low-loss value of 0.47 dB/km (at $\lambda = 1.18$ μm) [234]. This work opened the long wavelength window of optical communication for the first time. Furthermore, in 1979, another group at NTT Laboratories made an extremely low-loss fiber with a loss value of 0.2 dB/km [235], which is the closest to the theoretical loss limit of high-silica optical fibers at 1.55 μm.

In short, the key feature of the MCVD method is to combine the soot process and the CVD method by replacing the furnace with a flame torch. This makes it possible to use GeO_2 doping, $SiCl_4$ and $GeCl_4$, and a high deposition rate as well as to minimize the contamination from the transition metals through the use of the vapor-phase reaction.

Here, it is worthwhile to explain a little about the soot process, which is a fundamental process for high-silica glass systems. As mentioned, its origins can be traced to work undertaken in the early 1930s by J. F. Hyde [212]. He invented a flame hydrolysis process with the intention of producing silica glass at a relatively low temperature and, if desired, with a high degree of purity [212]. Also, he described in his patent the way in which the soot layer is formed on the mandrel by the deposition of fine glass particles that are synthesized by flame hydrolysis and how this soot layer is then vitrified into transparent silica particles by the oxy-hydrogen burner. Figure 1.11 is a schematic diagram of the soot process invented in 1934 by J. F. Hyde. The soot layer 58, which is deposited onto the mandrel 53 by the deposition burner 56 whose structure is shown in Figure 1.11(a), is vitrified (becomes transparent) by the vitrifying burner 57. With this process, it is possible to prepare very high purity silica glass at a relatively low temperature, that is, lower than the melting temperature of silica glass, because this is a sintering process

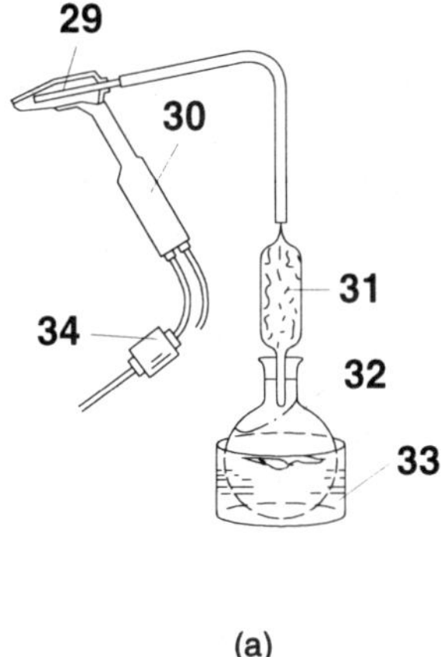

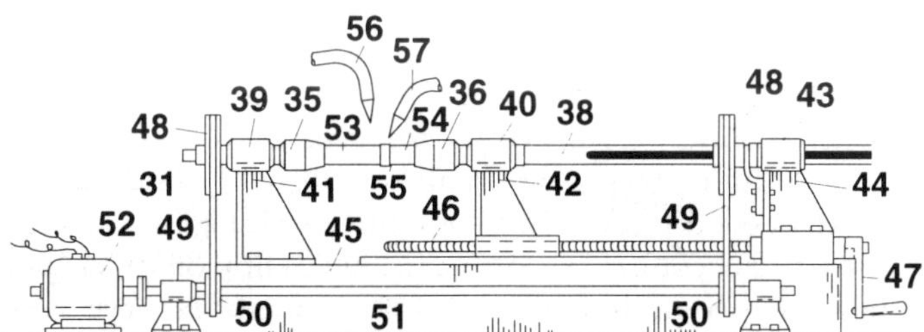

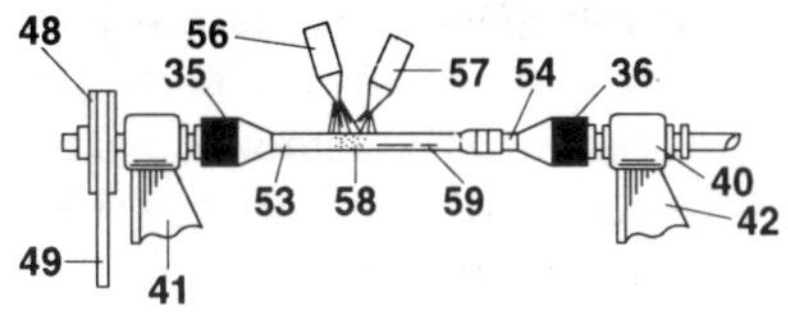

Figure 1.11 Schematic diagram of the soot process invented in 1934 by J. F. Hyde [212]. The soot layer 58 deposited onto the mandrel 53 by the deposition burner 56 whose structure is shown in (a) is vitrified (becomes transparent) by the vitrifying burner 57.

rather than a melting process. In addition, the glass particles are synthesized by the vapor phase reaction in the flame from the raw material purified through evaporation. In short, the advantage of the soot process is the potentiality of preparing high-purity silica glass at a temperature lower than its melting point. There then followed work by Nordbery [213] on silica containing TiO_2, which was originally intended as a glass with a low expansion coefficient. In both cases, silica or doped silica particles were formed by the introduction of silicon tetrachloride vapor and dopant chloride species into a gas burner flame (flame hydrolysis). These particles were collected and vitrified in a manner designed to produce an optical quality glass. Some 30 years later, this process was applied by Keck and Schultz to fabricate low-loss optical fibers from doped high-silica glasses. They invented the ISD method in 1970 [211].

In terms of the history of optical fiber fabrication methods, while high-silica fibers with impressive optical losses were being successfully fabricated with the MCVD method, there was other work undertaken that was aimed at developing a method for the mass production of high-silica optical fibers because, despite the fact that the MCVD method is the most widely used process in the world for fabricating low-loss high-silica fibers, it has a disadvantage in terms of manufacturing cost. This relates particularly to achievable blank sizes, deposition rates, and starting tube requirements. Typical MCVD blank sizes in use at that time yielded 5 km of fiber, the glass deposition rates were usually 0.1 to 0.3 gr/min, and high-quality starting tubes were needed [233,236]. In addition, such high-quality tubes were, in general, expensive and accounted for a significant fraction of MCVD preform costs. With this in mind, two representative fabrication methods were developed: the *outside vapor phase oxidation* (OVPO) or *outside vapor deposition* (OVD) method [236–238] and the *vapor-phase axial deposition* (VAD) method [239–241].

Figure 1.12 shows an outline of the OVPO method [238]. Fine glass particles, synthesized from raw vapor materials in a burner flame, are deposited on a rotating and traversing target rod. Layer by layer, a cylindrical porous glass preform is built up [Figure 1.12(a)]. After the glass layer deposition is completed, the porous preform is slipped off the target rod and then zone sintered (1,500 to 1,600°C) into a bubble-free and transparent glass blank by passing it through a furnace hot zone [Figure 1.12(b)]. The center hole remains in this blank but disappears when it is drawn into fiber at a much higher temperature (1,800 to 2,100°C), as seen in Figure 1.12(c). This OVPO method has several advantages in terms of manufacturing cost. Its effective deposition rate is about 2 gr/min and blank sizes yielding fibers several tens of kilometers in length can be easily made, which makes it more economical than the MCVD method. In addition, if desired, even the cladding layer can be formed without using a high-quality silica tube. Furthermore, blank sizes have now been improved to yield fibers that are several hundred kilometers in length.

Another fabrication method for the mass production of high-silica fibers, the VAD method, was developed in 1977 by a group at NTT Laboratories. Figure 1.13

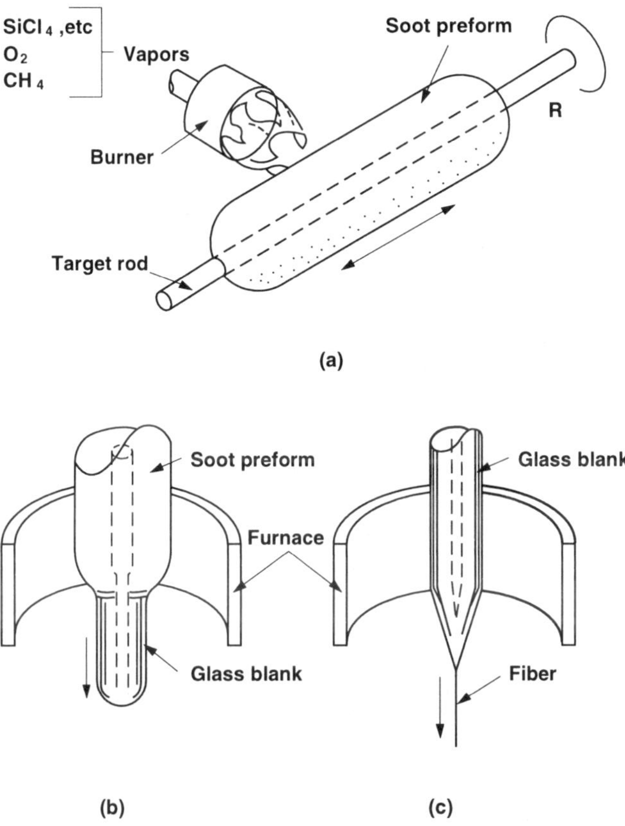

Figure 1.12 Outline of the OVPO method [238]: (a) soot deposition, (b) sintering, and (c) fiber drawing.

shows the basic principle for making fiber preforms by the VAD method. Raw halide materials are oxidized into fine glass particles (i.e., soot) by flame hydrolysis. Unlike the OVPO process, the core glass soot and sometimes the cladding glass soot are simultaneously deposited in an axial direction onto the end of the rotating fused silica target rod. As the rodlike porous preform grows devoid of a central hole, it is slowly retracted through a graphite resistance furnace, where it is consolidated (sintered) into a transparent preform by zone sintering. Occasionally the growth and consolidation of a porous preform are carried out separately. The development of the VAD method has its origins not only in the doping technique of the soot process but also in the axial deposition technique of the *direct deposition* process, which uses a plasma torch [224–227], an oxy-hydrogen flame [222], and a CO_2 laser [223] as melting heat sources. Furthermore, the direct deposition method originated from the so-called *Verneuil* method, where vitreous silica is made by passing crushed quartz particles through a Verneuil torch configuration [242].

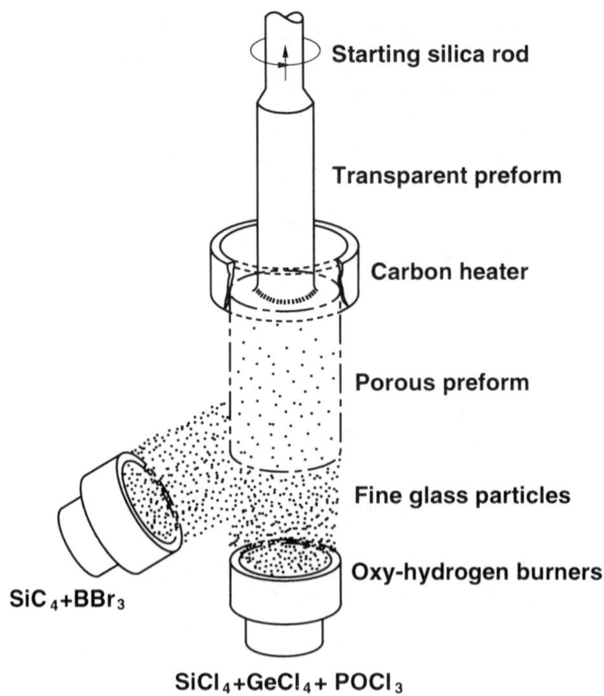

Figure 1.13 Basic principle for making fiber preforms by the VAD method [239–241].

However, there is a serious problem with the direct deposition method when it comes to preparing doped silica glasses, namely, that the blank surface temperature needed to achieve smooth glass deposition is so high that dopants such as GeO_2, P_2O_5, and B_2O_3 are mostly vaporized. The only dopants that can be used in this process are Al_2O_3 and TiO_2, which have higher boiling points than silica glass. Therefore, in short, the key feature of the VAD method is the axial deposition of fine glass particles consisting of GeO_2-, P_2O_5- or B_2O_3-doped high-silica glasses synthesized by flame hydrolysis and the subsequent consolidation of the porous preform made of these particles.

This VAD method offers several important advantages in relation to both mass production and fiber performance. First, since the VAD process involves essentially the same soot process as the OVPO process, the effective glass deposition rate is high, which offers the potential for mass production. Second, the VAD method involves an axial deposition process that makes the continuous fabrication of a fiber preform possible. Third, if the fabrication conditions can be successfully kept constant throughout the preform preparation, the preform will have good glass uniformity and, as a result, excellent attenuation reproducibility. In addition, VAD fibers have no dip in the center of their refractive-index profile, as commonly observed in MCVD fibers. These features of the VAD method, related to fiber performance, were confirmed by the attainment of ultimately low-loss optical fibers in 1980 [243].

Figure 1.14 shows a historical review of optical transmission losses achieved with high-silica glass fibers between 1970 and 1980. Figure 1.15 shows the contribution of various mechanisms to the spectral loss of ultimately low-loss optical fibers prepared in 1980 by the VAD method [243]. Although the transmission losses due to Rayleigh scattering and *ultraviolet* (UV) and *infrared* (IR) absorption are composition-dependent, they are all intrinsic loss mechanisms that can never be eliminated. The loss spectrum of the ultimately low-loss VAD fiber comprises virtually the only intrinsic attenuation loss of GeO_2-doped silica fiber. In fact, the difference between the loss value for the ultimately low-loss VAD fiber and the total loss estimated from the intrinsic loss factors is merely 0.03 dB/km, regardless of the wavelength. This is comparable to the margin of measurement error. This loss mechanism is commonly called "waveguide imperfection loss," derived from small undulations along the fiber axis. Another important extrinsic loss mechanism is impurity absorption. OH vibrational absorption, arising from OH- ions in the glass, is responsible for loss peaks at 0.95, 1.24, and 1.39 μm. However, in this loss spectrum the OH absorption loss peak is barely visible even in the vicinity of 1.39 μm. From this spectrum, it can be estimated that the OH concentration in this fiber is about 0.8 ppb, assuming that the relation between the loss increase at 1.39 μm and the OH content is 65 dB/km/ppm [244]. The advent of the ultimately

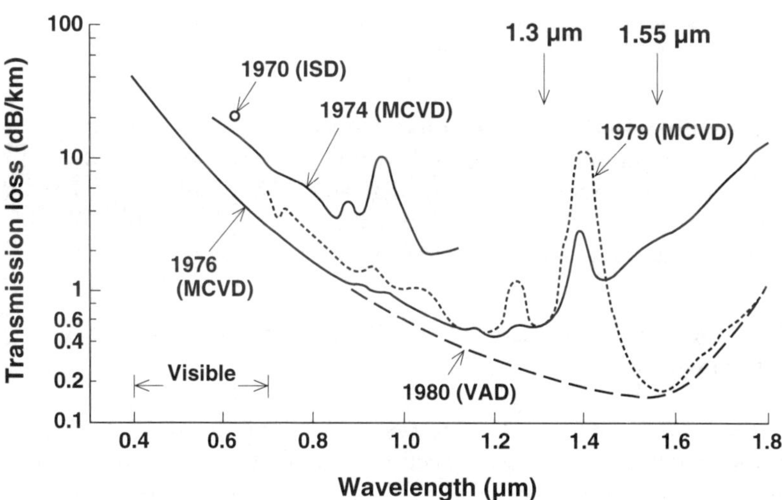

Figure 1.14 Historical review of optical transmission losses attained by high-silica glass fibers between 1970 and 1980.

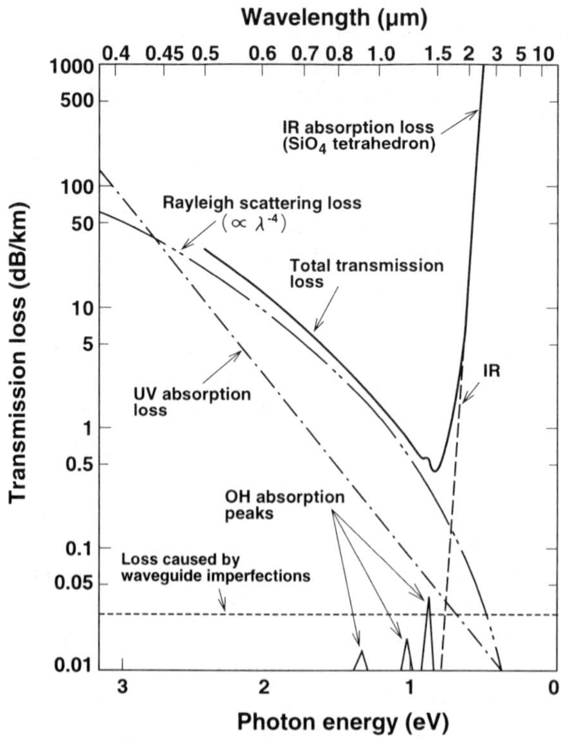

Figure 1.15 Contribution of various mechanisms to the spectral loss of ultimately low-loss optical fibers prepared in 1980 by the VAD method [243].

low-loss optical fibers ended research on low-loss optical fibers with high-silica glass systems and the development of a low-loss fiber fabrication method. Therefore, since 1980, the developmental trend in high-silica fiber fabrication methods has shifted toward lower cost fiber fabrication as well as the installation of high-silica fiber cables in actual transmission systems for business use.

1.4.4 Alternating Growth of Passive Fiber Technology and Active Fiber Technology

Since the development of ultimately low-loss optical fiber, research into fiber fabrication technology has moved in two directions: one focused on establishing a mass production method for preparing high-silica fibers at much lower cost and the other on the investigation of materials with ultra-low-loss properties with a view to developing fiber fabrication methods. As an example of work along the first of these two routes, the multiflame technique was developed as part of the VAD method to achieve large-volume production and a higher deposition rate for fiber preforms [245,246]. In addition, in relation to the OVD method, mechanical fiber performance has been improved, allowing for the production of various types of OVD fiber to commence [247,248]. Similar improvements were also achieved with the MCVD [249] and *plasma chemical vapor deposition* (PCVD) [250] methods, which enabled the levels of technology to be raised to the production stage. These goals had largely been accomplished by around 1987 [251–253].

With regard to the latter route involving the development of ultra-low-loss materials and fibers, the strongest research emphasis was placed on fluoride-based glass systems, especially on a ZrF_4-based mixture (ZrF_4-BaF_2-LaF_2-AlF_3-NaF, abbreviated as ZBLAN), whose total intrinsic loss was estimated to be close to 0.01 dB/km [254–256]. These research programs were initiated by proposals made in the late 1970s by several authors whose aim was to reduce the intrinsic attenuation to 0.01 dB/km or even lower in the 2- to 10-μm wavelength range [257–259]. Ideally, these ultra-low-loss optical fibers would enable communications systems to operate over several thousand kilometers without amplifiers or repeaters. In 1987, as a result of several years of investigation into materials and fiber fabrication methods, ZrF_4-based fluoride fiber with a minimum loss of less than 1 dB/km was successfully prepared by a group at NTT [260], and eventually in 1992 the minimum loss of 0.45 dB/km was achieved by a group at British Telecom [261,262]. However, they all confronted several difficulties such as the local crystallization of fluoride glass materials, the purification of raw materials, the preparation of long-length fibers, and long-term fiber reliability. Of these problems, the local crystallization of fluoride glass materials was the most important to overcome. To this end, a new process was developed in 1989 for fabricating ZrF_4-based fluoride glass preforms that used the CVD of fluoride materials [263]. In 1992, this process was used to prepare ZBLAN glass film on the inner surface of a ZBLYAN (ZrF_4-BaF_2-LaF_2-YF_3-AlF_3-NaF) support tube [264,265]. However, the use of this CVD process led to another problem regarding the difference in the deposition rates of fluoride materials such as ZrF_4, BaF_2, LaF_2, AlF_3, and NaF [264,265]. Consequently, despite intensive research on IR materials [266,267], until 1993 almost all research groups had concluded that there was little point in continuing their work on the fabrication of ultra-low-loss fibers with nonsilica glass materials.

The reason for their lack of success in fabricating ultra-low-loss fibers with nonsilica glass materials would seem to be that materials whose losses in the IR region are estimated to be lower than that for silica (SiO_2) glass must have a smaller phonon energy than that of silica glass, because optical loss in the IR region is determined by the phonon energy of the vibrations of the constituent atoms. However, a smaller phonon energy indicates a weaker bonding energy between atoms, and a weaker bonding energy leads to the easier movement of atoms and faster crystallization even at lower temperatures. Figure 1.16 outlines the difficulties involved in fabricating ultra-low-loss fibers with IR nonsilica materials based on the prior explanation. In short, materials with lower losses than that of silica glass are

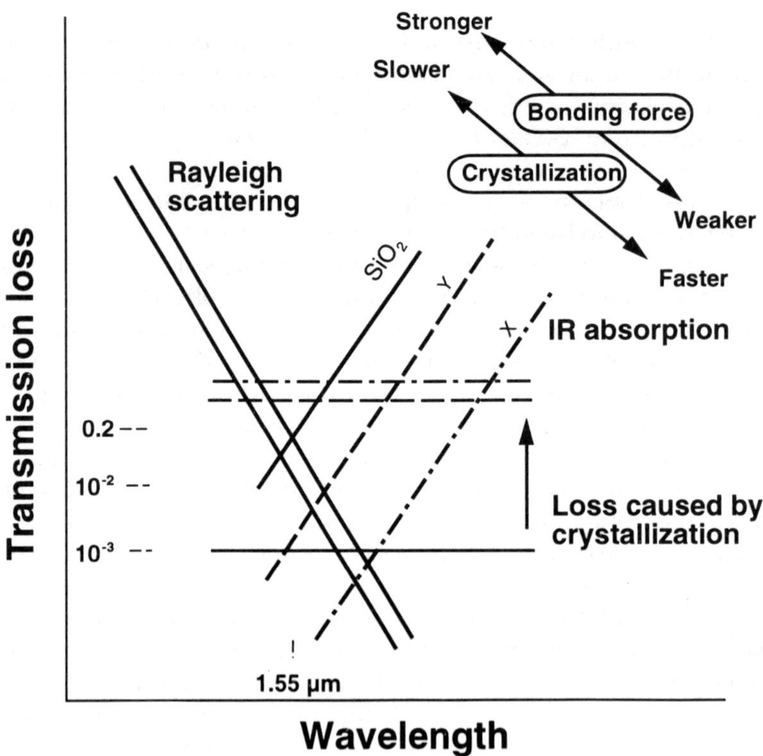

Figure 1.16 Difficulties involved in fabricating ultra-low-loss fibers with IR materials.

easier to crystallize than silica glass; and therefore, such materials have to have a higher scattering loss than silica glass. This is so even if their losses, estimated only from the infrared absorption caused by multiphonon and Rayleigh scattering losses, are lower than that for silica glass. As a result of various investigations into these infrared materials, we recognize that it is necessary to estimate the total transmission loss of fibers, which consists of the contribution of Rayleigh scattering, infrared absorption, and the scattering loss due to material crystallization or waveguide imperfections. This is the minimum requirement and shows that the traditional method of estimating the total transmission loss of fibers, from only the Rayleigh scattering contribution and infrared absorption, is incorrect [254–259]. Consequently, we believe that silica glass is the best material with which to achieve the lowest transmission loss for a long fiber, because silica glass is the only possible material that can be used to draw long fibers from a large volume fiber preform without any crystallization. Various studies on fiber materials over many years have clarified and confirmed this conclusion regarding materials for the fabrication of long practical fibers. One of the authors of this book stated in 1987 in the preface of a previous work [268]:

Although the authors are pessimistic with regard to the realization of an ultra-low-loss fiber of less than 0.01 dB/km, we believe that this book will assist in clarifying both the important and the significant factors in the manufacture of low-loss fibers.

1987 was the most exciting period for research on ultra-low-loss fibers using infrared materials such as fluoride glasses, chalcogenide glasses, and halide crystals.

Scanning the whole developmental history of glass fibers once more, we find an interesting pattern of the alternating growth of passive fibers for signal transmission and active fibers for lasing or amplification. Figure 1.17 outlines the developmental history of passive fibers for signal transmission and active fibers for lasing or amplification and highlights several facts. Particularly in the 1960s, research on active fibers was ahead of that on passive fibers. But, actually, in the early 1960s, neodymium-doped multicomponent glass fibers were prepared and high-gain fiber amplifiers were constructed using these neodymium-doped fibers [118–126], following the demonstration of ruby rod amplifiers in 1962 [13]. On the other hand, research on passive fibers for signal transmission made no great progress during this period, except for the important suggestion made by Kao and Hockham [184], because the loss of the glass fibers that had been prepared was still too high for application to signal transmission and few people believed that low-loss fibers could be achieved. In fact, the research on glass fibers confronted several difficulties and the optical loss of multicomponent glass fiber, which was a promising material during that period, was around 100 dB/km even in 1969 [209], as described in the previous section. In short, in the 1960s glass fiber was thought to be suitable for active devices because, although its optical loss was high, it was easy to dope with active ions, and, moreover, there were many promising candidates for signal transmission such as lenses, gas lenses, mirrors, hollows, and free beams.

The situation changed dramatically in 1970 as a result of two important developments: the realization of 20-dB/km silica glass fibers [210] and the successful room-temperature cw-operation of semiconductor laser diodes using a double-hetero junction structure [269]. These two achievements led to increasing efforts in research on low-loss silica glass fibers and a reduction in the attention paid to active fibers such as fiber lasers or amplifiers. In fact, in 1969, Dieter Ross stated in his book that [270]:

Amplification by stimulated emission is used nowadays in laser-oscillators for the production of coherent light. In comparison, the use of laser materials for light amplification is of less practical importance and has been investigated experimentally in detail only for helium-neon and ruby.

In 1972, a representative opinion was given in a paper published by R. Kompfner [271]:

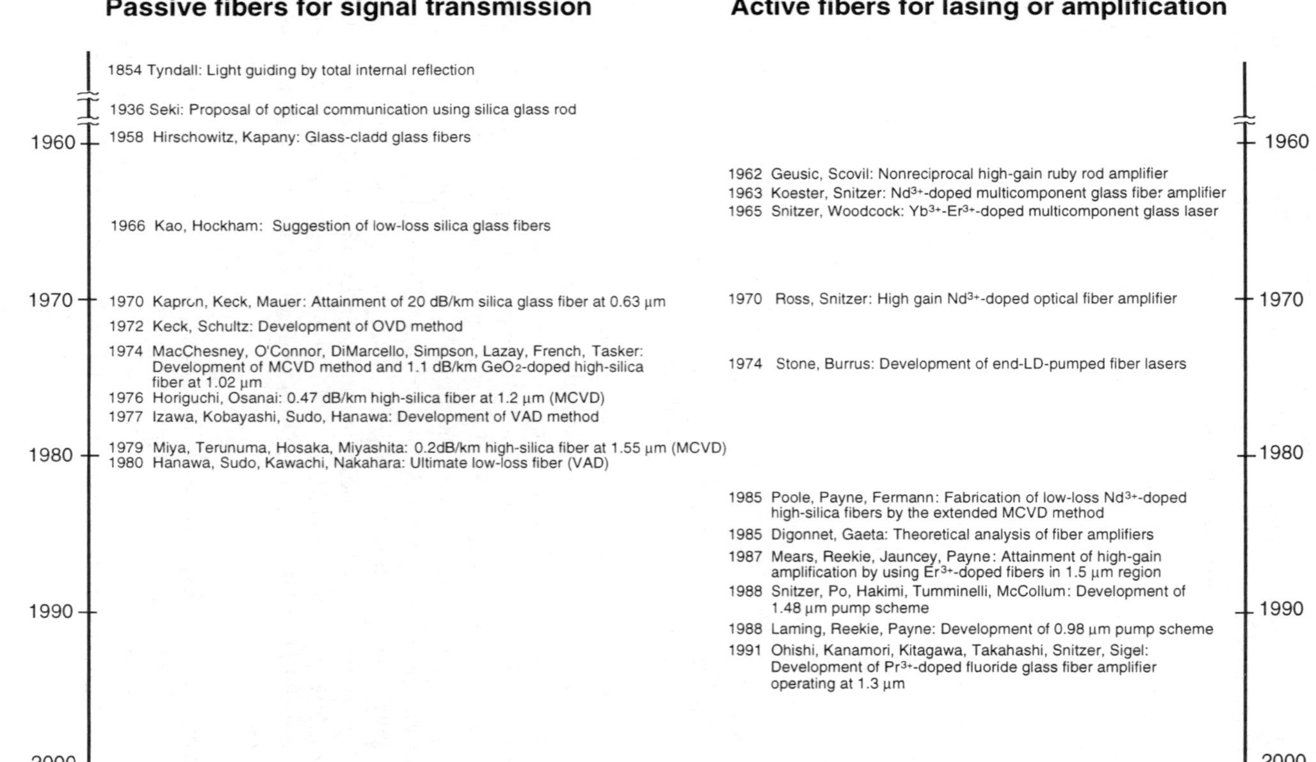

Figure 1.17 Brief summary of the developmental history of passive fibers for signal transmission and active fibers for lasing or amplification.

So far, laser amplifiers have not found many uses nor have they been proposed for any practical communication systems. This is undoubtedly due to the fact that most lasers have little gain; moreover, the more gain they have, the more susceptible they are to instability because of even minor reflections in the transmission path. Until laser amplifiers are invented that incorporate unilateral gain or, even better, have gain in the forward direction and loss in the backward direction, we do not expect them to play an important part in communications systems.

This opinion, which illustrates the little attention being paid to amplifiers, was a reflection of the then-current state of research where the main effort had shifted and was focused on developing low-loss fibers and highly reliable laser diodes. This is in part due to the poor characteristics of optical amplifiers at that time, but this is quite unlike the opinion of optical amplifiers that prevailed in the early 1960s. For example, in 1962, J. E. Geusic and H. E. D. Scovil, who made the first attempt to construct an optical amplifier with a ruby, stated that [272]:

> Since at optical frequencies the maser is the only available amplifier which at present preserves amplitude and phase information, it may well have to provide extremely high stable gains of between 30 dB and 60 dB in many applications, as for example in an optical communications system.

This pioneering statement agrees with that made by Schawlow in 1963 [273], because both statements show the authors' great expectations for optical communication systems and point the way to current amplifier development.

The 1970s saw the development of both low-loss, long-length, high-silica fibers and highly reliable semiconductor laser diodes with a lifetime of longer than a million hours. Consequently, the poor amplification characteristics and lack of technical breakthroughs moved research away from active fibers. This situation is clearly illustrated by the fact that there were few publications and few important advances in the field of active fibers between the early 1970s and the mid-1980s.

However, interestingly enough, in 1985, after the technology for fabricating passive fibers for signal transmission had been well established, the key technologies for fabricating active fibers were developed [151]—15 years after the last attempt at developing fiber amplifiers. However, this 15 years was not lost time for active fiber technology because the technology developed from 1970 to 1980, with regard to low-loss fibers and reliable laser diodes, was not only useful for optical signal transmission but also for the development of efficient fiber amplifiers. Actually, an extension of the well-established MCVD method made it possible to fabricate active fibers with excellent characteristics. In addition, the laser technology, which was developed in the period from 1970 to 1985, enabled the efficient pumping of active ions. Consequently, it can be said that active fibers and fiber amplifiers were reborn in 1985 [151] and in 1987 [127]. Since then, fiber amplifiers, especially *erbium-doped fiber amplifiers* (EDFAs), have become one of the most exciting new developments in the realm of optical communications and promise to revolutionize the field. It is widely said that no single subject has received more attention than erbium-doped fiber amplifiers.

Another interesting aspect of the alternating growth of technology lies in research on fluoride-based glass fiber. As described earlier, despite great research efforts on infrared materials including fluoride glasses, until 1993 almost all research groups had decided to terminate their research programs on the fabrication of ultra-low-loss fibers with fluoride-based glass materials because the lower phonon energy glasses have a weaker bonding energy, resulting in easier crystallization and an increase in scattering losses, although the lower phonon energy glasses were estimated to have a lower infrared absorption loss due to multiphonon vibration. Very fortunately, however, recent investigations have shown that, in principle, host materials with lower phonon energy are more suitable for active ions, in particular, for active ions with intermediate metastable energy levels such as Pr^{3+} ions and Tm^{3+} ions, than host materials with higher phonon energy because material with a lower phonon energy provides a lower multiphonon (nonradiative) relaxation rate resulting in a higher quantum efficiency of the transitions [274–278], as evidenced by the successful development of praseodymium (Pr^{3+})-doped fiber amplifiers operating in the 1.3-μm region [279,280].

Furthermore, it also emerged that fluoride glasses and multicomponent glasses, which were less successful as passive fibers for signal transmission, have several advantages over high-silica glasses as a host glass for active ions. First, these glass systems allow rare-earth ions to be doped into their glass networks at high concentrations because these systems have an ionic bonding structure rather than a covenant bonding structure and, hence, have more sites for rare-earth ion doping [281–286]. Second, this property of having many sites gives a smaller Stark split of the energy levels of rare-earth ions in glasses and, hence, provides a broader fluorescence spectrum than that for high-silica glass hosts [275,276,287–289]. As described in detail in Chapter 2, these features of high-concentration doping and broader fluorescence spectra are useful for practical fiber amplifiers. In fact, broadband EDFAs have been constructed using fluoride glass hosts [290–293] and will play an important role in WDM transmission systems.

In addition to 1.3-μm amplifiers (PDFAs) and broadband 1.5-μm amplifiers (B-EDFAs), which are constructed using a fluoride glass host, other applications of fluoride glass and multicomponent glass hosts include fiber amplifiers that are able to operate in other wavelength regions, such as 0.8, 1.4, and 1.65 μm [294–300]. Fluoride glass has a lower phonon energy than silica glass and, hence, has a smaller number of quenching levels due to multiphonon relaxation than silica glass, which indicates the possibility of operation at many wavelengths.

Finally, it is useful to briefly compare the historical progress made on optical and electronic devices, which have both been developed using the material theory based on quantum mechanics and quantum electronics. The historical progress made on optical fibers and electronic active devices is selected here to represent optical and electronic devices in general. Figure 1.18 shows the reduction in fiber loss as a function of time, which clearly indicates the historical progress of fiber technology [252]. In the course of their developmental history, optical glasses have shown an approximate 10^{-8} reduction in optical loss. Optical fiber, in particular, has shown a 10^{-4} reduction in transmission loss in only the last three decades. This progress is quite amazing and is far greater than almost all predictions. At the same time, the advent of low-loss fibers has initiated numerous changes in communication systems, networks, and equipments. Such revolutionary changes continue to be made.

Electronic devices, especially active elements such as transistors, integrated circuits, and large-scale integration have shown equal or even more remarkable

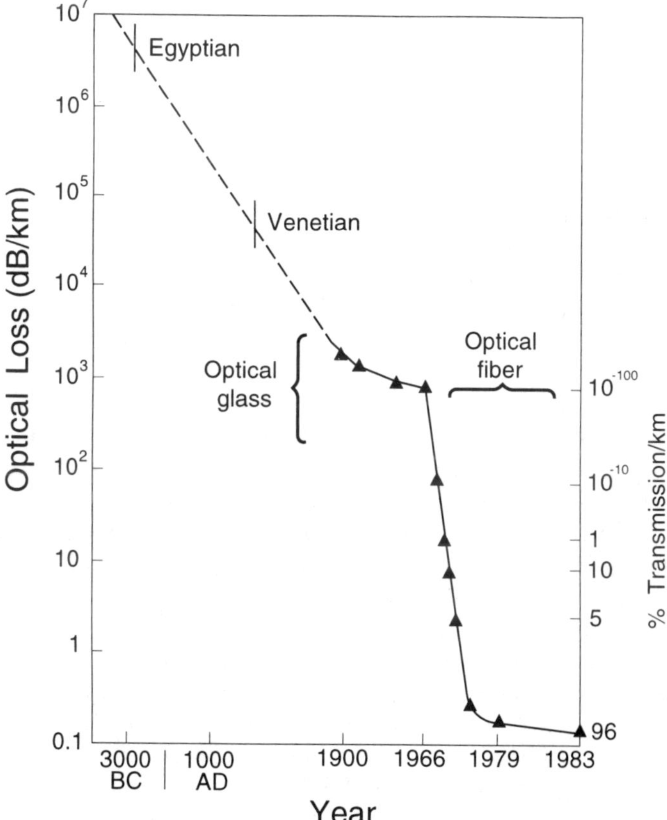

Figure 1.18 Reduction in fiber loss as a function of time, which clearly indicates the historical progress of fiber technology [252].

progress. Figure 1.19 depicts the progress in electronic devices using as an example the size of one active element from a vacuum tube to a *very large scale integration* (VLSI) over about a hundred years. In 1904 J. A. Fleming invented a diode vacuum tube (''valve'') that operated as a rectifier, and in 1906 Lee de Forest invented a triode vacuum tube (''audion'') for signal amplification. The size of these vacuum tubes ranged from several tens of centimeters for the largest to less than a centimeter for the smallest. On December 23, 1947, the great breakthrough came when J. Bardeen and W. H. Brattain demonstrated the operation of the point-contact transistor to the leaders of the research department at Bell Laboratories, as described in Brattain's famous notebook entry for Christmas Eve 1947 [301,302]. The device size was about 1 mm, although an emitter and collector were placed in close proximity (separation ~ 0.005 to 0.025 cm) on the top surface of a germanium block. It is widely recognized that the modern concept of an integrated circuit was

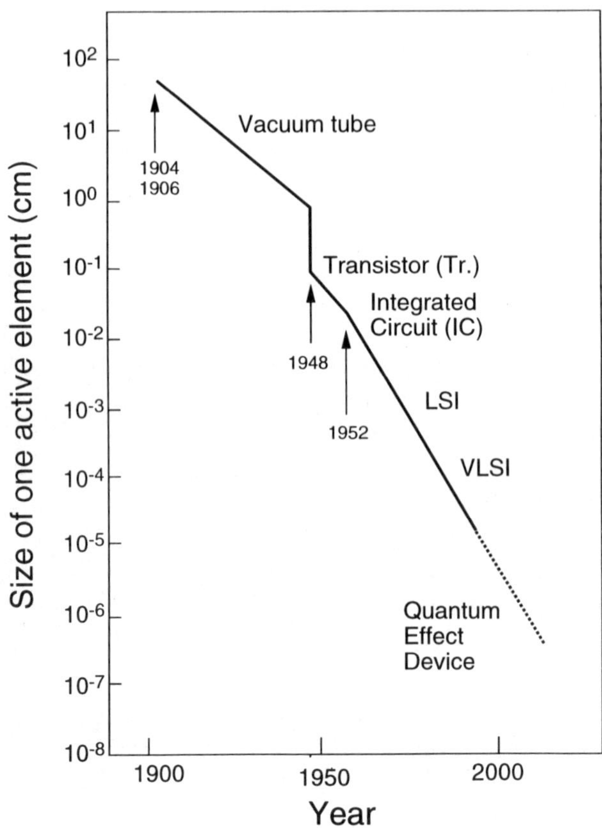

Figure 1.19 Progress on electronic devices using as an example the size of one active element from a vacuum tube to a VLSI over about a hundred years.

proposed in 1952, when G. W. A. Dummer proposed the idea in the closing paragraphs of an invited paper on radar component reliability presented at the annual electronic components symposium in Washington, DC [303]. This proposal initiated the work on integrated circuits, which led to the development of the VLSI. Of course, there were many important inventions and numerous investigations before VLSIs were developed [303]. In the course of the developmental history from the vacuum tube to the VLSI, active electronic elements have achieved an approximate reduction of 10^{-7} in size. In particular, semiconductor devices have achieved a 10^{-4} reduction in size in just the last four decades, as shown in Figure 1.19.

This progress in electronic devices is also quite outstanding and has outstripped almost all predictions. At the same time, the advent of transistors, ICs, LSIs, and VLSIs has led to the development of many familiar objects, such as televisions, computers, and radios, over the last several decades; and these electronic devices have wrought remarkable changes in our lives as well as in industry. Such revolutionary changes continue to be made today. As mentioned in the previous sections, these electronic devices could not have been achieved without quantum mechanics. Therefore, it can be clearly said that the successful consequences of quantum mechanics are all around us. In this sense, quantum mechanics is undoubtedly the most successful theory yet devised by the human mind.

To conclude this chapter, we list the keys to the achievement of low-loss fibers.

The first key was to use high-silica glasses as fiber materials. High-silica glasses are composed of SiO_2, GeO_2, B_2O_3, and P_2O_5, for example, and the SiO_2 content is about 80% or more. The important feature of high-silica glasses is glass uniformity because it leads to small scattering losses.

The second key was to develop the core/cladding structure, especially one with a small refractive-index difference. This structure is commonly called "weakly waveguiding," which is the key to achieving optical fibers with low-loss characteristics as it minimizes the scattering loss of the core/cladding boundary.

The third key was to use the vapor phase reaction process to synthesize high-silica glasses, especially from raw glass materials such as $SiCl_4$, $GeCl_4$, BCl_3, and $POCl_2$, which are all capable of purification via an evaporation process. The evaporation purification process makes possible achieving a significant reduction in transition metal contamination in the raw glass material. The vapor phase reaction can be used to synthesize high-silica glasses without transition metal contamination because this reaction does not require a crucible in which to synthesize the glass materials.

The fourth key was to apply the soot process to fiber preform fabrication, which enabled the GeO_2 to be doped into SiO_2, which slightly increased the refractive-index of SiO_2 glass, thus forming a weakly waveguiding structure in high-silica fibers. GeO_2 is the best material for effecting this slight change in the refractive-index of SiO_2 while minimizing the scattering loss increase.

These important developments have enabled us to prepare low-loss fibers from high-silica glass with good material uniformity, a smooth core/cladding boundary with a small refractive-index difference provided by GeO_2 doping, and very high purity glass material with an extremely low level of contamination by transition metals.

Finally, the keys to the construction of efficient amplifiers were as follows:

1. The first key was to use efficient transitions of rare-earth ions such as erbium (Er^{3+}), that is, high quantum efficiency with small or no nonradiative transition probability.
2. The second key was to use laser light with a very narrow spectral width as the pumping source, leading to very efficient pumping for the narrow absorption band of rare-earth ions.
3. The third key was to find energy levels and wavelengths for efficient pumping such as 1.48 and 0.98 μm.
4. The fourth key was to develop a fabrication method for low-loss, high-NA, and rare-earth-doped fibers by extending the well-established methods used for fabricating high-silica fibers.

References

[1] Einstein, A., "Johannes Kepler," *Ideas and Opinions*, New York: Crown, 1954, p. 266.

[2] Maiman, T. H., "Optical and Microwave-Optical Experiments in Ruby," *Phys. Rev. Lett.*, Vol. 4, 1960, pp. 564–566.

[3] Maiman, T. H., "Stimulated Optical Radiation in Ruby," *Nature*, Vol. 187, 1960, pp. 493–494.

[4] Sir Isaac Newton, *Opticks*, New York: Dover, 1952.

[5] Wolfgang von Goethe, Johann (transl. by L. Eastlake), *Theory of Colors*, Cambridge, MA: MIT Press, 1994.

[6] Maxwell, J. C., *A Treatise on Electricity and Magnetism, Vol. 1, Vol. 2*, New York: Dover, 1954.

[7] Einstein, A., "Zur Quantentheorie der Strahlung," *Phys. Z.*, Vol. 18, March 1917, pp. 121–128.

[8] The first publication is *Mitteilungen der Physikalischen Gesellschaft Zurich*, Nr. 18, 1916.

[9] Gordon, J. P., H. J. Zeiger, and C. H. Townes, "Molecular Microwave Oscillator and New Hyperfine Structure in the Microwave Spectrum of NH_3," *Phys. Rev.*, Vol. 95, 1954, pp. 282–284.

[10] Basov, G., and A. M. Prokhorov, JETP USSR, Vol. 27, 1954, pp. 431–438; Doklady Akad. Nauk USSR, Vol. 101, 1954, p. 47.

[11] Purcell, E. M., and R. V. Pound, "A Nuclear Spin System at Negative Temperature," *Phys. Rev.*, Vol. 81, 1951, pp. 279–280.

[12] Townes, C. H., Columbia Radiation Lab. Quart. Progr. Rep., Dec. 1951.

[13] Schawlow, A. L., and C. H. Townes, "Infrared and Optical Maser," *Phys. Rev.*, Vol. 112, 1958, pp. 1940–1949.

[14] Brombery, J. L., "The Birth of the Laser," *Phys. Today*, Oct. 1988, pp. 26–33.

[15] Storokin, P. P., and J. Stevenson, "Stimulated Infrared Emission from Trivalent Uranium," *Phys. Rev. Lett.*, Vol. 5, Dec. 1960, pp. 557–559.

[16] Maiman, T. H., "Stimulated Optical Emission in Fluorescent Solids. I. Theoretical Consideration," *Phys. Rev.*, Vol. 123, 1961, pp. 1145–1150.

[17] Maiman, T. H., R. H. Hoskins, I. J. D'Haenens, C. H. Asawa, and V. Evtuhov, "Stimulated Optical Emission in Fluorescent Solids. II. Spectroscopy and Stimulated Emission in Ruby," *Phys. Rev.*, Vol. 123, 1961, pp. 1151–1157.

[18] Schawlow, A. L., "Advances in Optical Masers," *Sci. American*, Vol. 209, 1963, pp. 34–45.

[19] Yariv, A., and J. P. Gordon, "The Laser," *Proc. IEEE*, Jan. 1963, pp. 4–29.

[20] Marcuse, D., *Engineering Quantum Electrodynamics*, Harcourt, Brace & World, Inc., 1970.

[21] Saleh, B. E. A., and M. C. Teich, *Fundamentals of Photonics*, New York: John Wiley & Sons, Inc., 1991.

[22] Copernicus, Nicholas (transl. and commentary by Edward Rosen), *On the Revolutions*, Baltimore: The Johns Hopkins University Press, 1992.

[23] Sir Isaac Newton (Motte's translation, revised by F. Cajori), *Principia: Mathematical Principles of Natural Philosophy and His Systems of the World*, Berkeley, CA: University of California Press, 1962.

[24] Galilei, Galileo, *Dialogue Concerning the Two Chief World Systems—Ptolemaic & Copernican* (translated by Stillman Drake, foreword by Albert Einstein), Berkeley, CA: University of California Press, 1967.

[25] Einstein, A., "Zur Elektrodynamik bewegter Korper," *Ann. Physik*, Vol. 17, 1905, pp. 891–921.

[26] Einstein, A., "Ist die Tragheit eines Korpers von seinem Energieinhalt abhangig?" *Ann. Physik*, Vol. 18, 1905, pp. 639–641.

[27] Einstein, A., H. A. Lorentz, H. Weyl, and H. Minkowski, *The Principle of Relativity*, New York: Dover, 1952.

[28] Planck, M., "Ueber das Gesetz der Energieverteilung im Normalspectrum," *Ann. Physik*, Vol. 4, 1901, pp. 553–563.

[29] Rutherford, E., "The Scattering of a and b Particles by Matter and the Structure of the Atom," *Phil. Mag.*, Vol. 21, 1911, pp. 669–688.

[30] Bohr, N., "On the Constitution of Atoms and Molecules," *Phil. Mag.*, Vol. 26, 1913, pp. 1–25.

[31] Bohr, N., "On the Constitution of Atoms and Molecules: Part II-Systems Containing Only a Single Nucleus," *Phil. Mag.*, Vol. 26, 1913, pp. 476–503.

[32] Bohr, N., "On the Constitution of Atoms and Molecules: Part III-Systems Containing Several Nucleii," *Phil. Mag.*, Vol. 26, 1913, pp. 857–875.

[33] Bohr, N., *Atomic Theory and the Description of Nature*, Cambridge, England: Cambridge University Press, 1934.

[34] Heisenberg, W., "Uber quantentheoretische Umdeutung kinematischer und mechanischer Beziehungen," *Z. Physik*, Vol. 33, 1925, pp. 879–893.

[35] Heisenberg, W., "Uber den anschaulichen Inhalt der Quantentheoretischen Kinematik und Mechanik," *Z. Physik*, Vol. 43, 1927, pp. 172–198.

[36] Schrödinger, E., "Quantisierung als Eigenwertproblem," *Ann. Physik*, Vol. 79, 1926, pp. 361–376.

[37] Schrödinger, E., "Quantisierung als Eigenwertproblem," *Ann. Physik*, Vol. 79, 1926, pp. 489–526.

[38] Schrödinger, E., "Uber das Verhatnis der Heisenberg-Born-Jordanschen Quantenmechnik zu der meinen," *Ann. Physik*, Vol. 79, 1926, pp. 734–756.

[39] Schrödinger, E., "Quantisierung als Eigenwertproblem," *Ann. Physik*, Vol. 80, 1926, pp. 437–490.

[40] Schrödinger, E., "Quantisierung als Eigenwertproblem," *Ann. Physik*, Vol. 81, 1926, pp. 109–139.

[41] Heisenberg, W. (transl. by C. Eckart and F. C. Hoyt), *The Principles of the Quantum Theory*, New York: Dover, 1949.

[42] Selleri, Franco, *Quantum Paradoxes and Physical Reality*, Dordrecht: Kluwer Academic Publishers, 1990.

[43] Peat, F. David, *Einstein's Moon: Bell's Theorem and the Curious Quest for Quantum Reality*, Chicago: Contemporary Books, 1990.

[44] Bell, J. S., *Speakable and Unspeakable in Quantum Mechanics*, Cambridge: Cambridge University Press, 1993.

[45] Whitaker, A., *Einstein, Bohr and the Quantum Dilemma*, Cambridge, England: Cambridge University Press, 1996.

[46] Herbert, N., *Quantum Reality*, New York: John Brockman, 1985.

[47] Sachs, M., *Einstein versus Bohr: The Continuing Controversies in Physics*, Chicago: Open Court, 1988.

[48] Gell-mann, M., *The Quark and the Jaguar: Adventures in the Simple and the Complex*, London: Abacus, 1994.

[49] Einstein, A., B. Podolsky, and N. Rosen, "Can Quantum-Mechanics Description of Reality Be Considered Complete?" *Phys. Rev.*, Vol. 47, 1935, pp. 777–780.

[50] Schrödinger, E., "Die Gegenwartige Situation in der Quantenmechanik," *Naturwissenschaften*, Vol. 23, 1935, pp. 807–812.

[51] Schrödinger, E., "Die Gegenwartige Situation in der Quantenmechanik," *Naturwissenschaften*, Vol. 23, 1935, pp. 823–828.

[52] Schrödinger, E., "Die Gegenwartige Situation in der Quantenmechanik," *Naturwissenschaften*, Vol. 23, 1935, pp. 844–849.

[53] Schrödinger, E., "Discussion of Probability Relations between Separated Systems," *Proc. Cambridge Philos. Soc.*, Vol. 31, 1935, pp. 555–563.

[54] Schrödinger, E., "Probability Relations between Separated Systems," *Proc. Cambridge Philos. Soc.*, Vol. 32, 1936, pp. 446–452.

[55] Schrödinger, E. (transl. by J. D. Trimmer), "The Present Situation in Quantum Mechnics: A Translation of Schrödinger's 'CAT PARADOX' Paper," *Proc. Amer. Phil. Soc.*, Vol. 124, 1980, pp. 323–338.

[56] Bohr, N., "Quantum Mechnics and Physical Reality," *Nature*, Vol. 136, 1935, p. 65.

[57] Bohr, N., "Can Quantum-Mechnical Description of Physical Reality be Considered Complete?" *Phys. Rev.*, Vol. 48, 1935, pp. 696–702.

[58] Schrödinger, E., "Discussion of Probability Relation betweeen Separated System," *Proc. Cambridge Philos. Soc.*, Vol. 31, 1935, pp. 555–562.

[59] Zukav, G., *The Dancing Wu Li Masters*, New York: Quill William Morrow, 1979.

[60] Einstein, A., "Autobiographical Notes," in *Albert Einstein, Philosopher-Scientist*, P. A. Schilpp (ed.), New York: Open Court, 1995, p. 85.

[61] Bell, J. S., "On the Einstein Podolsky Rosen Paradox," *Physics*, Vol. 1, 1964, pp. 195–200.

[62] Clauser, J. F., M. A. Horne, A. Shimony, and R. H. Holt, "Proposed Experiment to Test Local Hidden-Variable Theories," *Phys. Rev. Lett.*, Vol. 23, 1969, pp. 880–884.

[63] Freedman, S. J., and J. F. Clauser, "Experimental Test of Local Hidden-Variable Theories," *Phys. Rev. Lett.*, Vol. 28, 1972, pp. 938–941.

[64] Aspect, A., P. Grangier, and G. Roger, "Experimental Tests of Realistic Local Theories via Bell's Theorem," *Phys. Rev. Lett.*, Vol. 47, 1981, pp. 460–463.

[65] Strapp, H. P., "Bell's Theorem and World Process," *IL NUOVO CIMENTO*, Vol. 29 B, Oct. 1975, pp. 270–276.

[66] Strapp, H. P., "Are Superluminal Connections Necessary?" *IL NUOVO CIMENTO*, Vol. 40 B, 1977, pp. 191–205.

[67] Clauser J. F., and A. Shimony, "Bell's Theorem: Experimental Tests and Implications," *Rep. Prog. Phys.*, Vol. 41, 1978, pp. 1881–1927.

[68] Bell, J. S., "On the Impossible Pilot Wave," *Found. Phys.*, Vol. 12, 1982, pp. 989–999.

[69] Bell, J. S., " Against Measurement," *Phys. World*, Vol. 3, 1990, pp. 33–40.

[70] d'Espagnat, B., "Nonseparability and the Tentative Description of Reality," *Phys. Reports*, Vol. 110, 1984, 203–264.

[71] Bohm, D., *Wholeness and the Implicate Order*, New York: Routledge, 1992.

[72] Bohm, D., and B. J. Hiley, *The Undivided Universe: An Ontological Interpretation of Quantum Theory*, New York: Routledge, 1993.

[73] Hiley, B. J., and F. D. Peat (eds.), *Quantum Implications: Essays in Honour of David Bohm*, London: Routledge and Kegan Paul, 1987.

[74] Bohm, D. J., and B. J. Hiley, "On the Intuitive Understanding of Nonlocality as Implied by Quantum Theory," *Found. Phys.*, Vol. 5, 1975, pp. 93–109.

[75] Bohm, D. J., and B. J. Hiley, "Nonlocality in Quantum Theory Understood in Terms of Einstein's Nonlinear Field Approach," *Found. Phys.*, Vol. 11, 1981, pp. 529–546.

[76] Bohm, D. J., and B. J. Hiley, "I. Non-relativistic Particle Systems," *Phys. Report*, Vol. 144, 1987, pp. 323–348.

[77] Bohm, D. J., and B. J. Hiley, "II. A Causal Interpretation of Quantum Fields," *Phys. Report*, Vol. 144, 1987, pp. 349–375.

[78] Stroke, Henry (ed.), *The Physical Review: The First Hundred Years*, AIP Press, 1995, Chap. 14.

[79] Isham, C. J., *Lectures on Quantum Theory—Mathematical and Structural Foundations*, London: Imperial College Press, 1995.

[80] Selleri, Franco (ed.), *Wave-Particle Duality*, New York: Plenum Press, 1992.

[81] Mehra, J. (ed.), *The Physicist's Conception of Nature*, Dordrecht: D. Reidel Publishing Company, 1987.

[82] Wheeler, J. A., and W. H. Zurek (eds.), *Quantum Theory and Measurement*, Princeton, NJ: Princeton University Press, 1983.

[83] Schwarz, J. H. (ed.), *Superstrings: The First 15 Years of Superstring Theory, Vol. 1, Vol. 2*, Singapore: World Scientific, 1985.

[84] Kaku, M., *Introduction to Superstrings*, New York: Springer-Verlag, 1988.

[85] Kalara, S., and D. V. Nanopoulos (eds.), *Blackholes, Membranes, Wormholes and Superstrings*, Singapore: World Scientific, 1993.

[86] Taylor, J. G., P. C. Bressloff, and A. Restuccia, *Finite Superstrings*, Singapore: World Scientific, 1992.

[87] Schwarz, J. H., "Superstring Theory," *Phys. Report*, Vol. 89, 1982, pp. 223–322.

[88] Waldrop, M. M., "String as a Theory of Everything," *Science*, Vol. 20, 1985, pp. 1251–1253.

[89] Green, M. B., "Unification of Forces and Particles in Superstring Therories," *Nature*, Vol. 314, 1985, pp. 409–414.

[90] Green, N. B., "Superstrings," *Sci. American*, Vol. 255, 1986, pp. 44–56.

[91] Green, N. B., R. Lengo, S. Randibar-Daemi, E. Sezgin, and A. Strominger (eds.), *Superstrings '89: Proceedings of the Trieste Spring School 3–14 April 1989*, Singapore: World Scientific, 1990.

[92] Davies, P. C. W., and J. Brown, *Superstrings: A Theory of Everything?*, Cambridge, England: Cambridge University Press, 1988.

[93] Peat, F. D., *Supersttrings and the Search for the Theory of Everything*, Chicago: Contemporary Books, 1988.

[94] Kaku, M., *Beyond Einstein: The Cosmic Quest for the Theory of the Universe*, Doubleday, NY: Anchor Books, 1987, p. 49.

[95] Everett, H., III, "Relative State Formulation of Quantum Mechanics," *Rev. Mod. Phys.*, Vol. 29, 1957, pp. 454–462.

[96] Dewitt, B. S., and N. Graham (eds.), *The Many-World Interpretation of Quantum Mechanics*, Princeton, NJ: Princeton University Press, 1973.

[97] Griffiths, R. B., "Consistent Histories and the Interpretation of Quantum Mechanics," *J. Stat. Phys.*, Vol. 36, 1984, pp. 219–272.

[98] Griffiths, R. B., "Consistent Interpretation of Quantum Mechanics Using Quantum Trajectories," *Phys. Rev. Lett.*, Vol. 70, 1993, pp. 2201–2204.

[99] Gell-mann, M., and J. B. Hartle, "Quantum Mechanics in the Light of Quantum Cosmology," *Proc. 3rd Int. Symp. Fundations of Quantum Mechanics, Tokyo*, 1989, pp. 321–343.

[100] Bohr, N., *Niels Bohr Collected Works, Vol. 5: The Emergence of Quantum Mechnics, 1924–1926*, Amsterdam: North-Holland.

[101] Bohr, N., *Atomic Theory and the Description of Nature (The Philosophical Writings of Niels Bohr, Vol. I)*, Woodbridge, CT: Ox Bow, 1987.

[102] Bohr, N., *Essays 1932–1957 on Atomic Physics and Human Knowledge (The Philosophical Writings of Niels Bohr, Vol. II)*, Woodbridge, CT: Ox Bow, 1987.

[103] Bohr, N., *Essays 1958–1962 on Atomic Physics and Human Knowledge (The Philosophical Writings of Niels Bohr, Vol. III)*, Woodbridge, CT: Ox Bow, 1987.

[104] Folse, H. J., *The Philosophy of Niels Bohr: The Framework of Complementarity*, Amsterdam: North-Holland, 1985.

[105] Heisenberg, W., *Physics and Philosophy: The Revolution in Modern Science*, New York: Harper Torchbook, 1958.

[106] Heisenberg, W., *Philosophical Problem of Quantum Physics*, Woodbridge, CT: Ox Bow, 1979.

[107] Heisenberg, W., M. Born, E. Schrödinger, and P. Auger (transl. by M. Goodman and J. W. Binns), *On the Modern Physics*, New York: Collier Books, 1962.

[108] Dirac, P. A. M., "The Development of Quantum Mechanics," in *Directions in Physics*, H. Hora and J. R. Shepanski (eds.), Sidney: Wiley, 1976.

[109] Davies, P. C. W. (ed.), *The New Physics*, Cambridge, England: Cambridge Univesity Press, 1989.

[110] Davies, P. C. W., *God and the New Physics*, New York: Penguin Books, 1983.

[111] Wilber, K., *The Holographic Paradigm and Other Paradox*, Boston, MA: Shambhala, 1985.

[112] Lockwood, M., *Mind, Brain, & the Quantum*, Cambridge, MA: Blackwell Publishers Inc., 1989.

[113] Stapp, H. P., *Mind, Matter, and Quantum Mechanics*, New York: Springer-Verlag, 1993.

[114] Peat, F. D., *Synchronicity*, Toronto: Bantam Books, 1987.

[115] Capra, F., *The Tao of Physics: An Exploration of the Parallels Between Modern Physics and Eastern Mysticism*, 3rd ed., Updated, Boston, MA: Shambhala, 1991.

[116] Polkinghorne, J., *Beyond Science*, Cambridge: Cambridge University Press, 1996.

[117] Flowers, B. S. (ed.), *Joseph Campbell: The Power of Myth with Bill Moyers*, New York: Doubleday, 1988, p. 37.

[118] Koester, C. J., and E. Snitzer, "Fiber Laser as a Light Amplifier," *J. Opt. Soc. Amer.*, Vol. 53, No. 4, 1963, p. 515.

[119] Koester, C. J., and E. Snitzer, "Amplification in a Fiber Laser," *Appl. Optics*, Vol. 3, Oct. 1964, pp. 1182–1186.

[120] Snitzer, E., and R. Woodcock, "Yb^{3+}-Er^{3+} Glass Laser," *Appl. Phys. Lett.*, Vol. 6, Feb. 1965, pp. 45–46.

[121] Snitzer, E., "Glass Lasers," *Appl. Optics*, Vol. 5, Oct. 1966, pp. 1487–1499.

[122] Koester, C. J., "9A4-Laser Action by Enhanced Total Internal Reflection," *IEEE J. Quantum Electron.*, Vol. QE-2, Sept. 1966, pp. 580–584.

[123] Holst, G. C., and E. Snitzer, "Detection with a Fiber Laser Pre-Amplifier at 1.06 μm," *IEEE J. Quantum Electron.*, Vol. QE-5, 1969, p. 319.

[124] Holst, G. C., E. Snitzer, and R. Wallace, "High-Coherence High-Power Laser System at 1.0621 μm," *IEEE J. Quantum Electron.*, Vol. QE-6, 1970, p.361.

[125] Ross, B., and E. Snitzer, "Optical Amplification of 1.06-μm $InAs_{1-x}P_x$ Injection-Laser Emission," *IEEE J. Quantum Electron.*, Vol. QE-6, 1970, pp. 361–366.

[126] Snitzer, E., "Lasers and Glass Technology," *Ceramic Bull.*, Vol. 52, 1973, pp. 516–525.

[127] Mears, R. J., L. Reekie, I. M. Jauncey, and D. N. Payne, "High-Gain Rare-Earth-Doped Fiber Amplifier at 1.54 μm," *Tech. Dig. Conf. on Optical Fiber Communication/Int. Conf. Integrated Optics and Optical Fiber Communication (OFC/IOOC'87)*, Reno, Nevada, Feb. 1987, Paper W12.

[128] Mears, R. J., L. Reekie, I. M. Jauncey, and D.N. Payne, "Low-Noise Erbium-Doped Fibre Amplifier Operating at 1.54 μm," *Electron. Lett.*, Vol. 23, Sept. 1987, pp. 1026–1028.

[129] Geusic, J. E., and H. E. D. Scovil, "A Unidirectional Traveling-Wave Optical Maser," *Bell System Tech. J.*, Vol. 41, 1962, pp. 1371–1397.

[130] DeGrasse, R. W., E. O. Schulz-DuBois, and H. E. D. Scovil, "The Three-Level Solid-State Traveling-Wave Maser," *Bell System Tech. J.*, Vol. 38, 1959, pp. 305–334.

[131] Stone, J., and C. A. Burrus, "Neodymium-Doped Fiber Lasers: Room Temperature CW Operation with an Injection Laser Pump," *Appl. Opt.*, Vol. 16, 1974, pp. 1256–1258.

[132] Rosenkrantz, L. J., "GaAs Diode-Pumped Nd:YAG Lasers," *J. Appl. Phys.*, Vol. 43, 1972, pp. 4603–4605.

[133] Chesler, R. B., and D. A. Draegert, "Miniature Diode-Pumped Nd:YAlG Lasers," *Appl. Phys. Lett.*, Vol. 23, 1973, pp. 235–236.

[134] Stone, J., and C. B. Burrus, "Neodymium-Doped Silica Lasers in End-Pumped Fiber Geometry," *Appl. Phys. Lett.*, Vol. 23, 1973, pp. 388–389.

[135] Koester, C. J., and D. A. LaMarre, "Optimizing the Parameters for an End-Pumped Laser," *J. Opt. Soc. Amer.*, Vol. 52, 1962, p. 595.

[136] Burrus, C. A., and J. Stone, "Single-crystal Fiber Optical Devices: a Nd:YAG Fiber Laser," *Appl. Phys. Lett.*, Vol. 26, 1975, pp. 318–320.

[137] Stone, J., C. A. Burrus, and A. G. Dentai, "Nd:YAG Single-Crystal Fiber Laser: Room-Temperature CW Operation Using a Single LED As an End Pump," *Appl. Phys. Lett.*, Vol. 29, 1976, pp. 37–39.

[138] Burrus, C. A., J. Stone, and A. G. Dentai, "Room-Temperature 1.3 μm CW Operation of a Glass-clad Nd:YAG Single-crystal Fiber Laser End Pumped With a Single LED," *Electron. Lett.*, Vol. 12, 1976, pp. 600–602.

[139] Stone, J., and C. A. Burrus, "Self-Contained LED-Pumped Single-Crystal Nd:YAG Fiber Laser," *Fiber and Intergrated Optics*, Vol. 2, 1979, pp. 19–46.

[140] Fejer, M., R. L. Byer, R. Feigelson, and W. Kway, "Growth and Characterization of Single Crystal Refractory Oxide Fibers," *Advances in Infrared Fibers II*, Vol. 320, :SPIE, 1982, pp. 50–55.

[141] Fejer, M. M., J. L. Nightingale, G. A. Magel, and R. L. Byer, "Laser Assisted Growth of Optical Quality Single Crystal Fibers," *Processing of Guided Wave Optoelectronic Materials*, Vol. 460, :SPIE, 1984, pp. 26–32.

[142] Fejer, M. M., J. L. Nightingale, G. M. Magel, and R. L. Byer, "Laser-Heated Minitaure Pedestal Growth Apparatus for Single-Crystal Optical Fibers," *Rev. Sci. Instrum.*, Vol. 55, 1984, pp.1791–1796.

[143] Feigelson, R. S., W. L. Kway, and R. K. Route, "Single-Crystal Fibers by the Laser-Heated Pedestal Growth Method," *Optical Engrg.*, Vol. 24, 1985, pp.1102–1107.

[144] Nightingale, J. L., and R. L. Byer, "A Guided Wave Monolithic Resonator Ruby Fiber Laser," *Opt. Comm.*, Vol. 56, 1985, pp. 41–45.

[145] Digonnet, M. J. F., C. J. Gaeta, and H. J. Shaw, "1.064- and 1.32-μm Nd:YAG Single Crystal Fiber Lasers," *IEEE J. Lightwave Tech.*, Vol. LT-4, 1986, pp. 454–460.

[146] Digonnet, M. J. F., C. J. Gaeta, D. O'Meara, and H. J. Shaw, "Clad Nd:YAG Fibers for Laser Applications," *IEEE J. Lightwave Tech.*, Vol. LT-5, 1987, pp. 642–646.

[147] Digonnet, M. J. F., and C. J. Gaeta, "Theoretical Analysis of Optical Fiber Laser Amplifiers and Oscillators," *Appl. Opt.*, Vol. 24, 1985, pp. 333–342.

[148] Digonnet, M. J. F., "Closed-Form Expressions for the Gain in Three- and Four-Level Laser Fibers," *IEEE J. Quantum Electron.*, Vol. 26, 1990, pp. 1788–1796.

[149] MacChesney, J. B., P. B. O'Connor, F. V. Marcello, J. R. Simpson, and P. D. Lazay, "Preparation of Low Loss Optical Fibers Using Simultaneous Vapor Phase Deposition and Fusion," *Proc. Int. Congress on Glass*, Vol. 6, 1974, pp. 40–45.

[150] French, W. G., J. B. MacChesney, P. B. O'Connor, and G. W. Tasker, "Optical Waveguides with Very Low Losses," *Bell System Tech. J.*, Vol. 53, 1974, pp. 951–954.

[151] Poole, S. B., D. N. Payne, and M. E. Fermann, "Fabrication of Low-Loss Optical Fibres Containing Rare-Earth Ions," *Electron. Lett.* Vol. 21, Aug. 1985, pp. 737–738.

[152] Poole, S. B., D. N. Payne, R. J. Mears, M. E. Fermann, and R. I. Laming, "Fabrication and Characterization of Low-Loss Optical Fibers Containing Rare-Earth Ions," *J. Lightwave Tech.*, Vol. LT-4, July 1986, pp. 870–876.

[153] Snitzer, E., H. Po, F. Hakimi, R. Tumminelli, and B. C. McCollum, "Erbium Fiber Laser Amplifier at 1.55 μm with Pump at 1.49 μm and Yb Sensitized Er Oscillator," *Technol Digest of OFC'88*, 1988, pp. 447–450.

[154] Suyama, M., K. Nakamura, S. Kashiwa, and H. Kuwahara, "14.4-dB Gain of Erbium-Doped Fiber Amplifier Pumped by 1.490-μm Laser Diode," *Tech. Digest of OFC'89*, 1989, paper PD2.

[155] Nakazawa, M., Y. Kimura, and K. Suzuki, "Soliton Amplification and Transmission with an Er^{3+}-Doped Fiber Repeater Pumped by InGaAsP Laser Diodes," *Tech. Digest of OFC'89*, 1989, paper PD6.

[156] Laming, R. I., L. Reekie, D. N. Payne, P. L. Scrivener, F. Fontana, and A. Righetti, "Optimal Pumping of Erbium-Doped-Fibre Optical Amplifiers," *Tech. Digest of ECOC '88*, Part 2, 1988, pp. 25–28.

[157] Laming, R. I., M. C. Farries, P. R. Morkel, L. Reekie, D. N. Payne, P. L. Scrivener, F. Fontana, and A. Righetti, "Efficient Pump Wavelengths of Erbium-Doped Fibre Optical Amplifier," *Electron. Lett.*, Vol. 25, Jan. 1989, pp. 12–14.

[158] Laming, R. I., V. Shah, L. Curtis, R. S. Vodhanel, F. J. Favire, W. L. Barnes, J. D. Minelly, D. P. Bour, and E. J. Tarbox, "Highly Efficient 978 nm Diode Pumped Erbium-Doped Fibre Amplifier with 24 dB Gain," *Tech. Digest of IOOC '89*, July 1989, Vol. 5, paper 20PDA-4.

[159] Uehara, S., M. Horiguchi, T. Takeshita, M. Okayasu, M. Yamada, M. Shimizu, O. Kogure, and K. Oe, "0.98-μm InGaAs Strained Quantum Well Lasers for Erbium-Doped Fiber Optical Amplifiers," *Tech. Digest of IOOC '89*, July 1989, Vol. 5, paper 20PDB-11.

[160] Bell, A. G., "On the Production and Reproduction of Sound by Light," *Proc. Amer. Assoc. Advancement of Science*, Vol. 29, 1881, pp. 115–136.

[161] Bell, A. G., "On Selenium and the Photophone," *The Electrician*, Vol. 5, 1880, pp. 214–220.

[162] Seki, H., "Improvement of Optical Transmission Systems," Japanese Patent No. 125946, 1936.

[163] Ross, D., *Lasers: Light Amplifiers and Oscillators*, New York: Academic Press, 1969.

[164] Gloge, D., "Optical Waveguide Transmission," *Proc. IEEE*, Vol. 58, 1970, pp. 1513–1522.

[165] Miller, S. E., "Optical Communications Research Program," *Science*, Vol. 170, 1970, pp. 685–695.

[166] Kompfner, R., "Optics at Bell Laboratories-Optical Communications," *Appl. Opt.*, Vol. 11, 1972, pp. 2412–2425.

[167] Kompfner, R., "Optical Communications," *Science*, Vol. 150, 1965, pp. 149–155.

[168] Miller, S. E., and L. C. Tillotson, "Optical Transmission Research," *Appl. Opt.*, Vol. 5, 1966, pp. 1538–1549.

[169] Hogg, D. C., "On the Spectrum of Optical Waves Propagated Through the Atmosphere," *Bell System Tech. J.*, Vol. 42, 1963, pp. 2967–2969.

[170] Subramanian, M., and J. A. Collinson, "Modulation of Laser Beams by Atmospheric Turbulence," *Bell System Tech. J.*, Vol. 44, 1965, pp. 543–546.

[171] Hogg, D. C., "Scattering and Attenuation Due to Snow at Optical Wavelengths," *Nature*, Vol. 203, 1964, p. 396.

[172] Chu, T. S., "On the Wavelength Dependence of the Spectrum of Laser Beams Traveling Through the Atmosphere," *Appl. Opt.*, Vol. 6, 1967, pp. 163–164.

[173] Chu, T. S., and D. C. Hogg, "Effect of Precipitation on Propagation at 0.63, 3.5, and 10.6 Microns," *Bell System Tech. J.*, Vol. 47, 1968, pp. 723–759.

[174] Karbowiak, A. E., "Guided Wave Propogation in Submillimetric Region," *Proc. IRE*, Vol. 41, 1958, pp. 1706–1711.

[175] Marcatili, E. A. J., and R. A. Schmeltzer, "Hollow Metallic and Dielectric Wave-Guides for Long-Distance Optical Transmission and Lasers," *Bell System Tech. J.*, Vol. 43, 1964, pp. 1783–1809.

[176] Eaglesfield, C. C., "Optical Pipeline, a Tentative Assessment," *Proc. IEE (London)*, Vol. 109B, 1962, pp. 26–32.

[177] Prochazka, M., J. Pachman, and J. Muzik, "Experimental Investigation of a Pipeline for Optical Communication," *Electron. Lett.*, Vol. 3, 1967, pp. 73–74.

[178] Lamm, H., "Biegsame Optische Gerate," *Z. Instrumentenk*, Vol. 50, 1930, pp. 579–581.

[179] Tyndall, John, "On Some Phenomena Connected with the Motion of Liquids," *Proc. Royal Institution of Great Britain*, Vol. 1, 1854, pp. 446–448.

[180] Hirschowitz, B. I., L. E. Curtiss, C. W. Peters, and H. M. Pollard, "Demonstration of a New Gastroscope, The Fiberscope," *Gastroenterology*, Vol. 35, 1938, pp. 50–53.

[181] Kapany, N. S., "Fiber Optics. VI. Image Quality and Optical Insulation," *J. Opt. Soc. Amer.*, Vol. 49, 1959, pp. 779–787.

[182] Kapany, N. S., *Fiber Optics: Principle and Application*, New York: Academic Press, 1967.

[183] Okoshi, T., *Optical Fibers*, New York: Academic Press, 1982, pp. 10–11.

[184] Kao, K. C., and G. A. Hockham, "Dielectric-fibre Surface Waveguides for Optical Frequencies," *Proc. IEE*, Vol. 113, 1966, pp. 1151–1158.

[185] Gloge, D., "Weakly Guiding Fibers," *Appl. Opt.*, Vol. 10, 1971, pp. 2252–2258.

[186] Kapany, F. P., D. B. Keck, and R. D. Maurer, "Radiation Losses in Glass Optical Waveguides," *Technical Digest of Conference on Trunk Telecommunications by Guided Waves*, 29 September–2 October, 1970, pp. 148–153.

[187] Kapany, F. P., D. B. Keck, and R. D. Maurer, "Radiation Losses in Glass Optical Waveguides," *Appl. Phys. Lett.*, Vol. 17, 1970, pp. 423–425.

[188] Snitzer, E., "Cylindrical Dielectric Waveguide Modes," *J. Opt. Soc. Amer.*, Vol. 51, 1961, pp. 491–498.

[189] Snitzer, E., and H. Osterberg, "Observed Dielectric Waveguide Modes in the Visible Spectrum," *J. Opt. Soc. Amer.*, Vol. 51, 1961, pp. 499–505.

[190] Goubau, G., and J. R. Christian, "A New Waveguide for Millimeter Waves," *Proc. Army Sci. Conf. U. S. Military Acad.*, West Point, New York, 1959, pp. 291–303.

[191] Mink, J. W., "Experimental Investigations with an Iris Beam Waveguide," *IEEE Trans. Microwave Theory Tech.*, Vol. MTT-17, 1969, pp. 48–49.

[192] Goubau, G., and F. Schwering, "On the Guided Propagation of Electromagnetic Wave Beams," *IRE Trans. on Antennas and Propagation*, Vol. AP-9, 1961, pp. 248–256.

[193] Boyd, G. D., and J. P. Gordon, "Confocal Multimode Resonator for Millimeter Through Optical Wavelength Masers," *Bell System Tech. J.*, Vol. 40, 1961, pp. 489–508.

[194] Goubau, G., and J. R. Christian, "Loss Measurements with a Beam Waveguide for Long-Distance Transmission at Optical Frequency," *Proc. IEEE (Correspondence)*, Vol. 52, 1964, p. 1739.

[195] DeLange, O. E., "Losses Suffered by Coherent Light Redirected and Refocused Many Times in an Enclosed Medium," *Bell System Tech. J.*, Vol. 44, 1965, pp. 283–302.

[196] Gloge, D., "Experiments with an Underground Lens Waveguide," *Bell System Tech. J.*, Vol. 46, 1967, pp. 721–735.

[197] Gloge, D., "Pulse Shuttling in a Half-Mile Optical Lens Waveguide," *Bell System Tech. J.*, Vol. 47, 1968, pp. 767–782.

[198] Berreman, D. W., "A Lens or Light Guide Using Convectively Distorted Thermal Gradients in Gases," *Bell System Tech. J.*, Vol. 43, 1964, pp. 1469–1475.

[199] Berreman, D. W., "A Gas Lens Using Unlike Counter-Flowing Gases," *Bell System Tech. J.*, Vol. 43, 1964, pp. 1727–1735.

[200] Marcuse, D., and S. E. Miller, "Analysis of a Tubular Gas Lens," *Bell System Tech. J.*, Vol. 43, 1964, pp. 1759–1782.

[201] Miller, S. E., "Alternating-Gradient Focusing and Related Properties of Conventional Convergent Lens Focusing," *Bell System Tech. J.*, Vol. 43, 1964, pp. 1741–1758.

[202] Beck, A. C., "Gas Mixture Lens Measurements," *Bell System Tech. J.*, Vol. 43, 1964, pp. 1818–1820.

[203] Gloge, D., "Deformation of Gas Lenses by Gravity," *Bell System Tech. J.*, Vol. 46, 1967, pp. 357–365.

[204] Kaiser, P., "An Improved Thermal Gas Lens for Optical Beam Waveguides," *Bell System Tech. J.*, Vol. 49, 1970, pp. 137–153.

[205] Nishizawa, J., and S. Sasaki, Japan Patent Application 39-64040.

[206] Kao, K. C., and G. A. Hockham, "Dielectric Fiber Surface Waveguide for Optical Frequencies," *Proc. IEE*, Vol. 113, July 1966, pp. 1151–1158.

[207] Kao, K. C., and T., W. Davies, "Spectrophotometric Studies of Ultra Low Loss Optical Glasses I: Single Beam Method," *J. Sci. Instrum. (J. Phys. E)* Ser. 2, Vol. 1, 1968, pp. 1063–1068.

[208] Kao, K. C., and T. W. Davies, "Spectrophotometric Studies of Ultra Low Loss Optical Glasses II: Double Beam Method," *J. Sci. Instrum. (J. Phys. E)* Ser. 2, Vol. 2, 1969, pp. 331–335.

[209] Uchida, T., M. Furukawa, I. Kitano, K. Koizumi, and H. Matsamura, "A Light-Focusing Guide," *IEEE J. Quantum Electron.*, Vol. QE-5, 1969, p. 331.

[210] Kapron, F. P., D. B. Keck and R. D. Maurer, "Radiation Losses in Glass Optical Waveguides," *Conf. Trunk Telecommunications by Guided Wave*, 29 September–2 October 1970, pp.148–153.

[211] Keck, D. B., and P. C. Schultz, "Method of Producing Optical Waveguide Fibers," U.S. Patent 3.711.262 (Filed May 11, 1970).

[212] Hyde, J. F., "Method of Making a Transparent Article of Silica," U.S. Patent 2.272.342 (Filed Aug. 27, 1934; Patented Feb. 10, 1942).

[213] Nordberg, M. E., "Glass Having an Expansion Lower than that of Silica," U.S. Patent 2.326.059 (Filed Apr. 22, 1939; Patented Aug. 3, 1943).

[214] Kao, K. C., P. B. Dyott, and A. W. Snyder, "Design and Analysis of an Optical Fiber Waveguide for Communication," *Conf. Trunk Telecommunications by Guided Wave*, 29 September–2 October 1970, pp. 211–217.

[215] Kaiser, P., E. A. J. Marcatili, and S. E. Miller, "A New Optcal Fiber," *Bell System Tech. J.*, Vol. 52, 1973, pp. 265–269.

[216] Miyashita, T., T. Edahiro, S. Takahashi, M. Horiguchi, and K. Masuno, "Eccentric-Core Glass Optical Waveguide," *J. Appl. Phys.*, Vol. 45, 1974, pp. 808–809.

[217] Van Uitert, L. G., D. A. Pinnow, J. C. Williams, T. C. Rich, R. E. Jaeger, and W.H. Grodkiewicz, "Borosilicate Glasses for Fiber Optical Waveguides," *Mat. Res. Bull.*, Vol. 8, 1973, pp. 469–476.

[218] French, W. G., A. D. Pearson, G. W. Tasker, and J. B. MacChesney, "Low-Loss Fused Silica Optical Waveguide with Borosilicate Cladding," *Appl. Phys. Lett.*, Vol. 23, 1973, pp. 338–339.

[219] MacChesney, J. B., R. E. Jaeger, D. A. Pinnow, F. W. Ostermayer, T. C. Rich, and L. G. Van Uitert, "Low-Loss Fused Silica Optical Waveguide with Borosilicate Cladding," *Appl. Phys. Lett.*, Vol. 23, 1973, pp. 340–341.

[220] Schultz, P. C., "Preparation of Very Low Loss Optical Waveguides," *Amer. Ceramics Soc. Bull.*, Vol. 52, April 1973, p. 308.

[221] Payne, D. N., and W. A. Gambling, "New Silica-Based Low-Loss Optical Fiber," *Electron. Lett.*, Vol. 10, 1974, pp. 289–290.

[222] Kobayashi, S., H. Nakagome, N. Shimizu, H. Tsuchiya, and T. Izawa, "Low Loss Optical Glass Fiber with Al_2O_3-SiO_2 Core," *Electron. Lett.*, Vol .10, Oct. 1974, pp. 410–411.

[223] Kobayashi, S., S. Sudo, T. Miyashita, and T. Izawa, "Preparation of Silica Glass Using a CO_2 Laser," *Appl. Opt.*, Vol. 14, 1975, pp. 2817–2818.

[224] Nassau, K., J. W. Shiever, and T. Krause, "Preparation and Properties of Fused Silica Containing Alumina," *J. Amer. Ceramics Soc.*, Vol. 58, Oct. 1975, p. 461.

[225] Audsley, A., and R. K. Bayliss, "The Induced Plasma Torch as a High-Temperature Chemical Reactor, I. Oxidation of Silicon Tetrachloride," *J. Appl. Chem.*, Vol. 19, Feb. 1969, pp. 33–38.

[226] Nassau, K. C., T. C. Rich, and J. W. Schiever, "Low Loss Fused Silica Made by the Plasma Torch," *Appl. Opt.*, Vol. 13, April 1974, pp. 744–745.

[227] Nassau, K., and J. W. Schiever, "Plasma Torch Preparation of High Purity, Low OH Content Fused Silica," *Amer. Ceram. Soc. Bull.*, Vol. 54, 1975, pp. 1004–1011.

[228] MacChesney, J. B., P. B. O'Connor, F. V. DiMarcello, J. R. Simpson, and P. D. Lazay, "Preparation of Low Loss Optical Fibers Using Simultaneous Vapor Phase Deposition and Fusion," *Proc. Int. Congress on Glass*, Vol. 6, 1974, pp. 40–45.

[229] French, W. G., J. B. MacChesney, P. B. O'Connor, and G. W. Tasker, "Optical Waveguides with Very Low Losses," *Bell System Tech. J.*, Vol. 53, May–June 1974, pp. 951–954.

[230] MacChesney, J. B., and P. B. O'Connor, "Optical Fiber Fabrication and Resulting Production," U.S. Patent 4.217.027 (Filed Aug. 29, 1977; Patented Aug. 12, 1980).

[231] MacChesney, J. B., "Materials and Progresses for Preform Fabrication—Modified Chemical Vapor Phase Deposition and Plasma Chemical Vapor Deposition," *Proc. IEEE*, Vol. 68, 1980, pp. 1181–1184.

[232] MacChesney, J. B., P. B. O'Connor, J. R. Simpson, and F. V. DiMarcello, "Multimode Optical Waveguide Having a Vapor Deposition Core of Germania Doped Bolosilicate Glass," *Amer. Ceramic Soc. Bull.*, Vol. 52, July 1973, p. 704.

[233] French, W. G., R. E. Jaeger, J. B. MacChesney, S. R. Nagel, K. Nassau, and A. D. Pearson, "Fiber Preform Preparation," in *Opical Fiber Telecommunication*, S. E. Miller and A. G. Chynoweth (eds.), New York: Academic Press, 1979.

[234] Horiguchi, M., and H. Osanai, "Spectral Losses of Low-OH Content Optical Fibers," *Electron. Lett.* Vol.12, 1976, pp. 310–312.

[235] Miya, T., Y. Terunuma, T. Hosaka, and T. Miyashita, "Ultimate Low-Loss Single-Mode Fiber at 1.55 μm," *Electron. Lett.*, Vol.15, Feb. 1979, pp. 106–108.

[236] Schultz, P. C., "Vapor Phase Materials and Processes for Glass Optical Waveguides," *Fiber Optics*, B. Bendow and S. S. Mitra (eds.), New York and London: Plenum Press, 1979.

[237] Keck, D. B., and P. C. Schultz, "Method of Forming Optical Waveguide Fibers," U.S. Patent, 3.737.292 (Filed Jan. 3, 1972; Patented June 5, 1973).

[238] Schultz, P. C., "Fabrication of Optical Waveguide by the Outside Vapor Deposition Process," *Proc. IEEE*, Vol. 68. 1980, pp. 1187–1190.

[239] Izawa, T., S. Kobayashi, S. Sudo, and F. Hanawa, "Continuous Fabrication of High Silica Fiber Preform," *Technical Digest of IOOC'77* (Tokyo, Japan), July 1977, Paper C1-1.

[240] Izawa, T. Miyashita, and F. Hanawa, "Continuous Optical Fiber Preform Fabrication Method," U.S. Patent, 4.062.665 (Filed Apr. 5, 1977; Patented Dec. 13, 1977).

[241] Izawa, T., S. Sudo, and F. Hanawa, "Continuous Fabrication Process for High Silica Fiber Preforms," *Trans. IECE Japan*, Vol. E62, 1979, pp. 779–785.

[242] Verneuil, A., "Reproduction Artificielle du Rubis par Fusion," *Nature (Paris)* Vol. 32, 1904, p. 77.

[243] Hanawa, F., S. Sudo, M. Kawachi, and M. Nakahara, "Fabrication of Completely OH-Free VAD Fiber," *Electron. Lett.*, Vol. 16, 1981, pp. 699–700.

[244] Kaiser, P., A. R. Tynes, H. W. Astle, A. D. Pearson, W. G. French, R. E. Jaeger, and A. H. Cherin, "Spectral Losses of Unclad Vitreous Silica and Soda-Lime-Silicate Fibers," *J. Opt. Soc. Amer.*, Vol. 63, Sept. 1973, pp. 1141–1148.

[245] Suda, H., S. Shibata, and M. Nakahara, "Double-Flame VAD Process for High-Rate Optical Fiber Preform Fabrication," *Electron. Lett.*, Vol. 21, 1985, pp. 29–30.

[246] Murata, H., "Recent Developments in Vapor Phase Axial Deposition," *IEEE J. Lightwave Tech.*, Vol. LT-4, 1986, pp. 1026–1033.

[247] Vandewoestine, R. V., and A. J. Morrow, "Developments in Optical Waveguide Fabrication by the Outside Vapor Deposition Process," *IEEE J. Lightwave Tech.*, Vol. LT-4, Aug. 1986, pp. 1020–1025.

[248] Croft, T. D., J. E. Ritter, and V. A. Bhagavatula, "Low-Loss Dispersion-Shifted Single-Mode Fiber Manufactured by the OVD Process," *IEEE J. Lightwave Tech.*, Vol. LT-3, Oct. 1985, pp. 931–934.

[249] Jablonowski, D. P., "Fiber Manufacture at AT&T with the MCVD Process," *IEEE J. Lightwave Tech.*, Vol. LT-4, Aug. 1986, pp. 1016–1019.

[250] Lydtin, H., "PCVD: A Technique Suitable for Large-Scale Fabrication of Optical Fibers," *IEEE J. Lightwave Tech.*, Vol. LT-4, Aug. 1986, pp. 1034–1038.

[251] Izawa, T., and S. Sudo, "Optical Fibers: Materials and Fabrication," Dordrecht: D. Reidel Publishing Company, 1987.

[252] Nagel, S. R., "Optical Fiber—the Expanding Medium," *IEEE Comm. Mag.*, Vol. 25, Apr. 1987, pp. 33–43.

[253] Nagel, S. R., "R&D Directions for Optical Fiber," *IEEE LTS.*, Nov. 1992, pp. 26–34.

[254] Shibata, S., M. Horiguchi, K. Jinguji, S. Mitachi, T. Kanamori, and T. Manabe, "Prediction of Loss Minima in Infra-Red Optical Fibers," *Electron. Lett.*, Vol. 17, 1981, pp. 775–777.

[255] France, P. W., S. F. Carter, M. W. Moore, and C. R. Day, "Progress in Fluoride Fibers for Optical Communications," *Br. Telecom. Tech. J.*, Vol. 5, 1987, pp. 28–44.

[256] France, P. W., M. G. Drexhage, J. M. Parker, M. W. Moore, S. F. Carter, and J. V. Wright, *Fluoride Glass Optical Fibers*, London: Blackie, 1990, p. 65.

[257] Pinnow, D. A., A. L. Gentile, A. G. Standlee, and A. J. Timper, "Polycrystalline Fiber Optical Waveguides for Infrared Transmission," *Appl. Phys. Lett.*, Vol. 33, 1978, pp. 28–29.

[258] Van Uitert, L. G., and S. H. Wemple, "$ZnCl_2$ Glass: A Potential Ultralow-Loss Optical Fiber Material," *Appl. Phys. Lett.*, Vol. 33, 1978, pp. 57–59.

[259] Goodman, C. H. L., "Devices and Materials for 4 μm-Band Fiber-Optical Communication," *IEEE J. Solid-State and Electron Devices*, Vol. 2, Sept. 1978, pp. 129–137.

[260] Kanamori, T., and S. Sakaguchi, "Preparation of Elevated NA Fluoride Optical Fibers," *Japan. J. Appl. Phys.*, Vol. 25, 1986, pp. L468–470.

[261] Szebesta, D., S. T. Davey, J. R. Williams, and M. W. Moore, "OH Absorption in the Low Loss Window of ZBLAN(P) Glass Fiber," *Tech. Digest of 8th Int. Symp. Halide Glass* (Sept. 22–24, 1992, France), 1992, pp. 26–32.

[262] Szebesta, D., S. T. Davey, J. R. Williams, and M. W. Moore, "OH Absorption in the Low Loss Window of ZBLAN(P) Glass Fiber," *J. Non-Cryst. Solids*, Vol. 161, 1993, pp. 18–22.

[263] Fujiura, K., Y. Ohishi, and S. Takahashi, "New Process for Fluoride Glass Preform Fabrication Using Chemical Vapor Deposition," *Japan. J. Appl. Phys.*, Vol. 28, 1989, pp. L2236–L2238.

[264] Fujiura, K., Y. Nishida, H. Sato, S. Sugawara, K. Kobayashi, Y. Terunuma, and S. Takahashi, "Plasma-Enhanced Chemical Vapor Deposition of ZrF_4-Based Fluoride Glasses," *Tech. Digest of 8th Int. Symp. Halide Glass* (Sept. 22–24, 1992, France), 1992, pp. 16–20.

[265] Fujiura, K., Y. Nishida, H. Sato, S. Sugawara, K. Kobayashi, Y. Terunuma, and S. Takahashi, "Plasma-Enhanced Chemical Vapor Deposition of ZrF_4-Based Fluoride Glasses," *J. Non-Cryst. Solids*, Vol. 161, 1993, pp. 14–17.

[266] Day, C. R., P. W. France, S. F. Carter, M. W. Moore, and J. R. Williams, "Fluoride Fibers for Optical Transmission," *Opt. Quantum Electron.*, Vol. 22, 1990, pp. 259–277.

[267] Aggarwal, I. D., and G. Lu, *Fluoride Glass Fiber Optics*, New York: Academic Press, 1991.

[268] Izawa, T., and S. Sudo, *Optical Fibers: Materials and Fabrication*, Dordrecht: D. Reidel Publishing Company, 1987.

[269] Hayashi, I., P. B. Panish, P. W. Foy, and S. Sumski, "Junction Lasers Which Operate Continuously at Room Temperature," *Appl. Phys. Lett.*, Vol. 17, 1970, pp. 109–110.

[270] Ross, D., *Lasers: Light Amplifiers and Oscillators*, New York: Academic Press, 1969, p. 68.

[271] Kompfner, R., "Optics at Bell Laboratories-Optical Communications," *Appl. Opt.*, Vol. 11, 1972, p. 2421.

[272] Geusic, J. E., and H. E. D. Scovil, "A Unidirectional Traveling-Wave Optical Maser," *Bell Syst. Tech. J.*, Vol. 41, 1962, p. 1371.

[273] Schawlow, A. L., "Advances in Optical Masers," *Sci. American*, Vol. 209, 1963, pp. 34–45.

[274] Layne, C. B., W. H. Lowdermilk, and M. J. Weber, "Multiphonon Relaxation of Rare-Earth Ions in Oxide Glasses," *Phys. Rev. B*, Vol. 16, 1977, pp. 10–20.

[275] France, P. W., and M. C. Brierley, "Fluoride Glass Fibre Lasers and Amplifiers," *Fiber Laser Sources and Amplifiers*, Vol. 1171, :SPIE, 1989, pp. 65–71.

[276] Miniscalco, W. J., L. J. Andrews, B. A. Thompson, T. Wei, and B. T. Hall, "The Effect of Glass Composition on the Performance of Er^{3+} Fiber Amplifiers," *Fiber Laser Sources and Amplifiers*, Vol. 1171, :SPIE, 1989, pp. 93–102.

[277] Miniscalco, W. J., "Erbium-Doped Glasses for Fiber Amplifiers at 1500 nm," *IEEE J. Lightwave Tech.*, Vol. 9, Feb. 1991, pp. 234–250.

[278] Monerie, M., "Status of Fluoride Fiber Lasers," *Fiber Laser Sources and Amplifiers III*, Vol. 1581, :SPIE, 1991, pp. 2–13.

[279] Ohishi, Y., T. Kanamori, T. Kitagawa, S. Takahashi, E. Snitzer, and G. H. Sigel, Jr., "Pr^{3+}-doped Fluoride Fiber Amplifier Operating at 1.31 μm," *Tech. Dig. Conf. Optical Fiber Communication (OFC'91)*, 1991, Paper PD2.

[280] Ohishi, Y., T. Kanamori, M. Shimizu, M. Yamada, Y. Terunuma, J. Temmyo, M. Wada, and S. Sudo, "Praseodymium-Doped Fiber Amplifiers at 1.3 μm," *IEICE Trans. Comm.*, Vol. E77-B, Apr. 1994, pp. 421–440.

[281] Lea, K. R., M. J. M. Leask, and W. P. Wolf, "The Raising of Angular Momentum Degeneracy of *f*-Electron Terms By Cubic Crystal Fields," *J. Phys. Chem. Solids.* Vol. 23, 1962, pp. 1381–1405.

[282] Gruber, J. B., J. R. Henderson, M. Muramoto, K. Rajnak, and J. G. Conway, "Energy Levels of Single-Crystal Erbium Oxide," *J. Chem. Phys.*, Vol. 45, July 1966, pp. 477–482.

[283] Robinson, C. C., and J. R. Fournier, "Co-Ordination of Yb^{3+} in Phosphate, Silicate, and Germanate Glasses," *J. Phys. Chem. Solids.* Vol. 31, 1970, pp. 895–904.

[284] Robinson, C. C., "Multiple Sites for Er^{3+} in Alkali Silicate Glasses (I). The Principal Sixfold Coordinated Site of Er^{3+} in Silicate Glass," *J. Non-Crystalline Solids*, Vol. 15, 1974, pp. 1–10.

[285] Robinson, C. C., "Multiple Sites for Er^{3+} in Alkali Silicate Glasses (II). Evidence of Four Sites for Er^{3+}," *J. Non-Crystalline Solids*, Vol. 15, 1974, pp. 11–29.

[286] Weber, M. J., "Fluorescence and Glass Lasers," *J. Non-Crystalline Solids*, Vol. 47, 1982, pp. 117–134.

[287] Arai, K., H. Namikawa, K. Kumata, and T. Honda, "Aluminum or Phosphorus Co-Doping Effects on the Fluorescence and Structural Properties of Neodymium-Doped Silica Glass," *J. Appl. Phys.*, Vol. 59, May 1986, pp. 3430–3436.

[288] Millar, C. A., M. C. Brierley, and P. W. France, "Optical Amplification in an Erbium-Doped Fluorozirconate Fibre Between 1480 nm and 1600 nm," *Tech. Digest of ECOC '88*, 1988, pp. 66–69.

[289] Zemon, S., G. Lambert, W. J. Miniscalco, and L. J. Andrew, "Characterization of Er^{3+}-Doped Glasses by Fluorescence Line Narrowing," *Fiber Laser Sources and Amplifiers*, Vol. 1171, :SPIE, 1989, pp. 219–236.

[290] Clesca, B., D. Ronarc'h, D. Bayart, Y. Sorel, L. Hamon, M. Guibert, J. L. Beylat, J. F. Kerdiles, and M. Semenkoff, "Gain Flatness Comparison Between Erbium-Doped Fluoride and Silica Fiber Amplifiers With Wavelength-Multiplexed Signals," *IEEE Photonics Tech. Lett.*, Vol. 6, Apr. 1994, pp. 509–512.

[291] Clesca, B., D. Bayart, L. Hamon, J. L. Beylat, C. Cœurjolly, and L. Berthelon, "Over 25 nm, 16-Wavelength-Multiplexed Signal Amplification Through Fluoride-Based Fibre-Amplifier Cascade," *Electron. Lett.*, Vol. 30, Mar. 1994, pp. 586–588.

[292] Clesca, B., D. Bayart, C. Cœurjolly, L. Berthelon, L. Hamon, and J. L. Beylat, "Over 25 nm, 16 Wavelength-Multiplexed Signal Amplification Through Four Fluoride-Based Fibre-Amplifier Cascade and 440 km Standard Fiber," *Tech. Digest of OFC '94*, 1994, Paper PD20.

[293] Yamada, M., T. Kanamori, Y. Terunuma, K. Oikawa, M. Shimizu, and S. Sudo, "Fluoride-Based Erbium-Doped Fiber Amplifier with Inherently Flat Gain Spectrum," *IEEE Photon. Tech. Lett.*, Vol. 8, July 1996, pp. 882–884.

[294] Moneric, M., "Status of Fluoride Fiber Lasers," *Tech. Digest of SPIE*, Vol. 1581, Sept. 1991, pp. 2–13.

[295] Dye, S. P., and M. Fake, "Fully Engineered 800-nm Thulium-Doped Fluoride-Fiber Amplifier," *Tech. Digest of OFC '95*, Feb. 1995, Paper WD4.

[296] Sakamoto, T., M. Shimizu, T. Kanamori, Y. Ohishi, Y. Terunuma, M. Yamada, and S. Sudo, "Tm-Ho-Doped ZBLYAN Fiber Amplifiers for 1.4 μm-band," *Tech. Digest of OEC '92*, July 1994, Paper 13B2-2.

[297] Komukai, T., T. Yamamoto, T. Sugawa, and T. Miyajima, "1.47 μm Band Tm^{3+} Doped Fluoride Fiber Amplifier Using a 1.064 μm Upconversion Pumping Scheme," *Electron. Lett.*, Vol. 29, Jan. 1993, pp. 110–112.

[298] Sakamoto, T., M. Shimizu, M. Yamada, T. Kanamori, Y. Ohishi, Y. Terunuma, and S. Sudo, "35 dB Gain Tm-Doped ZBLYAN Fiber Amplifier Operating at 1.65 μm," *Tech. Digest of CLEO/Pacific Rim'95*, July 1995, Paper PD2.12.

[299] Barnes, W. L., and J. E. Townsend, "Highly Tunable and Efficient Diode Pumped Operation of Tm^{3+} Doped Fiber Laser," *Electron. Lett.*, Vol. 26, May 1990, pp. 746–747.

[300] Sankawa, I., H. Izumita, S. Furukawa, and K. Ishihara, "An Optical Fiber Amplifier for Wide-Band Wavelength Range Around 1.65 μm," *IEEE Photon. Tech. Lett.*, Vol. 2, 1990, pp. 422–424.

[301] Bardeen, J., and W. H. Brattain, "The Transistor, A Semi-Conductor Triode," *Phys. Rev.*, Vol. 74, 1948, pp. 230–231.

[302] Schockley, W., "The Path to the Conception of the Junction Transistor," *IEEE Trans. Electron. Devices*, Vol. ED-23, 1976, pp. 597–620.

[303] Wolff, M. F., "The Genesis of the Integrated Circuit," *IEEE Spectrum*, Vol. 13, 1976, pp. 45–53.

CHAPTER 2

Outline of Optical Fiber Amplifiers

S. Sudo

Most of the fundamental ideas of science are essentially simple and may, as a rule, be expressed in a language comprehensible to everyone. To follow up these ideas demands the knowledge of a highly refined technique of investigation. Mathematics as a tool of reasoning is necessary if we wish to draw conclusions which may be compared with experiments [1].

2.1 APPLICATION SYSTEMS AND REQUIREMENTS FOR OPTICAL FIBER AMPLIFIERS

The development of optical fiber amplifiers is having a most exciting effect on the topology of optical communication systems and will allow information technology to be accessed by a greater portion of the population. Figure 2.1 is a diagram of a conventional optical fiber transmission system with electrically regenerative repeaters. In this system, optical signals, which weaken as they are transmitted through fibers, are converted to electrical signals in the repeaters, hence making it possible for the signal to be totally recovered by electronic circuits with 3R functions such as reshaping, retiming, and regeneration. The optical signal regenerated by this modulating semiconductor laser is then relaunched into the fiber for transmission. In general, these electrically regenerative repeaters require high-speed electronics, which become increasingly complex and problematic in terms of equipment size as the bit rate increases. In addition, since the reliability required for systems consisting a multichain repeater is extremely high, this tends to make these repeaters very expensive once they exceed a few gigabits per second.

Figure 2.2 shows three types of transmission systems with fiber amplifiers. Figure 2.2(a) shows a long-distance transmission system with fiber amplifier repeat-

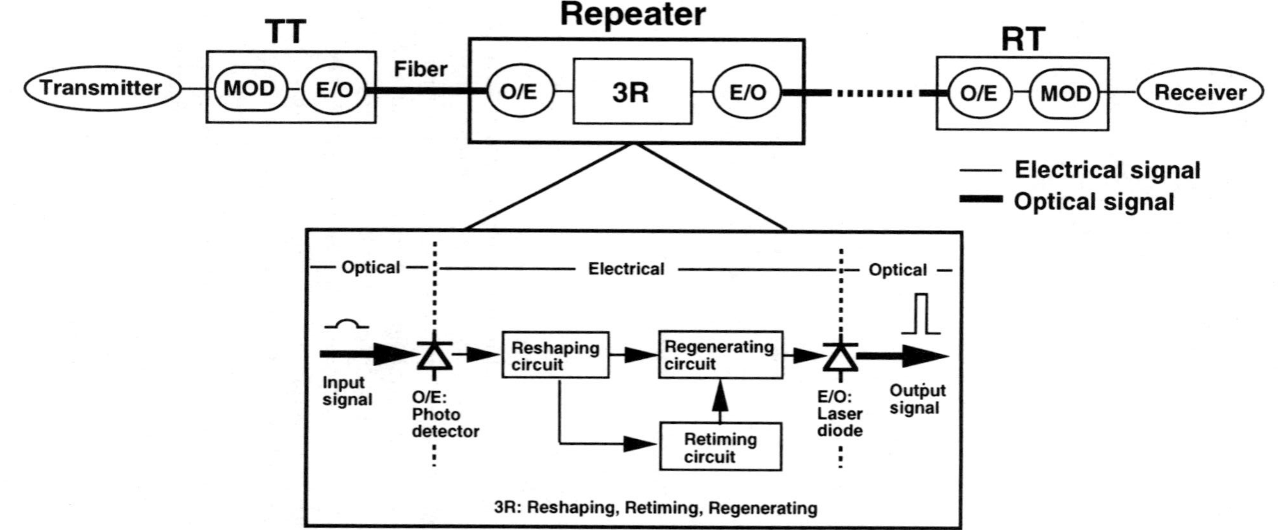

Figure 2.1 An example of conventional optical fiber transmission with electrical repeaters. TT is a transmitter terminal, RT is a receiver terminal, MOD is an electrical signal modulator, and E/O and O/E are electrical-to-optical and optical-to-electrical signal coverters, respectively. 3R indicates repetitive signal regeneration, which involves reshaping, retiming, and regeneration and operates in the electrical domain.

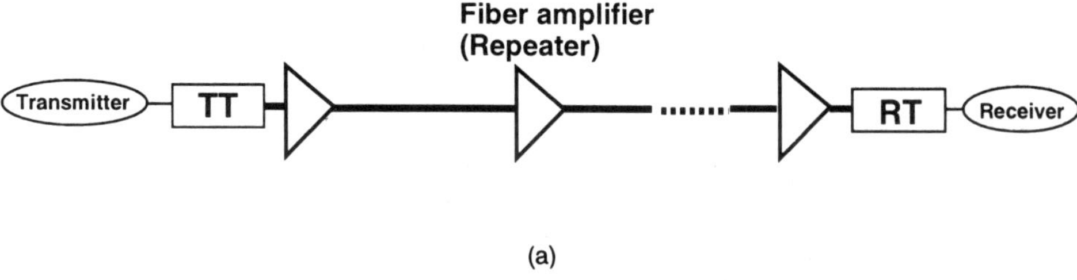

(a)

(b)

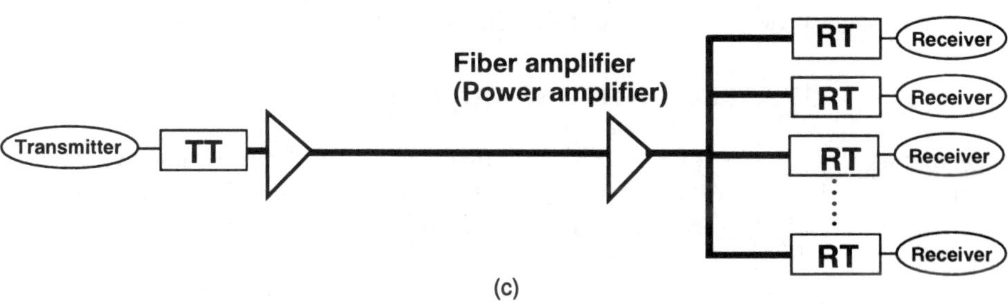

(c)

Figure 2.2 Three types of transmission systems with fiber amplifier: (a) a long-distance transmission system with fiber amplifier repeaters, (b) a long-span transmission using fiber amplifiers as a postamplifier and a preamplifier, and (c) an information distribution system using a fiber amplifier as a power amplifier.

ers, where optical signals transmitting along the line are amplified 10 to 1,000 times in amplitude with fiber amplifiers. Figure. 2.2(b) shows a long-span transmission using fiber amplifiers as a postamplifier and a preamplifier. Here, the optical signals are amplified up to maximum level by a fiber prior to long-span transmission through a long-span fiber. Figure 2.2(c) shows an information distribution system that uses a fiber amplifier as a power amplifier. Here, the visual signals are amplified before being distributed to several arms. In these application systems, the fiber amplifier repeaters do not require high-speed electronics and so the initial development investment is low, whereas electrically regenerative repeaters require high-speed electronics. This is because the fiber amplifier reshapes the transmission signals completely in the optical domain. This means that the optical gain of fiber amplifiers does not change with bit rate (that is, bit rate transparency is achieved) and makes fiber amplifier repeaters compatible with higher-speed systems. This feature allows system manufacturers to use the same fiber amplifier repeaters for different types of systems with different bit rates with no significant modification to the repeaters themselves. In fact, in the three application systems shown in Figure 2.2, the basic requirements for the fiber amplifiers are almost the same except for the specific characteristics required for each system. For example, in the long-distance transmission system shown in Figure 2.2(a), high-gain and low-noise characteristics are important, whereas the high-power characteristics are very important for use in the long-span transmission and information distribution systems shown in Figure 2.2(b,c).

Today, there are a variety of applications for optical amplifiers in addition to the aforementioned standard application systems that force a variety of requirements. In short, these include high gain, high power, low noise, broadband, and high reliability. In addition, low cost and compactness are required. Unfortunately, however, no one amplifier currently meets all these requirements. Nevertheless, many significant advances have been made very rapidly. The details of these advances and key issues are described in the following sections.

2.2 OUTLINE OF RARE-EARTH IONS AND AMPLIFICATION IN FIBERS

Before embarking on a detailed description of technical advances and key issues, we provide a brief description of host glasses, rare-earth ions, and amplification in fibers in order to assist readers to understand the individual technologies of fiber amplifiers.

Therefore, we shall begin by describing glass-forming elements as well as rare-earth ions. Figure 2.3 shows the periodic table designating the rare-earth series, main glass formers, main glass modifiers, and intermediate elements [2,3]. Main glass formers, main glass modifiers, and intermediate elements are designated for oxide glass systems and fluoride glass systems. Of course, other materials will form glasses, for example, chalcogenides, halides, nitrates, and some organic compounds. In this chapter, we restrict ourselves to oxide and fluoride glass systems.

2.2.1 Host Glasses

What is a glass? In general, when a liquid material is cooled down gradually, it will change to crystal, expelling the specific latent heat at a characteristic temperature,

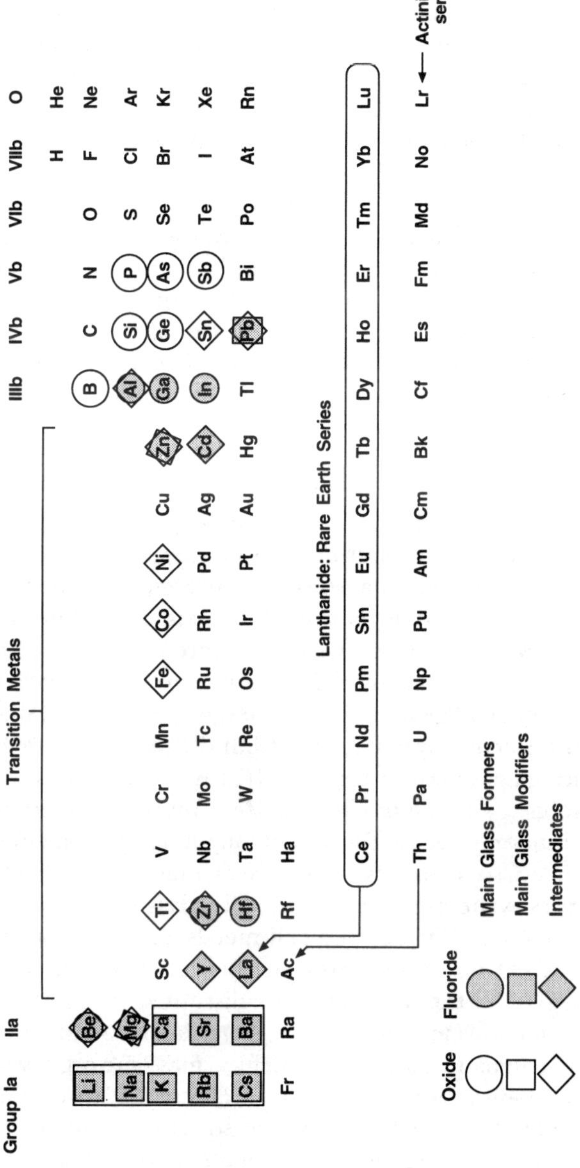

Figure 2.3 Periodic table designating rare-earth series, main glass formers, main glass modifiers, and intermediate elements [2,3].

that is, the melting temperature. However, some special materials do not crystallize below the melting temperature and their viscosity increases to such a high value that they are solid with a disordered arrangement of atoms, quite like that in the liquid state. A super cooled liquid is commonly called glass and, hence, is unstable from a thermodynamic standpoint because the enthalpy of the glassy state is higher than that of a crystal state. Therefore, glasses have a short-range order and form a three-dimensional matrix but lack the uniformity, symmetry, and structure of a crystalline material and have no long-range periodicity. Glass is in a state somewhere between that of a crystalline solid and a liquid.

Some oxides have the ability to form glasses by themselves, called *glass network formers*. SiO_2, GeO_2, B_2O_3, P_2O_5, As_2O_3, and Sb_2O_3 are known as glass network formers. Network formers can form glass not only by themselves but also by mixing with other network formers in any mixing ratio. They are all capable of forming a three-dimensional network with oxygen, thus providing the very strong covalent bonds that give glasses their characteristic properties. The requirements for forming stable oxide glasses were presented by Zachariasen in 1932 [4]. The conditions necessary for forming a glass with energy comparable to that of the crystalline form are as follows: (1) an oxygen atom must not be linked to more than two atoms; (2) the number of oxygen atoms surrounding atoms must be small; (3) the oxygen polyhedra must share corners with each other, not edges or faces; and (4) at least three corners in each oxygen polyhedron must be shared.

Figure 2.4 shows two-dimensional structure analogies for (a) crystal silica and (b) silica glass, where the black dots indicate silicon atoms and the circles indicate oxygen atoms. In reality, silicon atoms have one more bond with an oxygen atom and form a three-dimensional structure. The atomic arrangement of silica glass is thought to be a random arrangement of rigid oxygen tetrahedra (SiO_4^{-4}) that are formed based on the silicon-oxygen covalent bonds. As you can see, the atomic arrangement of silica glass shown in Figure 2.4(b) is very similar to the crystalline form of silica (crystobalite) shown in Figure 2.4(a), but the structure of silica glass, being slightly more random, leads to the lack of long-range periodicity or symmetry.

On the other hand, some oxides, which are commonly called network modifiers, cannot form a glass by themselves but can form a glass as a mixture with network formers to a certain content. Oxides of alkali metals and alkali-earth metals such as Li_2O, Na_2O, K_2O, Rb_2O, Cs_2O, CaO, SrO, and BaO are typical network modifiers. The silicon-oxygen glass network structure is disrupted by the introduction of network modifiers; these ions tend to be from group Ia or IIa of the periodic table. Figure 2.5 shows a schematic diagram of a silica glass matrix disrupted by the addition of network modifier ions. As shown in Figure 2.5, when these elements are added to the oxide glass, the silicon-oxygen structure is opened up, lowering the density of the glass, weakening the bond strength, and lowering both the fusion temperature and the viscosity of the glass. Since each tetrahedron can still be linked to at least three others, the random network will still be present. Some of the silicon-oxygen bridging bonds will have been broken, however, thus increasing the

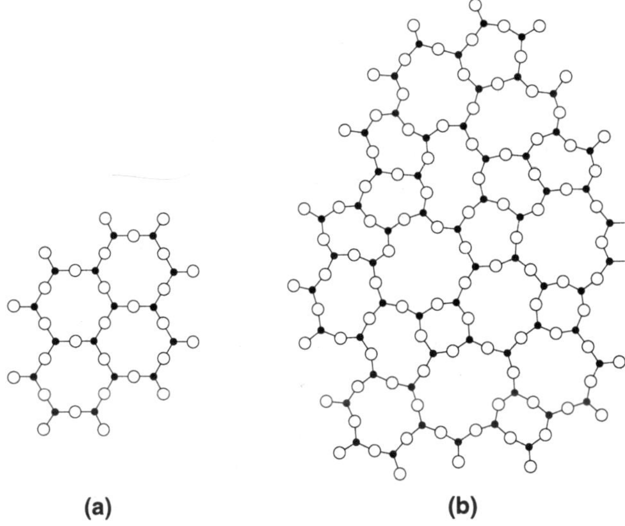

Figure 2.4 Two-dimensional structure analogies for (a) crystal silica and (b) silica glass, where the black dots indicate silicon atoms and the circles indicate oxygen atoms [4].

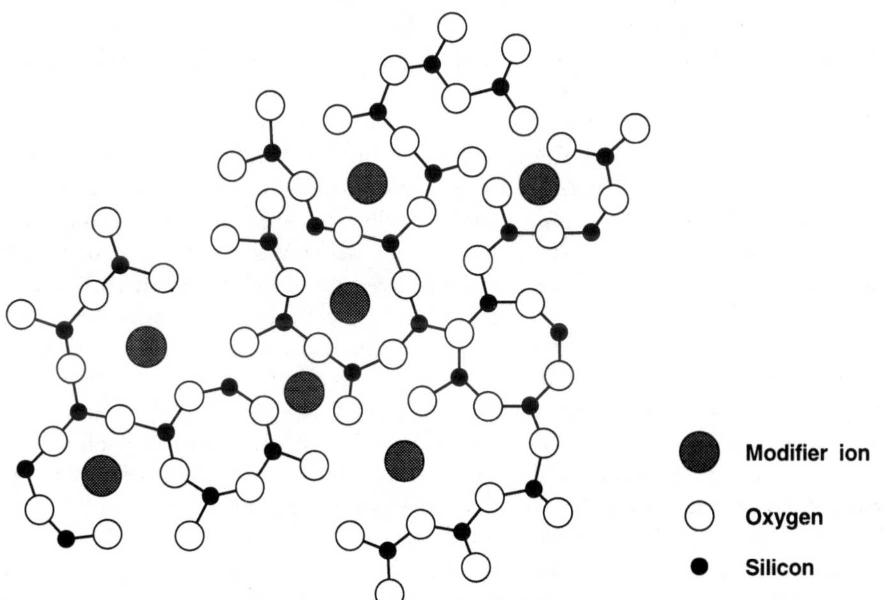

Figure 2.5 Schematic diagram of the silica glass matrix disrupted by the addition of network modifier ions.

number of nonbridging oxygens present; the spare negative charges on the oxygen ions are used to compensate for the positive charge of the metal ions. The addition of network modifiers is important as it allows glass to be processed at workable temperatures. Pure silica requires very high temperatures for working, whereas the chemical durability of glass containing alkali metal oxides is comparatively poor because the alkali ions in the glass dissolve in water.

In addition, TiO_2, SnO_2, Al_2O_3, ZrO_2, Fe_2O_3, BeO, MgO, NiO, ZnO, CoO, FeO, and PbO have characteristics that fall between those of network formers and network modifiers. These are called *intermediate oxides*. Most transition metal oxides in oxide glass introduce strong absorption in the visible and near-infrared wavelength region. Cr ions at a concentration of 1 ppb, for example, introduce a loss of about 1 dB/km at 1.0 to 1.8 μm. Therefore, transition metals should be removed from the fiber materials in order to make low-loss optical fibers. In contrast, fluoride glasses based on zirconium fluoride (ZrF_4) and indium fluoride (InF_3) provide an interesting alternative to oxides glasses as a rare-earth ion host. In the periodic table shown in Figure 2.3, the elements used to form fluoride glasses are designated by the meshed circles and squares. The zirconium fluoride glass system was first discovered at the University of Rennes, in France, in 1974 [5], although its glass structure is not well understood even today. The major problem with understanding the structure of fluoride glasses comes from the lack of a fixed structural unit. However, if the classical concepts of glass networks developed by Zacharisen, previously, are followed, the glass network model can be extended to fluoride glasses [6]. Figure 2.6 shows a two-dimensional schematic representation of a ZBLA glass [6,7]. Here Zr and Al are thought to be network formers, La to be an intermediate (in this case playing the role of a former), and Ba and Na to be network modifiers. The network of connecting fluoride ions is outlined in the diagram, where T indicates a terminal, E indicates an edge, and C indicates a corner. They are all based on the F^- ion, which has a very high electron affinity (3.4 eV) and a high electronegativity (4.0) [8]. With the major glass-forming component of fluoride glasses shown in Figure 2.3, the metal atoms have a low electronegativity; for example, the electronegativity of Zr is 1.4 and that of Al is 1.5 [8]. Therefore, with highly electronegative fluorine, atoms will form bonds that exhibit a high degree of ionicity, indicating that with relatively high atomic number cations, the closed shell configuration of F^- is relatively undeformed and it may be possible to obtain a good representation of heavy-metal fluoride bonding by considering spherically symmetric interatomic interactions. This is not to say that the bonding of heavy-metal fluoride glasses has no covalent characteristics but rather that a good first-order representation of their interatomic potentials can simply be ionic [9]. As described later, this key characteristic of bonding in fluoride glasses provides two important features. One is that metal atoms in fluoride glasses, especially La, can easily be substituted by an alternative rare-earth ion without significantly affecting the properties of the glass. The other is that glasses composed of heavy-metal fluoride exhibit lower fundamental phonon energies, together with the fact that the fluoride ion

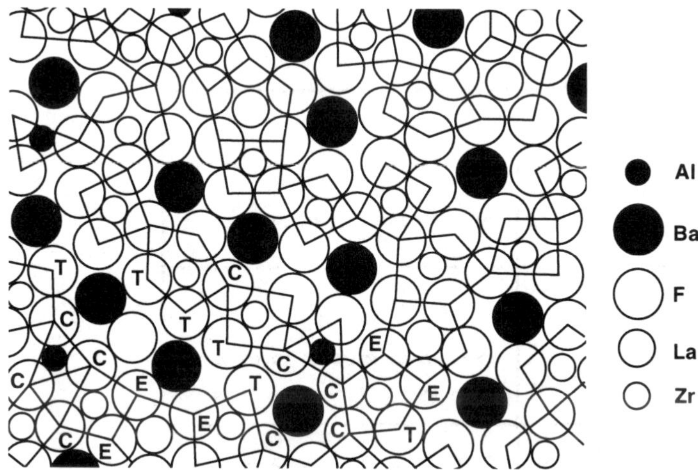

Figure 2.6 Two-dimensional schematic representation of a ZBLA glass [6,7]. Here Zr and Al are thought to be network formers, La to be an intermediate (in this case playing the role of a former), and Ba and Na to be network modifiers. The network of connecting fluoride ions is outlined in the diagram, where T indicates a terminal, E indicates an edge, and C indicates a corner.

is only singly charged. This property leads to lower nonradiative decay rates in fluoride glasses [10] as well as to the lower losses in the infrared regions described in Chapter 1.

2.2.2 Rare-Earth Ions and Their Transitions

As shown in Figure 2.3, the rare-earth elements are a series of 15 transition metals, beginning with lanthanum (atomic number 57) and ending with lutetium (atomic number 71), together with scandium (atomic number 21) and yttrium (atomic number 39). The electronic configuration of all the elements in the series is very similar and is based on the gradual filling of the 4f subshell along the series. Figure 2.7 shows the electronic configuration of rare-earth elements from cerium to ytterbium as neutral atoms that have been lased in crystalline and gassy hosts [11,12]. All the elements have the electronic configuration of $1s^2$, $2s^2$, $2p^6$, $3s^2$, $3p^6$, $3d^{10}$, $4s^2$, $4p^6$, $4d^{10}$, $4f^x$, $5s^2$, $5p^6$, and $6s^2$. These orbits represent the closed shells except for the $4f^x$ level, $x = 2, \ldots, 14$. The trivalent (3+) ionization preferentially removes the electrons of $6s^2$ and $4f^1$, and the electronic configuration for these ions is that of the xenon structure plus a certain number (1–13) of 4f electrons. For example, Ce can be expressed by (Xe) $4f^1$ and Pr is (Xe) $4f^2$. Therefore, the electronic state of these rare-earth ions is determined by the electronic configuration of the $4f^x$ level, and the observed IR and visible optical spectra of trivalent rare-earth ions are a consequence of transitions between 4f states. However, since the

| Atomic Element Number | | K | L | M | N | O | P |
| | | 1 | 2 | 3 | 4 | 5 | 6 |
		s	s p	s p d	s p d f	s p d f	s p d f
58	Ce	2	2 6	2 6 10	2 6 10 2	2 6	2
59	Pr	2	2 6	2 6 10	2 6 10 3	2 6	2
60	Nd	2	2 6	2 6 10	2 6 10 4	2 6	2
61	Pm	2	2 6	2 6 10	2 6 10 5	2 6	2
62	Sm	2	2 6	2 6 10	2 6 10 6	2 6	2
63	Eu	2	2 6	2 6 10	2 6 10 7	2 6	2
64	Gd	2	2 6	2 6 10	2 6 10 9	2 6	2
65	Dy	2	2 6	2 6 10	2 6 10 10	2 6	2
66	Ho	2	2 6	2 6 10	2 6 10 12	2 6	2
69	Tm	2	2 6	2 6 10	2 6 10 13	2 6	2
70	Yb	2	2 6	2 6 10	2 6 10 14	2 6	2

Trivalent ions lose $(6S)^2$ and $(4F)^1$

Figure 2.7 Electronic configuration of rare-earth elements from cerium to ytterbium as neutral atoms [11,12].

electrons of $5s^2$ and $5p^6$ shield $4f^x$ electrons, the effects of the environment on $4f^x$ electron are reduced. Figure 2.8 shows the radial parts of the 5s, 5p, and 4f wavefunctions for Ce^{3+} and also shows the shielding configuration of the 5s- and 5p-level electrons for 4f-level electrons [13].

The electronic energy levels of the rare-earth ions are determined by the electronic configuration of the $4f^x$ level and the effects of the environment on the $4f^x$ electrons, although such effects would not be strong. In general, the Schrödinger equations for many-electron atoms are extremely complicated because all the electrons interact with each other in various ways and there are contributions to the energy levels. In such many-electron atoms, there are mainly two interactions that we consider: (1) coulombic interaction between electrons, including spin correlation, and (2) spin-orbit coupling. The former is an electrostatic interaction, and the latter is a magnetic interaction. Therefore, the Hamiltonian of a many-electron system can be described as

$$H = H_c + H_{el\text{-}el} + H_{so} \tag{2.1}$$

where H_c is the Hamiltonian for the central field, $H_{el\text{-}el}$ is the Hamiltonian for the electrostatic interaction between electrons, and H_{so} is the Hamiltonian for the spin-orbit coupling. In addition, the 4f electron of rare-earth ions in condensed matter experiences interactions and effects from the structural environment in addition

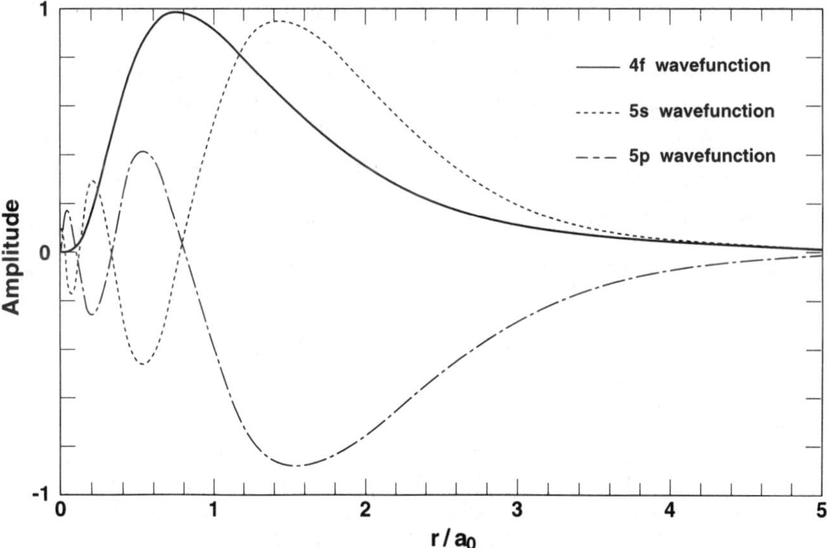

Figure 2.8 Radial parts of the 5s, 5p, and 4f wavefunctions for Ce^{3+} and the shielding configuration of 5s-level electrons and 5p-level electrons for 4f-level electrons [13].

to those with other 4f electrons. Therefore, the Hamiltonian can be written for an individual rare-earth ion and decomposed as

$$H = H_c + H_{el\text{-}el} + H_{so} \qquad \text{[Free ion]}$$
$$+ V_{ion\text{-}static\ lattice} + V_{ion\text{-}dynamic\ lattice} \qquad \text{[Structure]}$$
$$+ V_{electromagnetic} + V_{ion\text{-}ion} \qquad (2.2)$$

Here, $H_{free\ ion}$ is the Hamiltonian of the ion in complete isolation, $V_{ion\text{-}static\ lattice}$ and $V_{ion\text{-}dynamic\ lattice}$ contain the static and dynamic interactions of the ion with the host, $V_{electromagnetic}$ indicates the interaction of the ion with the electromagnetic field, and $V_{ion\text{-}ion}$ describes the interaction between rare-earth ions. Taking these interactions in atoms into account, we can describe the energy levels for rare-earth ions. Figure 2.9 shows a portion of the energy levels for praseodymium ions with two 4f-electrons (f indicates the f-orbit in the central field approximation) except for electrons in closed shells. This electron configuration can be described as (Xe) $4f^2$. The two 4f-electrons in the central field interact electrostatically (coulombic interaction between electron and spin correlation), and as a result their energy level splits into seven levels with different energies such as ^1S, ^3P, ^1D, ^3F, ^1G, ^3H, and ^1I [14]. This does not reflect the actual order of these energy levels, which have to be reordered based on Fund's law. Then, due to spin-orbit coupling (magnetic interaction), the energy levels of ^3H and ^3F that have parallel spin ($\uparrow\uparrow$, $S = 1$) split into three levels such as $^3H_6$, $^3H_5$, $^3H_4$ and $^3F_4$, $^3F_3$, $^3F_2$, whereas the energy level of ^1G does not split because of the paired spins ($\uparrow\downarrow$, $S = 0$). This paired-spin arrangement is called a "singlet" state; in the parallel case, the two spins add together to give a nonzero total spin and the resulting state is called a "triplet."

Here it is worthwhile to say a little about the expression "term symbol." Expressions such as $^3P_0$, $^3P_1$, and $^3P_2$ are called *term symbols*, which provide three pieces of information:

1. The letter (for example, S, P, and D in the examples) indicates the *total orbital angular momentum L*. The code for converting the value of L into a letter is designated by

$$L = 0,\ 1,\ 2,\ 3,\ 4,\ 5,\ 6,\ 7,\ 8,\ 9,\ 10 \ldots$$
$$S,\ P,\ D,\ F,\ G,\ H,\ I,\ K,\ L,\ M,\ N \ldots$$

   The codes are the same as for the s, p, d, f, . . . designations of orbitals.
2. The left superscript in the term symbol (for example, the 3 in $^3P_2$) shows the *multiplicity of the term*. This is given by the total spin S, and the multiplicity is $2S + 1$. For example, in the parallel case $S - 1/2 + 1/2 = 1$, so that $2S + 1 = 3$, that is, it is a *triplet*.

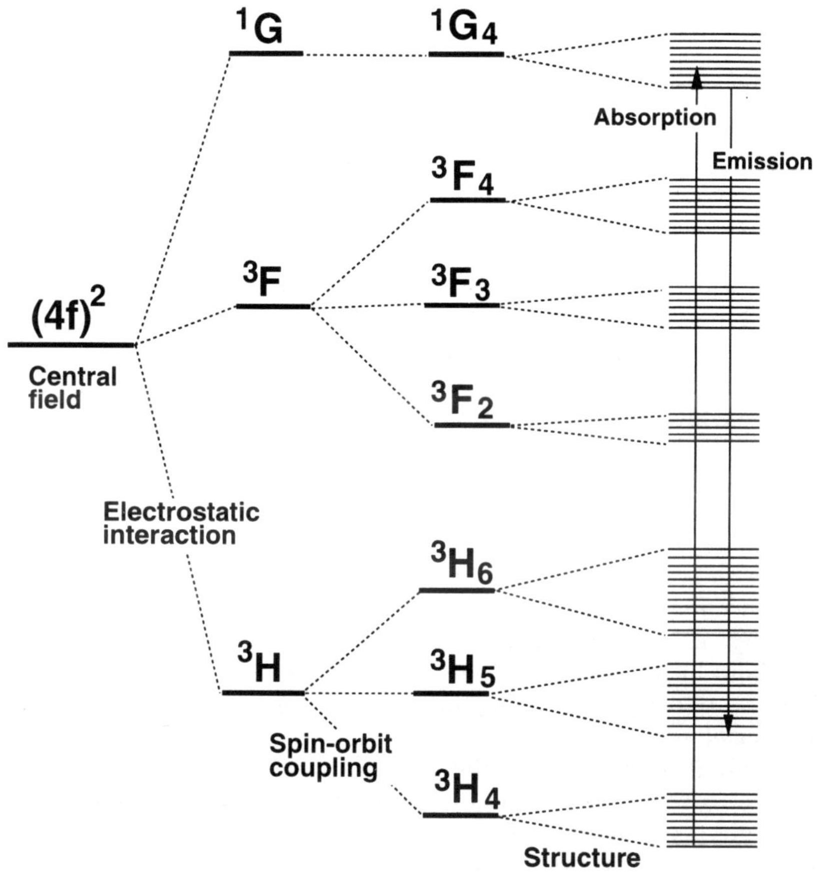

Figure 2.9 A portion of the energy levels for praseodymium ions with two 4f-electrons (f indicates the f-orbit in the central field approximation) except for electrons in closed shells.

3. The right subscript in the term symbol (for example, the 2 in $^3P_2$) is the value of the *total angular momentum quantum number J* and is given by

$$J = L + S, L + S - 1, \ldots, |L - S|$$

In the case of ^3P, $L = 1$, $S = 1$, so that $J = 2$, 1, and 0. Therefore, ^3P splits into $^3P_2$, $^3P_1$, and $^3P_0$. So, the term symbol is expressed as $^{2S+1}L_J$.

Another important illustration in Figure 2.9 is the effect of the structural environment in condensed matter. When a many-electron atom is placed in an external field of potential energy V as shown in (2.2) and such as a magnetic field or an electric field, the energy levels are split as shown in Figure 2.9. In general, when a magnetic field is applied to ions, the energy levels split into levels whose number depends on the value of the total angular momentum quantum number J. This splitting is called the *Zeeman effect*. On the other hand, when the electric field is applied to ions, the splitting is called the *Stark effect*. In glass materials, the electrostatic interaction between the ions and the lattice is dominant, so the energy level splitting is the Stark effect. The number of splits caused by the Stark effect is $2J+1$. In addition, the magnitude of the energy split depends on the magnitude of local electrical fields, which varies depending on their location in the condensed matter. Therefore, the Stark split in glass materials varies depending on the location of the ions and the field magnitude. This variation in Stark split energy provides the spectral band in the absorption and fluorescence characteristics of rare-earth ions. Consequently, different condensed materials such as silica glass, fluoride glass, and crystals provide individually different spectral characteristics of absorption and fluorescence of rare-earth ions because of their different Stark split energy levels.

Actual transitions between energy levels, that is, the absorption and emission of photons in rare-earth ions, take place between different energy levels, specifically, between the different Stark split energy levels shown in Figure 2.9. This provides a variety absorption and fluorescence wavelengths. Figure 2.10 is a diagram showing all the rare-earth ion energy levels and laser transitions ever reported [12]. The numbers shown in the figure indicate the approximate wavelengths of the rare-earth ion laser transitions, spanning the range from 0.172 to 5.15 μm. In total, 70 transitions exhibit laser operation.

Although the wavelengths of the rare-earth ion laser transitions concentrate most heavily in the visible and near-infrared regions, the important wavelength range for optical communication is from 0.8 to 1.6 μm, in which optical fibers exhibit low optical attenuation. Figure 2.11 shows the optical attenuation (loss) of high-silica fibers and the fluorescence wavelength band of the more important rare-earth ions (doped in glassy materials) in the 0.8- to 1.6-μm region. Each rare-earth ion has its own fluorescence wavelength band, which originates from the different Stark split energy levels of rare-earth ions placed in different locations in glassy materials. These fluorescence bands, which provide spectral bands for amplification, play a very important role in actual fiber amplifiers for use in optical fiber transmission.

2.2.3 Outline of Amplification in Rare-Earth-Doped Fiber

Figure 2.12 shows the basic configuration and the principle of optical signal amplification in rare-earth-doped fibers. As explained in Section 1.3, the basic configuration of the optical fiber amplifier is incredibly simple: it amplifies optical signals in rare-earth-doped fiber using the stimulated emission of optically excited rare-earth ions in the fiber core. However, it was about 30 years after the first attempt

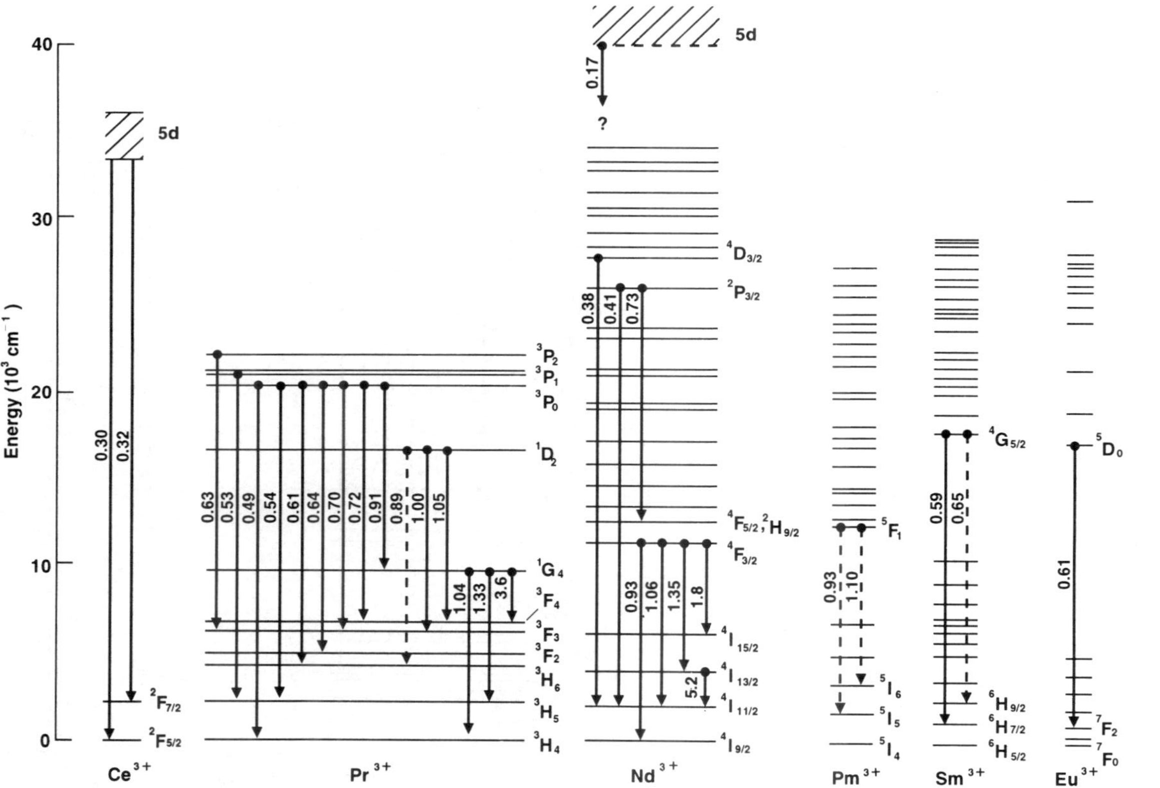

Figure 2.10 A diagram showing all the rare-earth ion energy levels and laser transitions ever reported [12]. The numbers indicate the approximate wavelengths of the rare-earth ion laser transitions, spanning the range from 0.172 to 5.15 μm. In total, 70 transitions exhibit laser operation.

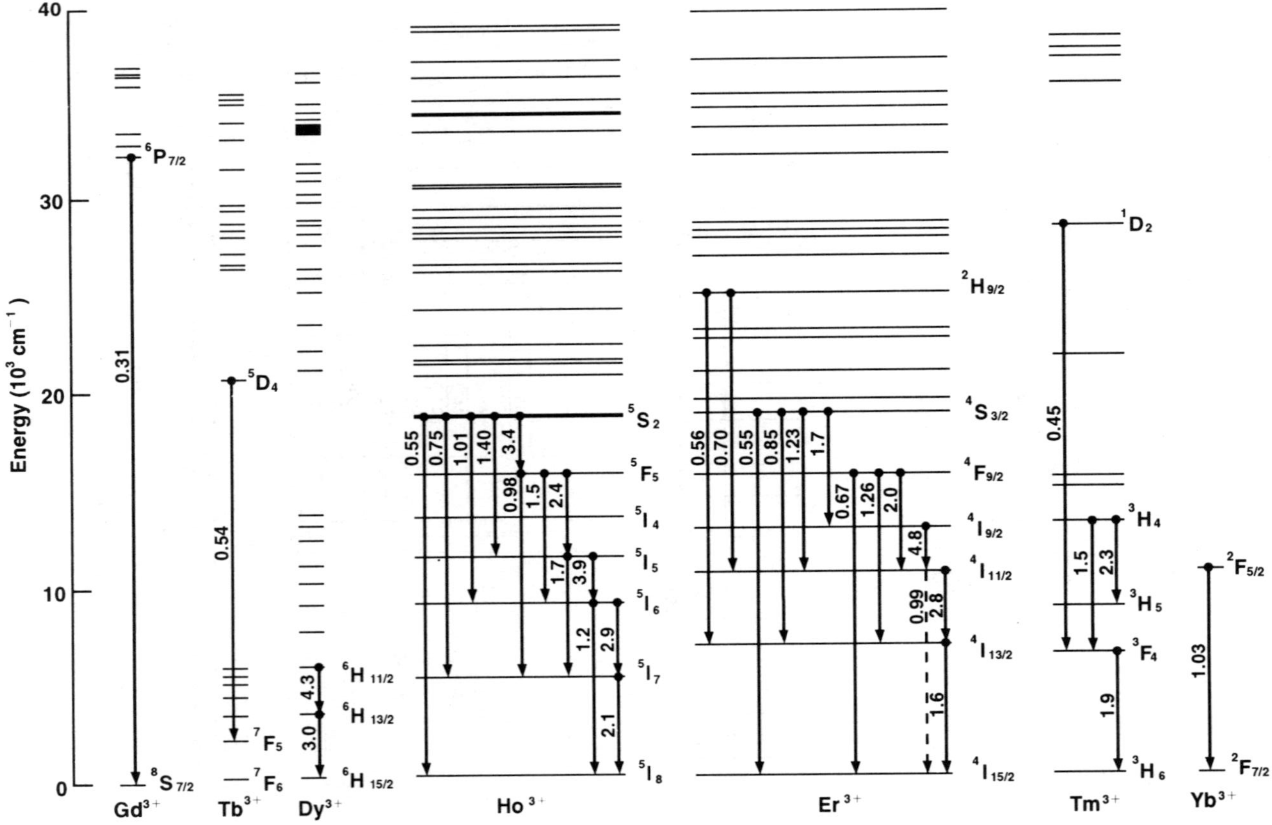

Figure 2.10 (continued).

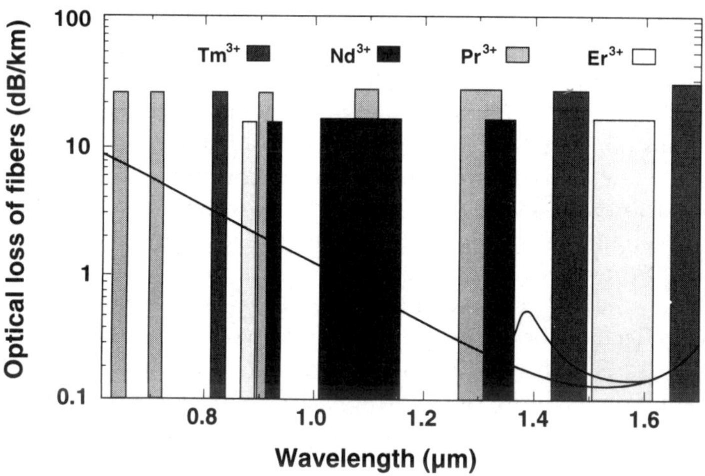

Figure 2.11 Optical attenuation (loss) of high-silica fibers and the fluorescence wavelength band of the main rare-earth ions (doped in glassy materials) in the 0.8- to 1.6-μm region.

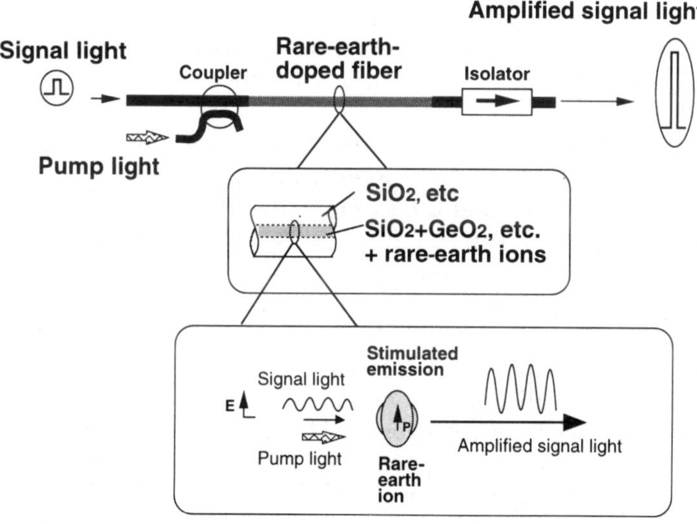

Figure 2.12 Basic configuration and principle of optical signal amplification in rare-earth-doped fibers.

to construct optical fiber amplifiers in the 1960s that efficient and low-noise fiber amplifiers were developed in 1987, as described in detail in Section 1.3.

In order to achieve efficient amplification in rare-earth-doped fiber, requirements should be met. The first is that rare-earth ions should be able to operate in a radiative transition with high quantum efficiency. This should have no quenching effect of ions, a low nonradiative transition caused by multiphonon relaxation, and a high optical intensity in the core part. The high optical intensity in the core increases the stimulated emission rate because the induced rate is proportional to the intensity of the electromagnetic field, whereas the spontaneous emission rate is independent of it [15]. The second requirement is that efficient pumping should be realized, reflecting the high-level pumping power and effective absorption of pump light by rare-earth ions. The third is that the fiber host should exhibit low loss, including low scattering loss and low absorption loss even for fibers with a high refractive index and rare-earth doping.

The rare-earth-doped fiber illustrated in Figure 2.12 consists of a core part containing SiO_2 and GeO_2, for example, and rare-earth ions and a cladding region consisting mainly of SiO_2. The structure of this rare-earth-doped fiber is almost the same as that of fibers prepared for optical transmission, except for the rare-earth ions doped in the core. In addition, rare-earth-doped fibers for amplification should also satisfy a set of requirements. They should exhibit low scattering and absorption losses, preferably have a low phonon energy for a low nonradiative transition, and also have a strongly waveguiding structure as a result of a high numerical aperture and single-mode operation. This is interesting in that it is different from the structure of fibers for transmission use, which is weakly waveguiding so that it can achieve ultimately low-loss characteristics by minimizing the scattering losses, as described in Section 1.4.

We next outline the characteristics of erbium ions for 1.5-μm amplification and praseodymium ions for 1.3-μm amplification as a background to a discussion of recent progress and key issues. Those ions are also of particular importance for practical fiber amplifiers. Figure 2.13 outlines the characteristics of erbium and praseodymium ions. Erbium ions are pumped at 980 nm and 1.48 μm. The signal is amplified by the transition from $^4I_{13/2}$ to $^4I_{15/2}$. This is a three-level transition, and its quantum efficiency is very high. However, there are two main degradation processes that affect amplification. One is pump *excited-state absorption* (ESA) with 980-nm pumping [16–19], and the other is cooperative upconversion with 1.48-μm pumping [20,21]. Specifically, 980-pump ESA appears during high-power operation, and cooperative upconversion appears when the erbium ion concentration is high.

On the other hand, praseodymium ions amplify the 1.3-μm signal by the transition from $^1G_4$ to $^3H_5$, which is basically a four-level transition [22]. However, the biggest problem with praseodymium is the nonradiative transition from $^1G_4$ to $^3F_4$ due to multiphonon relaxation [22,23]. Consequently, praseodymium needs a fiber host glass with low phonon energy such as ZrF_4-based fluoride glasses, InF_3-based fluoride glasses, and chalcogenide glasses. In addition, the gain characteristics

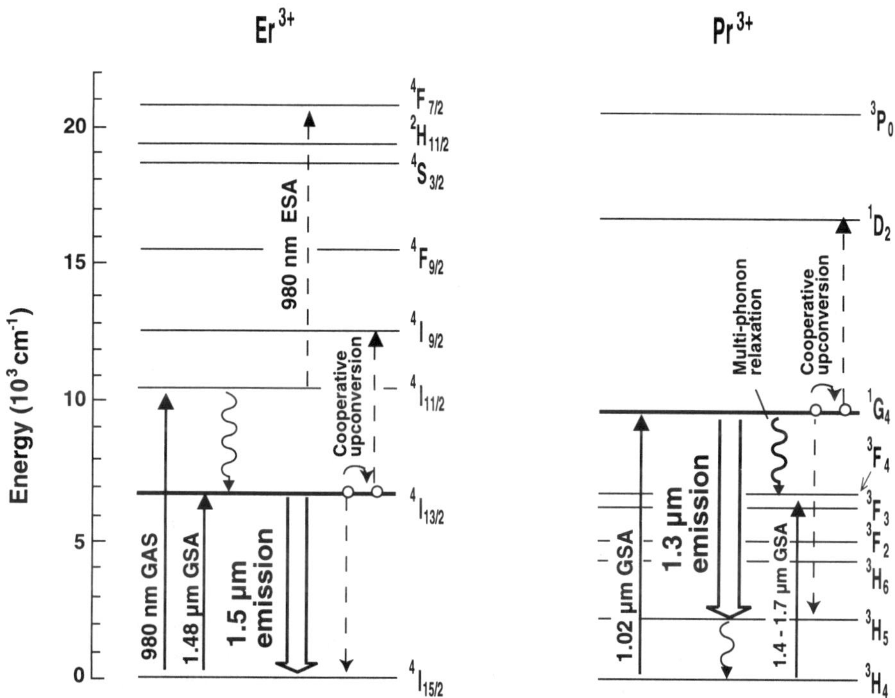

Figure 2.13 Characteristics of erbium and praseodymium ions.

at 1.3 μm are degraded by cooperative upconversion, ESA peaking at 1.38 μm, and *ground state absorption* (GSA) with a peak wavelength of 1.44 μm [23–25].

Next, we will briefly describe the fluorescence and signal amplification obtained using erbium at 1.5 μm and praseodymium at 1.3 μm. Figure 2.14 shows the fluorescence spectra of erbium ions doped in three different glass hosts: fluoride glass (ZBLYAN: ZrF_4-BaF_2-LaF_3-YF_3-AlF_3-NaF), SiO_2-Al_2O_3 glass, and SiO_2 glass. These fluorescence spectra result from the energy level transition from $^4I_{13/2}$ to $^4I_{15/2}$. This figure shows that, of the three hosts, fluoride glass results in the widest fluorescence spectrum for erbium ions. The second widest fluorescence spectrum is obtained with the SiO_2-Al_2O_3 glass host, in which Er^{3+} ions and Al_2O_3 are codoped as active ions and a dopant, respectively. The narrowest fluorescence spectrum is obtained with the SiO_2 glass host. These spectral widths originated from the Stark splits described for Pr^{3+} ions in Figure 2.9. That is, the energy level of erbium ions doped in glassy materials has Stark splits whose widths largely depend on the atomic structure around the erbium ions, in particular the electronic structure. Consequently the different host materials produce Stark splits of erbium ions with different widths and, hence, fluorescence spectra with different widths. In addition,

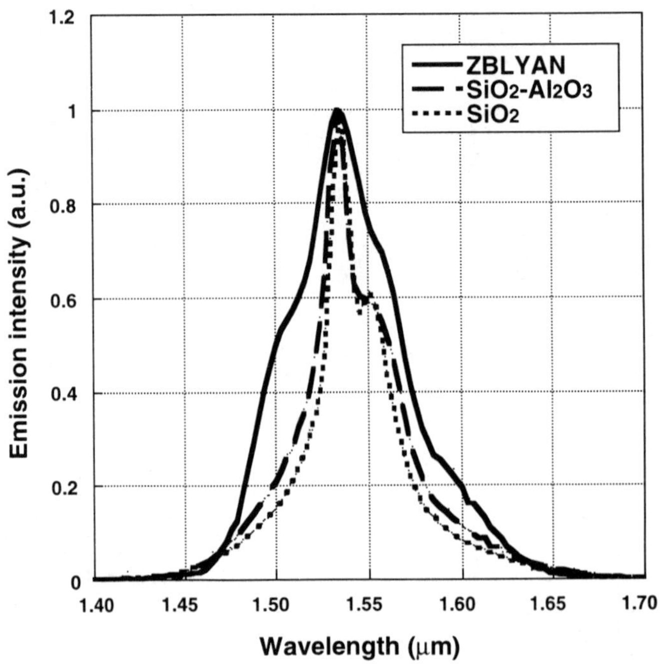

Figure 2.14 Fluorescence spectra of erbium ions doped in three different glass hosts: fluoride glass (ZBLYAN: ZrF_4-BaF_2-LaF_3-YF_3-AlF_3-NaF), SiO_2-Al_2O_3 glass, and SiO_2 glass.

each Stark split has some degree of shift and broadening caused by the local electrical field and the thermal fluctuation of surrounding atoms. In short, the spectral width of the fluorescence of rare-earth ions is determined by the Stark split width, or inhomogeneous broadening, and by homogeneous broadening. However, the mechanisms and the reasons for the different spectral widths with different hosts are rather complicated and so will be explained in detail in Sections 2.3 and 2.7.

Figure 2.15 shows the gain spectra of an EDFA with a fluoride (ZBLAN) glass host. The configuration of this EDFA is the same as that shown in Figure 2.12. The amplifier exhibits a high and flat gain of over 30 dB in the 1.53- to 1.56-μm range. The maximum gain reaches 40 dB. The maximum gain difference is about 6 dB in the 30-nm wavelength range at an input power level of -40 dBm. With an input level of -20 dBm, the gain difference for 30 nm becomes less than 2 dB, which is much smaller than that for EDFA with a silica glass host.

The fluorescence spectra of Pr^{3+} ions doped in fluoride glass (fluorozirconate: ZBLAN; ZrF_4-BaF_2-LaF_3-AlF_3-NaF) and chalcogenide glass (sulfide: AsS and GeS) are shown in Figure 2.16. Of course, here the fluorescence is produced by the energy level transition of praseodymium ions from $^1G_4$ to $^3H_5$. The fluorescence

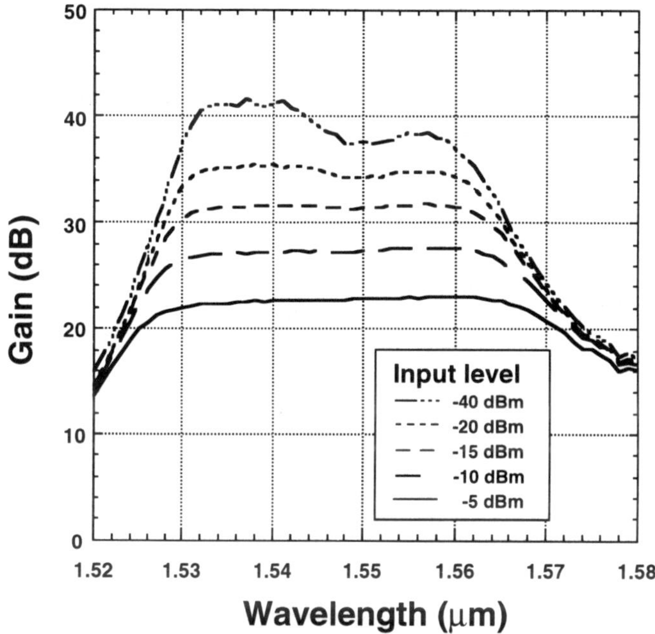

Figure 2.15 Gain spectra of an EDFA with a fluoride (ZBLAN) glass host.

spectra have peaks at 1.32 μm for fluoride glass and at 1.34 μm for chalcogenide glass. The widths of the fluorescence spectra are about 90 nm (full width at half maximum: FWHM) for ZBLAN and GeS and 76 nm for AsS. Interestingly, the fluorescence peak wavelengths for chalcogenide glass hosts are shifted toward a longer wavelength. This is in part because of the difference between the atomic bonding structures of fluoride and chalcogenide glasses surrounding rare-earth ions.

Figure 2.17 shows the gain spectra of a *praseodymium-* (Pr^{3+}) *doped fiber amplifier* (PDFA) with a fluoride (ZBLAN) glass host. The maximum gain reaches 42 dB for a small signal input of -30 dBm. The gain peak wavelength is around 1.30 μm, although the fluorescence spectra have peaks at 1.32 μm, which is possible because the amplified signal is absorbed in the wavelength region longer than 1.31 μm by the GSA due to the transition from $^3H_4$ to $^3F_3$, whose peak wavelength is 1.44 μm [23–25].

Finally, it is worth saying a little more about why Pr^{3+} ions need a low phonon energy glass host in order to achieve high gain amplification at 1.3 μm. As mentioned, the biggest problem in PDFAs is the nonradiative transition from $^1G_4$ to $^3F_4$ due to multiphonon relaxation [22,23]. Basically, the nonradiative decay rate W_{nr} can be expressed by

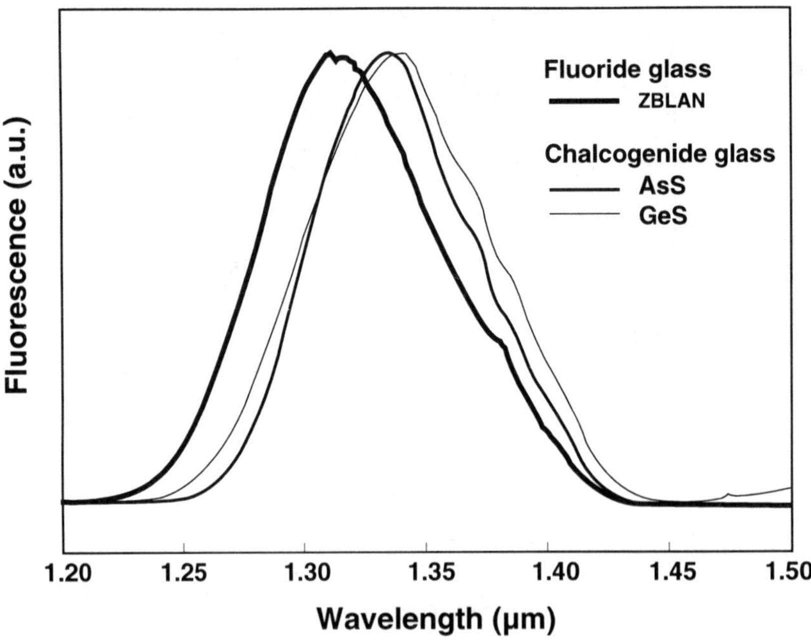

Figure 2.16 Fluorescence spectra of Pr^{3+} ions doped in fluoride glass (fluorozirconate: ZBLAN) and chalcogenide glass (sulfide: AsS and GeS).

$$W_{nr} = W_0 \exp(-\alpha \Delta E) \tag{2.3}$$

where W_0 is an experimental parameter corresponding to the decay rate for a zero energy gap and zero phonon emission, ΔE is the energy gap between transition levels, and α is a multiphonon relaxation parameter that depends on the phonon energy, energy gap, and electron phonon coupling strength [26–28]. Figure 2.18 shows the energy gap dependence of the nonradiative decay rate W_{nr} for three types of host glasses; silicate glass (oxide glass), fluorozirconate glass (fluoride glass), and sulfide (chalcogenide glass) [26–28]. As can be seen in the figure, since the energy gap between $^1G_4$ to $^3F_4$ of Pr^{3+} is about 2,900 cm^{-1}, the nonradiative decay rate varies greatly for the different host glasses. For example, the difference between silicate and fluoride is larger than 10^2 and the difference between silicate and sulfide is around 10^4. These large differences originate mainly from the difference between the phonon energies of the three types of host glass, which are 1,000 cm^{-1} for silicate, 500 cm^{-1} for fluorozirconate, and 300 cm^{-1} for sulfide. When we use silicate glass as a host for Pr^{3+} ions, we cannot obtain any gain in the 1.3-μm region because Pr^{3+} loses its energy through multiphonon relaxation instead of the stimulated emission for the amplification of signals. The ratio of the radiative transitions to the total of all transitions (radiative transitions plus nonradiative transitions) is commonly called the *radiative quantum efficiency*. The ratio of the

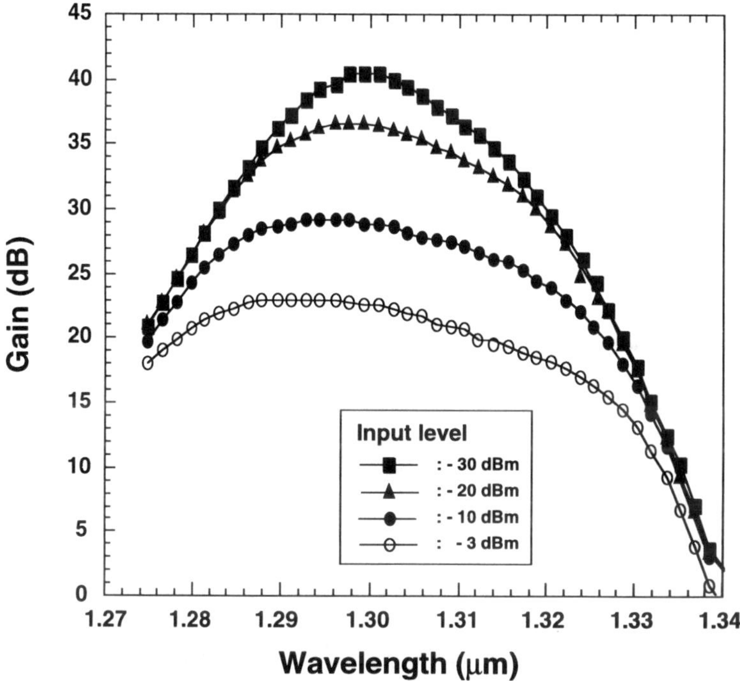

Figure 2.17 Gain spectra of a PDFA with a fluoride (ZBLAN) glass host.

radiative transitions at a certain wavelength to the total of radiative transitions is commonly called the *branching ratio*. Therefore, in short, a host glass with a smaller phonon energy provides a greater PDFA gain.

Here, once more, the three key requirements for achieving high gain amplification are as follows:

1. The first is to achieve a transition with high radiative quantum efficiency. This indicates no quenching effect of ions, a low nonradiative transition caused by multiphonon relaxation, and a high optical intensity in the core part.
2. The second is to realize efficient pumping, meaning a high-level pump power and the effective absorption wavelength of pump light by rare-earth ions.
3. The third is a low-loss fiber host. That is, it should exhibit low scattering and absorption losses at the operation wavelength.

2.3 KEY ISSUES FOR ERBIUM-DOPED FIBER AMPLIFIERS

2.3.1 Broadband

Broadband gain characteristics, which include a wide gain band and gain flattening, are currently receiving the greatest attention in connection with research on *wave-*

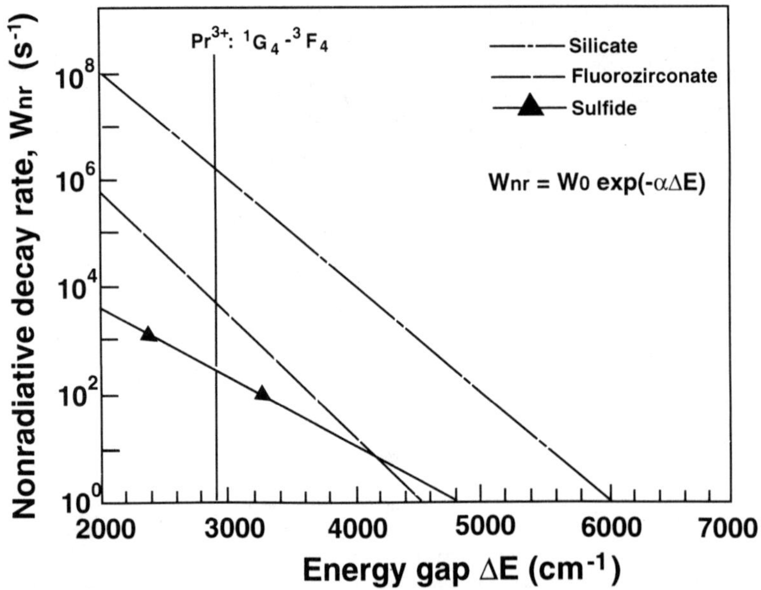

Figure 2.18 Energy gap dependence of nonradiative decay rate W_{nr} for three types of host glasses: silicate glass (oxide glass), fluorozirconate glass (fluoride glass), and sulfide (chalcogenide glass) [26–28].

length division multiplexing (WDM) systems. Several methods have already been proposed for achieving broadband gain characteristics, including the alumina-codoping technique [29], the filter method [30,31], and the use of fluoride host glasses [32–34]. Of these schemes, the use of fluoride host glasses has received the most attention recently, while the alumina-codoping technique for a silica-based EDFA can provide excellent broadband and low-noise characteristics [35,36]. This is because wider and flatter fluorescence spectra can be obtained with erbium ions in a fluoride glass host than in other host glasses, as shown in Figures 2.14 and 2.15.

Figure 2.19 compares the erbium ion *amplified spontaneous emission* (ASE) spectra of silica-based (alumina-codoped) and fluoride-based glass hosts measured at room temperature. The fluoride-based host glass is ZBLAN (ZrF_4-BaF_2-LaF_3-AlF_3-NaF), and the silica-based host glass is 4% alumina-codoped high-silica glass. Erbium with a fluoride host has a wider and flatter fluorescence spectrum in the 1.5-μm region. (The reason why erbium ions in a fluoride host have a wider fluorescence spectrum than in a silica host glass will be described later.) Figure 2.20 shows another example of the small signal gain spectra for a fluoride-based EDFA [37]. The ZBLAN fiber has a core diameter of 2.45 μm and an Er-doping density of 1,000 ppm. In this figure, the open and filled squares indicate the gain spectra for

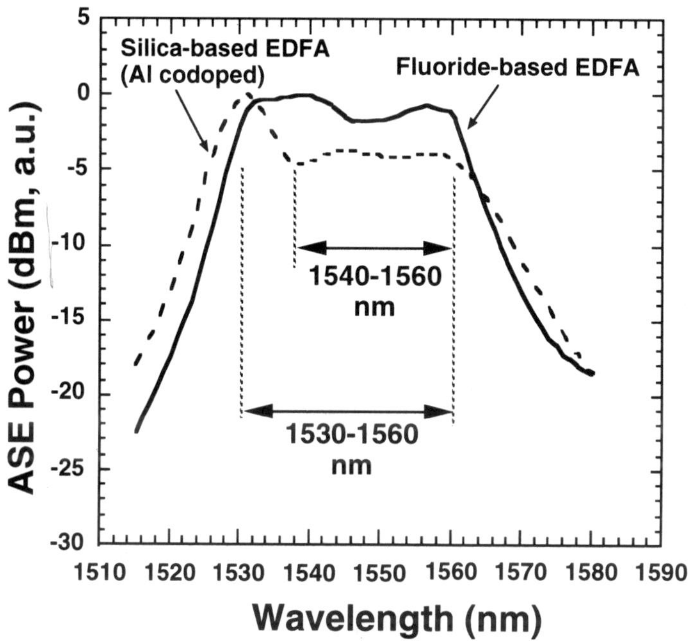

Figure 2.19 Erbium ion ASE spectra of silica-based (alumina-codoped) and fluoride-based glass hosts measured at room temperature.

pump powers of 64 and 30 mW, respectively, obtained using a 7.0-m-long fiber. The triangles indicate the gain spectrum with a pump power of 80 mW and a 9.5-m-long fiber. The signal input level is −40 dBm.

From the results of various investigations on the gain spectral width and flatness of EDFA [29–42], it can be summarized at present that a fluoride-based EDFA should be used when a wide wavelength range of over 20 nm is needed for WDM systems; in contrast an alumina-codoped silica-based EDFA is more suitable when the operation range is less than 15 or 20 nm and low-noise amplification is required. Therefore, the technical choice depends on the required operation wavelength range and noise figure.

Figure 2.21 shows experimental results for a four-EDFA cascade amplification of 16 channel signals [38–40] with (a) a fluoride-based EDFA and (b) a silica-based EDFA. The 16 channel signals with 1.6-nm spacing are allocated in the 1,533.7- to 1,558.2-nm wavelength range (25.5 nm). This figure shows that throughout the four cascade amplification the fluoride EDFA exhibits a gain excursion of only 4.5 dB compared with 23 dB for the silica-based EDFA. The gain difference along the amplification band increases in cascade amplification systems.

A further approach to gain flattening is a hybrid method that involves connecting *erbium-doped fibers* (EDF) with different host glasses, including the connection

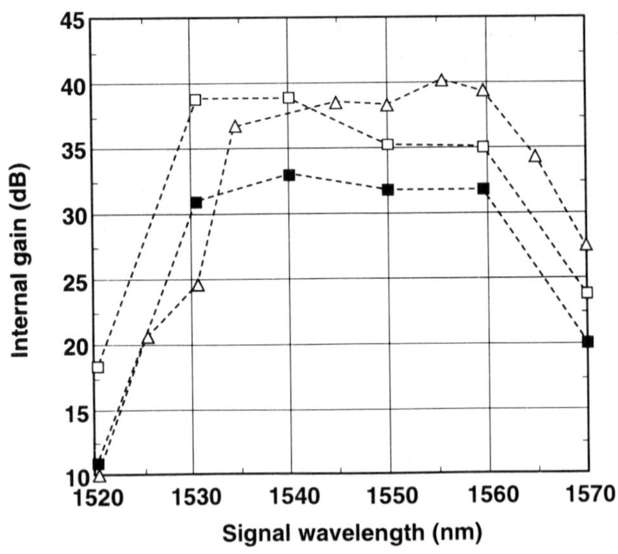

Figure 2.20 Another example of the small signal gain spectra for a fluoride-based EDFA [37].

of fluoride-based EDF with silica-based EDF [43,44], alumina-codoped EDF with silica-based EDF [35], and multicomponent-based EDF with silica-based EDF [45].

Here, it is useful to explain briefly why the fluoride-based EDFA has a wider gain (or fluorescence) spectrum than the silica-based EDFA. Figure 2.22 is a schematic diagram showing in detail the structure of Er^{3+} ion energy levels, which involve a Stark split, inhomogeneous broadening, and homogeneous broadening [46–73]. As shown in Figure 2.9, the energy levels of rare-earth ions are split by a coulombic interaction between electrons, including spin correlation and spin-orbit coupling, and eventually by the Stark effect due to crystal field and local coordination. The magnitude of the Stark split ranges from 200 to 400 cm^{-1} for Er^{3+}-doped fluoride, fluorophosphate, and silicate glasses [46,68,69,71–73]. However, the magnitude of the Stark split is slightly different in each ion because of the site-to-site difference in the crystal field surrounding the rare-earth ions. Therefore, the Stark level has an apparent broadening originating from the site-to-site difference in the Stark level of the individual ions. This broadening is commonly called *inhomogeneous broadening* and originates in the material structure [70]. In addition, the individual Stark levels fluctuate and broaden as a result of the fluctuation of the crystal fields caused by thermal atomic motions. This broadening due to thermal fluctuation is commonly called *homogeneous broadening* [60]. The magnitudes of inhomogeneous and homogeneous broadening are about 27 to 60 cm^{-1} [48,52,61,73] and 8 to 49 cm^{-1} [47–49,52,73], respectively. The energy spacing ΔE between adjacent Stark sublevels ranges from 20 to 80 cm^{-1} and its average value is 50 cm^{-1}. In a state of

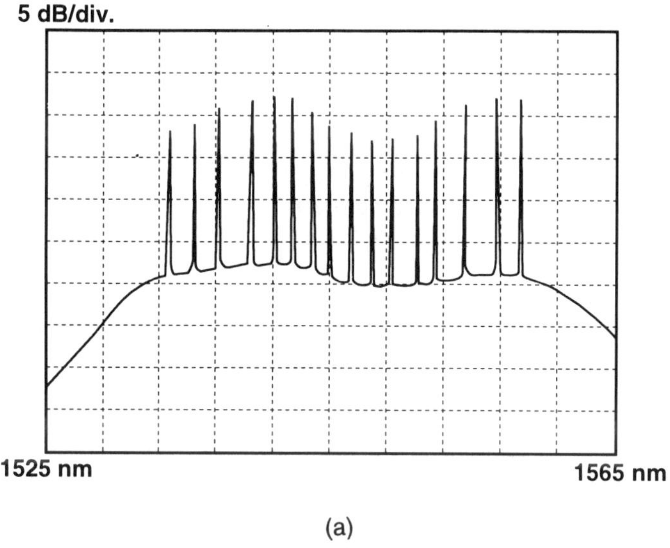

(a)

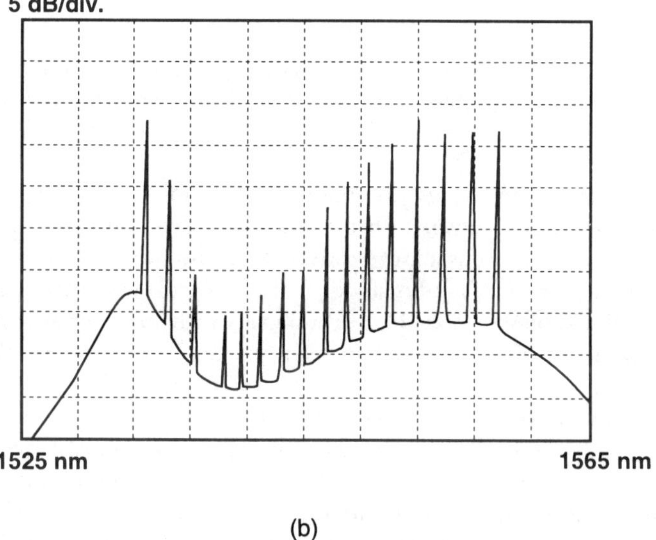

(b)

Figure 2.21 Experimental results for a four-EDFA cascade amplification of 16 channel signals [38–40] with (a) a fluoride-based EDFA and (b) a silica-based EDFA.

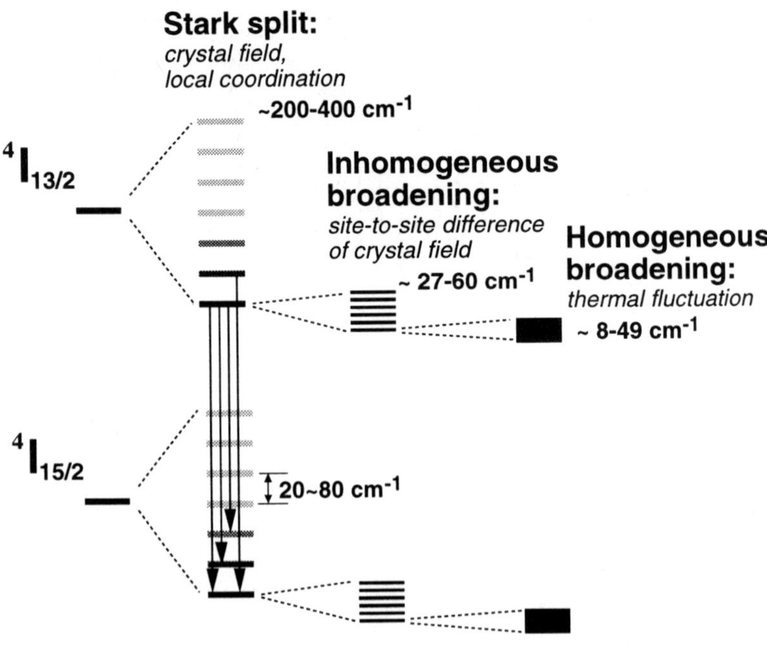

Figure 2.22 Schematic diagram showing details of the structure of Er^{3+} ion energy levels that involve a Stark split, inhomogeneous broadening, and homogeneous broadening [46–73].

thermal equilibrium, the energy population of Er^{3+} ions has a Boltzmann distribution. Since the ratio between two adjacent Stark sublevel populations is $\exp(-\Delta E/kT) = 0.78$, there is a large difference in the populations of those sublevels. In Figure 2.22, the color intensity of Stark sublevels indicates such population difference in the sublevels; that is, the sublevels with lower energy have a higher population of ions. The radiation transitions are indicated by the arrows in Figure 2.22.

Figure 2.23 shows the wavelength assignment of Er^{3+} ion fluorescence (radiation transitions) from the $^4I_{13/2}$ level to the $^4I_{15/2}$ level for an Al_2O_3-SiO_2 glass host: (a) shows the Stark levels and related fluorescence transition, and (b) shows the spectral composition of the room-temperature fluorescence for the Al_2O_3-SiO_2 glass host [68,69,72]. As shown in Figure 2.23(b), since the fluorescence lines F1 to F12 are closely spaced and, moreover, individual lines exhibit homogeneous broadening and inhomogeneous broadening, these line have a strong spectral overlap and form the complete fluorescence spectrum of Er^{3+} ions in the Al_2O_3-SiO_2 glass host. In fact, when erbium ions are incorporated into crystal hosts, the fluorescence exhibits different spectral characteristics from that of glassy hosts. Figure 2.24 shows the fluorescence spectra of Er^{3+} ions incorporated into $LiNbO_3$ crystal. Compared

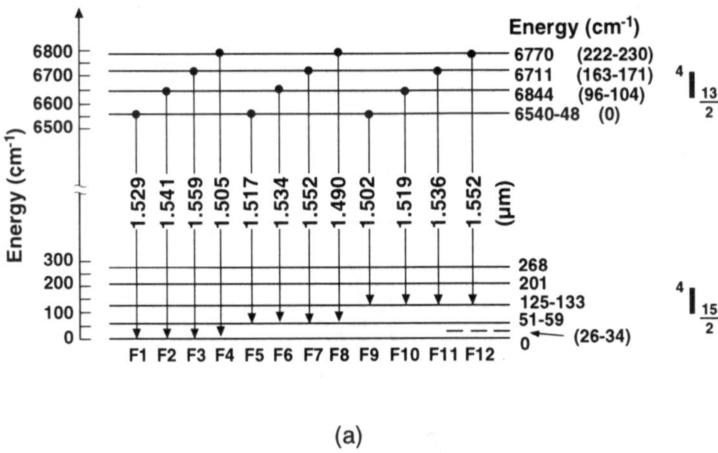

(a)

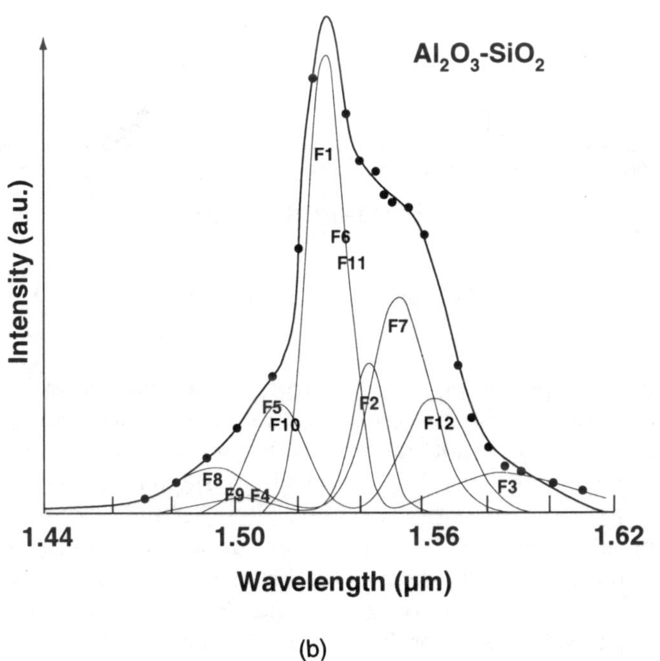

(b)

Figure 2.23 Wavelength assignment of Er^{3+} ion fluorescence (radiation transitions) from the $^4I_{13/2}$ level to $^4I_{15/2}$ level for an Al_2O_3-SiO_2 glass host: (a) the Stark levels and related fluorescence transition and (b) the spectral composition of the room-temperature fluorescence for the Al_2O_3-SiO_2 glass host [68,69,72].

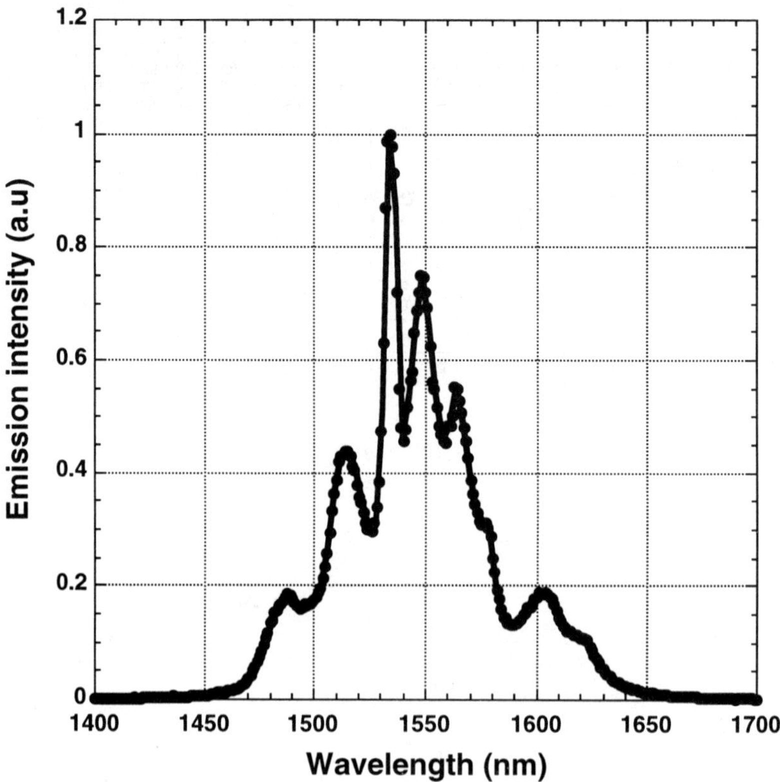

Figure 2.24 Fluorescence spectra of Er^{3+} ions incorporated into $LiNbO_3$ crystal.

with the fluorescence spectra for Er^{3+} ions incorporated into a glassy host as shown in Figures 2.14 and 2.23, the curve of the fluorescence spectrum for Er^{3+} ions incorporated into $LiNbO_3$ is not smooth and has several dips that are attributed to slight inhomogeneous broadening and possibly to slight homogeneous broadening as well.

Finally, by measuring the fluorescence spectra at a low temperature (4.2K), we can obtain additional useful data that helps us to understand the differences between the spectra of different host glasses and also to explain why the fluoride-based EDFA has a wider gain (or fluorescence) spectrum than the silica-based EDFA. Figure 2.25 compares fluorescence spectra measured at 4.2K for three host glasses [46]. From the fluorescence spectra shown in Figure 2.14, which are measured at room temperature, we expect erbium ions in fluoride-based glasses to have a wider Stark split than those in silica-based glasses because of their wider fluorescence spectrum. However, from the fluorescence spectra shown in Figure 2.25, which are measured at low temperature, it is found that erbium ions

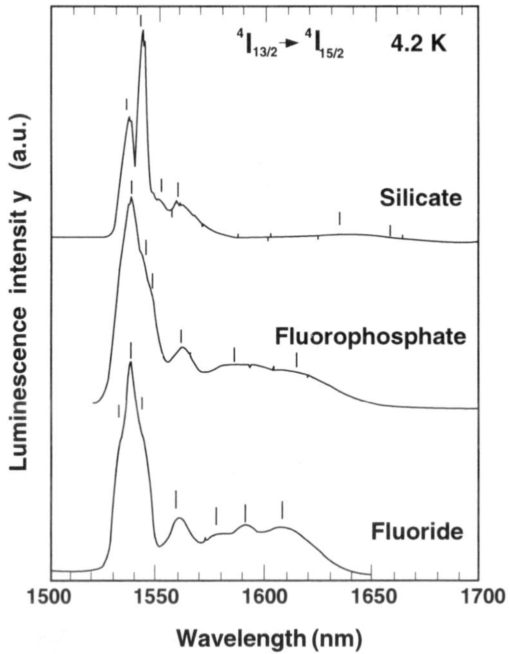

Figure 2.25 Comparison of fluorescence spectra measured at 4.2K for three host glasses [46].

in silica-based glass have a wider Stark split, ranging from 1.535 to 1.660 μm, than those in fluoride-based glasses, ranging from 1.533 to 1.610 μm.

Therefore, from these fluorescence spectra and other experimental results [47–73], we can determine why the fluoride-based EDFA has a wider gain (or fluorescence) spectrum than the silica-based EDFA in the 1.5-μm region. Some of the possible reasons can be summarized as follows. Erbium ions in fluoride-based glasses have a narrower Stark split, wider inhomogeneous broadening, and almost identical homogeneous broadening. These factors lead to a dense and packed spectrum in the 1.5-μm region, which is comparatively wide and flat.

2.3.2 High Gain

A high gain is of vital importance for optical amplifiers. The keys to achieving high gain amplification by optical fiber amplifiers are as follows: (1) a high optical intensity in the core of, for example, over 2%-Δn [74]; (2) low-loss and/or low-phonon-energy fiber hosts such as silica-based glasses [75] and fluoride-based glasses [32]; (3) efficient pumping in terms of pump wavelength [76,77], pump power, and configuration; (4) a suitable active ion content such as 10 to 1,000 ppm [78];

(5) an effective device configuration including, for example, a midway isolator and/or filter [79–82] and tilt splicing; and (6) control of the active ion energy level lifetime by codoping [83,84].

Figure 2.26 shows the highest EDFA gain (54 dB) ever reported [85–87]. The key features for achieving this 54-dB gain are the midway isolator, which can reduce backward ASE and Rayleigh backscattering and help prevent gain degradation, the low-loss host fiber, and in particular, the low Rayleigh scattering. Figure 2.27 is a schematic diagram of the midway isolator configuration [80]. Figure 2.28 shows the optimum position for the midway isolator in terms of high-gain and low-noise figure [82]. From the experimental results and theoretical calculations, for a forward pumping scheme the optimum range for the isolator position is determined as 0.15 to 0.45 of the total EDFA length from the signal-input end [80–87]. A configuration that incorporates the fiber of the optimal length and low reflection splicing is also important in achieving high gain. A theoretical estimation revealed that the upper limit of achievable gain in a EDFA is 57 to 70 dB, which is limited by gain degradation due to internal Rayleigh backscattering [88].

It is worth explaining a little about other processes of gain degradation in EDFAs. In addition to the degradation process caused by backward ASE and Rayleigh backscattering, there are several other types associated with the pumping scheme. Figure 2.29 shows several well-known gain degradation processes associated with pumping schemes in EDFAs [16–21,89–95]. The Er^{3+} ion absorption spectrum is shown on the right in the figure [95]. With the 980-nm pumping scheme, ESA appears for the 980-nm pump light. This is called pump ESA and is particularly evident during high-power operation [16–19,89,90]. The cooperative upconversion process at 1.48 μm appears particularly in EDFAs with a high Er^{3+} ion concentration because this process is caused by direct interaction between ions [20,21]. This

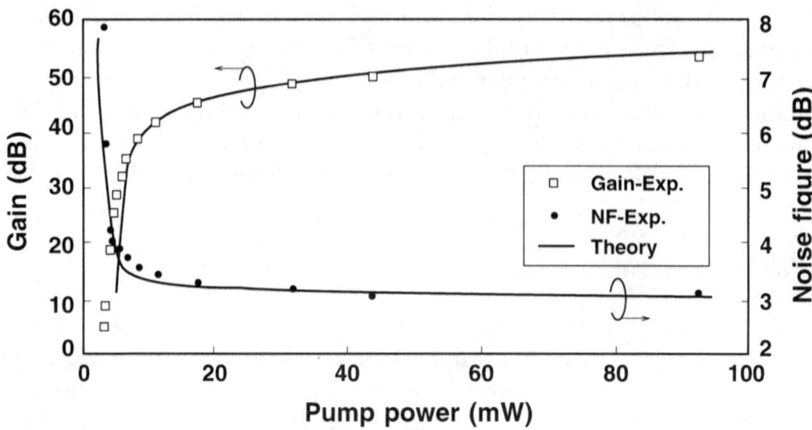

Figure 2.26 Highest EDFA gain (54 dB) ever reported [85–87].

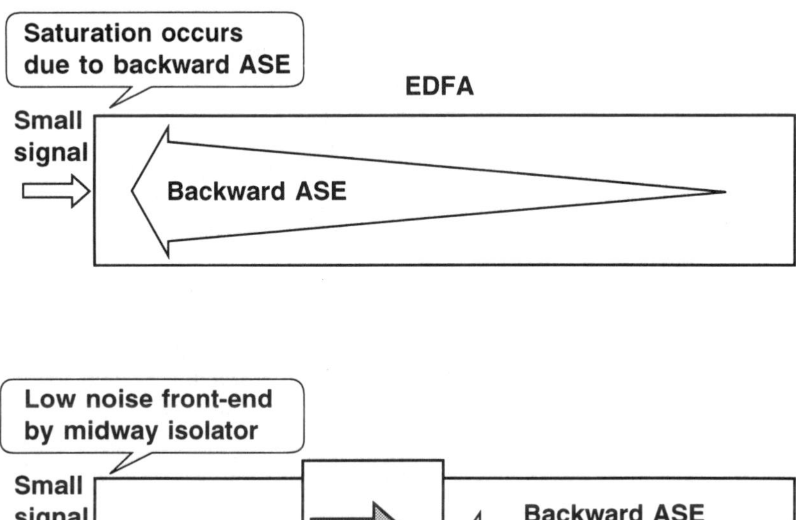

Figure 2.27 Schematic diagram of midway isolator configuration [80].

indicates that this is a dipole coupling process rather than a radiative process. The third type of gain degradation process is strong pump ESA at 800 nm, which makes the 800-nm pumping scheme difficult to employ [91–94]. Therefore, it is commonly recognized that a 1.48-μm pumping scheme is more suitable for high-power operation and a 980-nm pumping scheme is more suitable for low-noise operation, of course, taking into account the reliability of currently available pumping lasers.

2.3.3 Low Noise

The noise factors in fiber amplifiers have three origins: the first is signal-ASE beat noise and shot noise [96–98], and the second is multireflection in active fiber, which results in phase noise and intensity noise [99–101]. Figure 2.30 shows the interferometric noise from multiple reflections: (a) is a schematic diagram of a fiber transmission system with multiple reflection points, and (b) shows measured *bit error rate* (BER) curves versus R at 1 Gbps for two reflection points [99–101]. These two factors have already been characterized clearly and overcome. The key issue today, however, is the third factor, which is gain fading or S/N fluctuation

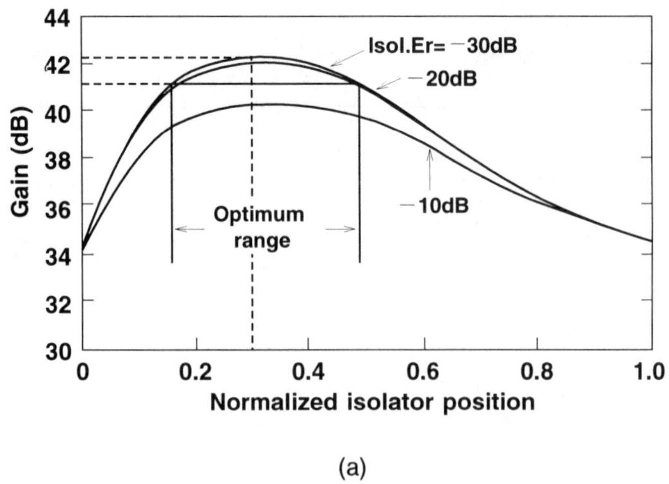

(a)

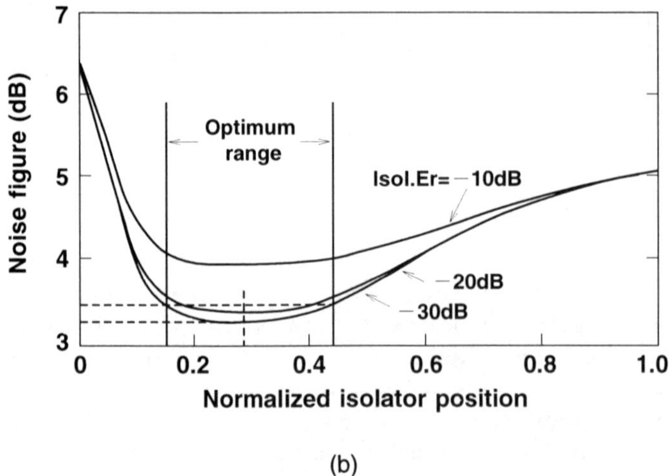

(b)

Figure 2.28 Optimum position for the midway isolator in terms of high-gain and low-noise figure [82]: isolator position dependence of (a) gain and (b) noise figure.

caused by *polarization hole burning* (PHB), which is of particular importance with regard to long-haul transmission. This polarization-dependence effect in long-haul optically amplified systems was observed for the first time by M. G. Taylor in 1993 [102]. BER degradation itself in long-distance (4,500 km) optical amplifier systems was reported by S. Yamamoto in January 1993 [103].

What is PHB in optically amplified systems? PHB is an anisotropic saturation that is created by a polarized saturating signal launched into an erbium-doped

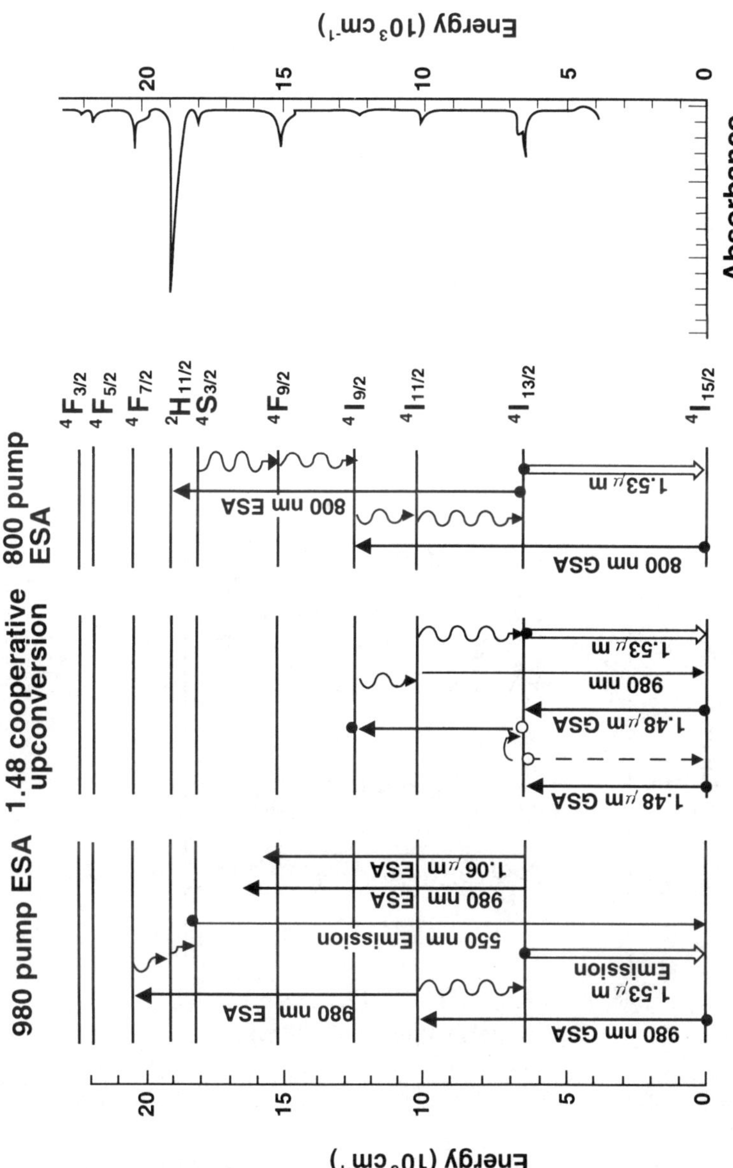

Figure 2.29 Several gain degradation processes associated with pumping schemes in EDFAs [16–21,89–90].

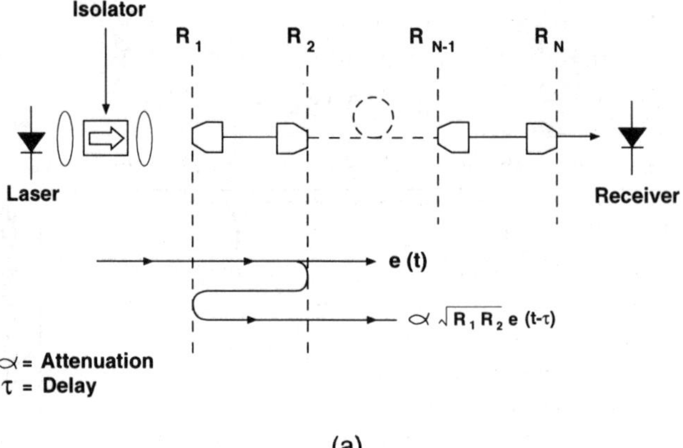

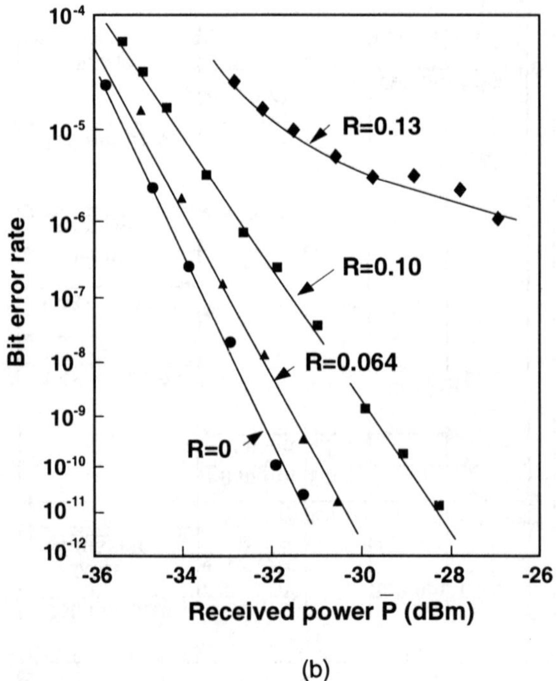

Figure 2.30 Interferometric noise from multiple reflections: (a) schematic diagram of a fiber transmission system with multiple reflection points and (b) measured BER curves versus R at 1 Gbps for two reflection points [99–101].

fiber. In Figure 2.31, S_i and S_o are saturating signals, and P_i and P_o are small probe signals that have an angle ϕ to the saturating signal [104–108]. As shown in the figure, when the probe has an orthogonal polarization to the saturating signal, the probe gain is maximum, and when it is parallel, the probe gain becomes minimum, although the gain change is very small. θ equal to zero indicates circular-polarization and 90 degrees indicates crosspolarization.

Next, why does PHB arise? Figure 2.32 shows the mechanisms of PHB and *polarization-dependent gain* (PDG). PHB arises from the randomly distributed orientation of erbium ions in a glass matrix and the selective de-excitation of those ions by a polarized signal, resulting in the anisotropic saturation of the signal gain. Consequently, the amplifier gain is slightly lower in the plane of a linearly polarized signal than in a perpendicular plane [104–109]. However, the birefringence in an EDFA has an averaging effect that reduces the measured PHB dependence on the launched polarization of the saturating signal. A similar effect, the polarization dependence of fluorescence or luminescence from rare-earth ions, is already known [110,111]. Figure 2.33 shows experimental evidence of the stimulated emission cross-section anisotropy of rare-earth ions, that is, the selective de-excitation of rare-earth ions [110]. This experiment was performed in 1983 by D. W. Hall and M. J. Weber.

In contrast, PDG is derived from selective excitation by the absorption of the polarized pump light [106,112–114]. Erbium ions with a randomly distributed orientation in a glass matrix are selectively excited by the absorption of the polarized pump. Those excited erbium ions whose major axes of s-ellipsoid are collinear with

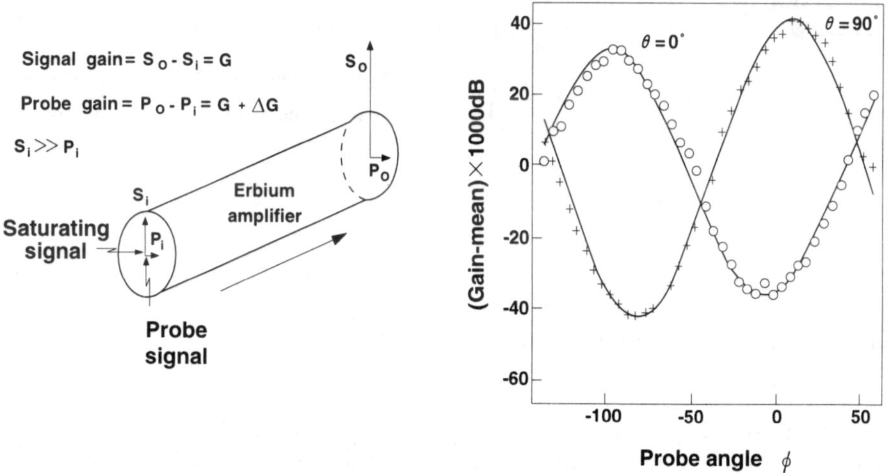

Figure 2.31 Explanation of PHB in optically amplified systems [104–108]. S_i and S_o are saturating signals, and P_i and P_o are small probe signals with an angle ϕ to the saturating signals.

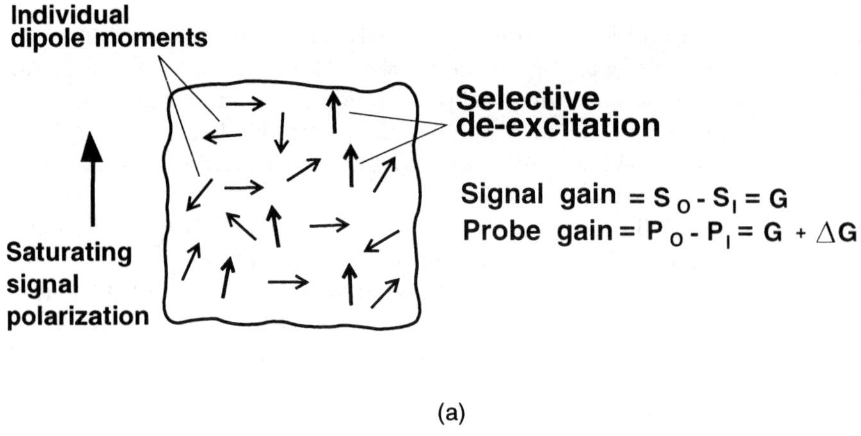

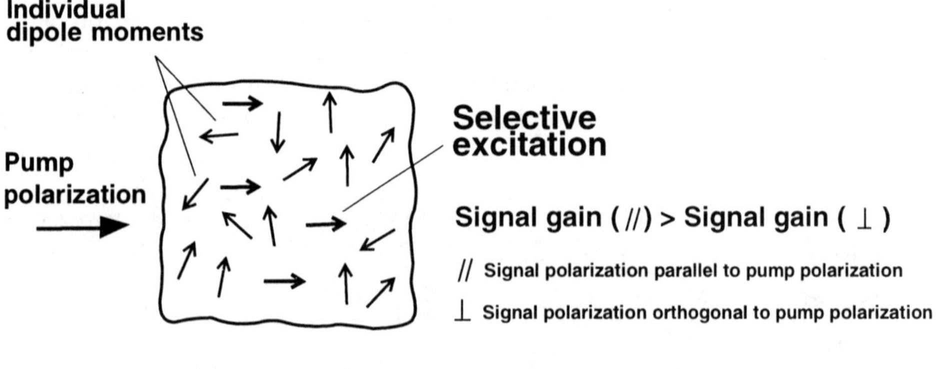

Figure 2.32 Mechanisms of (a) PHB and (b) PDG [104–114].

an incident linearly polarized optical signal are preferentially stimulated to emit photons. Therefore, the gain of a signal parallel to the pump light polarization is larger than that of a signal with orthogonal polarization.

Recently, this PHB problem has almost been solved by the introduction of a polarization scrambler [115–119], as shown in Figure 2.34, although the problem

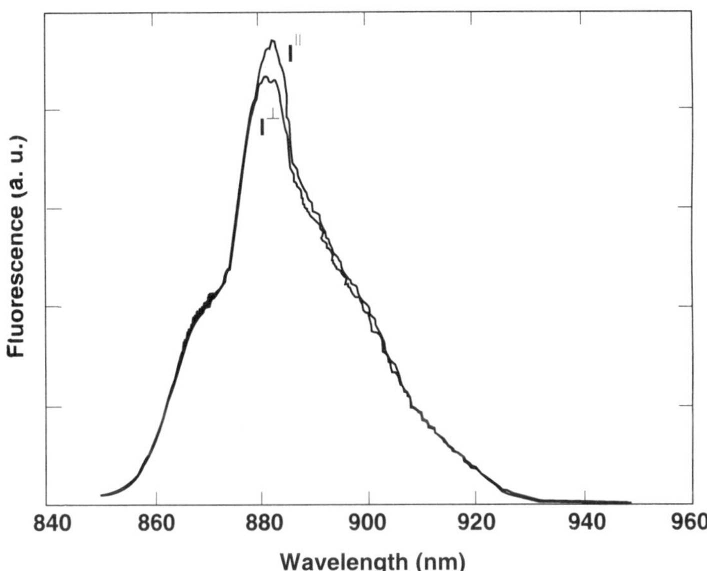

Figure 2.33 Experimental evidence of the stimulated emission cross-section anisotropy of rare-earth ions [110].

of timing jitter in the received signal, which is caused by a manifestation of the *polarization-mode dispersion* (PMD), still remains [120]. The flight time of the light in a fiber depends on the state of polarization of the launched signal. A polarization scrambler varies the polarization state between the fast and slow axes created by PMD. Thus, the received data are periodically displaced in time at harmonics of the scrambling frequency. The jitter was observed to be as large as 50 ps for the 5-Gbps and 9,000-km system [120]. In addition, it should be mentioned that PHB in Er-doped fiber is observed for a saturating signal with circular polarization [121].

2.3.4 High Power

Nowadays, high-power operation, in which low-noise signals are amplified to a high output power, is the key to the practical application of optical amplifiers, which means the application of fiber amplifiers as booster amplifiers. There are various methods for achieving high-power fiber amplifiers. These include: (1) high-power pumping, (2) strong absorption level pumping, (3) backward pumping, (4) the use of a cascade configuration, and (5) efficient power coupling. With these schemes, we should be able to achieve a high-power fiber amplifier.

Figure 2.35 shows an example of a high-power EDFA pumped by high-power 980-nm laser diodes [122]. This figure shows that the highest output power is

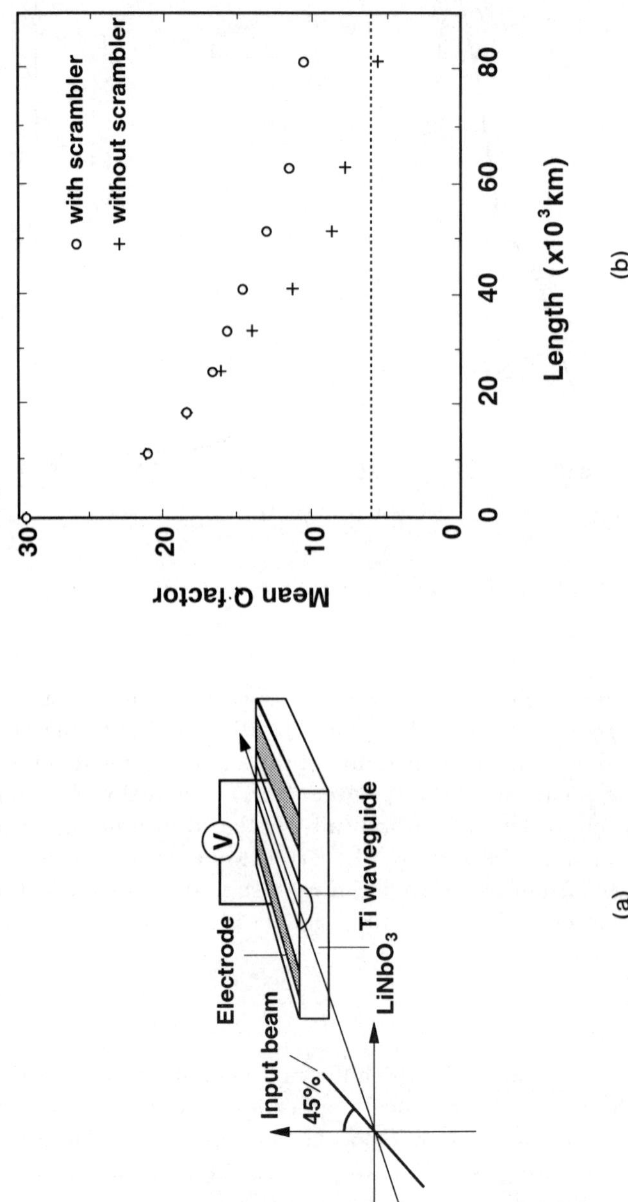

Figure 2.34 (a) Schematic diagram of a polarization scrambler consisting of a LiNbO$_3$ modulator and (b) the resulting improvement in transmission quality [116].

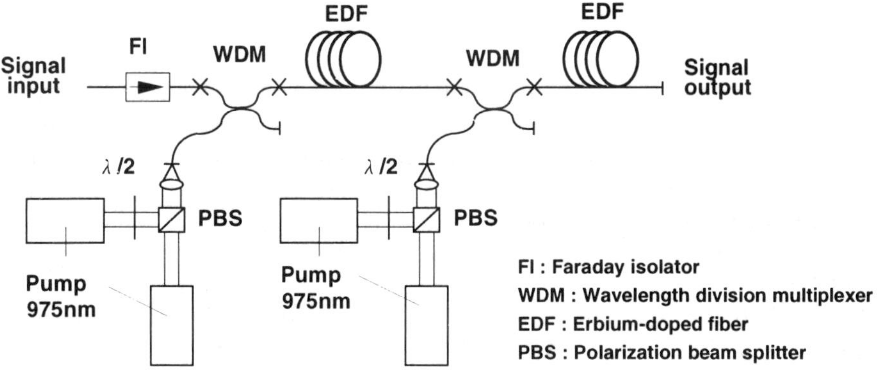

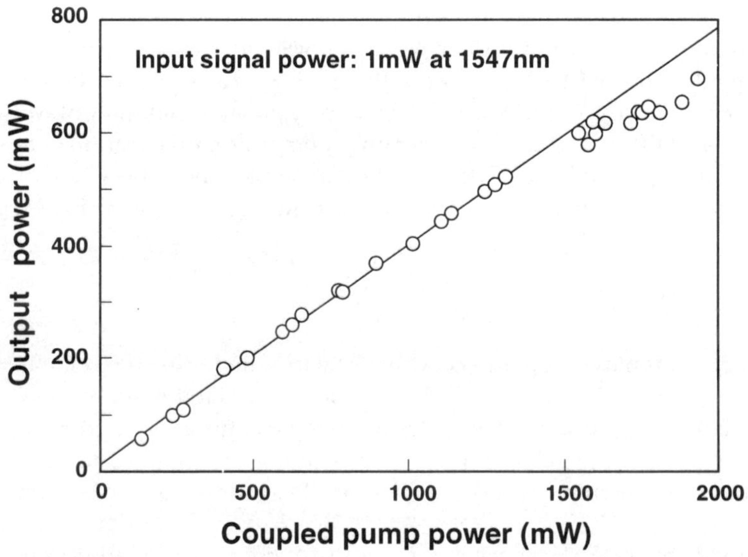

Figure 2.35 Example of a high-power EDFA pumped by high-power 980-nm laser diodes [122].

700 mW for a pump power of about 2W. The signal power level is 1 mW. The key features of this setup are 980-nm high-power pumping, a two-cascade configuration, and a forward pumping scheme. In general, a backward pumping scheme is more commonly used than a forward scheme for achieving high-power pumping. However, as mentioned, ESA appears during high-power operation with a 980-nm pumping scheme. This is called pump ESA [16,17]. Therefore, in this setup, a forward pumping scheme might be used in order to suppress the pump ESA at 980 nm. In fact, it can be seen in Figure 2.35 that pump ESA starts at a pump power of 1.5W.

Figure 2.36 shows another example of a high-power EDFA consisting of Er^{3+}/Yb^{3+}-codoped fiber and a Nd^{3+}:YLF pump laser [123,124]. With regard to energy transfer from Yb^{3+} to Er^{3+}, that is, the Yb^{3+}-sensitizer technique, there have been various experiments since the early stages of laser development [125–138]. In 1995, a high output power of +31.8 dBm (1.5W) was achieved by using a single diode-pumped, cladding-pumped Nd^{3+}-doped fiber laser [139–141]. In 1996, the highest output power reached 4.04W (36 dBm) for a total pump power of 12.6W using a three-stage Er^{3+}/Yb^{3+}-codoped fiber amplifier [142]. By contrast, the highest output power ever achieved with the 1,480-nm pumping scheme was +25.2 dBm using four *semi-multi-quantum-well* (SMQW) laser diodes [143].

Current areas of concern are how to achieve high-power, low-noise, broadband operation and high reliability in EDFAs. However, there are several technical choices to be made in order to realize these goals. For example, possible pumping wavelengths are 0.98, 1.047 (1.06), and 1.48 μm. Possible pumping configurations are backward, forward, and bidirectional. The fiber host material might be silica- or fluoride-based. These choices must be made while constructing the desired EDFA.

2.3.5 Reliability

The reliability of a fiber amplifier depends on various factors, including the reliability of the fiber, pump laser, other components, and the constructed module itself. These reliabilities depend on such parameters as temperature, humidity, and γ-ray irradiation. Of these factors, this section focuses on recent topics relating to fiber reliability and does not mention the reliability of the pump laser or other components.

Table 2.1 shows the EDFA gain changes due to temperature for different pump wavelengths [144,145]. For example, the data show that the temperature coefficient of the 1.48-μm pumping scheme is −0.07 dB/°C, which is about 20 times that for the 0.98-μm pumping scheme (−0.004 dB/°C), also shown in the table. Note that with a 1.48-μm pumping scheme the thermal excitation kT easily degrades the population inversion, leading to a gain change at 1.55 μm.

The next concern regarding fiber reliability is the loss increase due to H_2 diffusion [146–152]. Figure 2.37 shows the predicted long-term loss increases at

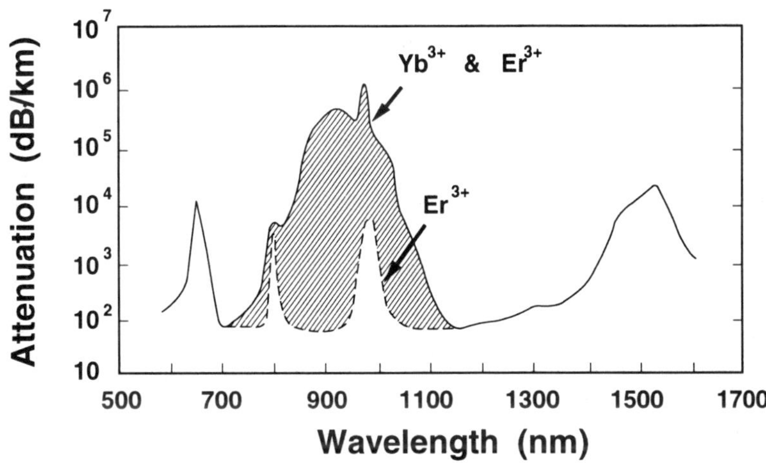

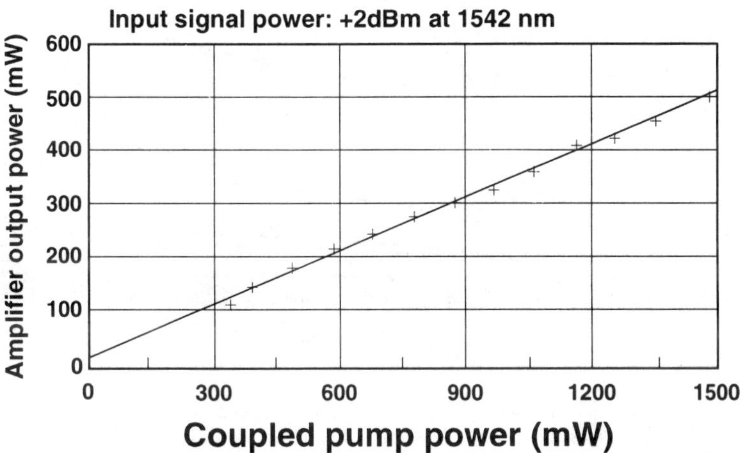

Figure 2.36 Another example of a high-power EDFA consisting of Er^{3+}/Yb^{3+}-codoped fiber and a Nd^{3+}:YLF pump laser [123,124].

Table 2.1
EDFA Gain Changes Due to Temperature for Different Pump Wavelengths

λp	Temperature Range		Gain Change	Temp. Coeff.
0.5145 μm	295K	77K	7 dB	−0.03 dB/°C
0.66 μm	+40°C	0°C	≈0	≈0
0.8 μm	300K	77K	6 dB	−0.03 dB/°C
0.98 μm	+80°C	−20°C	0.4dB	−0.004 dB/°C
1.48 μm	300K	77K	24 dB	−0.11 db/°C
1.48 μm	+85°C	−20°C	7 dB	−0.07 dB/°C

From: [144,145].

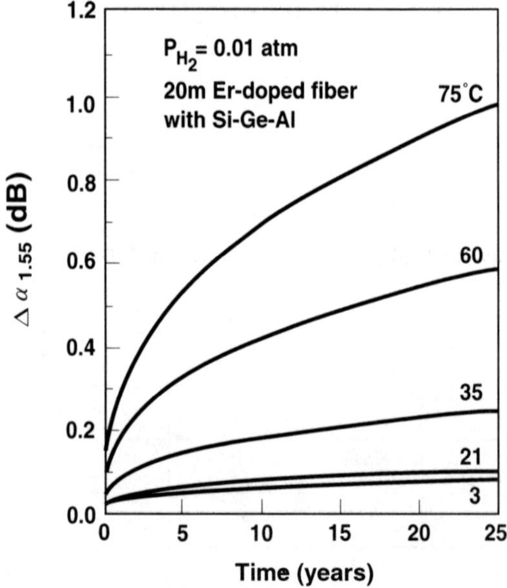

Figure 2.37 Predicted long-term loss increases at 1.55 μm for a 20-m-long erbium fiber doped with 18% GeO$_2$ and 1% Al$_2$O$_3$; the assumed H$_2$ pressure is 0.01 atm [150].

1.55 μm for a 20-m-long erbium fiber doped with 18% GeO$_2$ and 1% Al$_2$O$_3$; the assumed H$_2$ pressure is 0.01 atm [150]. This estimation was carried out on the basis of data obtained from loss increase experiments performed at 92°, 123°, 153°, 180°, and 225°C with a H$_2$ pressure of 0.012 atm. Other experiments have shown that hermetic coating techniques such as carbon coating effectively prevent any increase in the loss of erbium-doped fiber at 1.55 μm [146,151]. The loss increase due to γ-ray irradiation was investigated to estimate environmental effects on EDFA performance in a way similar to that used to investigate the loss increase due to H_2 diffusion [148,153,154]. Figure 2.38 shows the relationship between gain degradation and the total dose of γ-ray irradiation [154]. The γ-ray exposure tests for erbium-doped fibers indicated that the loss increase and gain degradation at 1.55 μm are negligible under realistic conditions for as long as 25 years [154].

Another key concern today regarding the reliability of fiber amplifiers is the reliability of the fluoride fiber host. Figure 2.39 shows experimental results and the predicted reliability of the fluoride fiber host [155,156]. The experimental data shown in the figure were obtained from a dynamic fatigue test on fluoride fibers wound around a drum 8 cm in diameter. The environmental temperature was kept at a constant 80°C. The humidity and tensile strength were varied. The result indicates that 350-MPa ZBLAN fluoride fiber has a lifetime of longer than 25 years even at 80°C and 80% humidity.

2.4 KEY ISSUES REGARDING PRASEODYMIUM-DOPED FIBER AMPLIFIERS

PDFA is expected to be employed in 1.3-μm optical transmission systems that use the standard 1.3-μm single-mode fiber most commonly installed in terrestrial systems

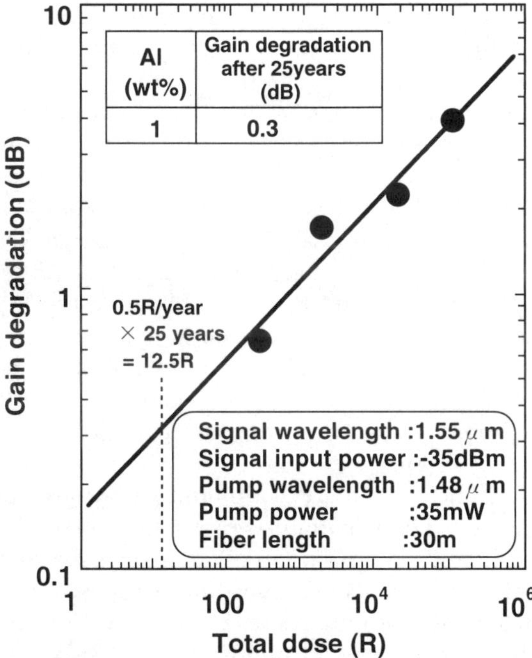

Figure 2.38 Relationship between gain degradation and the total dose of γ-ray irradiation [154].

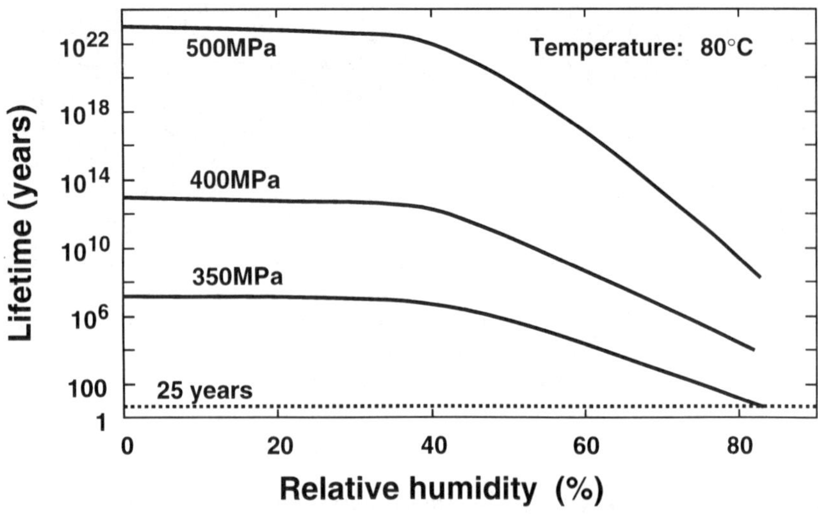

Figure 2.39 Experimental results and the predicted reliability of the fluoride fiber host [155,156].

[22–25]. Recently, digital and analog optical transmission experiments using PDFAs were successfully carried out [157–161], confirming that PDFAs can greatly improve the performance of 1.3-μm optical communications systems. This section summarizes the recent progress on PDFAs.

Figure 2.40 is a schematic diagram of a low-noise and high-power PDFA module with a cascade configuration [162,163], which has two amplification units. Each unit consists of a Pr^{3+}-*doped fluoride fiber* (PDF), a Nd-YLF laser, a WDM fiber coupler, and an isolator. The PDF has a Pr^{3+}-doped ZrF_4-BaF_2-LaF_3-YF_3-AlF_3-LiF-PbF_2 glass core and a ZrF_4-HfF_4-BaF_2-LaF_3-YF_3-AlF_3-LiF-NaF glass cladding. The Δn of the PDF is 2.5% and the scattering loss at 1.20 μm is 0.02 dB/m. The PDFs are pumped by a Nd-YLF laser operating at 1.047 μm. The output power from the pigtail fiber of the laser is about 550 mW. An optical isolator with an isolation of over 60 dB is inserted between the two amplification units and prevents any interaction between the two Nd-YLF lasers. In addition, 14-m- and 22-m-long PDFs are used in the first and second amplification units, respectively, to achieve a low *noise figure* (NF) and high-power operation.

The gain spectrum and NF of this PDFA module at various input signal powers were shown in Figure 2.17. The maximum signal gain is 40.6 dB at 1.30 μm, and the 3-dB down spectral bandwidth is about 17 nm from 1.292 to 1.309 μm for an input signal power of −30 dBm. The bandwidth increases with increasing input power as a result of the gain saturation effect, and hence the bandwidth for an input power of −3 dBm increases to 34 nm. Figure 2.41 shows the output power and NF of the PDFA module as a function of input power. This NF was measured by the optical method and found to be 4.8 dB. The output power increases with

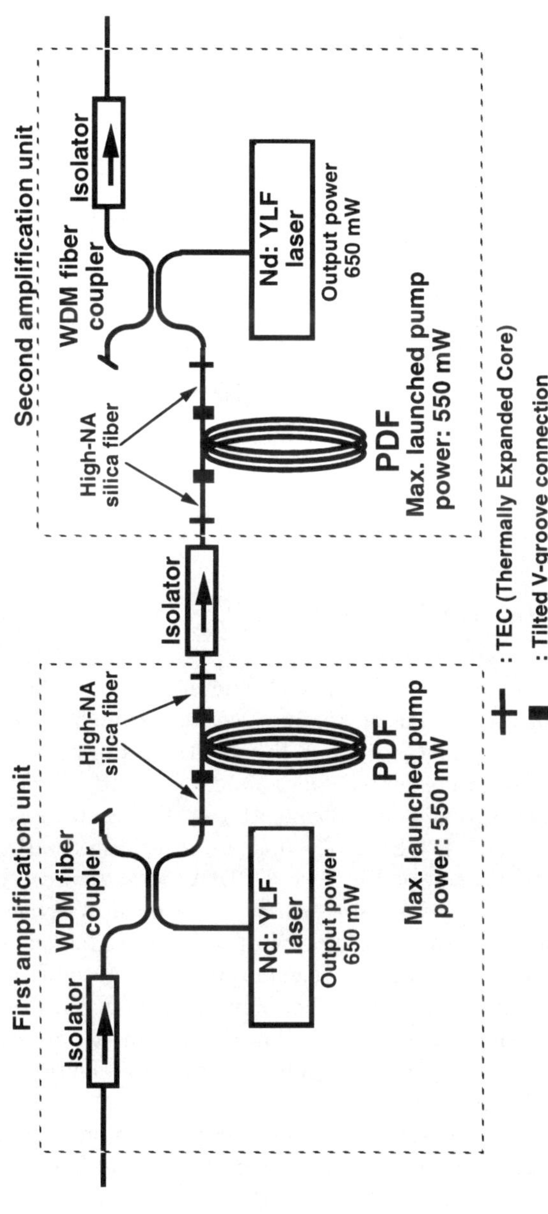

Figure 2.40 Schematic diagram of a low-noise and high-power PDFA module with a cascade configuration [162,163], which has two amplification units.

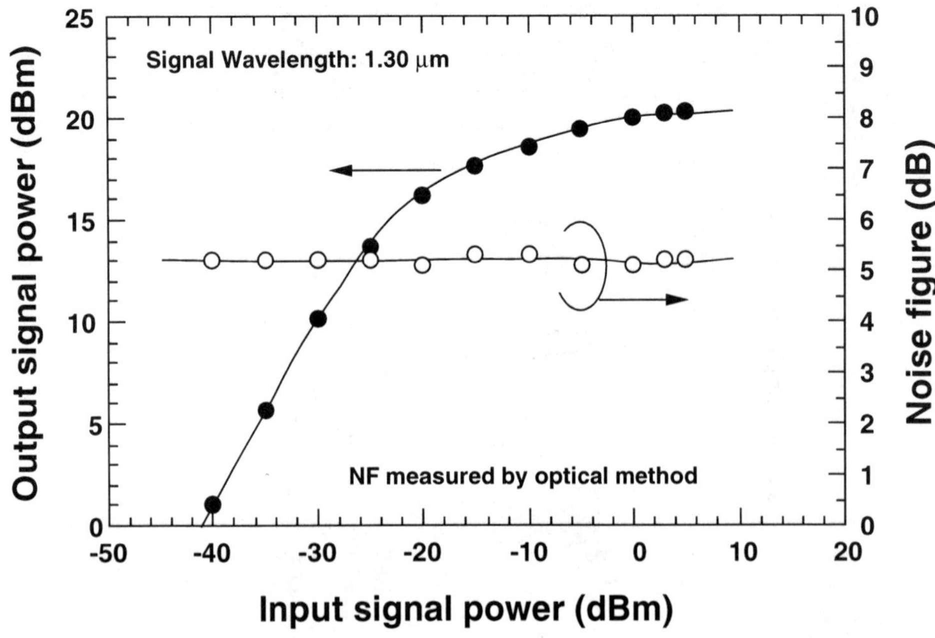

Figure 2.41 Output power and noise figure of the PDFA module as a function of input power.

increasing input power without any increase in the noise figure. An output signal power of 20.1 dBm was achieved at an input power of 0 dBm.

As described in Section 2.2.3, a much higher pump power is required from a PDFA than from an EDFA in order to obtain a high gain and a high signal output power because the quantum efficiency of the 1.3-μm $^1G_4$-$^3H_5$ transition of Pr^{3+} in ZrF_4-based fluoride glass is only 2.2% while that of the 1.5-μm transition of Er^{3+} is almost 100%. Therefore, a key to improving PDFA performance is to increase the quantum efficiency of 1.3-μm amplification. The most effective method for increasing the gain coefficient is to use low-phonon energy glass as the host fiber for Pr^{3+} ions because the multiphonon relaxation rate is inversely related to the number of phonons required to bridge the energy gap between the $^1G_4$ and $^3F_4$ levels. This means that host glasses with a lower phonon energy can provide a longer lifetime of the $^1G_4$ level and the higher quantum efficiency of the 1.3-μm $^1G_4$-$^3H_5$ transition.

There have been several attempts to apply low-phonon energy glasses such as In F_3-based fluoride glass and chalcogenide glasses to Pr^{3+} host fibers [23,164–168]. Figure 2.42 shows the calculated core radius dependence of the gain coefficient for Pr^{3+}-doped InF_3-based fluoride (IBSPZ: InF_3-BaF_2-SrF_2-PbF_2-ZnF_2) and La-Ga-S fibers compared with that of Pr^{3+}-doped ZBLAN [23]. A Δn of 3.7% is assumed in the calculations. The results shown in the figure indicate that gain

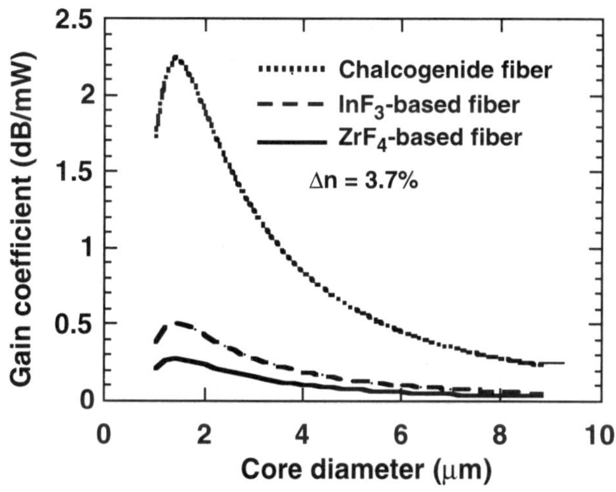

Figure 2.42 Calculated core radius dependence of the gain coefficient for Pr^{3+}-doped InF_3-based fluoride (IBSPZ: InF_3-BaF_2-SrF_2-PbF_2-ZnF_2) and La-Ga-S fibers compared with that of Pr^{3+}-doped ZBLAN [23].

coefficients of more than 0.4 dB/mW and more than 2 dB/mW can be expected with InF_3-based fluoride fiber and La-Ga-S chalcogenide fiber, respectively. Since the highest gain coefficient ever achieved using ZrF_4-based fluoride fiber is 0.24 dB/mW [23], its value can be increased from 2 to 10 times using these new hosts. Therefore, the development of these new hosts is the next major target of PDFA research.

Typical EDFA and PDFA performance is summarized and compared in Table 2.2 [169–177]. As indicated there is no great difference between EDFA and PDFA performance except with regard to the gain coefficient. The gain coefficient, that is, a significant figure of merit for fiber amplifiers, is more than 10 times larger for the EDFA than for the PDFA. As mentioned, this originates in the low quantum efficiency of the 1.3-μm $^1G_4$-$^3H_5$ transition of Pr^{3+} in ZrF_4-based fluoride glasses.

2.5 OTHER WAVELENGTH AMPLIFIERS

In addition to 1.3-μm amplifiers (PDFAs) and 1.5-μm amplifiers (EDFAs), other fiber amplifiers have been developed that are able to operate in other wavelength regions, such as 0.8, 1.4, and 1.65 μm [178–192]. Figure 2.43 shows various active ions and their fluorescence wavelengths [178]. As illustrated, there are more fluorescence wavelength regions for a fluoride glass host than for a silica glass host. As can be seen, fluoride glass has a lower phonon energy than silica glass and, hence, has a smaller number of quenching levels caused by multiphonon relaxation than silica glass.

Table 2.2
Summary of Typical EDFA and PDFA Performance

| | EDFA (1.5 μm) | | PDFA (1.3 μm) | |
	Top Data	Module	Top Data	Module
Gain (dB)	54 (with isolator)	~40	38.2 (internal)	42
Output power (dBm)	24.6	~24	24	13.0
Bandwidth (nm)	40 (Si-Al-P)	40 (Si-Al-P)	38	21
Noise figure (dB)	3.1 (with isolator)	~ 4 (0.98 pump)	3.4	4
Gain coefficient (dB/mW)	11.0	~ 5	0.24	0.4

From: [169–177].

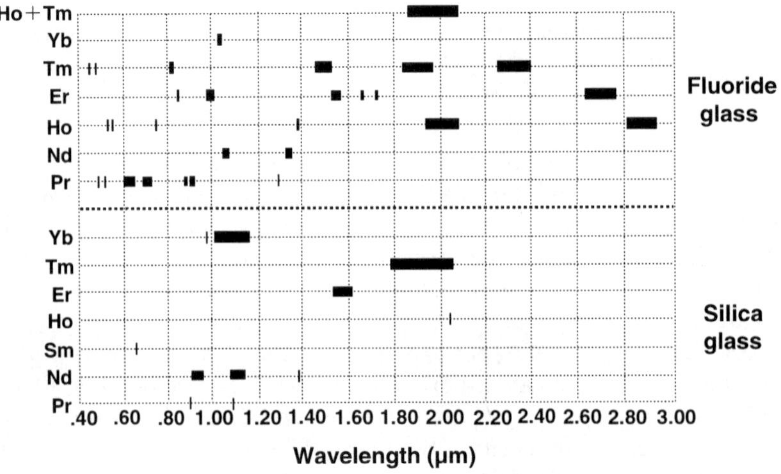

Figure 2.43 Various active ions and their fluorescence wavelengths [178].

Of these possible amplifiers, we first describe an attempt to construct a fiber amplifier operating at 1.4 μm. Figure 2.44 shows the energy diagram of Tm^{3+} ions and amplification in the 1.4-μm region [180,181,187,188]. 1.4-μm amplification is based on the four-level transition from $^3H_4$ to $^3F_4$. We call this 1.4-μm amplifier a *thulium-doped fiber amplifier* (TDFA). However, the 1.4-μm-band amplification is limited by the fact that the lifetime of the upper level, $^3H_4$, is longer than that of the lower level, $^3F_4$, which is a so-called *self-terminating system*. Therefore, since it is difficult to form a population inversion between $^3H_4$ and $^3F_4$, the lower level should be depopulated in order to achieve a high gain. There are two schemes proposed

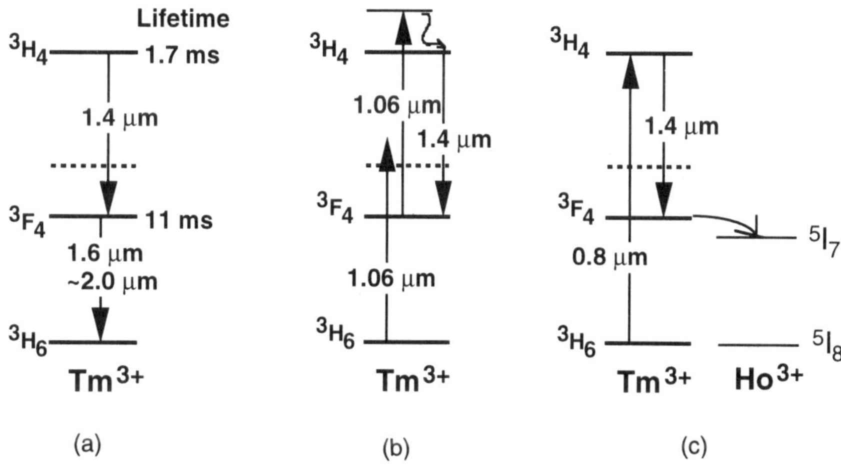

Figure 2.44 Energy diagram of Tm^{3+} ions and amplification in the 1.4-μm region [180]: (a) energy diagram, (b) upconversion pumping, and (c) the codoping of acceptor ions.

for this depopulation of the $^3F_4$ level: upconversion pumping and codoping of acceptor ions, as shown in Figure 2.44(b,c). With upconversion pumping using a Nd:YAG laser, Tm ions excited in the lower $^3F_4$ level are excited to the upper $^3H_4$ level, which results in the depopulation of the $^3F_4$ level of the Tm^{3+} ions as well as excitation to the $^3H_4$ level [187,188]. With codoping, the Tm^{3+} ions excited in the $^3F_4$ level are depopulated through an energy transfer to the acceptor ions [180,181]. The Ho^{3+} ion is one of the most efficient acceptor ions because it shortens the lifetime of the $^3F_4$ level of Tm^{3+} and does not shorten the lifetime of the $^3H_4$ level [180]. A high-power commercial LD operating at 0.8 μm can be used as a pump source with codoping. Figure 2.45 shows gain spectra of the upconversion TDFA, Tm-Ho-codoped TDFA, and Tm-doped TDFA [189]. The active fiber for upconversion TDFA is ZBLYAN fiber with 2,000 ppm of Tm^{3+} in its core, and the fiber for Tm-Ho-codoped TDFA is ZBLYAN fiber with 500 ppm of Tm^{3+} and 10,000 ppm of Ho^{3+} in its core. It is found in Figure 2.45 that the upconversion TDFA and the Tm-Ho-codoped TDFA have wide gain spectra at 1.4 μm, which can cover the entire 1.4-μm band.

Next, we describe an attempt to construct a fiber amplifier operating at 1.65 μm [190–192]. Tm^{3+} ions are being considered for use as active ions operating at 1.65 μm [191,192]. However, the problem with Tm^{3+} ions is that the high-gain region lies in the 1.75- to 2.0-μm wavelength range and ASE occurs easily in this region. Therefore, there is a significant deterioration in the amplification at 1.65 μm because of this ASE. In order to suppress the ASE, the cladding of the fiber is doped with Tb^{3+} ions because Tb^{3+} ions have a large absorption from 1.75 to 2.0 μm. Tm^{3+} ions are doped into the core of the fiber as active ions.

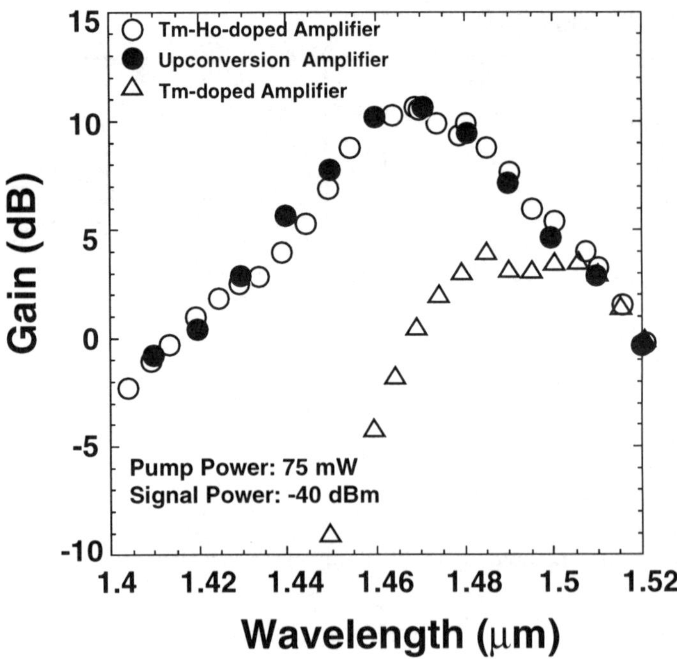

Figure 2.45 Gain spectra of upconversion TDFA, Tm-Ho-codoped TDFA, and simple Tm-doped TDFA [189]. The host is fluoride glass fiber (ZBLYAN: ZrF_4-BaF_2-LaF_3-YF_3-AlF_3-NaF).

Figure 2.46 shows energy diagrams of Tm^{3+} and Tb^{3+} ions and the energy transfer from Tm^{3+} ions (in core) to Tb^{3+} ions (in cladding) [190]. The host is fluoride glass fiber (ZBLYAN: ZrF_4-BaF_2-LaF_3-YF_3-AlF_3-NaF). The parameters of the tested ZBLYAN fibers are as follows: (1) Tm^{3+} ions at 2,000 ppm are doped in the core, and Tb^{3+} ions at 4,000 ppm are doped in the cladding; (2) the core diameter is 1.8 μm; and (3) the refractive index difference of the fiber Δn is 3.7%. Two doped fibers with an incorporated isolator were used for the amplification experiment. One of the ZBLYAN fibers was 13-m long and the other was 10-m long. Figure 2.47 shows the small signal gain characteristic of a thulium-doped fiber amplifier when the signal wavelength is 1.658 μm [190]. A high gain of 35 dB is obtained with a pump power of 140 mW. The gain efficiency was estimated to be 0.75 dB/mW up to a pump power of 100 mW. This gain characteristic indicates that Tb ions effectively suppress lasing as well as ASE in the high-gain region from 1.75 to 2.0 μm. This TDFA operating at 1.65 μm appears to be applicable to an optical transmission line monitoring system operating in the 1.65-μm band, which is indispensable for the construction of 1.3- and 1.5-μm transmission systems [193].

In addition to these TDFAs, various attempts have been made to construct fiber amplifiers and fiber lasers operating in the 0.4- to 3.5-μm wavelength region

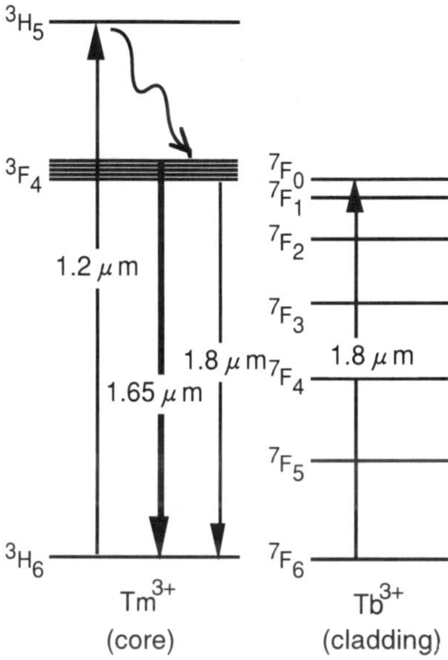

Figure 2.46 Energy diagrams of Tm^{3+} and Tb^{3+} ions and the energy transfer from Tm^{3+} ions (in core) to Tb^{3+} ions (in cladding) [190]. The host is fluoride glass fiber (ZBLYAN: ZrF_4-BaF_2-LaF_3-YF_3-AlF_3-NaF).

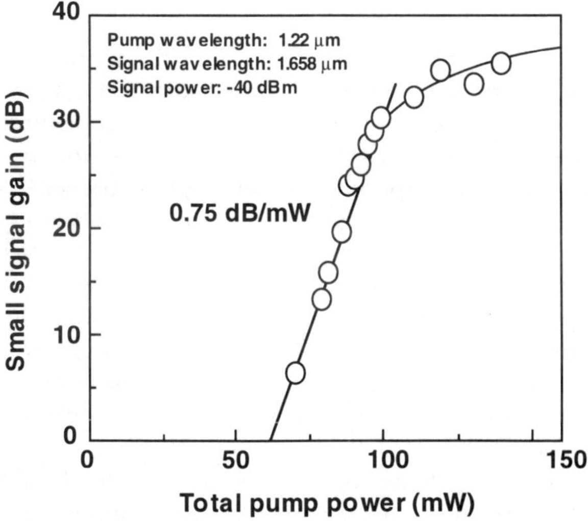

Figure 2.47 Small signal gain characteristic of a thulium-doped fiber amplifier when the signal wavelength is 1.658 μm [190].

by using rare-earth ions. These attempts regarding rare-earth ions and operation wavelengths are as follows: Er^{3+} ion: 0.4697 μm [194], 0.579 μm [195], 1.6 μm [196–200], 2.7 μm [201], 3.5 μm [202–203]; Pr^{3+} ion: 0.492 μm [204], 0.6328 μm [205-206], 1.048 μm [207]; Tm^{3+} ion: 0.48 μm [208–210], 0.68 μm [211], 0.81 μm [212-216], 1.9 μm [217–219], 2.3 μm [220]; Nd^{3+} ion: 0.603 μm [221], 0.938 μm [222], 1.047 μm [223], 1.06 μm [224–248]; Yb^{3+} ion: 1.015 μm [249]; Sm^{3+} ion; 0.651 μm [250]; Ho^{3+} ion: 2.08 μm [251]. However, these fiber amplifiers will not be described in this book.

2.6 KEY ISSUES REGARDING FABRICATION TECHNOLOGIES AND MATERIAL STRUCTURES

2.6.1 Fabrication Processes of Rare-Earth-Doped Fibers

Next, we will describe the key issues regarding the fabrication technology of rare-earth-doped fibers and their material structure, which has strong interactions with various characteristics of fiber amplifiers such as gain, bandwidth, and noise. Thus, we shall begin by reviewing the fabrication processes for rare-earth-doped fibers, focusing mainly on the fabrication process for silica-based rare-earth-doped fibers via the soot process briefly described in Chapter 1. A detailed description of fiber fabrication processes and materials will be given in Chapter 4, which reviews various fabrication methods and materials for transmission fibers as well as for rare-earth-doped fibers.

We will explain the fabrication of rare-earth-doped fibers with the VAD method as a representative of a fabrication method that uses the soot process. Figure 2.48 shows techniques for doping erbium ions into silica-based glasses with the VAD method. As shown in the figure, these rare-earth-doping techniques are categorized into two main processes in which erbium ions are doped into silica glass either during or after the deposition of fine glass particles. As shown in Figure 2.48(a), with the doping during deposition process, erbium ions are doped into silica glasses during glass particle deposition to form a porous preform, and then the porous preform with erbium oxides is consolidated into a transparent erbium-ion-doped fiber preform. In contrast, with the doping after deposition process, erbium ions are doped into the silica glass after soot preform preparation as shown in Figure 2.48(b), and then the porous preform with erbium oxides is consolidated into a transparent erbium-ion-doped fiber preform. We use the expressions "soot" and "glass particle" with almost the same meaning. In more detail, this process of doping after deposition employs two methods for incorporating erbium compounds into porous preforms as shown in Figure 2.48(b): vapor-phase impregnation of the erbium compound and solution (liquid-phase) impregnation of the erbium compound. In both cases, the transparent preform prepared by these processes contains a certain amount of erbium ions. However, it is evident

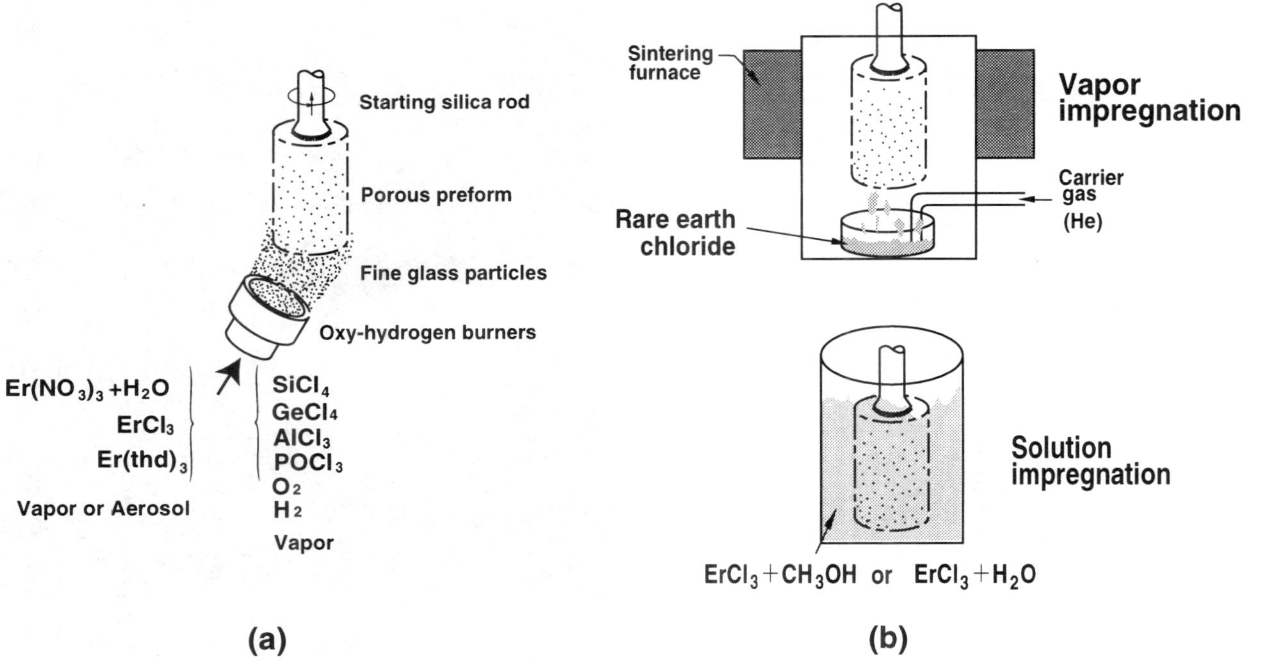

Figure 2.48 Techniques for doping erbium ions into silica-based glasses with the VAD method: (a) doping during deposition and (b) doping after deposition.

from the differences in the mechanisms of glass particle formation with erbium oxides that the microstructures of glass particles with erbium oxides and a transparent preform with erbium ion doping differ slightly depending on whether the doping is performed during or after the deposition process.

Figure 2.49 shows a photograph of porous preform preparation with the VAD method. Fine glass particles (soot) are synthesized in an oxy-hydrogen flame using "flame-hydrolysis," and they are deposited onto the end of the starting silica rod to form a porous preform. The starting silica rod is rotated and pulled upward to form the porous preform rod as seen in the figure. When erbium ions are doped into a porous preform, that is, doping during deposition, erbium compounds such as $Er(NO_3)_3$, $ErCl_3$, or $Er(thd)_3$ are fed into the flame mixed with silica glass raw materials such as $SiCl_4$, $GeCl_4$, and $AlCl_3$. In contrast, when doping after deposition, only silica glass raw materials are fed into the flame to form a porous preform with erbium ions.

The most important aspect of this porous preform preparation process lies in synthesizing the glass particles in the flame. Figure 2.50 shows the saturated vapor pressures for various oxides at high temperature [252,253]. It can be seen in the figure that the saturated vapor pressures of SiO_2 and TiO_2 are several orders of magnitude smaller than the same values for GeO_2 and B_2O_3 in the 1,300° to

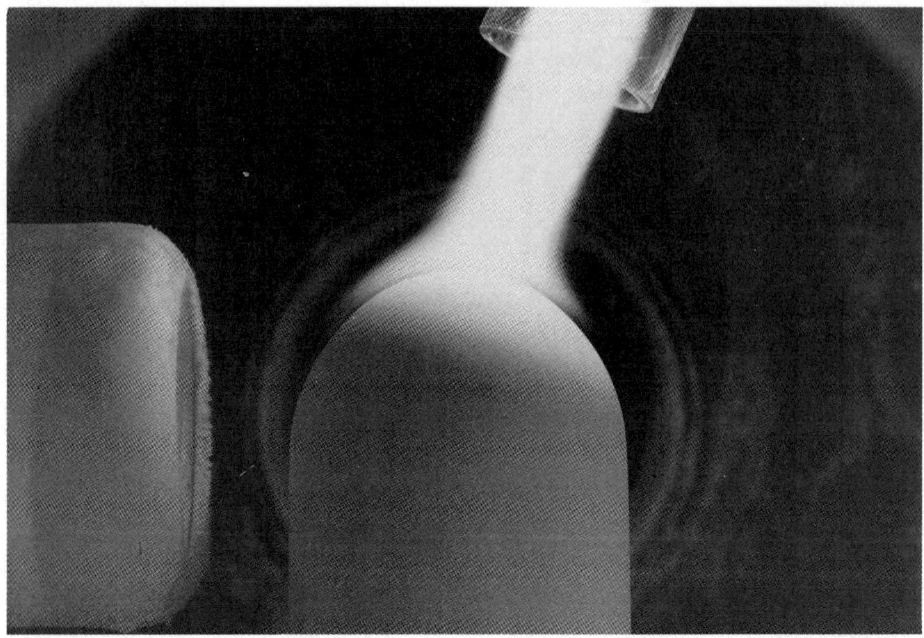

Figure 2.49 Photograph of porous preform preparation with the VAD method.

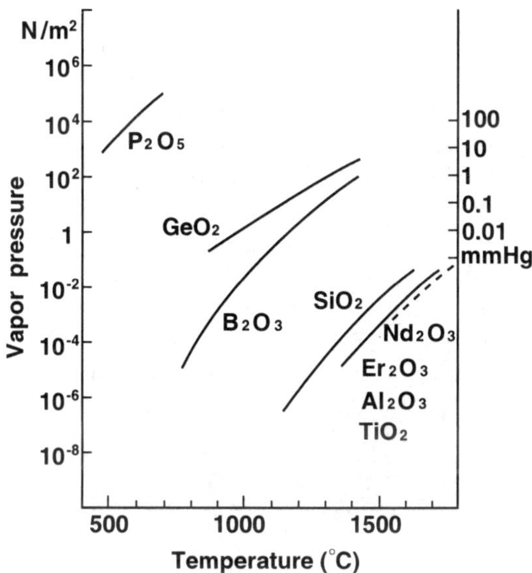

Figure 2.50 Saturated vapor pressures for various oxides at high temperature [252,253].

1,500°C temperature range. The value of P_2O_5 is, on the other hand, several orders of magnitude larger than that for GeO_2. Oxides with higher vapor pressures at a high temperature will tend to remain in the vapor state. For example, oxides whose curves are at higher temperatures than that for SiO_2, that is, GeO_2, B_2O_3, and P_2O_5, are more likely to remain in a vapor state in the flame and not form fine glass particles (solid state). On the other hand, oxides with a lower vapor pressure than SiO_2, that is, Er_2O_3, Al_2O_3, and Nd_2O_3, solidify easily to form fine glass particles in the flame. That means that oxide materials such as Er_2O_3, Al_2O_3, and Nd_2O_3 become solid before SiO_2 or at almost the same time because the vapor pressure differences between SiO_2 and those oxides are very small, as seen in Figure 2.50. Therefore, we can construct a model of the material structure of glass particles, in particular, of erbium-doped silica glass particles, which is of most interest to us.

Figure 2.51 is a photograph of fine glass particles synthesized by a flame hydrolysis reaction with the VAD process. Particles range in size from 500Å to 2,000Å and are almost spherical. The particle size is changed by changing such synthesis conditions as the flame temperature and the spatial concentration of raw materials. In general, these particles become smaller with increasing reaction temperature and increasing spatial concentration [254].

Figure 2.52 shows a SEM photograph of a porous preform growing surface. It can be seen that fine glass particles connect with each other to form open networks.

Returning to the structure of fine glass particles, on the basis of the vapor pressure theory, we can construct models of silica and erbium glass particles as

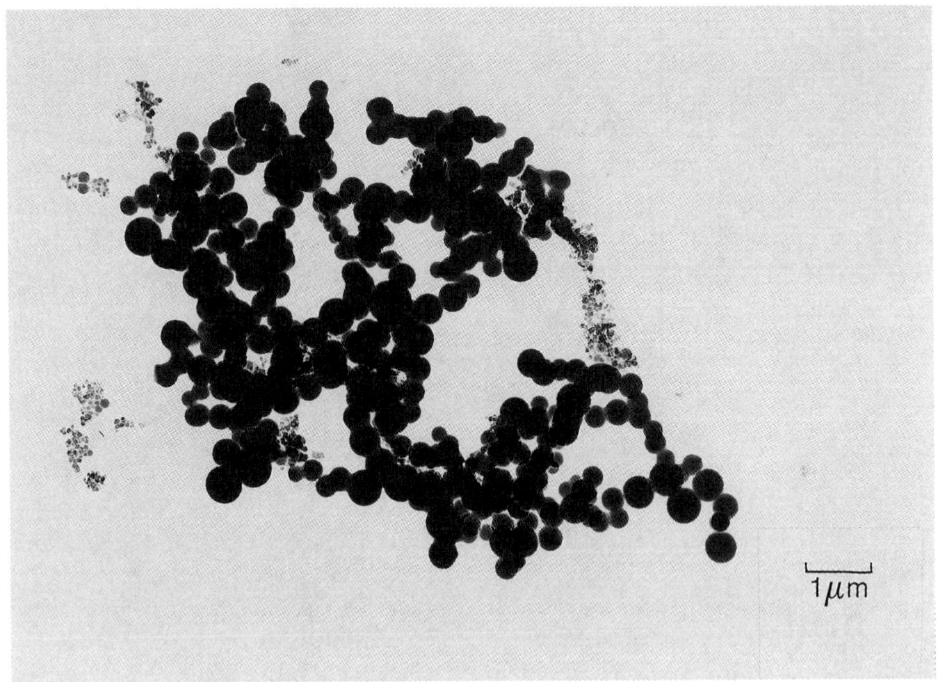

Figure 2.51 Photograph of fine glass particles synthesized by a flame hydrolysis reaction with the VAD process.

shown in Figure 2.53. When we prepare glass particles from a mixture of SiO_2 and Er_2O_3 by the doping during deposition process, they have the structure shown in Figure 2.53(a) because Er_2O_3 becomes solid before or at almost the same time as SiO_2 since Er_2O_3 and SiO_2 have almost the same vapor pressure as shown in Figure 2.50. Therefore, we have Er_2O_3 at the center of the SiO_2 particles if Er_2O_3 becomes solid a little before SiO_2, or we have glass particles with a dissolved SiO_2-Er_2O_3 structure when Er_2O_3 and SiO_2 become solid at almost the same time.

In contrast, when we dope erbium after deposition, we have glass particles with the structure shown in Figure 2.53(b). In this doping after deposition process, the glass particles have Er_2O_3 on their surface not inside, or Er_2O_3 forms independent particles because SiO_2 particles are already in existence. Consequently, these independent Er_2O_3 particles are likely to become clusters in glasses. This means we can prepare a fiber preform with uniformly doped erbium ions more easily by doping during deposition than after deposition because even if the glass particles shown in Figure 2.53 are placed in a high-temperature (around 1,500°C) consolidation process, it is difficult to melt the independent Er_2O_3 into SiO_2 glass networks and to form a dissolved SiO_2-Er_2O_3 structure. The Er^{3+} ions of the dissolved SiO_2-Er_2O_3 structure can act as active ions for 1.5-μm amplification without quenching.

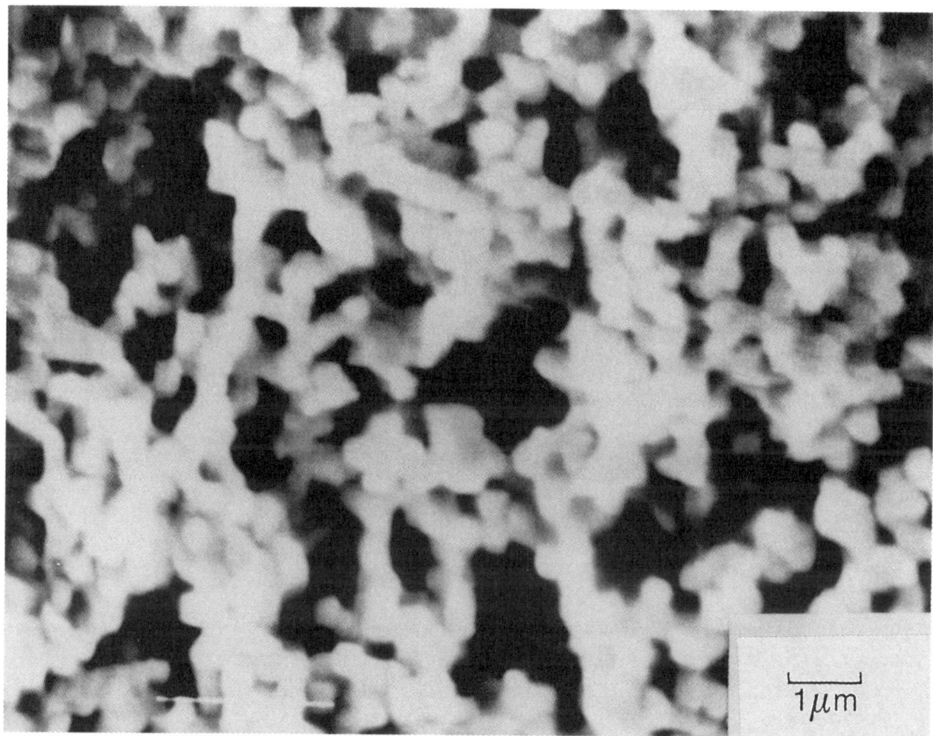

Figure 2.52 SEM photograph of a VAD porous preform growing surface.

The models of Er_2O_3 and SiO_2 glass structures shown in Figure 2.53 are valid for the two other soot-process fiber-fabrication methods: the MCVD and OVD methods. The glass particle structure with the OVD method is exactly the same as that with the VAD method. With the MCVD method, when we feed the erbium compounds and the glass raw materials such as $SiCl_4$ and $GeCl_4$ together into the silica tube as shown in Figure 1.5, the synthesized glass particles have the structure shown in Figure 2.53(a) because this erbium doping process is the aforementioned doping during deposition process. On the other hand, with the erbium doping after deposition process for the MCVD method, the solution doping technique has been developed [255,256]. With this technique, an unsintered (porous) layer of SiO_2 is first deposited inside a silica tube, and then this SiO_2 layer is doped by filling the tube with an aqueous rare-earth (erbium) chloride solution. After the solution is soaked and then drained completely, the impregnated layer is dried at a high temperature in a mixed chlorine/oxygen flow because dehydration will reduce the OH contamination. Eventually, the impregnated layer is consolidated into a transparent glass layer and the tube is collapsed to make a fiber preform.

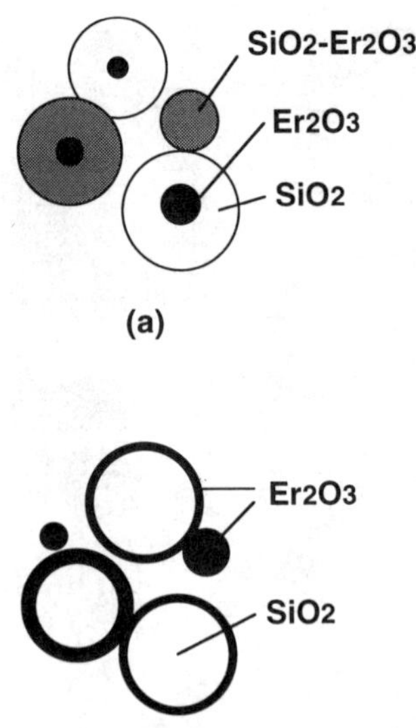

Figure 2.53 Structural models of silica and erbium glass particles: (a) Er-doping during soot deposition and (b) Er-doping after soot deposition.

In this doping after deposition process with the MCVD method, erbium oxides form independent fine particles similar to those shown in Figure 2.53(b). If there is a slight difference between this MCVD method and the VAD and OVD methods, the deposited SiO_2 glass particles are different sizes. With the VAD and OVD methods, the SiO_2 glass particles are synthesized in the oxy-hydrogen via "flame hydrolysis." In contrast, in the MCVD method, the SiO_2 glass particles are synthesized by the oxidation process inside the silica tube. In general, it is known that glass particles synthesized by the oxidation process are much smaller than glass particles synthesized by flame hydrolysis where there is a rich water vapor [257]. This size difference in deposited SiO_2 particles causes some difference in the glass uniformity of a SiO_2-Er_2O_3 mixture. It could be estimated that in the doping after deposition process uniform SiO_2-Er_2O_3 glass is more easily prepared than with the VAD method or the OVD method because of its small deposited SiO_2 glass particles.

Based on this discussion, we can say the following regarding rare-earth-doped fiber preform fabrication. First we can prepare a fiber preform with uniformly doped erbium ions more easily by doping during deposition than after deposition because, with the doping during deposition process, dissolved SiO_2-Er_2O_3 glass particles and glass particles with Er_2O_3 at their center are formed and, in turn, easily form a uniformly dissolved SiO_2-Er_2O_3 structure. Second, in the doping after deposition process, it is easier to prepare uniform SiO_2-Er_2O_3 with the MCVD method than with the VAD or OVD methods because of the small size of the deposited SiO_2 glass particles.

2.6.2 Material Structures of Rare-Earth-Doped Glasses and Their Effects on Amplification Characteristics

Now, we will take a closer look at rare-earth-doped glasses and describe their fine material (atomic) structures, which interestingly relate to the broadband characteristics of EDFAs and the gain characteristics of fiber amplifiers.

Figure 2.54 shows erbium ions in pure silica glass [66,258]. When we introduce an erbium ion into a SiO_2 glass network, each tetrahedron consisting of one Si and four oxygens (O) moves along the direction indicated by the arrows as shown in the figure. As a result, an "octahedral" complex is constructed from undistorted glass-forming tetrahedra of SiO_2. The expansion of the central void by the translation of each tetrahedron in the directions indicated by the arrows results in the formation of a trigonal field in the void (Er position). Because of this movement, two oxygens act as one in terms of charge, and as a result the coordination balance is achieved in the vicinity of erbium ion because of the 3+ erbium and the three

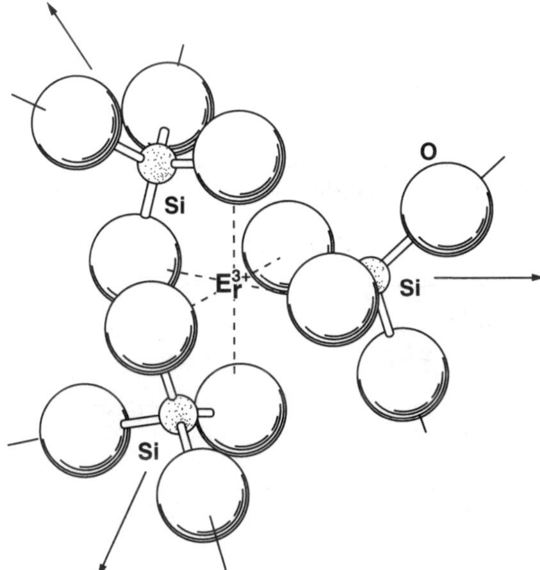

Figure 2.54 Erbium ions in pure silica glass [66,258]. Each tetrahedron consisting of one Si and four oxygens (O) moves in the direction indicated by the arrows. As a result, an "octahedral" complex is constructed from undistorted glass-forming tetrahedra of SiO_2. The expansion of the central void by the translation of each tetrahedron in the directions indicated by the arrows results in the formation of a trigonal field in the void (Er position). Because of this movement, two oxygens act as one in terms of charge and consequently the coordination balance is achieved in the vicinity of the erbium ion because of the 3+ erbium and the three −1 oxygens.

–1 oxygens. However, this structural distortion by the translation of each SiO_2 tetrahedron causes a nephelauxetic shift (a distortion of the electronic charge cloud of the erbium ions), which eventually leads to a large Stark split of the energy levels in the silica-based host, as explained in Section 2.3.1 and shown in Figure 2.25.

This structural distortion is very closely related to the alumina codoping technique and the concentration effect of the EDFA. In 1986–1987, K. Arai et al. constructed a comprehensive model of the coordination structure and atomic configuration for rare-earth-doped SiO_2 glass [259–261]. This model is shown in Figures 2.55 and 2.56. As mentioned, when we introduce rare-earth ions into silica

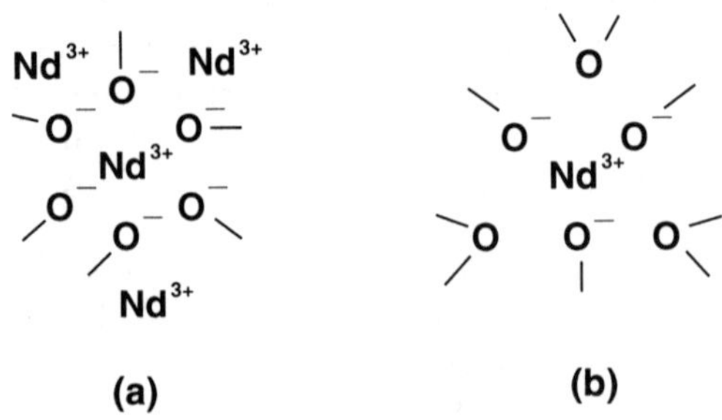

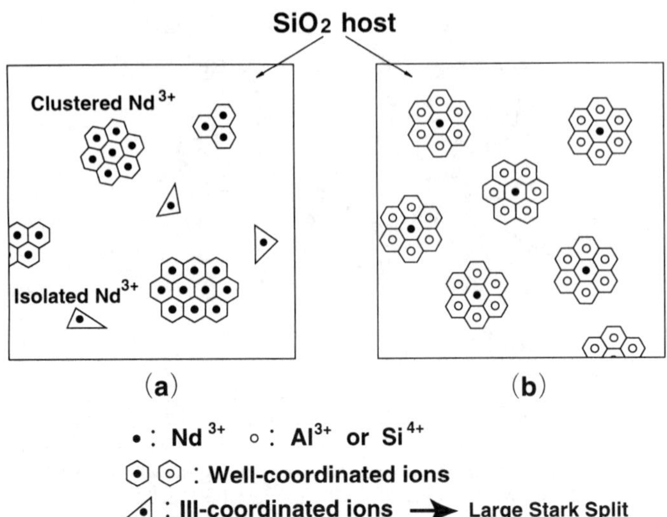

Figure 2.55 Model of coordination and atomic configuration for Nd^{3+}-doped SiO_2 glass constructed in 1986 by K. Arai et al. [259]: (a) clustering Nd^{3+} ions and (b) isolated Nd^{3+} ion.

Figure 2.56 Model of coordination and atomic configuration for Nd^{3+}-doped SiO_2 glass and effect of Al-codoping constructed in 1986 by K. Arai et al. [260,261]: (a) Nd-doped SiO_2 and (b) Nd-Al-codoped SiO_2.

glass, Nd^{3+}, into this model, the SiO_2 structure is distorted as shown in Figure 2.55(b). These ions are called "ill-coordinated ions" and are depicted by the triangular symbols in Figure 2.56. Nd^{3+} ions collect around an isolated Nd ion, hence achieving the coordination balance, which we call a cluster as shown in Figures 2.55(a) and 2.56(a). In contrast, when we codope SiO_2 glass with Al_2O_3 and rare-earth ions, charge compensation is achieved by the alumina surrounding an isolated Nd, as shown in Figure 2.56(b). We call these ions "well-coordinated ions," which means we should be able to dope rare-earth ions at a high concentration using alumina-codoping techniques without forming clusters, therefore avoiding any degradation of the gain coefficient [259–264]. In addition, these well-coordinated ions surrounded by alumina should have a small nephelauxetic shift (a small distortion of the electronic charge cloud of the erbium ion). This results in a small Stark split of the energy levels in this alumina-codoping silica glass host, which eventually leads to the flat gain spectrum achieved by alumina-codoping techniques, as shown in Figure 2.19. By contrast, when there is no alumina codoping, the gain degradation occurs easily at a content level of over a few 100 ppm [265].

When we introduce rare-earth ions into multicomponent glasses and fluoride glasses, charge compensation is readily achieved and the coordination is well balanced in the vicinity of rare-earth ions. Figure 2.57 shows the charge compensation and coordination balance in rare-earth-ion-doped multicomponent glass [64,65],

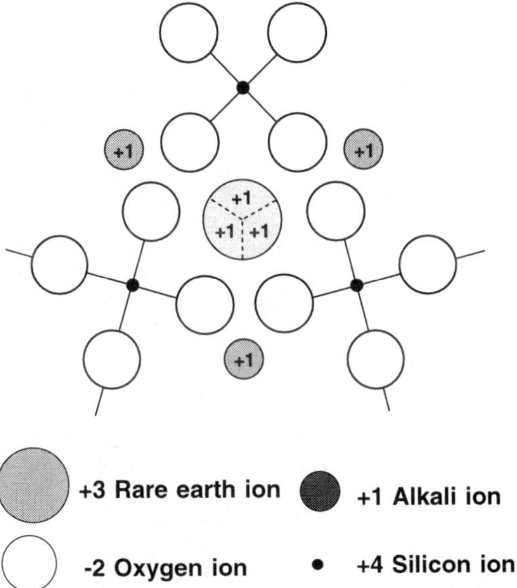

Figure 2.57 Charge compensation and coordination balance in rare-earth-ion-doped multicomponent glass [64,65], where the +3 charge of the rare-earth ion and the three +1 charges of the three alkali ions surrounding the rare-earth ion are well balanced with the six −1 charges of the six oxygen ions.

where the +3 charge of the rare-earth ion and the three +1 charges of three alkali ions surrounding the rare-earth ion is well balanced with the six −1 charges of the six oxygen ions. Therefore, we can dope rare-earth ions very heavily in multicomponent glass systems. The charge compensation is almost the same in the ZrF_3-based fluoride glass system shown in Figure 2.58 [266]. We can achieve a very high concentration without forming clusters in both glass systems. In addition, both glass systems can realize the well-balanced coordination surrounding rare-earth ions, which should have a small nephelauxetic shift (a small distortion of the electronic charge cloud of the erbium ions). This results in only a small Stark split in the energy levels in this alumina-codoped host, which eventually leads to the flat and broadband gain spectrum achieved by the fluoride-glass host shown in Figures 2.19 and 2.20.

Briefly summarizing the fabrication and materials, we arrive at the following features and advantages. First, in terms of fabrication, the doping during deposition process is superior to the after deposition process with regard to the uniform doping of rare-earth ions, which means "without forming a cluster." With regard to host material, a silica glass host has the advantage of extremely high gain. The advantage of alumina-codoped silica glass is its high gain and broadband gain spectrum. Multicomponent glass is suitable for a short-length amplifier because of the possibility of very high concentration. Fluoride glass has the advantage of a broadband and flat gain spectral characteristics.

2.7 RECENT TOPICS ON AMPLIFIED SYSTEMS

Various fiber amplification systems and system applications have already been investigated for employment in long, intermediate, and short transmission systems, WDM

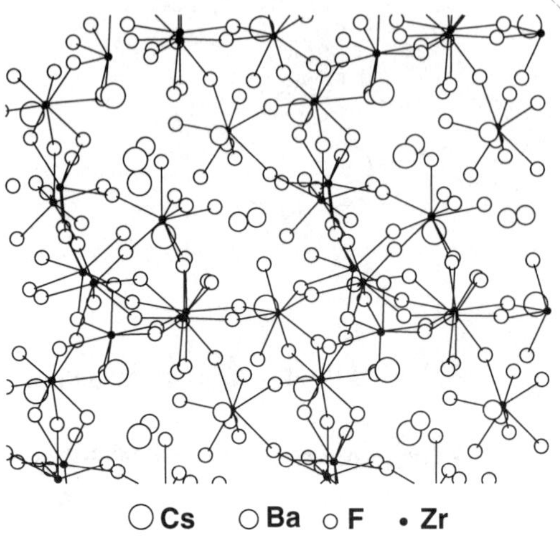

Figure 2.58 Glass structure of the ZrF_3-based fluoride glass system [266].

systems, CATV transmission systems, ultra high speed transmission systems, and soliton transmission systems. Some have already been practically installed.

Finally, we describe several recent topics with regard to transmission systems that employ fiber amplifiers. Recent amplified system experiments can mainly be divided into five groups. The first is long-distance transmission systems for NRZ signals [267–272] or solitons [273,274] using many fiber amplifiers. The second is unrepeated-long-span transmission [275,276] using remotely pumped EDFAs [277]. The third is WDM systems [278–281]. The fourth is ultra high speed transmission systems [282-284]. The fifth is CATV transmission systems [284,285]. Furthermore, combined transmission systems have been investigated, such as combinations of WDM and long distance [286,287] and WDM and soliton systems [288]. In 1996, three terabit-per-second transmission systems were demonstrated using a 55-wavelength by 20-Gbps WDM transmission [289], a 50-channel by 20-Gbps transmission [290], and a 10-channel by 100-Gbps transmission [291].

Optical networking is another topic of interest in the field of optical communications. WDM add/drop experiments and various other types of networking experiments are currently being energetically attempted [292–304]. Moreover, worldwide optical service networks are being rapidly constructed via undersea cables [305–315].

In terms of 1.3-μm systems, a 40-channel AM-VSB video signal transmission and a 10-Gbps digital transmission have been achieved through the use of PDFA modules [157–163].

2.7.1 Unrepeated Long-Span Transmission

Of these various experiments, first we review an unrepeated long-span transmission that employs remotely pumped EDFAs [275,276]. Figure 2.59 shows the unrepeated system configuration, (a) without and (b) with remotely pumped amplifiers. The major advantages of unrepeated systems are their reliability and low cost. The most obvious application of optical amplifiers in unrepeated systems is to amplify the signal before it is launched into the line. In Figure 2.59(a), the postamplifier provides a power budget improvement that is directly equal to the amplification gain. By contrast, at the other end of the line, the optical preamplifier improves the signal-to-noise ratio before the signal reaches the receiver photodiode. However, the improvement obtained using postamplification is limited by the maximum power that can be launched into the fiber, and this maximum power limitation originates in the fiber nonlinearity induced by a combination of the Kerr effect and chromatic dispersion. On the other hand, preamplification at the receiver side has intrinsic limitations due to noise. The remotely pumped EDFA was developed in 1989 to overcome these limitations [277]. This has recently been made practical by the realization of powerful pump lasers operating at 1.48 μm. As shown in Figure 2.59(b), the basic idea of remote amplification is to inject the pump power

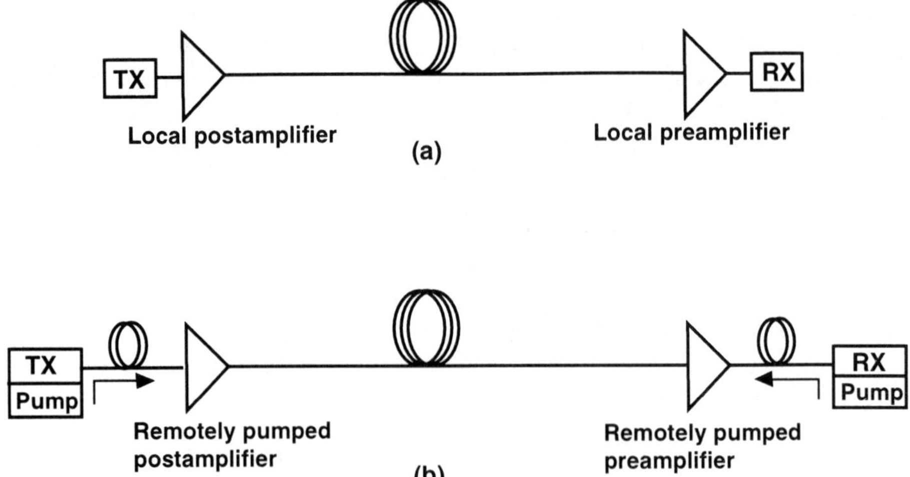

Figure 2.59 Unrepeated system configuration: (a) without and (b) with remotely pumped amplifiers.

into the fiber from the terminals in order to generate gain in the erbium-doped fiber sections that are located 50 to 100 km away from the terminals. Very long-span unrepeated systems of over 500 km have already been demonstrated that use this remote amplification technique [275,276]. Figure 2.60 is a schematic diagram of a demonstration of a 529-km-span unrepeated transmission [275]. As shown in the figure, two remotely pumped EDFAs were used: on the transmitter side as a booster amplifier and on the receiver side as a preamplifer or an in-line amplifier. These EDFAs are remotely pumped at a wavelength of 1.48 μm at which the transmission fibers have a lower transmission loss than at 0.98 μm or other pumping wavelengths.

2.7.2 Long-Distance Transmission

In long-distance transmission systems, the maximum power that can be launched into the fiber is limited by the fiber nonlinearity that is induced by a combination of the Kerr effect and chromatic dispersion, and the minimum power at the receiver side is limited by the signal-to-noise ratio, as shown schematically in Figure 2.61. Therefore, the maximum power ranges from 0 to 10 dBm, and the minimum power ranges from −10 to −20 dBm. With regard to long-distance transmission systems, an impressive transmission experiment was performed in 1993 by a joint group from KDD and AT&T [267]. They achieved a 10-Gbps, 9,040-km *intensity modulated and direct detection* (IM-DD) transmission using 274 EDFA repeaters. Figure 2.62 is a schematic diagram of the experimental setup for a 10-Gbps, 9,040-km IM-DD

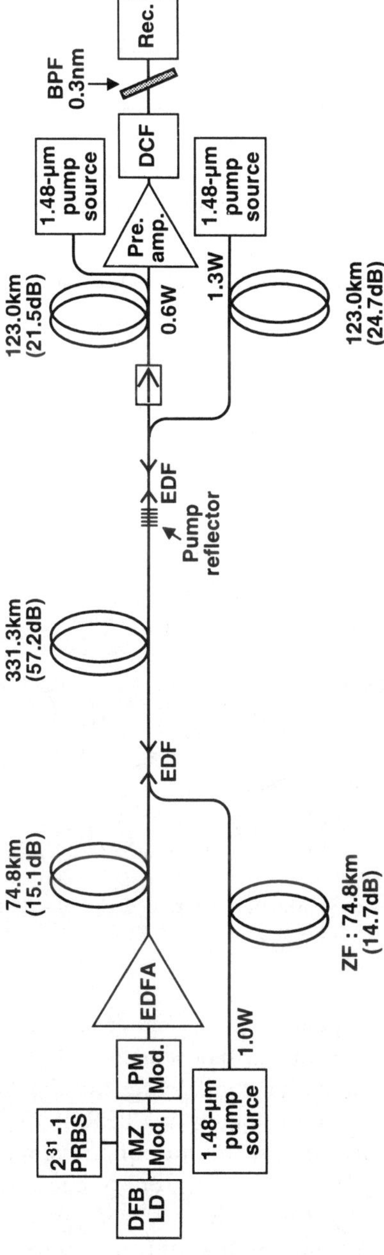

Figure 2.60 Schematic diagram of a demonstration of a 529-km-span unrepeated transmission [275].

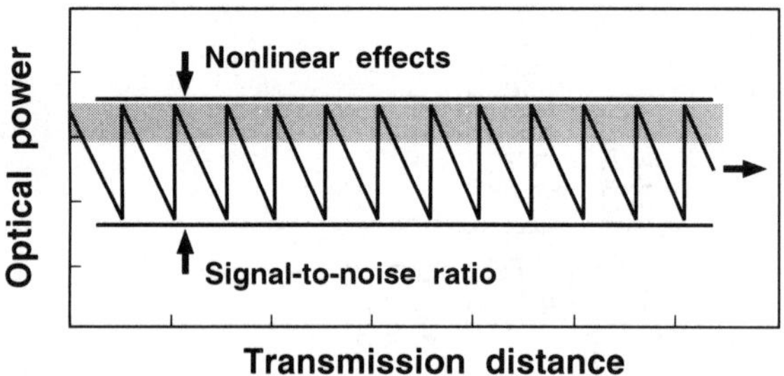

Figure 2.61 Schematic diagram of optical power evolution and repeater span limitation.

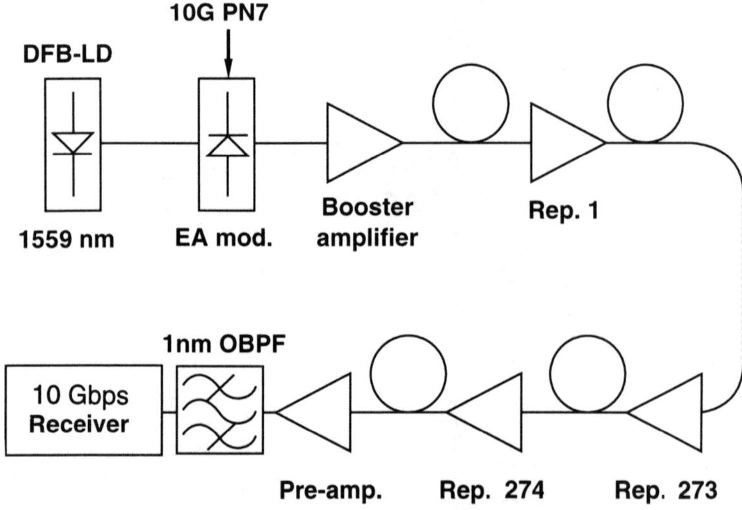

Figure 2.62 Schematic diagram of the experimental setup for a 10-Gbps, 9,040-km IM-DD transmission using 274 EDFA repeaters [267].

transmission using 274 EDFA repeaters [267]. Figure 2.63 shows the measured BER versus the optical preamplifier input signal power [267]. In this figure, the solid and open circles denote measured data for the baseline and the 9,040-km transmission, respectively, revealing that the power penalty at a BER of 10^{-9} due to the 9,040-km transmission was about 4.6 dB. The maximum optical signal power to the optical preamplifier is reported to be limited to −20 dBm by the restriction of the photo-current of the PIN-PD in the optical receiver.

An important feature in long-distance transmission systems is the gain and noise filtering effect in cascaded amplifiers [267–272]. Amplifiers, especially silica-based EDFAs, have a wavelength-dependent gain that reaches its maximum value

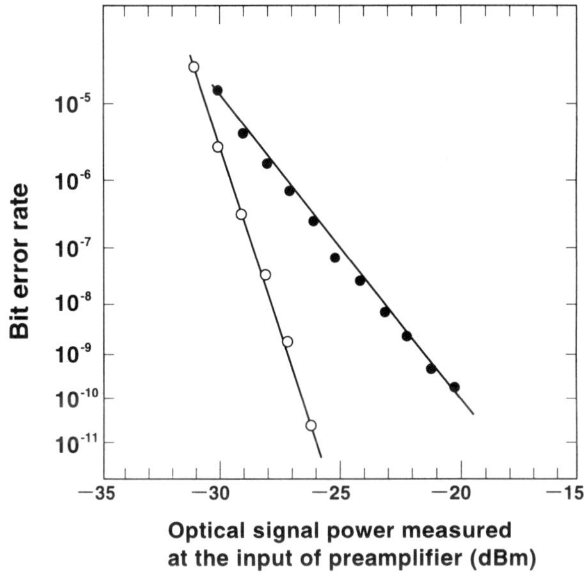

Figure 2.63 Measured BER versus optical preamplifier input signal power [267]. Solid and open circles denote measured data for the baseline and the 9,040-km transmission, respectively.

around 1,532.5 or 1,558.0 nm depending on the doped fiber length and the span loss, which induces a self-filtering effect. This self-filtering effect leads to a great reduction in the available system bandwidth, so the signal wavelength must be close to the maximum gain wavelength. This bandwidth reduction is a system constraint, but it is also very beneficial because it filters the noise generated by in-line amplifiers and eliminates the need for additional filters. Figure 2.64 shows the gain and noise filtering effect in cascaded amplifiers at the end of a 9,000-km link [271]. In this figure, the solid curve indicates the simulated optical spectrum of an amplifier designed for long-wavelength filtering, and the dashed curve indicates the simulated optical spectrum of an amplifier designed for short-wavelength filtering. The doped fiber lengths are 13m for the short-wavelength filtering amplifier and 20m for the long-wavelength filtering amplifier. The gain peak wavelengths are 1,532.5 nm for short-wavelength filtering and 1,558.5 nm for long-wavelength filtering. However, even though there is a noise filtering effect as a result of the gain filtering, in long-distance and cascaded amplifier transmissions, the ASE power increases with the transmission distance. Figure 2.65 shows the output power evolution of cascade amplifiers gain filtering at 1,532.5 nm [270]. In this experiment, the input power always remains around the gain peak at 1,532.5 nm and the total power is fixed. Since each EDFA adds ASE power and the total power is fixed, gain cannot compensate exactly for the transmission loss. This leads to a slow decay of signal power, as

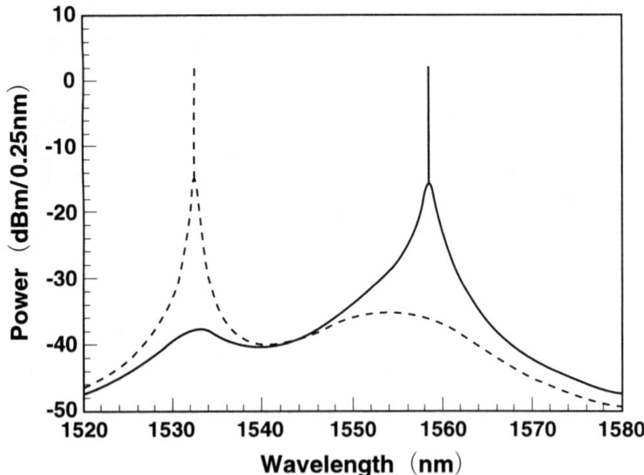

Figure 2.64 Gain and noise filtering effect in cascaded amplifiers at the end of a 9,000-km link [271]. The solid curve indicates the simulated optical spectrum of an amplifier designed for long-wavelength filtering, and the dashed curve indicates the simulated optical spectra of an amplifier designed for short-wavelength filtering.

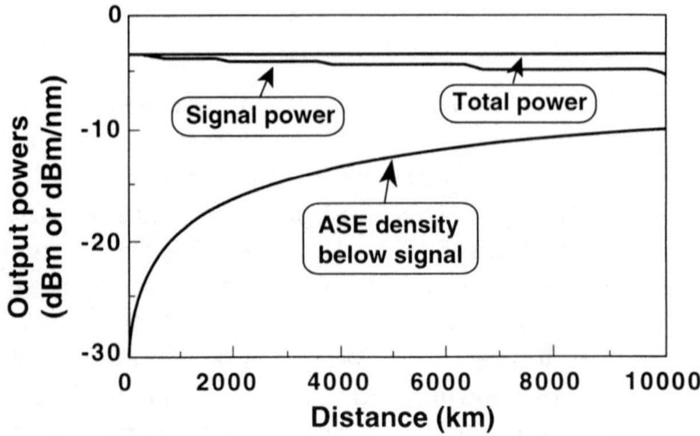

Figure 2.65 Output power evolution of cascade amplifier gain filtering at 1,532.5 nm [270].

shown in Figure 2.65. However, as seen in the figure, the ratio of signal power to ASE power is estimated to be nearly 5 dB even after 10,000-km transmission. The BER floor due to the ASE accumulation with a 5-Gbps signal would be lower than 10^{-20}. It is also evident that gain filtering leads to almost the same output power evolution as that with an additional Lorentzian filter [270].

2.7.3 WDM, Long-Distance, and/or High-Speed Transmission

Another recent area of interest in relation to amplified systems is the combination of WDM and a long transmission distance using either *non-return-to-zero* (NRZ) signals or solitons [286,288–316]. The transmission of many WDM channels over long distances can be limited by the nonlinear interaction between channels. Recent developments have made it possible to suppress these nonlinear interactions by tailoring or managing the chromatic dispersion of the transmission fibers and operating the system at low channel power [286,287]. Figure 2.66 shows the optical spectra of eight 5-Gbps WDM channels: (a) before and (b) after propagation over 8,000 km [286]. This transmission experiment was performed using an eight-wavelength transmitter and a 1,000-km EDFA chain in a circulating loop. The amplifier chain was constructed using 20 amplifier spans with 45-km lengths of negative dispersion fiber and two amplifier spans with 45-km lengths of positive dispersion fiber (that is, $l_0 = 1,300$ nm). This means that the large negative disper-

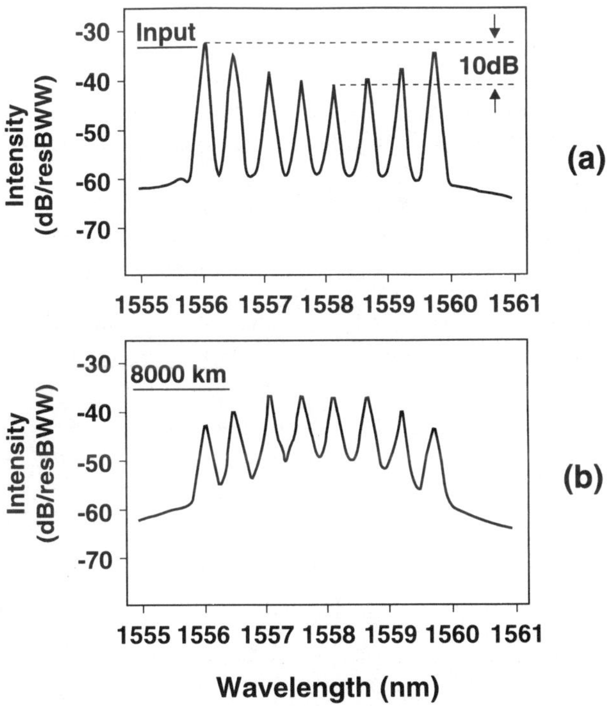

Figure 2.66 Optical spectra of eight 5-Gbps WDM channels: (a) before and (b) after propagation over 8,000 km [286].

sion in the transmission fiber was compensated for with 90 km of standard single-mode fiber with a large positive dispersion. Each 1,480-nm pumped EDFA had a small signal gain of about 15 dB and a noise figure of 5.2 dB at the operating wavelength. The total power launched into the transmission spans (for all channels) was +2.3 dBm; thus, the path-averaged power for any individual channel was only about 90 μW. Consequently, it was reported that no *four-wave mixing* (FWM) products were observed in the optical spectrum, demonstrating the effectiveness of the dispersion management scheme and the low-power-level operation.

Figure 2.67 shows the relative sideband powers or mixing efficiencies caused by the FWM effect for two channels as functions of channel separation $\delta\lambda$, for various values of chromatic dispersion D [293]. FWM, which is the dominant nonlinear effect in long-haul multichannel transmissions employing dispersion-shifted fibers, causes channel crosstalk. It is important to note in Figure 2.67 that a very small amount of dispersion, for example 1 to 2 ps/nm-km, disrupts the FWM phase-matching condition and hence significantly reduces the FWM effects.

Another idea has been proposed for reducing the FWM effects and the resultant channel crosstalk, which is called "unequally spaced channels" [317–320]. This technique employs an unequal channel spacing for WDM transmission in which the FWM effect suppressed by shifting its phase-matching conditions between channels. The use of this unequal channel spacing allows the fiber input power to be increased [317–320]. Actually, it allows a fivefold increase in the bit rate per

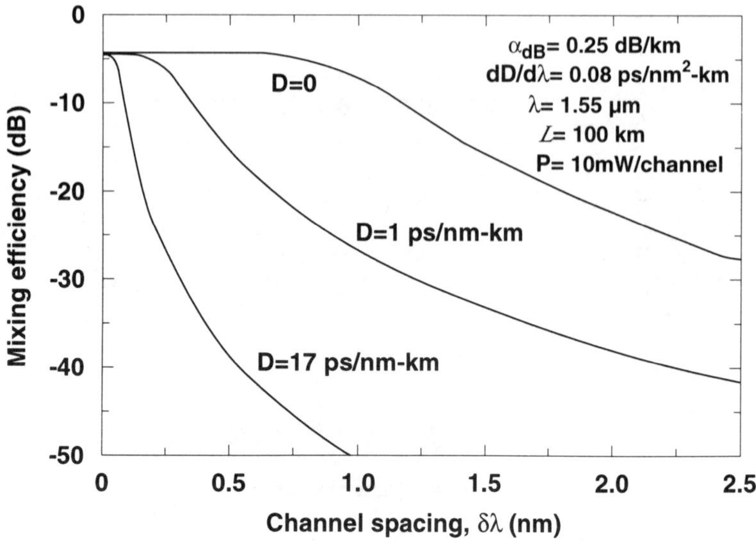

Figure 2.67 Relative sideband powers or mixing efficiencies caused by the FWM effect for two channels as functions of channel separation $\delta\lambda$ for various values of chromatic dispersion D [293].

channel (and therefore in the system capacity) or an increase in the system length because of the higher permissible input power.

In addition to this NRZ signal WDM transmission over long distance, a soliton WDM transmission experiment has been performed over 10 Mm using an 8- by 2.5-Gbps soliton signal [288]. A dispersion management scheme was used for soliton WDM transmission similar to that used for the NRZ WDM long-distance transmission. The 10-Mm soliton transmission line consisted of 54 amplifiers (noise figure 4.8 dB at 10 dB gain), spaced every 45 km. Each fiber span consisted of appropriate lengths of standard single-mode fiber (l_0 = 1,300 nm) with positive dispersion and dispersion-shifted fiber with negative dispersion in order to achieve the desired path-average dispersion of − 0.6 ps/nm-km at the mean signal wavelength (1,557.4 nm). At the Conference on Optical Fiber Communication, 1996, a major breakthrough was announced in the WDM field: Three terabit-per-second transmission systems were demonstrated that used a 55-wavelength by 20-Gbps WDM transmission [289], a 50-channel by 20-Gbps WDM transmission [290], and a 10-channel by 100-Gbps WDM transmission [291]. Figure 2.68 shows terabit-per-second WDM spectra (a) before (0 km) and (b) after (40 km) dispersion-shifted fiber transmission [291]. The figure also shows 10 channels allocated in the 1,533.6- to 1,562.0-nm wavelength range. The amplifiers used in this experiment were fluoride-based EDFAs—to be more exact, two-stage hybrid-type amplifiers consisting of silica-based amplifiers pumped at 980 nm as the first stage for low-noise amplification and a fluoride-based amplifier pumped at 1,480 nm as a second stage for wide-band amplification. Details of this hybrid-type EDFA are described in Chapter 5.

In connection with the available amplifiers, we can summarize the capabilities of WDM and *time-division multiplexing* (TDM) for increasing the aggregate transmission capacity in NRZ and soliton systems and the corresponding physical/technological sources of performance limitation [321]. Figure 2.69 summarizes the

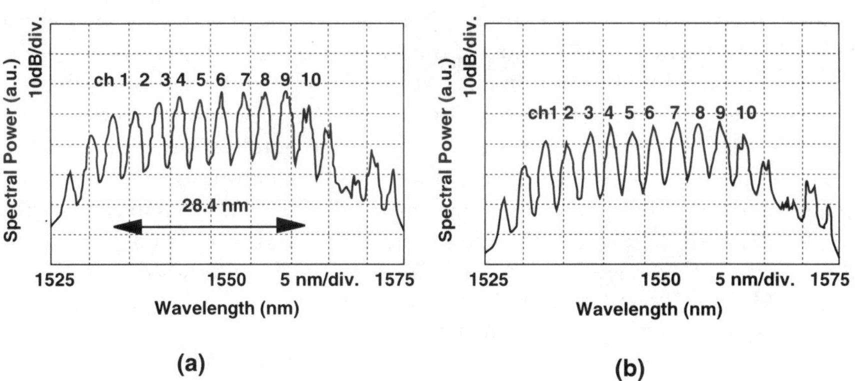

Figure 2.68 Terabit-per-second WDM spectra (a) before (0-km) and (b) after (40-km) dispersion-shifted fiber transmission [291].

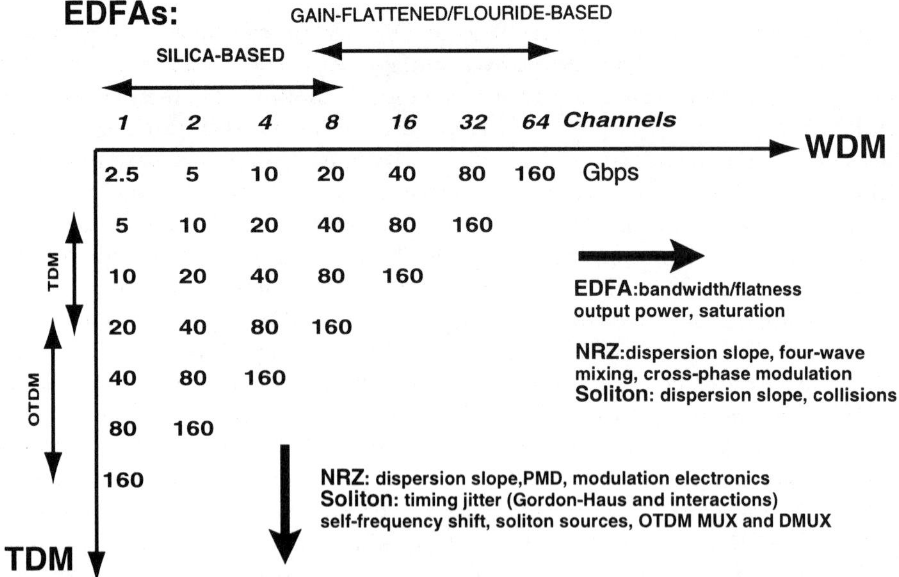

Figure 2.69 Summary of capabilities of WDM and TDM for increasing the aggregate transmission capacity in NRZ and soliton systems, and the corresponding physical/technological sources of performance limitation [293].

capabilities. As seen in Figure 2.68, when we increase the WDM channel number, we can increase the aggregate transmission capacity from 2.5 to 160 Gbps for 64-channel multiplexing. At the same time, along with this WDM increase, we need a wider bandwidth and flatter gain spectrum amplifier as well as higher output power. This suggests that for the multiplexing of more than eight channels we need gain-flattened silica-based EDFAs or fluoride EDFAs. In addition, the following limitations would arise: in NRZ systems, the fiber dispersion slope, FWM in a fiber, and cross-phase modulation in a fiber; and in soliton systems, the fiber dispersion slope and collisions between solitons in a fiber. On the other hand, when we increase the TDM bit rate, we can increase the aggregate transmission capacity of one channel from 2.5 to 160 Gbps. However, the bit rate would be limited by the following: in NRZ systems, the fiber dispersion slope, PMD (explained briefly in Section 2.3.3), and the speed of the modulation electronics; and in soliton systems, the timing jitter of propagating solitons (Gordon–Haus and interactions), the self-frequency shift due to self-phase modulation, soliton sources, and the multiplexing and demultiplexing scheme for optical TDM.

2.7.4 Optical Networking

The key feature of a WDM network is its "scalability" against node number and geographic span, which means WDM signals can be transmitted end-to-end without

O/E E/O conversion. Today, various types of WDM networking experiments are being energetically undertaken throughout the world. They are concerned with improving the large aggregate transmission capacity and all-optical networking [292–304]. For example, a 4- by 2.5-Gbps WDM transmission system experiment with optical add/drop multiplexing was performed over 3,711 km on installed submarine cables [298]. There are several research programs and consortiums working in the field including AON (Consortium on Wideband All-Optical Networks), MONET (Multiwavelength Optical Network Project), NTONC (National Transparent Optical Network Technology), and ONTC (Optical Networks Technology Consortium Project) in North America. In Europe, groups such as Race R2039 Atmos (ATM Optical Switching), and COBRA (Coherent Optical Systems Implemented for Business Traffic Routing and Access), are also conducting research.

Moreover, worldwide optical service networks are rapidly being constructed that employ undersea cables [305–315]. Since the deployment of the first deep-water repeated system in 1985, undersea lightwave technology has provided wideband digital connectivity among 60 nations and will connect 90 nations by the end of 1997, as shown in Figure 2.70 [305]. In addition, the following undersea networks are currently under construction or in their final stages of planning: the TAT-12/13 Cable Network [311], the FLAG Cable System [312], the TPC-5 Cable Network [313], the Asia Pacific Cable Network [314], and Africa ONE (the Africa Optical Network) [315]. All of them are third-generation transoceanic lightwave systems that are being deployed and operate at 5.0 Gbps per fiber pair and use EDFAs in repeaters to boost signals. These networks are described in [322,323]. Among the future plans described in these journals, the Africa ONE optical network is the biggest, which will encircle the entire continent of Africa with an undersea fiber optic ring network [306,315]. Figure 2.71 shows the Africa ONE optical network, which will employ a combination of WDM and *synchronous digital hierarchy* (SDH) to achieve network robustness. Africa ONE, which is a 40,000-km trunk and branch network landing in approximately 40 countries, is planned to be ready for service in 1999 [315].

2.7.5 1.3-μm Transmission

The last topic reviewed here is a transmission experiment using a PDFA operating at 1.3 μm. The PDFA module has been applied to experimental 40-channel AM-VSB video signal transmission and 10-Gbps digital transmission [160–163]. From a CATV experiment it was found that the degradation in picture quality monitored on a television set, before and after PDFA amplification, was imperceptible in subjective tests such as the measured *carrier-to-noise ratio* (CNR) and three types of distortion, *composite second-order* (CSO), *composite triple beat* (CTB), and *crossmodulation* (XM) for a 40-channel AM-VSB video signal transmission with and without a PDFA. In the digital transmission, a 10-Gbps digital signal was successfully transmitted more than

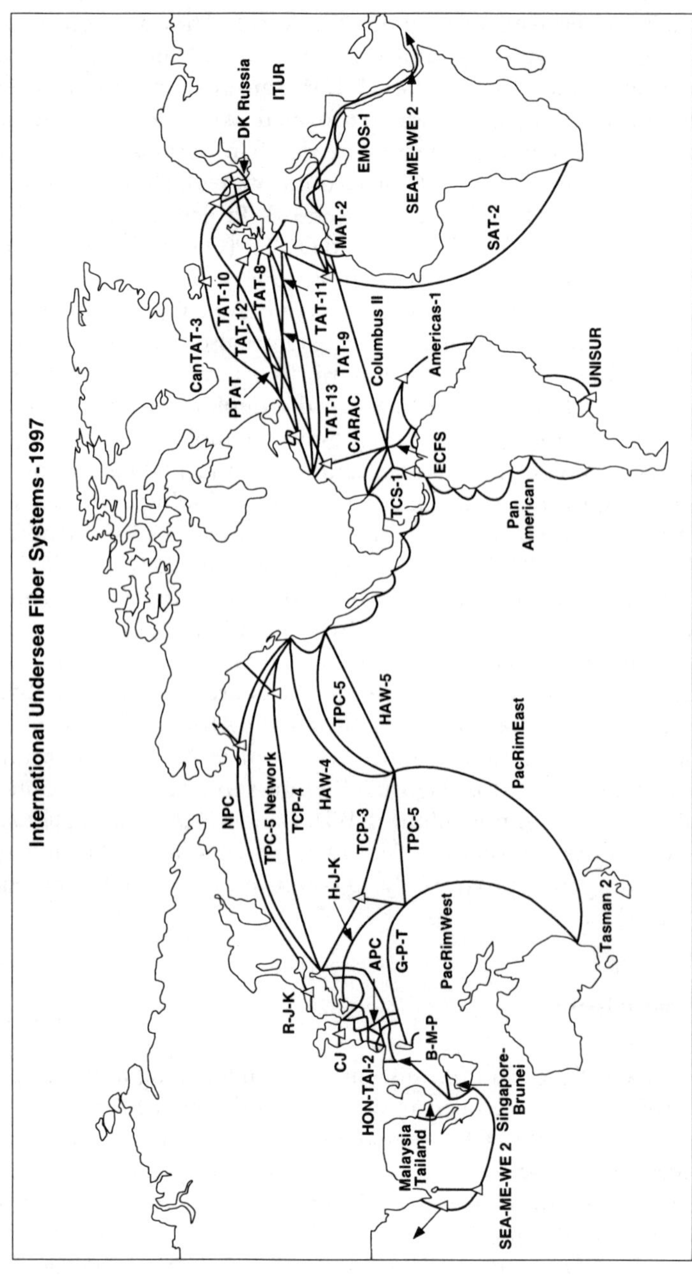

Figure 2.70 Installed and planned undersea fiber systems, which will connect 90 nations by the end of 1997 [305].

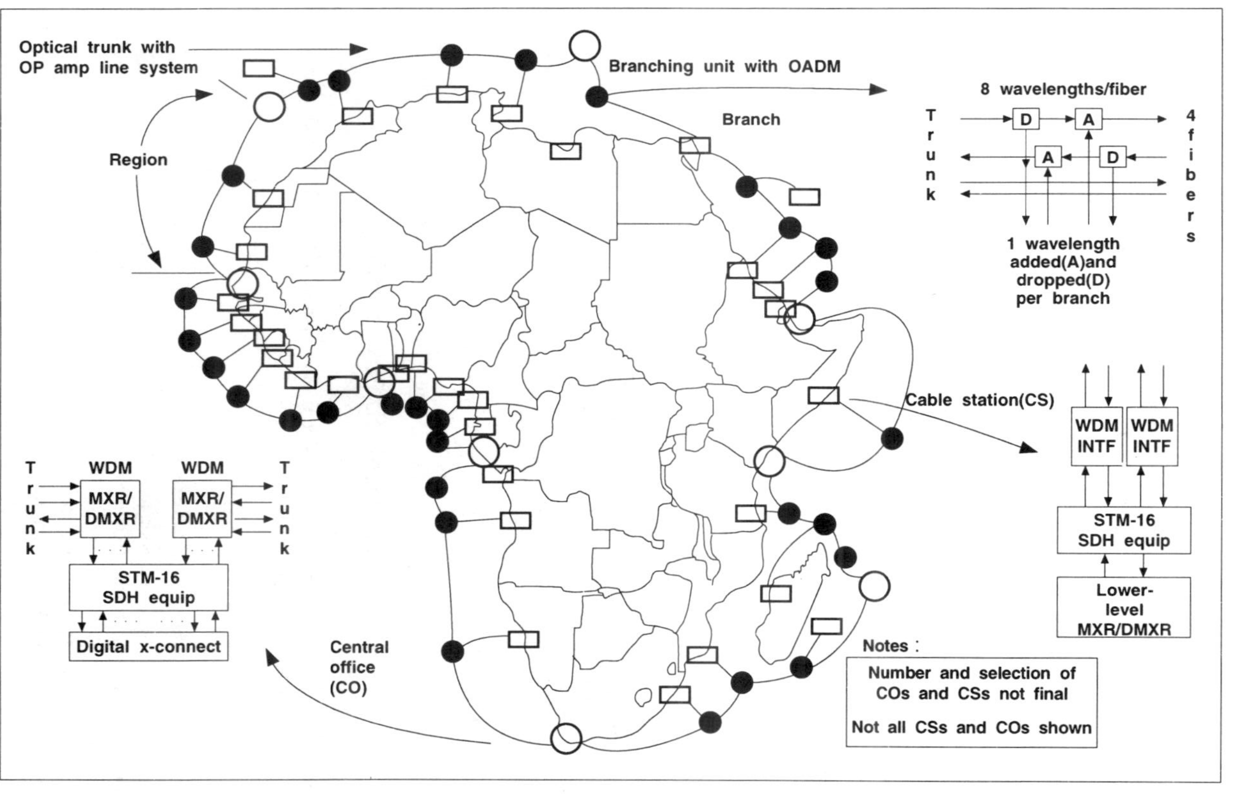

Figure 2.71 Africa ONE optical network, which encircles the entire continent of Africa with an undersea fiber optic ring network [306,315]. This network will employ a combination of WDM and SDH to achieve network robustness. Africa ONE, which is a 40,000-km trunk and branch network landing in approximately 40 countries, is planned to be ready for service in 1999.

110 km with a PDFA. Figure 2.72 shows the BER performance for a 10-Gbps digital signal transmission with a PDFA module used as a booster amplifier [162,163]. The transmitter was a strained *multi-quantum-well* (MQW) *distributed feedback* (DFB) laser operating at 1.315 μm. The laser was directly modulated using an NRZ pseudorandom signal with a pattern length of $2^{23} - 1$, and the output power was −1 dBm. An optical signal from the optical transmitter was fed to a 110.8-km standard single-mode fiber through a PDFA. The transmitted signal was received by an APD with an average received power of −25.4 dBm.

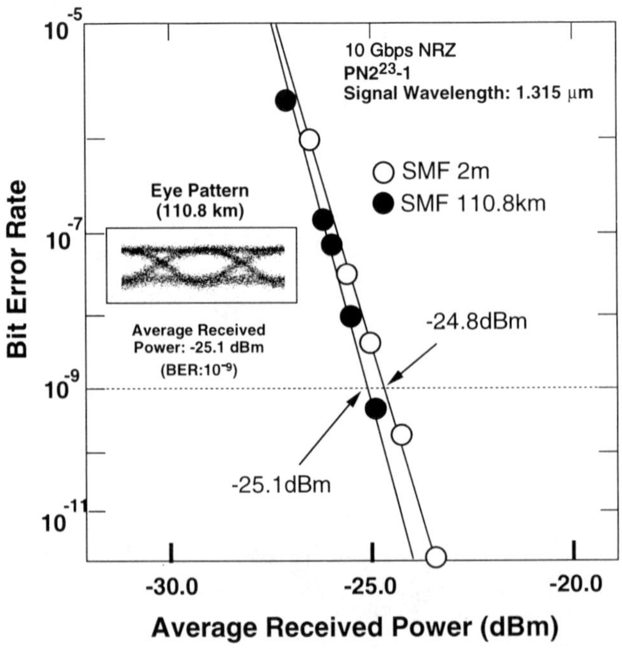

Figure 2.72 BER performance for a 10-Gbps digital signal transmission with a PDFA module used as a booster amplifier [162,163].

References

[1] Einstein, A., and L. Infeld, *The Evolution of Physics*, New York: Touchstone, 1938.

[2] Craig-Ryan, S. P., and B. J. Ainslie, "Glass Structure and Fabrication Techniques," in *Optical Fiber Lasers and Amplifiers*, P. W. France (ed.), Glasgow: Blackie, 1991, pp. 50–78.

[3] Izawa, T., and S. Sudo, *Optical Fibers: Materials and Fabrication*, Dordrecht: D. Reidel, 1987.

[4] Zacharisen, W. H., "The Atomic Arrangement in Glass," *J. Amer. Chem. Soc.*, Vol. 54, 1932, pp. 3841–3851.

[5] Poulain, M., and J. Lucas, "Verres Fluores au Tetrafluorure de Proprietes Optiques D'Un Verre Dope au Nd^{3+}," *Mat. Res. Bull.*, Vol. 10, 1975, pp. 243–246.

[6] Poulain, M., "Glass Systems and Structure," in *Fluoride Glasses*, A. E. Comyns (ed.), New York: Wiley, 1989, pp. 11–48.

[7] France, P. W., and M. C. Brierley, "Fluoride Fiber Lasers and Amplifiers," in *Optical Fiber Lasers and Amplifiers*, P. W. France (ed.), Glasgow: Blackie, 1991, pp. 183–211.

[8] Pauling, L., *General Chemistry*, New York: Dover, 1970, p. 182.

[9] Simmons, J. H., C. J. Simmons, and R. Ochoa, "Fluoride Glass Structure," in *Fluoride Glass Fiber Optics*, I. D. Aggarwal and G. Lu (eds.), New York: Academic Press, 1991, pp. 37–84.

[10] Layne, C. B., W. H. Lowdermilk, and M. J. Weber, "Multiphonon Relaxation of Rare-Earth Ions in Oxide glasses," *Phys. Rev. B*, Vol. 16, pp. 10–20.

[11] Snitzer, E., "Rare-Earth-Doped Fiber Lasers," *Tech. Digest of OFC'92*, Tutorial Session, 1992, pp. 417–484.

[12] Caird, J. A., and S. A. Payne, "Crystal Paramagnetic Ion Lasers," in *CRC Handbook of Laser Science and Technology, Supplement 1: Lasers*, M. J. Weber (ed.), Boca Raton, FL: CRC Press, 1991, pp. 3–100.

[13] Miniscalco, W. J., "Materials Aspects of Optical Fiber Amplifiers," *Tech. Digest of OFC'93*, Tutorial Session, 1993, pp. 115–173.

[14] Herzberg, G., *Atomic Spectra and Atomic Structure*, New York: Dover, 1944, pp. 120–151.

[15] Yariv, A., *Optical Electronics*, 3rd ed., New York: Holt, Rinehart and Winston, 1985, p. 130.

[16] Quimby, R. S., W. J. Miniscalco, and B. Thompson, "Excited State Absorption at 980 nm in Erbium Doped Glass," *Fiber Laser Sources and Amplifiers*, Vol. 1581, : SPIE, 1991, pp. 72–79.

[17] Quimby, R. S., "Output Saturation in a 980-nm Pumped Erbium-Doped Fiber Amplifier," *Appl. Optics*, Vol. 30, June 1991, pp. 2546–2552.

[18] Quimby, R. S., W. J. Miniscalco, and B. Thompson, "Excited State Absorption at 980 nm in Erbium-Doped Silica Glass," *Tech. Digest of OAA '92*, Vol. 17, June 1992, pp. 67–70.

[19] Quimby, R. S., W. J. Miniscalco, and B. Thompson, "Upconversion and 980-nm Excited-State Absorption in Erbium-Doped Glass," *Fiber Laser Sources and Amplifiers*, Vol. 1789, : SPIE, 1992, pp. 50–57.

[20] Ainslie, B. J., S. P. Craig-Ryan, S. T. Davey, J. R. Armitage, C. G. Atkins, and R. Wyatt, "Optical Analysis of Erbium Doped Fibres for Efficient Lasers and Amplifiers," *Tech. Digest of IOOC '89*, Vol. 3, July 1989, pp. 22–23.

[21] Blixta, P., J. Nilsson, T. Carlnäs, and B. Jaskorzynska, "Amplification Reduction in Fiber Amplifiers Due to Up-Conversion at Er^{3+}-Concentrations Below 1000 ppm," *Tech. Digest of OAA '91*, July 1991, pp. 52–55.

[22] Ohishi, Y., T. Kanamori, T. Kitagawa, S. Takahashi, E. Snitzer, and G. H. Sigel, Jr., "Pr^{3+}-Doped Fluoride Fiber Amplifier Operating at 1.31 μm," *Tech. Digest of OFC'91*, 1991, Paper PD2.

[23] Ohishi, Y., T. Kanamori, M. Shimizu, M. Yamada, Y. Terunuma, J. Temmyo, M. Wada, and S. Sudo, "Praseodymium-Doped Fiber Amplifiers at 1.3 μm," *IEICE Trans. Comm.*, Vol. E77-B, Apr. 1994, pp. 421–440.

[24] Pederson, B., W. J. Miniscalco, and S. Zemon, "Analysis of Pr^{3+} and Nd^{3+}-Doped Fiber Amplifiers at 1300nm," *Fiber Laser Sources and Amplifiers*, Vol. 1789, : SPIE, 1992, pp. 191–200.

[25] Pedersen, B., W. J. Miniscalco, and R. S. Quimby, "Optimization of Pr^{3+}:ZBLAN Fiber Amplifiers," *IEEE Photon. Tech. Lett.*, Vol. 4, May 1992, pp. 446–448.

[26] Miyakawa, T., and D. L. Dexter, "Phonon Sidebands, Multiphonon Relaxation of Excited States, and Phonon-Assisted Energy Transfer between Ions in Solid," *Phys. Rev. B*, Vol. 1, 1970, pp. 2961–2969.

[27] Layne, C. B., W. H. Lowdermilk, and M. J. Weber, "Multiphonon Relaxation of Rare-Earth Ions in Oxide Glasses," *Phys. Rev. B*, Vol. 16, 1977, pp. 10–20.

[28] Toratani, H., T. Izumitani, and H. Kuroda, "Compositional Dependence of Nonradiative Decay Rate In Nd Laser Glasses," *J. Non-Crystalline Solids*, Vol. 52, 1982, pp. 303–313.

[29] Ainslie, B. J., S. P. Craig, S. T. Davey, and B. Wakefield, "The Fabrication, Assessment and Optical Properties of High-Concentration Nd^{3+}- and Er^{3+}-Doped Silica-Based Fibres," *Mat. Lett.*, Vol. 6, Mar. 1988, pp. 139–144.

[30] Tachibana, M., R. I. Laming, P. R. Morkel, and D. N. Payne, "Erbium-Doped Fiber Amplifier with Flattened Gain Spectrum," *IEEE Photon. Tech. Lett.*, Vol. 3, Feb. 1991, pp. 118–120.

[31] Georges, E. S., "EDFA Gain Equalization with Fiber Filter for WDM Systems," *Tech. Digest of OAA '95*, 1995, Paper ThD5-1.

[32] Millar, C. A., M.C. Brierley, and P.W. France, "Optical Amplification in an Erbium-Doped Fluorozirconate Fibre Between 1480 nm and 1600 nm," *Tech. Digest of EOCO '88*, 1988, pp. 66–69.

[33] Miniscalco, W. J., and L. J. Andrews, "Fiber Optic Lasers and Amplifiers for the Near-Infrared," *Mat. Sci. Forum*, Vols. 32–33, 1988, pp. 501–510.

[34] Yamada, M., T. Kanamori, Y. Terunuma, K. Oikawa, M. Shimizu, and S. Sudo, "Fluoride-Based Erbium-Doped Fiber Amplifier with Inherently Flat Gain Spectrum," *IEEE Photon. Tech. Lett.*, Vol. 8, July 1996, pp. 882–884.

[35] Kashiwada, T., M. Shigematsu, M. Kakui, M. Onishi, and M. Nishimura, "Gain-Flattened Optical-Fiber Amplifiers with a Hybrid Er-Doped-Fiber Configuration for WDM Transimission," *Tech. Digest of OFC '95*, Feb. 1995, pp. 77–78.

[36] Bayart, D., B. Clesca, L. Hamon, and J. L. Beylat, "Experimental Investigation of the Gain Flatness Characteristics for 1.55 μm Erbium-Doped Fluoride Fiber Amplifiers," *IEEE Photon. Tech. Lett.*, Vol. 6, May 1994, pp. 613–615.

[37] Clesca, B., D. Ronarc'h, D. Bayart, Y. Sorel, L. Hamon, M. Guibert, J. L. Beylat, J. F. Kerdiles, and M. Semenkoff, "Gain Flatness Comparison Between Erbium-Doped Fluoride and Silica Fiber Amplifiers With Wavelength-Multiplexed Signals," *IEEE Photon. Tech. Lett.*, Vol. 6, Apr. 1994, pp. 509–512.

[38] Clesca, B., D. Bayart, L. Hamon, J. L. Beylat, C. Cœurjolly, and L. Berthelon, "Over 25 nm, 16-Wavelength-Multiplexed Signal Amplification Through Fluoride-Based Fibre-Amplifier Cascade," *Electron. Lett.*, Vol. 30, Mar. 1994, pp. 586–588.

[39] Clesca, B., D. Bayart, C. Cœurjolly, L. Berthelon, L. Hamon, and J. L. Beylat, "Over 25 nm, 16 Wavelength-Multiplexed Signal Amplification Through Four Fluoride-Based Fibre-Amplifier Cascade and 440 km Standard Fiber," *Tech. Digest of OFC '94*, 1994, Paper PD20.

[40] Beylat, J. L., D. Bayart, B. Clesca, and B. M. Desthieux, "Flat-Gain Fiber Optical Amplifiers for WDM Networks," *Tech. Digest of ECOC'95*, 1995, Paper Th.L.1.7.

[41] Goldstein, E. L., L. Eskildsen, V. da Silva, M. Andrejco, and Y. Silberberg, "Inhomogeneous Broadened Fiber-Amplifier Cascades for Transparent Multiwavelength Lightwave Networks," *L. Lightwave Tech.*, Vol. 13, 1995, pp. 782–790.

[42] Willner, A. E., and S. M. Hwang, "Transmission of Many WDM Channels Through a Cascade of EDFA's in Long-Distance Links and Ring Networks," *IEEE J. Lightwave Tech.*, Vol. 13, 1995, pp. 802–816.

[43] Semenkoff, M., M. Guibert, J. F. Kerdiles, and Y. Sorel, "High Power, High Gain Optical Fibre Amplifier for Multiwavelength Transmission Systems," *Electron. Lett.*, Vol. 30, Aug. 1994, pp. 1411–1413.

[44] Yamada, M., H. Ono, T. Kanamori, T. Sakamoto, Y. Ohishi, and S. Sudo, "A Low-Noise and Gain-Flattened Amplifier Composed of a Silica-Based and a Fluoride-Based Er^{3+}-Doped Fiber Amplifier in a Cascade Configuration," *IEEE Photon. Tech. Lett.*, Vol. 8, Mar. 1996.

[45] Yamada, M., M. Shimizu, Y. Ohishi, M. Horiguchi, S. Sudo, and A. Shimizu, "Flattening the Gain Spectrum of an Erbium-Doped Fibre Amplifier by Connecting an Er^{3+}-Doped SiO_2-Al_2O_3 Fibre and an Er^{3+}-Doped Multicomponent Fibre," *Electron. Lett.*, Vol. 30, Aug. 1994, pp. 1762–1764.

[46] Zemon, S., G. Lambert, W. J. Miniscalco, and L. J. Andrew, "Characterization of Er^{3+}-Doped Glasses by Fluorescence Line Narrowing," *Fiber Laser Sources and Amplifiers*, Vol. 1171, : SPIE, 1989, pp. 219–236.

[47] Zemon, S., G. Lambert, W. J. Miniscalco, and B. A. Thompson, "Homogeneous Linewidths in Er^{3+}-Doped Galsses Measured by Resonance Fluorescence Line Narrowing," *Fiber Laser Sources and Amplifiers III*, Vol. 1581, : SPIE, 1991, pp. 91–100.

[48] Zyskind, J. L., E. Desurvire, J. W. Sulhoff, and D. J. Di Giovanni, "Determination of Homogeneous Linewidth by Spectral Gain Hole-Burning in an Erbium-Doped Fiber Amplifier with GeO_2: SiO_2 Core," *IEEE Photon. Tech. Lett.*, Vol. 2, Dec. 1990, pp. 869–871.

[49] Desurvire, E., J. L. Zyskind, and J. R. Simpson, "Spectral Gain Hole-Burning at 1.53 μm in Erbium-Doped Fiber Aplifiers," *IEEE Photon. Tech. Lett.*, Vol. 2, Apr. 1990, pp. 246–248.

[50] Giles, C. R., and D. DiGiovanni, "Spectral Dependence of Gain and Noise in Erbium-Doped Fiber Amplifiers," *IEEE Photon. Tech. Lett.*, Vol. 2, Nov. 1990, pp. 797–800.

[51] Desurvire, E., and J. R. Simpson, "Evaluation of $^4I_{15/2}$ and $^4I_{13/2}$ Stark-Level Energies in Erbium-Doped Aluminosilicate Glass Fibers," *Optics Lett.*, Vol. 15, May 1990, pp. 547–549.

[52] Zyskind, J. L., E. Desurvire, J. W. Sulhoff, and D. J. DiGiovanni, "Spectral Gain Hole-Burning in an Erbium-Doped Fiber Amplifier with GeO_2:SiO_2 Core," *Tech. Digest of OAA '90*, MD4, 1990, pp. 507–509.

[53] Okuno, T., K. Tanaka, K. Koyama, M. Namiki, and T. Suemoto, "Homogeneous Line Width of Praseodymium Ions in Various Inorganic Materials," *J. Luminescence*, Vol. 58, 1994, pp. 184–187.

[54] Macfarlane, R. M., and R. M. Shelby, "Homogeneous Line Broadening of Optical Transitions of Ions and Molecules in Glasses," *J. Luminescence*, Vol. 36, 1987, pp. 179–207.

[55] Yen, W. M., and R. T. Brundage, "Fluorescence Line Narrowing in Inorganic Glasses: Linewidth Measurements," *J. Luminescence*, Vol. 36, 1987, pp. 209–220.

[56] Brundage, R. T., and W. M. Yen, "Low-Temperature Homogeneous Linewidths of Yb^{3+} in Inorganic Glasses," *Phys. Rev. B*, Vol. 33, Mar. 1986, pp. 4436–4438.

[57] Morgan, J. R., and M. A. El-Sayed, "Temperature Dependence of the Homogeneous Linewidth of the $^5D_0$-$^7F_0$ Transition of Eu^{3+} in Amorphous Hosts at High Temperatures," *Chem. Phys. Lett.*, Vol. 84, Dec. 1981, pp. 213–216.

[58] Hegarty, J., and W. M. Yen, "Optical Homogeneous Linewidths of Pr^+ in BeF_2 and GeO_2 Glasses," *Phys. Rev. Lett.*, Vol. 43, Oct. 1979, pp. 1126–1130.

[59] Brawer, S. A., and M. J. Weber, "Observation of Fluorescence Line Narrowing, Hole Burning, and Ion-Ion Energy Transfer in Neodymium Laser Glass," *Appl. Phys. Lett.*, Vol. 35, July 1979, pp. 31–33.

[60] Selzer, P. M., D. L. Huber, D. S. Hamilton, W. M. Yen, and M. J. Weber, "Anomalous Fluorescence Linewidth Behavior in Eu^{3+}-Doped Silicate Glass," *Phys. Rev.Lett.*, Vol. 36, Apr. 1976, pp. 813–816.

[61] Laming, R. I., L. Reekie, P. R. Morkel, and D. N. Payne, "Multichannel Crosstalk and Pump Noise Characteristics of Er^{3+}-Doped Fibre Amplifier Pumped at 980 nm," *Electron. Lett.*, Vol. 30, Mar. 1989, pp. 455–456.

[62] Weber, M. J., "Fluorescence and Glass Lasers," *J. Non-Crystalline Solids*, Vol. 47, 1992, pp. 117–134.

[63] Brecher, C., L. A. Riseberg, and M. J. Weber, "Line-Narrowed Fluorescence Spectra and Site-Dependent Transition Probabilites of Nd^{3+} in Oxide and Fluoride Glasses," *Phys. Rev.B*, Vol. 18, Nov. 1978, pp. 5700–5811.

[64] Robinson, C. C., "Multiple Sites for Er^{3+} in Alkali Silicate Glasses (I). The Principal Sixfold coordinated Site of Er^{3+} in Silicate Glass," *J. Non-Crystalline Solids*, Vol. 15, 1974, pp. 1–10.

[65] Robinson, C. C., "Multiple Sites for Er^{3+} in Alkali Silicate Glasses (II). Evidence of Four Sites for Er^{3+}," *J. Non-Crystalline Solids*, Vol. 15, 1974, pp. 11–29.

[66] Robinson, C. C., and J. R. Fournier, "Co-Ordination of Yb^{3+} in Phosphate, Silicate, and Germanate Glasses," *J. Phys. Chem. Solids*, Vol. 31, 1970, pp. 895–904.

[67] Lea, K. R., M. J. Leask, and W. P. Wolf, "The Raising of Angular Momentum Degeneracy of *f*-Electron Terms by Cubic Crystal Fields," *J. Phys. Chem. Solids*, Vol. 23, 1962, pp. 1381–1405.

[68] Gruber, J. B., J. R. Henderson, M. Muramoto, K. Rajnak, and J. G. Conway, "Energy Levels of Single-Crystal Erbium Oxide," *J.Chem. Phys.*, Vol. 45, July 1966, pp. 477–482.

[69] Gruber, J. B., W. F. Krupke, and J. M. Poindexter, "Crystal-Field Splitting of Trivalent Thulium and Erbium J. Levels in Yttrium Oxide," *J. Chem. Phys.*, Vol. 41, Dec. 1964, pp. 3363–3377.

[70] McCumber, D. E., and M. D. Sturge, "Linewidth and Temperature Shift of the R Line in Ruby," *J. Appl. Phys.*, Vol. 34, June 1963, pp. 1682–1684.

[71] Desurvire, E., and J. R. Simpson, "Evaluation of $^4I_{15/2}$ and $^4I_{13/2}$ Stark-Level Energies in Erbium-Doped Aluminosilicate Glass Fibers," *Optics Lett.*, Vol. 15, May 1990, pp. 547–549.

[72] Desurvire, E., "Erbium-Doped Fiber Amplifiers: Basic Physics and Characteristics," in *Rare Earth Doped Fiber Lasers and Amplifiers*, M. J. F. Digonnet (ed.), New York: Marcel Dekker, 1993.

[73] Miniscalco, W. J., "Optical and Electronic Properties of Rare Earth Ions in Glasses," in *Rare Earth Doped Fiber Lasers and Amplifiers*, M. J. F. Digonnet (ed.), New York: Marcel Dekker, 1993.

[74] Shimizu, M., M. Yamada, M. Horiguchi, T. Takeshita, and M. Okayasu, "Erbium-Doped Fibre Amplifiers with an Extremely High Gain Coefficient of 11.0 dB/mW," *Electron. Lett.*, Vol. 26, Sept. 1990, pp. 1641–1643.

[75] Mears, R. J., L. Reekie, I. M. Jauncey, and D. N. Payne, "High-Gain Rare-Earth-Doped Fiber Amplifier at 1.54 μm," *Tech. Digest Conf. Optical Fiber Communication/Int. Conf. Integrated Optics and Optical Fiber Communication (OFC/IOOC'87)*, Reno, Nevada, Feb. 1987, Paper W12.

[76] Laming, R. I., L. Reekie, D. N. Payne, P. L. Scrivener, F. Fontana, and A. Righetti, "Optimal Pumping of Erbium-Doped-Fibre Optical Amplifiers," *Tech. Digest of ECOC '88*, Part 2, 1988, pp. 25–28.

[77] Laming, R. I., V. Shah, L. Curtis, R. S. Vodhanel, F. J. Favire, W. L. Barnes, J. D. Minelly, D. P. Bour, and E. J. Tarbox, "Highly Efficient 978 nm Diode Pumped Erbium-Doped Fibre Amplifier with 24 dB Gain," *Tech. Digest of IOOC '89*, Vol. 5, July 1989, Paper 20PDA-4.

[78] Shimizu, M., M. Yamada, M. Horiguchi, and E. Sugita, "Concentration Effect on Optical Amplification Characteristics of Er-Doped Silica Single-Mode Fibers," *IEEE Photon. Tech. Lett.*, Vol. 2, Jan. 1990, pp. 43–45.

[79] Masuda, H., and A. Takada, "High Gain Two-Stage Amplification with Erbium-Doped FIbre Amplifier," *Electron. Lett.*, Vol. 26, May 1990, pp. 661–662.

[80] Yamashita, S., and T. Okoshi, "Improvement of Erbium-Doped Fiber Amplifier Characteristics by Insertion of a 'Midway' Isolator," *Tech. Digest of OEC '92*, July 1992, pp. 286–287.

[81] Povlsen, J. H., A. Bjarklev, O. Lumholt, H. V. Pommer, K. Rottwitt, and T. Rasmussen, "Optimizing Gain and Noise Performance of EDFA's with Insertion of a Filter or an Isolator," *Fiber Laser Sources and Amplifiers*, Vol. 1581, : SPIE, 1991, pp. 107–113.

[82] Zervas, M. N., R. I. Laming, and D. N. Payne, "Efficient Erbium-Doped Fibre Amplifier with an Integral Isolator," *Tech. Digest of OAA '92*, 1992, pp. 162–165.

[83] Sakamoto, T., M. Shimizu, T. Kanamori, Y. Ohishi, Y. Terunuma, M. Yamada, and S. Sudo; "Tm-Ho-Doped ZBLYAN Fiber Amplifiers for 1.4 μm-Band," *Tech. Digest of OEC '94*, July 1994, pp. 38–39.

[84] Sakamoto, T., M. Shimizu, M. Yamada, T. Kanamori, Y. Ohishi, Y. Terunuma, and S. Sudo, "35 dB Gain Tm-Doped ZBLAN Fiber Amplifier Operating at 1.65 μm," *Tech. Digest CLEO/Pacific Rim '95*, July 1995, Paper PD2.12.

[85] Laming, R. I., M. N. Zervas, and D. N. Payne, "Erbium-Doped Fiber Amplifier with 54 dB Gain and 3.1 dB Noise Figure," *IEEE Photon. Tech. Lett.*, Vol. 4, Dec. 1992, pp. 1345–1347.

[86] Laming, R. I., M. N. Zervas, and D. N. Payne, "54 dB gain Quantum-Noise-Limited Erbium-Doped Fibre Amplifier," *Tech. Digest of ECOC'92*, Vol. 1, 1992, pp. 89–92.

[87] Zervas, M. N., R. I. Laming, and D. N. Payne, "Efficient Erbium-Doped Fiber Amplifiers Incorporating an Optical Isolator," *IEEE J. Quantum Electron.*, Vol. 31, Mar. 1995, pp. 472–480.

[88] Hansen, S. L., K. Dybdal, and C. C. Larsen, "Gain Limit in Erbium-Doped Fiber Amplifiers due to Internal Rayleigh Backscattering," *IEEE Photon. Tech. Lett.*, Vol. 4, June 1992, pp. 559–561.

[89] Sceats, M. G., P. A. Krug, G. R. Atkins, S. C. Guy, and S. B. Poole, "Non-Linear Excited State Absorption in Er^{3+}-Doped Fibre with High-Power 980 nm Pumping," *Tech. Digest of OAA '91*, Vol. 13, July 1991, pp. 48–51.

[90] Krug, P. A., M. G. Sceats, G. R. Atkins, S. C. Guy, and S. B. Poole, "Intermediate Excited-State Absorption in Erbium-Doped Fiber Strongly Pumped at 980 nm," *Optics Lett.*, Vol. 16, Dec. 1991, pp. 1976–1978.

[91] Laming, R. I., S. B. Poole, and E. J. Tarbox, "Pump Excited-State Absorption in Erbium-Doped Fibres," *Optics Lett.*, Vol. 13, Dec. 1988, pp. 1084–1086.

[92] Atkins, C. G., J. R. Armitage, R. Wyatt, B. J. Ainslie, and S. P. Craig-Ryan, "Pump Excited State Absorption in Er^{3+} Doped Optical Fibres," *Optics Comm.*, Vol. 73, Oct. 1989, pp. 217–222.

[93] Zemon, S., G. Lambert, W. J. Miniscalco, R. W. Davies, B. T. Hall, R. C. Folweiler, T. Wei, L. J.

Andrews, and M. P. Singh, "Excited State Cross Section for Er-Doped Glasses," *Fiber Laser Sources and Amplifiers*, Vol. 1373, : SPIE, 1990, pp. 21–32.

[94] Zemon, S., B. Pedersen, G. Lambert, W. J. Miniscalco, L. J. Andrews, R. W. Davies, and T. Wei, "Excited-State Absorption Cross Section in the 800-nm Band for Er-Doped, Al/P-Silica Fibers: Measurements and Amplifier Modeling," *IEEE Photon. Tech. Lett.*, Vol. 3, July 1991, pp. 621–624.

[95] Stokowski, S. E., "Glass Laser," in *CRC Handbook of Laser Science and Technology, Vol. I, Lasers and Masers*, M. J. Weber (ed.), Boca Raton, FL: CRC Press, 1987.

[96] Laming, R. I., P. R. Morkel, D. N. Payne, and L. Reekie, "Noise in Erbium-Doped Fibre Amplifiers," *Tech. Digest of EOCO '88*, Part 1, 1988, pp. 54–57.

[97] Olshansky, R., "Noise Figure for Erbium-Doped Optical Fibre Amplifiers," *Electron. Lett.*, Vol. 24, Oct. 1988, pp. 1363–1365.

[98] Yamada, M., M. Shimizu, M. Okayasu, T. Takeshita, M. Horiguchi, Y. Tachikawa, and E. Sugita, "Noise Characteristics of Er^{3+}-Doped Fiber Amplifiers Pumped by 0.98 and 1.48 μm Laser Diodes," *IEEE Photon. Tech. Lett.*, Vol. 2, Mar. 1990, pp. 205–207.

[99] Gimlett, J. L., J. Young, R. E. Spicer, and N. K. Cheung, "Degradations in Gbit/s DFB Laser Transmission Systems due to Phase-to-Intensity Noise Conversion by Multiple Reflection Points," *Electron. Lett.*, Vol. 24, Mar. 1988, pp. 406–408.

[100] Gimlett, J. L., M. Z. Iqbal, L. Curtis, N. K. Cheung, A. Righetti, F. Fontana, and G. Grasso, "Impact of Multiple Reflection Noise in Gbit/s Lightwave Systems with Optical Fibre Amplifiers," *Electron. Lett.*, Vol. 25, Sept. 1989, pp. 1393–1394.

[101] Gimlett, J. L., and N. K. Cheung, "Effects of Phase-to-Intensity Noise Conversion by Multiple Reflections on Gigabit-per-Second DFB Laser Transmission Systems," *IEEE J. Lightwave Tech.*, Vol. 7, June 1989, pp. 888–895.

[102] Taylor, M. G., "Observation of New Polarisation Dependence Effect in Long Haul Optically Amplified System," *Tech. Digest of OFC/IOOC'93*, 1993, pp. 25–28.

[103] Yamamoto, S., N. Edagawa, H. Taga, Y. Yoshida, and H. Wakabayashi, "Observation of BER Degradation Due to Fading in Long-Distance Optical Amplifier Systems," *Electron. Lett.*, Vol. 29, Jan. 1993, pp. 209–210.

[104] Bergano, N. S., V. J. Mazurczyk, and C. R. Davidson, "Polarization Hole-Burning in Erbium-Doped Fiber-Amplifier Transmission Systems," *Tech. Digest of ECOC'94*, 1994, pp. 621–628.

[105] Mazurczyk, V. J., and J. L. Zyskind, "Polarization Hole Burning in Erbium Doped-Fiber Amplifiers," *Tech. Digest of LEOS'93*, CPD-26, Oct. 1993.

[106] Mazurczyk, V. J., and J. L. Zyskind, "Polarization Dependent Gain in Erbium Doped-Fiber Amplifiers," *IEEE Photon. Tech. Lett.*, Vol. 6, May 1994, pp. 616–618.

[107] Bergano, N. S., "Time Dynamics of Polarization Hole Burning in an EDFA," *Tech. Digest of OFC'94*, Vol. 4, 1994, pp. 305–306.

[108] Mazurczyk, V. J., and C. D. Poole, "The Effect of Birefringence on Polarization Hole Burning in Erbium Doped Fiber Amplifiers," *Tech. Digest of OAA'94*, Aug. 1994, pp. 77–79.

[109] Giles, C. R., "Suppression of Polarisation Holeburning-Induced Gain Anisotropy in Reflective EDFAs," *Electron. Lett.*, Vol. 30, June 1994, pp. 976–977.

[110] Hall, D. W., and M. J. Weber, "Polarized Fluorescence Line Narrowing Measurements of Nd Laser Glasses: Evidence of Stimulated Emission Cross Section Anisotropy," *Appl. Phys. Lett.*, Jan. 1983, pp. 157–159.

[111] Lebedev, V. P., and A. K. Przhevuskii, "Polarized Luminescence of Rare-Earth Activated Glasses," *Sov. Phys. Solid State*, Vol. 19, Aug. 1977, pp. 1389–1391.

[112] Greer, E. J., D. J. Lewis, and W. M. Macauley, "Polarisation Dependent Gain in Erbium-Doped Fibre Amplifiers," *Electron. Lett.*, Vol. 30, 1994, pp. 46–47.

[113] Keys, R. W., S. J. Wilson, M. Healy, S. R. Baker, A. Robinson, and J. E. Righton, "Polarization-Dependent Gain in Erbium-Doped Fibers" *Tech. Digest of OFC'94*, Vol. 4, 1994, pp. 306–307.

[114] Leners, R., T. Gerorges, P. L. François, and G. Stéphan, "Analytic Model of Polarization Dependent Gain in Erbium Doped Fibre Amplifiers," *Tech. Digest of OAA'94*, Aug. 1994, Paper PD1-1-4.

[115] Bergano, N. S., V. J. Mazurczyk, and C. R. Davidson, "Polarization Scrambling Improves SNR in a Chain of Erbium-Doped Fiber-Amplifiers," *Tech. Digest of OFC'94*, 1994, Paper ThR2.

[116] Letellier, V., G. Bassier, P. Marmier, R. Morin, R. Uhel, and J. Artur, "Polarization Scrambling in 5 Gbit/s 8100 km EDFA Based System," *Electron. Lett.*, Vol. 70, Mar. 1994, pp. 589–590.

[117] Jensen, N. G., T. Ono, and K. Fukuchi, "Elimination of Polarization Dependent Gain Using Polarization Scrambling," *Tech. Digest of OAA '94*, Vol. 14, 1994, pp. 53–55.

[118] Heismann, F., D. A. Gray, B. H. Lee, and R. W. Smith, "Electrooptic Polarization Scramblers for Optically Amplified Long-Haul Transmission Systems," *Tech. Digest of ECOC'94*, Vol. 2, 1994, pp. 629–632.

[119] Heismann, F., "Fast Polarization Controllers and Scramblers for High-Speed Lightwave Systems," *Tech. Digest of OFC'95*, 1995, Paper TuD3.

[120] Bergano, N. S., and C. R. Davidson, "Polarization Scrambling-Induced Timing Jitter in Optical Amplifier Systems," *Tech. Digest of OFC'95*, 1995, Paper WG3.

[121] Mazurczyk, V. J., R. H. Stolen, and C. D. Poole, "Observation of Polarization Hole Burning in Er-Doped Fiber for Circular Polarization of Saturating Signal," *Tech. Digest of OFC'95*, 1995, Paper TuJ7.

[122] Livas, J. C., S. R. Chinn, E. S. Kintzer, J. N. Walpole, C. A. Wang, and L. J. Missaggia, "High-Power Erbium-Doped Fibre Amplifier with 975 nm Tapered-Gain-Region Laser Pumps," *Electron. Lett.*, Vol. 30, June 1994, pp. 1054–1055.

[123] Grubb, S. G., W. H. Humer, R. S. Cannon, S. W. Vendetta, K. L. Sweeney, P. A. Leilabady, M. R. Keur, J. G. Kwasgroch, T. C. Munks, and D. W. Anthon, "+24.6 dBm Output Power Er/Yb Codoped Optical Amplifier Pumped by Diode-Pumped Nd: YLF Laser," *Electron. Lett.*, Vol. 28, June 1992, pp. 1275–1276.

[124] Barnes, W. L., S. B. Poole, J. E. Townsend, L. Reekie, D. J. Taylor, and D. N. Payne, "Er^{3+}-Yb^{3+} and Er^{3+} Fibre Laser," *J. Lightwave Tech.*, Vol. 7, Oct. 1989, pp. 1461–1465.

[125] Snitzer, E., and R. Woodcock, "Yb^{3+}-Er^{3+} Glass Laser," *Applied Physics Lett.*, Vol. 6, Feb. 1965, pp. 45–46.

[126] Gapontsev, Y. P., S. M. Matitsin, and A. A. Isineev, "Channels of Energy Losses in Erbium Laser Glasses in the Stimulated Emission Process," *Optics Comm.*, Vol. 46, July 1983, pp. 226–230.

[127] Hanna, D. C., A. Kazer, and D. P. Shepherd, "A 1.54 μm Er Glass Laser Pumped by a 1.064 μm Nd:YAG Laser," *Optics Comm.*, Vol. 63, Sept. 1987, pp. 417–420.

[128] Townsend, J. E., S. B. Poole, and D. N. Payne, "Solution-Doping Technique for Fabrication of Rare-Earth-Doped Optical Fibres," *Electron. Lett.*, Vol. 23, Mar. 1987, pp. 26–27.

[129] Maker, G. T., and A. I. Ferguson, "1.56 μm Yb-Sensitised Er Fibre Laser Pumped by Diode-Pumped Nd:YAG and Nd: YLF Lasers," *Electron. Lett.*, Vol. 24, Sept. 1988, pp. 1160–1162.

[130] Fermann, M. E., D. C. Hanna, D. P. Shepherd, P. J. Suni, and J. E. Townsend, "Efficient Operation of an Yb-Sensitised Er Fibre Laser at 1.56 μm," *Electron. Lett.*, Vol. 24, Sept. 1988, pp. 1135–1136.

[131] Grubb, S. G., R. S. Cannon, T. W. Windhorn, S. W. Vendetta, P. A. Leilabady, D. W. Anthon, K. L. Sweeney, W. L. Barnes, E. R. Taylor, and J. E. Townsend, "High-Power Sensitized Erbium Optical Fiber Amplifier," *Tech. Digest of OFC '91*, 1991, pp. 31–34.

[132] Grubb, S. G., T. H. Windhorn, W. F. Humer, K. L. Sweeney, P. A. Leilabady, J. E. Townsend, K. P. Jedrzejewski, and W. L. Barnes, "+21 dBm Erbium Power Amplifier Pumped by a Diode-Pumped Nd: YLF Laser," *Tech. Digest of OAA '91*, July 1991, pp. 345–348.

[133] Townsend, J. E., W. L. Barnes, K. P. Jedrzejewski, and S. G. Grubb, "Yb^{3+} Sensitised Er^{3+} Doped Silica Optical Fiber with Ultrahigh Transfer Efficiency and Gain," *Electron. Lett.*, Vol. 27, Oct. 1991, pp. 1958–1959.

[134] Minelly, J. D., R. I. Laming, J. E. Townsend, W. L. Barnes, E. R. Taylor, K. P. Jedrzejewski, and D. N. Payne, "High Gain Fiber Power Amplifier Tandem-Pumped by a 3-W Multistripe Diode," *Tech. Digest of OFC '92*, 1992, pp. 32–33.

[135] Minelly, J. D., W. L. Barnes, R. I. Laming, P. R. Morkel, J. E. Townsend, S. G. Grubb, and D. N. Payne, "Er^{3+}/Yb^{3+} Co-Doped Power Amplifier Pumped by a 1W Diode Array," *Tech. Digest of OAA'92*, June 1992, pp. 5–8.

[136] Grubb, S. G., W. F. Humer, R. S. Cannon, T. H. Windhorn, S. W. Vendetta, K. L. Sweeney, P. A. Leilabady, W. L. Barnes, K. P. Jedrzejewski, and J. E. Townsend, "+21 dBm Erbium Power Amplifier Pumped by a Diode-Pumped Nd: YLF Laser," *IEEE Photon. Tech. Lett.*, Vol. 4, June 1992, pp. 553–555.

[137] Grubb, S. G., P. A. Leilabady, and D. E. Frymyer, "Solid-State Laser Pumping of 1.5-μm Optical Amplifiers and Sources for Lightwave Video Transmission," *J. Lightwave Tech.*, Vol. 11, Jan. 1993, pp. 27–33.

[138] Minelly, J. D., W. L. Barnes, R. I. Laming, P. R. Morkel, J. E. Townsend, S. G. Grubb, and D. N. Payne, "Diode-Array Pumping of Er^{3+}/Yb^{3+} Co-Doped Fiber Lasers and Amplifiers," *IEEE Photon. Tech. Lett.*, Vol. 5, Mar. 1993, pp. 301–303.

[139] Grubb, S. G., "High-Power Diode-Pumped Fiber Lasers and Amplifiers," *Tech. Digest of OFC '95*, Feb. 1995, Paper TuJ1.

[140] Po, H., E. Snitzer, R. Tumminelli, L. Zenteno, F. Hakimi, N. M. Cho, and T. Haw, "Double Clad High Brightness Nd Fiber Laser Pumped by GaAlAs Phased Array," *Techical Digest of OSA '89*, 1989, pp. 260–263.

[141] Zenteno, L., "High-Power Double-Clad Fiber Lasers," *J. Lightwave Tech.*, Vol. 11, No. 9, Sept. 1993, pp. 1435–1446.

[142] Grubb, S. G., D. J. DiGiovanni, J. R. Simpson, W. Y. Cheung, S. Sanders, D. F. Welch, and Ben Rockney, "Ultrahigh Power Diode-Pumped 1.5-μm Fiber Amplifiers," *Tech. Digest of OFC '96*, Feb. 1996, Paper TuG4.

[143] Bousselet, P., R. Meilleur, A. Coquelin, P. Garabedian, and J. L. Beylat, "+25.2-dBm Output Power from an Er-Doped Fiber Amplifier with 1.48-μm SMQW Laser-Diode Modules," *Tech. Digest of OFC '95*, Feb. 1995, Paper TuJ2.

[144] Kagi, N., A. Oyobe, and K. Nakamura, "Temperature Dependence of the Gain in Erbium-Doped Fibers," *IEEE J. Lightwave Tech.*, Vol. 9, Feb. 1991, pp. 261–265.

[145] Suyama, M., R. I. Laming, and D. N. Payne, "Temperature Variation of Gain in a 1480nm-Pumped Erbium-Doped Fibre Amplifier," *Tech. Digest of OSA '90*, 1990, pp. 510–513.

[146] Lemaire, P. J., H. A. Watson, D. J. DiGiovanni, and K. L. Walker, "Hydrogen-Induced Loss Increases in Hermetic and Nonhermetic Erbium-Doped Amplifier Fibers," *Tech. Digest of OFC/IOOC '93*, Vol. 4, 1993, pp. 53–54.

[147] Lemaire, P. J., H. A. Watson, D. J. DiGiovanni, and K. L. Walker, "Prediction of Long-Term Hydrogen-Induced Loss Increases in Er-Doped Amplifier Fibers," *IEEE Photon. Tech. Lett.*, Vol. 5, Feb. 1993, pp. 214–217.

[148] Fukuda, C., Y. Koyano, T. Kashiwada, M. Onishi, Y. Chigusa, H. Kanamori, and S. Okamoto, "Hydrogen and Radiation Resistance of Erbium-Doped Fibers," *Tech. Digest of OFC '94*, Vol. 4, Feb. 1994, pp. 304–305.

[149] Marcerou, J. F., H. Février, P. M. Gabla, and J. Auge, "Sensitivity of Erbium-Doped Fibers to Hydrogen: Implications for Long-Term System Performance," *Tech. Digest of OFC '94*, Vol. 4, Feb. 1994, pp. 302–304.

[150] Lemaire, P. J., D. P Monroe, and H. A. Watson, "Hydrogen-Induced-Loss Increases in Erbium-Doped Amplifier Fibers: Revised Predictions," *Tech. Digest of OFC '94*, Vol. 4, Feb. 1994, pp. 301–302.

[151] Arai, S., A. Oyobe, A. Umeda, and K. Kokura, "Long-Term Reliability of Carbon-Coated Erbium-Doped Amplifier Fiber," *Tech. Digest of OEC '94*, July 1994, pp. 42–43.

[152] Larsen, C. C., and B. Pálsdóttir, "Hydrogen Immune Er-Doped Optical Fibres and Amplifiers," *Electron. Lett.*, Vol. 30, Aug. 1994, pp. 1414–1416.

[153] Simpson, J. R., M. M. Broer, D. J. DiGiovanni, K. W. Quoi, and S. G. Kosinski, "Ionizing and Optical Radiation-Induced Degradation of Erbium-Doped-Fiber Amplifiers," *Tech. Digest of OFC/IOOC '93*, Vol. 4, 1993, pp. 52–53.

[154] Fukuda, C., Y. Chigusa, T. Kashiwada, M. Onishi, and H. Kanamori, "Effect of γ-Ray Irradiation on Erbium-Doped Fibers," *Tech. Digest of OEC '94*, July 1994, pp. 44–45.

[155] Fujiura, K., K. Hoshino, T. Kanamori, Y. Nishida, Y. Ohishi, and S. Sudo, "Reliability of Fluoride Fibers for Use in Fiber Amplifiers," *Tech. Digest of OAA'95,* July 1995, pp. 104–107.

[156] Fujiura, K., K. Hoshino, T. Kanamori, Y. Nishida, Y. Ohishi, and S. Sudo, "Reliability of Fluoride Fibers for Use in Fiber Amplifiers," *IEEE Photon. Tech. Lett.,* to appear.

[157] Suda, H., N. Tomita, S. Furukawa, F. Yamamoto, K. Kimura, M. Shimizu, M. Yamada and Y. Ohishi, "Application of LD-pumped Pr-Doped Fluoride Fiber Amplifier to Digital and Analog Signal Transmission," *Tech. Digest of OAA'93,* July 1993, Paper PD8.

[158] Yamada, M., and M. Shimizu: "Progress toward the Telecommunication Application of 1.3-μm Praseodymium-Doped Fiber Amplifiers," *Tech. Digest of OAA'94,* Aug. 1994, Paper WC1.

[159] Whitley, T., R. Lobbett, R. Wyatt, and D. Szebesta, "5 Gbps Transmission Over 100 km of Optical Fiber Using a Directly Modulated DFB Laser and an Engineered 1.3 Micron Pr^{3+}-Doped Fluoride Fiber Power Amplifier," *Tech. Digest of OAA'94,* Aug. 1994, Paper WB1.

[160] Kikushima, K., M. Yamada, M. Shimizu, and J. Temmyo, "Distortion and Noise Properties of a Praseodymium-Doped Fluoride Fiber Amplifier in 1.3μm AM-SCM Video Transmission Systems," *IEEE Photon. Tech. Lett.,* Vol. 6, 1994, pp. 440–442.

[161] Yoshinaga, H., M. Yamada, and M. Shimizu, "Improved Performance of Pr^{3+}-Doped Fluoride Fiber Amplifier in Subcarrier Multiplexed Multichannel AM-VSB Video Signal Transmission," *Electron. Lett.,* Vol. 30, 1994, pp. 2042–2043.

[162] Yamada, M., M. Shimizu, T. Kanamori, Y. Ohishi, Y. Terunuma, K. Oikawa, H. Yoshinaga, K. Kikushima, Y. Miyamoto, and S. Sudo, "High-Power, Low-Noise Praseodymium-Doped Fluoride-Fiber Amplifiers," *IEEE Photon. Tech. Lett.,* Vol. 7, Aug. 1995, pp. 869–871.

[163] Shimizu, M., M. Yamada, T. Kanamori, Y. Ohishi, Y. Terunuma, H. Yoshinaga, and Y. Miyamoto, "High-Power, Low-Noise Praseodymium-Doped Fluoride-Fiber Amplifiers," *Tech. Digest of OFC'95,* Feb. 1995, Paper WD5.

[164] Ohishi, Y., T. Kanamori, T. Nishi, Y. Nishida, and S. Takahashi, "Quantum Efficiency at 1.3 μm of Pr^{3+} in Infrared Glasses," *Tech. Digest of XVI Int. Congress on Glass,* Oct. 1992, pp. 73–78.

[165] Becker, P. C., M. M. Broer, V. C. Lambrecht, A. J. Bruce and C. Nykolak, "Pr^{3+}:La-Ga-S Glass: A Promising Material for 1.3 μm Fiber Amplification," *Tech. Digest of OAA'92,* July 1992, Paper PDP5.

[166] Nishida, Y., Y. Ohishi, T. Kanamori, Y. Terunuma, K. Kobayashi, and S. Sudo, "Preparation of Pr^{3+}-Doped InF$_3$-Based Fluoride Single-Mode Fibers and First Demonstration of 1.3 μm Amplification," *Tech. Digest of ECOC'93,* Vol. 1, Sept. 1993, pp. 34–37.

[167] Ohishi, Y., A. Mori, T. Kanamori, K. Fujiura, and S. Sudo, "Fabrication of Praseodymium-Doped Arsenic Sulfide Chalcogenide Fiber for 1.3-μm Fiber Amplifiers," *Appl. Phys. Lett.,* Vol. 65, July 1994, pp. 13–15.

[168] Yanagida, H., K. Itoh, E. Ishikawa, H. Aoki, and H. Toratani, "26 dB Amplification at 1.31 μm in a Novel Pr^{3+}-Doped InF$_3$/GaF$_3$-Based Fiber," *Tech. Digest of OFC'95,* Feb. 1995, Paper PD2.

[169] Contained within a technical report from AMACO.

[170] Shimizu, M., M. Horiguchi, M. Yamada, I. Nishi, J. Noda, T. Takeshita, M. Okayasu, S. Uehara, and E. Sugita, "Compact and Highly Efficient Fiber Amplifier Modules Pumped by a 0.98-mm Laser Diode," *IEEE J. Lightwave Tech.,* Vol. LT9, 1991, pp. 291–296.

[171] Shimizu, M., M. Yamada, M. Horiguchi, T. Takeshita, and M. Okayasu, "Erbium-Doped Fibre Amplifiers with an Extremely High Gain Coefficient of 11.0 dB/mW," *Electron. Lett.,* Vol. 26, 1990, pp. 1641–1642.

[172] Whitley, T., R. Wyatt, D. Szebesta, S. Davey, and J. R. Williams, "Quarter Watt Output at 1.3 mm from a Praseodymium Doped Fluoride Fibre Amplifier Pumped with a Diode-Pumped Nd:YLF Laser," *Tech. Digest of OAA'92,* July 1992, Paper PDP4.

[173] Yamada, M., M. Shimizu, Y. Ohishi, T. Kanamori, M. Horiguchi, S. Takahashi, J. Temmyo, and M. Wada, "15.1-dB-Gain Pr^{3+}-Doped Fluoride Fiber Amplifier Pumped by High Power Laser Diode Modules," *Tech. Digest of ECOC'92,* Sept. 1992, pp. 49–52.

[174] Shimizu, M., T. Kanamori, J. Temmyo, M. Wada, M. Yamada, Y. Terunuma, Y. Ohishi, and

S. Sudo, "28.3 dB Gain 1.3 μm-Band Pr-Doped Fluoride Fiber Amplifier Module Pumped by 1.017 μm InGaAs-LD's," *IEEE Photon. Tech. Lett.*, Vol. 5, pp. 654–657, 1993.

[175] Yamada, M., M. Shimizu, Y. Ohishi, J. Temmyo, M. Wada, T. Kanamori, Y. Terunuma, and S. Sudo, "Gain Characteristics of Pr^{3+}-Doped Fluoride Fiber Amplifier at 1.3 μm," *Tech. Digest of Tech. Group*, OQE92-134, IEICE Japan, 1992, pp. 57–62.

[176] Yamada, M., M. Shimizu, Y. Ohishi, J. Temmyo, M. Wada, T. Kanamori, and S. Sudo, "One-LD-Pumped Pr^{3+}-Doped Fluoride Fiber Amplifier Module with a Signal Gain of 23 dB," *Tech. Digest of OAA'93*, July 1993, pp. 240–243.

[177] Ohishi, Y., T. Kanamori, Y. Terunuma, M. Shimizu, M. Yamada, and S. Sudo, "Investigation of Efficient Pump Scheme for Pr^{3+}-Doped Fluoride Fiber Amplifier," *IEEE Photon. Tech. Lett.*, Vol. 6, Feb. 1994, pp. 195–198.

[178] Moneric, M., "Status of Fluoride Fiber Lasers," *Tech. Digest of SPIE*, Vol. 1581, Sept. 1991, pp. 2–13.

[179] Dye, S. P., and M. Fake, "Fully Engineered 800-nm Thulium-Doped Fluoride-Fiber Amplifier," *Tech. Digest of OFC'95*, Feb. 1995, Paper WD4.

[180] Sakamoto, T., M. Shimizu, T. Kanamori, Y. Ohishi, Y. Terunuma, M. Yamada, and S. Sudo, "Tm-Ho-Doped ZBLYAN Fiber Amplifiers for 1.4 μm-band," *Tech. Digest of OEC'92*, July 1994, Paper 13B2-2.

[181] Sakamoto, T., M. Shimizu, T. Kanamori, Y. Terunuma, Y. Ohishi, M. Yamada, and S. Sudo, "1.4 μm-Band Gain Characteristics of a Tm-Ho-Doped ZBLYAN Fiber Amplifier Pumped in the 0.8-μm Band," *IEEE Photon. Tech. Lett.*, Vol. 7, Sept. 1995, pp. 983–985.

[182] Percival, R. M., D. Szebesta, and J. R. Williams, "Highly Efficient 1.064 μm Upconversion Pumped 1.47 μm Thulium Doped Fluoride Fibre Laser," *Electron. Lett.*, Vol. 30, June 1994, pp. 1057–1058.

[183] Percival, R. M., and J. R. Williams, "Highly Efficient 1.064 μm Upconversion Pumped 1.47 μm Thulium Doped Fluoride Fibre Amplifier," *Electron. Lett.*, Vol. 30, Sept. 1994, pp. 1684–1685.

[184] Miyajima, Y., T. Komukai, and T. Sugawa, "1-W CW Tm-Doped Fluoride Fibre Laser at 1.47 μm," *Electron. Lett.*, Vol. 29, Apr. 1993, pp. 660–661.

[185] Percival, R. M., D. Szebesta, and S. T. Davey, "Thulium Doped Terbium Sensitised CW Fluoride Fiber Laser Operating on the 1.47 μm Transition," *Electron. Lett.*, Vol. 29, June 1993, pp. 1054–1056.

[186] Percival, R. M., D. Szebesta, and S. T. Davey, "Highly Efficient CW Cascade Operation of 1.47 and 1.83 μm Transition in Tn-Doped Fluoride Fiber Laser," *Electron. Lett.*, Vol. 24, Sept. 1992, pp. 1866–1868.

[187] Komukai, T., T. Yamamoto, T. Sugawa, and T. Miyajima, "1.47 μm Band Tm^{3+} Doped Fluoride Fiber Amplifier Using a 1.064 μm Upconversion Pumping Scheme," *Electron. Lett.*, Vol. 29, Jan. 1993, pp. 110–112.

[188] Komukai, T., T. Yamamoto, T. Sugawa, and Y. Miyajima, "Upconversion Pumped Thulium-Doped Fluoride Fiber Amplifier and Laser Operating at 1.47 μm," *IEEE J. Quantum Electron.*, Vol. 31, Nov. 1995, pp. 1880–1889.

[189] Sakamoto, T., M. Yamada, M. Shimizu, T. Kanamori, Y. Terunuma, Y. Ohishi, and S. Sudo, "Thulium-Doped Fluoride Fiber Amplifiers for 1.4 μm and 1.6 μm Operation," *Tech. Digest of OAA'96*, 1996, Paper ThC1.

[190] Sakamoto, T., M. Shimizu, M. Yamada, T. Kanamori, Y. Ohishi, Y. Terunuma, and S. Sudo, "35 dB Gain Tm-Doped ZBLYAN Fiber Amplifier Operating at 1.65 μm," *Tech. Digest of CLEO/Pacific Rim'95*, July 1995, Paper PD2.12.

[191] Barnes, W. L., and J. E. Townsend, "Highly Tunable and Efficient Diode Pumped Operation of Tm^{3+} Doped Fiber Laser," *Electron. Lett.*, Vol. 26, May 1990, pp. 746–747.

[192] Sankawa, I., H. Izumita, S. Furukawa, and K. Ishihara, "An Optical Fiber Amplifier for Wide-Band Wavelength Range Around 1.65 μm," *IEEE Photon. Tech. Lett.*, Vol. 2, 1990, pp. 422–424.

[193] Takasugi, H., N. Tomita, J. Nakano, and N. Atobe, "Design of a 1.65-μm-Band Optical Time-Domain Reflectometer," *IEEE J. Lightwave Tech.*, Vol. 11, 1993, pp. 1743–1747.

[194] Hebert, T., R. Wannemacher, W. Lenth, and M. Macfarlane, "Blue and Green cw Upconversion Lasing in Er:YLiF$_4$," *Appl. Phys. Lett.*, Vol. 57, Oct. 1990, pp. 1727–1729.

[195] Chen, Y., and F. Auzel, "Spatial Domains in Avalanche Pumped Er:ZBLAN Fibre," *Electron. Lett.*, Vol. 30, Feb. 1994, pp. 323–324.

[196] Ghisler, C., M. Pollnau, G. Bunea, M. Bunea, W. Lüthy, and H. P. Weber, "Up-Conversion Cascade Laser at 1.7 μm with Simultaneous 2.7 μm Lasing in Erbium ZBLAN Fibre," *Electron. Lett.*, Vol. 31, Mar. 1995, pp. 373–374.

[197] Sikai, S., T. Fukuoka, and T. Tohi, "Application of an Er^{3+}-Doped Fibre Amplifier to an OTDR Operating at 1.6 μm," *Electron. Lett.*, Vol. 30, July 1994, pp. 1225–1226.

[198] Massicott, J. F., R. Wyatt, and B. J. Ainslie, "Low Noise Operation of Er^{3+} Doped Silica Fibre Amplifier Around 1.6 μm," *Electron. Lett.*, Vol. 28, Sept. 1992, pp. 1924–1925.

[199] Frerichs, Ch., and T. Tauermann, "Q-switched Operation of Laser Diode Pumped Erbium-Doped Fluorozirconate Fibre Laser Operating at 2.7 μm," *Electron. Lett.*, Vol. 30, Apr. 1994, pp. 706–707.

[200] Massicott, J. F., J. R. Armitage, R. Wyatt, B. J. Ainslie, and S. P. Draig-Ryan, "High Gain Broadbnad 1.6 μm Er^{3+} Doped Silica Fibre Amplifier," *Electron. Lett.*, Vol. 26, Sept. 1990, pp. 1645–1646.

[201] Brierley, M. C., and P. W. France, "Continuous Wave Lasing at 2.7 μm in an Erbium-Doped Fluorozirconate Fibre," *Electron. Lett.*, Vol. 24, July 1988, pp. 935–937.

[202] Pinto, J. F., G. H. Rosenblatt, and L. Esterowitz, "Continuous-Wave Laser Action in Er^{3+}:YLF at 3.41 μm," *Electron. Lett.*, Vol. 30, Sept. 1994, pp. 1596–1598.

[203] Többen, H., "Room Temperature CW Fibre Laser at 3.5 μm in Er^{3+}-Doped ZBLAN Glass," *Electron. Lett.*, Vol. 28, July 1992, pp. 1361–1362.

[204] Zhao, Y., and S. Poole, "Efficient Blue Pr^{3+}-Doped Fluoride Fibre Upconversion Laser," *Electron. Lett.*, Vol. 30, June 1994, pp. 967–968.

[205] Petreski, B. P., M. M. Murphy, S. F. Collins, and D. J. Booth, "Amplification in Pr^{3+}-Doped Fluorozirconate Optical Fibre at 632.8 nm," *Electron. Lett.*, Vol. 29, Aug. 1993, pp. 1421–1423.

[206] Koch, M. E., A. W. Kueny, and W. E. Case, "Photon Avalanche Upconversion Laser at 644 nm," *Appl. Phys. Lett.*, Vol. 56, Mar. 1990, pp. 1083–1085.

[207] Shi, Y., C. V. Poulse, M. Sejka, M. Ibsen, and O. Poulsen, "Tunable Pr^{3+}-Doped Silica-Based Fibre Laser," *Electron. Lett.*, Vol. 29, Aug. 1993, pp. 1426–1427.

[208] Millar, C. A., S. R. Mallinson, B. J. Ainslie, and S. P. Craig, "Photochromic Behaviour of Thulium-doped Silica Optical Fibres," *Electron. Lett.*, Vol. 24, May 1988, pp. 590–591.

[209] Allain, J. Y., and M. Monerie, Tunable CW Lasing around 0.82, 1.48, 1.88 and 2.35 μm in Thulium-Doped Fluorozirconate Fiber," *Electron. Lett.*, Vol. 25, Nov. 1989, pp. 1660–1662.

[210] Grubb, S. G., K. W. Bennett, R. S. Cannon, and W. F. Humer, "CW Room-Temperature Blue Upconversion Fibre Laser," *Electron. Lett.*, Vol. 28, June 1992, pp. 1243–1244.

[211] Tohmon, G., J. Ohya, and T. Uno, "Upconversion Blue Fiber Laser Pumped by 680 nm and 1.1 μm Light," *Tech. Digest of OEC '94*, July 1994, Paper 14E-5.

[212] Dye, S. P., and M. Fake, "Fully Engineered 800-nm Thulium-Doped Fluoride-Fiber Amplifier," *Tech. Digest of OFC'95*, Feb. 1995, Paper WD4.

[213] Percival, R. M., D. Szebesta, J. R. Williams, R. D. T. Lauder, A. C. Tropper, and D. C. Hanna, "Diode Pumped Operation of Thulium Doped Fluoride Fibre Amplifier Suitable for First Window Systems," *Electron. Lett.*, Vol. 30, Sept. 1994, pp. 1598–1599.

[214] Dennis, M. L., J. W. Dixon, and I. Aggarwal, "High Power Upconversion Lasing at 810 nm, in Tm:ZBLAN Fibre," *Electron. Lett.*, Vol. 30, Jan. 1994, pp. 136–137.

[215] Dennis, M. L., J. W. Dixon, and I. Aggarwal, "Upconversion-Pumped Thulium-Fiber Laser at 810 nm," *Tech. Digest of OFC'94.*, 1994, Paper WK10.

[216] Carter, J. N., R. G. Smart, D. C. Hanna, and A. C. Tropper, "Lasing and Amplification in the 0.8 μm Region in Thulium Doped Fluorozirconate Fibres," *Electron. Lett.*, Vol. 26, Oct. 1990, pp. 1759–1761.

[217] Percival, R. M., D. Szebesta, C. P. Seltzer, S. D. Perrin, S. T. Davey, and M. Louka, "A 1.6 μm Pumped 1.9-μm Thulium-Doped Fluoride Fiber Laser and Amplifier of Very High Efficiency," *IEEE Photon. Tech. Lett.*, Vol. 31, Mar. 1995, pp. 489–493.

[218] Oh, K., A. Kilian, Q. Zhang, T. F. Morse, and P. M. Weber, "Broadband Superfluorescent Emission of the $^3H_4$-> $^3H_6$ Transition in a Tm-Doped Multicomponent Silicate Fiber," *Optics Lett.,* Vol. 19, Aug. 1994, pp. 1131–1133.

[219] Yamamoto, T., Y. Miyajima, and T. Komukai, "1.9 μm Tm-Doped Silica Fibre Laser Pumped at 1.57 μm," *Electron. Lett.,* Vol. 30, Feb. 1994, pp. 220–221.

[220] Esterowitz, L., R. Allen, and I. Aggarwal, "Pulsed Laser Emission at 2.3 μm in a Thulium-Doped Fluorozirconate Fibre," *Electron. Lett.,* Vol. 24, Aug. 1988, p. 1104.

[221] Lenth, W., and R. M. Macfarlane, "Excitation Mechanisms for upconversion laser," *J. Luminescence,* Vol. 45, 1990, pp. 346–350.

[222] Reekie, L., I. M. Jauncey, S. B. Poole, and D. N. Payne, "Diode-Laser-Pumped Nd^{3+}-Doped Fibre Laser Operating at 938 nm," *Electron. Lett.,* Vol. 23, Aug. 1987, pp. 884–885.

[223] Pollak, T. M., W. F. Wing, R. J. Grasso, E. P. Chicklis, and H. P. Jenssen, "CW Laser Operation of Nd:YLF," *IEEE J. Quantum Electron.,* Vol. QE-18, Feb. 1982, pp. 159–163.

[224] Zenteno, L., "High-Power Double-Clad Fiber Lasers," *J. Lightwave Tech.,* Vol. 11, Sept. 1993, pp. 1435–1446.

[225] Po, H., J. D. Cao, B. M. Laliberte, R. A. Minns, R. F. Robinson, B. H. Rockney, R. R. Tricca, and Y. H. Zhang, "High Power Neodymium-Doped Single Transverse Mode Fibre Laser," *Electron. Lett.,* Vol. 29, Aug. 1993, pp. 1500–1501.

[226] Minelly, J. D., P. R. Morkel, K. P. Jedrzejewski, E. R. Taylor, J. Wang, and D. N. Payne, "Nd^{3+}-doped Singlemode Fibre Superfluorescent Source with 320 mW Output Power," *Electron. Lett.,* Vol. 29, Sept. 1993, pp. 1613–1614.

[227] Loh, W. H., D. Atkinson, P. R. Morkel, R. Grey, A. J. Seeds, and D. N. Payne, "Diode-Pumped Selfstarting Passively Modelocked Neodymium-Doped Fibre Laser," *Electron. Lett.,* Vol. 29, Apr. 1993, pp. 808–810.

[228] Wang, J. S., D. P. Machewirth, F. Wu, E. M. Vogel, and E. Snitzer, "Neodymium-Doped Tellurite Single Mode Fiber Laser," *Tech. Digest of OSA,* 1994, Paper 6264.

[229] Morkel, P. R., E. M. Taylor, J. E. Townsend, and D. N. Payne, *Wavelength Stability of Nd^{3+}-Doped Fibre Fluorescent Sources,* : Optoelectronics Research Centre, 1991, pp. 75–79.

[230] Digonnet, M. J. F., K. Liu, and H. J. Shaw, "Characterization and Optimization of the Gain in Nd-Doped Single-Mode Fibers," *IEEE J. Quantum Electron.,* Vol. 26, June 1990, pp. 1105–1110.

[231] Phillips, M. W., A. I. Ferguson, and D. B. Patterson, "Diode-Pumped FM Modelocked Fibre Laser with Coupled Cavity Bandwidth Selection, " *Tech. Digest of ECOC'89,* 1989, Paper TuA2.1.

[232] Phillips, M. W., A. I. Ferguson, and D. C. Hanna, "Frequency-Modulation Mode Locking of a Nd^{3+}-Doped Fiber Laser," *Optics Lett.,* Vol. 14, Feb. 1989, pp. 219–221.

[233] Hakimi, F., H. Po, R. Tumminelli, B. C. McCollum, L. Zenteno, N. M. Cho, and E. Snitzer, "Glass Fiber Laser at 1.36 mm from SiO_2:Nd," *Optics Lett.,* Vol. 14, Oct. 1989, pp. 1060–1062.

[234] Liu, K., M. Digonnet, K. Fesler, B. Y. Kim, and H. J. Shaw, "Broadband Diode-Pumped Fibre Laser," *Electron. Lett.,* Vol. 24, July 1994, pp. 838–840.

[235] Liu, K., M. Digonnet, H. J. Shaw, B. J. Ainslie, and S. P. Craig, "10 mW Superfluorescent Single-Mode Fibre Source at 1060 nm," *Electron. Lett.,* Vol. 23, Nov. 1987, pp. 1320–1321.

[236] Brierley, M. C., and P. W. France, "Neodymium-Doped Fluoro-Zirconate Fibre Laser," *Electron. Lett.,* Vol. 23, July 1987, pp. 815–817.

[237] Shimizu, M., H. Suda, and M. Horiguchi, "High-Efficiency Nd-Doped Fibre Lasers using Direct-Coated Dielectric Mirrors," *Electron. Lett.,* Vol. 23, July 1987, pp. 768–769.

[238] Zürn, M., J. Voigt, E. Brinkmeyer, R. Ulrich, and S. B. Poole, "Line Narrowing and Spectral Hole Burning in Single-Mode Nd^{3+}-Fiber Lasers," *Optics Lett.,* Vol. 12, May 1987, pp. 316–318.

[239] Jauncey, I. M., L. Reekie, D. N. Payne, and C. J. Rowe, "Single Longitudinal Mode Operation of a Fibre Laser," *CLEO '87,* Paper ThT9.

[240] Alcock, I. P., A. I. Ferguson, D. C. Hanna, and A. C. Tropper, "Tunable, Continuous-Wave Neodymium-Doped Monomode-Fiber Laser Operating at 0.900–0.945 and 1.070–1.135 μm," *Optics Lett.,* Vol. 11, Nov. 1986, pp. 709–711.

[241] Jauncey, I. M., J. T. Lin, L. Reekie, and R. J. Mears, "Efficient Diode-Pumped CW and Q-Switched Single-Mode Fibre Laser," *Electron. Lett.*, Vol. 22, Feb. 1986, pp. 198–199.

[242] Alcock, I. P., A. I. Ferguson, D. C. Hanna, and A. C. Tropper, "Mode-Locking of a Neodymium-Doped Monomode Fibre Laser," *Electron. Lett.*, Vol. 22, Feb. 1986, pp. 268–269.

[243] Alcock, I. P., A. C. Tropper, A. I. Ferguson, and D. C. Hanna, "Q-Switched Operation of a Neodymium-Doped Monomode Fibre Laser," *Electron. Lett.*, Vol. 22, Jan. 1986, pp. 84–85.

[244] Shigyouchi, K., S. Konishi, K. Susa, and I. Matsuyama, "Radial Gradient Refractive-Index Glass Rods Prepared by a Sol-Gel Method," *Electron. Lett.*, Vol. 22, Jan. 1986, pp. 99–100.

[245] Poole, S. B., D. N. Payne, and M. E. Fermann, "Fabrication of Low-Loss Optical Fibres Containing Rare-Earth Ions," *Electron. Lett.*, Vol. 21, Aug. 1985, pp. 737–738.

[246] Mears, R. J., L. Reekie, S. B. Poole, and D. N. Payne, "Neodymium-Doped Silica Single-Mode Fibre Lasers," *Electron. Lett.*, Vol. 21, Aug. 1985, pp. 738–740.

[247] Uppal, J. S., T. P. S. Nathan, and D. D. Bhawalkar, "Performance Characteristics of a Nd:Glass Laser Amplifier from Fluorescence Emission Studies," *Opt. and Quantum Electron.*, Vol. 17, 1985, pp. 323–328.

[248] Stone, J., and C. A. Buruus, "Neodymium-Doped Silica Lasers in End-Pumped Fiber Geometry," *Appl. Phys. Lett.*, Vol. 23, Oct. 1973, pp. 388–389.

[249] Hanna, D. C., R. M. Pervical, I. R. Perry, R. G. Smart, P. J. Suni, and A. C. Tropper, "An Ytterbium-Doped Monomode Fiber Laser: Broadly Tunable Operation from 1.010 μm to 1.162 μm and Three-Level Operation at 0.974 nm," *J. Modern Opt.*, Vol. 37, 1990, pp. 517–525.

[250] Farries, M. C., P. R. Morkel, and J. E. Townsend, "Samarium^{3+}-Doped Glass Laser Operating at 651 nm," *Electron. Lett.*, Vol. 24, May 1988, pp. 709–711.

[251] Brierley, M. C., P. W. France, and C. A. Millar, "Lasing at 2.08 μm and 1.38 μm in a Holmium Doped Fluoro-Zirconate Fibre Laser," *Electron. Lett.*, Vol. 24, Apr. 1988, pp. 539–540.

[252] Samsonov, G. V., *The Oxide Handbook*, New York: IFI/Plenum, 1973, p. 178.

[253] Sudo, S., "Studies on the Vapor-Phase Axial Deposition Method for Optical Fiber Fabrication," Ph.D. thesis, University of Tokyo, 1982, p. 164.

[254] Dunning, W. J., "Theory of Crystal Nucleation from Vapor, Liquid and Solid Systems," in *Chemistry of the Solid State*, W. E. Garner (ed.), London: Butterworths, 1955.

[255] Townsend, J. E., S. B. Poole, and D. N. Payne, "Solution Doping Technique for Fabrication of Rare-Earth Doped Optical Fibers," *Electron. Lett.*, Vol. 23, 1987, pp. 329–331.

[256] Simpson, J. R., "Rare Earth Doped Fiber Fabrication: Techniques and Physical Properties," in *Rare Earth Doped Fiber Lasers and Amplifiers*, M. J. F. Digonnet (ed.), New York: Marcel Dekker, 1993, pp. 1–18.

[257] Izawa, T., and S. Sudo, *Optical Fibers: Materials and Fabrication*, Dordrecht: D. Reidel, 1987.

[258] Snitzer, E., "Lasers and Glass Technology," *Ceramic Bull.*, Vol. 52, 1973, pp. 516–525.

[259] Arai, K., H. Namikawa, Y. Ishii, H. Imai, H. Hosono, and Y. Abe, "Nature of Doping into Pure Silica Glass by Plasma Torch CVD," *J. Non-Crystalline Solids*, Vols. 95/96, 1987, pp. 609–616.

[260] Arai, K., "Preparation of Active Ion Doped Silica Glasses and Their Properties," *Ceramics*, Vol. 21, 1986, pp. 419–424.

[261] Arai, K., H. Namikawa, K. Kumata, and T. Honda, "Aluminum or Phosphorus Co-Doping Effects on the Fluorescence and Structural Properties of Neodymium-Doped Silica Glass," *J. Appl. Phys.*, Vol. 59, May 1986, pp. 3430–3436.

[262] Ainslie, B. J., S. P. Craig, R. Wyatt, and K. Moulding, "Optical and Structural Analysis of Neodymium-Doped Silica-Based Optical Fibre," *Mat. Lett.*, Vol. 8, July 1988, pp. 204–208.

[263] Ainslie, B. J., S. P. Craig, S. T. Davey, and B. Wakefield, "The Fabrication, Assessment and Optical Properties of High-Concentration Nd^{3+}-and Er^{3+}-Doped Silica-Based Fibres," *Mat. Lett.*, Vol. 6, Mar. 1988, pp. 139–144.

[264] Laming, R. I., J. E. Townsend, D. N. Payne, F. Meli, G. Grasso, and E. J. Tarbox, "High-Power Erbium-Doped-Fiber Amplifiers Operating in the Saturated Regime," *IEEE Photon. Tech. Lett.*, Vol. 3, Mar. 1991, pp. 253–255.

[265] Shimizu, M., M. Yamada, M. Horiguchi, and E. Sugita, "Concentration Effect on Optical Amplification Characteristics of Er-Doped Silica Single-Mode Fibers," *IEEE Photon. Tech. Lett.*, Vol. 2, Jan. 1990, pp. 43–45.

[266] Inoue, H., and I. Yasui, "F ion Conduction Mechanism in Glasses Based on ZrF_4," *Mat. Sci. Forum*, Vol. 19–20, 1987, pp. 161–170.

[267] Taga, H., N. Edagawa, H. Tanaka, M. Suzuki, S. Yamamoto, H. Wakabayashi, N. S. Bergano, C. R. Davidson, G. M. Hmsey, D. J. Kalmus, P. R. Trischitta, D. A. Gray, and R. L. Maybach, "10Gbit/s, 9,000 IM-DD Transmission Experiments Using 274 Er-doped Fiber Amplifier Repeaters," *Tech. Digest of OFC/IOOC 93*, Postdeadline Papers, 1993, Paper PD1.

[268] Edagawa, N., Y. Yoshida, H. Taga, S. Yamamoto, K. Mochizuki, and H. Wakabayashi, "904km, 1.2Gbit/s Non-Regenerative Optical Transmission Experiment Using 12 Er-Doped Fibre Amplifiers," *Tech. Digest of ECOC '89*, Paper PDA-8.

[269] Saito, S., T. Imai, and T. Ito, "An Over 2200-km Coherent Transmission Experiment at 2.5 Gb/s Using Erbium-Doped-Fiber In-Line Amplifier," *IEEE J. Lightwave Tech.*, Vol. 9, Feb. 1991, pp. 161–169.

[270] Blondel, J.-P., and J. F. Marcerou, "Erbium-doped Fiber Amplifier Behavior in Transoceanic Links," *Fiber Laser Sources and Amplifiers III*, Vol. 1581, : SPIE, 1991, pp. 218–233.

[271] Blondel, J.-P., and P. M. Gabla, "Gain in Filtering in Cascade Amplifier Systems," *Tech. Digest of OFC'94*, 1994, Paper ThC1.

[272] Blondel, J.-P., and P. M. Gabla, "Optical Amplification in Submarine Systems: New Concepts and Ultimate Limits," *Tech. Digest of IOOC'95*, 1995, Paper ThA2-1.

[273] Nakazawa, M., Y. Kimura, K. Suzuki, H. Kubota, K. Komukai, E. Yamada, T. Sugawa, E. Yoshida, T. Yamamoto, T. Imai, A. Sahara, H. Nakazawa, O. Yamauchi, and M. Umezawa, "Field Demonstration of Aoliton Transmission at 10 Gb/s Over 2,500 km and 20 Gb/s Over 1,000 km in the Tokyo Metropolitan Optical Network," *Tech. Digest of IOOC '95*, 1995, Paper PD2-1.

[274] Suzuki, M., I. Morita, S. Yamamoto, N. Edagawa, H. Taga, and S. Akiba, "Timing Jitter Reduction by Periodic Dispersion Compensation in Soliton Transmission," *Tech. Digest of OFC'95*, Postdeadline Papers, 1995, Paper PD20.

[275] Hansen, P. B., L. Eskildsen, S. G. Grubb, A. M. Vengsarkar, S. K. Korotky, T. A. Strasser, J. E. J. Alphonsus, J. J.Veselka, D. J. DiGiovanni, D. W. Peckham, E. C. Beck, D. Truxal, W. Y. Cheung, S. G. Kosinski, D. Gasper, P. F. Wysocki, V. L. da Silva, and J. R. Simpson, "2.488-Gb/s Unrepeatered Transmission over 529 km using Remotely Pumped Post- and Pre- Amplifiers, Forward Error Correction, and Dispersion Compensation," *Tech. Digest of OFC'95*, Postdeadline Papers, 1995, Paper PD25.

[276] Sian, S. S., O. Gutheron, M. S. Chaudhry, C. D. Stark, S. M. Webb, K. M.Guild, M. Mesic, J. M. Dryland, J. R. Chapman, A. R. Docker, E. Brandon, T. Barbier, P. Garabedian, and P. Bousselet, "511km at 2.5 Gbit/s and 531km at 622Mbit/s-Unrepeatered Transmission with Remote Pumped Amplifiers, Forward Error Correction and Dispersion Compensation," *Tech. Digest of OFC'95*, Postdeadline Papers, 1995, Paper PD26.

[277] Aida, K., S. Nishi, Y. Sato, K. Hagimoto, and K. Nakagawa, "1.8 Gb/s 310km Fiber Transmission without Outdoor Repeater Equipment Using a Remotely Pumped In-Line Er-Doped Fiber Amplifier in an IM/Direct-Detection System," *Tech. Digest of ECOC '89*, Paper PDA-7.

[278] Yoshida, S., S. Kuwano, M. Yamada, T. Kanamori, N. Takachio, and K. Iwashita, "A 10 Gbit/s x 10 Channel Transmission Experiment Over 600 km with 100 km Repeater Spacing Employing Unequal Channel Spacing and Cascaded Fluoride-Based Erbium Doped Fiber Amplifiers," *Tech. Digest of IOOC '95*, 1995, Paper PD2-3.

[279] Taylor, N. H., K. P. Jones, D. Simeonidou, M. S. Cahudhry, G. M. Long, and P. R. Morkel, "4x2.5 Gbit/s WDM Transmission, with Optical Add/Drop Multiplexing, Over 3711km on the Installed RIOJA Submarine Cable System," *Tech. Digest of OAA'95*, 1995, Paper PD7.

[280] Clesca, B., A. Jourdan, S. Artigaud, G. Da Loura, L. Hamon, G. Soulage, J. C. Jacquinot, P. Doussiére, P. de Vivie de Régie, D. Bayart, J. L. Beylat, M. Sotom, C. Jœrgensen, T. Durhuus, K.

Stubkjaer, M. Semenkoff, and M. Guibert, "A Joint RACE Experiment: 80 Gbit/s WDM Transmission Combining Wideband Optical Fibre Amplifiers with an In-Line 10 Gbit/s All-Optical Wavelength Converter," *Tech. Digest of ECOC '94*, Vol. 4, pp. 27–31, 1994.

[281] Tkach, R. W., A. H. Gnauck, A. R. Charapolyvy, R. M. Derosier, C. R. Giles, B. M. Nyman, G. A. Ferguson, J. W. Sulhoff, and J. L. Zyskind, "One-Third Terabit/s Transmission through 150 km of Dispersion-Managed Fiber," *Tech. Digest of ECOC '94*, Vol. 4, pp. 45–48, 1994.

[282] Kawanishi, S., H. Takara, T. Morioka, O. Kamatani, and M. Saruwatari, "200 Gbit/s, 100 km TDM Transmission using Supercontinuum Pulses with Prescaled PLL Timing Extraction and All-Optical Demultiplexing," *Tech. Digest of OFC'95*, Postdeadline Papers, 1995, Paper PD28.

[283] Nakazawa, M., E. Yoshida, E. Yamada, K. Suzuki, T. Kitoh, and M. Kawachi, "80 Gbit/s Soliton Data Transmission Over 500km with Unequal Amplitude Solitons for Timing Clock Extraction," *Electron. Lett.*, Vol. 30, Oct. 1994, pp. 1777–1778.

[284] Shimizu, M., M. Yamada, T. Kanamori, Y. Ohishi, Y. Terunuma, and S. Sudo, "High-Power, Low-Noise Praseodymium-Doped Fluoride-Fiber Amplifiers," *Tech. Digest of OFC'95*, 1995, Paper WD5.

[285] Kikushima, K., H. Yoshinaga, and M. Yamada, "Signal Crosstalk due to Fiber Nonlinearity in Wavelength Multiplexed SCM-AM-TV Transmission Systems," *Tech. Digest of OFC'95*, Postdeadline Papers, 1995, Paper PD24.

[286] Bergano, Neal S., C. R. Davidson, B. M. Nyman, S. G. Evangelides, J. M. Darcie, J. D. Evankow, P. C. Corbett, M. A. Mills, G. A. Ferguson, J. A. Nagel, J. L. Zyskind, J. W.Sulhoff, A. J. Lucero, and A. A. Klein, "40 Gb/s WDM Transmission of Eight 5 Gb/s Data Channels Over Transoceanic Distances using the Conventional NRZ Modulation Format," *Tech. Digest of OFC'95*, Postdeadline Papers, 1995, Paper PD19.

[287] Suzuki, M., I. Morita, S. Yamamoto, N. Edagawa, H. Taga, and S. Akiba, "Timing Jitter Reduction by Periodic Dispersion Compensation in Soliton Transmission," *Tech. Digest of OFC'95*, Postdeadline Papers, 1995, Paper PD20.

[288] Nyman, B. M., S. G. Evangelides, G. T. Harvey, L. F. Mollenauer, P. V. Mamyshev, M. L. Saylors, S. K. Korotky, U. Koren, V. Mizrahi, T. A. Strasser, J. J. Veselka, J. D. Evankow, A. Lucero, J. Nagel, J. Sulhoff, J. Zyskind, P. C. Corbett, M. A. Mills, and G. A. Ferguson, "Soliton WDM Transmission of 8 X 2.5 Gb/s, Error Free over 10 Mm," *Tech. Digest of OFC'95*, Postdeadline Papers, 1995, Paper PD21.

[289] Onaka, H., H. Miyata, G. Ishikawa, K. Otsuka, H. Ooi, Y. Kai, S. Kinosha, M. Seino, H. Nishimoto, and T. Chikama, "1.1 Tb/s WDM Transmission Over a 150 km 1.3 μm Zero-Dispersion Singlemode Fiber," *Tech. Digest of OFC'96*, San Jose, CA, Feb. 1996, Paper PD19.

[290] Gnauck, A. H., R. W. Tkach, F. Forghieri, R. M. Derosier, A. R. McCormick, A. R. Chraplyvy, J. L. Zyskind, J. W. Sulhoff, A. J. Lucero, Y. Sun, R. M. Jopson, and C. Wolf, "One Terabit/s Transmission Experiment," *Tech. Digest of OFC'96*, San Jose, CA, Feb. 1996, Paper PD20.

[291] Morioka, T., H. Takara, S. Kawanishi, O. Kamatani, K. Takiguchi, K. Uchiyama, M. Saruwatari, H. Takahashi, M. Yamada, T. Kanamori, and H. Ono, "100 Gbit/s x 10 Channel OTDM/WDM Transmission Using a Single Supercontinuum WDM Source," *Tech. Digest of OFC'96*, San Jose, CA, Feb. 1996, Paper PD21.

[292] Brackett, C. A., Dense Wavelength Division Multiplexing Networks: Principles and Applications," *IEEE J. Sel. Areas in Comm.*, Vol. 8, Aug. 1990, pp. 948–964.

[293] Li, T., "The Impact of Optical Amplifiers on Long-Distance Lightwave Telecommunications," *Proc. IEEE*, Vol. 81, Nov. 1993, pp. 1568–1579.

[294] Nosu, K., H. Toba, K. Inoue, and K. Oda, "100 Channel Optical FDM Technology and Its Applications to Optical FDM Channel-Based Networks," *IEEE J. Lightwave Tech.*, Vol. 11, May/June 1993, pp. 764–776.

[295] Alexander, S. B., R. S. Bondurant, D. Byrne, V. W. S. Chan, S. G. Finn, R. Gallager, B. S. Glance, H. A. Haus, P. Humblet, R. Jain, I. P. Kanimov, M. Karol, R. S. Kennedy, A. Kirby, H. Q. Le, A. A. M. Saleh, B. A. Schofield, J. H. Shapiro, N. K. Shankaranarayanan, R. E. Thomas, R. C. Williamson, and R. W. Wilson, "A Precompetitive Consortium on Wide-Band All-Optical Networks," *IEEE J. Lightwave Tech.*, Vol. 11, May/June, 1993, pp. 714–735.

[296] Chidgey, P. J., "Multi-Wavelength Transport Networks," *IEEE Comm. Mag.*, Dec. 1994, pp. 28–35.

[297] Green, P. E., Toward Customer-Usable All-Optical Networks," *IEEE Comm. Mag.*, Dec. 1994, pp. 44–49.

[298] Taylor, N. H., K. P. Jones, D. Simeonidou, M. S. Chaudhry, G. M. Long, and P. R. Morkel, "4x2.5 Gbit/s WDM Transmission, with Optical Add/Drop Multiplexing, over 3711 km on the Installed RIOJA Submarine Cable System," *Tech. Digest of OAA '95*, June 1995, PD7.

[299] Toba, H., K. Oda, and K. Inoue, "Wavelength Division Multiplexing Networks," *Tech. Digest of OAA '95*, June 1995, pp. 204–207.

[300] O'Mahony, M. J., D. Simeonidou, A. Yu, and J. Zhou, "The Design of a European Optical Network," *IEEE J. Lightwave Tech.*, Vol. 13, May 1995,

[301] Chan, V. W. S., "All-Optical Networks," *Sci. American*, Sept. 1995, pp. 72–76.

[302] Hill, A. M., and A. J. N. Houghton, "Optical Networking in the European ACTS Programme," *Tech. Digest of OFC'96*, San Jose, CA, Feb 1996, Paper ThI1.

[303] Hui, B., "Progress of Optical Networking and Technology in the U. S. A.," *Tech. Digest of OFC'96*, San Jose, CA, Feb 1996, Paper ThI2.

[304] Saleh, A. A. M., "Overview of the MONET, Multiwavelength Optical Networking, Program," *Tech. Digest of OFC'96*, San Jose, CA, Feb. 1996, Paper ThI3.

[305] Sipress, J. M., "Undersea Communications Technology," *AT&T Tech. J.*, Vol. 74, Jan./Feb. 1995, pp. 4–7.

[306] Zsakany, J. C., N. W. Marshall, J. M. Roberts, and D. G. Ross, "The Application of Undersea Cable Systems in Global Networking," *AT&T Tech. J.*, Vol. 74, Jan./Feb. 1995, pp. 8–15.

[307] Schesser, J., S. M. Abbott, R. L. Easton, and M. S. Stix, "Design Requirements for the Current Generation of Undersea Cable Systems," *AT&T Tech. J.*, Vol. 74, Jan./Feb. 1995, pp. 16–32.

[308] Mortenson, R. L., B. S. Jackson, S. Shapiro, and W. F. Sirocky, "Undersea Optically Amplified Repeatered Technology, Products, and Challenges," *AT&T Tech. J.*, Vol. 74, Jan./Feb. 1995, pp. 33–45.

[309] Stafford, E. K., J. Mariano, and M. M. Sanders, "Undersea Non-Repeatered Technologies, Challenges, and Products" *AT&T Tech. J.*, Vol. 74, Jan./Feb. 1995, pp. 47–59.

[310] Lynch, R. L., R. L. Maybach, and P. F. Walch, "Design and Development of Optically Amplified Undersea Systems," *AT&T Tech. J.*, Vol. 74, Jan./Feb. 1995, pp. 83–92.

[311] Trischitta, P., M. Colas, M. Green, G. Wuzniak, and J. Arena, "The TAT-12/13 Cable Network," *IEEE Comm. Mag.*, Vol. 34, Feb. 1996, pp. 24–28.

[312] Welsh, T., R. Smith, H. Azami, and R. Chrisner, "The FLAG Cable System," *IEEE Comm. Mag.*, Vol. 34, Feb. 1996, pp. 30–35.

[313] Barnett, W. C., H. Takahira, J. C. Baroni, and Y. Ogi, "The TPC-5 Cable Network," *IEEE Comm. Mag.*, Vol. 34, Feb. 1996, pp. 36–40.

[314] Gunderson, D. R., A. Lecroart, and K. Takekura, "The Asia Pacific Cable Network," *IEEE Comm. Mag.*, Vol. 34, Feb. 1996, pp. 42–48.

[315] Marra, W. C., and J. Schesser, "Africa ONE: The Africa Optical Network," *IEEE Comm. Mag.*, Vol. 34, Feb. 1996, pp. 50–57.

[316] Gnauck, A. H., "Ultra-High-Capacity Amplified WDM Lightwave Systems," *Tech. Digest of OAA '95*, June 1995, pp. 17–20.

[317] Forghieri, F., R. W. Tkach, A. R. Chraplyvy, and D. Marcuse, "Reduction of Four-Wave-Mixing Crosstalk in WDM Systems Using Unequally Spaced Channels," *OFC/IOOC'93*, San Jose, CA, 1993, Paper FC4.

[318] Forghieri, F., R. W. Tkach, A. R. Chraplyvy, and D. Marcuse, "Reduction of Four-Wave-Mixing Crosstalk in WDM Systems Using Unequally Spaced Channels," *IEEE Photon. Tech. Lett.*, Vol. 6, June, 1994, pp. 754–756.

[319] Forghieri, F., R. W. Tkach, and A. R. Chraplyvy, "WDM Systems with Unequally Spaced Channels," *IEEE J. Lightwave Tech.*, Vol. 13, May 1995, pp. 889–897.

[320] Forghieri, F., R. W. Tkach, and A. R. Chraplyvy, "Performance of WDM Systems with Unequal Channel Spacing to Suppress Four-Wave-Mixing," *Tech. Digest of ECOC'94*, Firenze, Italy, 1994, Vol. 2, pp. 741–744.

[321] Desurvire, F., J.-P. Hamaide, and E. Brun, "Soliton in Long-Haul Communications Systems," *Tech. Digest of OFC'96,* San Jose, CA, Feb 1996, Paper WC2.

[322] *AT&T Tech. J.,* Vol. 74, Jan./Feb. 1995.

[323] *IEEE Comm. Magazine,* Vol. 34, Feb. 1996.

Chapter 3

Rare-Earth Ions in Glasses and Transitions for Optical Amplification

Y. Ohishi

> Even for the physicist the description in plain language will be a criterion
> of the degree of understanding that has been reached.
> —Werner Heisenberg, *Physics and Philosophy*

3.1 INTRODUCTION

To know the spectroscopic properties of rare earths doped in glasses is the first step toward predicting the characteristics of rare-earth-doped fiber lasers and amplifiers. The characteristics of optical devices based on rare-earth-doped materials are determined by the optical properties of the rare-earth ion-material combinations used. If we can obtain relevant information on the optical properties of these combinations, it will be possible to accurately predict the performance of such devices as lasers and amplifiers. One of the most powerful tools for obtaining the necessary information is optical spectroscopy related to absorption and emission transitions. This chapter provides a basic introduction to the spectroscopy of rare-earth ions doped in glasses. The Judd–Ofelt theory is explained, starting with a simple overview of the electronic structure of isolated rare-earth ions in materials. This theory is powerful and important in that it can be used to obtain stimulated emission cross sections for laser transitions, to look for new laser transitions, and to assess parasitic transitions. This allows us to assess the suitability of new laser hosts and to understand the performance limitations imposed on active media. In addition, we describe other practical procedures for obtaining emission cross sections that can be conveniently employed when the Judd–Ofelt theory is difficult to use. We will describe nonradiative processes, including ion-ion interactions and the multiphonon relaxation process, which influence the quantum efficiency of rare-earth-doped devices and the doping concentrations of rare-earth ions in glasses. Absorptions and emissions due to $4f$-$4f$ transitions broaden homogeneously or inhomogeneously in glasses. The spectral broadening bandwidth depends on the glass composition. The spectral broadening effect determines not only the optical bandwidth of amplifiers but also affects other gain characteristics, such as gain saturation and signal crosstalk for multichannel signals. The basics of spectral broadening will be described here, mainly in terms of spectroscopy. Rate equations are used in analyzing amplifier performance. Finally, we will describe the development of rate equations for analyzing steady-state characteristics, such as pumping, population inversion, and gain saturation, in simplified four- and three-level system amplifiers.

3.2 THE CONFIGURATION OF THE $4f$ STATES IN CONDENSED MATERIALS

The rare earths consist of 14 elements from Ce (atomic number Z of 56) to Lu (Z = 71) in the periodic table. In condensed matter the trivalent (3+) state is the most stable for rare-earth ions, and most optical devices use trivalent ions. The electronic configuration of a trivalent rare-earth is given by

$$1s^2 2s^2 2p^6 3s^2 3p^6 3d^{10} 4s^2 4p^6 4d^{10} 4f^N 5s^2 5p^6 \qquad (N = 1, \ldots, 14)$$

The Hamiltonian can be written for an individual rare-earth ion and decomposed as

$$H = H_0 + H_{so} + V_{crys} \qquad (3.1)$$

where H_0 is the Halmitonian of the $4f$ electrons in rare-earth ions considering the electrostatic interaction and is given by the equation

$$H_0 = -\sum_{i=1} \frac{\hbar^2 \Delta_i^2}{2m} - \sum_{i=1} \frac{ze^2}{r_i} + \sum_{i=1} \frac{e^2}{r_{ij}} \qquad (3.2)$$

H_{so} is the spin-orbital interaction given by

$$H_{so} = \lambda(S,L)\mathbf{L} \cdot \mathbf{S} \qquad (3.3)$$

and V_{crys} is the potential energy due to the crystalline field around rare earths. The number of electrons in an ion is assumed to be N. The standard approach to H_0 is to employ a central field approximation in which each electron is assumed to move independently in a spherically symmetric potential formed by the nucleus and the average potential of all other electrons. The solutions to this problem can then be factored into the product of a radial and angular function. While the radial function depends on the details of the potential, the spherical symmetry ensures that the angular component is identical to that of a hydrogen atom and can be expressed as spherical harmonics. The solutions of the central field problem are products of one-electron states that are antisymmetric under the interchange of a pair of electrons, as required by the Pauli exclusion principle.

Because these solutions are constructed from hydrogenic states, total orbital angular momentum **L** and total spin **S** are "good" quantum numbers (that is, exact eigenvalues of the Hamiltonian). **L** and **S** are the vector sums of the orbital and spin quantum numbers for all the $4f$ electrons on the ion. Each $4f$ electron contributes an orbital quantum number of 3 and a spin of $1/2$. Total orbital angular momentum is specified by the letters S, P, D, F, G, H, I, K, ... to represent $L = 0, 1, 2, 3, 4, 5, 6, 7, \ldots$, respectively. Each $4f^N$ state splits into several ^{2S+1}L states due to the Coulomb interaction between the $4f$ electrons. Russell–Saunders coupling is used for the states of rare earths. In this scheme **L** and **S** are vectorially added to form the total angular momentum **J** and the states are labeled $^{2S+1}L_J$. The quantum numbers (L, S, J, and M, which is the azimuthal quantum number of J) ($J = |L - S|, |L - S| + 1, |L - S| + 2, \ldots, L + S, M = -J, -J + 1, -J + 2, \ldots, J$) define the terms of the configuration, all of which are degenerate in the central field approximation, as illustrated in Figure 3.1 [1]. Furthermore, each ^{2S+1}L state is split into several J states by the spin-orbital interaction, as shown in Figure 3.1. When the spin-orbital interaction is considered, L and S are no longer good quantum numbers, but J and M are good quantum numbers. In this case, an electronic state is expressed by a linear combination of Russell–Saunders states with the same J and M values as follows:

$$|4f^N[\alpha SL]J\rangle = \sum_{\alpha'S'L'} \alpha_J(\alpha SL; \alpha' S' L') |4f^N \alpha' S' L' J\rangle \qquad (3.4)$$

where α is a quantum number that distinguishes the $^{2S+1}L_J$ states with the same S, L, and J numbers. The quantum numbers within brackets are not good quantum numbers. The actual numbers α, S, and L on the left-hand side indicate the α, S, and L quantum numbers of a Russell–Saunders state that mostly contributes to forming the electronic state. And the electronic state is expressed by the Russell–Saunders state $^{2S+1}L_J$ for convenience. When the spin-orbital interaction H_{so} is considered, J is a good quantum number and the electronic state is identified by J. The $4f$ energy states of trivalent rare-earth ions are indicated by Figure 3.2 [2]. When the potential energy, V_{crys}, has spherical symmetry, the $^{2S+1}L_J$ state degenerates into a $2J + 1$ manifold. However, this degenerate is split by an external field with lower symmetry.

We assume here that the rare-earth concentration is sufficiently low and the interactions between rare earths can be ignored. Although the $4f$ electron of rare

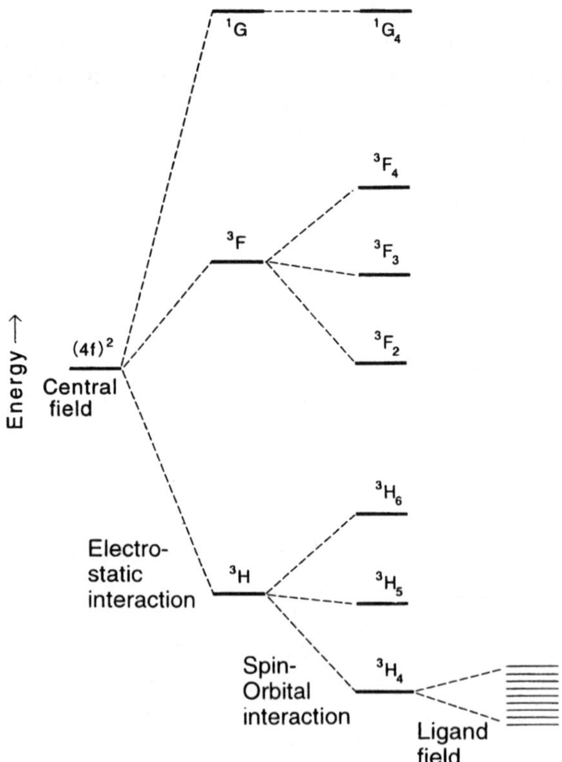

Figure 3.1 Energy level diagram illustrating the energy level splitting hierarchy resulting from various interactions.

earths is influenced by the electric field formed by the surrounding ions, which is called crystalline field (V_{crys}), as well as by Coulomb interactions in the ion, it almost maintains its own electronic state in the free ion state. This is because the spread of the $4f$ orbital is almost one half that of the $5s$ and $5p$ orbitals, as shown in Figure 3.3, and the $4f$ electrons are electrostatically shielded by the $5s$ and $5p$ electrons from the external electric field.

Now, we consider point charges with the electric charge Q_i as the nuclei of ions at a position $\mathbf{R}_i$ ($i = 1, 2, \ldots$). Then the potential energy for the electron due to the crystalline field formed by the point charges is given by

$$V_{\text{cry}} = \sum_i \sum_j \frac{-eQ}{|\mathbf{r}_i - \mathbf{R}_j|} \tag{3.5}$$

Furthermore, when $\mathbf{R}_j$ is expressed by $(R_j, \theta_j, \varphi_j)$ and the angle formed by $\mathbf{r}_i$ and $\mathbf{R}_j$ is Θ_{ij}, the crystalline field is expressed by

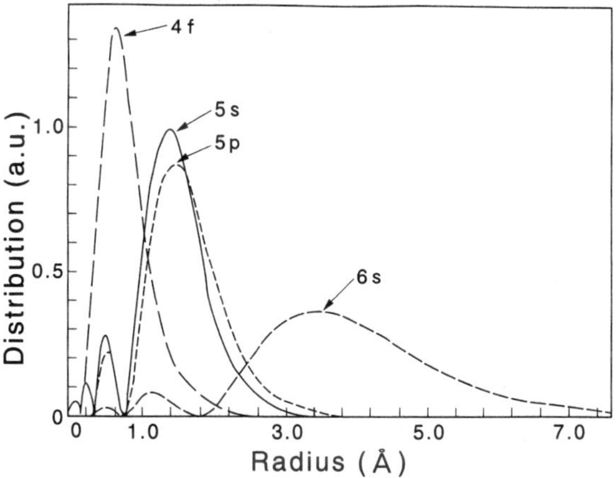

Figure 3.2 Spatial distributions for 4*f*, 5*s*, 5*p*, and 6*s* orbitals.

$$V_{\text{cry}} = \sum_i \sum_{t,p} r_i^t A_{tp} C_p^{(t)}(\theta_i, \varphi_i) = \sum_{t,p} A_{tp} D_p^{(t)} \tag{3.6}$$

where $1/|\mathbf{r}_i - \mathbf{R}_j|$ is expanded using the Legendre polynomial $P_k(\cos \Theta_{ij})$ as

$$\frac{1}{|\mathbf{r}_i - \mathbf{R}_j|} = \sum_{k=0} \frac{1}{R_j} \left(\frac{r_i}{R_j}\right)^k P_k(\cos \Theta_{ij}) \tag{3.7}$$

$$P_k(\cos \Theta_{ij}) = \frac{4\pi}{(2k+1)} \sum_{m=-h} Y_{km}(\theta_i, \varphi_j) Y_{km}^*(\theta_j, \varphi_i) \tag{3.8}$$

where

$$C_p^{(t)}(\theta_i, \varphi_i) = \left(\frac{4\pi}{(2k+1)}\right)^{1/2} Y_{tp}(\theta_i, \varphi_i) \tag{3.9}$$

$$A_{tp} = \left(\frac{4\pi}{(2k+1)}\right)^{1/2} \sum_j \frac{-eQ}{R_j^{t+1}} Y_{tp}^*(\theta_j, \varphi_i) \tag{3.10}$$

$$D_p^{(t)} = \sum_i r_i^t C_p^{(t)}(\theta_i, \varphi_i)$$

where $Y_{km}(\theta_i, \varphi_i)$ is a spherical function.

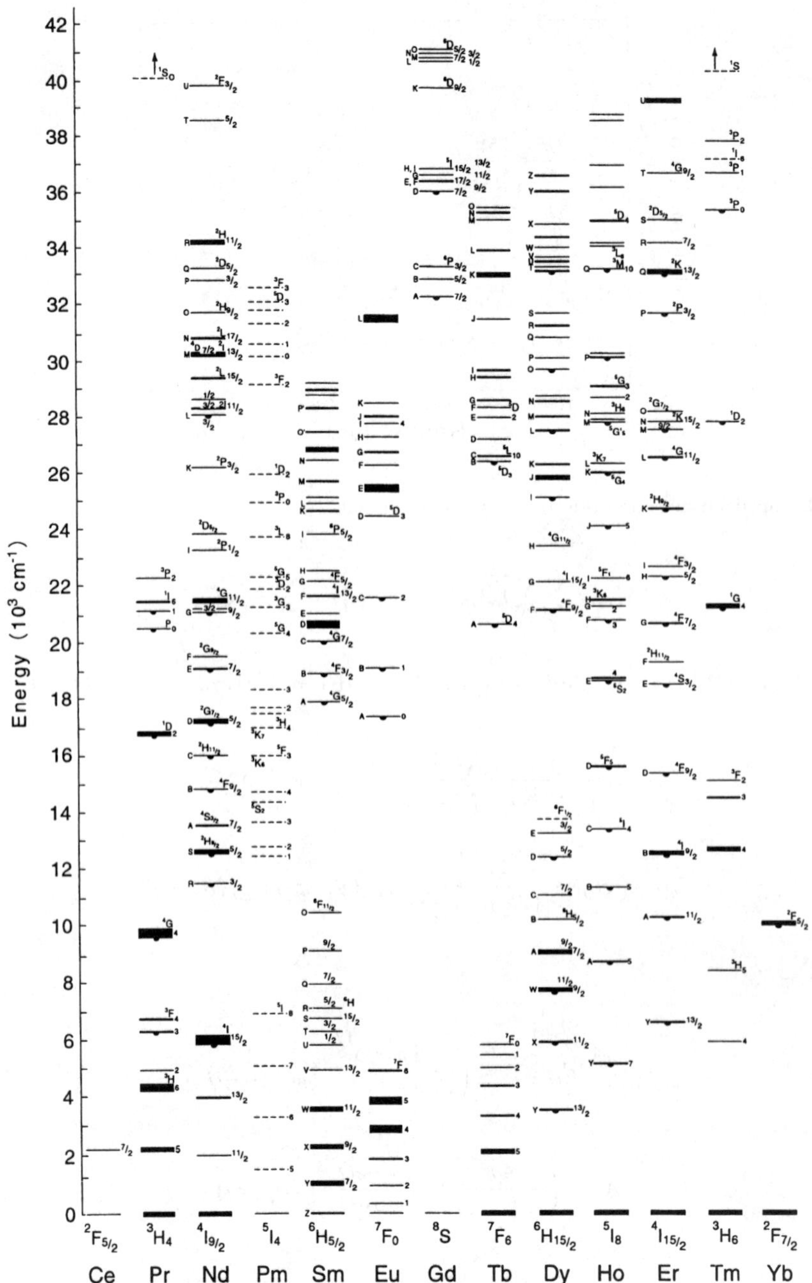

Figure 3.3 Energy level diagram for trivalent rare-earth ions.

When the strength of V_{crys} is much smaller than that of H_0 and H_{so}, V_{crys} can be dealt with via the perturbation theory. However, if the crystalline field is taken into consideration, J is no longer a good quantum number and the electronic state is identified by the irreducible expression of the point group. The electronic state that degenerates into the $(2J + 1)$ manifold in the free ion state is split as shown in Figure 3.1 and the split level is called the Stark level. In crystals or glasses, the $4f$ electron state of rare earths is split into Stark levels due to the crystalline field of the anions.

3.3 THE JUDD–OFELT THEORY FOR DETERMINING TRANSITION INTENSITIES

A knowledge of the transition intensities of $4f$-$4f$ transitions and of absorption and emission cross sections is the first step in investigating the performance of rare-earth-doped fiber lasers and amplifiers.

The spectral intensities of $4f$-$4f$ transitions of rare-earth ions can be obtained using the Judd–Ofelt theory [3,4]. However, since articles on fiber laser and amplifiers do not normally describe the Judd–Ofelt theory, it is explained in some detail here, citing [5].

3.3.1 The Judd–Ofelt Theory

The oscillator strength P of a component of an electric dipole transition from the ground state $|A\rangle$ to an excited state $|B\rangle$ is given by

$$P = \chi \left[\frac{8\pi^2 mc\sigma}{h} \right] \left| \langle A | D_q^{(1)} | B \rangle \right|^2 \tag{3.11}$$

where m is the electron mass, h Planck's constant, c the velocity of light, σ the energy of the transition in cm^{-1}, and χ the Lorentz field correction for the refractivity of the medium. The matrix elements of the electric dipole operator vanish between states of the same parity and so between states arising from the same configuration.

In the free ion approximation the states of the $4f^N$ configuration are taken as linear combinations of Russell–Saunders coupled states $|f^N \alpha SLJ\rangle$:

$$|f^N \alpha [SL]J\rangle = \sum_{\alpha,S,L} C_{\alpha SL} |f^N \alpha SLJ\rangle \tag{3.12}$$

where the $C_{\alpha SL}$ are the coupling coefficients transforming the Russel–Saunders states to intermediate coupled states. The symbol J is the total angular moment, which is a good quantum number; and S and L are the spin and orbital quantum numbers, respectively. The symbol α includes all other quantum numbers needed to specify the states. In these intermediate coupled states the quantum numbers S and L, although convenient for purposes of labeling, are not good and are enclosed in square brackets. To save space the wavefunction defined in (3.12) will frequently be written $|f^N\psi J\rangle$.

The matrix of the electric dipole operator, then, vanishes between these states. In order to force an electric dipole transition it is necessary to mix another configuration with opposite parity into the $4f^N$ configuration. Such mixing may be accomplished by the odd parity terms of the expansion of the crystal field potential

$$V_{\text{cry}} = \sum_{t,p} A_{tp} D_p^{(t)} \quad \text{with } t \text{ odd} \tag{3.13}$$

Considering the crystal field as a first-order perturbation and mixing in states of a higher energy opposite parity configuration $|nl\alpha''[S''L'']J''M''\rangle$ (which will be written $|\psi''\rangle$) we may write $|A\rangle$ and $|B\rangle$

$$|A\rangle = |f^N\psi JM\rangle + \sum_k \frac{\langle\psi''|\langle f^N\psi JM|V_{\text{cry}}|\psi''\rangle}{E(4f^N J) - E(\psi'')} \tag{3.14}$$

$$|B\rangle = |f^N\psi' J'M'\rangle + \sum_k \frac{\langle\psi''|\langle f^N\psi' J'M'|V_{\text{cry}}|\psi''\rangle}{E(4f^N J') - E(\psi'')} \tag{3.15}$$

where k stands for all the quantum numbers of the excited configuration. Then the dipole strength $D = e^2\langle A|D^{(1)}q|B\rangle^2$ of a transition from $|A\rangle$ to $|B\rangle$ is

$$D = \left[e\sum_{k,t,p} A_{tp} \left\{ \frac{\langle f^N\psi JM|D_q^{(1)}|\psi''\rangle\langle\psi''|D_q^{(t)}|f^N\psi' J'M'\rangle}{E(4f^N J') - E(\psi'')} \right. \right.$$
$$\left. \left. + \frac{\langle f^N\psi JM|D_q^{(t)}|\psi''\rangle\langle\psi''|D_q^{(1)}|f^N\psi' J'M'\rangle}{E(4f^N J) - E(\psi'')} \right\} \right]^2 \tag{3.16}$$

Now the problem is to reduce this perturbation expression to a usable form. We initially consider only the first half of (3.16). The treatment of the second half is analogous.

First we can express the operators $D_q^{(1)}$ and $D_q^{(t)}$ as $\sum_i r_i(C_q^{(1)})_i$ and $\sum_i r_i^t(C_p^{(t)})_i$, respectively, and so take out radial integrals. The expression then becomes

$$e\sum_{k,t,p} A_{tp}\langle f^N\psi JM|\sum_i (C_q^{(1)})_i|\psi''\rangle\langle\psi''|\sum_i (C_q^{(t)})_i|f^N\psi' J'M'\rangle$$
$$\cdot \langle 4f|r|nl\rangle\langle nl|r^t|4f\rangle[E(4f^N J') - E(\psi'')]^{-1} \tag{3.17}$$

where $\langle nl | r^k | n'l' \rangle$ is an abbreviation for

$$\int_0^\infty R(nl)\, r^k R(n'l')\, dr \tag{3.18}$$

and R/r is the radial part of the appropriate one electron wave function.

In these equations, $C^{(k)}$ are irreducible tensors, defined by Racah, which transform as spherical harmonics, having components

$$C_q^{(k)} = \left(\frac{4\pi}{2k+1}\right)^{1/2} Y_{kq} \tag{3.19}$$

where Y_{kq} is a spherical harmonic of rank k.

$\mathbf{C}^{(0)}$, then, has only one component: $C_0^{(0)} = (4\pi)^{1/2} Y_{00} = 1$ and is a scalar.

The position vector $\mathbf{r}$ is a tensor of rank 1 and so may be related to $\mathbf{C}^{(1)}$ as

$$r = rC^{(1)} \tag{3.20}$$

having components

$$r_{+1} = rC_1^{(1)} = r(4\pi/3)^{1/2}Y_{11} = -(2)^{-1/2}(x + iy) \tag{3.21a}$$

$$r_0 = rC_0^{(1)} = r(4\pi/3)^{1/2}Y_{10} = z \tag{3.21b}$$

$$r_{-1} = rC_{-1}^{(1)} = r(4\pi/3)^{1/2}Y_{1-1} = (2)^{-1/2}(x - iy) \tag{3.21c}$$

The electric dipole moment operator $\mathbf{P}$ is the sum over all electrons of the position vectors of these electrons

$$P = -e\sum_i r_i = -e\sum_i r_i(C^{(1)})_i \tag{3.22}$$

and is written $-eD^{(1)}$ with components $-eD_q^{(1)}$ where $q = 0, \pm1$.

Another operator we will encounter later is the crystal field operator V_{cry}, which, in general, is of rank t:

$$V_{\text{cry}} = \sum_{t,p} A_{tp} \sum_i r_i(C_p^{(t)})_i \tag{3.23}$$

It is written $\sum_{t,p} A_{tp} D_p^{(t)}$, where A_{tp} are the crystal field parameters.

One way of simplifying this expression is to use the procedure known as closure [6]. If the energy of the perturbing configuration is invariant with respect to α'', S'', L'', and M'', the following equation would be exact:

$$\sum_{\alpha'' S'' L'' J'' M''} \langle f^N \psi JM | \sum_i (C_q^{(1)})_i | \psi'' \rangle \langle \psi'' | \sum_i (C_q^{(t)})_i | f^N \psi' J' M' \rangle \tag{3.24}$$

$$= (-1)^{p+q+1}[\lambda] \begin{pmatrix} 1 & \lambda & t \\ q & -p-q & p \end{pmatrix} \langle f^N \psi JM | \sum_i ((C_q^{(1)} C_p^t)_{-p-q}^{(\lambda)})_i | f^N \psi' J' M' \rangle$$

The abbreviation $[k] = (2k + 1)$ has been introduced. The combined operator

$$(C_q^{(1)} C_p^{(t)})_{-p-q}^{(\lambda)} \tag{3.25}$$

may then be further simplified and (3.24) becomes

$$(-1)^{p+q}(-1)^{f+l}[\lambda][f][l] \begin{pmatrix} 1 & \lambda & t \\ q & -p-q & p \end{pmatrix} \begin{Bmatrix} 1 & t & \lambda \\ f & f & l \end{Bmatrix}$$

$$\cdot \langle f \| C^{(1)} \| l \rangle \langle l \| C^{(t)} \| f \rangle \langle f^N \psi JM | U_{-p-q}^{(\lambda)} | f^N \psi' J' M' \rangle \tag{3.26}$$

$U^{(\lambda)} = \Sigma u_i^{(\lambda)}$, where $u_i^{(\lambda)}$ is defined by $\langle nl \| u(l) \| n'l' \rangle = \delta(nn')\delta(ll')$ and cooperates within the configuration. Judd made the assumption that the energies of the excited configurations are independent of all quantum numbers except n and l; that is, that the excited configurations are completely degenerate. This is an admitted weak link in the theory. The final simplification is, however, probably the least justified. This involves equating $E(4f^N J) - E(\psi'')$ and $E(4f^N J') - E(\psi'')$ and replacing them by an average energy denominator $\Delta E(\psi'')$. This is clearly a rather bad approximation for certain rare earths where the energy of, say, the $4f^{N-1}5d$ configuration is not very much greater than that of the $J - J'$ transitions being considered. However, it does result in significant simplification. If the previously outlined procedures [of (3.16) to (3.18)] are applied to both halves of (3.15), the result for each half will differ only in the $3 - j$ symbol. Because of the symmetrical relation [3]

$$\begin{pmatrix} 1 & \lambda & t \\ q & -p-q & p \end{pmatrix} = (-1)^{1+\lambda+t} \begin{pmatrix} t & \lambda & 1 \\ p & -p-q & q \end{pmatrix} \tag{3.27}$$

the two halves are equal if λ is even and cancel each other out if λ is odd. Thus, by introducing the approximation of an average energy denominator we can remove all the terms involving odd λ. The $6 - j$ symbol restricts l to values of less than or equal to 6. Finally, then, we obtain

$$D = \left[e \sum_{p,t,\text{even}\,\lambda} (-1)^{p+q} A_{tp}[\lambda]\, \Xi(t, \lambda) \begin{pmatrix} 1 & \lambda & t \\ q & -p-q & p \end{pmatrix} \langle f^N \psi JM | U^{(\lambda)}_{-p-q} | f^N \psi' J' M' \rangle \right]^2$$

(3.28)

where

$$\Xi(t, \lambda) = 2\sum (-1)^{f+l} [f][l] \begin{Bmatrix} 1 & t & \lambda \\ f & l & f \end{Bmatrix} \langle f \| C^{(1)} \| l \rangle$$

$$\cdot \langle l \| C^{(t)} \| f \rangle \langle 4f | r | nl \rangle \langle nl | r^t | 4f \rangle \Delta E(\psi'')^{-1}$$

(3.29)

and there is an implicit summation over n and l of all configurations desired for mixing. The matrix element in (3.28) can now be reduced and the resulting expression for D inserted into (3.11) to give

$$P_{\text{E.D.}} = \chi \left[\frac{8\pi^2 mc\sigma}{h} \right] \left[\sum_{p,t,\text{even}\,\lambda} (-1)^{p+q} [\lambda] A_{tp} \begin{pmatrix} 1 & \lambda & t \\ q & -p-q & p \end{pmatrix} \begin{pmatrix} J & \lambda & J' \\ -M & -p-q & M' \end{pmatrix} \right.$$

$$\left. \cdot \Xi(t, \lambda) \langle f^N \psi J | U^{(\lambda)} | f^N \psi' J' \rangle \right]^2$$

(3.30)

This is the oscillator strength of a component of an electric dipole transition.

In the form (3.30) the theory is equipped to deal with transitions between individual Stark levels, whose wavefunctions are generally linear combinations of $[f^N \alpha [SL] JM\rangle$ states.

In solution, however, such transitions cannot usually be distinguished and it is convenient to use the sum of all the Stark levels of the ground state (assuming all are equally populated). At the same time we can determine the sum of the components of $D_q^{(1)}$ and the components of $D_p^{(t)}$, which is appropriate for isotropic light. The $3 - j$ symbols in (3.30) vanish and are thus replaced by a factor of $3 - 1(2J + 1) - 1(2t + 1) - 1$. Finally, the oscillator strength of an electric dipole transition is given by the equation

$$P_{\text{E.D.}} = \sum_{\lambda=2,4,6} \sigma \mathfrak{S}_\lambda \langle f^N \alpha [SL] J \| U^{(\lambda)} \| f^N \alpha' [S'L'] J' \rangle^2 (2J + 1)^{-1}$$

(3.31)

where

$$\mathfrak{S}_\lambda = \chi \left[\frac{8\pi^2 mc\sigma}{h} \right] [\lambda] \sum_{p,t} |A_{tp}|^2 \Xi^2(t, \lambda)(2J + 1)^{-1}$$

(3.32)

Many researchers have used an alternative notation for work involving the isotropic spectra of ions in crystals. Because the refractive index correction χ varies with wavelength, it is not incorporated in the Judd–Ofelt parameters but left explicitly outside the summation sign. In this case

$$P_{\text{E.D.}} = \chi \left[\frac{8\pi^2 mc\sigma}{h} \right] \sigma \sum_{\lambda=2,4,6} \Omega_\lambda \langle f^N \alpha [SL]J \| U^{(\lambda)} \| f^N \alpha' [S'L']J' \rangle^2 (2J+1)^{-1}$$

(3.33)

where

$$\Omega_\lambda = [\lambda] \sum_{p,t} |A_{tp}|^2 \, \Xi^2(t, \lambda)(2J+1)^{-1}$$

(3.34)

This is the well-known expression of the oscillator strength of an electric dipole transition.

The quantity analogous to the oscillator strength normally used in dealing with emission spectra is the spontaneous emission coefficient, $A_{\text{E.D.}}$

$$A_{\text{E.D.}} = \chi \left[\frac{64\pi^4 e^2}{3h} \right] \sigma^3 \sum_{\lambda=2,4,6} \Omega_\lambda \langle f^N \alpha [SL]J \| U^{(\lambda)} \| f^N \alpha' [S'L']J' \rangle^2 (2J+1)^{-1}$$

(3.35)

and Ω_λ is defined in (3.35). No distinction will be made between the Ω_λ values measured by means of absorption spectra and those obtained from emission spectra. The normal spectroscopic convention will be used when discussing individual transitions: the arrow connecting states goes from right to left for absorption and from left to right for emission. The reduced matrix element $\langle f^N \alpha [SL]J \| U^{(\lambda)} | f^N \alpha' [S'L']J' \rangle^2$ (frequently written $\Gamma^{(\lambda)}$) for transitions in rare earths are given in [7]. Table 3.1 tabulates the major reduced matrix elements for some rare earths [8].

3.3.2 Selection Rules From the Judd–Ofelt Theory

Several selection rules may be obtained from the Judd–Ofelt theory. The $6-j$ symbol in (3.29) gives $\lambda \leq 6$ (and so because λ is even, $\lambda = 2, 4, 6$); $t \leq 7$ (and so because t is odd, $t = 1, 3, 5, 7$); and $\Delta l = \pm 1$ (which means that only $l = d$ and g configurations may be mixed into $4f^N$). The $3-j$ symbol in (3.28) gives the selection rules in t with respect to λ: $\lambda = 2$, $t = 1, 3$; $\lambda = 4$, $t = 3, 5$; $\lambda = 6$, $t = 5, 7$. The Kronecker delta gives $\Delta S = 0$ and the $6-j$ symbol $|\Delta J| \leq \lambda$, $|\Delta L| \leq \lambda$, and if J or $J' = 0$, $|\Delta J|$ must be even, and $J = 0 - J' = 0$ is forbidden. Finally the $3-j$ symbol involving M and M' in (3.30) gives selection rules for M and M' (which in turn

Table 3.1

Squared Reduced Matrix Elements for Transitions from the Ground States of Rare-Earth Ions

Pr^{3+} (^{3}H$_4$)

	$[U(2)]^2$	$[U(4)]^2$	$[U(6)]^2$
^{3}H$_5$	0.110	0.202	0.611
^{3}H$_6$	0.000	0.033	0.140
^{3}F$_2$	0.509	0.403	0.118
^{3}F$_3$	0.065	0.347	0.698
^{3}F$_4$	0.019	0.050	0.485
^{1}G$_4$	0.001	0.007	0.027
^{1}D$_2$	0.003	0.017	0.052
^{3}P$_0$	0	0.173	0
^{3}P$_1$	0	0.171	0
^{1}I$_6$	0.009	0.052	0.024
^{3}P$_2$	0	0.036	0.136
^{1}S$_0$	0	0.007	0

Nd^{3+} (^{4}I$_{9/2}$)

	$[U(2)]^2$	$[U(4)]^2$	$[U(6)]^2$
^{4}I$_{11/2}$	0.019	0.107	1.165
^{4}I$_{13/2}$	0	0.014	0.456
^{4}I$_{15/2}$	0	0	0.045
^{4}F$_{3/2}$	0	0.229	0.055
^{4}F$_{5/2}$	0.001	0.237	0.397
^{2}H$_{9/2}$	0.009	0.008	0.115
^{4}F$_{7/2}$	0	0.003	0.235
^{4}S$_{3/2}$	0.001	0.042	0.425
^{4}F$_{9/2}$	0.001	0.009	0.042
^{2}H$_{11/2}$	0	0.003	0.010
^{4}G$_{5/2}$	0.898	0.409	0.036
^{4}G$_{7/2}$	0.076	0.185	0.031
^{2}K$_{13/2}$	0.007	0	0.031
^{4}G$_{7/2}$	0.055	0.157	0.055
^{4}G$_{9/2}$	0.005	0.061	0.041
^{2}K$_{15/2}$	0	0.005	0.014
^{2}G$_{9/2}$	0.001	0.015	0.014
^{2}D$_{3/2}$	0	0.019	0
^{4}G$_{11/2}$	0	0.005	0.008
^{2}P$_{1/2}$	0	0.037	0
^{2}D$_{5/2}$	0	0	0.002
^{2}P$_{3/2}$	0	0.001	0.001
^{4}D$_{3/2}$	0	0.196	0.017
^{4}D$_{5/2}$	0	0.057	0.028
^{2}I$_{11/2}$	0.005	0.015	0.003
^{4}D$_{1/2}$	0	0.258	0
^{2}L$_{15/2}$	0	0.025	0.010
^{4}I$_{13/2}$	0	0.001	0.002
^{4}D$_{7/2}$	0	0.004	0.008

Sm^{3+} (^{6}H$_{5/2}$)

	$[U(2)]^2$	$[U(4)]^2$	$[U(6)]^2$
^{6}H$_{7/2}$	0.206	0.196	0.095
^{6}H$_{9/2}$	0.026	0.140	0.327
^{6}H$_{11/2}$	0	0.024	0.265
^{6}H$_{13/2}$	0	0.001	0.066
^{6}F$_{1/2}$	0.194	0	0
^{6}H$_{15/2}$	0	0	0.004
^{6}F$_{3/2}$	0.144	0.136	0
^{6}F$_{5/2}$	0.033	0.284	0
^{6}F$_{7/2}$	0.002	0.143	0.430
^{6}F$_{9/2}$	0	0.021	0.341
^{6}F$_{11/2}$	0	0.001	0.052
^{4}G$_{5/2}$	0	0.001	0
^{4}M$_{15/2}$	0	0	0.031
^{4}I$_{11/2}$	0	0	0.011
^{4}I$_{13/2}$	0	0.003	0.023
^{6}P$_{5/2}$	0	0.026	0
^{4}L$_{13/2}$	0	0.008	0.010
^{6}P$_{3/2}$	0	0.168	0
^{6}P$_{7/2}$	0	0.002	0.075
^{4}D$_{3/2}$	0	0.025	0
^{4}D$_{7/2}$	0	0.001	0.037

give selection rules for the crystal field quantum numbers μ): $|\Delta M| = p + q$; so for σ polarized spectra ($q = 0$) $|\Delta M| = 3p$ and for π polarized spectra ($q = \pm 1$) $|\Delta M| = p \pm 1$. The value of p is of course determined by the particular point group considered: for example in D_{3h} symmetry the odd crystal field parameters are A_{33}, A_{53}, and A_{73}; thus $|\Delta M) = (\sigma)$ and $|\Delta M| = 2, 4(\pi)$ [9]. Summarizing,

Table 3.1
(continued)

Eu^{3+} ($^7F_0$)

	$[U(2)]^2$	$[U(4)]^2$	$[U(6)]^2$
$^7F_2$	0.137	0	0
$^7F_4$	0	0.140	0
$^7F_6$	0	0	0.145
$^5D_2$	0.001	0	0
$^5L_6$	0	0	0.016
$^5G_6$	0	0	0.004
$^5H_6$	0	0	0.006

Eu^{3+} ($^7F_1$)

	$[U(2)]^2$	$[U(4)]^2$	$[U(6)]^2$
$^7F_3$	0.209	0.128	0
$^7F_4$	0	0.174	0
$^7F_5$	0	0.119	0.054
$^7F_6$	0	0	0.377
$^5D_1$	0.003	0	0
$^5L_6$	0	0	0.009
$^5L_7$	0	0	0.018
$^5G_5$	0	0.001	0.010
$^5H_7$	0	0	0.007
$^5H_5$	0	0.002	0.007

Gd^{3+} ($^8S_{7/2}$)

	$[U(2)]^2$	$[U(4)]^2$	$[U(6)]^2$
$^6P_{7/2}$	0.001	0	0
$^6I_{7/2}$	0	0	0.004
$^6I_{9/2}$	0	0	0.010
$^6I_{17/2}$	0	0	0.021
$^6I_{11/2}$	0	0	0.018
$^6I_{13/2}$	0	0	0.024

$^6I_{15/2}$	0	0	0.027
$^6D_{9/2}$	0.006	0	0
$^6D_{7/2}$	0.004	0	0

Tb^{3+} ($^7F_6$)

	$[U(2)]^2$	$[U(4)]^2$	$[U(6)]^2$
$^7F_5$	0.538	0.642	0.118
$^7F_4$	0.089	0.516	0.265
$^7F_3$	0	0.232	0.413
$^7F_2$	0	0.048	0.470
$^7F_1$	0	0	0.376
$^7F_0$	0	0	0.144
$^5D_4$	0.001	0.001	0.001
$^5D_3$	0	0	0.001
$^5G_6$	0.002	0.005	0.012
$^5L_{10}$	0	0	0.059
$^5G_5$	0.001	0.002	0.014
$^5L_9$	0	0.002	0.047
$^5L_8$	0	0	0.024
$^5L_7$	0.001	0	0.012

Dy^{3+} ($^6H_{15/2}$)

	$[U(2)]^2$	$[U(4)]^2$	$[U(6)]^2$
$^6H_{13/2}$	0.246	0.414	0.682
$^6H_{11/2}$	0.092	0.037	0.641
$^6H_{9/2}$	0	0.018	0.199
$^6F_{11/2}$	0.939	0.829	0.205
$^6F_{9/2}$	0	0.574	0.721
$^6H_{7/2}$	0	0.001	0.039
$^6H_{5/2}$	0	0	0.003
$^6F_{7/2}$	0	0.136	0.715
$^6F_{5/2}$	0	0	0.345
$^6F_{3/2}$	0	0	0.061
$^4F_{9/2}$	0	0.005	0.030
$^4I_{15/2}$	0.007	0	0.065
$^4G_{11/2}$	0	0.015	0

$$\Delta l = \pm 1 \qquad \Delta s = 0 \qquad |\Delta L| \leq 6$$

$$|\Delta J| \leq 6 \quad \text{unless } J \text{ or } J' = 0 \text{ when } |\Delta J| = 2,\ 4,\ 6$$

$$|\Delta M| = p + q$$

Table 3.1

(continued)

	$[U(2)]^2$	$[U(4)]^2$	$[U(6)]^2$
$^4F_{7/2}$	0	0.077	0.026
$^4K_{17/2}$	0.011	0.005	0.094
$^4M_{19/2}$	0	0.017	0.102
$^6P_{5/2}$	0	0	0.070
$^6P_{7/2}$	0	0.522	0.013
$^6P_{3/2}$	0	0	0.110

Ho^{3+} ($^5I_8$)

	$[U(2)]^2$	$[U(4)]^2$	$[U(6)]^2$
$^5I_7$	0.025	0.134	1.522
$^5I_6$	0.008	0.039	0.692
$^5I_5$	0	0.010	0.094
$^5I_4$	0	0	0.008
$^5F_5$	0	0.425	0.569
$^5S_2$	0	0	0.227
$^5F_4$	0	0.239	0.707
$^5F_3$	0	0	0.346
$^5F_2$	0	0	0.192
$^3K_8$	0.021	0.033	0.158
$^5G_6$	1.520	0.841	0.141
$^5G_5$	0	0.534	0
$^5G_4$	0	0.031	0.036
$^3K_7$	0.006	0.005	0.034
$^3H_5$	0	0.079	0.161
$^3H_6$	0.215	0.118	0.003
$^3L_9$	0.019	0.005	0.154
$^3F_4$	0	0.126	0.005
$^3M_{10}$	0	0.070	0.081
$^5D_4$	0	0.304	0.049
$^3H_4$	0	0.263	0.004

Er^{3+} ($^4I_{15/2}$)

	$[U(2)]^2$	$[U(4)]^2$	$[U(6)]^2$
$^4I_{13/2}$	0.020	0.117	1.432
$^4I_{11/2}$	0.028	0	0.395
$^4I_{9/2}$	0	0.173	0.010
$^4F_{9/2}$	0	0.535	0.462
$^4S_{3/2}$	0	0	0.221
$^2H_{11/2}$	0.713	0.413	0.093
$^4F_{7/2}$	0.147	0.627	
$^4F_{5/2}$	0	0	0.223
$^4F_{3/2}$	0	0	0.127
$^2G_{9/2}$	0	0.019	0.226
$^4G_{11/2}$	0.918	0.526	0.117
$^4G_{9/2}$	0	0.242	0.124
$^2K_{15/2}$	0.022	0.004	0.076
$^2G_{7/2}$	0	0.017	0.116
$^2P_{3/2}$	0	0	0.017
$^2K_{13/2}$	0.003	0.003	0.015
$^4G_{5/2}$	0	0	0.003
$^4G_{7/2}$	0	0.033	0.003
$^4D_{7/2}$	0	0.892	0.029
$^2L_{17/2}$	0.005	0.066	0.033

Tm^{3+} ($^3H_6$)

	$[U(2)]^2$	$[U(4)]^2$	$[U(6)]^2$
$^3F_4$	0.537	0.726	0.238
$^3H_5$	0.107	0.231	0.638
$^3H_4$	0.237	0.109	0.595
$^3F_3$	0	0.316	0.841
$^3F_2$	0	0	0.258
$^1G_4$	0.048	0.075	0.013
$^1D_2$	0	0.316	0.093
$^1I_6$	0.011	0.039	0.013
$^3P_0$	0	0	0.076
$^3P_1$	0	0	0.124
$^3P_2$	0	0.265	0.022

3.3.3 Other Multiple Transitions

Similar to electric dipole radiation, $4f$-$4f$ transitions can absorb magnetic dipole and even higher electric multipole radiation (such as quadrupole and hexadecapole). Magnetic dipole transitions are parity allowed between states of f^N and subject to selection rules $\Delta\alpha = \Delta S = \Delta L = 0$, $\Delta J = 0, \pm 1$ in Russel–Saunders limit. The line strength for *magnetic dipole* (md) transitions between J manifolds is

$$S_{\text{md}}(\alpha J; \alpha' J') = \frac{eh^2}{16n^2 n^2 c^2} |\langle f^N \alpha [SL]J \| L + 2S \| f^N \alpha' [S'L']J' \rangle|^2 \qquad (3.36)$$

where e is the charge of an electron, the squared terms including $U^{(t)}$ are the matrix elements of the doubly reduced unit tensor $U^{(t)}$ for the rare-earth ion, m is the mass of an electron, c is the velocity of light, $\mathbf{L}$ is orbital angular momentum, $\mathbf{S}$ is spin angular momentum, and f^N is the electronic configuration of the rare earth.

The effect of the host's Stark field reduces the $(2J + 1)$-fold degeneracy of the free ion states. The resulting eigenstates are linear combinations of basis states $|f^N \alpha SLJ\rangle$ given by (3.12). The matrix elements of $L + 2S$ are given by

$$J' = J$$
$$\langle f^N \alpha SLJ \| L + 2S \| f^N \alpha' S'L'J \rangle = \delta(\alpha, \alpha')\delta(S, S')\delta(L, L')\beta$$
$$\cdot [(2J + 1)/4J(J + 1)]^{1/2}[S(S + 1) - L(L + 1) + 3J(J + 1)] \qquad (3.37)$$

$$j' = J - 1$$
$$\langle f^N \alpha SLJ \| L + 2S \| f^N \alpha' S'L'J - 1 \rangle \qquad (3.38)$$

$$= \delta(\alpha, \alpha')\delta(S, S')\delta(L, L')\beta \left\{ \frac{[(S + L + 1)^2 - J^2][J^2 - (L - S)^2]}{4J} \right\}^{1/2}$$

$$j' = J + 1 \qquad (3.39)$$
$$\langle f^N \alpha SLJ \| L + 2S \| f^N \alpha' S'L'J - 1 \rangle$$

$$= \delta(\alpha, \alpha')\delta(S, S')\delta(L, L')\beta \left\{ \frac{[(S + L + 1)^2 - (J + 1)^2][(J + 1)^2 - (L - S)^2]}{4(J + 1)} \right\}^{1/2}$$

where $\beta = eh/2mc$

Eigenfunctions for $4f$ states of Pr^{3+}, Er^{3+}, and Tm^{3+} [10–12] are tabulated in Table 3.2.

3.3.4 Experimental Procedure for Obtaining Ω Parameters

The total oscillator strength P of an absorption transition at frequency ν is given by

$$P(aJ; bJ') = \frac{8\pi^2 m\nu}{3h(2J + 1)} \left[\frac{(n^2 + 2)^2}{9n} S_{\text{ed}} + nS_{\text{md}} \right] \qquad (3.40)$$

Table 3.2
Eigenfunctions for Pr^{3+}, Er^{3+}, and Tm^{3+}

Pr^{3+}

$|\ [^3H_4]\ >=0.9878\ |\ ^3H_4\ >+0.1534\ |\ ^1G_4\ >-0.0282\ |\ ^3F_4\ >$

$|\ [^3H_5]\ >=1.0000\ |\ ^3H_5\ >$

$|\ [^3H_6]\ >=-0.9985\ |\ ^3H_6\ >+0.0541\ |\ ^1I_6\ >$

$|\ [^3F_2]\ >=0.9800\ |\ ^3F_2\ >+0.1475\ |\ ^1D_2\ >-0.0133\ |\ ^3P_2\ >$

$|\ [^3F_3]\ >=1.000\ |\ ^3F_3\ >$

$|\ [^3F_4]\ >=0.8544\ |\ ^3F_4\ >-0.5092\ |\ ^1G_4\ >+0.1035\ |\ ^3H_4\ >$

$|\ [^1G_4]\ >=0.8469\ |\ ^1G_4\ >+0.5188\ |\ ^3F_4\ >-0.1167\ |\ ^3H_4\ >$

$|\ [^1D_2]\ >=-0.9483\ |\ ^1D_2\ >+0.2823\ |\ ^3P_2\ >+0.1452\ |\ ^3F_2\ >$

$|\ [^3P_0]\ >=0.9962\ |\ ^3P_0\ >+0.0876\ |\ ^1S_0\ >$

$|\ [^1I_6]\ >=0.9985\ |\ ^1I_6\ >+0.0541\ |\ ^3H_6\ >$

$|\ [^3P_1]\ >=1.0000\ |\ ^3P_1\ >$

$|\ [^3P_2]\ >=0.9592\ |\ ^3P_2\ >+0.2812\ |\ ^1D_2\ >-0.0290\ |\ ^3F_2\ >$

$|\ [^1S_0]\ >=-0.9962\ |\ ^1S_0\ >+0.0876\ |\ ^3P_0\ >$

Er^{3+}

$|\ [^4I_{15/2}]\ >=0.9852\ |\ ^4I_{15/2}\ >-0.1708\ |\ ^2I_{15/2}\ >-0.0176\ |\ ^2L_{15/2}\ >$

$|\ [^4I_{13/2}]\ >=-0.9955\ |\ ^4I_{13/2}\ >-0.0319\ |\ ^2I_{13/2}\ >+0.0896\ |\ ^2K_{13/2}\ >$

$|\ [^4I_{11/2}]\ >=0.9125\ |\ ^4I_{11/2}\ >-0.1073\ |\ ^2H(11)_{11/2}\ >+0.3740\ |\ ^2H(21)_{11/2}\ >+0.0631\ |\ ^2I_{11/2}\ >$
$\qquad +0.1094\ |\ ^4G_{11/2}\ >$

$|\ [^4I_{9/2}]\ >=-0.7322\ |\ ^4I_{9/2}\ >+0.2765\ |\ ^2G(20)_{9/2}\ >-0.2204\ |\ ^2G(21)_{9/2}\ >+0.1953\ |\ ^2H(11)_{9/2}\ >$
$\qquad -0.4125\ |\ ^2H(21)_{9/2}\ >+0.3611\ |\ ^4F_{9/2}\ >+0.0116\ |\ ^4G_{9/2}\ >$

$|\ [^4F_{9/2}]\ >=0.7725\ |\ ^4F_{9/2}\ >+0.2882\ |\ ^2G(20)_{9/2}\ >-0.2158\ |\ ^2G(21)_{9/2}\ >-0.0012\ |\ ^2H(11)_{9/2}\ >$
$\qquad +0.0838\ |\ ^2H(21)_{9/2}\ >+0.0883\ |\ ^4G_{9/2}\ >+0.5087\ |\ ^4I_{9/2}\ >$

$|\ [^4S_{3/2}]\ >=0.8371\ |\ ^4S_{3/2}\ >-0.4196\ |\ ^2P_{3/2}\ >-0.2666\ |\ ^2D(20)_{3/2}\ >-0.0207\ |\ ^2D(21)_{3/2}\ >$
$\qquad +0.0415\ |\ ^4D_{3/2}\ >+0.2237\ |\ ^4F_{3/2}\ >$

$|\ [^2H_{11/2}]\ >=-0.6908\ |\ ^2H(21)_{11/2}\ >+0.1500\ |\ ^2H(11)_{11/2}\ >-0.0570\ |\ ^2I_{11/2}\ >-0.5962\ |\ ^4G_{11/2}\ >$
$\qquad +0.3762\ |\ ^4I_{11/2}\ >$

$|\ [^4F_{7/2}]\ >=0.9624\ |\ ^4F_{7/2}\ >+0.0499\ |\ ^2F(10)_{7/2}\ >+0.0542\ |\ ^2F(21)_{7/2}\ >+0.2096\ |\ ^2G(20)_{7/2}\ >$
$\qquad +0.0111\ |\ ^4D_{7/2}\ >+0.0294\ |\ ^4G_{7/2}\ >-0.1545\ |\ ^2G(21)_{7/2}\ >$

$|\ [^4F_{5/2}]\ >=0.9232\ |\ ^4F_{5/2}\ >-0.3483\ |\ ^2D(20)_{5/2}\ >+0.1284\ |\ ^2D(21)_{5/2}\ >+0.0402\ |\ ^2F(10)_{5/2}\ >$
$\qquad +0.0708\ |\ ^2F(21)_{5/2}\ >-0.0422\ |\ ^4D_{5/2}\ >-0.0370\ |\ ^4G_{5/2}\ >$

$|\ [^4F_{3/2}]\ >=0.08002\ |\ ^4F_{3/2}\ >+0.0702\ |\ ^2P_{3/2}\ >+0.4486\ |\ ^2D(20)_{3/2}\ >-0.0039\ |\ ^2D(21)_{3/2}\ >$
$\qquad +0.3918\ |\ ^4S_{3/2}\ >+0.0016\ |\ ^4D_{3/2}\ >$

$|\ [^2H_{9/2}]\ >=0.2706\ |\ ^2H(11)_{9/2}\ >-0.4032\ |\ ^2H(21)_{9/2}\ >+0.4397\ |\ ^2G(20)_{9/2}\ >-0.3920\ |\ ^2G(21)_{9/2}\ >$
$\qquad -0.4862\ |\ ^4F_{9/2}\ >+0.2450\ |\ ^4G_{9/2}\ >+0.3475\ |\ ^4I_{9/2}\ >$

$|\ [^4G_{11/2}]\ >=0.7771\ |\ ^4G_{11/2}\ >+0.3306\ |\ ^2H(11)_{11/2}\ >-0.56117\ |\ ^2H(21)_{11/2}\ >-0.0209\ |\ ^2I_{11/2}\ >$
$\qquad +0.1569\ |\ ^4I_{11/2}\ >$

$|\ [^2G_{9/2}]\ >=0.0126\ |\ ^2G(20)_{9/2}\ >+0.0545\ |\ ^2G(21)_{9/2}\ >+0.0997\ |\ ^2H(11)_{9/2}\ >-0.3806\ |\ ^2H(21)_{9/2}\ >+$
$\qquad 0.0209\ |\ ^4F_{9/2}\ >-0.8916\ |\ ^4G_{9/2}\ >+0.2161\ |\ ^4I_{9/2}\ >$

$|\ [^2K_{15/2}]\ >=0.9548\ |\ ^2K_{15/2}\ >+0.2440\ |\ ^2L_{15/2}\ >+0.1698\ |\ ^4I_{15/2}\ >$

$|\ [^2G_{7/2}]\ >=-0.5140\ |\ ^2G(20)_{7/2}\ >+0.4848\ |\ ^2G(21)_{7/2}\ >+0.1380\ |\ ^2F(10)_{7/2}\ >+0.1439\ |\ ^2F(21)_{7/2}\ >$
$\qquad +0.0560\ |\ ^4D_{7/2}\ >+0.1943\ |\ ^4F_{7/2}\ >-0.6482\ |\ ^4G_{7/2}\ >$

Table 3.2
(continued)

$| [^2P_{3/2}] > = 0.6049 \, | \, ^2P_{3/2} > +0.4720 \, | \, ^2D(20)_{3/2} > +0.1832 \, | \, ^2D(21)_{3/2} > +0.3380 \, | \, ^4S_{3/2} > -0.1770 \, | \, ^4D_{3/2} >$
      $+0.4819 \, | \, ^4F_{3/2} >$

$| [^2K_{13/2}] > = 0.9461 \, | \, ^2K_{13/2} > +0.3149 \, | \, ^2I_{13/2} > +0.0751 \, | \, ^4I_{13/2} >$

$| [^2D_{7/2}] > = 0.0442 \, | \, ^2F(10)_{7/2} > +0.0329 \, | \, ^2F(21)_{7/2} > +0.5270 \, | \, ^2G(20)_{7/2} > -0.3883 \, | \, ^2G(21)_{7/2} >$
      $+0.0354 \, | \, ^4D_{7/2} > -0.1589 \, | \, ^4F_{7/2} > -0.7362 \, | \, ^4G_{7/2} >$

$| [^2D_{5/2}] > = -0.7688 \, | \, ^2D(20)_{5/2} > +0.3789 \, | \, ^2D(21)_{5/2} > -0.0195 \, | \, ^2F(10)_{5/2} > +0.0266 \, | \, ^2F(21)_{5/2} >$
      $-0.3626 \, | \, ^4D_{5/2} > -0.3627 \, | \, ^4F_{5/2} > -0.0340 \, | \, ^4G_{5/2} >$

$| [^4G_{9/2}] > = 0.4900 \, | \, ^2G(20)_{9/2} > -0.3945 \, | \, ^2G(21)_{9/2} > -0.3011 \, | \, ^2H(11)_{9/2} > +0.5617 \, | \, ^2H(21)_{9/2} >$
      $-0.1879 \, | \, ^4F_{9/2} > -0.3552 \, | \, ^4G_{9/2} > -0.1912 \, | \, ^4I_{9/2} >$

$| [^3H_6] > = -0.9955 \, | \, ^3H_6 > -0.0943 \, | \, ^1I_6 >$

$| [^3F_4] > = 0.8079 \, | \, ^3F_4 > +0.5228 \, | \, ^1G_4 > -0.2718 \, | \, ^3H_4 >$

Tm^{3+}

$| [^3H_5] > = 1.0000 \, | \, ^3H_5 >$

$| [^3H_4] > = -0.7818 \, | \, ^3H_4 > -0.5022 \, | \, ^3F_4 > +0.3696 \, | \, ^1G_4$

$| [^3F_3] > = 1.0000 \, | \, ^3F_3 >$

$| [^3F_2] > = 0.8800 \, | \, ^3F_2 > -0.4550 \, | \, ^1D_2 > -0.1362 \, | \, ^3P_2 >$

$| [^1G_4] > = -0.7681 \, | \, ^1G_4 > +0.3083 \, | \, ^3F_4 > -0.5612 \, | \, ^3H_4 >$

$| [^1D_2] > = 0.6436 \, | \, ^1D_2 > +0.6327 \, | \, ^3P_2 > +0.4306 \, | \, ^3F_2 >$

$| [^1I_6] > = -0.9955 \, | \, ^1I_6 > +0.0943 \, | \, ^3H_6 >$

$| [^3P_0] > = 0.9762 \, | \, ^3P_0 > -0.2169 \, | \, ^1S_0 >$

$| [^3P_1] > = 1.0000 \, | \, ^3P_1 >$

$| [^3P_2] > = 0.7623 \, | \, ^3P_2 > -0.6155 \, | \, ^1D_2 > -0.2002 \, | \, ^3F_2 >$

$| [^1S_0] > = 0.9762 \, | \, ^1S_0 > +0.2169 \, | \, ^3P_0 >$

---

when using the line strengths for electric dipole and magnetic dipole transitions (S_{ed} and S_{md}, respectively), where

$$S_{ed}(\alpha J; \alpha' J') = \sum_{\lambda=2,4,6} \Omega_\lambda \langle f^N \alpha [SL]J | U^{(\lambda)} | f^N \alpha' [S'L']J' \rangle^2 \tag{3.41}$$

$$S_{md}(\alpha J; \alpha' J') = \beta^2 | \langle f^N \alpha [SL]J | L + 2S | f^N \alpha' [S'L']J' \rangle |^2 \tag{3.42}$$

and αJ and $\alpha' J'$ are quantum numbers indicating an excited energy level and the ground state, respectively.

Experimentally, the oscillator strength P can be obtained by

$$P = \frac{9mcn}{\pi e^2 (n^2 + 2)^2} \int \sigma(\nu) \, d\nu \tag{3.43}$$

where $\sigma(\nu)$ is the absorption cross section for an absorption transition.

The standard method for obtaining Ω parameters (Ω_2, Ω_4, and Ω_6) is to choose the values that minimize the root mean square deviation between the observed oscillator strengths (3.43) and those calculated by (3.40). The oscillator strength for any emission transition can be calculated using the Ω parameters obtained with (3.40). The spontaneous emission rate is given by

$$A(\alpha J;\, \alpha'J') = \frac{64\pi^4 \nu^3}{3hc^3(2J+1)}\left[\frac{n(n^2+2)^2}{9}S_{\mathrm{ed}} + n^3 S_{\mathrm{md}}\right] \qquad (3.44)$$

The peak-induced emission cross section is related to the radiative probability by the transition

$$\sigma(\lambda_p) = \frac{\lambda_p^4}{8\pi c n^2 \Delta\lambda_{\mathrm{eff}}} A(\alpha J;\, \alpha'J') \qquad (3.45)$$

where the effective linewidth $\Delta\lambda_{\mathrm{eff}}$ is used for glasses because the emission bands as well as absorption bands are characteristically asymmetrical.

The radiative lifetime of level J can be expressed in terms of spontaneous emission probabilities as

$$\tau_J^{-1} = \sum_{J'} A(J, J') \qquad (3.46)$$

where the summation is over all terminal levels J'. The fluorescence-branching ratio from level J and J'' is given by

$$\beta_{JJ'} = \frac{A(J, J')}{\displaystyle\sum_{J'} A(J, J')} \qquad (3.47)$$

The Judd–Ofelt analysis makes it possible to obtain spectroscopic properties, such as the stimulated emission transition cross section, the radiative lifetime, and the branching ratio, for any $4f$-$4f$ transition. Furthermore, the Judd–Ofelt analysis can be used to estimate excited state absorption and the probability of ion-ion interaction that causes energy transfer and fluorescence quenching phenomena.

3.4 OTHER PROCEDURES FOR OBTAINING EMISSION CROSS SECTIONS

When it is difficult to obtain the stimulated-emission cross section for a rare-earth ion transition using the Judd–Ofelt analysis, it can be calculated from the absorption cross section using the Einstein A and B coefficients for a two-level system [13]. This method has frequently been used to estimate cross sections of Er^{3+}. When generalized for the case in which the lower state (level 1) and upper state (level 2) are split into multiple components, the relationship becomes

$$g_1 \int \nu^2 \sigma_a(\nu) \, d\nu = g_2 \int \nu^2 \sigma_e(\nu) \, d\nu \tag{3.48}$$

where g_1 is the degeneracy of level 1, ν is the photon frequency, and σ_a and σ_e are the absorption and stimulated-emission cross sections, respectively. Equation (3.48), which is a more general form of the Ladenburg–Fuchtbauer relationship [14], is valid for rare-earth ions only if one of the two following conditions is met: (1) all components of the two levels must be equally populated, and (2) all the transitions must have the same strength regardless of the components involved. This method is very convenient. However, the relationship does not always provide accurate emission cross sections of rare earths because the manifold width of the $4f$ state often exceeds 300 cm^{-1} and the first condition is not satisfied at room temperature $(kT \sim 200 \text{ cm}^{-1})$. Furthermore, the transition strength is sensitive to the Stark levels [15].

The McCumber theory provides another method of transformation between the absorption and emission cross sections. The only assumption needed by the McCumber theory is that the time required to establish thermal distribution within each manifold be short compared with the lifetime of that manifold. The absorption and emission cross sections are then related by [16]

$$\sigma_e(\nu) = \sigma_a(\nu) \exp[(\epsilon - h\nu)/kT] \tag{3.49}$$

where ϵ is the temperature-dependent excitation energy. The physical interpretation of ϵ is as the net free energy required to excite a rare-earth ion from the ground state to an excited state at temperature T. At frequencies higher than ν_c $(\nu_c = \epsilon/h)$, the emission cross section is smaller than the absorption cross section and vice versa for $\nu < \nu_c$. Another important characteristic of (3.49) is that even a relative measurement of the spectrum of either cross section is sufficient to generate a relative spectrum of the other cross section. This is a feature clearly not available from the Einstein relation in (3.48).

In addition to providing relative cross-section spectra, the McCumber theory can also provide absolute values if the parameter ϵ can be evaluated. For this purpose a useful alternative definition is [16]

$$\frac{N_1}{N_2} = \exp(\epsilon/kT) \tag{3.50}$$

where N_i is the equilibrium population of the ith level at temperature T in the absence of optical pumping. If the positions of all the Stark components are known, ϵ is calculated according to

$$\frac{N_1}{N_2} = \frac{1 + \displaystyle\sum_{j=2} \exp(-E_{1j}/kT)}{\exp(-E_0/kT)\left[1 + \displaystyle\sum_{j=2} \exp(-E_{2j}/kT)\right]} \tag{3.51}$$

In this expression E_0 is the separation between the lowest component of each manifold, E_{ij} is the difference in energy between the jth and the lowest component of level i.

Next, we describe the applicability of McCumber theory to Er^{3+}-doped glasses. $\sigma_a(\nu)$ can be determined from absorption measurements of bulk samples with a known Er^{3+} concentration. From corrected emission spectra and measured lifetimes for the $^4I_{13/2}$ state, the experimental stimulated-emission cross sections can be obtained using the expression [16]

$$\frac{1}{\tau} = \frac{8\pi n^2}{c^2} \int \nu^2 \sigma_e(\nu)\ d\nu \tag{3.52}$$

where τ is the radiative lifetime and n is the refractive index.

The absorption cross sections are transformed into relative emission cross-section spectra with (3.49) and converted into absolute cross sections by means of (3.52). Figure 3.4 shows the shape of $\sigma_e(\nu)$, calculated in this way for a fluorophosphate glass [17]. There is very good agreement between the emission cross sections calculated by the McCumber theory and those determined experimentally.

Figure 3.5 compares stimulated-emission cross sections determined by three different scaling procedures [17]. The measured curve was scaled using the lifetime. The McCumber curve was scaled by estimating by (3.50) and (3.51), and the Einstein curve was calculated using (3.48). The McCumber curve is in good agreement with the measured cross-section spectrum, whereas the Einstein curve overestimates it. The McCumber theory is an advanced tool for analyzing cross sections for rare-earth-doped glasses.

3.5 ENERGY TRANSFER PHENOMENA BETWEEN RARE EARTHS

3.5.1 Formalism of Resonance Energy Transfer Between Rare Earths

The resonance transfer of energy between a donor (sensitizer) ion (S) and an acceptor (activator) ion (A) occurs when the energy differences between the ground and excited states of the S and the A systems are identical. If there is a suitable

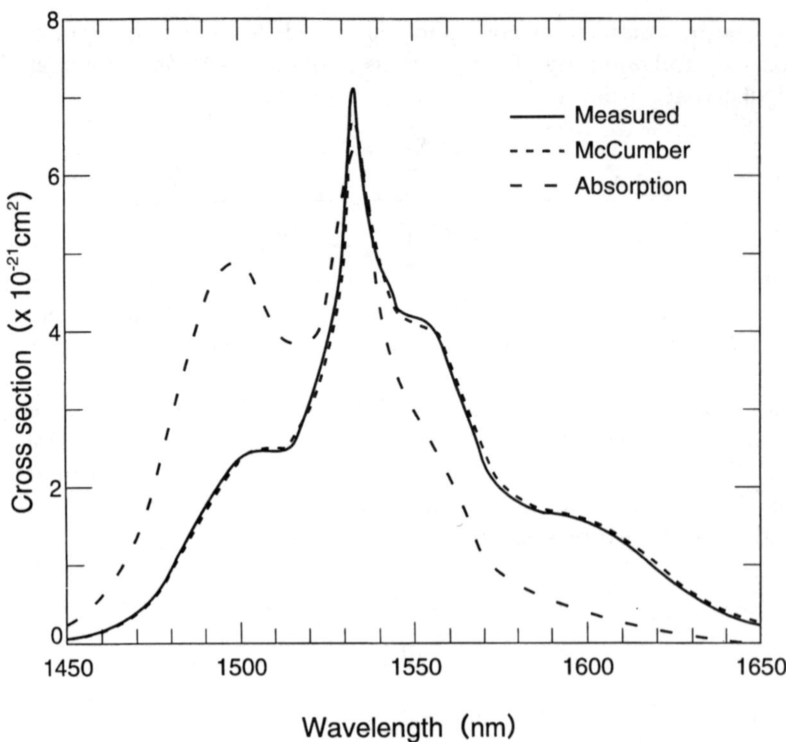

Figure 3.4 The measured stimulated-emission cross section and that calculated from the absorption cross section by the McCumber theory. Host glass: fluorophosphate glass.

interaction (such as an exchange interaction, a superexchange interaction, or a multipolar interaction) between the two electronic systems, the energy absorbed by S can be transferred to A.

The coupling of adjacent ions in such a case can arise via an exchange interaction if their wavefunctions overlap directly, via a superexchange interaction when intervening ions are involved, or via various electric or magnetic multipolar interactions.

Dexter [18] derived the following general expression for the probability of resonance transfer per unit time that energy will be transferred between the two ions:

$$P(R) = \left(\frac{2\pi}{\hbar}\right) |\langle \Psi_S^* \Psi_A | H_{SA} | \Psi_S \Psi_A^* \rangle|^2 \int g_S(E) g_A(E) \, dE \qquad (3.53)$$

where Ψ_S^* and Ψ_S, Ψ_A^* and Ψ_A are the excited and ground state wave-functions of the sensitizer S and the activator A ions, respectively, and where H_{SA} is the interaction

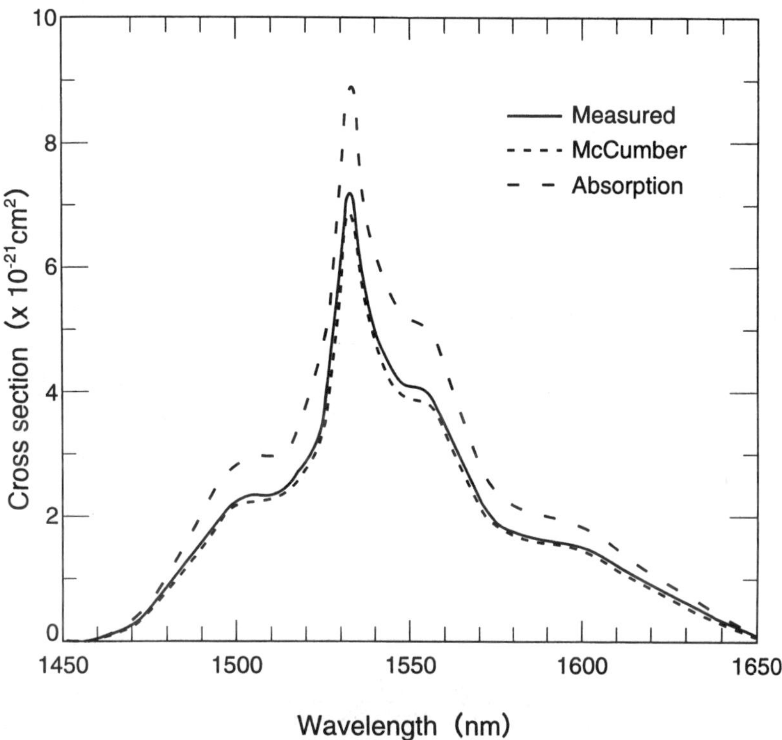

Figure 3.5 Comparison of stimulated-emission cross sections using the radiative lifetime, the Einstein relation, and the McCumber theory. Host glass: fluorophosphate glass.

Hamiltonian. The achieved overlap integral, $\bar{S} = \int g_S(E) g_A(E)\ dE$, is over the normalized emission band shape of the sensitizer and the absorption band of the activator in which the transitions $S - S^*$ and $A - A^*$ are represented by the line shape functions $g_S(E)$ and $g_A(E)$ each normalized in the sense

$$\int g(E)\ dE = 1 \tag{3.54}$$

The ion-ion interaction depends on the donor-acceptor distance for various mechanisms. The radial dependence of the ion-pair transfer rate is derived from the square of the matrix element of H_{SA} in (3.53).

Reviews on energy transfer theory with special emphasis on rare-earth ions have been published by Reisfeld [19], Riseberg and Weber [20], Watts [21], and Auzel [22]. Below is an outline of the expressions for the matrix elements and transfer probabilities of magnetic and electrostatic mechanisms that dominate the energy transfer phenomena for rare-earth-doped amplifier media [8].

3.5.1.1 Magnetic Dipole-Dipole Interaction

Another mechanism of paramagnetic ion coupling is the *magnetic dipole-dipole* (MDD) interaction. The Hamiltonian of this interaction is of the form

$$H_{\mathrm{MDD}} = \sum_{i,j} \left[\frac{\boldsymbol{\mu}_i \cdot \boldsymbol{\mu}_j}{R^3} - \frac{3(\boldsymbol{\mu}_i \cdot \mathbf{R})(\boldsymbol{\mu}_j \cdot \mathbf{R})}{R^5} \right] \tag{3.55}$$

where $\boldsymbol{\mu}_i = \mathbf{l}_i + 2\mathbf{s}_i$, and $\mathbf{l}_i$ and $\mathbf{s}_i$ are the orbital and spin operators for the ith electron of ions S and A, respectively. The selection rules ΔS, ΔL, $\Delta J = 0 \pm 1$ for transition between $4f^N$ states are relaxed by SLJ state admixing. The MDD interaction has the same long-range R^{-3} radial dependence as the electron dipole interaction, which will be dealt with below.

3.5.1.2 Electrostatic Interaction

The electrostatic interaction is represented by

$$H_{\mathrm{es}} = \sum_{i,j} \frac{e^2}{K |(\mathbf{r}_{S_i} - \mathbf{R} - \mathbf{r}_{A_j})|} \tag{3.56}$$

Here $\mathbf{r}_{S_i}$ and $\mathbf{r}_{A_j}$ are the coordinate vectors of electrons i and j belonging to ions S and A, respectively; $\mathbf{R}$ is the nuclear separation; and K is the dielectric constant. The various multipolar terms result from a power series expansion of the denominator. This expansion was expressed by Kushida [23] in terms of tensor operators. The leading terms are the *electric dipole-dipole* (EDD), *dipole-quadrupole* (EQD), and *quadrupole-quadrupole* (EQQ) interactions. These have radial a dependence of R^{-3}, R^{-4}, and R^{-5}, respectively.

In his calculation of the induced dipole-dipole and dipole-quadrupole processes of energy transfer, Kushida [23] made use of the Judd–Ofelt expression for the forced electric dipole transition probability in rare earths incorporated in solids. The transfer rates for dipole-dipole, dipole-quadrupole, and quadrupole-quadrupole processes in rare earths are of the forms

$$P_{\mathrm{SA}}^{(\mathrm{dd})} = \frac{1}{(2J_S + 1)(2J_A + 1)} \left(\frac{2}{3} \right) \left(\frac{2\pi}{\hbar} \right) \left(\frac{e^2}{R^3} \right)^2$$

$$\cdot \left[\sum_t \Omega_{tS} \langle J_S \| U^{(t)} \| J_S' \rangle^2 \right] \left[\sum_t \Omega_{tA} \langle J_A \| U^{(t)} \| J_A' \rangle^2 \right] \bar{S} \tag{3.57}$$

$$P_{SA}^{(dq)} = \frac{1}{(2J_S + 1)(2J_A + 1)} \left(\frac{2\pi}{\hbar}\right) \left(\frac{e^2}{R^4}\right)^2 \left[\sum_t \Omega_{tS} \langle J_S \| U^{(t)} \| J_S' \rangle^2\right] \langle 4f | r_A^2 | 4f \rangle^2$$

$$\cdot \langle f \| C^{(2)} \| f \rangle^2 \langle J_A \| U^{(2)} \| J_A' \rangle^2 \bar{S} \tag{3.58}$$

$$P_{SA}^{(dq)} = \frac{1}{(2J_S + 1)(2J_A + 1)} \left(\frac{14}{5}\right) \left(\frac{2\pi}{\hbar}\right) \left(\frac{e^2}{R^5}\right)^2 \langle 4f | r_S^2 | 4f \rangle^2 \langle 4f | r_A^2 | 4f \rangle^2$$

$$\cdot \langle f \| C^{(2)} \| f \rangle^4 \langle J_S \| U^{(2)} \| J_S' \rangle^2 \langle J_A \| U^{(2)} \| J_A' \rangle^2 \bar{S} \tag{3.59}$$

where the Ω values are the Judd–Ofelt intensity parameters; $\langle J \| U^{(t)} \| J' \rangle$ is the matrix element of the transition between the ground and excited state of the sensitizer and activator, respectively; $\bar{S}$ is the overlap integral; R is the interionic distance; $C^{(2)}$ is a numerical factor that depends on the orientation of the coordinate axis; and $\langle 4f \| r_A^2 \| 4f \rangle^2$ is the expectation value of the radial integral of the $4f$ orbital. The dependence of the transfer rate on S-A separation can be written as

$$P_{SA} = \frac{\alpha^{(6)}}{R^6} + \frac{\alpha^{(8)}}{R^8} + \frac{\alpha^{(10)}}{R^{10}} + \cdots \tag{3.60}$$

where $\alpha^{(6)}$, $\alpha^{(8)}$, and $\alpha^{(10)}$ are constants for dipole-dipole, dipole-quadrupole, and quadrupole-quadrupole transitions that may be evaluated from (3.57) through (3.59). The dipole-dipole process gives rise to the longest ion-ion interaction.

3.5.2 Concentration Quenching

Concentration quenching is a phenomenon of the quantum efficiency reduction of an emission transition of an ion with increasing concentration of that ion due to the ion-ion interactions. It can occur as a result of the energy transfer processes of cross relaxation and cooperative upconversion and has important implications for the performance of fiber amplifiers because it leads to a reduction in pump efficiency. The two mechanisms, cross relaxation and cooperative upconversion, are explained in the following subsections.

3.5.2.1 Cross Relaxation

A rare-earth ion in an excited state transfers part of its excitation to a neighboring ion in a cross-relaxation process as illustrated in Figure 3.6. This process occurs via the multipolar interactions discussed in Section 3.5.1. The cross-relaxation rate

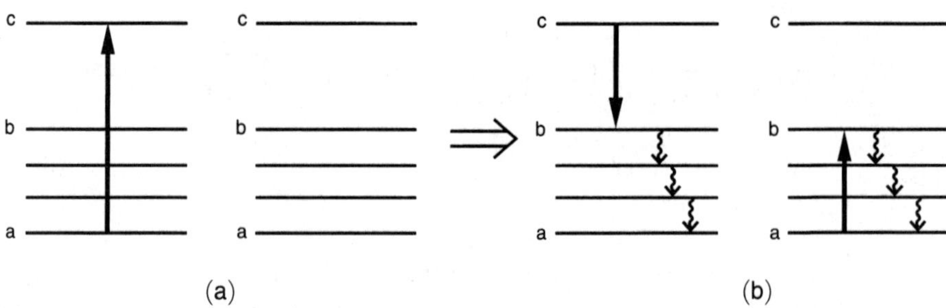

Figure 3.6 A schematic diagram of the cross-relaxation process: (a) excitation process and (b) energy transfer process.

becomes large at a high concentration. If an ion excited to the metastable **c** level interacts with a nearby ion in the ground state, the first ion transfers part of its energy to the second. Both ions occupy in the intermediate **b** state. When the energy gaps to the lower lying states are small, both ions decay nonradiatively to the ground state from the intermediate state. As a result, part of the excitation energy is converted into heat. The emission lifetime of a transition from the metastable state **c** shortens when there is cross relaxation among the ions. This quenching process will manifest itself as a nonexponential decay that is independent of pump power, since only one excited ion is required for cross relaxation. Measurements of emission lifetime dependence on concentration tell us the concentration range where there is no cross relaxation.

3.5.2.2 Cooperative Upconversion

The cooperative upconversion process is shown in Figure 3.7 and is the inverse of cross relaxation. When two ions are in an excited state [Figure 3.7(a)], one transfers

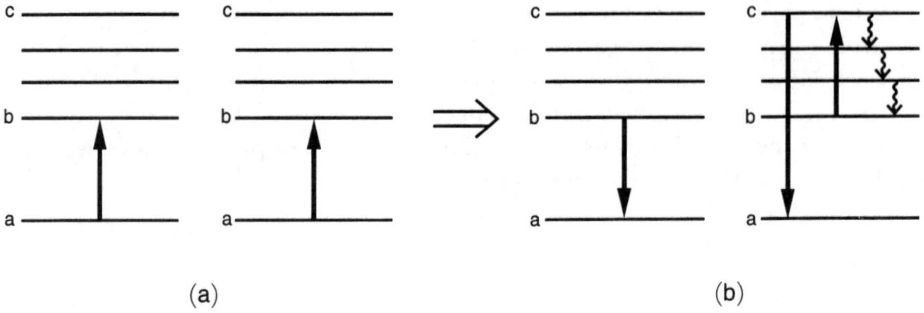

Figure 3.7 A schematic diagram of the cooperative upconversion process: (a) two-ion excitation process and (b) energy transfer process.

its energy to the other, leaving itself in the ground state and the other in a higher excited state [Figure 3.7(b)]. This is effectively an Auger process. Since cooperative upconversion requires two interacting ions to be in an excited state, it is not evident at low pumping levels. It appears at high pump powers, and nonexponential decay of emission from the excited state **b** can be observed. This behavior has been observed [24,25], and further evidence can be found in the report of the upconverted $^4I_{11/2}$ emission of erbium at 980 nm with pumping at 1,480 nm [26]. The pump power dependence of the upconversion mechanism has important consequences for three-level laser amplifiers since the quenching process is most deleterious at the high excited state populations required to achieve high single-pass gain and a good signal-to-noise ratio.

3.6 NONRADIATIVE RELAXATION BY THE MULTIPHONON EMISSION PROCESS

Nonradiative relaxation caused by the multiphonon emission process greatly influences the radiative quantum efficiency of the transition between the $4f$ states of rare-earth ions. Nonradiative relaxation between various $4f$ states occurs as a result of the simultaneous emission of several phonons that conserve the energy of the transitions. These multiphonon processes arise from the interaction of the electronic levels of the rare earth with the vibrations of the host lattice. The lattice vibrations are quantized as phonons having excitation energies determined by the masses of the constituent ions and the binding energies between the ions. The multiphonon process mechanism by which a nonresonant energy transfer between a donor and an acceptor ion may occur was first formulated by Orbach [27] and developed quantitatively by Miyakawa and Dexter [28] and Soules and Duke [29]. The theory of multiphonon process relaxation was further developed by Fong and his group [30]. Kiel [31] treated the problem by applying the time-dependent perturbation theory to higher orders. Multiphonon processes involve a combination of crystal field theory and lattice dynamics, and because of the tremendous difficulties in the rigorous calculation, suitable approximations are made in all quantitative calculations. The greatest barrier to finding a rigorous solution is the lack of detailed information regarding the frequency, polarization, symmetry, and propagation properties of the vibrations and the strength of the ion-phonon coupling constants. Riseberg and Moos [32] dealt with the problem of multiphonon relaxation phenomenologically. Here, we cite [33] for an understanding of the multiphonon relaxation process.

An isolated rare-earth ion can interact with its crystal field that varies with time because of lattice vibration. The interaction Hamiltonian [32] can be written as

$$H = V_0 + \sum_i V_i Q_i + \frac{1}{2} \sum_{i,j} V_{ij} Q_i Q_j + \cdots \tag{3.61}$$

where V_0 is the interaction with the static field. The remaining terms, involving the normal coordinates Q_i, represent the interaction with the vibrating lattice. The expression for $W(p)$ (the transition rate between two electronic levels), involving the emission of p phonons, can be obtained using either the pth-order time-dependent perturbation theory for the second term in (3.61) or the pth term in the Hamiltonian, in the first-order perturbation theory.

The phenomenological approach [32] using both mechanisms predicts that the multiphonon process transition rate has the form

$$W^{(p)} = W_0 \prod_i (n_i + 1)^{p_i} \tag{3.62}$$

Here, p_i is the number of phonons emitted with energy $\hbar\omega$, so $\Sigma p_i = p$. W_0 is the transition rate for spontaneous decay at $T = 0K$, and n_i is the Bose–Einstein occupation probability

$$n_i = [e^{\hbar\omega/kT} - 1]^{-1} \tag{3.63}$$

The 1 within brackets of (3.63) represents the spontaneous emission of phonons. If ΔE is the energy gap between two electronic levels, a number p_i of phonons, of energy $\hbar\omega_i$, is required for energy conservation; hence, the order of the process is determined by the condition

$$\sum p_i \hbar\omega = \Delta E \tag{3.64}$$

The critical feature in the temperature dependence of $W^{(p)}$ is the order of the process, which governs the relaxation. It has been found in practice that phonons of single frequency can be inserted in (3.64) for transitions in several crystals [28]. However, this is not always the case [34].

Decay occurs between groups of Stark levels of two J multiplets. If levels of the upper multiplet are designated by the subscript a and levels of the lower multiplet by the subscript b, then the decay rate from one of the upper levels is $\Sigma_b W_{ab}$. Since the levels within the multiplet are in thermal equilibrium, the combined rate is a Boltzmann average of the separate levels given by

$$W = \frac{\Sigma_a \Sigma_b g_a W_{ab}(T)\exp(-\Delta E/kT)}{\Sigma_a g_a \exp(-\Delta E/kT)} \tag{3.65}$$

where W_{ab} is the individual decay rate from upper multiplet level a to lower multiplet b, g_a is the degeneracy, and ΔE is the energy separation of the ath level from the upper level of the upper multiplet. A similar expression exists for the temperature dependence of the total radiative rate, where W_{ab} is replaced by $(W_{ab} + A_{ab})$.

The intrinsic temperature dependence of the multiphonon rate W_{ab} is expressed as follows using the single-frequency model:

$$W(T) = W_0\left[\frac{\exp(\hbar\omega/kT)}{\exp(\hbar\omega/kT) - 1}\right]^{p_i} = W_0[1 - \exp(-\hbar\omega/kT)]^{-p_i} \qquad (3.66)$$

The significance of (3.66) is in the possibility of establishing the order of the multiphonon decay process and the energies of the dominant phonons involved. The multiphonon relaxation rate decreases with decreasing temperature.

The dependence of the multiphonon rates on the energy gap, given by Miyahawa and Dexter [28] is of the form

$$W^{p_i} = W^0\exp(-\alpha\Delta E) \qquad (3.67)$$

where W^0 is the transition probability extrapolated to a zero energy gap, ΔE is the energy gap between two successive electronic levels, and α is expressed by

$$\alpha = (\hbar\omega)^{-1}\ln[p/g(n + 1)]^{-1} \qquad (3.68)$$

where g is the electron phonon coupling constant.

Equation (3.67) is obtained under the assumption that the electron-phonon coupling is weak (which is the case for rare-earth ions) and that the occupation number n is not much greater than unity. The exponential dependence of (3.67) is only approximate because both α and W^0 depend on ΔE.

In the phenomenological model the energy gap dependence of the multiphonon transition rate arises as a consequence of the convergence of the perturbation expansion [35]. When a single-frequency model is considered, the rate for the pth-order process can be written phenomenologically as

$$W^p = A\omega^p \qquad (3.69)$$

where A is a constant, $p = \Delta E/\hbar\omega$, and ϵ is a coupling constant whose magnitude determines the rapidity of the convergence of the terms in the perturbation expansion. We see that the multiphonon relaxation probability has an approximately exponential dependence on the energy gap between levels.

Figure 3.8 shows the multiphonon relaxation rate dependence on the energy gap of the J states of rare-earth ions in various host glasses [phosphate ($\hbar\omega = 1{,}200$ cm^{-1}), silicate ($\hbar\omega = 1{,}100$ cm^{-1}), tellurite ($\hbar\omega = 700$ cm^{-1}), and ZrF$_4$-based fluoride ($\hbar\omega = 560$ cm^{-1})] [36]. The multiphonon relaxation rate in silicate glass ($\hbar\omega = 1100$ cm^{-1}) is more than three orders of magnitude larger than that in ZrF$_4$-based fluoride glass, which is the smallest. Some transitions that are not laser active in silicate glass can be active in ZrF$_4$-based fluoride glass. Glasses with

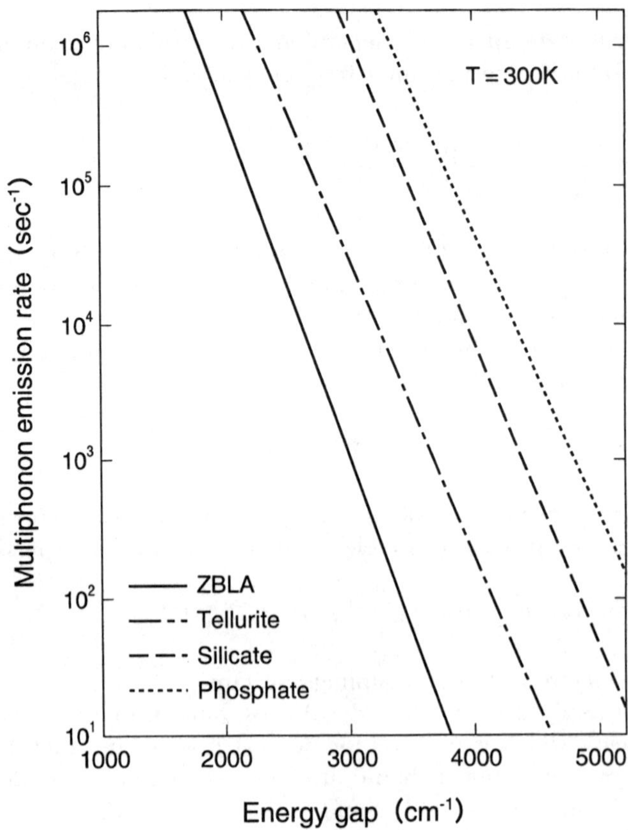

Figure 3.8 Multiphonon emission relaxation rate as a function of energy gap for several glasses.

lower energy phonons have the potential of increasing the range of laser emission wavelengths.

3.7 SPECTRAL BROADENING PHENOMENA

Homogeneous and inhomogeneous processes cause a broadening of the spectral line shape of transition of rare earths in glasses. These processes provide rare-earth-doped amplifier performance with additional features. The homogeneous broadening of transitions between Stark components of different J multiplets is caused by lifetime broadening that is dominated by phonon-induced transitions between the Stark components within a multiplet and phonon broadening caused by high-frequency lattice vibrations. The spectral linewidth of an optical transition (ij) can be given by the relation [37]

$$\Delta\omega_{ij} = A_i + A_j + 2/T_{2,ij} \tag{3.70}$$

where $A_{i,j}$ is the decay rate of the i or j level, and $T_{2,ij}$ is an effective dephasing time associated with phonon broadening. All rare earths with a homogeneously broadened linewidth experience the same broadening effects in the host. For a homogeneously broadened transition, a given wavelength will interact with all ions with equal probability. Thus any pump wavelength will produce the same gain spectrum and any signal wavelength can saturate the entire band. In a homogeneously broadened system, all the power can be extracted by a signal at any wavelength within the gain spectrum and signals at different wavelengths can interact with each other by perturbing the gain spectrum. This leads to high-efficiency lasers and power amplifiers but presents the possibility of cross-saturation for amplifiers in WDM systems.

Because of differences in terms of bonding to nearest neighbor ions and, in multicomponent glasses, in the types and distribution of more distant neighbor ions, the local fields at individual sites vary in strength and symmetry. This is the origin of inhomogeneous broadening and results in site-to-site differences in the energy levels and radiative and nonradiative transition probabilities of the laser ions. Variations in the stimulated emission cross section originate from differences in the center frequency ν_i, homogeneous linewidth $\Delta\nu_h$, and line strength S_i of the transition. These differences are illustrated in Figure 3.9 [38] for site i and all other sites j. Each of the three spectroscopic features may differ from site to site. The radiative and nonradiative decay rates and quantum efficiency may also change. Inhomogeneous linewidths are essentially independent of temperature. If inhomogeneous broadening dominates, individual sections of the gain spectrum act independently and can be individually addressed by photons of different wavelengths. For an inhomogeneously broadened transition, laser and power amplifier efficiencies will be lower but there is little interaction between signals in a WDM system.

Homogeneous linewidths at room temperature are obtained by extrapolating low-temperature linewidths. Figure 3.10 shows examples of polarized output spectra

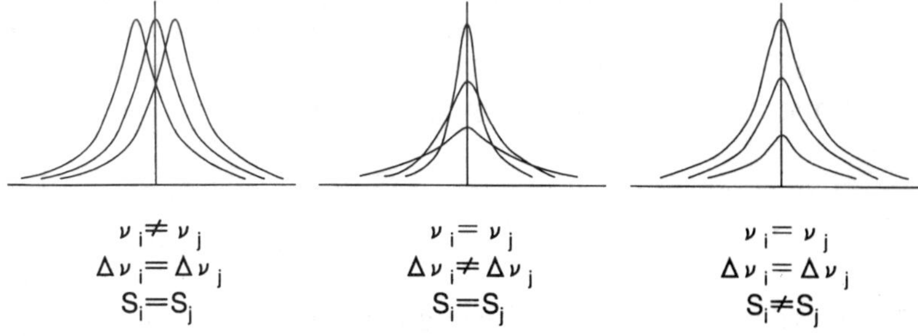

Figure 3.9 Differences in the center frequency ν, homogeneous linewidth $\Delta\nu$, and line strength S of optical transitions for ions in physically different sites i and j.

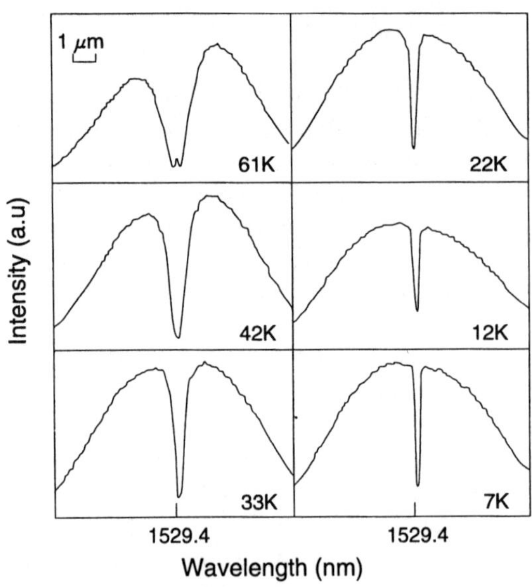

Figure 3.10 Polarized output spectra recorded in the polarization direction transverse to that of the output signal near 1.53 μm.

of an Al_2O_3-GeO_2-doped SiO_2 fiber, which were recorded in the polarization direction transverse to that of the output saturating signal near 1.53 μm [39]. The input signal power is 50 μW. The hole width $\Delta\lambda_{hole}$ decreases from 1.3 to 0.21 nm when the temperature decreases from 61K to 7K. A plot of $\Delta\lambda_{hole}$ against temperature is shown in Figure 3.11 [39]. The plot shows that in the 20-K to 77-K temperature range, $\Delta\lambda_{hole}$ follows a $T^{1.73}$ law, which agrees with the near quadratic law observed for the homogeneous linewidth $\Gamma_h(T)$ of most rare-earth inorganic glasses [40,41]. The break in the power dependence at lower temperatures (6K to 20K), $\Delta\lambda_{hole} \sim T^{0.3}$, is perhaps attributable to the existence of low-energy tunneling modes or so-called two-level system effects [41], which are known to contribute to a near linear component in the temperature dependence of Γ_h.

The homogeneous linewidth Γ^h is proportional to the hole width $\Delta\lambda_{hole}$ [37], that is,

$$\Gamma_h \sim \Delta\lambda_{hole} / (1 + (1 + P_s/P_{sat})^{1/2}) \tag{3.71}$$

where P_s is the signal power and P_{sat} the saturation power of the inhomogeneously broadened transitions, respectively. When the condition $P_s/P_{sat} \ll 1$ is satisfied, the homogeneous linewidth is given by $\Gamma_h \sim \Delta\lambda_{hole}/2$. The homogeneous linewidth at room temperature is found to be 11.5 nm from Figure 3.11.

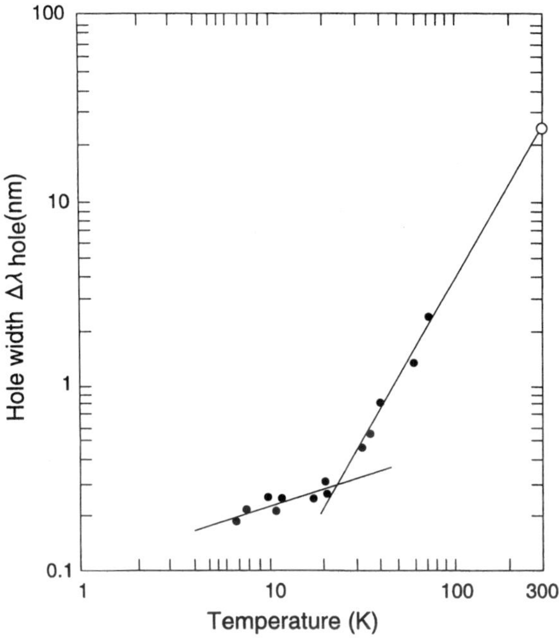

Figure 3.11 Relation of hole width $\Delta\lambda_{\text{hole}}$ and temperature. The point (o) indicates an evaluated room-temperature width of $\Delta\lambda_{\text{hole}} = 23$ nm.

Table 3.3 lists homogeneous and inhomogeneous linewidths at room temperature for Er^{3+} doped in several glasses [1]. The linewidths were obtained by cross modulation [42], hole burning [39,43], and fluorescence line narrowing [44,45]. An interesting byproduct of these investigations has been the observations for some Er^{3+}-doped glasses of significant departures from the expected T^2 dependence of

Table 3.3

Homogenous ($\Delta\nu_{\text{H}}$) and Inhomogenous ($\Delta\nu_{\text{IH}}$) Linewidths for Er^{3+} Doped in Glasses

Host	$\Delta\nu_{\text{H}}$ (cm^{-1})	$\Delta\nu_{\text{IH}}$ (cm^{-1})	Cite
GeO_2-doped silica	17	30	[42]
GeO_2-doped silica	17	34	[43]
GeO_2-doped silica	14	27	[44]
Al_2O_3-GeO_2-doped silica	49	49	[39]
Al_2O_3-P_2O_5-doped silica	8	60	[45]
Fluorophosphate (low fluorine)	7–10	60	[45]
Fluorophosphate (high fluorine)	20–35	60	[45]
Fluorozirconate	20–35	60	[45]

the homogeneous linewidth. For GeO_2-doped silica, Zyskind et al. found $m = 1.61$ for T^m at the peak (1,535 nm) and $m = 0.54$ at 1,550 nm [43]. The data reported by Zemon et al. for both an Al_2O_3-P_2O_5-doped silica and low-fluorine-content fluorophosphate were best fit by $m = 1.2$ [45]. At present there is no explanation for this behavior or for the difference in $\Delta\nu_H$ between Al_2O_3-GeO_2-doped and Al_2O_3-P_2O_5-doped silicas in Table 3.3.

Whether the stimulated emission for amplification is a homogeneous or inhomogeneous transition affects the gain saturation characteristics. A homogeneous line saturates in the following way

$$g_{homo} \propto \frac{1}{1 + I_s/I_{sat}} \tag{3.72}$$

while an inhomogeneous line saturates in the following way

$$g_{inhomo} \propto \frac{1}{\sqrt{1 + I_s/I_{sat}}} \tag{3.73}$$

where g_{homo} and g_{inhomo} are gain coefficients for homogeneous and inhomogeneous transitions, and I_s and I_{sat} are the signal and saturation intensity (see Section 3.8), respectively.

This difference between the linear and square-root dependence in the saturation denominator is the distinguishing difference between homogeneous and inhomogeneous saturation, as illustrated in Figure 3.12 [37]. The inhomogeneous gain seems to saturate somewhat more slowly at large intensities because the signal can draw energy from an increasingly wider range of packets as its intensity increases. That is, in an inhomogeneous line, the applied signal saturates not only the spectral packet with which it is in exact resonance but also other adjoining packets with which it has a weaker interaction.

3.8 THREE- AND FOUR-LEVEL AMPLIFIER SYSTEMS

Rare-earth-doped amplifiers are classified as either four- or three-level systems. Figures 3.13 and 3.14 show transitions for four- and three-level amplifier systems [37]. Amplifications due to the $^1G_4$-$^3H_5$ transition at 1.31 μm of Pr^{3+}, the $^4F_{3/2}$-$^4I_{13/2}$ transition at 1.34 μm of Nd^{3+}, and the $^3H_4$-$^3F_4$ transition at 1.47 μm of Tm^{3+} are classified as four-level system amplification. Amplifications due to the $^4I_{13/2}$-$^4I_{15/2}$ transition at 1.55 μm of Er^{3+} and the $^3F_4$-$^3H_6$ transition at 1.65 μm of Tm^{3+} are classified as three-level system amplification. In the four-level system, the final level of the stimulated-emission transition is a intermediate level, while that in the three-level system is the ground level. This aspect of the three-level system is a serious disadvantage, since more than half the rare-earth ions initially in the ground

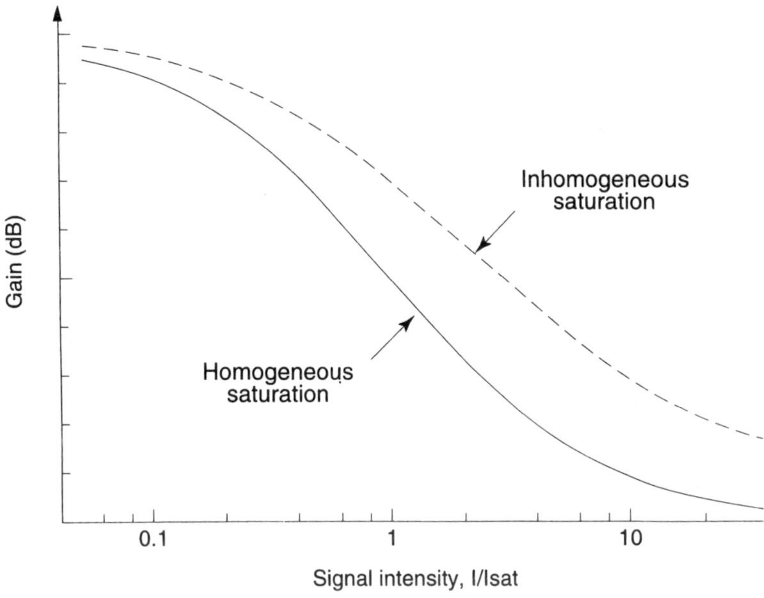

Figure 3.12 Difference in homogeneous and inhomogeneous saturation in response to a strong applied signal.

state must be pumped to the upper level, which is the initial level of the amplification transition, in order to obtain a stimulated transition. Therefore, three-level system amplifiers are usually not as efficient as four-level amplifiers. This section describes the difference between the four- and three-level system amplifiers.

3.8.1 Population Inversion

We begin by analyzing the rate equations so as to understand the population inversion condition. In order to analyze the four-level system we write the rate equation based on the model shown in Figure 3.13 and then solve it to obtain steady-state solutions. Assuming $W_{14} = W_{41} = W_p$ between levels 1 and 4, the rate equation for level 4 is

$$\frac{dN_4}{dt} = W_p(N_1 - N_4) - N_4/\tau_4 \tag{3.74}$$

where τ_4 is the total lifetime of level 4. The steady-state population of level 4 when $dN_4/dt = 0$ is given by

$$N_4 \approx W_p \tau_4 N_1 \tag{3.75}$$

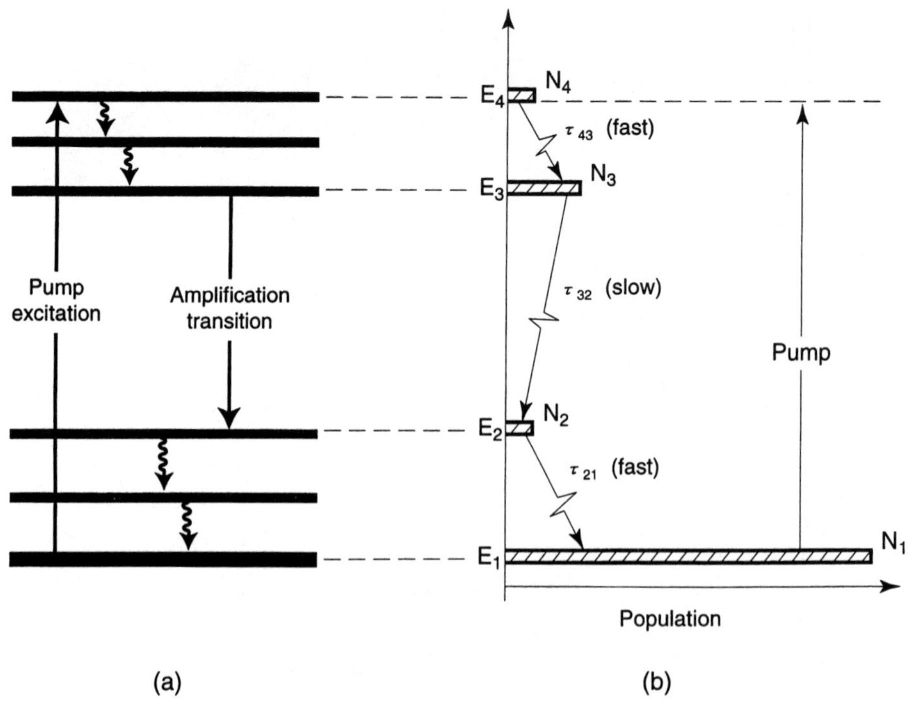

Figure 3.13 A four-level pumping model: (a) energy levels and transitions and (b) population of each level.

if $W_p\tau_4 \ll 1$. The rate equations for levels 3 and 2 are

$$\frac{dN_3}{dt} = \frac{N_4}{\tau_{43}} - \frac{N_3}{\tau_3} \tag{3.76}$$

$$\frac{dN_2}{dt} = \frac{N_4}{\tau_{42}} + \frac{N_3}{\tau_{32}} - \frac{N_2}{\tau_{21}} \tag{3.77}$$

Equation (3.76) gives the steady state ($dN_3/dt = 0$)

$$N_3 = \frac{\tau_3}{\tau_{43}} N_4 \tag{3.78}$$

In an ideal system the relaxation from level 4 to level 3 is very fast and level 3 has a long lifetime. Then N_3 is much larger than N_4 ($N_3 \gg N_4$). The population inversion $N_3 - N_2$ is given by

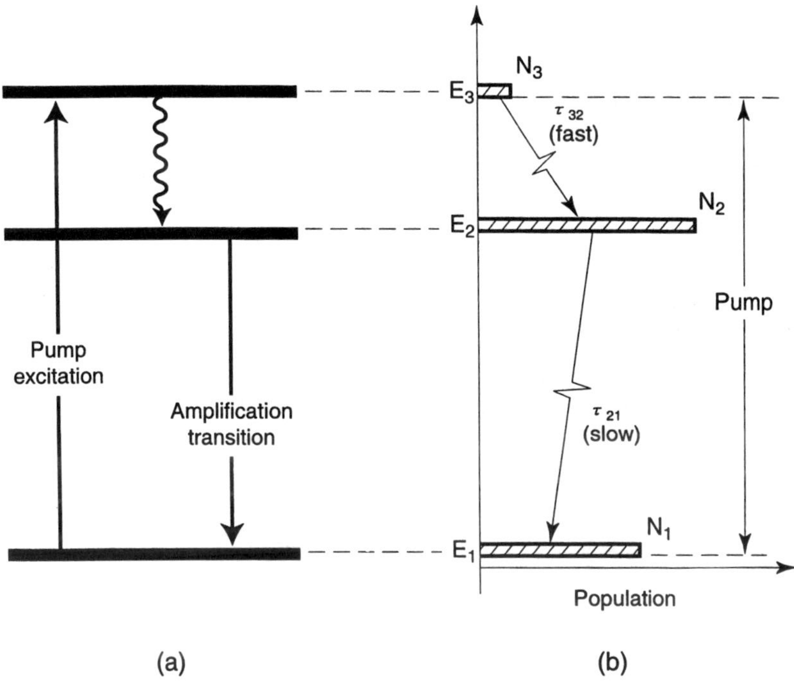

Figure 3.14 A three-level pumping model: (a) energy levels and transitions and (b) population of each level.

$$N_3 - N_2 = \frac{W_p \tau_{rad}}{1 + W_p \tau_{rad}} N \tag{3.79}$$

if $\tau_{21}/\tau_{32} \ll 1$, utilizing the ion conservation condition of $N = N_1 + N_2 + N_3 + N_4$, where τ_{rad} is the radiative lifetime of level 3 and N is the total ion number density.

With a three-level system and assuming that $W_{13} = W_{31} = W_p$ between levels 1 and 3, the rate equations for the two upper levels 2 and 3 are given by

$$\frac{dN_3}{dt} = W_p(N_1 - N_3) - \frac{N_3}{\tau_3} \tag{3.80}$$

$$\frac{dN_2}{dt} = \frac{N_3}{\tau_{32}} - \frac{N_2}{\tau_{21}} \tag{3.81}$$

The steady-state population inversion $N_2 - N_1$ is given by

$$N_2 - N_1 = \frac{(1 - \beta)\eta W_p \tau_{rad} - 1}{(1 + 2\beta)\eta W_p \tau_{rad} + 1} N \tag{3.82}$$

where η and β are given by

$$\eta = \frac{\tau_3 \tau_{21}}{\tau_{32} \tau_{\text{rad}}} \tag{3.83}$$

$$\beta = \frac{\tau_{32}}{\tau_{21}} \tag{3.84}$$

employing the usual conservation equation of $N = N_1 + N_2 + N_3$. Population inversion in the three-level system can be obtained only if $\beta < 1$ and occurs only when the pumping rate exceeds a threshold value given by

$$W_p \tau_{\text{rad}} \geq \frac{1}{\eta(1 - \beta)} \tag{3.85}$$

When $\eta \cong 1$ and $\beta \cong 0$, the population inversion is reduced to

$$N_2 - N_1 \approx \frac{W_p \tau_{\text{rad}} - 1}{W_p \tau_{\text{rad}} + 1} N \tag{3.86}$$

Figure 3.15 illustrates the difference in population inversion versus pumping rate for a three- and a four-level system [37]. If the parameters have the same values, a four-level system has a much lower pumping threshold than a three-

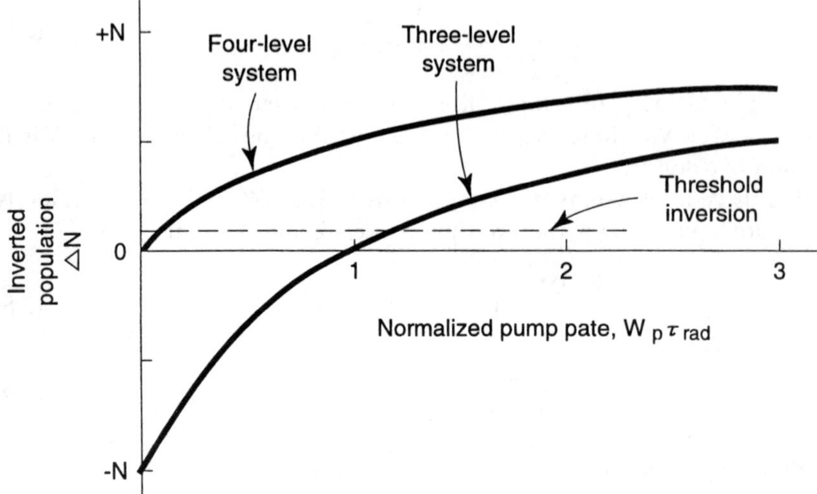

Figure 3.15 Population inversion versus pumping rate for four- and three-level systems.

level system. The population inversion of a four-level system increases linearly with pumping intensity W_p at lower pump levels. But it then approaches a limiting value N, when $W_p \tau_{rad} \gg 1$, as the ground state is depleted and a large fraction of the ions are lifted into the upper level. By contrast, the population inversion of the three-level system does not easily approach N even if the pumping intensity is greatly increased. It can be seen that a large population inversion is much easier for a four-level system than a three-level system.

3.8.2 Gain Saturation

The signal gain decreases as the input signal intensity is increased. This phenomenon is called gain saturation and it is an important aspect of optical amplifier performance. There is a difference in the gain saturation characteristics of three- and four-level systems.

The signal intensity I_s evolution along a fiber is given by the differential equation

$$\frac{dI_s}{dz} = g(I_s) I_s \tag{3.87}$$

where $g(I_s)$ is the gain coefficient and is expressed by

$$g(I_s) = \frac{g_0}{1 + I_s / I_{sat}} \tag{3.88}$$

In (3.88), g_0 is unsaturated gain or small signal gain and I_{sat} is called the saturation intensity, which is the intensity of a signal passing through an amplification medium that will saturate the gain coefficient down to half its unsaturated gain. With a four-level system, I_{sat} is given by

$$I_{sat} = \frac{h\nu_s}{\sigma_s \tau_3} \tag{3.89}$$

and for a three-level system

$$I_{sat} = \frac{h\nu_s}{\sigma_s}(W_p + 1/\tau_2) \tag{3.90}$$

where ν_s is the frequency of the signal light, σ_s the stimulated-emission cross section for the signal light, W_p the pumping rate, and τ_2 and τ_3 the lifetimes of the initial levels of amplification transition. It should be noticed that the saturation intensity depends only on the material parameters and not on the pumping rate, that is, the pumping power, while the saturation intensity for a three-level system depends on the pumping rate. This is because the depletion of the population inversion is caused only by the stimulated-emission transition caused by the signal light in a four-level system, while that for a three-level system is determined by the balance between the stimulated-emission transition and the pumping transition. The saturation intensity with a four-level system becomes larger for amplification media with a smaller stimulated-emission cross section and shorter lifetime.

Figures 3.16 and 3.17 show typical gain saturation characteristics of a PDFA as an example of a four-level system and those of an EDFA as an example of a three-level amplifier, respectively [46,47]. We can confirm from Figures 3.16 and 3.17 that the saturation intensity of a PDFA does not depend on the pumping power whereas that of an EDFA does.

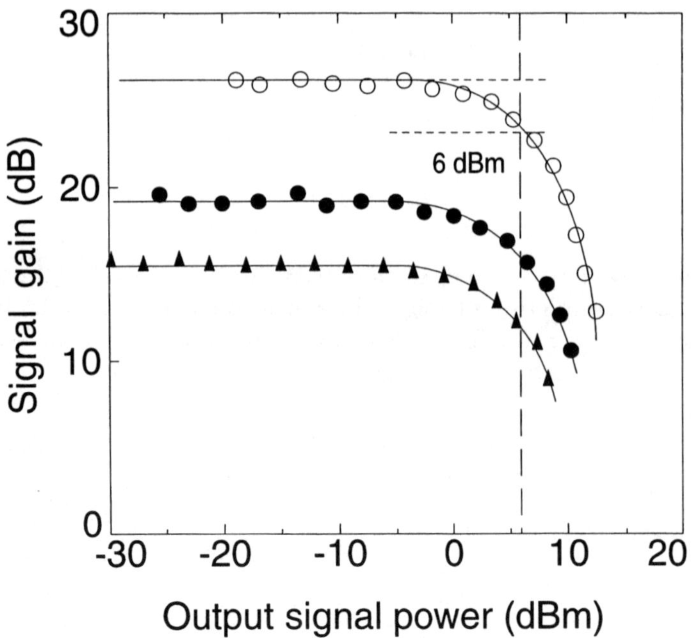

Figure 3.16 Gain saturation characteristics of a PDFA as an example of a four-level system. Signal wavelength: 1.30 μm.

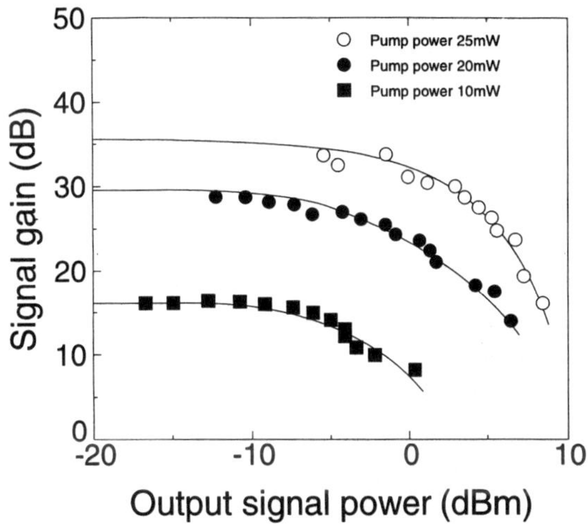

Figure 3.17 Gain saturation characteristics of an EDFA as an example of a three-level system. Signal wavelength: 1.536 μm.

References

[1] Miniscalco, W. J., "Optical and Electronic Properties of Rare Earths Ions in Glasses," in *Rare Earth Doped Fiber Lasers and Amplifiers*, M. J. F. Digonnet (ed.), New York: Marcel Dekker, Inc., 1993, pp. 19–133.

[2] Dieke, G. H., and H. M. Crosswhite, "The Spectra of the Doubly and Triply Ionized Rare Earths," *Appl. Opt.*, Vol. 2, 1963, pp. 675–686.

[3] Judd, B. R., "Optical Absorption Intensities of Rare-Earth Ions," *Phys. Rev.*, Vol. 127, 1962, pp. 750–761.

[4] Ofelt, G. S., "Intensities of Crystal Spectra of Rare-Earth Ions," *J. Chem. Phys.*, Vol. 37, 1962, pp. 511–520.

[5] Peacock, R. D., "The Intensities of Lanthanide f - f Transitions," *Structure and Bonding*, Vol. 22, New York: Springer-Verlag, 1975, pp. 83–122.

[6] Griffith, J. S., "Some Investigations in the Theory of Open-Shell Ions Part IV. The Basis of Intensity Theory," *Mol. Phys.*, Vol. 3, 1960, pp. 477–483.

[7] Carnall, W. T., H. Crosswhite, and H. M. Crosswhite, "Energy Level Structure and Transition Probabilities of the Trivalent Lanthanides in LaF_3," Argonne, IL: Argonne National Laboratory.

[8] Reisfeld, R., and C. K. Jorgensen, *Lasers and Excited States of Rare Earths*, New York: Springer-Verlag, 1977.

[9] Wybourne, B. G., *Spectroscopic Properties of Rare Earth*, New York: John Wiley, 1965.

[10] Weber, M. J., "Spontaneous Emission Probabilities and Quantum Efficiencies for Excited States of Pr^{3+} in LaF_3," *J. Chem. Phys.*, Vol. 48, 1968, pp. 4774–4780.

[11] Weber, M. J., "Probabilities for Radiative and Nonradiative Decay of Er^{3+} in LaF_3," *Phys. Rev.*, Vol. 157, 1967, pp. 262–272.

[12] Spector, N., R. Reisfeld, and L. Boehm, "Eigenstates and Radiative Transition Probabilities for $Tm^{3+}(4f^{12})$ in Phosphate and Tellurite Glasses," *Chem. Phys. Lett.*, Vol. 49, 1977, pp. 49–53.

[13] DiBartolo, B., *Optical Interactions in Solids*, New York: Wiley, 1968.

[14] Sandoe, J. N., P. H. Sarkies, and S. Parke, "Variation of Er^{3+} Cross Section for Stimulated Emission with Glass Composition," *J. Phys. D: Appl. Phys.*, Vol. 5, 1972, pp. 1788–1799.

[15] Zemon, S., G. Lambert, W. J. Miniscalco, L. J. Andrews, and B. T. Hall, "Characterization of Er^{3+}-Doped Glasses by Fluorescence Line Narrowing," *Proc. SPIE*, Vol. 1171, 1989, pp. 219–236.

[16] McCumber, D. E., "Theory of Phonon-Terminated Optical Masers," *Phys. Rev.*, Vol. 134, 1964, pp. A299–A306.

[17] Miniscalco, W. J., and R. S. Quimby, "General Procedure for the Analysis of Er^{3+} Cross Sections," *Opt. Lett.*, Vol. 16, 1991, pp. 258–260.

[18] Dexter, D. L., "A Theory of Sensitized Luminescence in Solids," *J. Chem. Phys.*, Vol. 21, 1953, pp. 836–850.

[19] Reisfeld, R., *Structure and Bonding*, Vol. 13, New York: Springer-Verlag, 1973.

[20] Riseberg, L. A., and M. J. Weber, "Relaxation Phenomena in Rare-Earth Luminescence," *Progress in Optics*, Vol. 14, E. Wolf (ed.), Amsterdam: North-Holland Publishing Co., 1975.

[21] Watts, R. K., *Optical Properties of Solids*, R. DiBartolo (ed.), New York: Plenum Press, 1975.

[22] Auzel, F. E., "Materials and Devices Using Double-Pumped Phosphors with Energy Transfer," *Proc. IEEE*, Vol. 61, 1973, pp. 758–786.

[23] Kushida, T., "Energy Transfer and Cooperative Optical Transitions in Rare-Earth-Doped Inorganic Materials. I. Transition Probability Calculation," "II. Comparison with Experiments," and "III. Dominant Transfer Mechanism," *J. Phys. Soc. Japan*, Vol. 34, 1973, pp. 1318–1326, 1327–1333, 1334–1337.

[24] Wyatt, R., "Spectroscopy of Rare-Earth-Doped Fibres," *Proc. SPIE*, Vol. 1171, 1989, pp. 54–64.

[25] Ainslie, B. J., S. P. Craig-Ryan, S. T. Davey, J. R. Armitage, C. G. Atkins, J. F. Massicott, and R. Wyatt, "Erbium Doped Fibres for Efficient Optical Amplifiers," *IEE Proc. Pt. J. Optoelectronics*, Vol. 137, pp. 205–208.

[26] Kagi, N., A. Oyobe, and K. Nakamura, "Efficient Optical Amplifier Using a Low-Concentration Erbium-Doped Fiber," *IEEE Photon. Tech. Lett.*, Vol. 2, 1990, pp. 559–561.

[27] Orbach, R., *Optical Properties of Ions in Crystals*, H. M. Crosswhite and H. W. Moos (eds.), New York: Wiley-Interscience Inc., 1967.

[28] Miyakawa, T., and D. L. Dexter, "Phonon Sidebands, Multiphonon Relaxation of Excited States, and Phonon-Assisted Energy Transfer Between Ions in Solids," *Phys. Rev. B*, Vol. 1, 1970, pp. 2961–2969.

[29] Soules, T. F., and C. B. Duke, "Resonant Energy Transfer Between Localized Electronic States in a Crystal," *Phys. Rev. B*, Vol. 3, 1971, pp. 262–274.

[30] Fond, F. K., S. L. Naberhuis, and M. M. Miller, "Theory of Radiationless Relaxation of Rare-Earth Ions in Crystals," *J. Chem. Phys.*, Vol. 56, 1972, pp. 4020–4027.

[31] Kiel, A., "Multiphonon Spontaneous Emission in Parametric Crystals," *Quantum Electronics*, P. Grivet and N. Bloemberger (eds.), New York: Columbia U. P., 1964, pp. 765–777.

[32] Riseberg, L. A., and H. W. Moos, "Multiphonon Orbit-Lattice Relaxation of Excited States of Rare-Earth Ions in Crystals," *Phys. Rev.*, Vol. 174, pp. 429–438.

[33] Reisfeld, R., "Radiative and Non-Radiative Transitions of Rare-Earth Ions in Glasses," *Structure and Bonding*, Vol. 22, New York: Springer-Verlag, 1975, pp. 123–175.

[34] Reed, E. D., Jr., and H. W. Moos, "Multiphonon Relaxation of Excited States of Rare-Earth Ions in YVO_4, $YAsO_4$, and YPO_4," *Phys. Rev. B*, Vol. 8, 1973, pp. 980–987.

[35] Riseberg, L. A., and M. J. Weber, "Relaxation Phenomena in Rare-Earth Luminescence," *Progress in Optics*, Vol. 14, E. Wolf (ed.), Amsterdam: North-Holland Publishing Co., 1975.

[36] Sibley, W. A., D. C. Yeh, M. Suscavage, and M. G. Drexhage, "Multiphonon Relaxation and Infrared to Visible Conversion of Er^{3+} and Yb^{3+} Ions in Barium-Thorium Fluoride Glass," *Materials Science Forum*, Vol. 19–20, 1987, pp. 567–572.

[37] Siegman, A. E., *Lasers*, Mill Valley, CA: Univ. Sci. Books, 1986.

[38] Weber, M. J., "Fluorescence and Glass Lasers," *J. Non-Crystal. Solids*, Vol. 47, 1982, pp. 117–134.

[39] Desurvire, E., J. L. Zyskind, and J. R. Simpson, "Spectral Gain Hole-Burning at 1.53 μm in Erbium-Doped Fiber Amplifiers," *IEEE Photon. Tech. Lett.*, Vol. 2, 1990, pp. 246–248.

[40] Macfarlane, R. M., and R. M. Shelby, "Homogeneous Line Broadening of Optical Transitions of Ions and Molecules in Glasses," *J. Luminescence*, Vol. 36, 1989, pp. 179–207.

[41] Yen, W. M., and R. T. Brundage, "Fluorescence Line Narrowing in Inorganic Glasses: Linewidth Measurements," *J. Luminescence*, Vol. 36, 1987, pp. 209–220.

[42] Laming, R. I., L. Reekie, P. R. Morkel, and D. N. Payne, "Multichannel Crosstalk and Pump Noise Characterisation of Er^{3+}-Doped Fibre Amplifier Pumped at 980 nm," *Electron. Lett.*, Vol. 25, 1989, pp. 455–456.

[43] Zyskind, J. L., E. Desurvire, J. W. Sulhoff, and D. J. Di Giovanni, "Determination of Homogeneous Linewidth by Spectral Gain Hole-Burning in an Erbium-Doped Fiber Amplifier with $GeO_2{:}SiO_2$ Core," *IEEE Photon. Tech. Lett.*, Vol. 2, 1990, pp. 869–871.

[44] Guy, S. C., R. A. Minasian, S. B. Poole, and M. G. Sceats, "Fluorescence Line Narrowing in Erbium-Doped Fiber Amplifiers," *Technical Digest of OFC'91*, Vol. 4, San Diego, CA: Optical Society of America, 1991, p. 194.

[45] Zemon, S., G. Lambert, W. Miniscalco, and B. A. Thompson, "Homogeneous Linewidths in Er^{3+}-Doped Glasses Measured by Resonance Fluorescence Line Narrowing," *Proc. SPIE*, Vol. 1581, 1991, pp. 91–100.

[46] Shimizu, M., T. Kanamori, J. Temmyo, M. Wada, M. Yamada, Y. Terunuma, Y. Ohishi, and S. Sudo, "28.3 dB gain 1.3 μm-band Pr-doped fluoride fiber amplifier module pumped by 1.017 μm InGaAs-LD's," *IEEE Ohoton. Technol. Lett.*, Vol. 5, 1993, pp. 654–657.

[47] Yamada, M., M. Shimizu, T. Takeshita, M. Okayasu, M. Horiguchi, S. Uehara, and E. Sugita, "Er^{3+}-doped fiber amplifier pumped by 0.98 μm laser diodes," *IEEE Photon. Technol. Lett.*, Vol. 1, 1989, pp. 422–424.

CHAPTER 4

Fiber Materials and Fabrications

K. Fujiura
T. Kanamori
S. Sudo

As in Mathematicks, so in Natural Philosophy, the Investigation of difficult Things by the Method of Analysis, ought ever to precede the Method of Composition. This Analysis consists in making Experiments and Observations, and in drawing general Conclusions from them by Induction, and admitting of no Objections against the Conclusion, but such as are taken from Experiments, or other certain Truths [1].

The keys to achieving low-loss fibers for long-distance transmission are the realization of low absorption, low scattering, and weak waveguiding in fibers. The first two requirements are common to all low-loss optical fibers, but the last applies uniquely to fibers used for long-distance lightwave transmission. This concept, called "weakly waveguiding," was introduced by D. Gloge in 1970 [2]. Many fabrication methods, various materials, and several fiber structures have been investigated with a view to meeting these requirements. Many investigations have made it clear that high-silica glass fiber with a SiO_2-GeO_2-based glass core and pure SiO_2 cladding are the best material and structure, respectively, for the fiber. In terms of fiber fabrication methods, those based on the "soot process" have recently been recognized as superior to any other. The others are the MCVD method, the VAD method, and the OVD method.

In contrast, however, the requirements for the active fibers used in fiber amplifiers and lasers are not the same as those for fibers used for long-distance transmission. Active fibers need a greater variety of materials and structures in order to realize a superior characteristics of amplification and lasing at a variety of operating wavelengths.

Table 4.1 summarizes the major fiber and fiber preform fabrication methods, fiber materials, and requirements for transmission fibers and active fibers. There

Table 4.1
Major Fiber and Fiber Preform Fabrication Methods, Fiber Materials,
and Requirements for Transmission Fibers and Active Fibers

Fabrication Process	Materials	Items	Requirements	
			Transmission	Active
Direct deposition	High-silica	Loss	●	Δ
Flame Laser Plasma	Multicomponent	Δn	○	●
Soot	Fluoride	Fiber length	●	Δ
MCVD VAD OVD	Chalcogenide	Reliability	●	○
PCVD Crucible	Crystal	Rare-earth solibility	—	●
Laser pedestal				

Importance: ● Major, ○ Average, Δ Minor

are various fiber fabrication processes such as the direct deposition process that are categorized into three types in terms of heat sources; the soot process, which includes the MCVD process, the VAD process, the OVD process, and the PCVD process; a fabrication process using a crucible; and a *laser-heated pedestal growth* (LHPG) method for crystal fibers. There are five major fiber materials: high-silica glass, multicomponent oxide glass, fluoride glass, chalcogenide glass, and crystals. The fibers that are currently used or being investigated all derive from a combination of these fabrication methods and materials. Furthermore, fibers used either for transmission or as amplifiers should satisfy requirements with regard to the following: optical loss, Δn, fiber length, and reliability and, in addition, rare-earth ion solubility for active fibers.

This chapter describes various host materials, glass and fiber characteristics, and a number of fiber fabrication methods including details of those for transmission fiber and rare-earth-doped fiber.

4.1 FIBER MATERIALS AND COMPOSITIONS

The essential amplification properties of active-ion-doped optical fiber depend strongly on the choice of active ions and fiber materials. There are many factors that influence the fabrication and characteristics of practical optical fibers for amplification. The most importance of these are as follows:

1. The homogeneous doping of active ions in host materials;
2. The phonon energy properties of materials;
3. The refractive index properties and the controllability of the refractive-index profile in fibers;
4. Low intrinsic loss and the potential for extrinsic loss reduction at operating wavelengths;
5. Precise control of the shape and size of the cross section and along the axial direction of fibers;
6. The high chemical and mechanical durability of materials.

The first five factors strongly affect the amplification characteristics of fibers. The clustering of active ions shortens the lifetime of radiative transitions. The phonon energy of host materials decides the radiative quantum efficiency. The transmission loss and refractive-index difference between the core and cladding (Δn) of fibers, which are related to factors 3 and 4, strongly influence the gain coefficient. Factor 6 is necessary for practical fiber fabrication and application. The poor chemical and mechanical durability of materials could be overcome by employing such engineering methods as the use of a closed metal vessel. However, in order to increase the productivity and reliability of fibers and lower the fabrication cost, it is very important that the fiber material itself has high chemical and mechanical durability.

Table 4.2 shows representative inorganic fiber materials with typical fiber compositions, the transparent wavelength regions where the transmission loss is less than 1,000 dB/km, and the minimum transmission loss for the fibers [3–19]. Oxide glasses, fluoride glasses, chalcogenide glasses, and alkali-halide crystals are the representative materials for low-loss optical fiber. Since fiber transmission loss has been reduced to below 1,000 dB/km, these materials have sufficient potential for short-distance applications.

These materials are classified into three groups: single crystalline, polycrystalline, and glasses.

Single crystalline materials have an essentially low intrinsic loss because Rayleigh scattering loss due to density fluctuation is very low. The active-ion doping, refractive-index control, and shape control of single crystalline materials are very difficult since doping the host with active ions or refractive-index modifiers sometimes prevents single crystal growth and as-grown single crystals have rough specific crystal surfaces. In addition, because of the low speed of crystal growth, single crystal fiber is difficult to mass produce.

By comparison, the speed of fiber fabrication with polycrystalline materials is high and there is little roughness on the surfaces and core/cladding interfaces of the fibers. However, polycrystalline materials exhibit a large loss as a result of light scattering at the grain boundary.

Glasses are made by cooling certain molten materials in such a manner that they do not crystallize but remain in an amorphous state, and their viscosity increases

Table 4.2
Fiber Materials

Materials	Systems	Typical Components	Transmitting Wavelengths* (μm)	Transmission Loss (dB/km)	
Oxide glasses[a]	Silica-based glasses	$SiO_2 + GeO_2$, F	0.25–2.4	0.15	(1.55 μm)
	Multicomponent glasses	$SiO_2 + Na_2 O + B_2 O_3 + \cdots$	0.42–1.9	3.4	(0.84 μm)
Fluoride glasses[b]	ZrF_4-based glasses	$ZrF_4 + BaF_2 + LaF_3 + \cdots$	0.24–4.4	0.7	(2.6 μm)
	AlF_3-based glasses	$AlF_3 + CaF_2 + BaF_2 + \cdots$	–4.0	6	(2.6 μm)
	InF_3-based glasses	$InF_3 + GaF_2 + ZnF_2 + \cdots$	–	73	(2.6 μm)
Chalcogenide glasses[c]	Sulfide glasses	Ge + S	0.81–5.1	148	(1.7 μm)
		As + S	0.92–6.6	24	(2.3 μm)
	Selenide glasses	As + Ge + Se	1.3 –9.5	182	(2.1 μm)
Halide crystals[d]	Single crystal	CsBr	–	400	(10.6 μm)
	Poly crystal	AgBr	–	70	(10.6 μm)
		TlBrl(KRS-5)	–	90	(10.6 μm)

*Loss less than 1000 dB/km. a, [3–8]; b, [9–13]; c, [14–16]; d, [17–19].

to such high values that they become solid. Glasses with high thermal stability permit the comparatively easy fabrication of fibers without the appearance of extrinsic scattering sources such as specific crystal surface roughness or grain boundaries. Rayleigh scattering loss, originating from the density and constituent fluctuations of noncrystalline materials, is larger than that of crystalline materials. However, the Rayleigh scattering loss is much smaller than the loss caused by grain boundary scattering in polycrystalline materials. Moreover, the refractive indexes of these glasses are easily controlled by changing the constituent ratio, which does not severely limit the glass formation conditions.

In the following subsections, we will describe the fundamental characteristics of representative fiber materials from the viewpoint of optical fiber materials for amplification.

4.1.1 Oxide Glass [20]

Some semimetal and semiconductor oxides have the ability to form glasses by themselves. They are called glass network formers. SiO_2, GeO_2, B_2O_3, P_2O_5, As_2O_3, and Sb_2O_3 are known glass network formers. Network formers can form glass not only by themselves but also when mixed with other network formers at any mixing ratio. The requirements to be satisfied for forming stable glass were described by Zachariasen [21]; that is, the conditions necessary for forming a glass with energy comparable to that needed for the crystalline form: (a) an oxygen atom must be not linked to more than two atoms; (b) the number of surrounding oxygen atoms must be small; (c) the oxygen polyhedrons must share corners, not edges or faces; and (d) at least three corners in each oxygen polyhedron must be shared. Some oxides, called network modifiers, cannot form a glass by themselves but can form glass when mixed with a given amount of network formers. Oxides of alkali metal and alkali-earth metal such as CaO, SrO, BaO, Li_2O, Na_2O, K_2O, Rb_2O, and Cs_2O are typical network modifiers. Metal oxides, such as TiO_2, SnO_2, Al_2O_3, ZrO_2, BeO, MgO, ZnO, and PbO, have characteristics that place them between network formers and network modifiers, and they are called intermediate oxides.

Physical properties of the oxide glass, such as the refractive index, thermal expansion coefficient, and glass transition temperature, can be changed easily by adding other network formers, network modifiers, or intermediate oxides.

Except for SiO_2, glass network formers are not necessarily very stable in the presence of moisture and some of them are toxic. The chemical durability of glass containing a large amount of alkali metal oxide is comparatively poor because the alkali ions in the glass dissolve in water. The kinds of oxide glass systems suitable for fabricating low-loss fibers are severely limited for these reasons.

Silica glass and multicomponent silicate glass are representative materials for low-loss optical fibers.

4.1.1.1 Silica Glass

Silica glass is one of the most suitable materials for low-loss optical fibers. SiO_2 can be made into glass easily without crystallizing, which means that it is easy to fabricate low-loss fibers without extrinsic scattering centers such as separated crystals. Silica glass has high transparency in the visible and near-IR wavelength region, and its refractive index can be changed easily by doping it with other oxides such as GeO_2, Al_2O_3, and TiO_2 or fluorine. The purification of these raw materials is easily accomplished. The chemical durability and mechanical strength of silica glass are extremely high. Furthermore, the raw materials are relatively cheap. On the other hand, silica glass has some disadvantages in terms of its use as optical fiber for amplification. It is difficult to dope a large amount of radiative ions such as rare-earth elements homogeneously. The phonon energy is high compared to other fiber materials. This greatly reduces the radiative quantum efficiency of doped ions under certain conditions. For example, amplification cannot be realized in the 1.3-μm wavelength band using Pr^{3+}-doped silica fibers for this reason.

4.1.1.2 Multicomponent Silicate Glass

Soda-lime-silicate is a typical glass composition for optical fibers because of its stable glassy state, relatively low dispersion, and low Rayleigh scattering loss. In this glass system, SiO_2, GeO_2, Na_2O, CaO, Li_2O, and MgO are used as components [8]. Compared with high silica glass, the chemical durability and mechanical strength of multicomponent silicate glass containing alkali metal oxides are poor. It is not easy to purify raw materials for alkali and alkali-earth metal oxides. However, multicomponent silicate glass has certain advantages. Though the phonon energy is almost the same as that of silica glass, it is easy to dope a large amount of rare-earth elements as radiative ions homogeneously. Moreover, many kinds of glass with a large variety of optical and thermal properties can be easily prepared at a comparatively low cost.

4.1.1.3 Nonsilicate Glass

GeO_2-based glass [22], Al_2O_3-CaO-based glass [23], and TeO_2-based glass [24] are fiber materials with relatively low phonon energy and fairly poor glass stability. GeO_2 glass has a high glass-forming tendency, but the chemical durability is very poor. The refractive index of GeO_2 glass can be controlled by doping it with Sb_2O_3. Al_2O_3-CaO-based oxide glasses, such as Al_2O_3-CaO-BaO-Y_2O_3, have a high refractive index of about 1.7; and TeO_2-based glasses, such as TeO_2-ZnO-BaO, have a very high refractive index of above 2. The chemical durability of both glasses is fairly high.

4.1.2 Halide Glass

Halide glasses are transparent in the IR region where oxide glasses have high transmission losses. Because of these low-phonon energy characteristics, halide glasses are considered to be promising host materials for active ions. There are many kinds of halide glass systems including fluoride glass, chloride glass, and mixed halide glass systems. Compared to oxide glass, these halide glasses are generally not stable with regard to crystallization and the effects of moisture. Except for some fluoride glass systems, halide glasses are too hygroscopic for practical use and some of them are toxic. Fluoride glass systems are divided into four groups: BeF_2-based glasses, AlF_3-based glasses, ZrF_4/HfF_4-based glasses, and InF_3/GaF_3-based glasses. BeF_2 easily forms glass on cooling from the molten state and essentially satisfies Zachariasen's glass-forming conditions as with SiO_2. The difference is that the ionic valence of BeF_2 is half that of SiO_2. Although BeF_2-based glasses have potentially attractive properties, they are unsuitable as practical fiber materials because of their highly toxic and hygroscopic nature. Compared with BeF_2-based glasses, AlF_3-based glasses, ZrF_4/HfF_4-based glasses, and InF_3/GaF_3-based glasses, which are called heavy-metal fluoride glasses, are much more stable against moisture, although they exhibit a fairly low glass-forming tendency. Although these fluoride systems can be fully vitrified when they are rapidly cooled from the molten state, they sometimes crystallize during a reheating process such as fiber drawing. Therefore, it is important to select a very stable glass system for fiber fabrication. Heavy-metal fluoride glasses have a refractive index of 1.4 to 1.6 and glass transition temperature of 250° to 450°C.

4.1.2.1 AlF₃-Based Glass

These glasses generally contain 30 to 50 mol% AlF_3 and 30 to 50 mol% alkali-earth metal fluoride as the primary constituents with another metal fluoride, such as ZrF_4, YF_3, LaF_3, and/or NaF, playing the role of the glass stabilizer. Typical glass compositions are $40AlF_3$-$22CaF_2$-$22BaF_2$-$16YF_3$ [25] and $30.2AlF_3$-$20.2CaF_2$-$10.6BaF_2$-$13.2SrF_2$-$3.5MgF_2$-$8.3YF_3$-$10.2ZrF_4$-$3.8NaF$ [26]. Constituents are given in mol%. Compared with other heavy-metal fluoride glasses, AlF_3-based glasses have a low refractive index, high glass-transition temperature, high chemical durability, and high phonon energy.

4.1.2.2 ZrF₄/HfF₄-Based Glass

These glasses are the most stable of the heavy-metal fluoride glasses. They consist of 40 to 60 mol% ZrF_4/HfF_4 and 20 to 40 mol% BaF_2 as the primary constituents together with another metal fluoride, such as AlF_3, YF_3, LaF_3, GdF_3, PbF_2, LiF, and/or NaF. Typical glass compositions are $53ZrF_4$-$20BaF_2$-$4LaF_3$-$3AlF_3$-$20NaF$ [27] and

47.5ZrF$_4$-23.5BaF$_2$-2.5LaF$_3$-2YF$_3$-4.5AlF$_3$-20NaF [10], which are known as ZBLAN and ZBLYAN, respectively. Acronyms are used based on the first letter of each component. ZrF$_4$/HfF$_4$-based glasses have a moderate refractive index, low glass-transition temperature, fairly low chemical durability, and moderate phonon energy.

4.1.2.3 InF$_3$/GaF$_3$-Based Glass

These glasses generally contain 30 to 50 mol% InF$_3$/GaF$_3$, 10 to 30 mol% ZnF$_2$, and 30 to 50 mol% alkali-earth metal fluoride/PbF$_2$ as their primary constituents along with another metal fluoride, such as YF$_3$, LaF$_3$, GdF$_3$, YbF$_3$, AlF$_3$, LiF, and/or NaF. Typical glass compositions are 34InF$_3$-6GaF$_3$-20ZnF$_2$-15BaF$_2$-20SrF$_2$-2GdF$_3$-1YF$_3$ [13] and 17InF$_3$-17GaF$_3$-19ZnF$_2$-43PbF$_2$-4LaF$_3$ [28]. InF$_3$/GaF$_3$-based glasses have the lowest phonon energy of the heavy-metal fluoride glasses. Glasses containing alkali-earth metal fluoride have a moderate refractive index, moderate glass transition temperature, and moderate chemical durability. Glasses containing PbF$_2$ have very high refractive indexes. As described previously, heavy-metal fluoride glasses contain a few mol% of rare-earth fluorides as the glass stabilizer, which means that it is very easy to dope a large amount of rare-earth elements as radiative ions homogeneously. This offers great advantages for fabricating fibers for optical amplifiers.

4.1.2.4 Chloride and Mixed Halide Glass

Typical chloride and mixed halide glass systems are ZnCl$_2$-based glasses and CdCl$_2$-based glasses. They have a low glass-forming tendency and a low glass-transition temperature below 200°C in addition to poor chemical durability. 33CdCl$_2$-17CdF$_2$-13BaF$_2$-34NaF-3KF [29] glasses have been reported as fiber materials for optical amplification.

4.1.3 Chalcogenide Glass

Chalcogenide glasses are solid solutions of metal sulfides, selenides, and tellurides such as GeS$_3$, As$_2$S$_3$, Sb$_2$S$_3$, As$_2$Se$_3$, Sb$_2$Se$_3$, and GeTe. They have a considerable glass-forming ability, wide transmission windows with a long IR cutoff wavelength, and very high refractive indexes of above 2, and they can be easily drawn into fibers at low temperature [14,30,31]. These properties offer great benefits as fiber materials for amplification. Normal chalcogenide glasses have some disadvantages. For example, it is difficult to dope radiative ions such as rare-earth elements homogeneously. Photostructural changes, which appear as a shift in the band gap absorption edge and refractive index changes, are easily induced by light with the band gap energy, indicating that chalcogenide glasses with low band gap energy, like selenide and telluride glasses, are not suitable for application to amplifiers operating in the optical communication band in the near-IR wavelength region.

Ga-La-S glasses have potentially attractive properties, although they have a low glass-forming tendency compared with normal chalcogenide glasses [32]. Ga-La-S glasses have high band gap energy, and rare-earth ions can be doped by substituting them for La.

4.1.4 Crystals

Metal halides such as KRS-5 (TlBrI) [19], AgCl [33], AgBr [18], KCl [34], CsBr [17], and CsI [35] are used as crystal fiber materials. They have wide transmission windows with a long cutoff wavelength. Alkali metal halides have low refractive indexes and poor stability against moisture. KRS-5, AgCl, and AgBr with high refractive indexes are fairly stable against moisture. Oxides such as sapphire [36] and YAG [37] are other crystal materials that are used for fiber fabrication. They have high refractive indexes and high chemical durability. From the viewpoint of fiber materials for optical amplifiers, however, crystal materials have a large disadvantage in that single-mode fiber fabrication techniques have not yet been well developed.

4.2 TRANSMISSION LOSS OF FIBER MATERIALS

One of the most important criteria in the selection of fiber host materials is that they are capable of being formed into fibers with sufficiently low loss in the pumping and signal wavelength regions for optical signal amplification. With optical amplifier host fibers, host materials are selected in terms of their amplification characteristics, which can be estimated from the optical properties of bulk samples as described in Chapter 3. We then make every effort to fabricate a rare-earth-doped fiber with low attenuation. Therefore, it is very useful to understand the physical properties of materials, as this allows us to estimate the difficulty of fabricating a host fiber with a required loss, prior to fiber fabrication. In addition, an analysis of the attenuation mechanism is indispensable if we are to improve the fiber fabrication process. The physical properties of host materials and their attenuation mechanisms have been fully investigated during the development of low-loss telecommunication media, which are described in various textbooks [20,38]. In general, optical attenuation consists of optical absorption and scattering, which are classified into intrinsic and extrinsic factors as shown in Figure 4.1. Figure 4.2 is a diagram showing the optical attenuation mechanisms in fibers. Of these mechanisms, electron transition absorption, weak absorption tail, multiphonon absorption, Rayleigh scattering, Brillouin scattering, Raman scattering, and scattering caused by homogeneous crystallization are classified as intrinsic loss factors. Extrinsic absorption losses originate

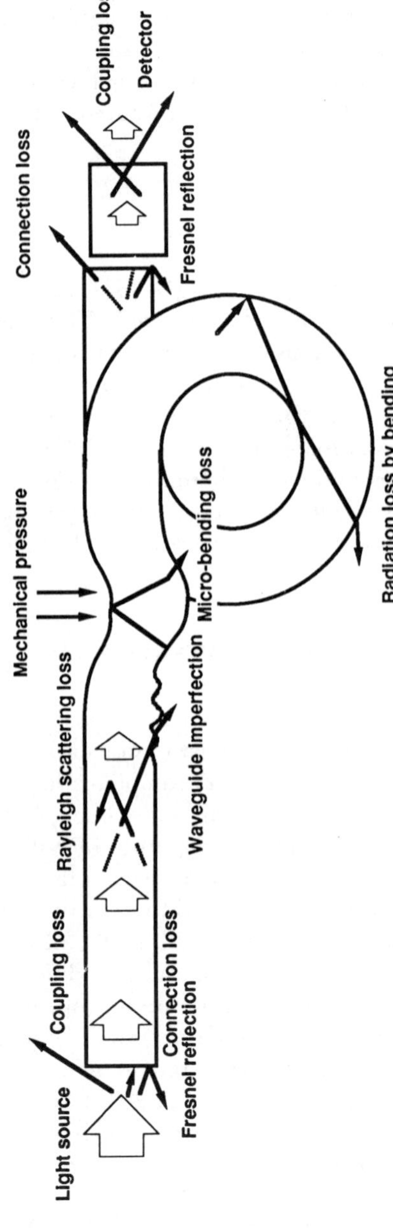

Figure 4.1 Schematic representation of the attenuation factors in optical fiber.

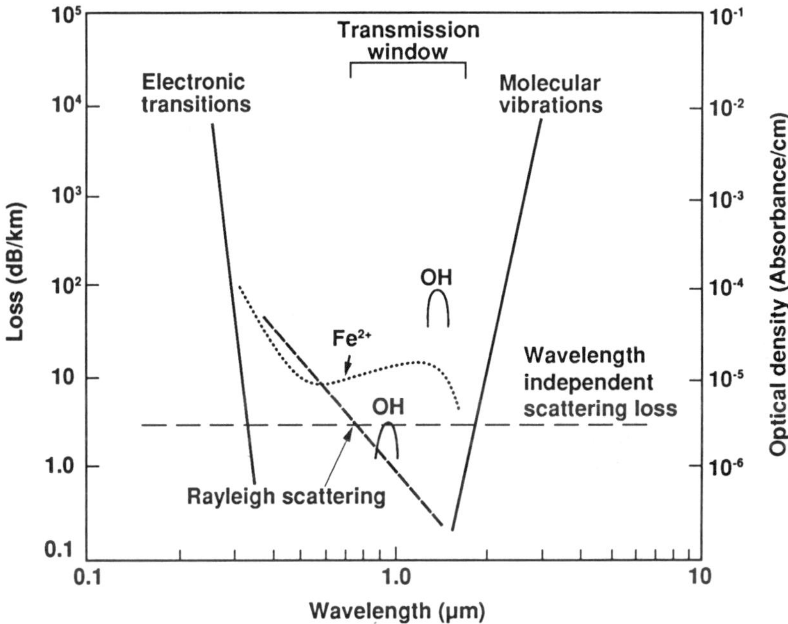

Figure 4.2 Schematic diagram of the attenuation mechanisms for glasses in the transparent regime.

from impurities, such as transition metal ions, rare-earth ions except doped ions, and OH ions. Extrinsic scattering losses are caused by such imperfections as bubbles, striae, microcrystals caused by heterogeneous nucleation, and core-cladding fluctuations, which mostly depend on fiber fabrication conditions. These major loss factors, guidelines for the selection of host glass materials for optical fiber amplifiers, and the keys to reducing the extrinsic losses will be discussed in the following subsections.

In this section, it is emphasized that scattering loss caused by homogeneous crystallization is classified as an intrinsic loss factor. Fortunately, with silica fibers, the scattering loss caused by crystallization is negligible [39] because silica glass is exceptionally stable against crystallization. The tendency of glass to crystallize, however, cannot be ignored and becomes important in terms of the host materials for optical fiber amplifiers for two reasons. One is the low solubility of rare-earth ions in some host materials. The other is that low phonon energy glass materials such as heavy-metal fluoride glasses and chalcogenide glasses make optical amplification possible at a number of wavelengths that was unavailable when using rare-earth-doped silica fibers, and in general, low phonon energy means high crystallization tendency. The effect of small particles in the core on fiber loss is described in this section, and crystallization theory and the properties of glasses are described in detail in the next section.

4.2.1 Intrinsic Loss Factors

4.2.1.1 Electron Transition Absorption

Optical absorption due to electronic transitions is observed in the UV region. On the basis of the band structure for crystalline solids, the absorption process consists of promoting electrons from the valence band to the conduction band, and there is a series of broad absorption bands originating from a quasi-continuum of electronic transition above the band gap ω_g. For insulating materials, the band gap ω_g, which is the lowest electronic transition, corresponds to frequencies somewhere in the UV to visible portions of the spectrum.

In real solids, absorption edges have been observed for photon energies smaller than the band gap ω_g. This weak tail is often referred to as the "Urbach tail" [40] and is caused by anharmonic interactions, for example, electron-phonon interactions. This enables absorption below the band gap to occur by virtual excitation of electrons across the gap, with the subsequent annihilation of one or more phonons in order to conserve energy. The Urbach tail is expected to fall off exponentially as a function of frequency below the band gap [41]. In general, the absorption coefficient A_u varies with temperature T and photon energy $h\omega$

$$\alpha = A_u \exp\, g(h\omega - h\omega_0) \tag{4.1}$$

where h is Planck's constant divided by 2π; k_B is Boltzmann's constant; and A, ω_0, and g are fitting parameters (with $g = \sigma/k_B T$ at high temperature, σ being nearly unity and weakly dependent on temperature). This empirical characterization was first enunciated by Urbach to describe his observations in alkali-halide crystals. Figure 4.3 shows the temperature dependence of absorption of an AgBr crystal in which the straight lines are the $1/kT$ slopes at the corresponding temperature [40]. Not only alkali halides, but also II-IV compounds, III-V semiconductors, and amorphous materials sometimes exhibit exponential tails. Table 4.3 shows the parameters of Urbach's rule for various materials [42–46]. The slope factor is very small. Therefore, the absorption loss tail decreases abruptly at longer wavelengths and does not seriously affect the total loss of materials.

For host glass materials, the electronic absorption tail contributes very little except with some chalcogenide glasses because the tail is relatively small compared with the Rayleigh scattering loss at wavelengths longer than 0.6 μm. As described in the following sections, the slope factor of the Urbach's tail is generally in the 0.03- to 0.07-eV range and the weak absorption tail factor is 0.16 to 1.6 eV for various materials. Exceptionally, some chalcogenide glasses have semiconductive properties and the electronic absorption tails longer than 1.0 μm are observed. Figure 4.4 shows the attenuation spectrum for $As_{38}Ge_5Se_{57}$ bulk glass and unclad fiber in the UV region [14]. As the wavelength increases, the rate of loss decrease slows at the wavelengths longer than 0.65 μm for $As_{40}S_{60}$ and 0.9 μm for $As_{38}Ge_5Se_{57}$.

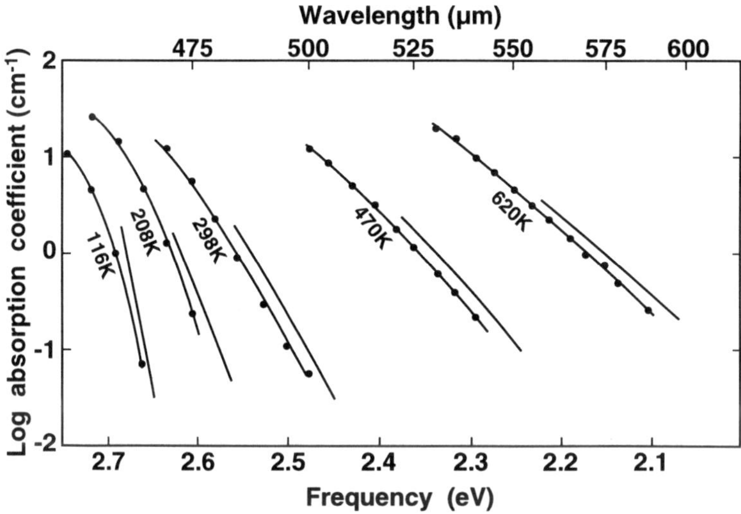

Figure 4.3 Temperature dependence of absorption of an AgBr crystal in which the straight lines are the $1/kT$ slopes at the corresponding temperatures [40].

Table 4.3
Urbach Tail Coefficients

Materials	$K(cm^{-1})$	g	w_0
KCl	8,696	4.2	213
KBr	6,077	4.2	166
NaCl	24,273	4.8	268
BaF_2	49,641	4.5	344
CaF_2	105,680	5.1	482
MgF_2	11,213	4.4	617
LiF	21,317	4.4	673
Al_2O_3	55,222	5.0	900
NaF	41,000	5.0	425

From: [42–46].

This phenomenon is attributed to a weak absorption tail, which will be described in the next section.

The prior discussion focused on the frequency dependence, which is analogous to that of multiphonon processes; however, the temperature dependence is not. This fact reveals that other mechanisms in addition to anharmonicity contribute to the Urbach tail. Dow reviewed this subject [40].

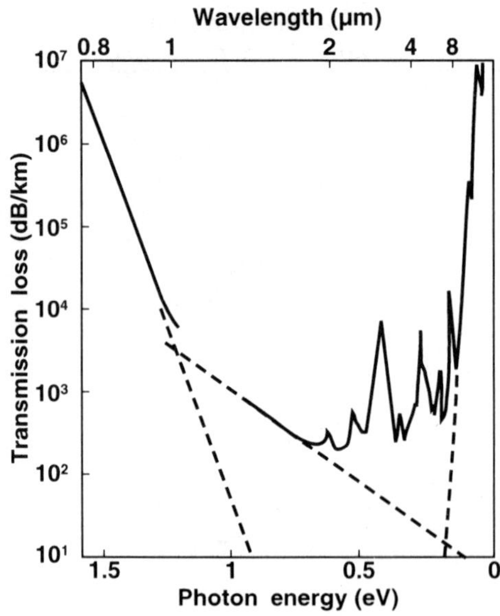

Figure 4.4 Transmission loss versus photon energy for $As_{38}Ge_5Se_{57}$ bulk glass and unclad fiber [14].

4.2.1.2 Weak Absorption Tail

For glasses, the absorption edge in the UV region has the shape shown in Figure 4.5 [38]. In this figure, part A corresponds to the electronic transition from the valence band to the conduction band that is broadened in part B by the processes described in the previous section. At low absorption levels, a tail C is often observed that is much flatter than the Urbach tail. This tail is very troublesome since it extends from the absorption edge and limits the maximum transparency of the glass in the window between the electronic transitions and vibrational modes.

The absorption tail C is called the weak absorption tail [47] and can be described as

$$\alpha = \alpha_0 \exp(\hbar\omega/E_u) \tag{4.2}$$

The weak absorption tail was first observed in amorphous semiconductors such as As_2S_3 and $Ge_{33}As_{11}Se_{55}$ [47]. These tails have also been observed in alkali-halide crystals BaF_2 and SiO_2 [48]. In most cases, the absorption constants follow Urbach's rule for $\alpha > 0.1$ cm^{-1}, and E_u is in the 0.03- to 0.07-eV range. For $\alpha < 0.1$ cm^{-1}, one observes another exponential part of absorption where the slope factor E_u is 0.16 to 1.6 eV for various materials, as shown in Table 4.4. This tail could influence the total loss of materials in the near-IR region where the scattering loss and phonon absorption loss are very small.

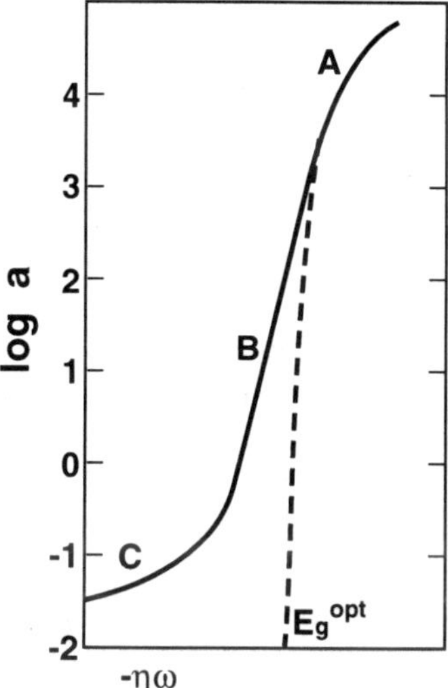

Figure 4.5 Absorption edge of amorphous solids. Part A corresponds to the electronic transition from the valence band, to the conduction band [38]. Part B is the Urbach edge. Part C is the absorption tail.

Table 4.4

Multiphonon Absorption Coefficients

Materials	$A\ (cm^{-1})$	$E_u\ (E_v)$	$h\omega\ (eV)$
KCl	1.26×10^{10}	0.033	7.76
KBr	2.4×10^{6}	0.033	–
SnO_2	1.5×10^{5}	0.037	3.75
KI	5.9×10^{9}	0.032	5.89
NaCl	1.2×10^{10}	0.034	8.03
$As_{40}S_{60}$	–	0.054	–
$As_{38}Ge_5Se_{60}$	–	0.050	–
$Ge_{20}S_{80}$	–	0.073	–

From: [20].

The origin of the weak absorption tail has not yet been fully clarified. It is thought that there are mechanisms in addition to electron-phonon interactions that influence the electronic edge of glass. For example, disordered solids possess a distribution of atomic configurations, each with a different set of electronic energy levels. Thus, the effective gap will be determined by the most probable atomic configuration, but absorption can occur below energy transitions. This broadening is due entirely to structural disorder, regardless of electron-phonon interactions

[41]. According to experimental results, the absorption coefficient of alkali-halide crystals can be changed by thermal treatment [20]. For example, the absorption coefficient increased for a KCl crystal heated to about 600K and then rapidly quenched. The quenched sample was again heated to 600K and then annealed. The coefficient returned to its initial value. These quenching and annealing effects have also been clearly observed in CsI crystals. In addition, it has been noted that the absorption coefficient is affected by treatment in chloride gas. Figure 4.6 shows the change in the absorption edge of an amorphous Ge film produced by annealing. One film had a rather broad edge after deposition, but it sharpened significantly after annealing [49,50]. Experiences with other amorphous materials are similar. With a-Se, purer samples give sharper edges until the slope corresponding to the Urbach tail is reached; then the absorption tail part C is observed [51]. Similar results were obtained for the chalcogenide glasses shown in Figure 4.7 [52]. These experimental results support the idea that the origin of the weak absorption tail cannot be attributed to impurities but to defects and dislocations. A weak absorption tail can be observed when the scattering loss, which has λ^{-4}-dependence, is comparatively low and the UV-absorption peak is at a longer wavelength. Chalcogenide glasses are typical materials satisfying this condition. Figure 4.8 shows the loss spectrum of a chalcogenide glass [53]. In this figure, weak absorption tails are clearly observed. Weak absorption tails can be observed in various materials via laser calorimetric measurements, as shown in Figure 4.9.

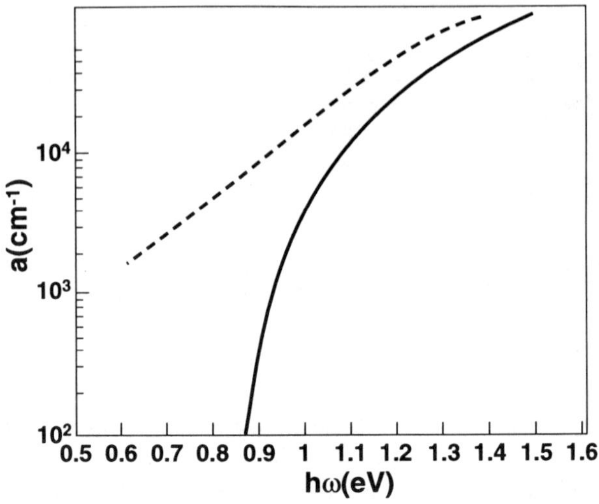

Figure 4.6 The change in the absorption edge of an amorphous Ge film produced by annealing (*After*: [49]). Dashed line: as deposited; solid line: after annealing.

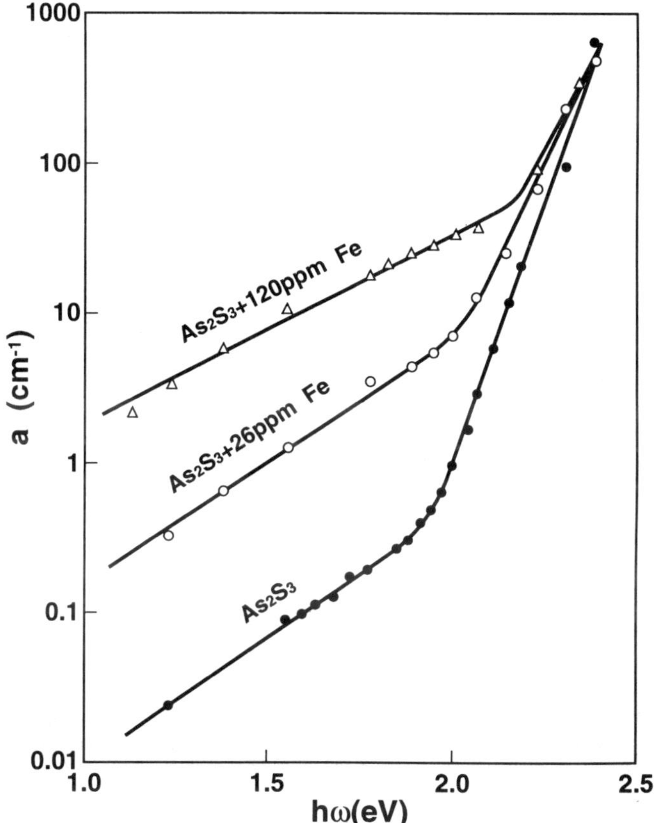

Figure 4.7 Absorption edge of amorphous As_2S_3 nominally pure and doped with Fe (*After:* [52]).

4.2.1.3 Multiphonon Absorption

The absorption spectra in the IR region are usually determined by phonon absorption, which is caused by the bending and stretching vibration of constituent molecules. It is recognized that the multiphonon process provides the intrinsic residual absorption in the shorter wavelength region rather than at the phonon absorption wavelength.

The phonon absorption frequency will be described with a simple model. With crystals that have two atoms per primitive cell, there are three acoustic branches and three optical branches to the phonon dispersion relation. Interaction between a photon and a phonon takes place at the crosspoint of the optical phonon branch, and the phonon dispersion relation is given by $\omega = ck$. The momentum of a photon

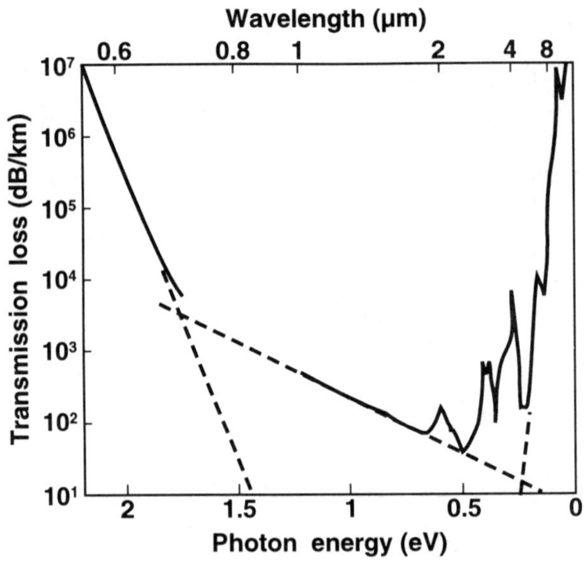

Figure 4.8 Transmission loss versus photon energy for $As_{40}S_{60}$ bulk glass and unclad fiber [53]. This spectrum shows a typical Urbach tail and a weak absorption tail.

is negligible compared to that of a phonon, and the crosspoint is at $k = 0$. The fundamental vibrational frequency ω_0 at the crosspoint is given by the following equation for the transverse mode:

$$\omega_0{}^2 = f\left(\frac{1}{M_1} - \frac{1}{M_2}\right) \tag{4.3}$$

where M_1, M_2 are the mass of atoms and f is the force constant. The light frequency ω_0 depends on the mass of two atoms and is absorbed resonantly by the crystal. At this frequency, the absorption coefficient is very large; the frequency ω_0 will be small when the masses of the two atoms, M_1 and M_2, are large and the binding force f between the two atoms is small. We can observe weak absorption peaks at shorter wavelengths that cannot be attributed to phonon absorption. It has been reported that the multiphonon process provides this residual absorption [38,54–57] in the shorter wavelength region. Multiphonon absorption results from the combined effect of two interrelated mechanisms: the electric moment associated with the distorted charge density of real materials to which light couples directly and the anharmonic interionic potential that stems from the interactions between electronic charge densities. These mechanisms originate from the nonlinearity of the effective dipole moment M and the ionic potential energy V for lattice displacements. In detail, the anharmonicity absorption is caused by multiphonon absorption proceed-

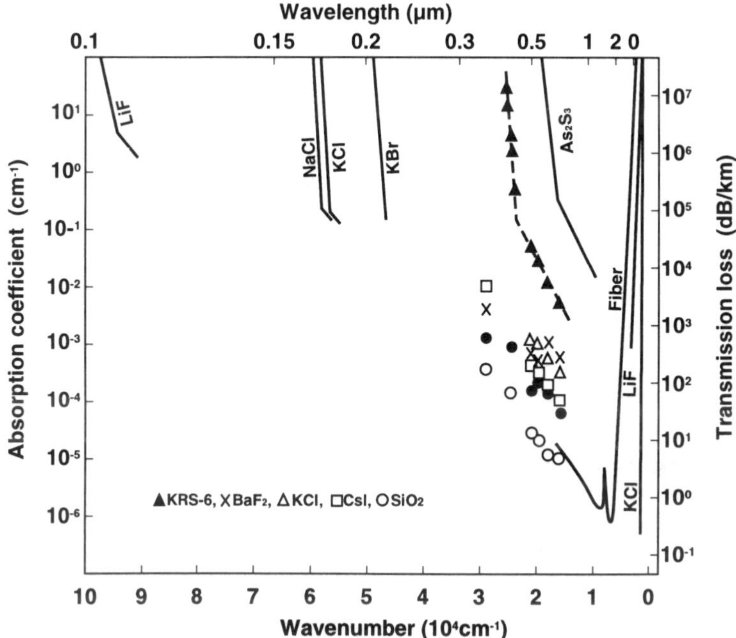

Figure 4.9 Absorption loss coefficients for low-loss materials in the visible region [48]. Solid lines show reported absorption coefficients in the UV and IR regions. KRS-6, BaF$_2$, KCl, CsI, SiO$_2$.

ing the virtual excitation of an IR-active phonon and its subsequent decay via anharmonic interactions. Figure 4.10 shows multiphonon absorption processes [58]. According to theoretical analysis by Bendow [59], the detailed spectral structure is formulated by taking account of the n-phonon density of state ρ_n as follows:

$$\alpha(\omega) = \frac{1}{N(\omega) + 1} \sum_n^\infty f_n \rho_n(\omega) \tag{4.4}$$

where $N(\omega)$ is the Bose–Einstein function and f_n possesses an exponential-like dependence on n, for example, $f_n \propto c^n$.

In real materials, multiphonon absorption is related to the electric moment of the material; in other words, it is related to the ionicity of the atomic bonding. With ionic crystals, a clear relation can be observed that agrees with (4.5). When the ionicity of a crystal is weak, nonlinearity of the optical interaction becomes strong and absorption spectra with a fine structure can be observed. Figure 4.11 shows the multiphonon absorption tails for various materials, and Table 4.5 [60] shows the values of K, γ, and ω_0. The ionic crystals shown in Figure 4.11 demonstrate a nearly exponential absorption as a function of frequencies $\omega/\omega_0 \geq 3$, where ω_0

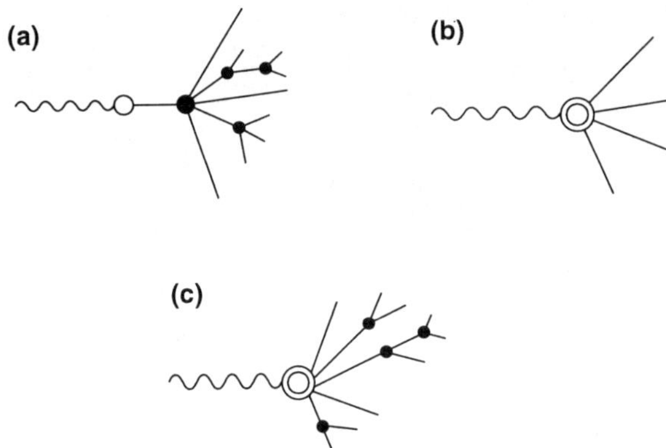

Figure 4.10 Multiphonon absorption processes: (a) phonon (wavy line) excites the fundamental mode, which subsequently decays anharmonically into other phonons; (b) photon excites many phonons directly via the nonlinear moment mechanism; (c) the process in (b), but now allowing for anharmonic decay of the primary phonons [58].

is the average optical phonon frequency. This dependency arises from the broad densities of phonon states, and large anharmonicities of the absorption coefficient are given by

$$A_{MP} = K \exp(-\gamma\omega/\omega_0) \tag{4.5}$$

where K, γ, and ω_0 are parameters characteristic of the material in question. The value of γ for various materials lies between 4 and 5, and the value of ω_0 is related [55] to the longitudinal optical phonon absorption wavenumber.

Spectra with a clear structure are observed for most semiconductors, although their overall dependence remains exponential-like. Figure 4.12 shows the IR absorption edge versus dimensionless frequency for selected semiconductors. The structure seen in Figure 4.12 is attributed to the structure in the phonon density of states. This structure persists but broadens somewhat at a higher temperature, at least for the more ionic semiconductors [41].

The transmission loss at wavelengths longer than those of the fundamental phonon absorption peaks can be represented by the equation [61,62]

$$n\alpha = K(\hbar\omega)\beta \tag{4.6}$$

where α is the absorption coefficient, $\omega/2\pi$ is the frequency of the far-IR radiation, and n is the refractive index. For most of the materials investigated, $\beta < 2$. The origin of this tail is the disorder-induced coupling of the radiation to vibrational modes. The values β [62] for SiO_2, GeO_2, B_2O_3, As_2Se_3, and As_2S_3 are all 2.0, and

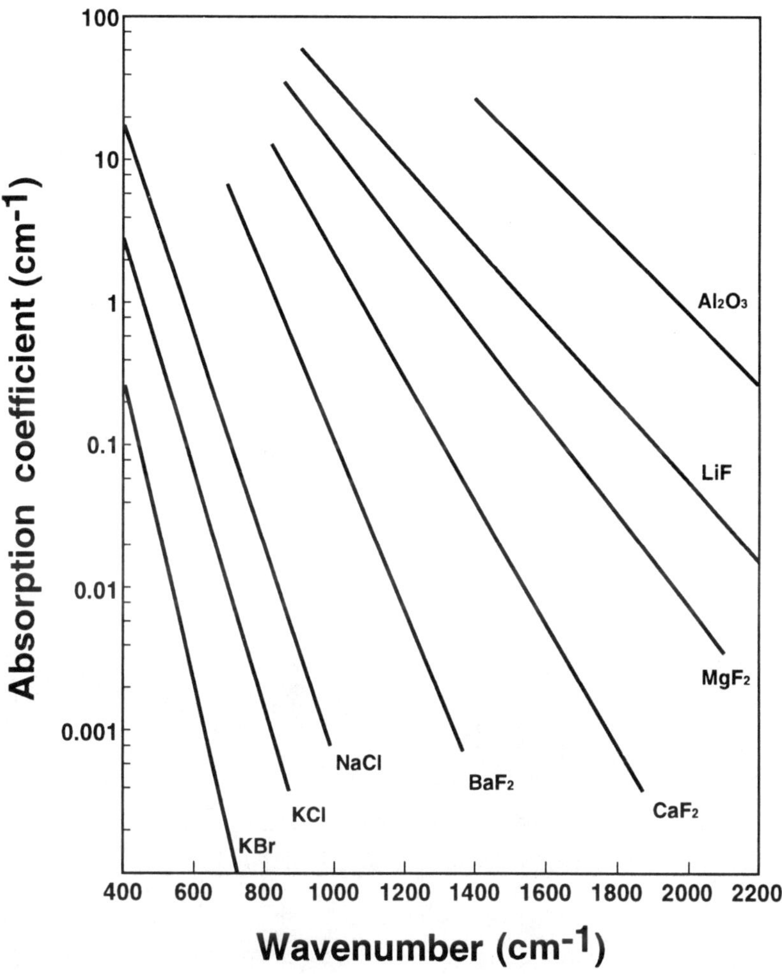

Figure 4.11 Absorption coefficient for crystalline materials resulting from multiphonon absorption [60].

the absorption loss decreases gradually with increasing wavelength. Therefore, low absorption loss cannot be expected in the far-IR region. Absorption loss spectra for various materials are shown in Figure 4.13 [62].

4.2.1.4 Scattering Loss

Raleigh Scattering, Brillouin Scattering, and Raman Scattering [20,63–65]

Scattering loss is one of the major factors of loss in the visible and near-IR region. It is well known that there are three kinds of scattering: Rayleigh scattering, Brillouin

Table 4.5

Results of Least Squares Fit of Absorption Coefficient Data to an Exponential and to the Power Law Dependence on Wavenumber

| | | Raw Data | $\beta = A\ exp\ [-\gamma(\omega/\omega_0)]$ | | Power | | | β scale. (Exponential Law) | |
| | ω | β | A | ω_0/γ | Law | $\omega_{1.0}$ | | $10.6\mu m$ | $5.3\ \mu m$ |
Material	Range (cm^{-1})	Range (cm^{-1})	(cm^{-1})	(cm^{-1})	n	(cm^{-1})	$\gamma\omega_{1.0}/\omega_0$	(cm^{-1})	
KBr	250–600	12–0.002	6,077	39.1	10.1	166	4.2	2×10^{-7}	8×10^{-18}
KCl	312–750	20–0.003	8,696	50.8	9.6	213	4.2	2×10^{-5}	7×10^{-13}
NaCl	356–945	44–0.001	24,273	56.0	10.5	268	4.8	1.1×10^{-3}	6×10^{-10}
BaF$_2$	1,250–800	1.3–0.004	49,641	75.9	13.3	344	4.5	0.19	0.8×10^{-6}
SrF$_2$	1,450–850	1.9–0.003	22,548	90.4	12.5	395	4.4	0.65	0.2×10^{-6}
CaF$_2$	1,600–1,000	2–0.004	105,680	93.6	13.6	482	5.1	4.36	1.8×10^{-4}
MgF$_2$	2,000–1,250	1.9–0.007	11,213	143.8	11.1	617	4.4		2.2×10^{-2}
LiF	2,500–1,450	1.5–0.002	21,317	153.2	12.4	673	4.4		0.93×10^{-1}
Al$_2$O$_3$	3,000–1,800	1.8–0.004	55,222	179.0	13.1	900	5.0		1.44

From: [60].

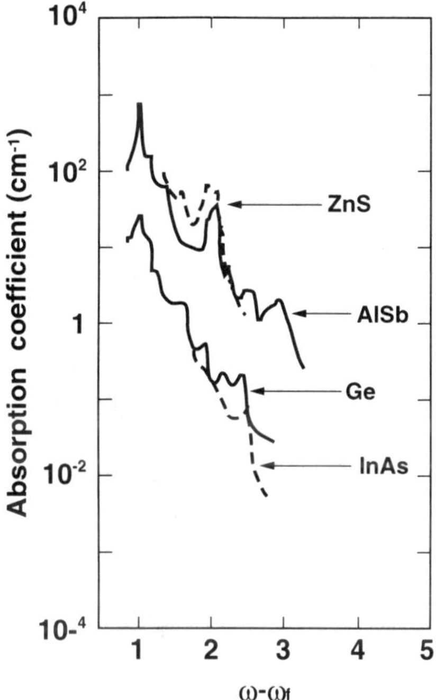

Figure 4.12 IR absorption edge versus dimensionless frequency for selected semiconductors [41].

scattering, and Raman scattering. These are caused by density and concentration fluctuation, acoustic phonon, and optical phonon interactions, respectively.

Light scattering in liquids and glasses is known to be due to microscopic variations in the local dielectric constant associated with the random molecular structure of these materials. If the spatial extent of the fluctuations $\Delta\epsilon(r)$ of permittivity is smaller than wavelength scale, then the scattered light intensity $I_s(\theta)$ at the angle θ to the incident direction of propagation takes on a particularly simple form that is independent of the shape of the scattering centers. In the Born approximation at a distance R from the scattering center this relationship is [66]

$$I_s(\theta) = \frac{I_0 L^3 \pi^2 (1 + \cos^2\theta)}{2\lambda^4 R^2} \left\langle \int \Delta\epsilon(r)\Delta\epsilon(0)\, d^3r \right\rangle \tag{4.7}$$

in which $R \gg L$, where L^3 is the volume of scattering material. The total scattered light power can be obtained by summing over all scattered angles; then the scattering loss coefficient is given by

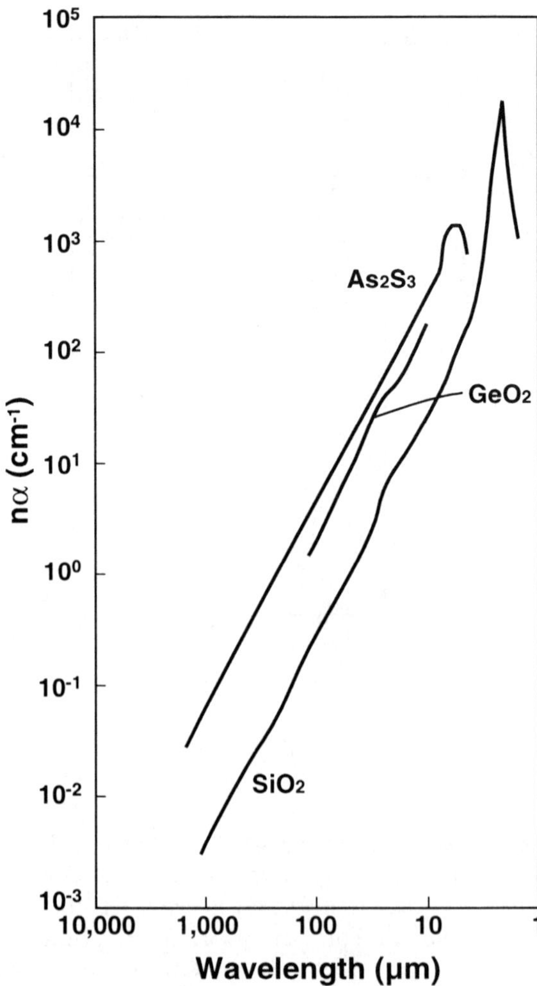

Figure 4.13 Product of refractive index and absorption coefficient for amorphous materials in the far-IR region [62].

$$\alpha = \frac{8\pi^3}{3\lambda^4}\langle(\Delta\epsilon)^2\rangle v_\epsilon \tag{4.8}$$

where v_ϵ is a correlation volume. The equation shows that the attenuation via scattering arises from any process that contributes to a fluctuation in the dielectric constant. The fluctuation of $\Delta\epsilon$ is a function of the conjugate thermodynamic macroscopic variables; temperature and entropy (T, S); pressure and volume (P, V); and (for multicomponent glasses with m constituents) chemical potential and concentration components (μ_i, c_i). In addition to these variables, microscopic rotational and vibrational mode momentum and displacement coordinates (P_i, Q_i) must be considered; then $\Delta\epsilon$ is written as

$$\langle(\Delta\epsilon)^2\rangle v_\epsilon = (\partial\epsilon/\partial P)^2\langle(\Delta P)^2\rangle v_p + (\partial\epsilon/\partial S)^2\langle(\Delta S)^2\rangle v_s$$

$$+ \sum_{i\neq m}(\epsilon'_i - \epsilon'_m)^2\langle(\Delta c_i)^2\rangle v_{ci} + \sum_i(\partial\epsilon/\partial Q_i)^2\langle(\Delta Q_i)^2\rangle v_{Qi} \qquad (4.9)$$

in which v_x $(x = P, S, c_i, Q_i)$ are correlation volumes

$$v_x = \frac{4\pi\int_0^\infty r^2\langle\Delta x(r)\Delta x(0)\rangle\,dr}{\langle\Delta x(0)\Delta x(0)\rangle} \qquad (4.10)$$

all presumed to be small on the scale of λ_3 [64]. In frequency space, the second and third (S and c_i) contribute quasi-statistically to the Rayleigh scattering peak. The first and fourth are Brillouin scattering and Raman scattering, respectively—both dynamic in origin. In addition to thermodynamic considerations concerning perturbations of Raleigh, Brillouin, concentration and Raman fluctuations, and dielectric dispersion, the following expressions are obtained where total loss $\alpha = \alpha_0$ at the vacuum wavelength $\lambda = \lambda_0$ of minimum refractive-index dispersion [64,65]:

$$(\alpha_0)_{\text{Density}} = 7.0 \times 10^{-6}E_0E_dE_1^2(1 - \Lambda)^2TK_T \quad \text{dB/km} \qquad (4.11)$$

$$(\alpha_0)_{\text{Conc}} = 0.0084E_0E_dE_1^2VF \quad \text{dB/km} \qquad (4.12)$$

$$(\alpha_0)_{\text{Raman}} = \frac{1.40E_0E_dE_1^2Vd^2\Lambda^2}{z_Rl^2m\omega}\coth(0.72\omega/T) \quad \text{dB/km} \qquad (4.13)$$

The various quantities involved are defined as follows. First, E_0, E_d, and E_1 are the three Sellmeier energies introduced by Wemple et al. [67,68] in terms of which the minimum refractive-index dispersion vacuum wavelength is [69]

$$\lambda_0 = 1.63(E_d/E_0^3E_1^2)^{1/4} \quad \mu\text{m} \qquad (4.14)$$

Physically, E_0 is the average electronic energy gap, while E_dE_0 and E_1^2 are proportional to the electronic and lattice oscillator strengths, respectively. It is presumed that one particular local motion (with a number z_R of such independent modes per molecular unit) involving a highly polarizable bond type of bond length l dominates both the bond-dependent density and Raman scattering mechanism [65]. Λ is a dimensionless measure of the density dependence of the average bond polarizability. For example, [64,65] concluded that lattice stuffing without production of nonbridging oxygen in silicate glasses causes a $(1 - \Lambda)$ parameter decrease. K_T is the static isothermal compressibility, and V is the molecular volume. Parameter m is the effective mass of the dominant Raman mode and has a frequency ω. F is a dimensionless parameter that is zero for single-component materials but, for multicomponent species, involves the molar fractions c_i and dielectric constants ϵ_i of various components i in the manner set out in [64]. Applying the prior theoretical results, numerical estimates have been reported for a number of materials including single-component oxide, halide, and chalcogenide glasses as well as a few multicomponent glasses and single-component halide crystals. The results are shown in Figure 4.14.

The intensity ratio of Rayleigh and Brillouin scattering (I_R/I_B), which is known as the Landau–Placzek ratio, is 10–45 for various glasses and 23.2 for fused silica glass. The density fluctuation in crystals is comparatively small, therefore Brillouin scattering is dominant in high-quality crystals. The intensity of synthesized silica crystal was 1/15 that of synthesized fused silica. Raman scattering also has λ^{-4} dependence. The Raman scattering intensity is usually very weak except for stimulated scattering, and it is 10^{-3} smaller than Rayleigh scattering [20].

Scattering Caused by Homogeneous Crystallization

Since glass is in a metastable phase under equilibrium conditions on the basis of thermodynamics, crystallization must proceed with homogeneous and heterogeneous nucleation. Although the rate of crystallization depends on glass structure and its bonding nature, even pure glass material without a nucleus crystallizes in time. The crystallization is accelerated by heating during fiber drawing to form scattering centers in the fibers. Every fiber fabrication technique necessarily contains a process in which glass is heated above its glass-transition temperature. Therefore, scattering loss caused by homogeneous crystallization should be classified as an intrinsic loss factor. This degree of crystallization will be described later. In this section, only the effect of scattering centers on optical fiber loss will be discussed.

A homogeneous medium containing nonabsorbing spherical particles is assumed as a simple model to deal with the scattering loss of optical fibers [70]. The model of assumed imperfections is shown in Figure 4.15. The homogeneous medium and the particles represent glass and microcrystals, respectively. When the particle number per unit volume n is so small that different scattered light rays from particles do not interfere with each other, the scattering extinction coefficient ϵ of this system is given by

$$\epsilon = n\pi a^2 Q \qquad (4.15)$$

where Q is the efficiency factor for scattering from each particle and a is the radius of the particles [71–72]. The scattering loss α (dB/km) for an optical fiber of this system is

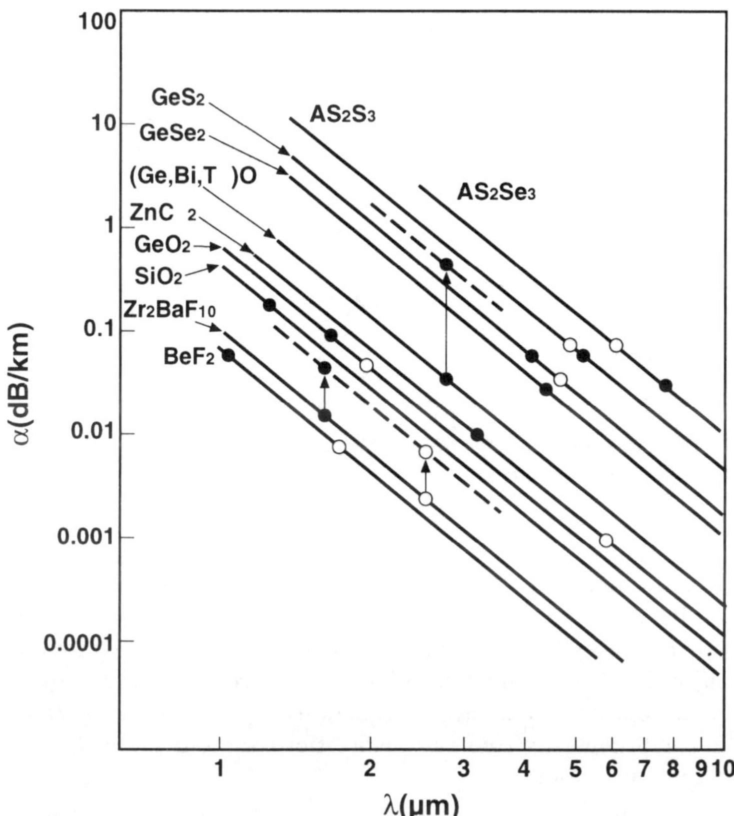

Figure 4.14 Plots of density fluctuation attenuation α (solid lines) as a function of vacuum wavelength λ for vitreous materials [65]. Solid circles indicate the points at which refractive index dispersion is minimum, and open circles denote the positions of minimum total attenuation in cases for which estimates can presently be made. The dashed lines are relevant for mixed component glasses with concentration fluctuation scattering contributions calculated in an ideal mixing approximation. The associated attenuation increments for these materials over the "density only" fluctuation estimates at minimum dispersion and (where available) minimum total attenuation are indicated by arrows.

$$\alpha = (10/\ln 10)\epsilon = (N/L)(10/\ln 10)(a/r)^2 Q \tag{4.16}$$

where L is the fiber length (km), r is the core radius, and N is the total number of particles in an optical fiber. Scattering loss per particle in the fiber $\hat{\alpha}$ (dB) can be defined as

$$\hat{\alpha} = (10/\ln 10)(a/r)^2 Q \tag{4.17}$$

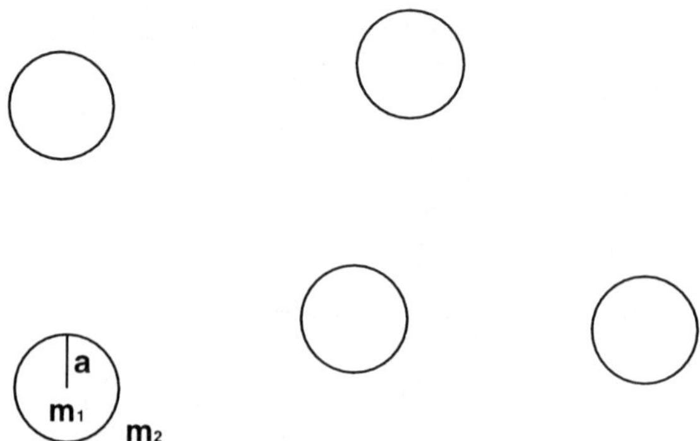

Figure 4.15 Model of glass containing imperfections. Circles and their surroundings represent imperfections and glass, respectively.

Therefore, scattering loss and its wavelength dependence can be evaluated by obtaining the efficiency factor Q, which for spherical particles can be obtained by solving Maxwell's equations (Mie theory). Efficiency factors have been reported for many relative indexes m [73], where $m = m_1/m_2$, where m_1 is the refractive index of the particle and m_2 is the refractive index of the surrounding medium. However, it is known that approximate analytical equations for calculating Q are described in restricted conditions. Approximate expressions for estimating the scattering characteristics are as follows:

Rayleigh Scattering. When the particle size is smaller than the wavelength λ, $x = 2\pi a m^2/\lambda < 1$. Then the efficiency factor is given by

$$Q = (8/3)[(m^2 - 1)/(m^2 + 2)]^2 x^4 \tag{4.18}$$

Rayleigh-Gans Scattering. When the relative refractive index is close to 1, $m - 1 < 1$; and when the phase shift ρ is small, $\rho = 2x(m - 1) < 1$. Then,

$$Q = (m - 1)^2\{(5/2) + 2x^2 - [\sin(4x)/4x] - 7[1 - \cot(4x)]/16x^2$$
$$+ (1/2x^2 - 2)[\gamma + \ln(4x) - C_i(4x)]\} \tag{4.19}$$

where $\gamma = 0.577$ is Euler's constant and

$$C_i(x) = -\int_x^\infty [\cos(u)/u]\, du \tag{4.20}$$

Geometrical Optics and Diffraction. When the relative refractive index is close to 1, $m - 1 < 1$; and when the particle size is larger than the wavelength, $x > 1$. Then

$$Q = 2 - 4\sin(\rho)/\rho + 4[1 - \cos(\rho)]/\rho^2 \tag{4.21}$$

where

$$\rho = 2x(m - 1) \tag{4.22}$$

The relative refractive indexes and size parameter affect the efficiency factor Q. Figure 4.16 shows an example of scattering efficiency factors calculated for ZrO_2 and ZrF_4 particles in fluoride glasses [70]. This figure clearly shows that the wavelength dependency of efficiency factor Q varies with the relative refractive indexes and particle size. In addition, the Rayleigh factor is a function of the particle numbers shown in Figure 4.17 [70]. In real fiber host materials, especially in multicomponent

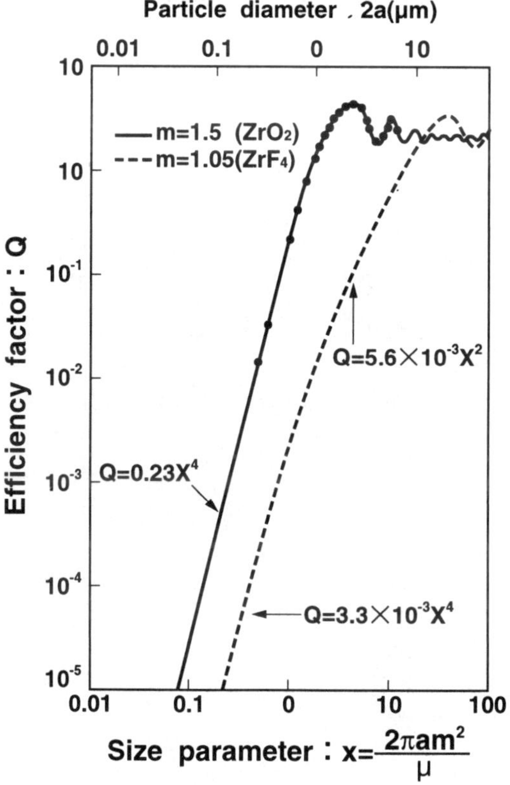

Figure 4.16 Scattering efficiency factors of ZrO_2 and ZrF_4 particles in ZrF_4-BaF_2-LaF_3-YF_3-AlF_3-LiF (ZBLYAL) glass as a function of $x = 2\pi a m_2/\lambda$ where a is the particle radius, m_2 is the refractive index of ZBLYAL, and λ is the wavelength of light [70]. The Q values represented by the solid circles are quoted from [71].

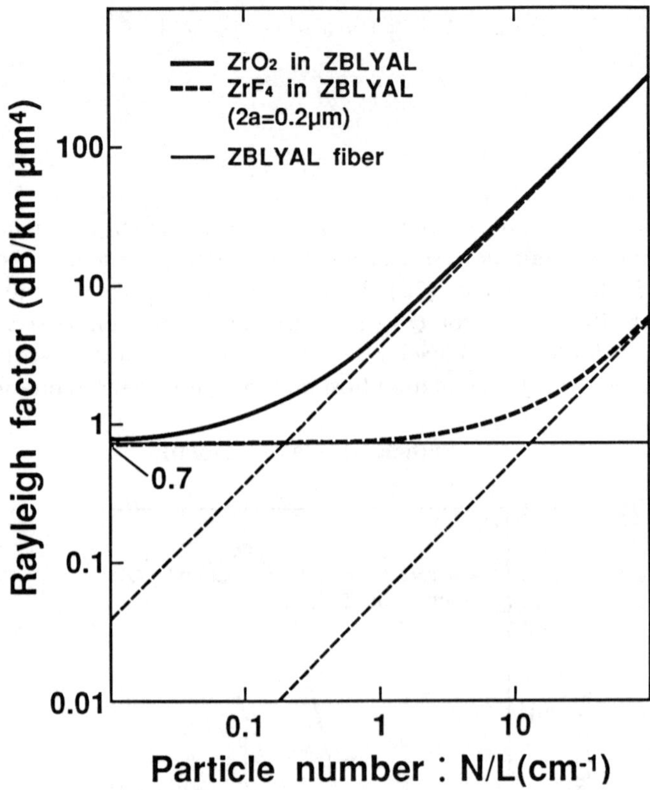

Figure 4.17 Rayleigh factor of a ZrF_4-BaF_2-LaF_3-YF_3-AlF_3-LiF optical fiber containing ZrO_2 and/or ZrF_4 particles 0.2 μm in diameter as a function of the number of particles per unit fiber length where the fiber core radius is 15 μm [70].

glasses, several kinds of crystals are formed with different growth rates, so the scattering losses might have various types of wavelength dependency including Rayleigh-like, Mie-like, and independent.

4.2.2 Extrinsic Loss Factors

Loss factors that are not associated with the fundamental material properties are categorized as extrinsic. These are generally associated with additional substances present in the glass compound that are not essential to the light-guiding properties of the fiber. They, and the associated losses, can in principle be removed with appropriate refinements to the fabrication process [74].

4.2.2.1 Impurity Absorption

Cation impurities comprise one class of extrinsic loss. They are composed of transition metals such as V, Cr, Ni, Mn, Cu, and Fe, all of which exhibit very strong absorption bands in the visible and near-IR region. Other main impurities are rare-earth ions other than the dopant that introduce absorption losses that are important in the near-IR region. Even if these impurities have no absorption bands at the pumping or signal wavelength, energy transfer from dopant to impurities causes a degradation in optical amplification performance. Therefore, all impurities must be reduced to concentration levels of the order of a few *parts per billion* (ppb) in order to reduce their loss contribution and interaction to negligible levels.

Cationic Impurities

The 3d transition metals and rare earths are used as coloring agents for glasses, and their optical absorption spectra in oxide glasses have been investigated [75,76]. Schultz clarified the practical absorptivities of 3d transition metals in silica glass in order to analyze the sources of absorption losses in silica fibers [77]. The absorption spectra of 3d transition metals in ZrF_4-based fluoride glasses have also been measured [78].

The 3d transition metal ions are in octahedral or tetrahedral symmetry with their nearest neighbor anions. Their free ion energy levels for octahedral or tetrahedral coordination are split by the action of the ligand field. And since the ligand field depends on the glass structure and bonding nature, the absorption spectra are affected by the host glasses. The absorption of 3d transition metal ions in silica glass is shown in Figure 4.18 [77]. With silica glasses, transition metal ions give rise to broad, intense absorption in the visible and near-IR regions. The quantity of each transition metal element giving 1-dB/km absorption loss is 2 ppb for Cr, 10 ppb for Mn, 20 ppb for Fe, 0.2 ppb for Co, and 2 ppb for both Ni and Cu. In the first stage of developing low-loss fibers, the transition metal ions were major loss origins. However, when the "soot process" is used, reducing the transition metal impurities is no longer a serious problem. In recent low-loss glasses, these ions cause no detectable absorption. The results of impurity reduction in silica fibers are described in the following section.

The absorption loss of 3d transition elements is more serious for heavy-metal fluoride glasses because fluoride glasses are generally prepared by the conventional melt casting method, which does not include an effective purification process such as the vaporization of the starting materials as used for silica fiber fabrication. The first quantitative estimation of the absorptivity of 3d transition metals and rare-earth elements in ZrF_4-based fluoride glasses was made by Ohishi et al. [78]. Figures 4.19 and 4.20 show the absorption loss spectra of 3d transition metal elements [78]. In recent fluoride fibers, concentrations of 3d impurities of less than 1 ppb [79]

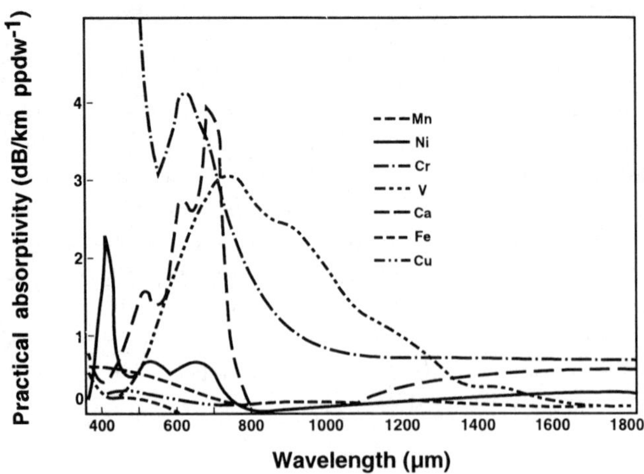

Figure 4.18 Absorption due to transition metal ions in silica glass (*After*: [77]).

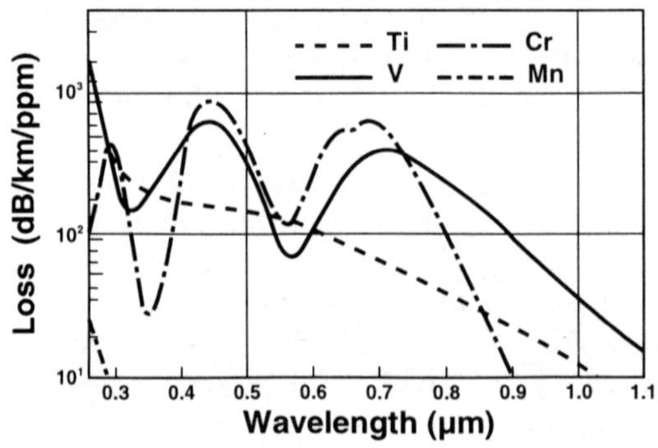

Figure 4.19 Calculated absorption loss spectra due to 1 ppm of transition metals [78].

have been achieved and such low levels have no effect on the optical amplification properties.

Rare-earth ions have unoccupied 4f shells. The internal transitions in 4f shells give rise to optical absorption. The major difference between 3d and 4f electrons is that 4f electrons exist in the internal shell that is screened from external fields by 5s and 5p electrons, while 3d electrons exist in the outer shell and are directly affected by external fields. Therefore, degenerate ^{2S+1}L terms of 4f electrons are mainly perturbed by the spin-orbit interaction even in condensed matter and *J*-multiplets give the energy levels of rare-earth ions in glasses like in free state. Since *J*-multiplets with a $4f^n$ electronic configuration arise in the internal shell of several hundreds to thousands cm^{-1}, optical absorption due to rare-earth ions occurs

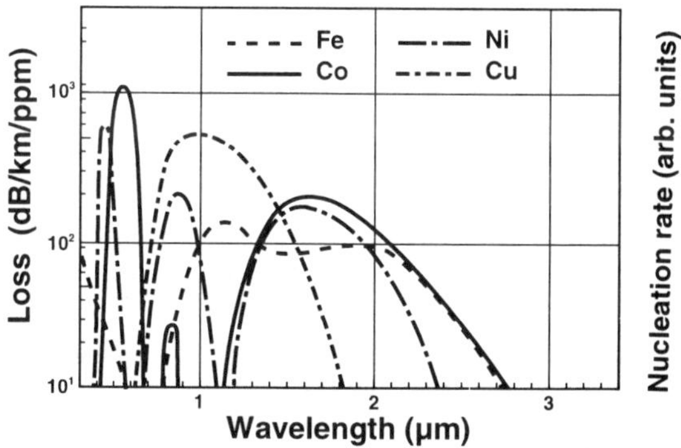

Figure 4.20 Calculated absorption loss spectra due to 1 ppm of transition metals [78].

throughout the visible to mid-IR region. Composition is not an important factor with regard to the optical absorption characteristics of rare-earth ions. The absorption properties of rare-earth ions are described in Chapter 3.

Anionic Impurities

One of the most difficult extrinsic loss factors to remove is the hydroxyl group. The fundamental stretching resonance of OH is centered at a wavelength between 2.7 and 3.0 μm. The OH vibrational mode is slightly anharmonic, which leads to oscillation at overtone. With silica glasses, the OH ion gives rise to strong absorption [80] at 2.72 μm. The overtones and the combination tones with Si-O-Si bending vibration mode are observed at 2.20, 1.90, 1.38, 1.24, 1.13, 0.95, and 0.88 μm. The OH ions of 1 ppm give rise to the absorption losses of 0.7 and 65 dB/km at 1.30 and 1.38 μm, respectively. On the other hand, the loss induced by OH at 1.6 μm is less than 0.1 dB/km. In general, the OH ion concentration should be reduced to less than 1 ppb when the fiber is used in the 1.0- to 1.6-μm region. The absorption peaks shift slightly when other oxides are used as the dopant. OH ions in GeO_2 glass give rise to an absorption peak at 2.86 μm. With GeO_2-doped silica glass, the peaks shift toward longer wavelength with increasing GeO_2. For fluoride glasses, the fundamental OH resonance is observed at around 2.9 μm [81], and other incidental absorption peaks are observed at 2.24 and 2.42 μm and are attributed to combined tones between ZrF_4 and stretching vibrations of free HF and OH, respectively.

In addition, polymeric molecular ions such as NH_4^+, CO_3^{2-}, SO_4^{2-}, and PO_4^{3-} have their fundamental vibration in the IR region. Since some of these influence the signal and pumping wavelengths, they must be reduced according to concentrations in the fibers.

4.2.2.2 Extrinsic Scattering

Extrinsic scattering loss factors in optical fibers caused by the imperfections in a waveguide, such as small bubbles and crystalline particles in the core glass and irregularities between the core glass and cladding layer, have no dependence on the wavelength; that is, when the fiber transmission loss can be expressed as

$$\text{Loss}(\lambda) = A + B/\lambda^2 + C/\lambda^4 \tag{4.23}$$

where A is the wavelength-independent loss due to waveguide imperfections, B is Mie scattering loss constant, and C is Rayleigh scattering loss constant. As described in Section 4.2.1.4, the scattering loss and its wavelength dependency vary with refractive index and crystal size formed in the fiber. The wavelength-independent loss is due to scattering centers whose size is much larger than a wavelength. When scattering center size is of the order of a wavelength, the scattering is Mie in character, exhibiting λ^{-2} dependence.

With pure silica fibers, extrinsic scattering loss has become negligibly small through the use of mature fiber fabrication technologies and their extreme stability against crystallization. However, the doping of rare-earth ions into silica fiber hosts causes rare-earth ions to cluster in the glasses [82,83]. This causes an increase in scattering loss [84]. The clustering can be suppressed by codoping Al_2O_3 and P_2O_5 since this can compensate for the electric charge around rare-earth ions. In contrast, the doping of rare-earth ions does not increase scattering loss since rare-earth ions are composed of fluoride glasses as network modifiers and their solubility is one order of magnitude higher than that of silica-based glasses. However, fluoride glasses are less stable than silica-based glasses and crystallize during fiber fabrication processes such as casting and fiber drawing. Furthermore, the core glass in a single-mode fiber undergoes considerably more thermal processing than multimode fibers, so Mie scattering and wavelength-independent loss are the dominant loss factors [85].

The irregularity of the core-cladding interface that occurs during fiber fabrication is one of the dominant loss factors. The fabrication of high-Δn fibers generally results in a scattering loss increase because a high-Δn requires core and cladding glasses whose compositions are quite different. In the fiber fabrication process, the viscosity difference between the core and cladding glasses causes a fluctuation in the core-cladding interface.

Other extrinsic loss factors are caused by inclusions, bubbles, and core-cladding boundary fluctuations. Since extrinsic scattering is closely related to the fiber fabrication technologies, other scattering centers are described in the following sections.

4.3 THERMAL PROPERTIES OF FIBER MATERIALS

As described in the previous section, the optical properties of rare-earth-doped fibers are associated with the physical and chemical properties of their host glasses, which originates in the chemical bonding and coordination number of the constituents. These basic factors determine the thermodynamic resistance to crystallization and nucleation and the crystal-growth rates. In addition, the solubility of rare-earth ions in the glass matrices, that is, thermodynamic stability of rare-earth-doped glasses, is influenced by the local structure in the vicinity of the rare-earth ions. Therefore, a stable glass system and a stable glass composition should be chosen in order to fabricate a low-loss host fiber without the rare-earth ions forming clusters in the practical concentration range. In the following subsection, the structure, bonding nature, and crystallization tendency of various host glasses will be described.

4.3.1 Glass Structure and Chemical Bonds

Prior to discussing the thermodynamic stability of host glasses from a macroscopic viewpoint, we would like to mention the solubility of rare-earth ions in host glasses. The solubility may be determined by the types of chemical bonds of the constituents. The classification of glass-forming substances based on the type of chemical bonding is useful in relation to the concentration limits below which the rare-earth ions are uniformly dispersed without any clustering as shown in Table 4.6 [86]. The glass structures are quite different in each class.

In this section, typical glasses with different chemical bonds and composed of silica, silicate, chalcogenide, and fluoride will mainly be described because these glasses have different structures and are important as optical amplifier hosts in terms of their unique optical properties.

Criteria relating to glass-forming ability and the chemical bond strength of constituents were first clarified by Sun [87]. Since the process of atomic

Table 4.6

Classification of Glass-Forming Materials by Type of Bonding

Bond Type	Examples
Covalent	Oxides (silicates, borates, phosphates, germanates, etc.)
	Chalcogenides
	Organic polymers
Ionic	Halides, nitrates, carbonates, sulfates
Hydrated ionic	Aqueous, solution of salts
Molecular	Organic liquids
Metallic	Splat-cooled alloys

rearrangement that causes crystallization may involve the breaking and reforming of interatomic bonds, one can expect a correlation between the strength of these bonds and the glass-forming ability. The stronger the bonds, the more sluggish the rearrangement process will be and hence the more readily a glass will be formed. Table 4.7 shows the calculated bond strengths and coordination numbers of oxides. The strength is calculated by dividing the energy required to dissociate oxide MO_x into its constituent gaseous atoms by the number of oxygen atoms surrounding the atom M in a crystal or glass. The results in Table 4.7 reveal that glass-forming oxides have single bond strength of less than 90 kcal/mol and the modifiers have a bond strength of less than 60 kcal/mol. Although this criterion can be applied only to a specific system, it is clearly demonstrated that glass-forming ability is strongly related to the bonding nature of the constituents.

The structure of covalently bonded glass has already been extensively studied and described in detail. The covalent bonds give rise to molecular chains and networks, which are favorable for glass formation. A typical example is vitreous silica. Another important factor is the coordination number of an atom in a glass or the number of other atoms of a specific kind that are bonded to it. Since glass is rigid, the coordination number of its atoms is relatively fixed as a function of time. The coordination number of an atom is important in determining its role in the glass as part of a three-dimensional network, a chain, a disrupter of networks and chains, or as a source of incipient crystallization. Therefore, considerable attention must be paid to determining the coordination numbers of atoms in glasses [86].

4.3.1.1 Silica and Silicate Glasses

The simplest of the silicates is pure silica (SiO_2); and the fundamental molecular structure is shown in Figure 4.21, where the silicon atom in the center is covalently bonded to four oxygen atoms. The oxygen atoms are arranged tetrahedrally around the silicon atom, and Zachariasen's postulate of a three-dimensional network without periodicity is formed by the union of these tetrahedra at their corners [21]. This structure is very similar to the crystalline form of silica, crystobalite; but the structure, being slightly more random, results in a lack of long-range periodicity or symmetry, as shown in Figures 4.22 and 4.23. The tetrahedron in Figure 4.21 can be attached to no, one, two, three, or four other tetrahedra by silicon-oxygen bonds at their corners depending on the concentration of the other oxides present.

Some ions, such as alkali or alkaline earth ions, disrupt the silicon-oxygen glass network as evidenced by the much lower viscosity of these glasses compared to pure silica. In general, as long as the number of A_2O_3 or AO units, A being a metallic ion, has less than a 1:1 ratio to the number of SiO_2 units, the basic glass structure can be maintained, since each tetrahedron can still be linked to at least three other tetrahedra. Some of the silicon-oxygen bridging bonds, however, will

Table 4.7
Calculated Bond Strengths and Coordination Numbers of Oxides

M in MO$_x$	*Valence*	*Dissociation Energy per MO$_x$ [Ed(kcal)]*	*Coordination No.*	*Single Bond Strength B$_{M-O}$ (kcal)*
Glass formers				
B	3	356	3	119
Si	4	424	4	106
Ge	4	431(?)	4	108
Al	3	402–317	4	101–79
B	3	356	4	89
P	5	442	4	88–111
V	5	449	4	90–112
As	5	349	4	70–87
Sb	5	339	4	68–85
Zr	4	485	6	81
Intermediates				
Ti	4	435	6	73
Zn	2	144	2	72
Pb	2	145	2	73
Al	3	317–402	6	53–67
Th	4	516	8	64
Be	3	250	4	63
Zr	4	485	8	61
Cd	2	119	2	60
Modifiers				
Sc	3	362	6	60
La	3	406	7	58
Y	3	399	8	50
Sn	4	278	6	46
Ga	3	267	6	45
In	3	259	6	43
Th	4	516	12	43
Pb	4	232	6	39
Mg	2	222	6	37
Li	1	144	4	36
Pb	2	145	4	36
Zn	2	144	4	36
Ba	2	260	8	33
Ca	2	257	8	32
Sr	2	256	8	32
Cd	2	119	4	30
Na	1	120	6	20
Cd	2	119	6	20
K	1	115	9	13
Rb	1	115	10	12
Hg	2	68	6	11
Cs	1	114	12	10

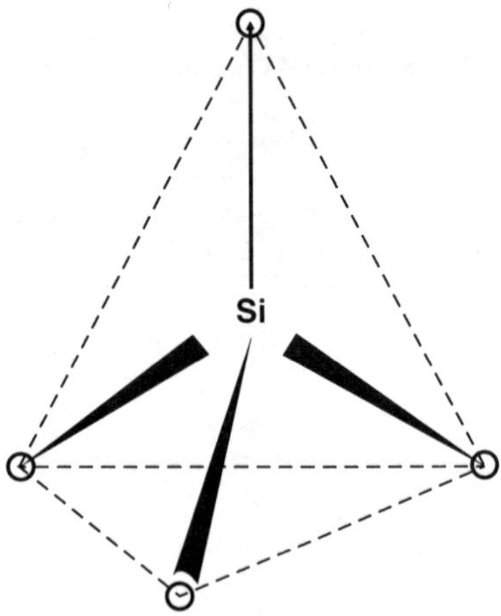

Figure 4.21 Schematic representations of the three-dimensional tetrahedral arrangement in fused silica glass.

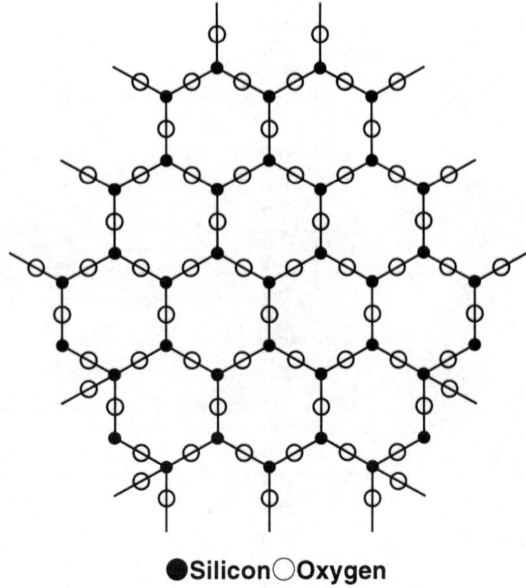

Figure 4.22 Diagram of the two-dimensional matrix of silicon and oxygen atoms, as found in crystobalite.

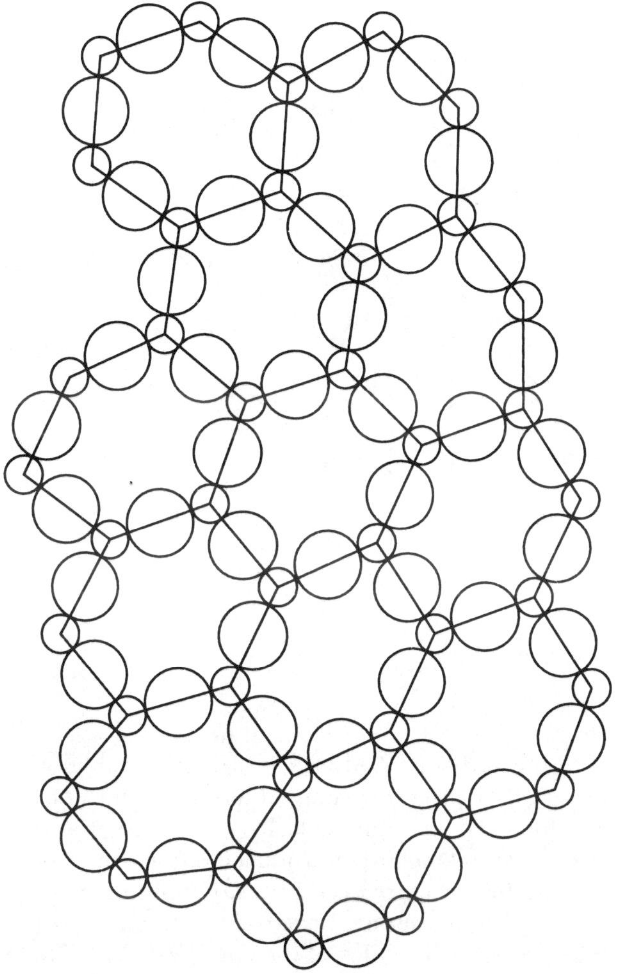

Figure 4.23 Schematic diagram of a two-dimensional structure of fused silica.

have been broken, thus increasing the number of nonbridging oxygens present. The structure in the vicinity of these nonbridging oxygen ions carries a negative charge, so neutrality is maintained locally in the structure. As an example of these glasses, the structure of Na_2O-SiO_2 glass is shown in Figure 4.24 [88]. Pure silica requires a very high temperature to make it workable. The addition of so-called network modifiers, therefore, is important because they reduce this temperature [89].

Other ions that can be substituted in a number of silicon sites are aluminum, phosphorus, boron, and germanium; and these ions are recognized as the most important in rare-earth-doped fiber amplifiers.

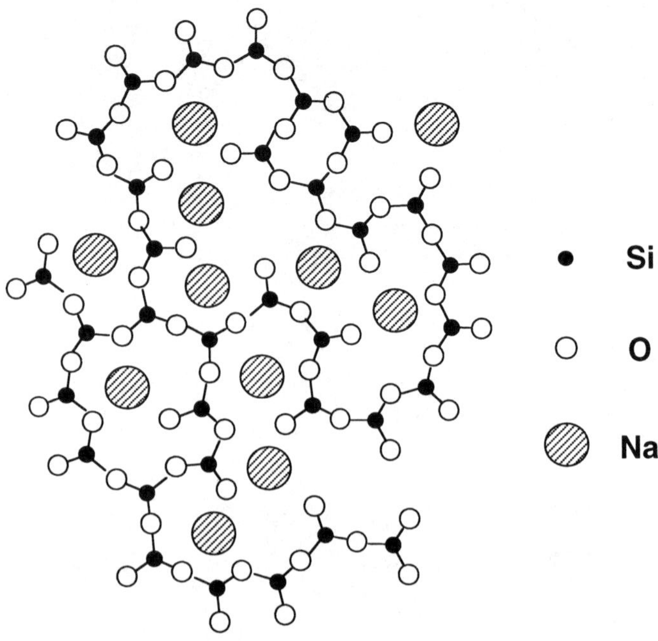

Figure 4.24 Structure of Na_2O-SiO_2 glass [88].

With regard to the addition of Al_2O_3 to silica, it is well known that Al^{3+} can occupy silicon sites if other positively charged atoms are introduced at the same time. For example, for each Al^{3+} ion replacing Si^{4+} in a tetrahedral position, a single N_a^{+} ion must be simultaneously introduced so that the total charge in the glass is neutralized and this structure need have no nonbridging oxygens. In fibers for telecommunications, aluminum has been used as an alternative to germanium in order to increase the refractive index of the core [89]. As will be seen later, the rare-earth ions are positively charged and, when introduced into an Al_2O_3-SiO_2 system, tend to congregate preferentially near aluminum sites [90]. This can be viewed as the aluminum charge deficiency being equilibrated by the rare-earth ions.

Phosphate glasses have phosphorus-oxygen tetrahedrons as their basic building blocks, as shown in Figure 4.25. In contrast to silica glasses, phosphorus has a double bond to one of its surrounding oxygen atoms. Thus, it seems likely that the structure of glassy P_2O_5 is a three-dimensional network of these phosphorus-oxygen tetrahedra, each tetrahedron being bonded to three rather than four other tetrahedra in silica glasses. Phosphorus is commonly used in conjunction with fluorine to dope silica-based optical fibers. This glass is used as a cladding material with the same refractive index as pure silica but with a viscosity that can be effectively lowered by the addition of a small amount of phosphorus [44].

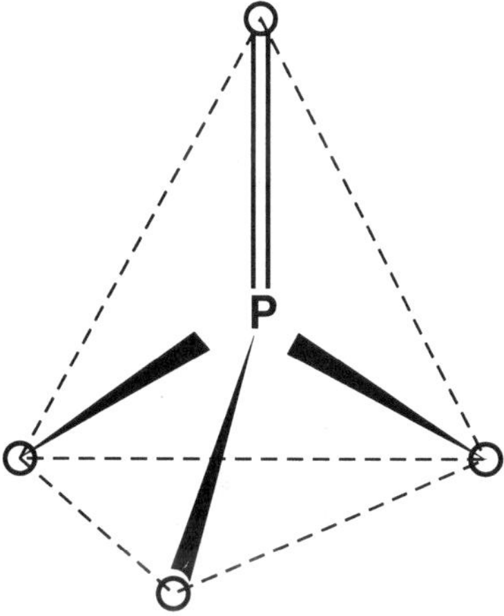

Figure 4.25 Diagram of a three-dimensional phosphorus and oxygen tetrahedron showing the double bond between one oxygen and the phosphorus.

Borate glasses are of interest because their structure and properties are quite different from those of silicates, and boric oxide is used extensively in borosilicate glasses. Since boron is trivalent, it forms only three bonds with neighboring oxygen atoms and is composed of a planar ring consisting of six alternating atoms of boron and oxygen. These rings are linked together in a three-dimensional network by boron-oxygen-boron bonds, as shown in Figure 4.26. The addition of B_2O_3 to pure silica reduces the refractive index and lowers the softening temperature. This glass has a very high phonon energy corresponding to an absorption edge at about 7 μm.

Germanium oxide glass consists of germanium-oxygen tetrahedra in a random network, and its properties are similar to those of silica glass. However, there is some evidence that the germanium coordination number changes from four to six as alkali oxide is added to GeO_2 [86,91]. In telecommunication fibers, germanium is commonly used to dope the silica glass during the core deposition because it raises the refractive index above that of the cladding glass and makes it possible to fabricate a wide range of guiding structures [92–93].

4.3.1.2 Chalcogenide Glass

Chalcogenide glasses are composed of Group V elements, particularly, arsenic and antimony combined with Group VI elements such as sulfur, selenium, and tellurium and sometimes with the addition of rare earths, halogens, and other elements. As rare-earth host glasses, they have attractive features such as low phonon energy and relatively high stability against crystallization. Figure 4.27 shows a boundary in the periodic table that separates these elements from the metals. The structures of

Molecular structure

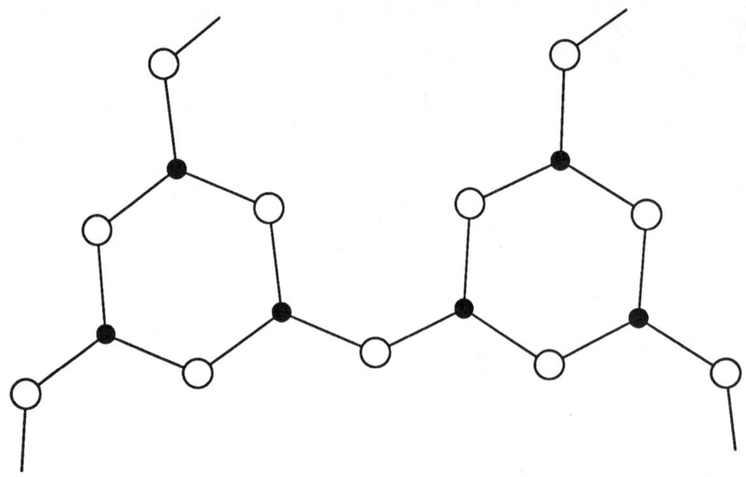

Figure 4.26 Two boroxyl groups linked by a shared oxygen. Closed circle: boron; open circle: oxygen.

Mg	Al	Si	P	S
Zn	Ga	Ge	As	Se
Cd	In	Sn$_{gray}$ / Sn$_{white}$	Sb	Te
Hg	Tl	Pb	Bi	Po

Figure 4.27 Borderline between metallic and nonmetallic structure in the periodic system.

elements to the right of the line are characterized by a low coordination number in contrast to the close-packed structures of the true metals. The lighter elements in particular crystallize with structures in which each atom has 8-N neighbors, N being the group number. Thus silicon and germanium have a diamond structure, characterized by a coordination number of 4. For all the various modifications of phosphorus and arsenic, it is believed that each atom is bonded to its three nearest neighbors whilst the solid and liquid forms of sulfur and selenium contain chains and/or rings, each atom having two nearest neighbors. In all these elements, the structure is predominantly determined by covalent interatomic bonds. As the atomic number increases there is a gradual transition to structures with a higher coordination number, and the increasingly metallic nature is generally attributed to this structure change [88].

In the chalcogenide glasses, arsenic sulfide glass has been studied extensively as optical fibers because of its IR transparency. The compounds As_2S_3 and As_2Se_3 form glasses easily with structures closely related to the structure of crystalline As_2S_3 (orpiment). The structure of crystalline As_2S_3 is shown in Figure 4.28 [94]. X-ray data [95,96] suggest that the orpiment layer structure of As_2S_3 glass retains short-range ordering but is heavily distorted. The layers are more widely separated and persist over a greater radial distance than intralayer correlations, with a greater range of bond length and angles. There is no signal from the 12-member ring structure in the crystal. The bonds are predominantly covalent [97].

4.3.1.3 Fluoride Glass

In contrast, the structure of fluoride glass has not been sufficiently studied. Some information about their structural characteristics has emerged, but much work

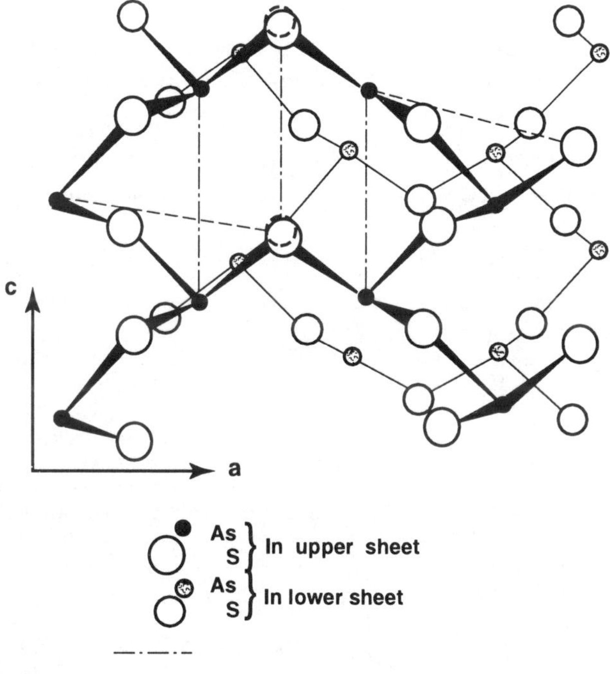

Figure 4.28 Structure of orpiment (As_2S_3) [94].

remains to be done. With regard to chemical bonding, the F^- anion is small and has a very high electron affinity (3.4 eV) that is a measure of its heat of formation and, hence, of the ability of the neutral F atom to acquire the missing electron needed to complete a $(1s)^2(2s)^2(2p)^6$ closed shell configuration. An indication of the character of a particular bond is given by the electronegativities of the two atoms involved that reflect both their electron affinity and ionization potential. With the major glass-forming components of fluoride glasses, it is clear that the metal atoms have a low electronegativity and, therefore, will form bonds exhibiting a high degree of ionicity with the highly electronegative fluorine atoms. This bonding nature determines the local structure and also the macro properties such as thermodynamic stability. This does not mean, however, that there is no covalency in the bonding of heavy-metal fluoride glasses but rather that a good first-order representation of their interatomic potentials can simply be ionic. In terms of the glass structure, an electric field around the F^- species cannot be screened until it is surrounded by positively charged particles because of the strong ionicity of the F^- ions. It is this character that allows the monovalent fluorine ion to form strong cation-anion-cation bridges, leading to the sluggish relaxation behavior near freezing point that favors glass formation. For systems where the bonding is at least partially ionic in character, the coordination polyhedra present are to a large extent determined by the relative sizes of their constituent ions, usually expressed as the radius ratio, r_A/r_X, in which A is a cation and X is an anion. The critical radius ratio $(r_A/r_X)_c$ is shown in Table 4.8. The effective radius ratio for a given pair of

Table 4.8

Coordination Number and Critical Radius Ratio for the Common Coordination Polyhedra

Coordination Number	Minimum Radius Ratio	Orbitals	Geometrical Structure
2		sp, dp	linear
3	0.155	sp^2, ds^2	trigonal planar
		p^3, d^2p	trigonal pyramid
4	0.225	sp^2d, p^2d^2	square planar
		sp^3, d^3s	tetrahedral
		d^4	tetragonal pyramid
5	0.155	sp^3d, d^3sp	trigonal bipyramid
6	0.414	d^2sp^3	octahedral
		d^4sp	trigonal prism
7	0.592		I atom above the face of an octahedron
8	0.645	d^4sp^3	dodecahedral square
	0.732		antiprism, cube
9	0.732		right triangular prism
12	1.00		cube, octahedron

ions is a function of the degree of ionicity of their bond, and small coordination numbers often are associated with a strong covalent bonding character because of the compatibility of the corresponding polyhedral shapes with common bond hybridization geometries. Conversely, with high coordination numbers, there is an increased degree of ionicity in the bonds and the different polyhedra are determined by ionic size and packing geometry. There is also an increased probability of both edge and face sharing between adjacent polyhedra since the required increase in strain energy is much less [98]. Figures 4.29 [99] and 4.30 [100] show structure models of HfF_4-BaF_2 glass and ZrF_4-BaF_2-LaF_3-AlF_3 glass. As shown in these figures, one interesting feature of fluoride glasses is that they already contain La^{3+} as an integral component, and this can easily be substituted by an alternative rare-earth ion without significantly affecting the properties of the glass [101]. The bonding nature described might allow various kinds of cation to be incorporated in fluoride glasses without clustering. This seems to be a great advantage with regard to rare-earth-doped host glass fibers.

4.3.2 Theory and Kinetics of Crystallization

As described in the previous sections, the background loss of current host fibers is mainly due to a light scattering including both intrinsic and extrinsic factors. Since the latter can be reduced by improving the fabrication process, we should pay careful attention to the intrinsic factors, especially scattering loss caused by homogeneous crystallization when choosing host materials, because a quantitative evaluation of the crystallization tendency is relatively difficult compared with evaluations of optical properties such as optical absorption and Rayleigh scattering. Moreover, crystallization deals the host fiber a fatal scattering loss. The tendency to crystallize is attributed to the structure of the glass and its bonding nature, which can be analyzed based on a thermodynamics and a kinetics. This section deals with the theory and the kinetic analysis of the homogenous crystallization of host glasses and how to evaluate the scattering loss from the basic crystallization properties of the bulk glass samples.

4.3.2.1 Evaluation of Stability of Bulk Glasses

$T_x - T_g$ Gap and Critical Cooling Rate

The first step when evaluating glass stability is to assess whether a sample is glassy or not. Here, there is no major problem in setting the limits of a vitreous area. The next step is thermal analysis using *differential thermal analysis* (DTA) or *differential scanning calorimetry* (DSC). Figure 4.31 [101] shows a schematic diagram of thermal analysis. Thermal analysis will confirm the occurrence of a glass transition at T_g and the onset of crystallization at T_x. T_m is the melting point of crystals formed at T_x. T_c is the peak temperature of the crystallization, and T_l is the liquidus tempera-

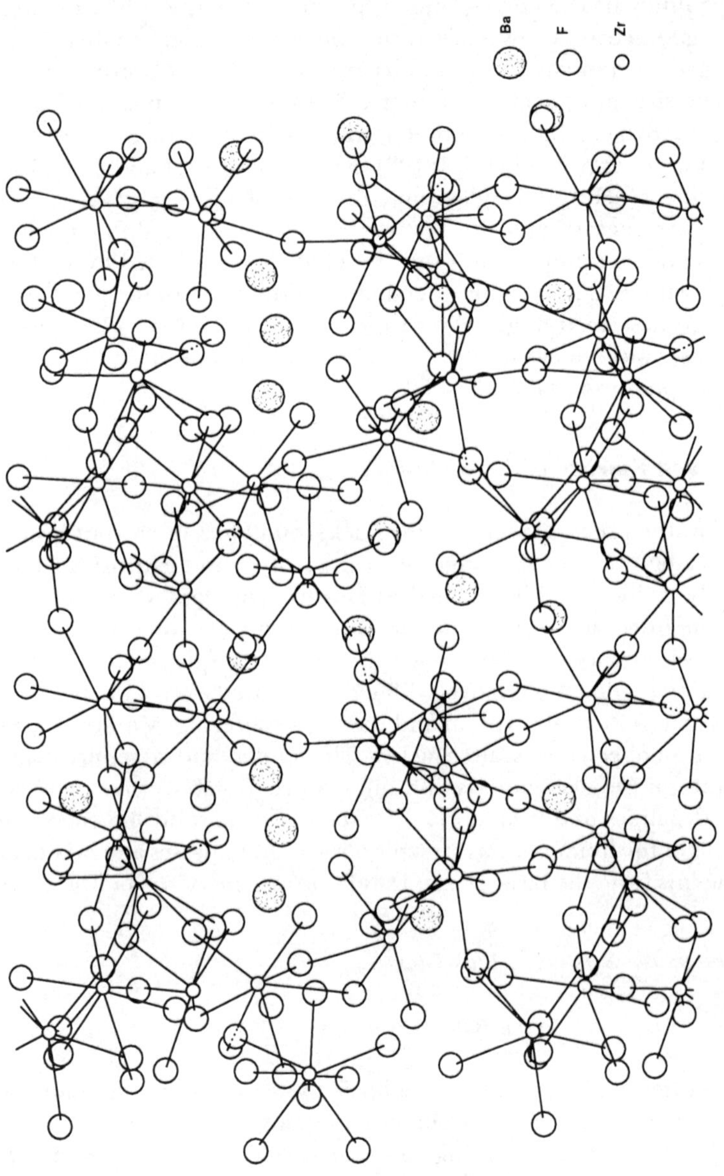

Figure 4.29 Structure model for $3HfF_4$-$2BaF_2$ glass obtained from MD simulation [99].

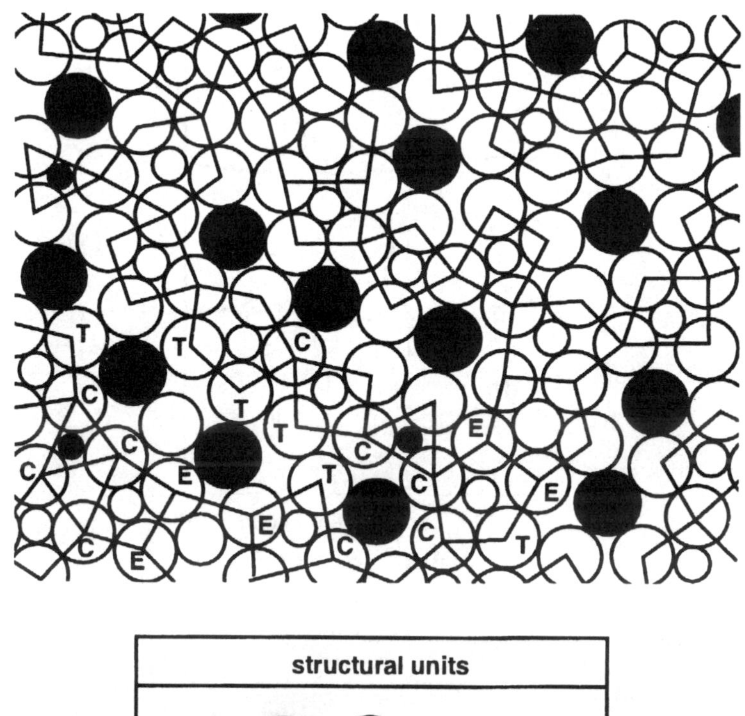

Figure 4.30 Schematic representation of the structure of a typical fluorozirconate glass [100]. There are three types of fluorine ions: terminal (T), edge (E), and corner (C).

ture. T_g, T_x, and T_m are the basic parameters for evaluating glass stability, in other words, glass-forming ability or the tendency of a glass to crystallize [102].

The problem lies in the definition of glass stability. In general, two criteria have been used to determine glass stability against crystallization. One is the critical cooling rate, which is defined as the minimum melt-cooling rate at which a glass forms without crystallization [103], and the other is the $T_x - T_g$ gap in the reheating process.

The critical cooling rate is measured by DSC or DTA. In a DSC trace for cooling a fluoride glass melt, the crystallization temperature tends to appear in the lower temperature region when the melt is cooled at a high cooling rate. Thus, supercooling, which refers to the difference between the liquidus temperature

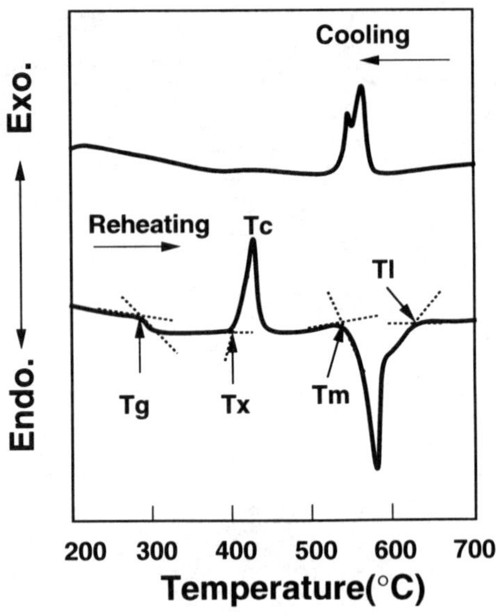

Figure 4.31 DSC data for a ZrF$_4$-based fluoride glass [101].

and the crystallization temperature, increases with increasing cooling rate. The relationship is given by

$$\ln R = \ln R_c - \frac{B}{\Delta T_c^2} \tag{4.24}$$

where R is the cooling rate, R_c is the critical cooling rate, B is a constant, and ΔT_c is the supercooling. Figure 4.32 shows a relationship between R and ΔT_c [10]. The critical cooling rate is obtained by extrapolation to $1/\Delta T_c = 0$ in Figure 4.32 and represents the glass-forming ability in the melt-cooling process. It is closely related to the homogeneity of the preform. Tables 4.9 and 4.10 [101] show critical cooling rates measured for host glasses. Since the critical cooling rate depends on the thermodynamic stability of a glass-melt, glass-melt composed of chemical bonds with high covalency tends to have a low critical cooling rate. In practical glass systems, silica glass and chalcogenide glass are thermodynamically more stable than fluoride glass. The $T_x - T_g$ gap and Hruby parameter (Hr) in (4.25) provide a crude guide to stability when it is reheated in relation to such factors as elongation and fiber drawing. These parameters can be simply and rapidly obtained from the DSC curves. For loss stable glasses, crystallization occurs rapidly when the sample is heated beyond the glass transition. Therefore, the difference of $T_x - T_g$ is correlated with the tendency to devitrify. Moreover, a thermodynamically stable glass

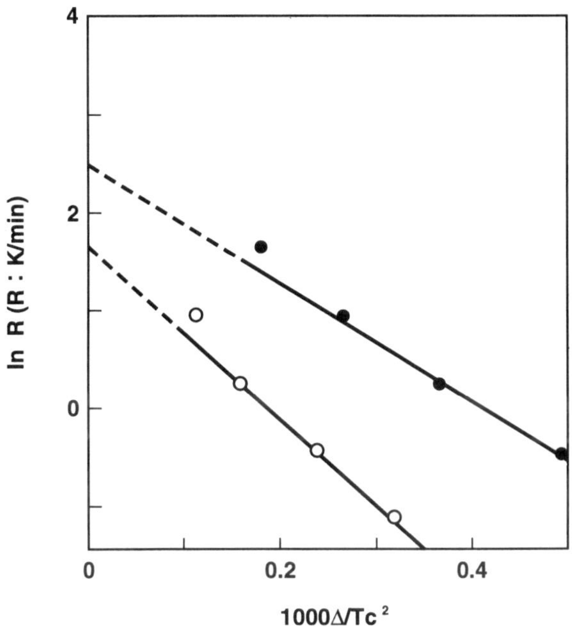

Figure 4.32 Plots of supercooling T_c with cooling rate R [10]. Solid circles: 47.5ZrF$_4$-23.5BaF$_2$-2.5LaF$_3$-2YF$_3$-4.5AlF$_3$-10LiF-10NaF; open circles: 48ZrF$_4$-23.5BaF$_2$-2.5LaF$_3$-3.5AlF$_3$-20NaF (mol%).

Table 4.9

Calculated Critical Cooling Rates for
Forming Various Oxides (1 cm^3) as
Glasses

Material	Cooling Rate (K/s)
P$_2$O$_5$	10^{-23}
B$_2$O$_3$	10^{-16}
GeO$_2$	10^{-11}
SiO$_2$	10^{-1}
As$_2$O$_3$	10^{7}

system has a stable liquid phase, which corresponds to a relatively low T_g. From this aspect, Hr parameter includes a $T_m - T_x$ term.

$$\text{Hr} = (T_x - T_g)/(T_m - T_x) \tag{4.25}$$

In addition to the $T_x - T_g$ and the Hr parameters, DSC curves may give us important information about nucleation and crystal growth rate. Figure 4.33 shows

Table 4.10
Typical Glass Compositions for Fluoride Optical Fiber and Their Critical Cooling Rates

	Composition (mol %)								
Glass	ZrF_4	BaF_2	LaF_3	GdF_3	YF_3	AlF_3	LiF	NaF	R_c (K/min)
ZBG	63	33		4					370
ZBLA	57	34	5			4			60
ZBGA	60	32		4		4			70
ZBLAN	51	20	4.5			4.5		20	0.7
ZBLALi	52	20	5			3	20		26
ZBLYAN	47.5	23.5	2.5		2	4.5		20	1.1
ZBLYALi	49	22	3		3	3	20		26

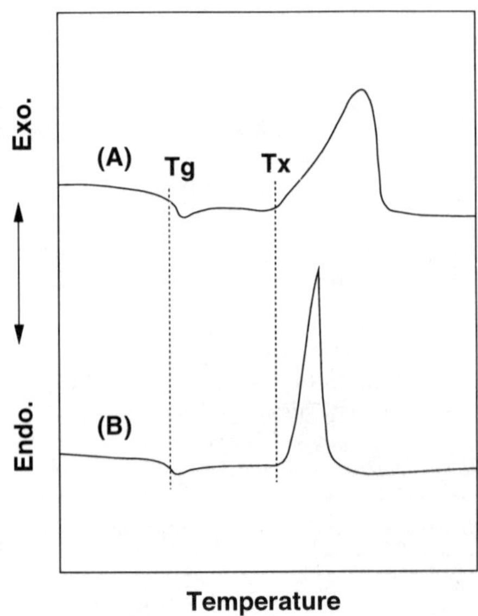

Figure 4.33 DSC curves for glasses with the same $T_x - T_g$.

a schematic diagram of DSC curves, where both curves indicate the same $T_x - T_g$ but the shapes of the exothermic peaks are different. Since crystallization rate is determined by the product of nucleation rate and crystal growth rate, a shape of a DSC peak is affected by temperature dependencies of nucleation and crystal growth rates. In the case of peak A, the broad exothermic peak may originate from crystallization that the nucleation rate is fast but the crystal growth rate is low. In this case, there are a large number of nuclei in the glass, but the volume of crystal gradually increases because of the low crystal growth rate. This gradual increase in the crystal volume leads to the broad exothermic peak. In the case of the peak B, the sharp exothermic peak may originate from crystallization that the nucleation rate is slow but the crystal growth rate is fast. In this case, once the nucleation occurs, a volume of crystal rapidly increases because of the high crystal growth rate. This rapid increase leads to the sharp exothermic peak. These crystallization processes reflect on a morphology of crystals; the former may form a large amount of small crystals that causes a continuous scattering in the fiber, and the latter may form a small amount of large crystals that causes discrete bright scattering centers in the fiber. In terms of an amplifier host fiber, the latter is preferable because the fiber with the length required for an amplifier host can be obtained between the bright scattering centers.

Figures 4.34 and 4.35 show the critical cooling rates and $T_x - T_g$ gaps of Pr-doped ZrF_4-based fluoride glasses against PbF_2 content [85]. This glass system is used as a core glass for 1.3-μm optical fiber amplifiers [85,104]. In these figures, it is revealed that the glass stability against crystallization becomes worse as the PbF_2 content is increased. Figure 4.36 shows transmission loss spectra of Pr^{3+}-doped fluoride glass single-mode fibers. The main loss factor of these fibers is wavelength-independent scattering caused by crystallization. This result suggests that an analysis of the crystallization properties of host glasses will enable us to know whether a specific glass system can be employed as an amplifier host fiber or not. The theory and a quantitative analysis of crystallization kinetics are provided in the following sections.

Nucleation and Crystal Growth Rates

The kinetics of nucleation and crystal growth from condensed phase has been discussed thoroughly elsewhere [105], and devitrification studies on host glasses

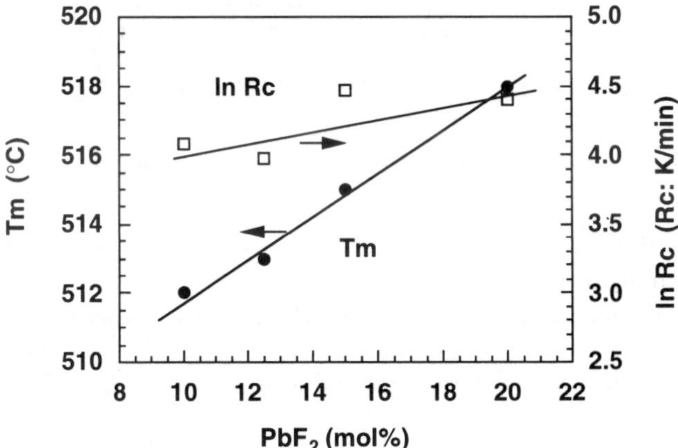

Figure 4.34 Effects of PbF_2 content of T_m and critical cooling rate [85].

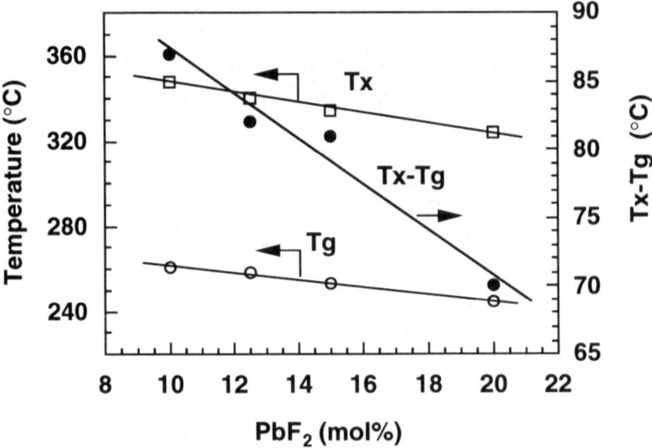

Figure 4.35 Effects of PbF$_2$ content on glass transition temperature and crystallization temperature [85].

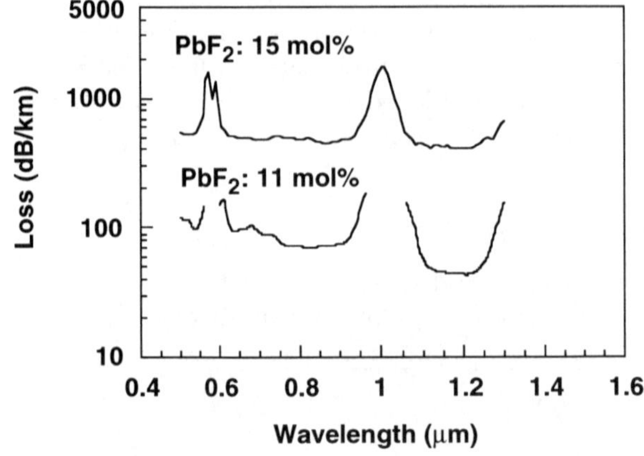

Figure 4.36 Transmission loss spectra of Pr^{3+}-doped single-mode fluoride glass fiber [85].

have been carried out according to Avrami's law [106,107] in which specific nucleation is assumed. In general, a nucleation temperature, at which the nucleation rate reaches its maximum, is close to a crystal growth temperature. Therefore, the nucleation and the growth can simultaneously take place in the transformation range. Crystallization kinetics was analyzed using a following nucleation and growth theory.

At first, it is assumed that crystallization occurs with interface-controlled growth because crystals formed during heating have a similar composition of the glass matrix. The linear crystal growth rate (u) independent of time and the composition of the growth crystal is the same as that of the supercooled liquid. When the nuclei are formed at the time t', a linear dimension of the growth particle is obtained as $u(t - t')$ in the time t. Then, the volume of growth particle (V_p) is given by

$$V_p = g u_x u_y u_z (t - t')^3 \tag{4.26}$$

where g is the shape factor. The nucleation rate I per unit volume and time is given by

$$I = \frac{dN_c}{dt} \tag{4.27}$$

where N_c is the number of critical nuclei.

If the impingement of growth crystals is ignored, total crystal phase volume (V_t) formed in the glass (V_0) at the limit can be expressed by

$$dV_t = V_p V_0 I dt' \tag{4.28}$$

where $V_0 I dt'$ is the number of nuclei formed in dt', which is equal to $V_0 dN_c$. Then

$$\frac{dV_t}{V_0} = dX_t = V_p \, dN_c \tag{4.29}$$

where X_t is the volume fraction of crystal phase at the time t.

According to the theory revised by Johnson and Mehl [107], the impingement of growing particles is taken into account by multiplying (4.29) by the volume fraction of residual glass

$$\frac{dV_t}{V_0} = dX_t = V_p dN_c (1 - X_t) \tag{4.30}$$

Then

$$
\begin{aligned}
1 - X_t &= \exp\left(-\int_{t'=0}^{t'=t} V_p \, dN_c\right) \\
&= \exp\left(-\int_{t'=0}^{t'=t} V_p I \, dt'\right) \\
&= \exp\left[-\int_{t'=0}^{t'=t} \{g u^n (t - t')^n\} I \, dt'\right] \\
&= \exp\left[\frac{-g u^n t^{n+1} I}{n + 1}\right]
\end{aligned}
\tag{4.31}
$$

where n is the time dimension of crystal growth. If the synthesized glass is inhomogeneous, which means that nuclei have been formed during the glass synthesis, the crystal growth may take place from their nuclei. In this case, the number of the nuclei do not vary remarkably, so

$$N_c = N_0 \tag{4.32}$$

$$1 - X_t = \exp\,(-V_p N_0) = \exp\,(-gu^n t^n N_0) \tag{4.33}$$

Comparing (4.33) with (4.31), dimension of time t is different between homogeneous and inhomogeneous nucleations. Therefore, an increase in the volume fraction of crystal can be separated into two parts: homogeneous crystallization and inhomogeneous crystallization, by analyzing the time dependence (in this case, it is assumed that growth crystal is unique).

If the growing particles are sphere, V_p is given by

$$V_p = \frac{4}{3}\pi u^3 (t - t')^3 \tag{4.34}$$

Then, $1 - X_t$ for homogeneous crystallization can be expressed from (4.31) and

$$-\ln(1 - X_t) = \frac{1}{3}\pi u^3 t^4 I \tag{4.35}$$

In contrast, $1 - X_t$ for inhomogeneous crystallization be expressed from (4.31) and (4.33) as

$$-\ln(1 - X_t) = \frac{4}{3}\pi u^3 t^3 N_0 \tag{4.36}$$

Since the synthesized glasses generally include inhomogeneity, the measured results of $-\ln(1 - X_t)$ might be obtained as the summation of (4.35) and (4.36), so

$$-\ln(1 - X_t) = \frac{1}{3}\pi u^3 t^4 I + \frac{4}{3}\pi u^3 t^3 N_0 \tag{4.37}$$

Therefore, in order to evaluate an intrinsic stability against crystallization, a part of the homogeneous crystallization must be extracted from the results. A nonlinear curve fitting of (4.37) was applied for the results, and then inhomogeneous crystallization was removed from the results to obtain (4.35). The critical volume fraction of glass is defined to be $X_t = 10^{-6}$. Then, in this study, crystallization time t_g corresponding to $X_t = 10^{-6}$ was used as a criterion of stability against crystallization.

In the previous discussion, the crystallization rate is formulated as a function of time, and crystal volume is expressed as the product of nucleation and crystal growth rates. To clarify the formation of scattering centers in optical fibers, temperature dependence of nucleation and growth rates must be evaluated. A DSC measurement could not analyze nucleation and growth rates separately because only crystal volume can be measured as an integration of exothermic peaks and nucleation rate cannot be detected by DSC.

Figure 4.37 [108] shows the schematic temperature dependence of nucleation and crystal growth rates. As shown in Figure 4.37, these curves do not overlap for a stable glass. Therefore, two-stage heat treatment in the two temperature regions of nucleation and crystal growth enables measuring the nucleation rate indirectly. Figure 4.38 shows the temperature diagram of two-stage heat treatment. In this method, glass samples are heated at T_1, at which nucleation is dominant for time

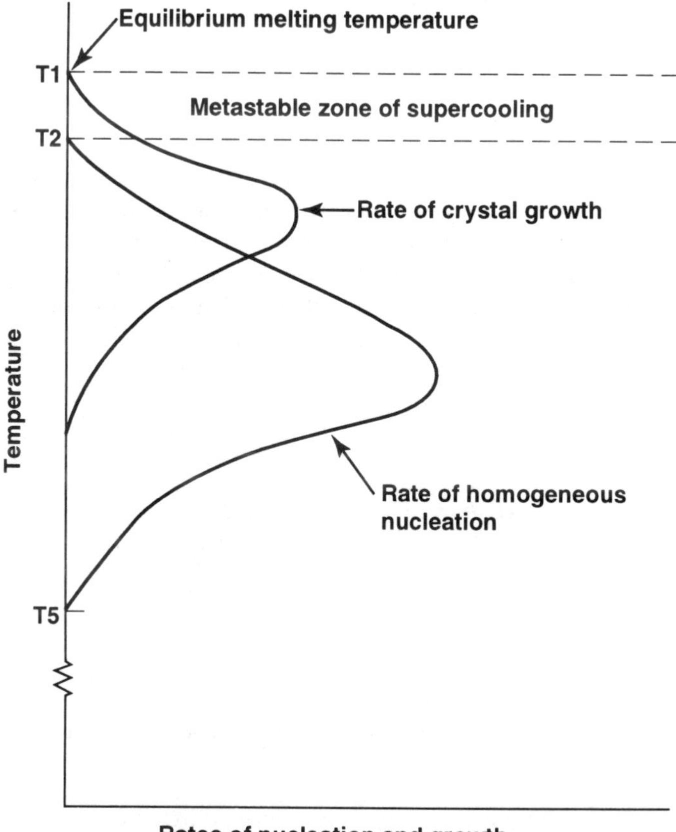

Figure 4.37 Effect of temperature on the rates of homogeneous nucleation and crystal growth [108].

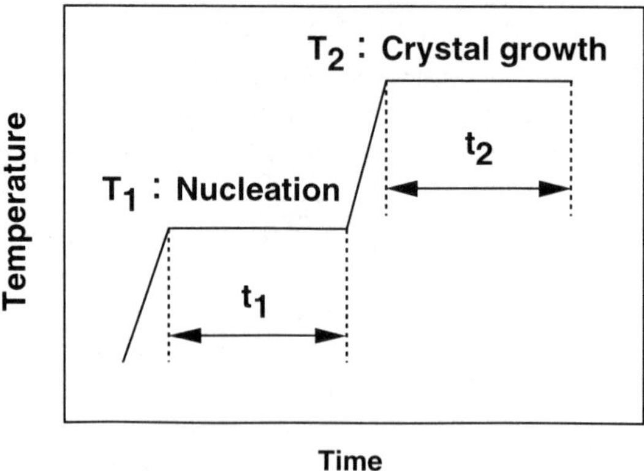

Figure 4.38 Temperature profile of two-stage heating technique.

t_1, and then heated at T_2, at which crystal growth is dominant for time t_2. During heat treatment at T_1, the amount of nuclei formed is the product of nucleation rate I_1 and time t_1. In the temperature region of crystal growth, crystal growth occurs from the nuclei formed at T_1. In general, nuclei are too small to affect a total crystal volume. Therefore, a number of nuclei cannot be measured by DSC. By using this two-stage heat treatment, however, we can estimate a relative nucleation rate. This is because glass with a large number of nuclei leads to a rapid increase in the crystal volume at T_2 rather than that with a small amount of nuclei. According to this discussion, the crystal volume fraction can be expressed as

$$-\ln(1 - X_t) = \int_{t'=0}^{t'=t_1} \frac{4}{3}\pi u_1^{3}(t - t')^{3} I_1 \, dt' + \frac{4}{3}\pi u_1^{3} N_0 t_1^{3}$$

$$+ \int_{t''=0}^{t''=t_2} \frac{4}{3}\pi u_2^{3}(t - t'')^{3} I_2 \, dt'' + \frac{4}{3}\pi u_2(I_1 t_1 + N_0) t_2^{3} \qquad (4.38)$$

Then

$$-\frac{3}{\pi}\ln(1 - X_t) = u_2^{3} I_2 t_2^{4} + 4u_2^{3}(I_1 t_1 + N_0) t_2^{3} + u_1^{3} I_1 t_1^{4} + 4u_1^{3} N_0 t_1^{3} \qquad (4.39)$$

First, in order to obtain the temperature dependence of the nucleation rate, the time dependence of X_t is measured at T_2 after heating the samples at various T_1 with constant dwelling time t_1. In this measurement, since u_1 is negligible, (4.39) is given by

$$-\frac{3}{\pi}\ln(1 - X_t) = u_2{}^3 I_2 t_2{}^4 + 4u_2{}^3 (I_1 t_1 + N_0) t_2{}^3 \tag{4.40}$$

By fitting (4.40) in the measured results, $4u_2{}^3(I_1 t_1 + N_0)$ can be obtained as a coefficient of $t_2{}^3$ term. Furthermore, $4u_2{}^3(I_1 t_1 + N_0)$ is measured for various nucleation time t_1 to obtain $u_2{}^3 I_1$ and $u_2{}^3 N_0$. Since u_2 is constant, $u_2{}^3 I_1$ corresponds to a relative nucleation rate at T_1. On the other hand, the temperature dependence of the crystal growth rate is also obtained by measuring the time dependence of X_c at various T_2 after heating the samples at constant T_1 and t_1. In this case, since $4u_2{}^3 (I_1 t_1 + N_0)$ is obtained as a coefficient of $t_2{}^3$ and $I_1 t_1 + N_0$ is constant, $4u_2{}^3(I_1 t_1 + N_0)$ corresponds to a relative crystal growth rate. Moreover, by measuring u_2 directly using, for example, an optical microscope, temperature dependence of I_1 can be obtained. After formulating the nucleation rate, crystal growth rate is measured according to (4.40).

As an example of crystallization analysis, PbF_2 containing ZrF_4-based fluoride glass, which has been used as a core of the PDFAs, is described based on the aforementioned theory. In this analysis, the following nucleation and crystal growth rate equations [109] were used in addition to the prior equations:

$$I = \frac{n_v kT}{3\pi\lambda^3 \eta}\exp\left\{-\frac{16\pi\sigma^3 V_m{}^2}{3kT} \cdot \frac{T_m{}^2}{\Delta H_f (T_m - T)^2}\right\} \tag{4.41}$$

$$u = \frac{fkT}{3\pi\lambda^2 \eta}\left[1 - \exp\left\{-\Delta H_f\left(\frac{1}{T} - \frac{1}{T_m}\right)\right\}\right] \tag{4.42}$$

where n_v is a number of atoms in a unit volume, λ is a jumping distance of atoms in a diffusion process, η is a viscosity of a glass, σ is a free energy of an interface in a unit surface area, V_m is a molar volume of a crystal phase, T_m is a melting point, ΔH_f is an enthalpy of fusion, and f is a effective site fraction on a nuclear surface. Figure 4.39 shows the temperature dependence of the relative nucleation rates. In this figure, open circles and the solid line indicate the measured results and the fitting curve according to (4.41). The relative nucleation rate can be obtained as

$$u_2{}^3 I_1 = 3.71 \times 10^{10} T \exp\left\{-\frac{8.76 \times 10^8}{T(787.6 - T)^2}\right\} \tag{4.43}$$

In this equation, T_m was measured using DSC prior to the analysis. This figure reveals that the temperature dependence of the nucleation rates seems to be small, and the nucleation indicates a maximum at a temperature of 20°C higher than T_g. Figure 4.40 shows the temperature dependence of the relative crystal growth rates. The fitting results of (4.42) are expressed as

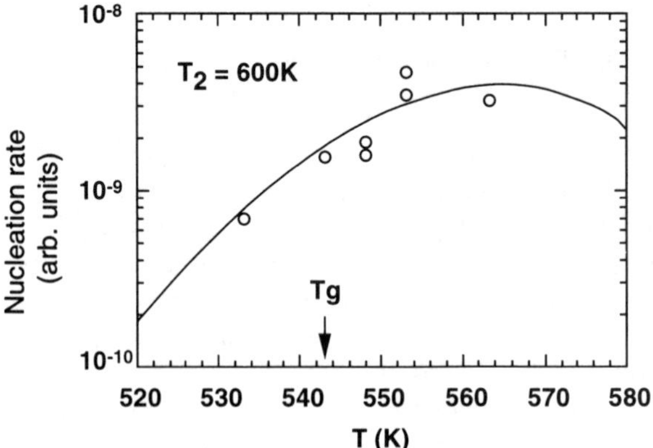

Figure 4.39 Temperature dependence of nucleation rate for ZrF$_4$-based fluoride glass measured using the two-stage heating technique.

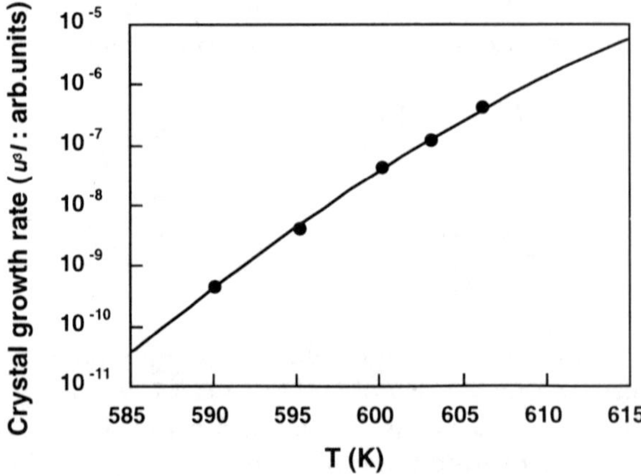

Figure 4.40 Temperature dependence of crystal growth rate for ZrF$_4$-based fluoride glass measured using the two-stage heating technique.

$$u_2^{3}I_1 = 2.60\left[1 - \exp\left\{-5.0\left(\frac{1}{T} - \frac{1}{787.6}\right)\right\}\right] \qquad (4.44)$$

Figure 4.41 shows the nucleation and crystal growth rates and the deformation temperature range of this glass system. It is clear that the nucleation and crystal growth simultaneously occur in the fiber drawing process so that a scattering loss caused by the crystallization might be sensitive to the fiber drawing condition. Therefore, this result suggests that a scattering loss of the fibers prepared using

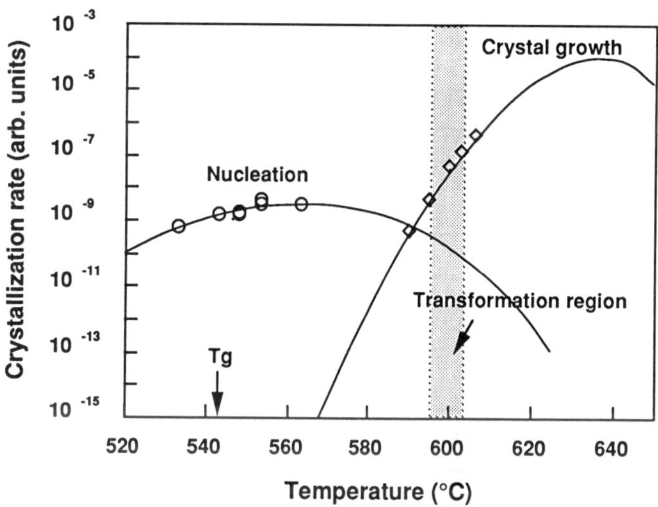

Figure 4.41 Temperature dependence of crystallization rate for ZrF_4-based fluoride glass.

this glass system may be determined by the intrinsic crystallization rate and may vary with their composition.

The representative parameter indicating the degree of the crystallization tendency is a crystallization time t_g, as previously described. Figure 4.42 shows loss spectra for the fibers fabricated using ZrF_4-based fluoride glasses with different PbF_2 contents. Since the transmission losses decrease with increasing t_g, it is clear

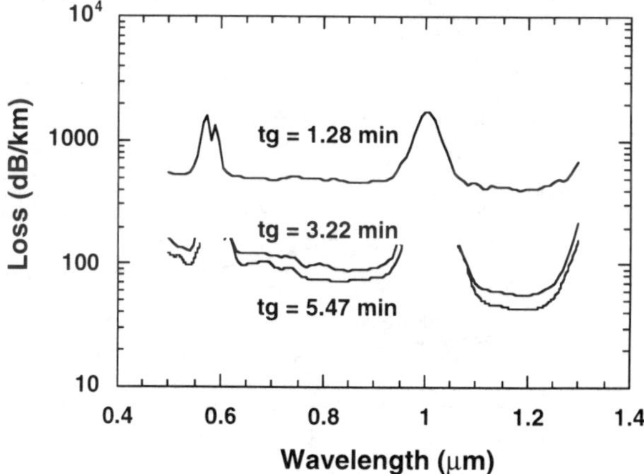

Figure 4.42 Loss spectra for fibers fabricated using ZrF_4-based fluoride glasses with different PbF_2 contents.

that the intrinsic crystallization tendency reflects the scattering loss in this glass system. In addition, the relationship between the transmission losses at 1.2 μm and t_g is shown in Figure 4.43. The spectrum analysis indicates that the dominant loss factor is a waveguide imperfection caused by crystallization, and the minimum loss of the fiber with this glass system may be 20 dB/km unless a fabrication method is developed.

Consequently, a potential of a specific glass system for an amplifier host can be evaluated using the analysis of the crystallization kinetics in terms of a scattering loss.

4.3.2.2 Evaluation of Scattering Loss due to Crystallization in Fiber Fabrication Process

As described in the following sections, thermal properties of those glasses might be the most important criteria in deciding which fiber fabrication techniques can be adopted. Except silica fibers, in general, a glass-melt is cast into a metallic mold to make a preform, and then the preform is drawn into fibers. This process is adopted for fluoride glasses and multicomponent silica-based glasses. With regard to these glass systems, casting and drawing are two major heat processes that determine a degree of crystallization. Therefore, it is of critical importance to be able to quantitatively evaluate the number and size of crystals during casting and fiber drawing and to relate these to a degree of an increase in scattering loss.

The basic theory of crystallization composed of nucleation and crystal growth is described by a crystallization model demonstrated by Drehman [110] in Section 4.3.2.1. In this section, however, Schneider and Staudt's model [111] is adopted

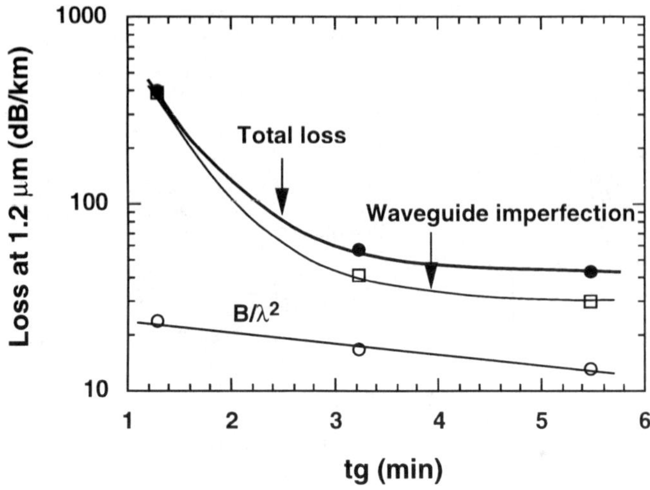

Figure 4.43 Relationship between the transmission losses at 1.2 μm and the crystallization time t_g.

because the calculated crystallization fraction has been demonstrated to fit the measured results [112–114]. The model is given by

$$I = a_0 T \exp\{-a_1/(T - a_3)\}\exp[-a_2/\{T(T - a_4)^2\}] \qquad (4.45)$$

$$K = b_0(b_1 - T)\exp\{-b_2/(t - b_3)\} \qquad (4.46)$$

where I is the initiation rate in $\mathrm{cm}^{-3} \cdot \mathrm{sec}^{-1}$, K is the growth rate in $\mathrm{mm} \cdot \mathrm{sec}^{-1}$, and $a_0 - a_4$ and $b_0 - b_3$ are constants. The parameters measured for ZrF_4-based fluoride glasses are listed in Table 4.11 [111]. While Drehman's model seems to exaggerate the crystallization, that of Hart et al. offers the lowest bound [109]. Thus, an evaluation based on Hart's model is made as a comparison. Their model is given by

$$I = c_0 T \exp\{-c_1/T\}\exp[-c_2/\{T(T - c_3)^2\}] \qquad (4.47)$$

$$K = d_0 T_0(b_1 - T)[1 - \exp\{-d_2(T - d_3)/T\}]$$
$$\exp[-2d_3/\{(T - d_4 + (T - d_4)^2 + 4d_5 T)^{1/2}\}] \qquad (4.48)$$

where c and d are constants. For ZrF_4-based fluoride glasses, numerics are e^{256}, 1.05×10^5, 3.7×10^9, and 850 for c_0, c_1, c_2, and c_3; and 0.0006, 850, 5, 110, 632, and 1.2 for d_0, d_1, d_2, d_3, d_4, and d_5, respectively [109,112].

To determine the degree of crystallization during melt-casting, temperature changes in the glass and mold regions are calculated first and then crystallization is calculated using the nucleation and growth rate equations. The temperature change in glass and mold regions are described by

$$\frac{\partial T_i}{\partial t} = \frac{k_i}{\rho_i c_i}\left(\frac{\partial^2 T_i}{\partial r^2} + \frac{1}{r}\frac{\partial T_i}{\partial r}\right) \qquad (4.49)$$

where T is the temperature, t is the time, k is the thermal conductivity, ρ is the density, c is the heat capacity, r is the radius, and the subscript indicates the glass or mold regions. The initial conditions are

Table 4.11
Parameters from Least Square Fits of Nucleation Data and Crystal Growth Data

	a_0	a_1	a_2	a_3	a_4
Nucleation	$2.2 \times 10^{55}\ \mathrm{cm}^{-3}\mathrm{sec}^{-1}$	1,317K	3.659×10^{10}K	483.2K	1,307K
Crystal growth	$40.08\ \mu\mathrm{m} \cdot \mathrm{sec}^{-1}$	720K	847.9K	530K	–

From: [111].

$$T = T_i \quad \text{at } t = 0 \tag{4.50}$$

for glass, mold, and ambient regions. The boundary conditions are given by

$$\frac{\partial T}{\partial r} = 0 \quad \text{at } r = 0 \tag{4.51}$$

$$\frac{\partial T_i}{\partial r} = \frac{\partial T_{i+1}}{\partial r} \tag{4.52}$$

$$T_i = T_{i+1} \quad \text{at } r = r_i \tag{4.53}$$

and

$$\frac{\partial T_i}{\partial r} = -h(T_i - T_0) \quad \text{at } r = r_i \tag{4.54}$$

where h is the heat transfer coefficient and T_0 is the ambient temperature [114].

In the fiber-drawing process, the temperature distribution in the preform neckdown region is given as a steady-state heat condition by [115]

$$V \, dt/dz = -(T - T_1)2h/(\rho cr) \tag{4.55}$$

$$F = 3\pi r^2 \mu \, dV/dz \tag{4.56}$$

where V is the speed, F is the drawing tension, T_1 is the furnace temperature, and μ is the glass viscosity. These simultaneous differential equations are numerically solved assuming that the furnace temperature profile and the viscosity are given by

$$T_1 = T_m \exp\{-(z - z_0)^2/g\} \tag{4.57}$$

and

$$\mu = \mu_0 \exp(B/T) \tag{4.58}$$

where T_m is the maximum temperature, z_0 and g are profile parameters, and μ_0 and B are constants. The preform is substantially heated until neckdown begins. This preheating stage strongly affects crystallization because of slow preform feed speed. For this region, the temperature is obtained by extrapolating the furnace temperature [115].

The number of crystals is calculated for a short dwell time, and their size is calculated in the next time segment. These values accumulate for the whole heat-treatment process. Such a calculation for crystallinity based on additivity is useful for a dynamic thermal process [116]. In the case of glasses prepared by vapor phase technique, crystallization only by fiber drawing must be taken into account.

The nucleation rate data from Drehman [110] and Hart et al. [109] are shown in Figure 4.44, where it can be seen that the nucleation rates are widely scattered and varied with the measurements, because the nucleation rate can only be measured by the indirect method such as the two-stage heating method and the nucleation undergoes the influence of impurities and the surface crystallization. In addition, nucleation and crystal growth of several kinds of crystals may occur simultaneously in the case of multicomponent glasses. Therefore, a glass sample with high purity must be used for the measurement and the growth crystal phase must be identified prior to the measurement. Figure 4.45 shows the crystal growth rates of ZBLAN measured by Hart [109] and Schneider [111], where both data are coincident, because a crystal growth can be directly measured and is insensitive to impurities. Using these data, the number and the size distribution of crystals can be estimated for several heating conditions based on the aforementioned theory. Figures 4.46 and 4.47 show the crystal size distribution in the core region for the fiber drawn from 15-mm preform and the extrinsic scattering loss spectra in the 1- to 3-μm wavelength range [114]. Figure 4.46 reveals that the crystals after casting exhibit a

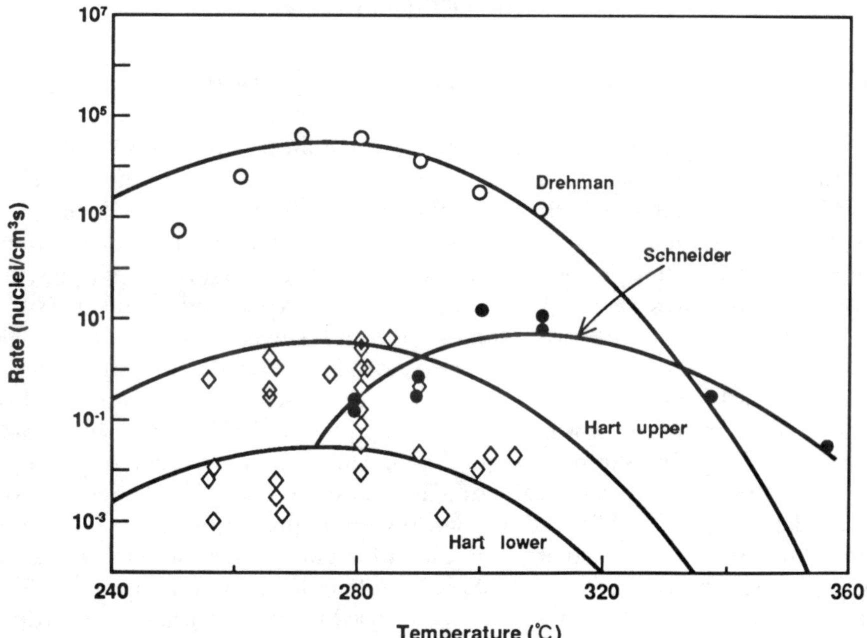

Figure 4.44 Temperature dependence of nucleation rates measured for ZBLAN glass.

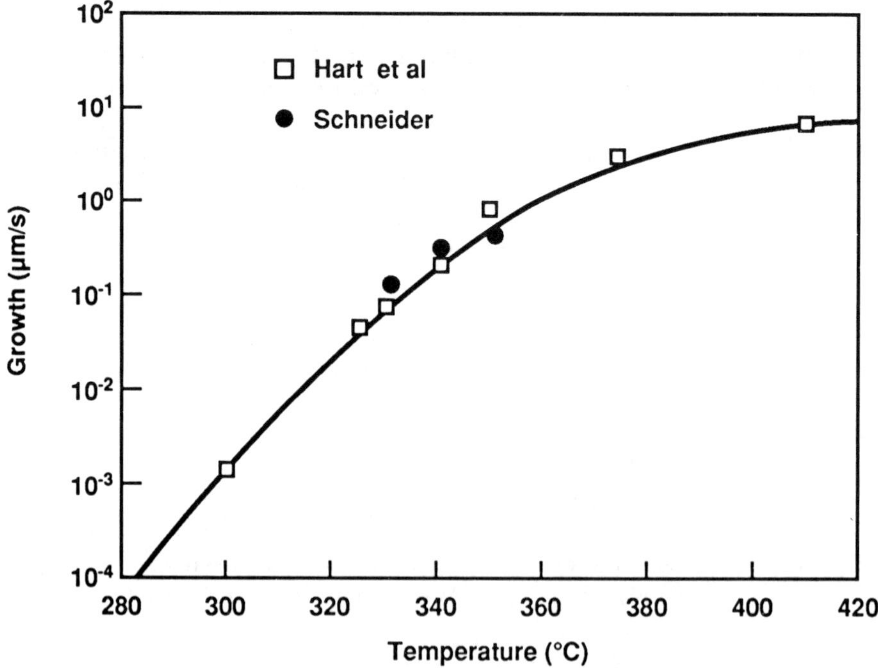

Figure 4.45 Temperature dependence of crystal growth rates measured for ZBLAN glass.

gradual distribution but that the distribution after drawing has a clear peak. This peak indicates growth during drawing of crystals that were present after casting. The peak value corresponds to the amount of growth during drawing. Figure 4.46 shows the scattering loss spectra in the 1- to 3-μm wavelength range for the fiber whose crystal distribution is shown in Figure 4.46. In this figure, m denotes the relative refractive index ($m = m_1/m_2$, where m_1 is the crystal index and m_2 is the matrix index). The inherent extrinsic scattering loss due to crystallization is strongly affected by m; and 80 dB/km and less than 10 dB/km at 1.5 μm are estimated for $m = 1.05$ and 1.01, respectively. Schneider et al. also calculated the scattering loss of ZrF_4-based fluoride fibers, which are fabricated by suction casting using molds with different diameters shown in Figure 4.48. The fabrication technique will be described in the following sections. Since the scattering loss strongly depends on the mold diameter, the stability of this glass system might not be stable enough to achieve a low-loss value of less than 0.1 dB/km with a fiber length which is required for long-distance transmission media. Since the length of an amplifier host fiber is relatively short, a small-diameter mold can be used for the fabrication. According to this scattering loss estimation, relatively low-loss ZrF_4-based fluoride fiber can be obtained by using the small-diameter mold. Therefore, ZrF_4-based fluoride glass is applicable to an amplifier host fiber.

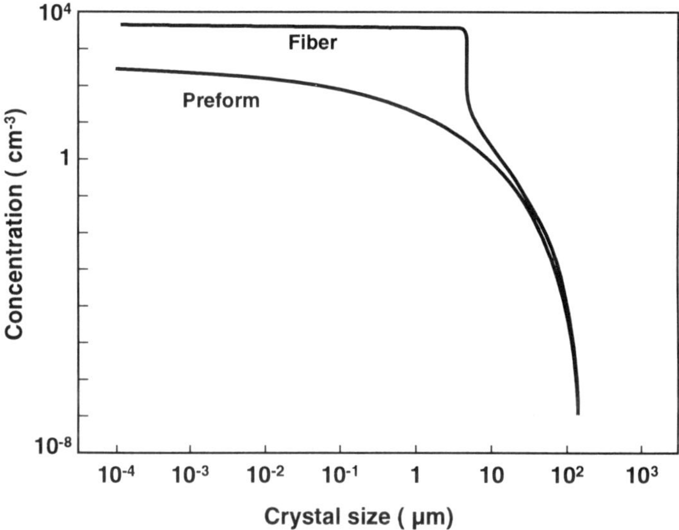

Figure 4.46 Crystal size distribution in a preform with 7-mm core and 15-mm cladding diameters and in a fiber drawn with a 125-μm diameter [114].

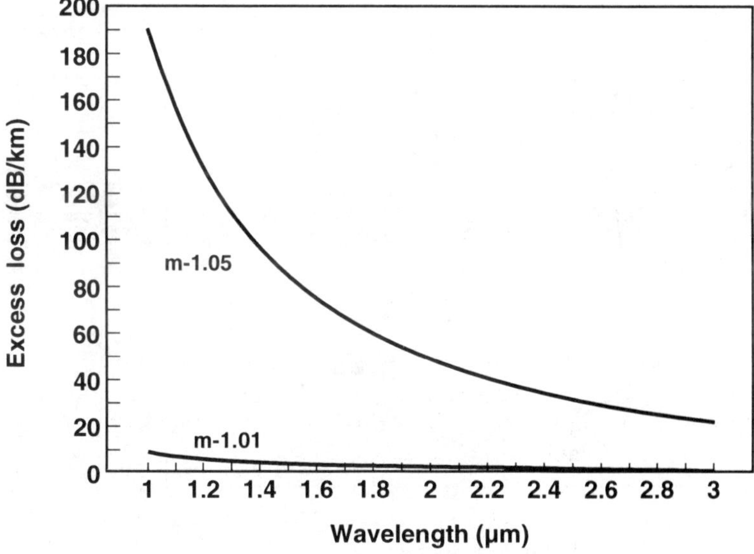

Figure 4.47 Extrinsic scattering loss spectra in the 1- to 3-μm wavelength range for the fiber whose crystal distribution is shown in Figure 4.46 [114].

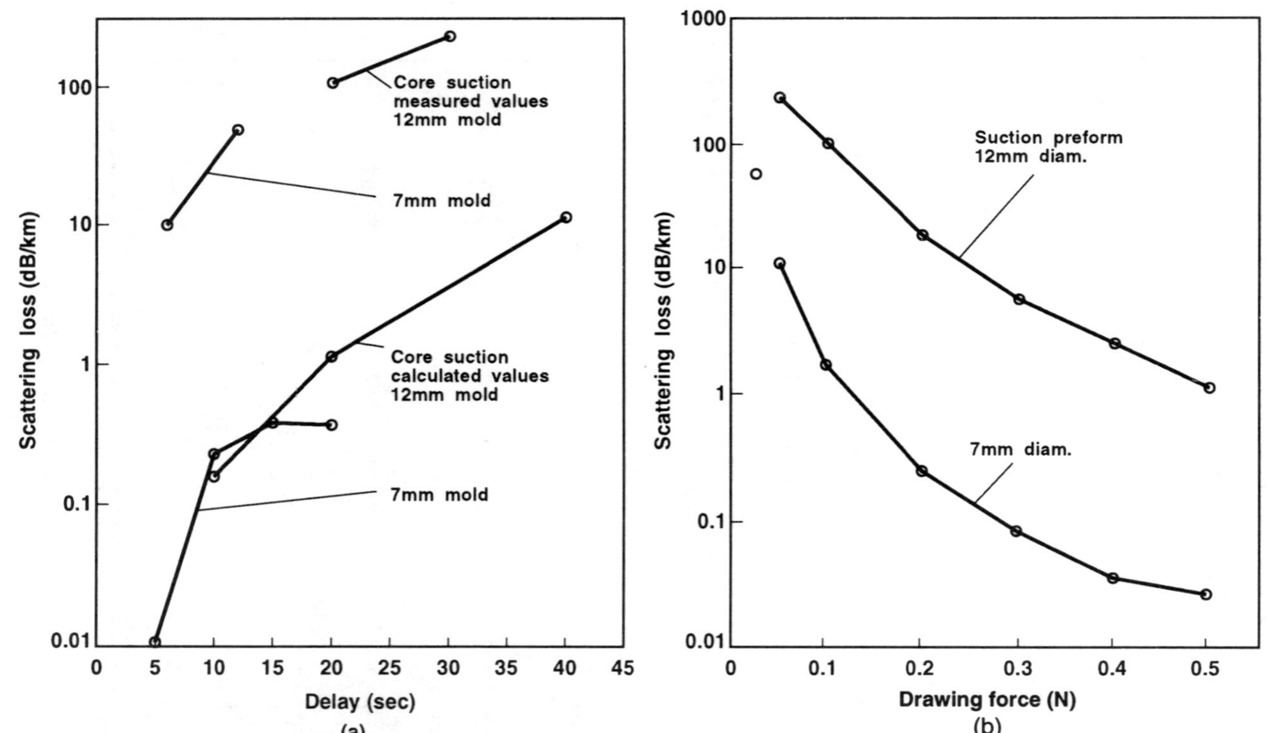

Figure 4.48 Calculated scattering losses at 2.5 μm resulting from (a) preform crystals and (b) preform and fiber crystals for 160 μm fiber with $\Delta n = 0.006$: (a) suction technique, 650°C melt and 220°C mold temperature, 75% core; (b) same preforms as (a) with 5 sec delay, drawing speed 11 cm/s, cone length three times preform radius [111].

Scattering loss due to crystallization has been thoroughly evaluated from ZrF_4-based fluoride glass fibers [111,112,114,117]. These works were done in order to clarify the possibility of achieving long-length and ultra-low-loss optical fibers, and the evaluation based on the previous equations revealed that it is impossible to realize long-length and low loss simultaneously. Although the evaluation of crystallization derived negative results in this case, this technique is very useful in judging the capability of a specific glass system as an amplifier host fiber and in analyzing a fiber fabrication processes.

4.4 HIGH-SILICA FIBER FABRICATION PROCESS AND RARE-EARTH DOPING

As mentioned earlier, the requirements for fibers used in fiber amplifiers and lasers operating as active devices are different from those for fibers used for long-distance transmission. Basically, active fibers need a greater variety of materials and structures in order to realize efficient amplification and lasing at various operating wavelengths.

However, the preparation of rare-earth-doped fibers is based on the method used for long-distance transmission fibers. Rare-earth-doped fibers are prepared by modifying standard fabrication processes so that rare-earth ions can be doped into silica glass. Therefore, we first describe the standard fabrication processes for transmission fibers and then move on to the rare-earth doping techniques used in individual fabrication processes.

4.4.1 Soot Process: Origin of High-Silica Glass Fabrication Process

Oxide glass fibers are commonly separated into two main groups: high-silica glasses, which consist of SiO_2 as the main component and one or more oxide (GeO_2, P_2O_5, B_2O_3) as a dopant to change the refractive index, and multicomponent glasses, which are commonly used as optical glasses in such devices as lenses and light guides. These glasses consist of one or more glass formers, for example, SiO_2, B_2O_3, or P_2O5, modified by such nonglass formers as Na_2O, CaO, or Al_2O_3. High-silica glass fibers are made by drawing fiber preforms, which have a similar structure, that is, core-cladding structure, to that of the desired fibers. In contrast, multicomponent glass fibers are prepared using a double crucible.

However, today, it is widely recognized that high-silica glass systems are the best materials for transmission fiber in terms of optical loss, fiber structure, fiber strength, and long-term reliability. They already occupy a prominent position as the glass material for transmission fiber use in the 0.6- to 1.7-μm wavelength region.

The high purity of high-silica glass results from the use of a fabrication process now known as the "soot process." This process is basically a vapor phase reaction. As mentioned in Section 1.4.3, the origin of the soot process for synthesizing high silica glasses can be traced to work undertaken in the early 1930s by J. F. Hyde [118]. He invented a flame hydrolysis process with the intention of producing silica glass at a relatively low temperature and, if desired, with a high degree of purity [118]. As shown in Figure 1.11(a,b), Hyde also described in his patent the way in which the soot layer is formed on the mandrel by the deposition of fine glass particles that are synthesized by flame hydrolysis and how this soot layer is then vitrified into transparent silica particles by the oxy-hydrogen burner, which is the origin of the OVD method described later. In short, the advantage of the soot process is the potential for preparing high-purity silica glass at a temperature lower than its melting point. There then followed work by M. E. Nordberg in 1939 on the preparation of doped silica glasses [119]. Nordberg prepared silica glass containing TiO_2, which was originally intended as a glass with a low expansion coefficient. In this process, silica or doped silica particles were formed by the introduction of silicon tetrachloride vapor and chloride species dopant into a gas burner flame (flame hydrolysis). These particles were collected and vitrified in a manner designed to produce doped silica glass with optical quality. Figure 4.49 shows a doped silica glass preparation process invented by M. E. Nordberg [119]. In Figure 4.49, the raw materials such as $SiCl_4$ and $TiCl_4$ are placed in containers

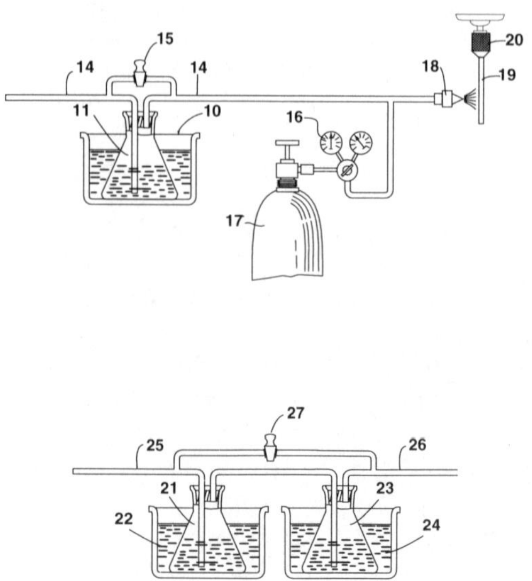

Figure 4.49 TiO_2-doped silica glass preparation process invented by M. E. Nordberg [119]. $SiCl_4$ is in a container (11 or 21) and $TiCl_4$ is in another container (23). In the preparation of TiO_2-doped glass, a mixture of $SiCl_4$ vapor and $TiCl_4$ vapor is supplied to a gas burner (18) and the SiO_2 glass particles containing TiO_2 are deposited onto a mandrel (19).

(indicated as 11, 21, and 23) and are transported by the carrier gas that bubbles in the containers. In the preparation of TiO_2-doped silica glass, a mixture of $SiCl_4$ vapor and $TiCl_4$ vapor is supplied to a gas burner (18) and the SiO_2 glass particles containing TiO_2 are deposited onto a mandrel (19).

Some 30 years later, this soot process of synthesizing TiO_2-doped silica glasses was applied by Keck and Schultz to fabricate low-loss optical fibers from doped high-silica glasses. They invented the ISD method in 1970 and achieved high-silica glass optical fibers with a low loss of 20 dB/km [120–122]. Until that time, TiO_2-doped silica glasses were used as the fiber core glass [120,121]. However, the disadvantage of TiO_2-SiO_2 glass is that the very small amount of Ti^{3+} ions formed in the prepared glass causes a very strong absorption in the visible and near-IR regions, although this absorption can be reduced to as low as 20 dB/km by annealing the fiber in an oxygen atmosphere for a few hours at 800° to 1,000°C [120]. However, it is very difficult to reduce this absorption below 20 dB/km, and in addition, annealing at a high temperature degrades the mechanical strength of the fiber as a result of silica crystallization of silica through alkaline metal contamination.

The real potential of the soot process was revealed when the MCVD method was developed in 1974 by a group at Bell Laboratories that succeeded in fabricating a 1.1-dB/km (at $\lambda = 1.06$ μm) fiber [123–125]. They succeeded in using the soot process to fabricate silica glasses containing GeO_2 whose vapor pressure is rather higher than that of SiO_2 and hence difficult to dope into silica glass, as will be explained later. GeO_2 is the best dopant material for changing (increasing) the refractive index of SiO_2 to a desired level to form the core/cladding structure in fibers, especially with regard to the "weakly waveguiding" structure. In addition, there is no crystallization and no extra absorption caused by GeO_2 doping because Ge atoms belong to group IV in the periodic table, the same as Si atoms, and hence GeO_2 has almost the same bonding structure as SiO_2. In this sense, GeO_2 can be called the best partner with which to form high-silica glasses. The development of this MCVD method gave a powerful impetus and stimulated the development and improvement of two other representative fabrication methods: OVPO or OVD method [126,127] and the VAD method [128–130]. These three methods will be described in detail.

Here, it is worthwhile to explain a little about the mechanisms that enable the soot process to produce high-purity (low-loss) doped silica glasses. There are two key features in the soot process that relate to the purity of the prepared glass and refractive-index controllability. First, the soot process is basically a vapor phase reaction in a flame. The glass particles are synthesized in the flame from raw vapor materials such as $SiCl_4$ and $GeCl_4$. As shown in Figure 4.49, the raw materials in the vapor phase are evaporated in a container and transported by a carrier gas to a flame formed by oxygen and hydrogen or methane gas. An important aspect of this purification process is the difference between the vapor pressures of metal halides, especially the difference between those of the raw glass materials and transition metal halides that cause absorption loss in the wavelength region of interest. Figure 4.50 shows the vapor pressures of the raw glass materials of metal halides [131,132]. From the data in Figure 4.50 it is clear that the large vapor difference between such raw glass materials such as $SiCl_4$ and $GeCl_4$ and such transition metals as $FeCl_3$ and VCl_4 makes it possible to greatly reduce transition metal impurities in the raw glass materials transported into the oxy-hydrogen flame. This very high level distillation of the starting halide materials enables the preparation of high-purity silica glass and the resultant low absorption loss fibers.

The second key feature of the soot process lies in the advantage of being able to prepare GeO_2-doped silica glass. Despite the great advantages of GeO_2 as a dopant material for SiO_2, the fabrication of fibers with a GeO_2-doped silica glass core was impossible before the MCVD method appeared in 1974. There is a clear reason for the difficulty in preparing GeO_2-doped silica glass, which will be described. In the oxy-hydrogen flame used to synthesize oxide glasses from raw metal halides, the synthesized oxides such as SiO_2 and GeO_2 are exposed to a very high temperature of over 2,000°C. Therefore, each individual oxide has a high

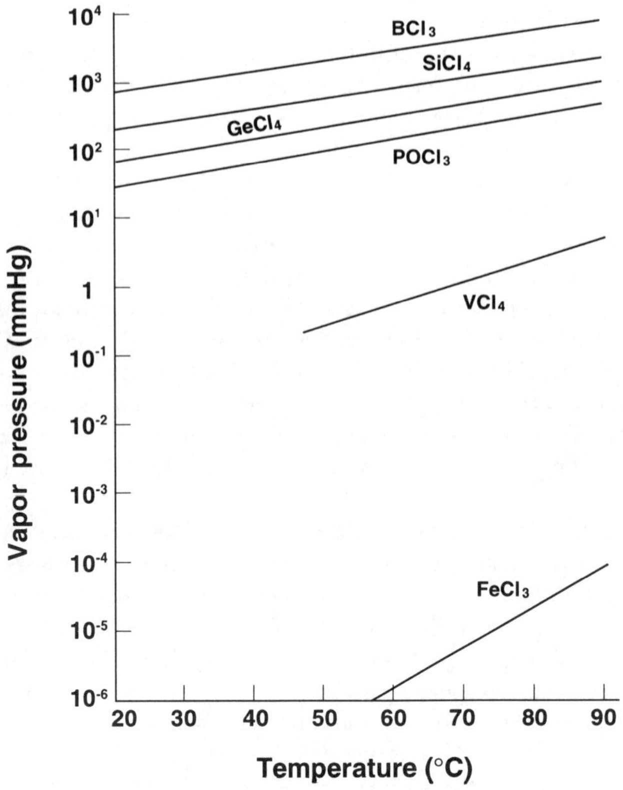

Figure 4.50 Vapor pressures of raw glass materials for metal halides [131,132].

vapor pressure. Figure 4.51 shows the saturated vapor pressures for various oxides of glass fiber compounds [133,134]. As mentioned in Section 2.6.1 and shown in this figure, the saturated vapor pressures of GeO_2 and B_2O_3 are several orders of magnitude higher than those for SiO_2, TiO_2, and Al_2O_3. In addition, the vapor pressure of P_2O_5 is several orders of magnitude higher than that for GeO_2 and B_2O_3. Oxides with higher vapor pressures at a high temperature will tend to remain in the vapor state. For example, GeO_2, B_2O_3, and P_2O_5, whose vapor pressure curves are located at higher temperatures than that for SiO_2, are more likely to remain vapor in the flame and not form fine glass particles (solid state). On the other hand, TiO_2 and Al_2O_3, which have a lower vapor pressure than SiO_2, solidify easily to form fine glass particles in the flame. This means TiO_2 and Al_2O_3 can be easily doped in SiO_2 glass to change the refractive index. On the other hand, GeO_2 and B_2O_3, which have a higher vapor pressure than SiO_2, are difficult to dope into SiO_2 glass because they are likely to remain vapor and to move away from the deposited SiO_2 glass. In addition, P_2O_5, which has a much higher vapor pressure than GeO_2 and B_2O_3, is much more likely to remain vapor and is therefore very difficult to dope into silica glass because of the great difference in oxide vapor pressure.

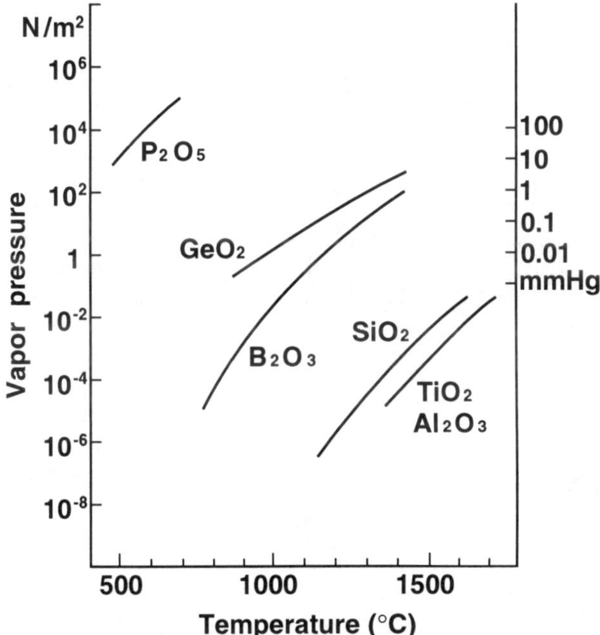

Figure 4.51 Saturated vapor pressures for various oxides of glass fiber compounds [133,134].

With the *direct deposition* (DD) process using an oxy-hydrogen flame [135], a plasma torch [136–139], or a CO_2 laser [140–142] as a heat source and, since the deposited SiO_2 glass is heated to a high temperature to obtain a transparent silica glass rather than a soot glass, GeO_2 or B_2O_3 easily evaporates from the surface because the temperature is too high for them to become solid. Figure 4.52 shows silica glass production by the DD process using the plasma torch as a heat source

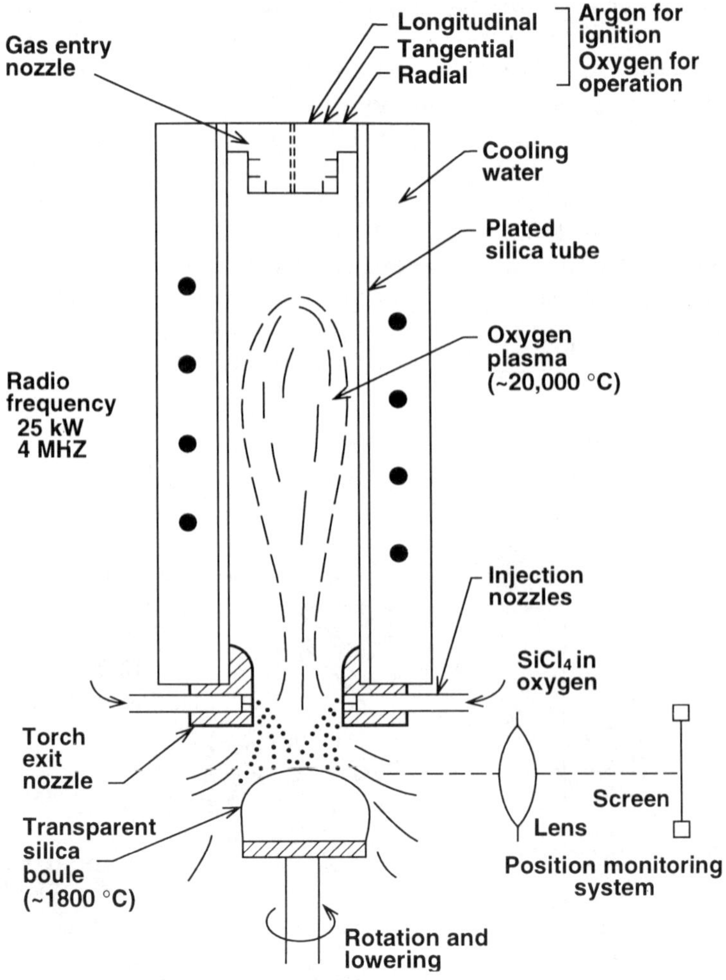

Figure 4.52 Silica glass production by the DD process using a plasma torch a heat source [138]. Because of the high temperature (1,800°C) of the silica glass surface and the resultant high vapor pressure, GeO_2, B_2O_3, and P_2O_5 are difficult to dope into a SiO_2 glass boule. Available dopants are Al_2O_3 and TiO_2.

[138]. In this case, a transparent silica glass boule is made directly from raw halide material ($SiCl_4$). This means that both soot deposition and sintering (becoming transparent) take place simultaneously. Therefore, the surface of the silica glass boule must be held at a temperature higher than 1,800°C to ensure that a smooth and transparent glass surface is obtained, which exhibits stable growth. The high temperature of the silica glass surface and the resultant high vapor pressure make it difficult to dope GeO_2 and B_2O_3 in a SiO_2 glass boule. Furthermore, P_2O_5 cannot be doped into silica glass in this process, which is why GeO_2-, B_2O_3-, and P_2O_5-doped silica glass cannot be prepared with the DD process.

The only possible dopants that can be used in the DD process are Al_2O_3 and TiO_2, which have a lower vapor pressure than silica glass, as shown in Figure 4.51. With regard to Al_2O_3-SiO_2 compositions, in 1974, Kobayashi et al. [135] successfully prepared a 10-dB/km single-mode fiber with Al_2O_3-SiO_2 glass core using an oxyhydrogen flame. In 1975 Nassau et al. [139] also prepared a silica glass boule containing alumina using a plasma torch. With Al_2O_3-SiO_2, however, it is difficult to prepare a doped silica glass with a high refractive index because of partial crystallization in the Al_2O_3 rich phase. As a result, the possible refractive index is limited to 1.468 or below. With the TiO_2-SiO_2 composition, as mentioned, it is difficult to prepare low-loss optical fiber because of the absorption loss caused by Ti^{3+} ions, although this system was the most commonly used glass system for fiber fabrication in the early stage [120–122]. Consequently, it became recognized in the mid-1970s that, with the DD method, it is difficult to prepare doped silica glass with a low loss and a sufficiently high refractive index for high-quality optical fiber fabrication.

In short, the key features of the soot process as a method for fabricating optical fibers are its ability to eliminate transition metal impurities and the possibility of using dopants that have a higher vapor pressure than SiO_2. First, fine glass particles (soot) are prepared by means of a vapor phase reaction at sufficiently low temperature for GeO_2, B_2O_3, or P_2O_5 to be doped into SiO_2 glass. These glass particles are then consolidated into transparent glass boules. This enables the preparation of GeO_2-, B_2O_3-, or P_2O_5-doped silica glass with a very high purity. Three representative fabrication methods—that is, the MCVD method, the OVD method, and the VAD method—are all based on this soot process, although they employ different ways of making fiber preforms. In the following subsections, we will describe these three representative methods of fabricating high-silica fiber preforms and their rare-earth doping schemes.

4.4.2 MCVD Process and Rare-Earth Doping

4.4.2.1 Outline of MCVD Process

The origin of the MCVD process, developed at Bell Laboratories in 1974, can be found in the CVD method, where the thin SiO_2 film formation process used in the semiconductor industry was adopted for optical fiber fabrication [143–145]. A relatively high refractive-index SiO_2-GeO_2 glassy film is deposited on the inside surface of a fused silica tube by means of the oxidation reaction between SiH_4 and GeH_4 vapor and O_2 gas. After a sufficient film thickness is obtained, the tube is collapsed so that the deposited layer forms a high-index core and the tube provides a lower index cladding. The CVD method produces fibers with losses of less than 10 dB/km, however, it suffers from two major disadvantages: a low deposition rate for the glass layer and OH ion contamination originating in the use of hydride reactors such as SiH_4 and GeH_4.

In 1970 Corning Glass Works lodged a patent for a new fiber preparation technique, the ISD method [120], where the technique for preparing high-silica glass through the soot process was applied to the fabrication of optical fibers. With the ISD method, the inside wall of a thick-walled fused silica tube is coated with a thin layer of TiO_2-doped SiO_2 core glass soot by aiming the soot stream from an oxy-hydrogen burner down through the tube hole. Then this soot layer is sintered to form a bubblefree glass film. Finally, the composite preform is drawn into a fiber to obtain a solid glass. The first low-loss TiO_2-doped SiO_2 core fiber was made using this ISD method [120,122]. However, there are also several disadvantages with this method related to fiber performance, as exemplified by the need for a 800° to 1,000°C heating treatment to reduce the absorption loss due to Ti^{3+} ions and the difficulty in the formation of a uniformly thick soot layer along the tube. The MCVD method, which originated in a combination of the earlier CVD process and the ISD process, was developed in order to overcome the aforementioned problems and to produce fibers with a low-OH content and high optical quality [123–125]. Figure 4.53 shows the basic principle for making fibers by the MCVD method. Here a silica glass tube is rotated on a glass-working lathe. Raw halide material vapors carried by oxygen gas are introduced into the silica tube, which is heated by an oxy-hydrogen flame from the outside to about 1,500° to 1,650°C. The oxy-hydrogen flame is moved repeatedly along the tube. During each traverse, the halide vapor materials are oxidized into fine glass particles and deposited on the inner surface of the silica tube. These particles are then immediately sintered into a transparent glass film. Multiglass layers are formed by traversing the flame repeatedly along the tube, which is then thermally collapsed to form a preform rod. This silica tube becomes the outer cladding (or called jacketing) and the deposited glass layers become the core and inner cladding of the fiber preform. By changing the $GeCl_4$ concentration in the halide vapors for each traverse, the refractive index of each glass layer can be controlled. Both the core and cladding glass are formed with this process. This preform is subsequently drawn into a fiber in a carbon furnace.

The MCVD method is an elegant fabrication process and hence has several important advantages mainly relating to fiber performance. First, since water vapor and OH-ion contamination can be completely avoided using this enclosed-space reaction process and using halide raw materials, the deposited glass and the resultant fiber core are essentially hydroxylfree. Therefore, it is easy to attain low-loss attenua-

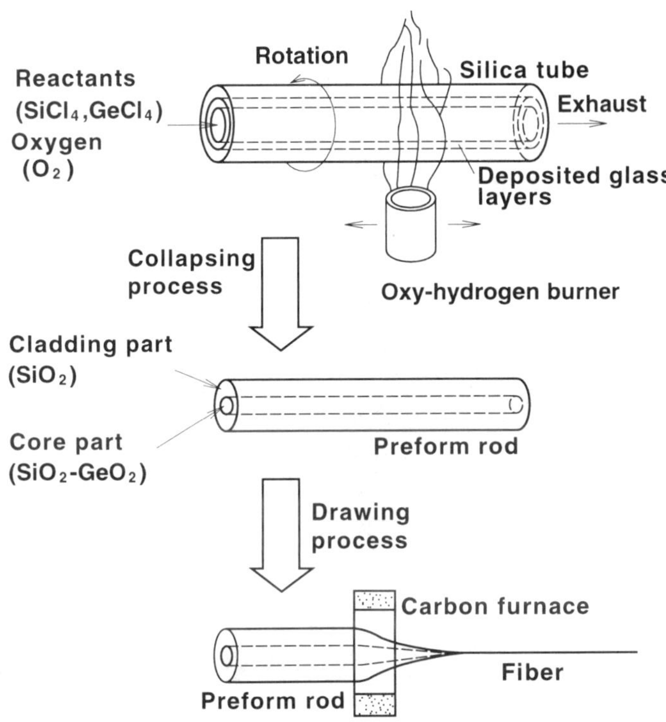

Figure 4.53 Basic principle for making fibers by the MCVD method.

tion. Although OH diffusion from the starting silica tube increases loss in ultra-low-loss fibers, increasing the barrier cladding layer thickness and decreasing the deposition temperature helps to minimize the diffusion of OH ions from the tube into the core glass [146,147]. Second, since the fine glass particles (soot layer) are first deposited and then the soot layer becomes transparent, GeO_2 is readily doped into the silica glass to form the fiber core. This GeO_2 doping leads to excellent waveguiding with very low scattering and radiation losses. Third, soot deposition followed immediately by consolidation using an oxy-hydrogen burner leads to a great improvement in the glass layer deposition rate compared with the CVD method. This improvement is realized as a result of the "thermophoretic effect" in glass particle deposition, which originates from the temperature gradient in the glass particle deposition region inside the silica tube [148]. In 1979, an extremely low-loss single-mode fiber was made with a loss value of 0.2 dB/km [5].

4.4.2.2 Setup for MCVD Process

The fundamental apparatus for the MCVD process consists of a glass-working lathe, a raw material supplier, and an exhaust gas treatment apparatus. Figure 4.54 shows

Figure 4.54 Photographs of MCVD apparatus: (a) a glass-working lathe and its set-up and (b) a silica tube heated by an oxy-hydrogen burner.

photographs of MCVD apparatus: (a) the glass-working lathe and its setup and (b) a silica tube heated by an oxy-hydrogen burner. Figure 4.55 shows a schematic diagram of the MCVD apparatus. In the glass-working lathe shown in Figure 4.54(a), the center of the two chucks should align coaxially so that the center difference is reduced to less than 0.05 mm in order to prevent the silica tube from deforming during the processes of deposition and collapse. A motor-driven carriage that carries oxy-hydrogen torches at a constant speed is added to the lathe. It is preferable to install a radiation thermometer and outer diameter monitor that move with the torches to measure the temperature and diameter of the heated zone. The heated zone can be kept at a constant temperature by controlling the flow rates of oxygen and hydrogen using the signal from the radiation thermometer.

Vapor mixtures of metal halides such as $SiCl_4$, $GeCl_4$, $POCl_3$, and BCl_3 and oxygen gas are fed into the silica tube by the gas supplying system, as shown in Figure 4.55. Hydrogen and oxygen gases are used for heating the silica tube. The main function of the gas supply system is to precisely control the supply rates of the vapor mixture and gases. The halide vapor mixing ratio should be maintained at the programmed value for each layer, and the combusting gases are controlled

Figure 4.54 (continued).

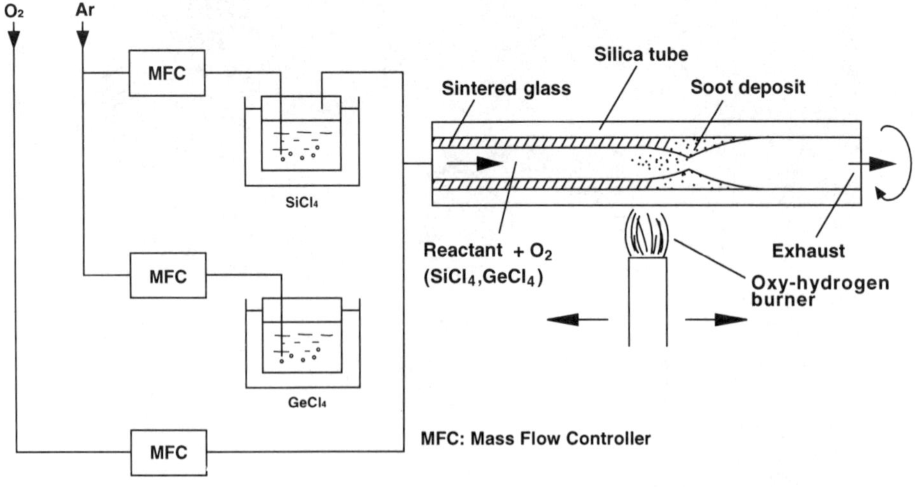

Figure 4.55 Schematic diagram of MCVD method.

via the signal from the thermometer measuring the surface temperature of the silica tube. The metal halides used in the soot process remain in the liquid state at less than 50°C (except for BCl_3), so they are vaporized by bubbling Ar or O_2 gas through them and are then carried into the silica tube together with oxygen, as shown in the figure. High-purity halides are placed in containers made of glass or stainless steel that are housed in constant-temperature baths to maintain the liquid halides at the desired temperature. The halide vapor flow rates are controlled by the flow rate of the carrier gas using thermal mass flow controllers. The flow rates of the halides, which are vapors at room temperature, such as BCl_3, are controlled by thermal mass flow controllers without using a carrier gas.

The reacted gas exhaust from the silica tube contains fine glass particles and chlorine gas. These materials must be removed before releasing the exhaust gas into the atmosphere by using a cleaning system that consists of a soot trapper, water shower, and flow control valve. Fine glass particles in the exhaust gas are first removed by the soot trapper. Next, Cl_2 and HCl are removed by the water shower along with neutralization using NaOH. The exhaust gas flow is controlled by a valve, so the pressure measured with the transducer can be kept at a constant value.

4.4.2.3 Details of MCVD Preform Fabrication

The outer diameter of a tube shrinks from an initial value of 14 mm to 9 mm after about 30 deposition passes at 1,600°C. The thickness of the tube wall increases as it shrinks and consequently the inner surface temperature is reduced. Sometimes it becomes difficult to continue the deposition even if the surface temperature is high enough. In order to avoid this problem and to keep the tube diameter at its initial value, a slight positive internal pressure is applied to the silica tube by adjusting a valve near its exhaust end. The outer diameter of the silica tube in the hot zone is monitored by a noncontact-measuring set that employs a deflecting He-Ne laser beam. The signal is fed back to the pressure-adjusting apparatus in order to maintain the outer diameter to the desired value.

There are many surface imperfections such as small scratches and bubbles on the inner surface of the tube. They would cause an increase in the scattering loss and a decrease in the strength of the produced fibers. The typical additional loss caused by these surface imperfections is estimated to be 0.5 to 1.5 dB/km [149]. Prior to glass deposition, the scratches and bubbles can be removed by prebaking the tube at a high temperature of about 1,600°C. At higher temperatures, the scattering center density decreases abruptly, however, a small number of centers remain. It appears that some of the scratches create new bubbles because they are covered with a glass layer before being removed by surface tension.

The hydroxyl contamination in the glass fibers may be attributed to two causes. One is the contamination due to hydrous and hydroxyl impurities in the raw material chlorides and water vapor in the carrier gas. The other is OH diffusion from the silica tube. The major hydrous impurity in $SiCl_4$ is trichlorosilane (SiHCl3), which can be removed by multistep distillation. Water vapor contained in the carrier gas for the chlorides, oxygen or argon is also an important cause of OH contamination. To remove this water vapor, it is very important not only to purify the gas itself but also to prevent air leaking from the tube connection. The rotating connector used at the raw material inlet of the silica tube and the plastic fittings are the major causes of these leaks.

Interestingly it is found that OH ions in the silica tube diffuse into the silica glass during the deposition process. The diffusion length is about 47 μm before collapse, which is in fairly good agreement with the expected value of 34 μm, calculated by taking a diffusion constant of 7.3×10^{-9} cm^2sec^{-1} at 1,600°C and $t = 400$ sec and by integrating the heating time during the fabrication [146]. Figure 4.56 shows OH ion distribution profiles that were obtained from the optical density at 2.73 μm for four performs with different thicknesses of deposited silica glass of 1.0, 1.5, 2.0, and 2.5 mm [146]. The distance from the inner surface of the deposited tube before collapse is also indicated at the top of the figure. It is clear from the data in this figure that buffer layers are needed to prevent the OH contamination of the silica tube. The OH absorption loss of single-mode fibers changes with the thickness of the deposited cladding layers because a fair proportion of the light power propagates in the cladding. It turns out that the ratio of cladding diameter to core diameter should be larger than 5 to obtain a low-loss fiber [146,147].

The transmission loss of MCVD fibers is strongly affected by the preform fabrication conditions, in particular, the glass deposition temperature. When this temperature is too low, the deposited glass particles cannot be sufficiently sintered

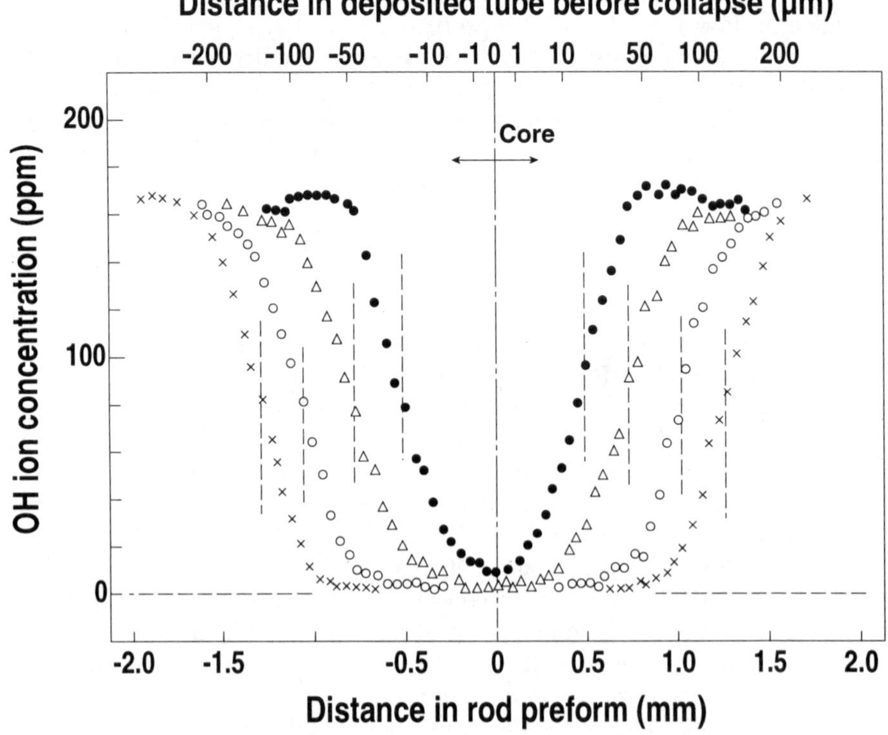

Figure 4.56 OH ion distribution profiles obtained from the optical density at 2.73 μm for four performs with different thicknesses of deposited silica glass of (a) 1.0 mm, (b) 1.5 mm, (c) 2.0 mm, and (d) 2.5 mm [146]. The distance from the inner surface of the deposited tube before collapse is also indicated at the top of the figure.

to form a transparent glass. In contrast, when it is too high, the silica tube shrinks too much for glass deposition to continue. Therefore, the deposition temperature is categorized by the following four values [149]:

1. Appropriate temperature T_{ap};
2. Intermediate temperature T_{im};
3. Critical temperature of bubble creation T_{cr};
4. Bubble creation temperature T_b.

T_b is the temperature at which the scattering center in the glass layers can be seen easily during the deposition process as shown in Figure 4.57(a). T_{cr} is the temperature at which the scattering center is barely seen. T_{ap} is the temperature where the scattering centers are completely eliminated, as shown in Figure 4.57(b). Just above T_{cr}, scattering centers still remain in the glass layers and the structural

Figure 4.57 Photographs of silica tubes during the deposition process [149]: (a) glass deposition at the bubble creation temperature T_b (1,500°C) and (b) glass deposition at the appropriate temperature T_{ap} (1,650°C).

scattering loss is clearly recognizable in the loss characteristics [149]. The surfaces of glasses formed at four different temperatures appear quite different. Figure 4.58 shows electron microscope photographs of glass surfaces formed at the temperatures of T_{ap} (1,650°C), T_{im} (1,615°C), T_{cr} (1,580°C), and T_b (1,500°C), respectively [149,20]. These temperatures vary slightly with the glass composition, and it is very important to deposit glass layers at the optimum temperature in order to fabricate low-loss fibers.

However, interestingly enough, the porous materials seen in Figure 4.58(d) should play an important role with regard to doping rare-earth ions in silica glass. This will explained in a subsequent section.

4.4.3 Rare-Earth Doping in MCVD Process

There are several differences between fibers for transmission and fibers for amplification in terms of structure, fiber material, and preform fabrication process. For example, with regard to fiber structure, as mentioned in Chapter 1, the "weakly waveguiding" structure is suitable for the transmission fiber that is required in order to have a very low loss over long distances [2]. The weakly waveguiding structure is effective for achieving low loss because it exhibits lower scattering loss in the fiber core and the core/cladding boundary than the "strongly waveguiding" structure. In contrast, a strongly waveguiding structure is needed for fiber used in amplification because the induced rate is proportional to the intensity of the electromagnetic field [150] that is created by the high-Δn and strong waveguiding in the fiber core.

Moreover, since fiber used for amplification need be only a few hundred meters or less in length, Al_2O_3 or other oxides can be used as dopants in addition to GeO_2. Although GeO_2 is the best dopant for transmission fiber, many oxides such as GeO_2, B_2O_3, P_2O_5, and Al_2O_3 are good candidates for amplification fibers. Furthermore, the fabrication process must be changed because the raw materials of rare-earth ions have a very low vapor pressure at room temperature and hence a high evaporation point of several hundred degrees centigrade. This makes it difficult to transport the rare-earth raw materials to a desired reaction region. Consequently, we have to develop a new method that makes it possible to enable rare-earth raw materials to be transported to a glass deposition region or porous silica glass body. Figure 4.59 shows the vapor pressures of reactant halides used for making rare-earth-doped high-silica glasses [151–155]. Halide raw materials of silicon and dopants such as germanium, phosphorus, aluminum, and boron are introduced into the reaction stream as vapors carried by oxygen or inert gas at a temperature near 30°C. The halide compounds of rare-earth ions are, however, generally less volatile than the chlorides that are commonly used as glass-forming raw materials. Therefore, they require volatilizing and need delivery temperatures of a few hundred degrees as shown in Figure 4.59. This requirement has stimulated the development of vapor and liquid phase handling methods.

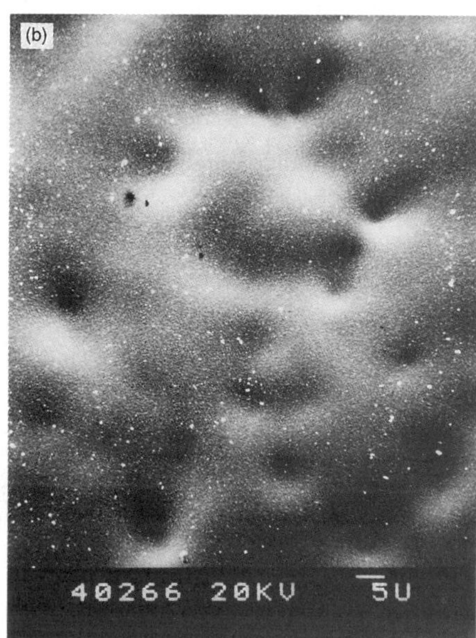

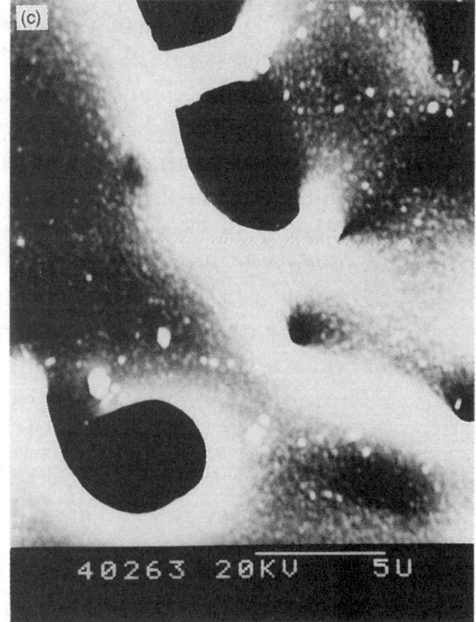

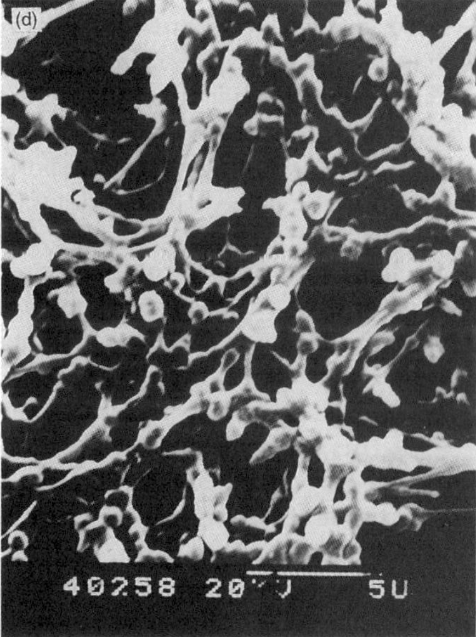

Figure 4.58 Electron microscope photographs of glass surfaces formed at the different glass deposition temperatures of (a) T_{ap} (1,650°C), (b) T_{im} (1,615°C), (c) T_{cr} (1,580°C), and (d) T_b (1,500°C) [149,20].

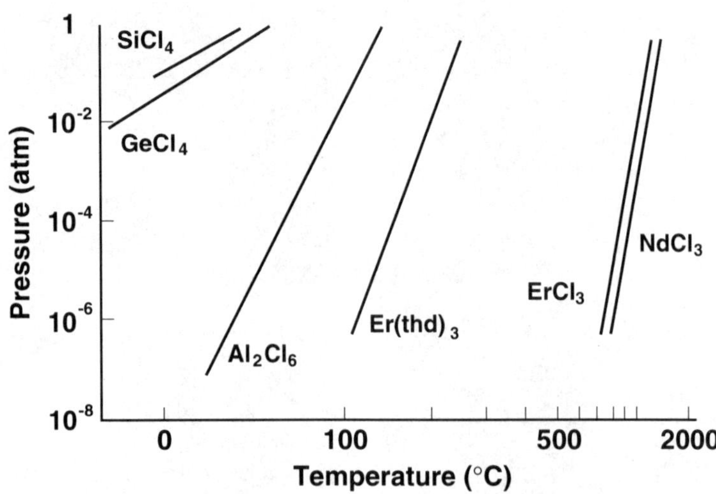

Figure 4.59 Vapor pressures of reactant halides for making rare-earth-doped high-silica glasses [151–155].

Various methods have been investigated for doping rare-earth ions into high-silica glass with the MCVD process. As described in Chapter 2, these rare-earth-doping methods can be categorized by two types: (1) rare-earth ions are doped into silica glass during the deposition of fine glass particles and (2) rare-earth ions are doped into silica glass after the fine glass particles have been deposited. With the MCVD process, there are five different techniques for supplying the rare-earth materials when doping during deposition. These are heated fit, heated source, heated source injector, aerosol delivery, and chelate delivery [154–159]. Three of these—heated source, heated source injector, and chelate delivery—are shown in Figure 4.60. With these techniques, the cladding layer is first deposited in the usual way, and during core deposition the dopant chamber is heated to ≈900° to 1,000°C to increase the vapor pressure of the rare-earth halide. The vapor is incorporated with the main reactants and included in the deposited core layers. Eventually, the tube is collapsed to form a preform and fiber is drawn in the usual way. Al$_2$O$_3$ can be added to the core host glass by transporting Al$_2$Cl$_6$ or AlCl$_3$ to the reaction zone, as shown in Figure 4.60(b,c). Some work has been reported that suggests aerosols are useful for incorporating a wide range of involatile species into the core glass [159].

On the other hand, the latter type of doping scheme used with the MCVD method, that is, doping after deposition, is shown in Figure 4.61 [160–162]. Although this method was first reported by Stone and Burrus in 1973 [163], it was not used again until 1987 [160]. The first stage of the process is almost the same as that for producing a standard MCVD preform except that the core material is

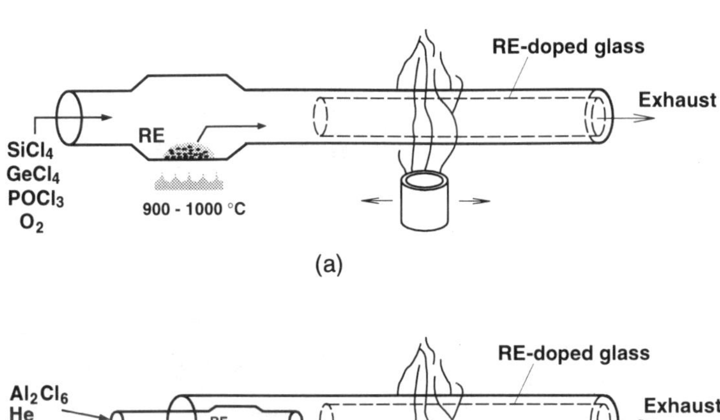

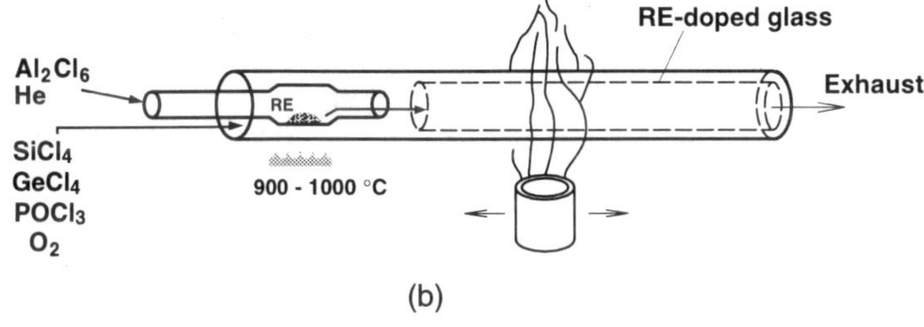

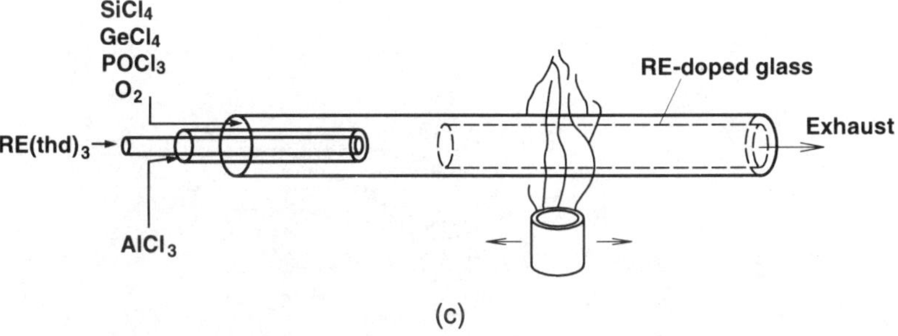

Figure 4.60 Rare-earth doping for the MCVD method. Techniques for doping during deposition [153–158]: (a) heated source, (b) heated source injector, and (c) chelate delivery.

deposited at a lower temperature than normal, which prevents the complete sintering of the fine glass particles and leads to the formation of a porous glass layer. This glass layer, deposited at a lower temperature, is almost the same as the glass layer deposited at T_b (1,500°C), as shown in Figure 4.58(d). Obviously the halide oxidation temperature must be exceeded in order for the reaction to take place; thus with low-viscosity core glass compositions, for example, heavily P_2O_5-doped silica, it is difficult to prevent complete sintering [156]. This potential problem can be overcome and a porous core layer achieved by reversing the traverse direction of the core and subsequently lightly sintering the layer in the absence of reactants

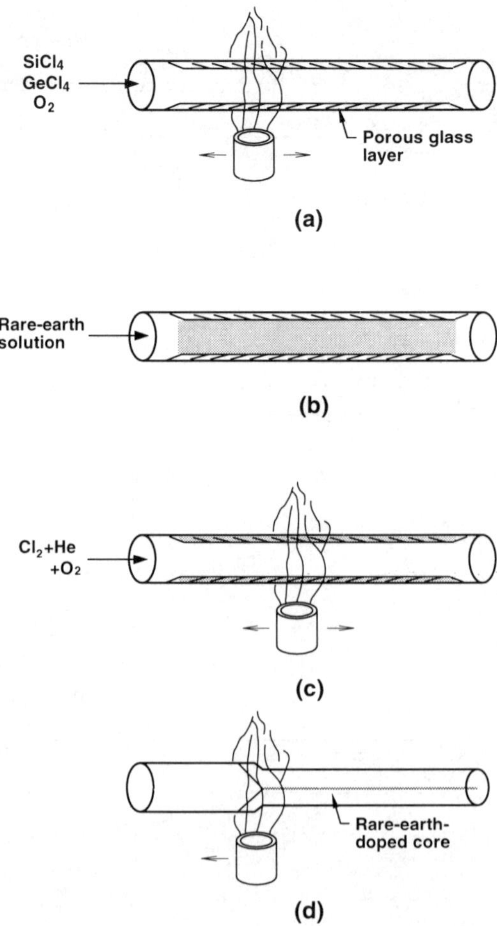

Figure 4.61 Rare-earth doping for the MCVD method. Techniques for doping after deposition: (a) porous layer deposition, (b) solution impregnation, (c) deposit drying, and (d) preform collapse [160–162].

[156]. Then, a solution of rare earth to be incorporated is introduced into the tube and allowed to soak into the pores [as seen in Figure 4.58(d)] in the porous glass layer for approximately one hour. Various salts in either aqueous or alcoholic solutions have been used including halides, nitrates, and sulphates. The liquid is subsequently expelled from the tube and the porous layer blown dry. Then the porous layer is carefully sintered in a Cl_2-O_2 atmosphere to completely dry the material as shown in Figure 4.61(c) and the tube is collapsed in the usual way. Slight modifications to the process include the deposition of a pure silica core layer

instead of GeO_2-P_2O_5-doped silica and soaking the layer in an Al^{3+} rare-earth-ion solution to make an Al_2O_3-SiO_2 host glass.

Since they have already been described in Chapter 2, we will make no mention here of the structural differences between the rare-earth-doped fine glass particles prepared by doping during or after deposition or of the influence of such structural differences on the bandwidth characteristics of fiber amplifiers.

4.4.4 VAD Process

4.4.4.1 Outline of VAD Process

In 1977, the VAD method was developed by a group from NTT Ibaraki Electrical Communication Laboratory, mainly for the mass production of high-quality optical fibers [128–130]. The VAD method was developed by combining the doping technique from the soot process and the axial deposition technique from the DD process shown in Figure 4.4, based on the concepts of fabrication process simplification and fiber preform mass production. Actually, at the beginning, this method was called the "vapor-phase Verneuil method" [164] rather than the "vapor-phase axial deposition method." The use of "Verneuil" originated in the so-called "Verneuil method" for vitreous-silica production in which vitreous silica is formed in the axial direction by passing crushed quartz particles through a Verneuil torch configuration [165].

Therefore, the key feature in the VAD method is the axial deposition of fine glass particles. Figure 4.62 shows the basic principle for making fiber preforms by the VAD method: raw halide materials, blown from a torch, oxidize into fine glass particles or soot by flame hydrolysis. The soot both the core glass and cladding glass is deposited simultaneously in an axial direction onto the end of a rotating fused-silica target rod. As the rodlike porous preform grows devoid of a central hole, unlike with the OVD method, the preform is slowly retracted through a graphite resistance furnace, where it is consolidated into a transparent preform by zone sintering. Occasionally, the porous preform preparation process and the consolidation process using a furnace are separated for the sake of setup convenience.

The adaptation of the VAD process mentioned previously offered the possibility of several important advantages [128–130] related to both mass production and fiber performance. First, the VAD process has a high glass-deposition rate and hence the potential for application to mass production. Second, if the fabrication conditions could be successfully stabilized throughout preform preparation, the preform would have good uniformity with regard to glass material and hence an excellent attenuation reproducibility. Third, VAD fibers have no dip in the center of their refractive-index profile, as commonly observed in MCVD fibers. This was the initial evaluation of the VAD method in 1977.

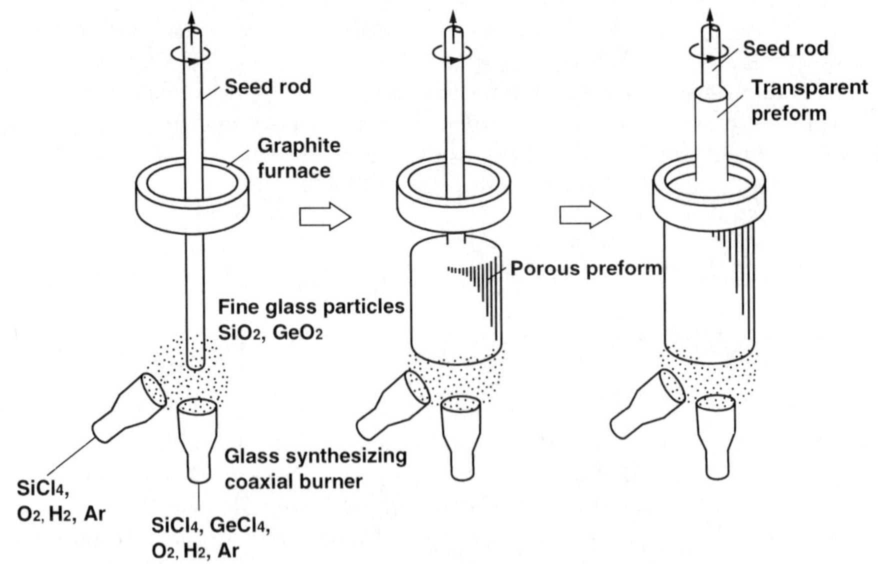

Figure 4.62 Basic principle for making fiber preforms by the VAD method.

On the other hand, the VAD process posed the following difficulties. First, it is difficult to prepare a rodlike porous preform, made only of fine glass particles in the axial direction. Second, it is difficult to completely consolidate porous preforms and effectively reduce the OH-ion impurities in the preform resulting from the flame hydrolysis products. Third, it is difficult to control the glass composition gradient accurately (and thus, the refractive-index profile) across a blank core, and so it is hard to obtain wide-bandwidth graded-index fibers because of the index profile in the spatial domain, unlike with the OVD and MCVD processes. Fourth, it is difficult to make a single-mode fiber preform with a large core/cladding ratio. In the next section, we will describe in detail the technical problems with the VAD method, which can be summarized as follows:

1. Porous preform preparation in an axial direction;
2. Complete consolidation to make a transparent preform;
3. Dehydration to reduce OH contamination;
4. Refractive-index profile control;
5. Single-mode fiber preform preparation.

4.4.4.2 VAD Apparatus

Figure 4.63 is a schematic diagram of the simplest VAD preform fabrication apparatus, which basically consists of the following components [134,2]:

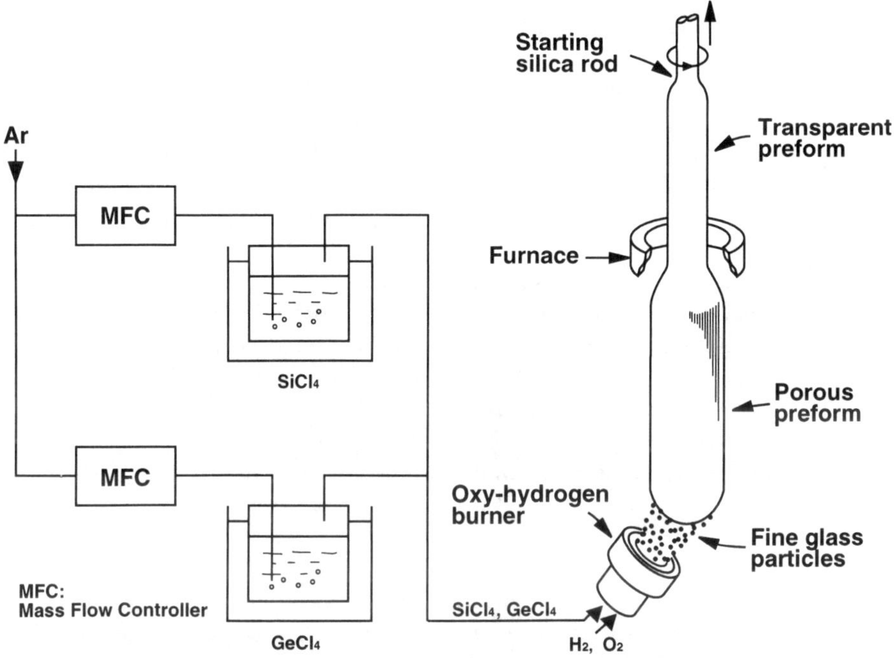

Figure 4.63 Schematic diagram of the simplest VAD preform fabrication apparatus.

1. A system for supplying raw materials and flame combustion gases;
2. A porous preform preparation chamber;
3. A heater or furnace for consolidation;
4. A pulling mechanism and its positional control system;
5. An exhaust system.

Raw halide materials such as $SiCl_4$ and $GeCl_4$ are carried by Ar gas from the raw material and gas supply system. $SiCl_4$ and $GeCl_4$ remain in the liquid state at less than 50°C, so they are vaporized by bubbling the Ar gas through the liquid halides and then carried into the burner together with the carrier Ar gas (at all moderate temperatures less than 50°C). The flow rates are controlled by a thermal *mass flow controller* (MFC) and kept at a constant value with a fluctuation of less than ±1%. The actual flow rate M for the $SiCl_4$ and $GeCl_4$ carried by the Ar gas is given by the following equation for temperatures below boiling point

$$M = mp/(760 - P) \tag{4.59}$$

where p is the halide vapor pressure and m is the Ar carrier gas flow rate. Furthermore, in this distillation, transition-metal impurities such as $FeCl_3$, VCl_4, and $CuCl_2$

in $SiCl_4$ and $GeCl_4$ can be dramatically reduced to a very low level, as previously mentioned.

Fine glass particles are synthesized by the flame hydrolysis reaction of these raw halide vapors blown from the burner. The rodlike porous preform is grown by the deposition of these particles onto the end of the target silica rod in the direction of its rotating axis. The increasing end surface temperature of the porous preform formed is measured with a pyrometer to control the index profile in the core preform as detailed later. The porous preform is continuously consolidated by an electric furnace. The transparent preform is pulled upward by the pulling mechanism rotating around its axis, while the pulling speed is controlled so as to keep the growing end of the porous preform in a fixed position, measured with the monitoring system. In addition, the undeposited glass particles and H_2O and HCl gases caused by the flame hydrolysis are removed by the exhaust system to maintain a uniform porous preform diameter. Figure 4.64 is a photograph of VAD apparatus. When the porous preform preparation process in an axial direction and the consolidation process using a furnace are separated for the sake of setup convenience, the furnace is a separate piece of equipment with an up and down mechanism.

One of the key pieces of apparatus for the VAD method is the burner for synthesizing fine glass particles and forming porous preforms. Figure 4.65 shows the concentric nozzle VAD burner for synthesizing fine glass particles and forming porous preforms [134,166–169]. These burners or torches are usually made of silica glass and basically have four concentric nozzles. The halide material vapor mixture flows from nozzle I. An inert, such as argon, flows from nozzle II. Hydrogen gas for flame combustion flows from nozzle III. Oxygen flows from nozzle IV. A typical example of the gas supply conditions for graded-index fiber fabrication is as follows. Halide materials from nozzle I: Ar gas flowing at 100 cc/min saturated with 40°C $SiCl_4$, and Ar gas flowing at 100 cc/min saturated with 30°C $GeCl_4$. Inert gas from nozzle II: 1,000 cc/min of Ar gas. Combustion gas from nozzle III: 3,000 cc/min of H_2 gas. Auxiliary gas from nozzle IV: 4,000 cc/min of O_2 gas. The main reason for using this torch is to blow the inert gas from nozzle II because this plays an important role in adjusting the flame temperature as well as in preventing fine glass particles from being deposited on the torch end. Furthermore, several burner structures have been investigated based on this fundamental structure, but these are much more complex.

4.4.4.3 Porous Preform Preparation

Techniques for preparing porous preforms in an axial direction are not only fundamental keys for fiber preform fabrication by the VAD method but also of extreme importance in relation to both the transmission characteristics of the resultant optical fibers and the axial uniformity of the preform diameter. It seems to be

Figure 4.64 Photograph of VAD apparatus.

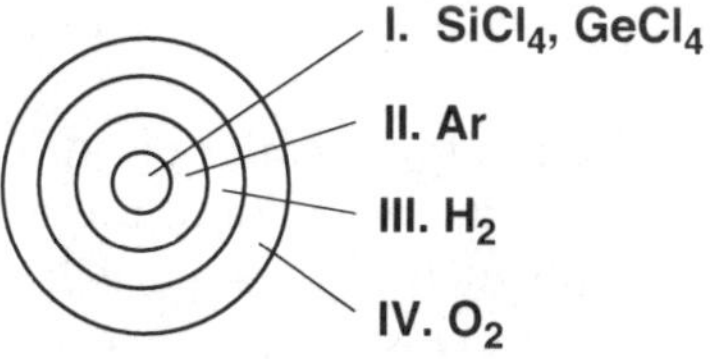

Figure 4.65 Concentric nozzle VAD burner for synthesizing fine glass particles and forming porous preforms [134].

technically difficult, however, to form a porous rodlike body composed only of fine glass particles in the axial direction. Specifically, it is only in the early stage of porous preform preparation that the target rod has an influence on the preform dimensions and on the shape of the growing surface shape, while in a steady state, the dimensions and shape may be determined by a complicated interaction among various preparation factors, such as the fine particle stream, the growing surface temperature, and the starting material supply rate.

The VAD porous preform preparation process is comprised of four main processes—(1) synthesis of fine glass particles; (2) particle deposition; (3) porous preform preparation in an axial direction; and (4) size control—and a technique for simultaneously forming a porous cladding and core that leads to a technique for preparing a porous preform from which to fabricate single-mode fiber. Oxide products such as SiO_2 and GeO_2 are synthesized by the flame hydrolysis reaction of starting halide materials such as $SiCl_4$ and $GeCl_4$ carried from the supply system. Such flame hydrolysis reactions are expressed as

$$SiCl_4 + 2H_2O \rightarrow SiO_2 + 4HCl \tag{4.60}$$

$$GeCl_4 + 2H_2O \rightarrow GeO_2 + 4HCl \tag{4.61}$$

It has been found by modeling hydrolysis reactions in an electric furnace that the reactions in (4.60) and (4.61) begin taking place in the 800° to 900°C temperature range and are complete at temperatures above 1,000°C [170]. The reaction temperature range found in the model experiment is 200° to 300°C lower than in the oxidation reaction of the same halide materials [171–173].

In the hydrolysis reaction, the SiO_2 product is not formed directly from $SiCl_4$ but rather by passing through the $Si(OH)_4$ state where it is synthesized by the reaction of $SiCl_4$ with H_2O molecules [170]. Therefore, the practical activation energy for the SiO_2 product formation becomes lower. Furthermore, since in the actual flame hydrolysis reaction the existence of radical O, H, and OH ions in the oxy-hydrogen flame seems to promote the oxide product formation reaction [174,175] and intermediate products such as $SiHCl_3$ are formed, this latter reaction takes place at lower temperatures and with higher reaction speeds [170]. At the same time, in the oxidation reaction, a higher temperature is required for the formation of oxide products because they are formed directly from halide materials [171–173].

Oxide products synthesized in the manner given in (4.60) and (4.61) solidify as fine glass particles and are then deposited onto the surface of the silica rod end. Figure 4.66 shows a photograph of fine glass particles synthesized by the flame hydrolysis reaction in the VAD process. They range in size from 500Å to 2,000Å and have an almost spherical shape. The particle size is dependent on such synthesis conditions as the flame temperature and spatial concentration of raw materials. In general, these particles become smaller with increasing reaction temperature and

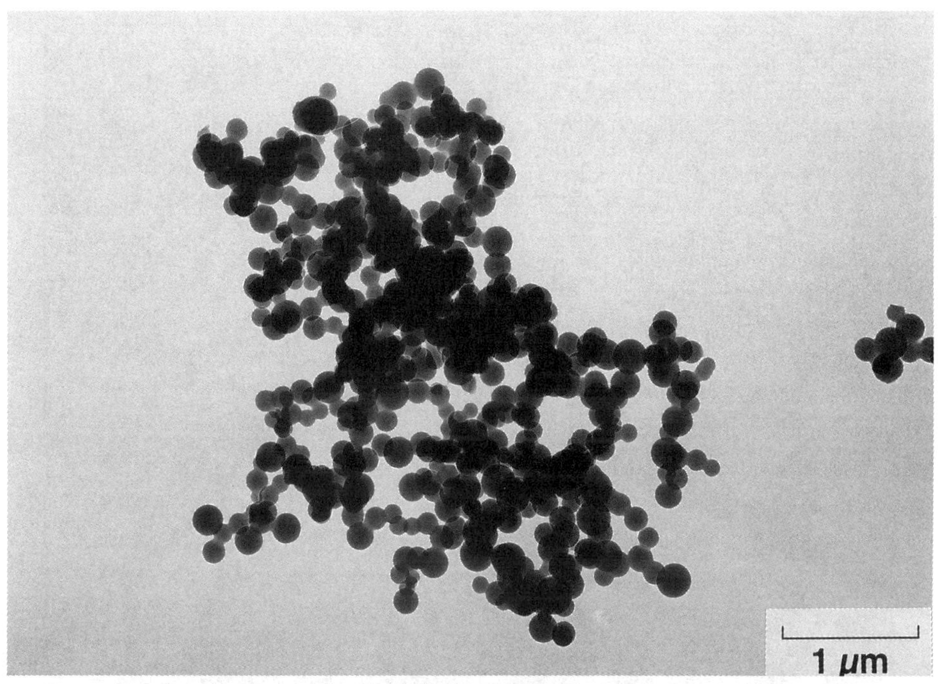

Figure 4.66 Photograph of the fine glass particles synthesized by the flame hydrolysis reaction in the VAD process.

increasing spatial concentration, Figure 4.67 shows a photograph of the process of synthesis and deposition of fine glass particles. It can be seen that fine glass particles are synthesized in the oxy-hydrogen flame blown from the torch and then deposited onto the growing surface to form the porous preform and that H_2O and HCl released by the reaction and residual fine glass particles not attached to the growing surface are carried away from the deposition region by the exhaust port.

Figure 4.68 shows a SEM photograph of a porous preform growing surface. It can be seen that fine glass particles connect together to form open networks. Such connections of fine glass particles generate mechanical strength in the porous preform and thus make it possible to prepare a rodlike porous preform in the axial direction. Under ordinary conditions, the average density of these porous preforms is 0.2 to 0.5 g/cm^3, which is 1/10 to 1/4 times that of silica glass. The average mechanical strength of the porous preform is 0.5 to 1.0 kg/cm^2 for a cross-sectional area. This strength value enables raising vertically a preform several kilograms in weight.

In order to fabricate high-quality fiber preforms, it is important to prepare rodlike porous preforms with a uniform outer diameter. The stability of the growing

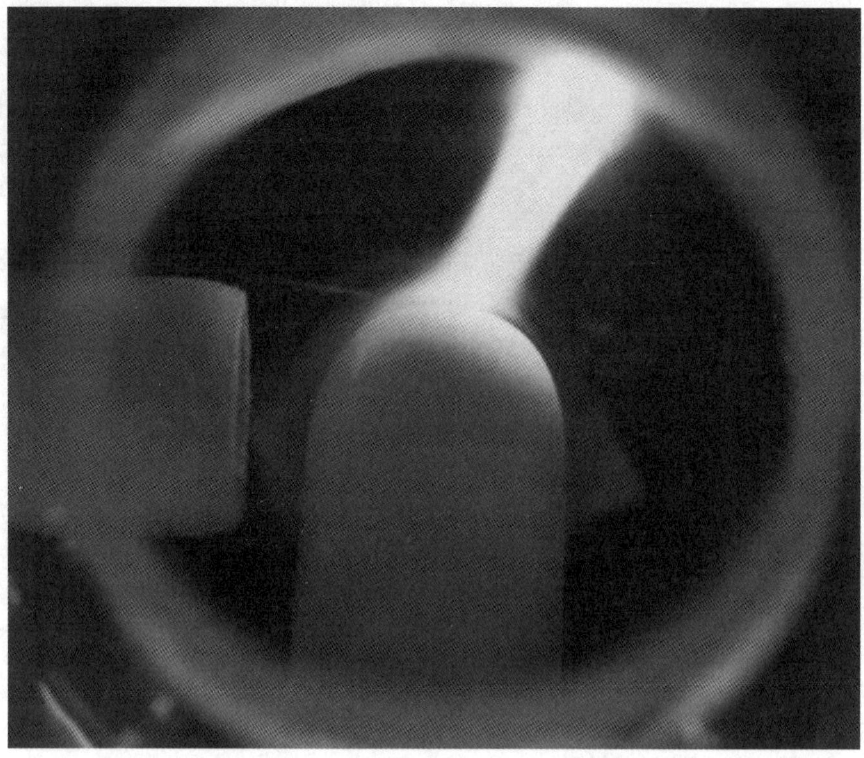

Figure 4.67 Photograph of the process of synthesis and deposition of fine glass particles.

surface shape and the uniformity in outer diameter do not depend on the starting silica rod but are almost completely dominated by various deposition conditions such as the flame temperature, vapor supply rate, exhaust gas velocity, preform pulling speed, and gas flow surrounding the porous preform growing end surface [166–169]. The fluctuations in those factors should be eliminated during the preparation of porous preforms. In addition, to keep the growing end of the porous preform at a fixed position, the pulling speed is controlled by a feedback mechanism that uses a positional signal from a video camera. Figure 4.69 shows a photograph of a porous VAD preform prepared using this stabilization technique. This figure shows that the porous preform has a uniform diameter.

4.4.4.4 Simultaneous Cladding and Single-Mode Fiber Preform Preparation

In order to reduce the loss in VAD optical fiber, it is important to prepare a cladding layer with a lower refractive index on the surface of the porous preform at the

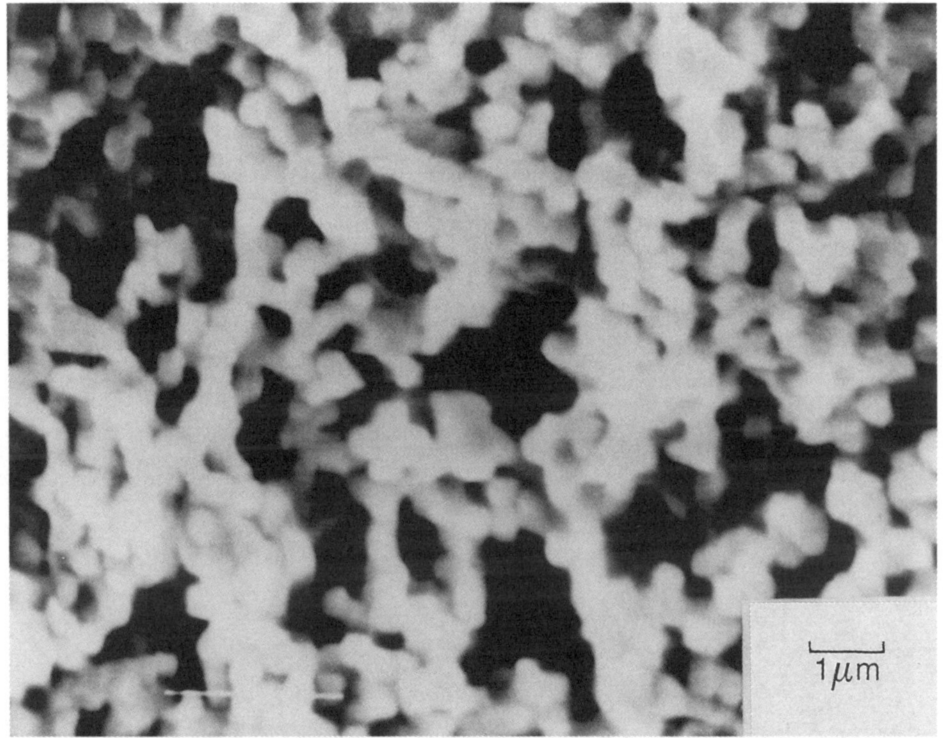

Figure 4.68 SEM photograph of a porous preform growing surface.

same time as the porous core preform is prepared. Figure 4.70 shows a photograph of this simultaneous cladding formation; a SiO_2-B_2O_3 layer of fine glass particles is deposited as a cladding on the periphery of a porous SiO_2-GeO_2 glass preform already prepared using another burner. This cladding technique has an extremely positive effect in term of loss reduction in graded-index fibers, step-index fibers, and high NA fibers. However, a thicker cladding layer is needed to reduce the loss in single-mode fibers.

In order to fabricate a single-mode fiber preform by the VAD method, a porous preform with a small diameter must first be prepared for the core. A thick cladding layer then is formed on this slender rodlike porous preform using a second torch. In addition to the simultaneous cladding technique, a technique for preparing a small-diameter porous preform is absolutely essential for achieving low-loss single-mode fibers with the VAD method. However, it is rather difficult to fabricate a porous preform of less than 30 mm in diameter using the process described previously with a concentric nozzle VAD burner. Therefore, a burner

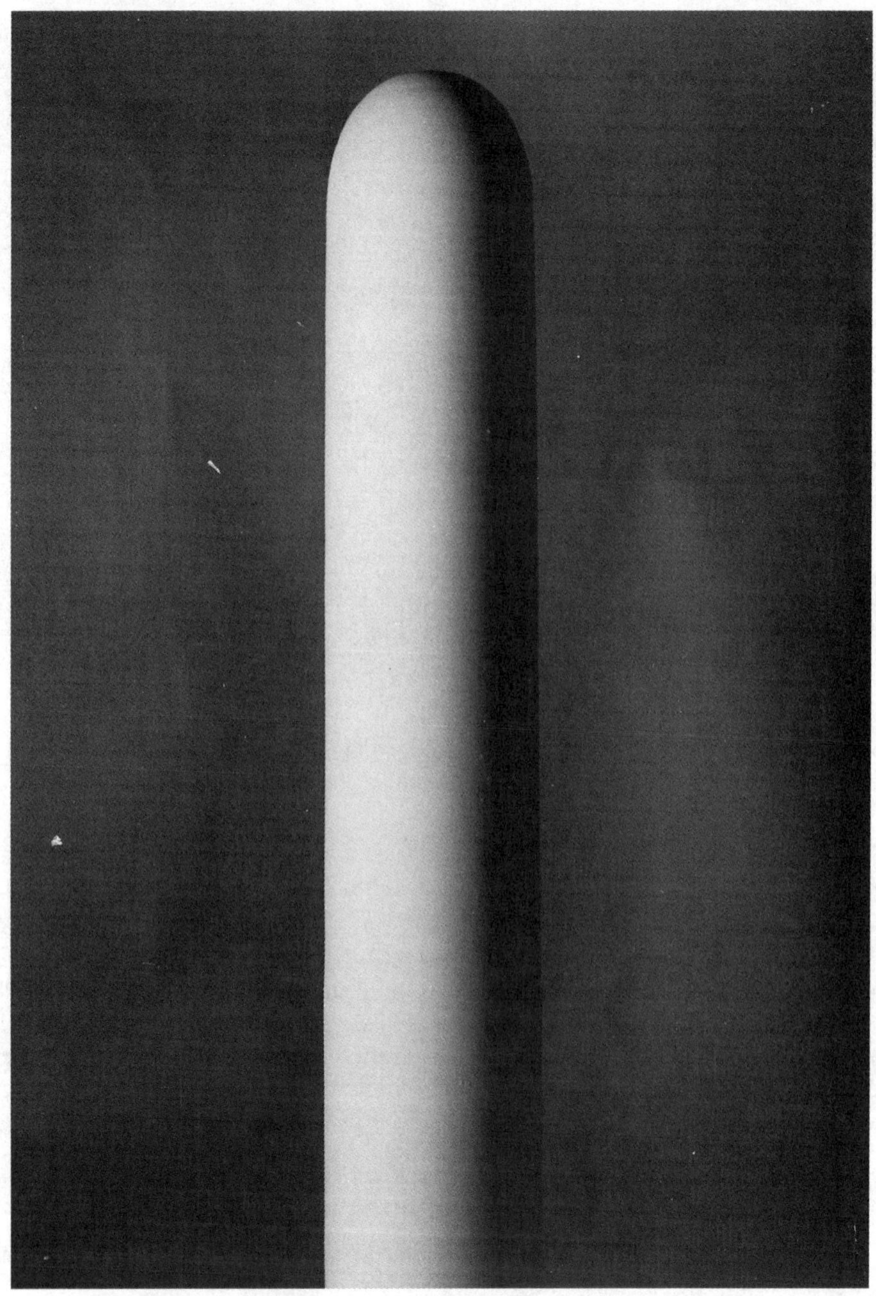

Figure 4.69 Photograph of porous preform fabricated by VAD.

Figure 4.70 Photograph showing simultaneous cladding formation; a SiO_2-B_2O_3 layer of fine glass particles is deposited as a cladding on the surface of a SiO_2-GeO_2 glass porous preform already prepared using another burner.

suitable for fabricating small-diameter core glass for single-mode fibers has been developed [166–169]. The key feature of this burner is that it has a nozzle for blowing halide materials that is set at a position away from the center of the burner flame. Thereby, the core diameter is sufficiently reduced to enable the deposition of a thick cladding layer on the surface of the porous core rod. Figure 4.71 shows the preparation of a porous preform for single-mode fiber [166–169]: (a) schematic diagram and (b) a photograph of a single-mode porous preform being prepared. As you can see in this figure, the core torch is located at an angle with respect to the axial direction of the preform, and the exhaust port is also located near the periphery of the porous core body. The fine glass particle stream from the core torch is blown only through a small arc of the flame, and glass particles are deposited only on the top of the porous core body growing surface. Next, a thick porous glass cladding is formed on the periphery of this core porous body, as seen in Figure 4.71.

4.4.4.5 Dehydration and Consolidation

The VAD method consists of two major processes: the process whereby fine glass particles are synthesized and deposited to form a porous preform and a process

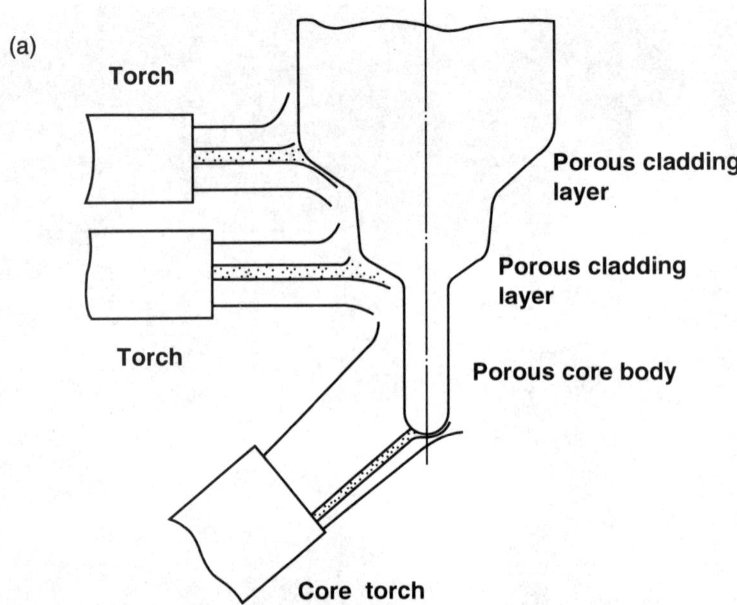

Figure 4.71 Preparation of a porous preform for single-mode fiber [166–169]: (a) schematic diagram and (b) a photograph of a single-mode porous preform being prepared.

for its dehydration-consolidation, as shown in Figure 4.62. The dehydration and consolidation techniques are of essential importance in terms of reducing optical fiber loss.

The dehydration-consolidation process involves two steps: in one the porous glass preform is consolidated into a bubblefree transparent glass preform by zone sintering using an electric furnace; the other is the dehydration process, usually included in the consolidation process, where a large amount of OH ions and H_2O molecules are removed from the in porous preforms. In the following subsections, we describe the consolidation process, with emphasis on process analysis, based on the elemental closed-pore model. This clarifies the roles of both the gas environment and the speed with which the temperature increases in the furnace. Then, we describe the dehydration techniques in detail, that is, the principle of the dehydration process using $SOCl_2$ and Cl_2 gases, and technical keys to achieving complete dehydration.

Consolidation Technique

The achievement of a high degree of uniform transparency in the preform, which is important in relation to transmission loss characteristics for a drawn fiber, was

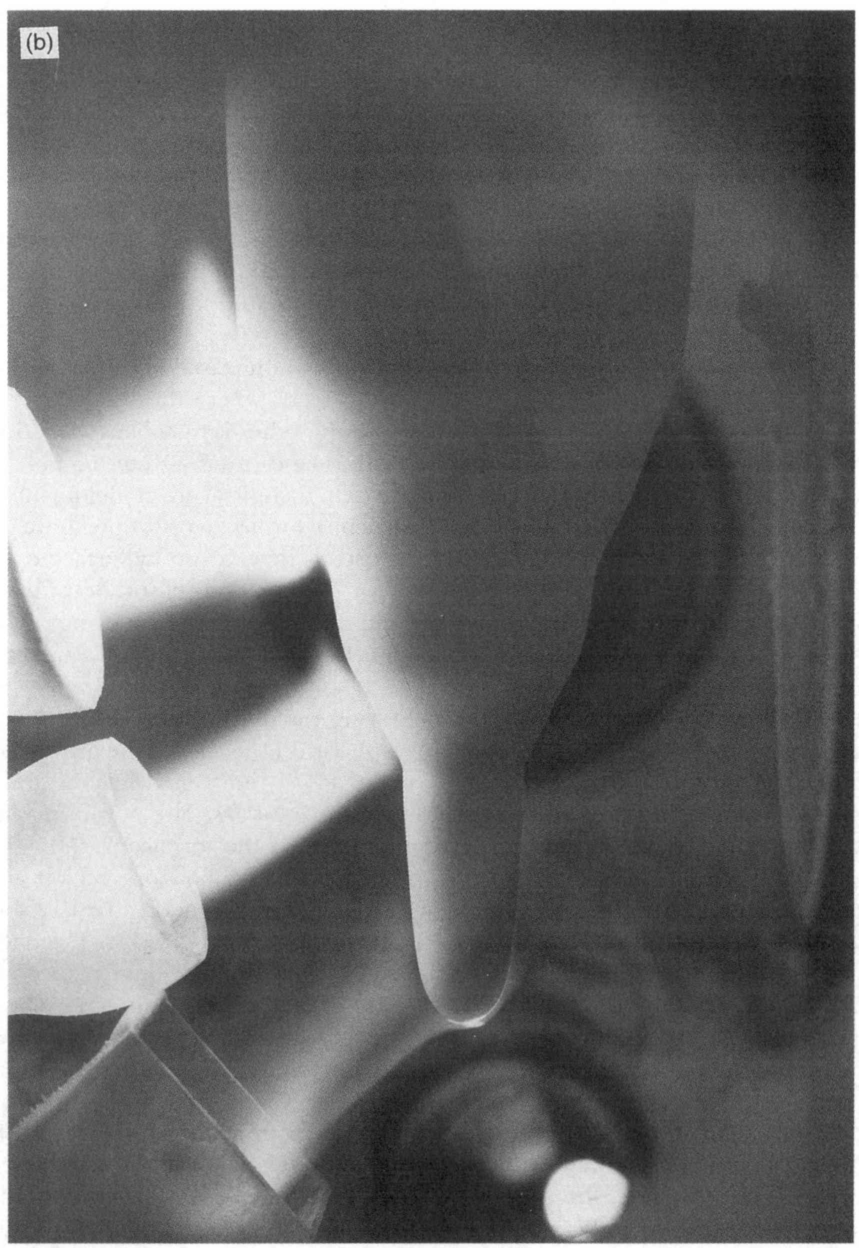

Figure 4.71 (continued).

one of the main problems to be solved in the development of the VAD. In the early stages, nitrogen or argon gas was used as the environment gas in the consolidation furnace [134,20]. However, it was difficult to obtain a transparent bubblefree preform; but by using a helium atmosphere, transparent bubblefree preforms were easily obtained. Accordingly, when we started using a helium atmosphere, it seemed that the problem of the consolidation process had been solved experimentally. However, the problem of bubbles remaining in the preform has arisen again along with the development of dehydration techniques and high-speed porous preform production. Therefore, it is necessary to undertake a more detailed investigation in order to find a complete solution to this consolidation process problem so as to obtain high-quality preforms.

The sintering of a low-density porous preform is comprised of two steps [176]. The first step is the densification of the porous glass, where the transition from open-pore state to closed-pore state occurs, and the other is to collapse the closed pores. The first step was dealt with by Sherer in relation to the sintering behavior of OVPO porous preforms [177]. In his approach, a cubic array model formed by intersecting cylinders was used as the microstructure for low-density open-pore glass preforms; the cubic array was densified by a viscous flow driven by surface energy reduction. He described an elegant scheme for the porous preform densification model. However, it is necessary to analyze the latter process of sintering, that is, closed-pore collapse, which is more important for obtaining a highly transparent bubblefree preform.

Therefore, the closed-pore behavior during the VAD porous preform zone-sintering process should be analyzed with emphasis on the role of the gas environment in the furnace [178]. In the experiment [178], inert gas (helium and/or argon) was blown into the furnace as an environmental gas. The porous preform was consolidated in the furnace. The temperature of the furnace was 1,500° to 1,600°C at the center, and the temperature gradient in the hot zone region along its axis was about 140°C/cm; the pulling speed was 40 mm/min. Figure 4.72 shows a photograph of the porous and transparent preform by zone sintering. The diameters of the porous and transparent preforms were about 60 and 25 mm, respectively. Next, a cross section of the preform, in the transitional region from a porous to a transparent state, was observed using a scanning electron microscope. Specimens were prepared by cutting the preform along the axial direction of the transitional region.

Figure 4.73 shows SEM photographs of a preform taken to show each state between the porous state and the transparent state. Their locations are shown in Figure 4.74. In this case, the porous preform was sintered in a helium gas atmosphere. Figure 4.73(a) indicates the initial porous state, where fine glass particles (0.05 to 0.2 μm in diameter) form an open network; (b) shows the first sintering state, where particles begin to melt into each other to form open pores. Gases existing in the open pores can move freely through pores. Figure 4.73(c) shows the second sintering state, where the pores are isolated from each other. At this

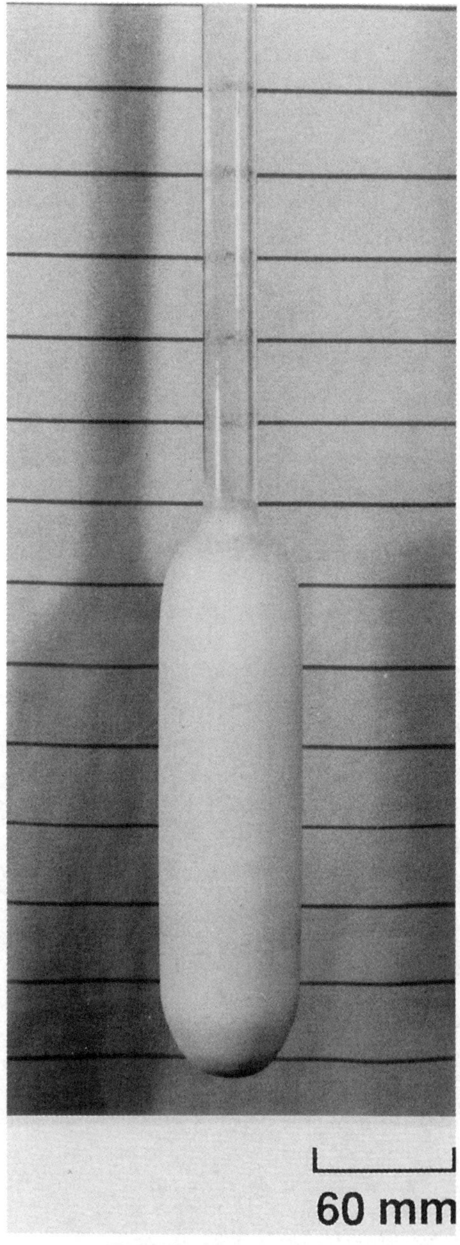

Figure 4.72 Photograph of a porous and transparent preform fabrication by zone-sintering.

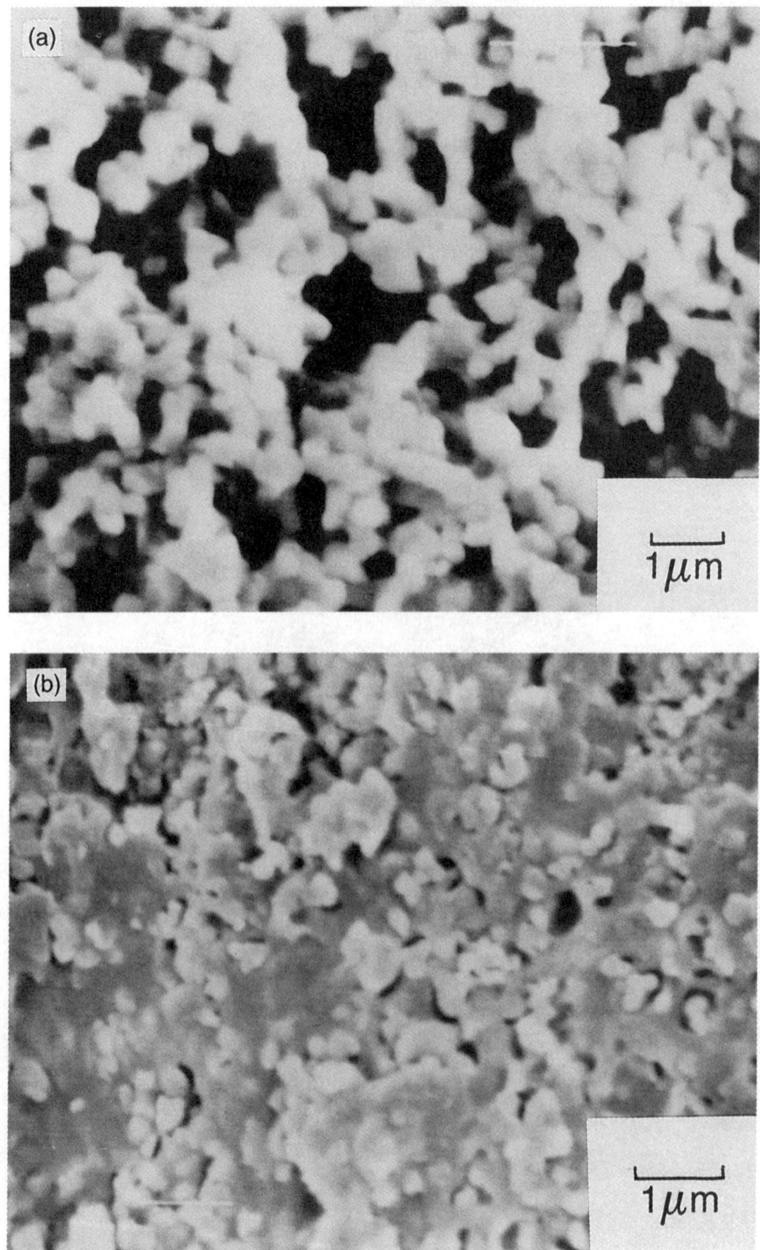

Figure 4.73 SEM photographs of a preform sintered in a helium gas atmosphere showing each state from porous to transparent [134,178]: (a) initial porous state, (b) open-pore state, (c) closed-pore state, (d) closed-spherical pore state, (e) bubblefree transparent glass.

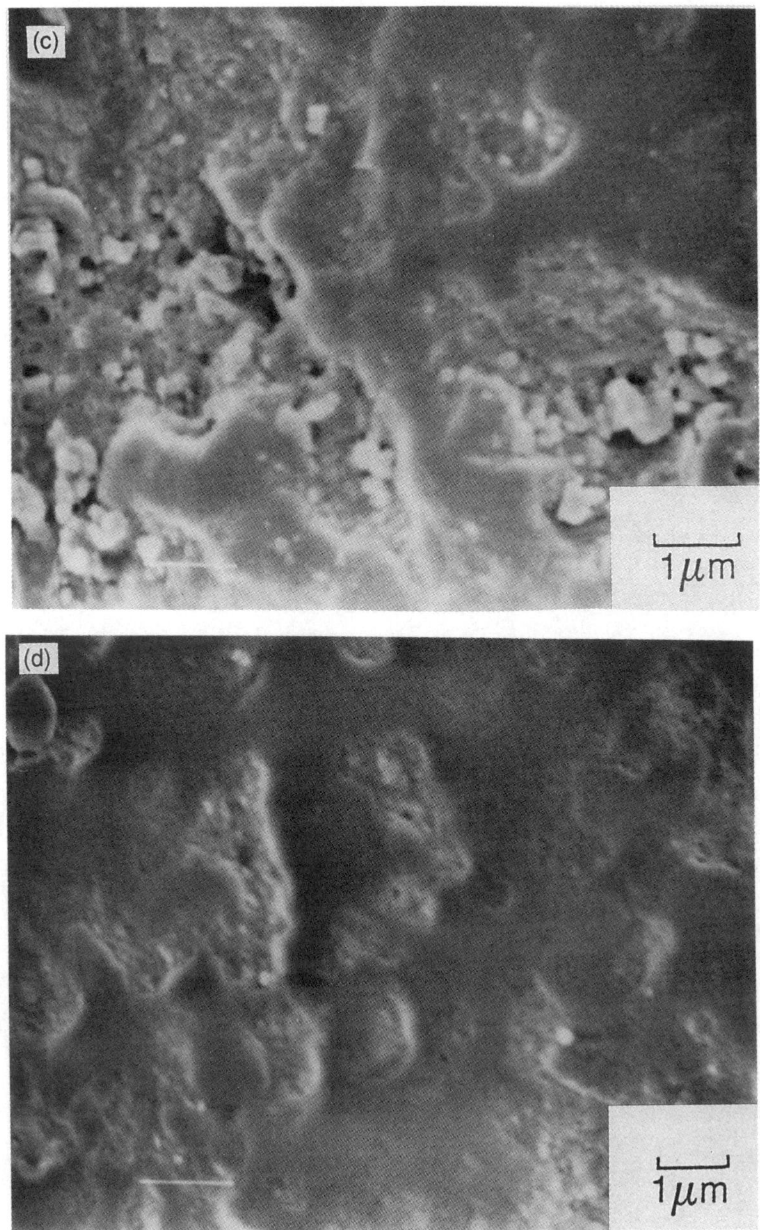

Figure 4.73 (continued).

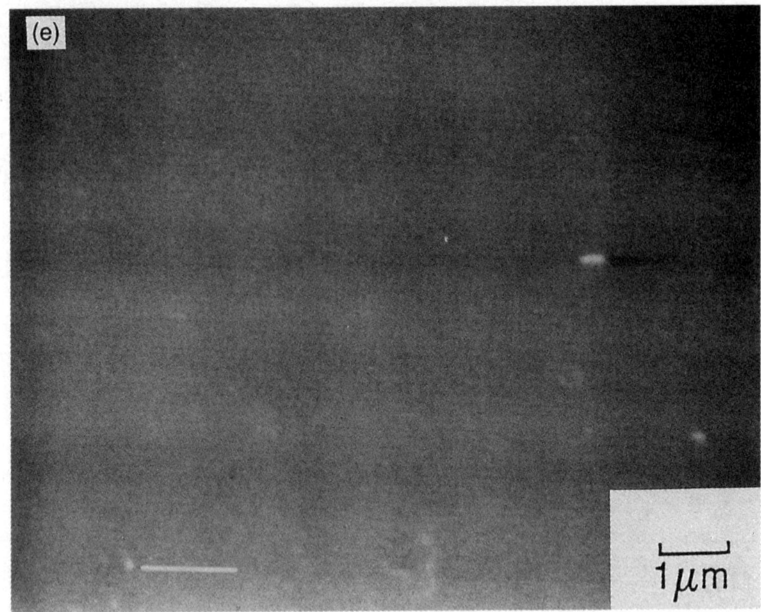

Figure 4.73 (continued).

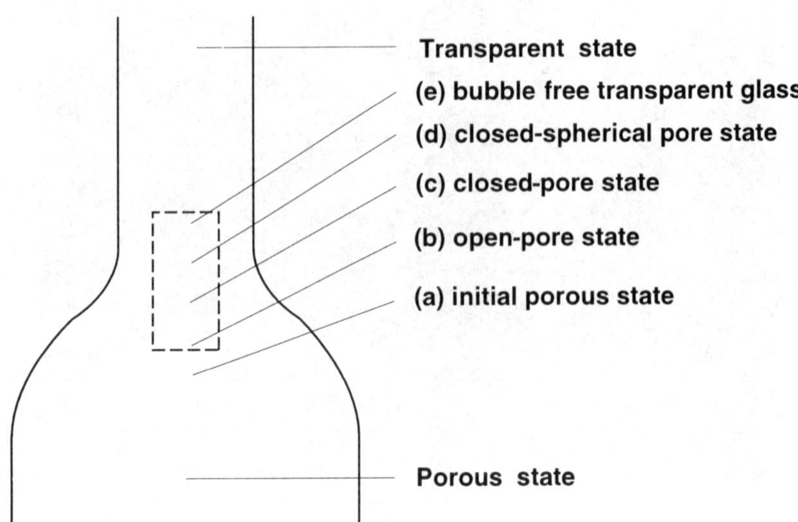

Figure 4.74 Illustration showing each position in the preform corresponding to Figure 4.73. The broken-line outlines the specimen dimension.

stage, the preform becomes transparent and the gas is confined in each pore. Porous preform densification has almost finished in this stage. Figure 4.73(d) shows the closed-spherical pore state; the closed-pore shapes in the sintered glass are changed because of the low viscosity and surface tension. The final state in the sintering process is shown in Figure 4.73(e), where pores have collapsed and the preform has become bubblefree transparent glass.

Figure 4.75 shows SEM photographs of a preform sintered in an argon gas atmosphere at each point illustrated in Figure 4.76. Figure 4.75(a) shows the open-pore state that corresponds to Figure 4.73(b). Figure 4.75(b) indicates the closed-pore state. The behavior of the closed pores differs from that in the preform sintered in a helium gas atmosphere in that the pores are larger than 1 μm in diameter but tend to expand. Figure 4.75(c,d) shows the closed pores in the transparent glass and the expanded pores, respectively.

Sintering porous preforms in different gas atmospheres lead to preforms with different optical qualities. Table 4.12 shows the glass quality of preforms sintered under five different atmospheric gas conditions. The He to Ar ratios are: (a) 100:0, (b) 80:20, (c) 60:20, (d) 40:60, (e) 20:80, and (f) 0:100 (in mole percent). The table shows that a He gas content of more than 60 mol% is necessary to sinter a porous preform into a bubblefree transparent preform.

A Model for the Final Sintering Stage

The above SEM observations indicate that the initial sintering states, from the as-deposited porous state to the closed-pore state, show a similar trend regardless of whether sintering is performed in a He or a Ar gas atmosphere. The initial sintering state corresponds to the shrinkage process for an as-deposited porous preform. This has been dealt with theoretically by Sherer [177]. The final sintering states in the two atmospheres (He and Ar), however, show quite different behavior. The preform sintered in the Ar gas atmosphere contains closed pores, hardly any of which have collapsed, as shown in Figure 4.75(c,d). On the other hand, the preform sintered in the He gas atmosphere contains no closed pores since they collapsed completely during the process. In the following sections, we offer some considerations on the final sintering stage based on a model [134,178]. In this model, the difference between the permeabilities of environmental gases in silica glass plays an important role in the final sintering stage.

The behavior of the closed pores in the final stage of the sintering process seems to depend on the balance between the rate at which the environmental gas permeates into the surrounding glass and the pore expansion rate during a temperature increase. The simple model shown in Figure 4.77 has been developed [134,178], where a closed spherical pore filled with inert gas (He and/or Ar) is surrounded by a spherical glass body. The glass wall thickness is L and the pore volume is V. On the basis of this pore model, pore expansion or shrinkage can be described as follows.

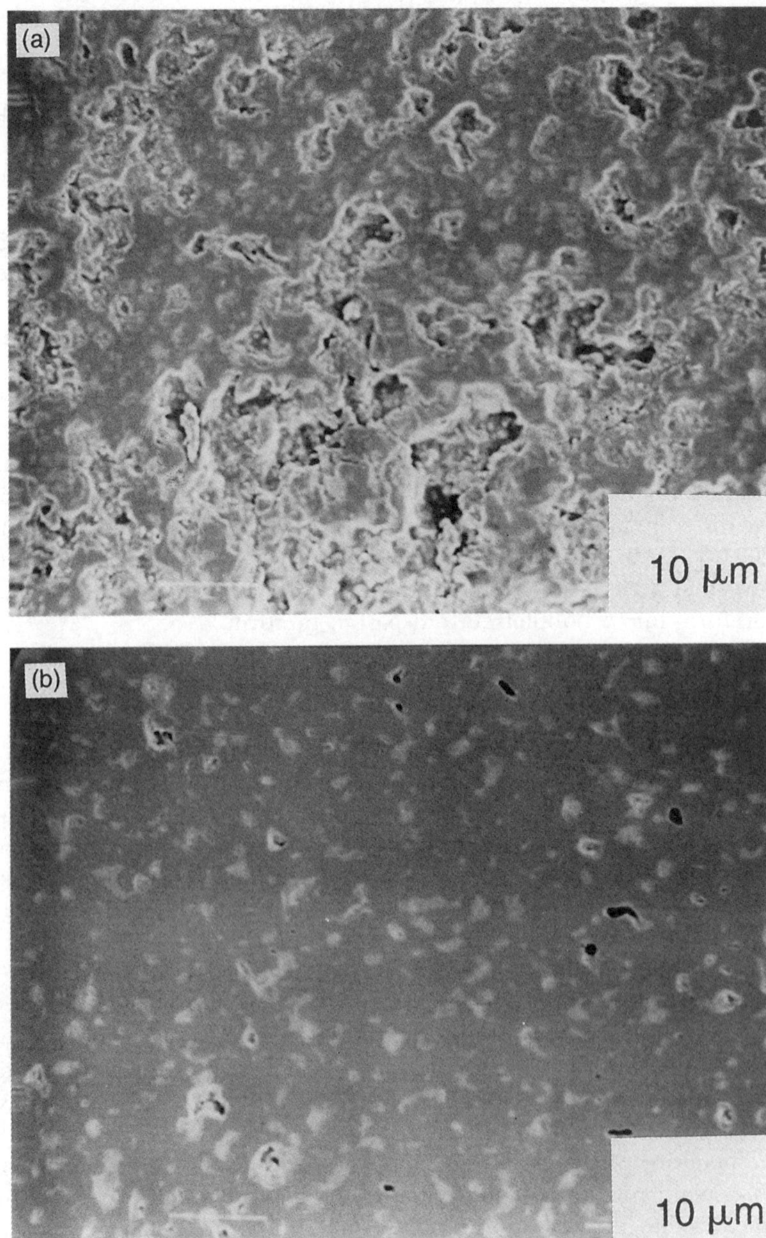

Figure 4.75 SEM photographs of a preform sintered in an argon gas atmosphere showing each state from porous to transparent [134,178]: (a) open-pore state, (b) closed-pore state, (c) closed-spherical pore state, and (d) expanded pores.

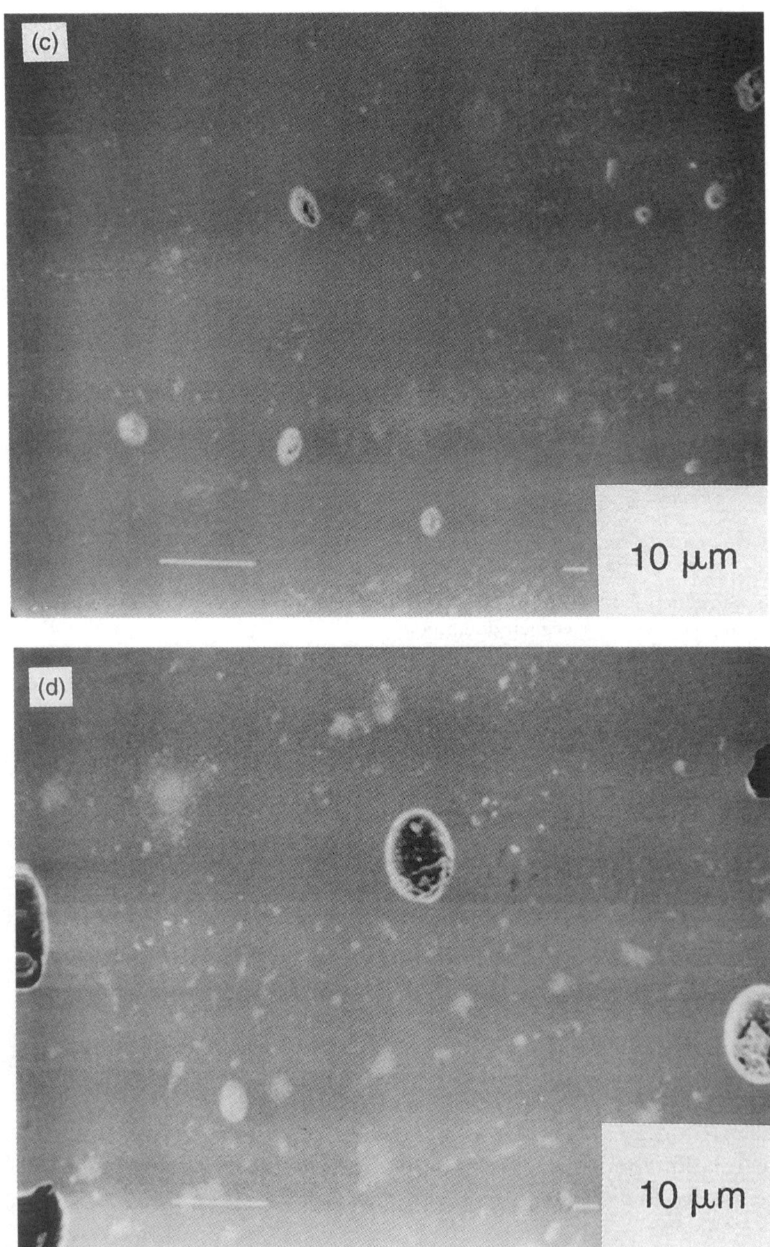

Figure 4.75 (continued).

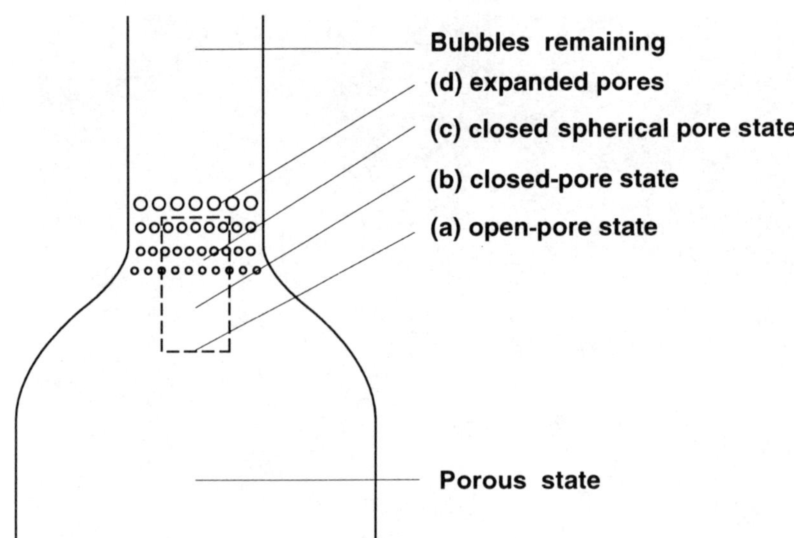

Figure 4.76 Illustration showing each position in the preform corresponding to Figure 4.75. The broken-line outlines the specimen dimension.

Table 4.12
Glass Quality of the Preform Sintered Under Five
Different Atmospheric Gas Conditions

He:Ar Ratio (mol %)	Glass Quality
100:0	Bubblefree
80:20	Bubblefree
60:40	Bubblefree
40:60	Few/some bubbles remaining
20:80	Bubbles remaining
0:100	Bubbles remaining

From: [134,178].

The relation between pore volume and surrounding temperature can be written by Boyle–Shalle's law

$$PV = nRT \tag{4.62}$$

where P is the pressure inside the pore, n is the mole number for the gas inside the pore, and R is the gas constant. The small pore volume increment ΔV caused by a slight temperature increase ΔT can be obtained by differentiating (4.62)

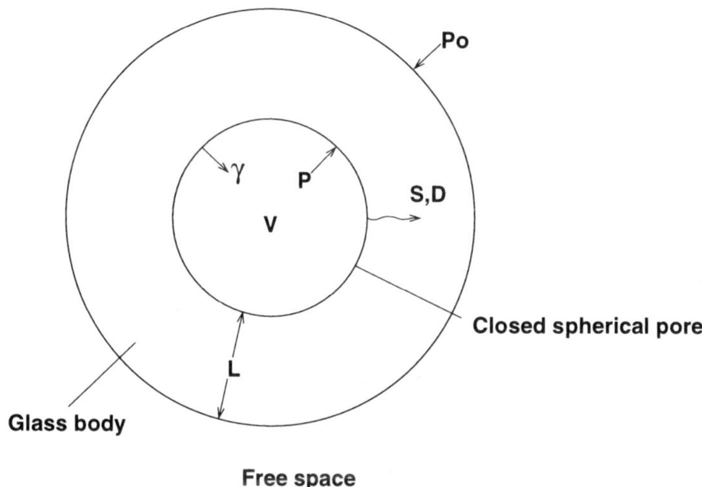

Figure 4.77 A model of a closed-spherical pore in the final sintering stage [134,178].

$$\Delta PV + P\Delta V = \Delta nRT + nR\Delta T \tag{4.63}$$

The sign of $\Delta V/\Delta T$, as determined from (4.63), dominates the pore behavior with increasing temperature. The gas pressure inside the pore P can be related to the gas pressure in the surrounding glass, P_0, by (4.64)

$$P - P_0 = 2\gamma\pi \, (3V/4\pi)^{-1/3} \tag{4.64}$$

where γ is glass surface tension. The small change in Δn in (4.63) represents a mole number variation caused by gas diffusion into the glass wall. Δn can roughly be expressed by

$$\Delta n = -JA\Delta t/N_0 \tag{4.65}$$

where J is the gas diffusion density $(\text{cm}^3/\text{cm}^2 \cdot \text{sec})$, A is the pore inner surface area $(= (36\pi)^{1/3}V^{2/3})$, Δt is the time required for the temperature to increase from T to $T+\Delta T$, and N_0 is Avogadro's number. The diffusion density J is approximately given by [179]

$$J = DS \, (P - P_0)/L \tag{4.66}$$

where D is the diffusion coefficient of the gas in the glass, S is the gas solubility in glass, and L is the glass wall thickness. Substituting (4.64) through (4.66) into (4.63) gives

$$\frac{\Delta V}{\Delta T} = \frac{R(n - \beta T)}{P_0 + \frac{4}{3}\left(\frac{4\pi}{3}\right)^{1/3} \gamma V^{-1/3}} \tag{4.67}$$

where

$$n = \frac{V}{RT}\left[P_0 + 2\gamma\left(\frac{4\pi}{3}\right)^{1/3} V^{-1/3}\right] \tag{4.68}$$

$$\beta = \frac{8\pi}{N_0}\left(\frac{3}{4\pi}\right)^{1/3} \frac{\gamma DSV^{1/3}}{CL} \tag{4.69}$$

$$C = \Delta T/\Delta t \tag{4.70}$$

which indicates the speed of the temperature increase. It is obvious from (4.67) that the sign of $\Delta V/\Delta T$ changes if $n > \beta T$ or $n < \beta T$. When $n = \beta T$, the pore diameter is given as the critical value

$$d_c = \left(\frac{3}{4\pi}V_c\right)^{1/3} = -2\left(\frac{36}{\pi^2}\right)^{1/3} \frac{\gamma}{P_0}$$

$$+ \sqrt{4\left(\frac{36}{\pi^2}\right)^{2/3}\left(\frac{\gamma}{P_0}\right)^2 + 48\left(\frac{3}{4\pi}\right)^{1/3} \frac{\gamma RT^2}{P_0 N_0} \frac{DS}{CL}} \tag{4.71}$$

A pore with a diameter larger than d_c expands with increasing temperature, while a pore with a diameter less than d_c tends to shrink.

It must be noted here that the aforementioned analysis is rather qualitative because the dynamic motion of the pore expansion or shrinkage is disregarded. In order to analyze the dynamic motion of the pore behavior, a more complicated consideration is necessary. It can be said that glass viscosity does not play an important role in determining the critical pore diameter d_c. This may be explained by the fact that a pore of the critical diameter will maintain stationary, even if the temperature is increased.

Results and Discussion of Consolidation Conditions

Substituting the following numerical values into (4.71)

$$P_0 = 1 \text{ (atom)}$$
$$\gamma = 3.0 \times 10^{-4} \text{ (atom} \cdot \text{cm) [180]}$$
$$R = 82 \text{ (cm}^2 \cdot \text{atom} \cdot \text{K}^{-1} \cdot \text{mol}^{-1})$$
$$T = 1600 \text{ (K)} \tag{4.72}$$

we obtain the critical pore diameter

$$d_c = -0.545 \times 10^{-3} + (0.297 \times 10^{-6} + 3.09 \times 10^2 K/CL)^{1/2} \qquad (4.73)$$

where $K = DS$ is the gas permeability in glass. The critical diameter d_c increases with increasing gas permeability K and decreases with the speed of the temperature increase C and wall thickness L.

Under these experimental conditions, C was about 1 K/s, estimated from the temperature gradient (140°C/cm) in the furnace and the preform pulling speed (4 mm/min). The wall thickness L was about 0.1 cm from the SEM photograph at the transition, although the situation in the transition zone shown in Figure 4.73 is more complicated than the simple case shown in Figure 4.77. Permeabilities for He and Ar in silica glass, using data reported by Perkins [179], were found to be

$$K_{He} = 8.32 \times 10^{-7} \ (cc(STP) \cdot cm^{-1} \cdot sec^{-1} \cdot atm^{-1} \cdot K^{-1})$$
$$K_{Ar} = 2.27 \times 10^{-11} (cc(STP) \cdot cm^{-1} \cdot sec^{-1} \cdot atm^{-1} \cdot K^{-1}) \qquad (4.74)$$

Substituting these data into (4.8), the critical diameters for He and Ar result in

$$d_{c,He} = 500 \ \mu m$$
$$d_{c,Ar} = 0.6 \mu m \qquad (4.75)$$

This means that the Ar gas-filled pores, with a pore diameter larger than 0.6 μm, do not shrink with the increasing temperature. This observation shown in Figure 4.75 indicates a fairly good agreement with the estimated critical diameter scheme. With an He gas atmosphere, on the other hand, closed pores with a diameter smaller than 500 μm are rarely formed in the usual densification process during the initial sintering process. Figure 4.78 shows the relation between gas permeability and critical diameter increase with increasing gas permeability. A transparent bubblefree preform is easily obtained in an atmosphere containing a more highly permeable gas.

Obviously, from (4.71) or (4.73), the critical diameter d_c increases with a decrease in the speed of the temperature increase C, which result indicates that with zone sintering, a transparent preform is more easily obtained when the temperature gradient in the hot-zone region along its axis decreases and also when the pulling speed decreases. With uniform sintering, a transparent preform is more easily obtained when the temperature increase is slow.

Table 4.12 shows some examples of transparency for preforms sintered under different gas conditions. The results indicate that it is necessary to maintain a sintering atmosphere with an He gas content of more than 60% to obtain bubblefree preforms. This result can be explained as follows, based on the above discussion.

Gas permeability K_m for an He-Ar mixed gas is presented as

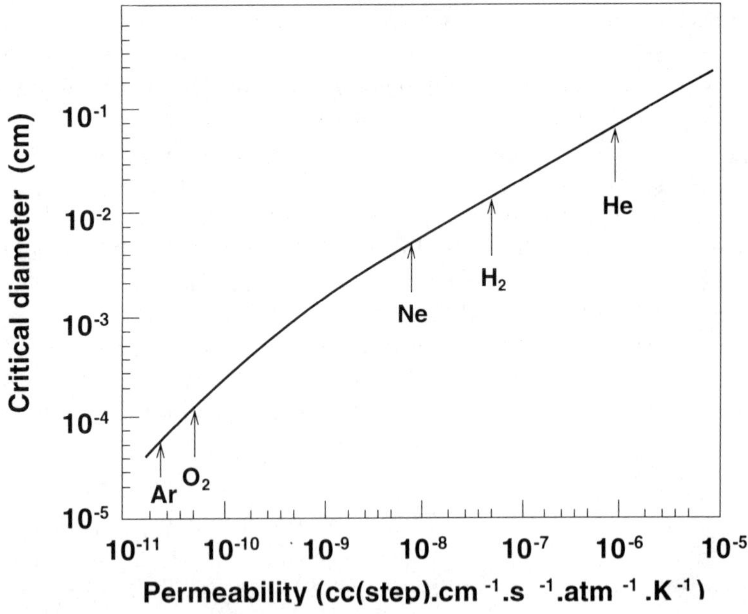

Figure 4.78 Relation between gas permeability and the critical diameter increase with increasing gas permeability [134,178].

$$K_m = K_{He}M_{He} + K_{Ar}M_{Ar} \tag{4.76}$$

where M_{He} and M_{Ar} are the mixing ratio of He and Ar, respectively. The critical pore diameter, at which a pore starts to shrink, is given by substituting (4.76) into (4.73), giving

$$d_{cm} = -0.545 \times 10^{-3} + (0.297 \times 10^2 \, K_m/CL)^{1/2} \tag{4.77}$$

This equation indicates that a closed pore, with a diameter d of less than d_{cm}, starts to shrink until most of the He gas has diffused out of the pore into the surrounding glass. As a result, a closed pore containing residual Ar gas has a diameter given by $dM_{Ar}^{-1/3}$. So, if this pore diameter is smaller than the critical diameter $d_{c,Ar}$, that is, if the initial pore diameter is smaller than $d_{c,Ar}M_{Ar}^{-1/3}$, the closed pore tends to diminish. When $M_{Ar} = 0.4$, the value of $d_{c,Ar}M_{Ar}^{-1/3}$ is about 0.8 μm and d_{cm} is about 120 μm. This discussion shows that a pore with a diameter smaller than 0.8 μm will finally disappear. On the other hand, a pore with a diameter between 120 and 0.8 μm will only shrink until the He gas has diffused out and the pore does not disappear. Experimental results, shown in Table 4.12, indicate that actual preforms mostly contain closed pores with diameters smaller than 0.8 μm, although the discussion here is based on a rather oversimplified consideration.

It seems worthwhile to note finally that this discussion regarding the mixed gas consolidation is of practical interest in connection with recent developments on the dehydration technique for the porous preforms. Mixtures containing halide gas are now used during the sintering process to eliminate OH ions and H_2O molecules from the preforms.

It must also be noted here that the critical diameter depends not only on gas permeability K but also on the speed of the temperature increase C. It was found that a bubblefree preform can be obtained, even in a pure Ar gas atmosphere, with a C value smaller than $0.1°C/s$.

These results are not only useful for determining the sintering conditions for porous preforms and designing the furnace structure for the sintering process in the VAD method but also applicable to the OVD sintering process.

Dehydration Principle

It is widely recognized that the harmonic stretching vibration for OH ions gives rise to one of the principal losses in low-loss optical fibers [181]. Loss reduction has been pursued by reducing the OH ions in optical fiber to achieve low loss in the long 1.1- to 1.8-μm wavelength band because the first and second overtones and related vibrations for OH absorption at 2.73 μm appear at wavelengths of 1.39, 1.24, and 0.95 μm [79].

The VAD porous preform consists of fine glass particles, 500Å to 2,000Å in diameter, as shown in Figure 4.66. Since these fine glass particles are synthesized as a result of a flame hydrolysis reaction, they contain a large amount of OH ions and water molecules in and/or on the particle surface, which then contaminate the transparent preform during consolidation. Accordingly, in order to reduce loss, especially in the long-wavelength region, it is necessary to find an effective way to remove residual OH ions and water molecules.

Figure 4.79 shows the IR reflectance spectrum for VAD fine glass particles. It is difficult to quantitatively determine the OH concentration from this diffused reflectance spectrum. It can, however, be seen that there is a strong absorption peak at 2.73 μm due to OH ions trapped near Si-sites and a broad absorption peak near 3 μm related to the water molecules. Particle structures containing absorbed water molecules and OH ions are shown in Figure 4.80, which is estimated from the observed data. There have been several investigations into removing OH ions and H_2O molecules from the glass surface from the viewpoint of surface chemistry [182–184]. As shown in Figure 4.81, on the basis of experimental results and previous work [182–184], the behavior of OH ions and H_2O molecules on the glass particle surface can be qualitatively interpreted as follows. First, when fine glass particles containing OH ions and H_2O molecules are heated in a dry atmosphere, the physically absorbed H_2O molecules are readily removed at 150°C and then part of the chemically bonded Si-OH is dehydrated at about 400°C. However, as the

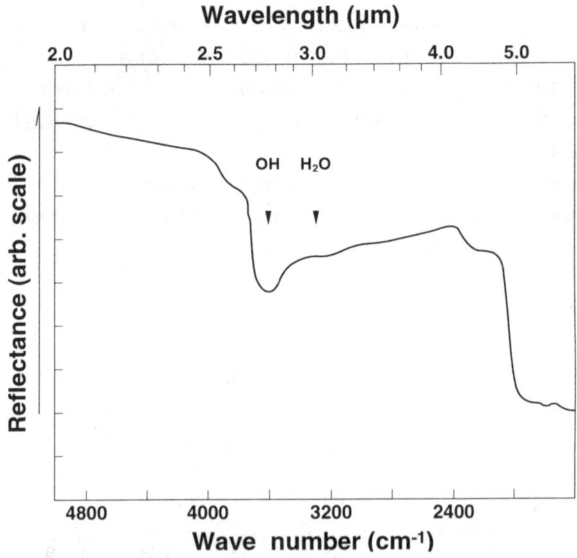

Figure 4.79 Infrared reflectance spectrum for VAD fine glass particles.

figure shows, some Si-OH bonds remain at the particle surface, as isolated OH ions, even at an elevated temperature of about 800°C. OH ions in the particles diffuse out according to the OH ion concentration gradient, and the diffused-out OH ions condense to water molecules. Even so, isolated OH ions remain on the particle surface. This means that porous preform consolidation in a dry atmosphere results in the presence of some amount of residual OH ions in the transparent preform (OH content: experimentally 5 to 30 ppm). Considering that one or two ions remain in a 100-$\overset{\circ}{A}{}^2$ area on the surface of a particle with an average diameter of 0.1 μm [134], it can be estimated that residual OH ions amount to 30 ppm, which is comparable to the prior experimental value. A chemical treatment using a reactive reagent is necessary in order to reduce these isolated OH ions.

For the investigation, thionyl chloride was mainly used as the dehydration reagent because the chemical properties of $SOCl_2$ and its dehydration effect on silica glass surface have been well studied and $SOCl_2$ reactivity as a dehydration reagent is very effective. Chemical reactions in $SOCl_2$ with H_2O and OH ions are expressed by [185]

$$SOCl_2 + H_2O \rightarrow SO_2 + 2HCl \tag{4.78}$$

$$SOCl_2 + Si\text{-}OH \rightarrow SO_2 + HCl + Si\text{-}Cl \tag{4.79}$$

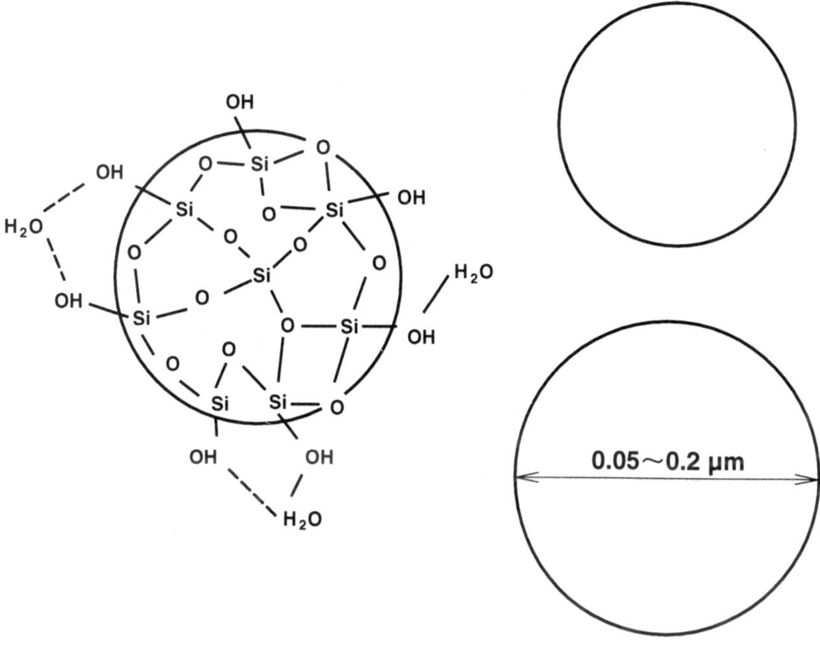

Figure 4.80 Schematic illustration of VAD fine glass particles containing H_2O molecules and OH ions.

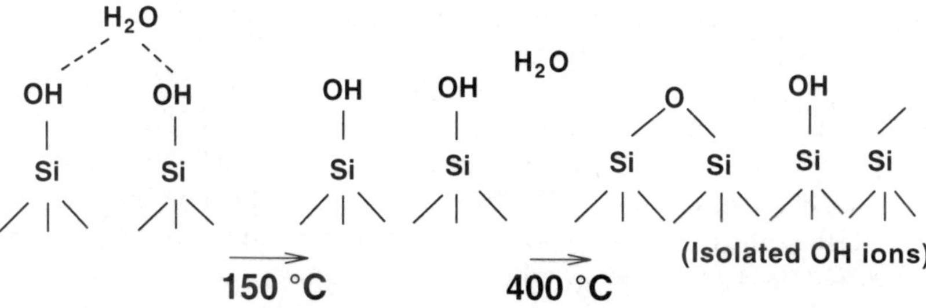

Figure 4.81 Diagram showing the behavior of H_2O molecules and OH ions on the fine glass particle surface.

This chemical treatment is essentially a halogenating process, where an isolated OH ion is replaced by Cl-, Br-, or F-ions. The halogenating process shown in (4.78) and (4.79) generates Si-Cl bonds. However, absorption loss due to such Si-Cl bonds has no serious influence on the fiber loss in the wavelength region of present interest (that is, 0.6- to 2.0-μm wavelength range) because the fundamental absorption peak due to the Si-Cl bonds is around a 25-μm wavelength. Thus, chemical dehydration using a halogenating process has an effect on the reduction of residual OH-ion content. It has also been determined that Cl_2 gas has the same effect as $SOCl_2$ on reducing OH content. We can use Cl_2 gas as a chemical reagent for soot preform dehydration with the same effectiveness as $SOCl_2$.

Figure 4.82 shows IR transmission spectra for consolidated preform samples with and without dehydration [185]. The absorption peak at 2.73 μm for the sample without dehydration is from the residual OH ions trapped near Si-sites. This residual OH-ion content is estimated to be about 30 ppm. The preform consolidated in a dry atmosphere has this level of OH contamination. In contrast, a preform consolidated with $SOCl_2$ has no OH peak at 2.73 μm. This confirms that dehydration using a chemical reaction is very useful for reducing OH content in a VAD preform.

Dehydration Conditions

The chemical dehydration reaction is sensitive to the furnace temperature. Figure 4.83 shows the relation between the dehydration temperature and the residual-ion content in consolidated preforms [185–187]. In the experiment, only the dehydration temperature was changed using $SOCl_2$ as a dehydration reagent and a uniform-heating furnace. The dehydration treatment was carried out for 2 hr. The figure shows that the residual OH-ion content decreases steeply in the vicinity of 700°C and about 1,200°C. An OH-ion content of about 30 ppm remains in the preform treated at temperatures lower than 700°C, and about 0.5 ppm remains in preforms treated in the 700° to 1,200°C temperature range. Finally, it was possible to reduce the OH-ion content to less than 0.1 ppm by treating the porous preform at temperatures higher than 1,200°C.

This experimental result can be explained as follows. First, a critical point appears at 700°C in Figure 4.83, where the residual OH content decreases sharply and is limited by the chemical reaction between OH ions and $SOCl_2$ on the particle surface [188,189] rather than OH diffusion because the OH diffusion length $(4DT)^{1/2}$ in silica glass, calculated using the diffusion constant for OH ions in silica glass at 600°C ($D = 10^{-11}$ cm^2/s) and dehydration time, is about 505 μm, which is larger than the particle size. This indicates that most of the OH ions existing inside the particles easily diffuse out to the particle surface. The porous preform starts to shrink at about 1,200°C. However, the preforms contain many pores until consolidation is complete, as can be seen in Figure 4.83, and water vapor migrates easily into the open pores and recombines with or adheres to the particle surface as a result of the recombination process from Si-Cl to Si-OH. This means that dehydrated particle surfaces are recombined with OH ions or H_2O molecules during the consolidation period, when water vapor is contained in the environment. It also means that it is necessary to maintain a water-vaporfree atmosphere, even during the consolidation period. Gases, such as helium or oxygen, used during consolidation usually contain more than 1-ppm water. To prevent further contamination during the consolidation process, a dehydration reagent such as chlorine or thionyl chlo-

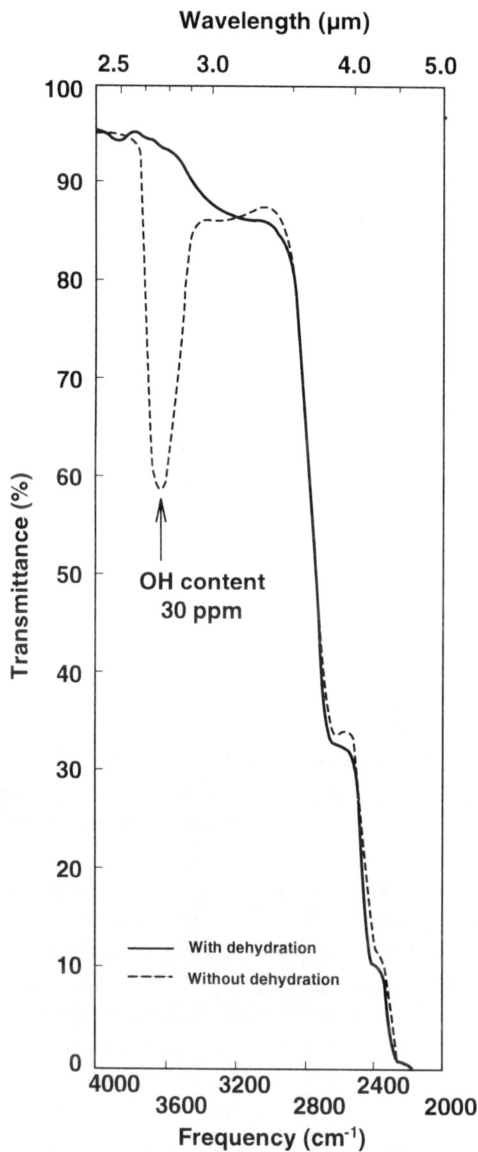

Figure 4.82 Infrared transmission spectra of consolidated preform samples with and without dehydration [185].

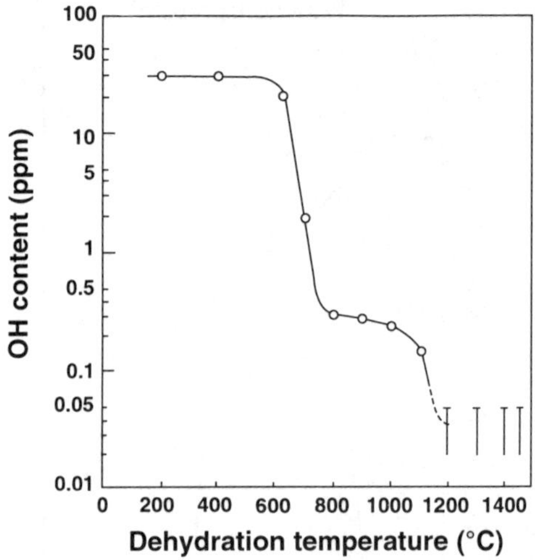

Figure 4.83 Relation between the dehydration temperature and the residual-ion content in consolidated preforms [185–187].

ride vapor should flow into the furnace throughout the dehydration period and consolidation process.

Another important dehydration condition is the partial pressure of the dehydration reagent (Cl_2 gas) in the consolidation atmosphere. Figure 4.84 shows the relation between the partial pressure of the dehydration reagent (Cl_2 gas) in the atmosphere and the residual OH-ion content of consolidated preforms [185–187]. This figure shows that the residual OH-ion content decreases as the Cl_2 gas partial pressure is increased. It is, however, necessary to choose an adequate partial pressure, taking into account the critical diameter given by (4.77). The result shown in Figure 4.84 and the limit to the consolidation suggest that the appropriate partial pressure range is 10 to 100 mmHg.

Other important dehydration conditions are the dehydration time and type of dehydration reagent. With regard to dehydration time, experimental investigations at a dehydration temperature of 770°C and 10-mmHg $SOCl_2$ partial pressure, revealed that 2 hr is necessary to reduce the residual OH content to 0.01 ppm or below [134]. The result suggests that the dehydration time is not determined by the OH-ion diffusion time from the particle volume but by the speed of the $SOCl_2$ reaction with OH ions on the particle surface. The dehydration time can be shortened using $SOCl_2$ gas at a higher partial pressure.

With regard to dehydration reagents, halogenating regents such as $SOBr_2$ (brominating reagent) and $C_2Cl_2F_2$ (fluorinating reagent) appear to be suitable rather than chlorinating reagents such as $SOCl_2$ or Cl_2. However, from dehydration

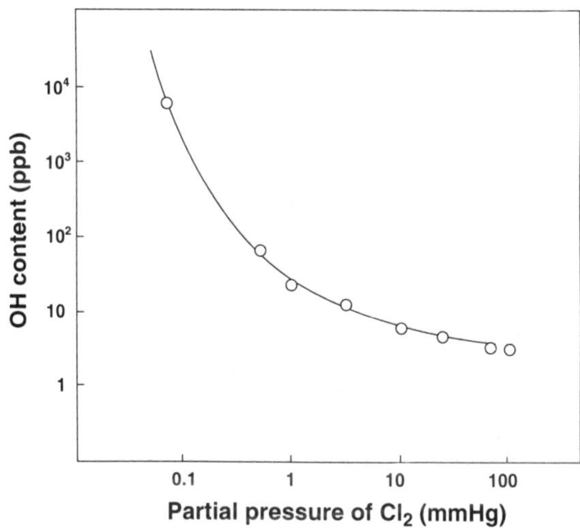

Figure 4.84 Relation between the partial pressure of the dehydration reagent (Cl$_2$ gas) in the atmosphere and the residual OH-ion content [185–187].

experiments undertaken using SOBr$_2$ (brominating reagent), it is revealed that the amount of residual OH-ions contained in a transparent preform dehydrated with SOBr$_2$ is one order of magnitude larger than in that dehydrated with SOCl$_2$ [134]. This result means that the reactivity of Cl-ion with Si-OH is stronger than that of Br-ions with Si-OH, understood in terms of the difference between the electronegativity of Cl atoms and Br atoms defined by L. Pauling [190].

When a fluorinating reagent with a strong dehydration activity is used, the silica glass particles themselves are fluorinated and vaporized. Therefore, chemical dehydration treatments using a fluorinating reagent are essentially difficult to employ, especially with wet porous preforms. However, fluorination is used to form glass with a lower refractive index than pure silica and to terminate dangling bonds to prevent the effects of hydrogen permeation [191].

In addition to the complete dehydration of a consolidated preform, the cladding thickness is an important parameter in terms of achieving ultra-low-loss fibers, especially with single-mode fibers. The investigation indicated that the core/cladding diameter ratio needs to be about 1/10 to achieve low-loss single-mode fibers. In 1980, a preform was dehydrated and consolidated under conditions in which the dehydration temperature was higher than 1,200°C, the Cl$_2$ gas partial pressure was 100 mmHg, and the dehydration time was 2 hr. Completely OH-free fibers can be prepared using an all-VAD-synthesized fiber [192]. Its loss curve is shown in Figures 1.14 and 1.15, and the residual OH-ion content is less than 1 ppb.

4.4.4.6 Refractive-Index Profile Control

The key feature of the VAD method is that a porous preform is prepared in the axial direction. This allows refractive-index profile control in the spatial domain, which is not the case with the MCVD and OVD methods.

In the early stages of the development of the VAD process, the profile formation concept was constructed on the basis that raw material vapors for different compositions were blown from different torch nozzles, mixed spatially with each other by diffusion, and overlapped to form a spatial dopant distribution on the growing surface of the porous preform [128–130]. This idea was thought to explain profile formation. However, the fact that graded-index profiles could be obtained with a single torch and a single raw material vapor nozzle could not be explained simply by this mixing effect idea [193].

Precise experiments on the properties of SiO_2-GeO_2 particle deposition in flame hydrolysis reactions made it clear that the GeO_2 concentration of a deposited particle depends strongly on the substrate temperature during deposition. There is, however, some difficulty in fully clarifying the relation of these deposition properties to practical behavior occurring during the flame hydrolysis reaction and synthesized glass particle deposition.

Dopant Concentration

Figure 4.85 shows a photograph of a flame in which fine glass particles are synthesized. In the VAD process, the vapor phase reaction in the flame and on the porous preform growing surface has a direct influence on index-profile formation. Therefore, a fine glass particle deposition experiment was conducted to clarify the mechanism underlying index-profile formation in the VAD process. Figure 4.86 shows the experimental arrangement for fine glass particle deposition [194]. The key to the success of this experimental setup is to control the surface temperature of the substrate tube independently of the flame temperature by employing a cooling gas supply. In this experiment, a vapor-phase binary mixture of $SiCl_4$-$GeCl_4$ (10 mol%), $SiCl_4$-BBr_3 (8 mol%), or $SiCl_4$-$TiCl_4$ (4 mol%) was fed into an oxyhydrogen torch at a rate of about 160 cc/min, with He gas (flow rate: 100 cc/min) as a carrier. The torch was composed of the coaxial silica tubes shown in Figure 4.65. The halide gas mixture was fed into the central tube and the combustion gases (H_2 and O_2) into the outer tubes. Oxide glass particles synthesized in the flame were deposited on a silica substrate tube (10×9 mm). The distance between the silica substrate tube and the end of the torch was about 30 mm. The surface temperature of the tube was maintained at a desired value between 200° and 800°C by introducing cooling gas into the tube. The temperature was monitored using a two-dimensional optical pyrometer.

After glass deposition under various conditions, the dopant concentrations in the fine glass particles deposited on the substrate were characterized by IR

Figure 4.85 Photograph of a flame in which fine glass particles are synthesized.

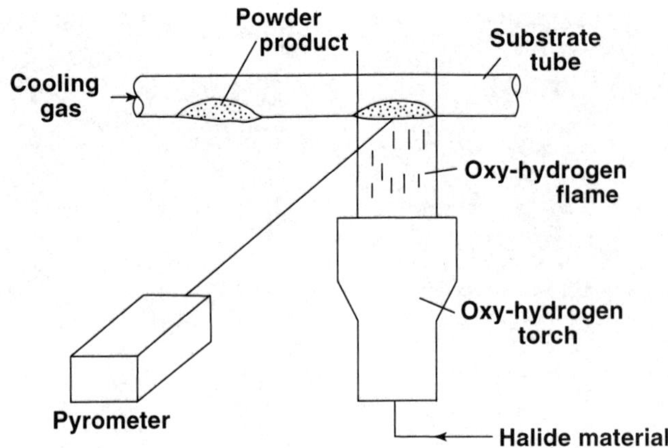

Figure 4.86 Experimental arrangement for fine glass particle deposition [194].

spectrum measurements to clarify the relationship between the glass deposition temperature and the dopant concentration. For example, the IR spectra for SiO_2-GeO_2 glass have strong absorption peaks at 870 and 660 cm^{-1}, which are attributed to the vibration of Ge-O-Ge and Si-O-Ge chains, respectively [195]. This suggests that there is a structure that connects SiO_2 and GeO_2 other than the simple oxides of Si and Ge. The absorption peaks at 1,380, 920, and 950 cm^{-1} are attributed to the vibration of B-O-B, Si-O-B, and Si-O-Ti chains, respectively [196,197]. The relative dopant concentrations of fine glass particles prepared at different substrate temperatures can be determined from these absorption peaks. Whether the phase of the fine glass particles is crystalline or amorphous is determined from x-ray diffraction patterns [198]. Sharp diffraction peaks that are attributed to the crystalline state of GeO_2 and B_2O_3 are observed for the samples prepared at lower temperatures.

These experimental results confirmed that the SiO_2-B_2O_3 and SiO_2-GeO_2 systems are similar in terms of deposition properties. The concentration and crystalline structures show a considerable dependence on the substrate temperature. By contrast, the TiO_2 concentration was found to have no significant dependence on the substrate temperature in the SiO_2-TiO_2 system. Figures 4.87 to 4.89 schematically illustrate the substrate temperature dependence of the GeO_2, B_2O_3, and TiO_2 concentrations, respectively, as evaluated from IR and x-ray data [134,191].

As shown in Figure 4.87, the GeO_2 component in the SiO_2-GeO_2 system was deposited with a crystalline structure at substrate temperatures below 400°C but assumed a noncrystalline form dissolved in the SiO_2 glass network at about 500°C. The crystalline GeO_2 concentration decreases with increasing substrate temperature. The concentration of dissolved GeO_2 increases with increases in the substrate temperature from 500° to 800°C.

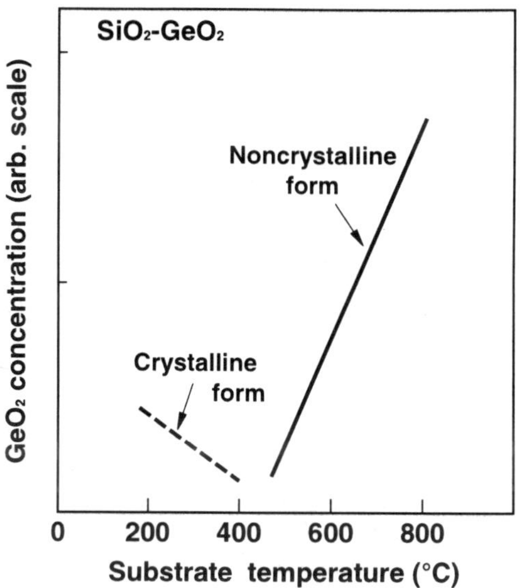

Figure 4.87 Substrate temperature dependence of GeO_2 concentration in SiO_2-GeO_2 glass particles [134,194].

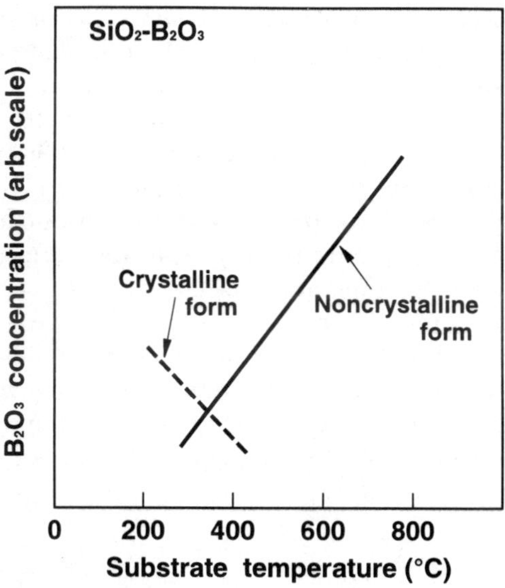

Figure 4.88 Substrate temperature dependence of B_2O_3 concentration in SiO_2-B_2O_3 glass particles [134,194].

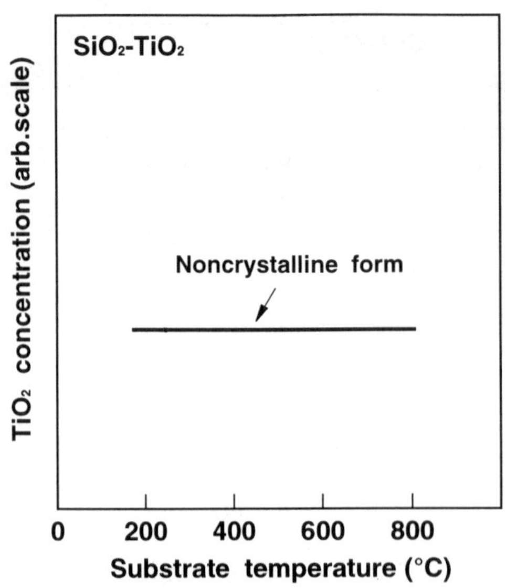

Figure 4.89 Substrate temperature dependence of TiO_2 concentration in SiO_2-TiO_2 glass particles [134,194].

The low-temperature phase of the B_2O_3 component, formed in the low substrate temperature region below 500°C (Figure 4.88), is a crystalline state combined with some water in an impurity-hydrate form. The crystalline B_2O_3 concentration decreases with increasing substrate temperature. The dissolved B_2O_3 concentration increases with an increase in substrate temperature from 200° to 800°C. The data in Figures 4.87 and 4.88 indicate that the SiO_2-B_2O_3 and SiO_2-GeO_2 particles are not completely formed in the flame. The dependence of the GeO_2 and B_2O_3 concentrations and their crystalline structures on the substrate temperature suggests that GeO_2 and B_2O_3 components solidify near the surface of the substrate, depending on the substrate temperature. GeO_2 and B_2O_3 seem to remain in the vapor state in the oxy-hydrogen flame.

On the other hand, the deposition properties of the SiO_2-TiO_2 system are in sharp contrast to those of the SiO_2-GeO_2 and SiO_2-B_2O_3 systems. The TiO_2 component is deposited with a noncrystalline structure dissolved in SiO_2 glass, regardless of the substrate temperature, and the TiO_2 concentration does not depend significantly on the substrate temperature.

Fine Glass Particle Formation Mechanism

It is useful to say a little about the vapor phase reaction in the VAD process, that is, the mechanisms by which fine glass particles of GeO_2, B_2O_3, and TiO_2 are formed and their differences [134,20]. As shown in Figure 4.51, the vapor pressures of GeO_2 and B_2O_3 at high temperatures of about 1,200° to 1,500°C are several orders of magnitude larger than the values for SiO_2 and TiO_2. This should have an important effect on the deposition properties of high-silica glass particles in a hydrolysis reaction.

In fact, from another investigation on the formation properties of oxide particles in the flame [134,199], it was found that each raw halide gas such as $SiCl_4$, $GeCl_4$, BBr_3, and $TiCl_4$ has a threshold flow rate below which the oxide product does not form solid particles in the flame but remains in a vapor state. The order of this threshold flow rate is dominated by the saturated vapor pressure values shown in Figure 4.51 for the oxide product at a high temperature. In addition, a model experiment with the flame hydrolysis reaction has shown that the dissolved SiO_2-GeO_2 particles are formed by the reaction of vapor phase GeO_2 with raw $SiCl_2$ gas on the SiO_2 particles [134].

Therefore, in the SiO_2-GeO_2 system, only SiO_2 forms solid particles in the flame, while the GeO_2 product remains in the vapor state because of its large saturated vapor pressure. This vapor-phase GeO_2 will be carried to the substrate (actually, the porous preform) surface and be cooled there. When the substrate temperature is sufficiently low, the vapor-phase GeO_2 will solidify in crystalline form on the substrate. On the other hand, when the substrate temperature is higher, the vapor-phase GeO_2 will not be cooled sufficiently on the substrate for it to be deposited in the crystalline state. At higher substrate temperatures, above 500°C, the GeO_2 component can remain on the substrate, forming an interconnected structure with SiO_2. This interconnected structure is formed by the reaction of GeO_2 vapor with raw $SiCl_2$ gas on the surface of the SiO_2 particles already formed in the flame. As a result, dissolved SiO_2-GeO_2 particles are formed. The formation of a GeO_2 component interconnected with SiO_2 depends largely on both the reaction temperature (that is, the substrate temperature) and environmental gas composition. Accordingly, the GeO_2 concentration in the SiO_2-GeO_2 glass particles deposited on the substrate surface (that is, the porous preform growing surface) is dependent on the substrate temperature, as shown in Figure 4.87. The formation mechanism of the SiO_2-B_2O_3 system is almost the same as that of the SiO_2-GeO_2 system.

On the other hand, in the SiO_2-TiO_2 system, TiO_2 vapor produced in the flame will solidify immediately and form an interconnected structure with coexisting SiO_2 vapor because of the low saturated vapor pressures in the flame. The formation of SiO_2-TiO_2 particles is, therefore, completed in the flame. This leads to the fact that the TiO_2 concentration and its crystalline structure do not depend on the substrate temperature.

Profile Formation Mechanism and Profile Control

The fine glass particle formation mechanism, as described, plays a dominant role in forming the refractive-index profile (that is, the dopant concentration distribution) in the VAD process. In particular, in the SiO_2-GeO_2 system, the refractive-index profile in the preforms depends largely on the temperature distribution on the growing surface of the porous preform shown in Figure 4.67. Figure 4.90 shows a typical surface-temperature distribution for the growing end of a porous preform [134,200]. The temperature distribution along the center line of the preform is indicated on the left in the figure, from which it can be seen that the temperature reaches a maximum of about 650°C at the center and decreases gradually to 300°C in the radial direction. The dopant concentration distribution in the prepared preforms is determined by this temperature distribution.

Thus, on the basis of these results for the vapor-phase reaction during flame hydrolysis, the formation mechanism of the refractive-index profile in the VAD process can be explained as follows. When the porous preform is grown by the deposition of fine glass particles, as shown in Figure 4.63, its growing surface usually exhibits a parabolic temperature distribution, where the center region has a high temperature that gradually decreases in the radial direction, as shown in Figure 4.90. The substrate temperature dependence of the GeO_2 concentration plays an important role in the formation of the index-profile formation. That is, when the porous perform is grown at surface temperatures ranging from 500° to 800°C using the SiO_2-GeO_2 system with the deposition property shown in Figure 4.87, the GeO_2 concentration is high at the center and gradually decreases to zero in the radial direction and the GeO_2 concentration gradient in the core porous preform is formed according to the surface-temperature gradient. In this way, a graded-index profile is realized in the SiO_2-GeO_2 system. In the SiO_2-TiO_2 system, however, since the TiO_2 concentration has no substrate temperature dependence, the core porous preform will have a steplike TiO_2 concentration; moreover, the surface-temperature gradient will be the same as in the SiO_2-GeO_2 system. This steplike TiO_2 concentration has been confirmed experimentally.

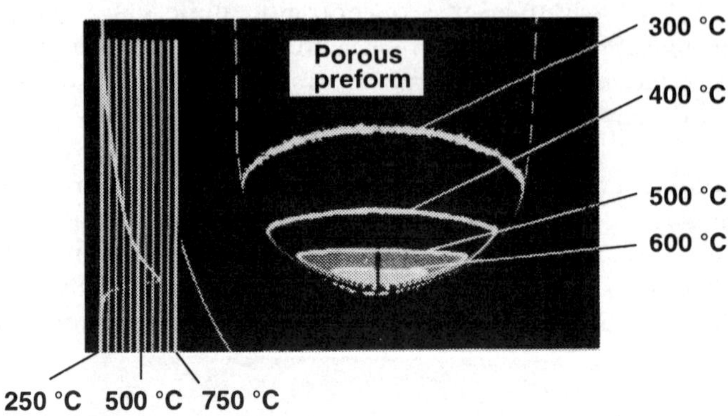

Figure 4.90 Typical surface-temperature distribution of the growing end of porous preform [134,200].

Refractive-Index Profile Control

In addition to its dependence on substrate temperature, the GeO_2 concentration in the SiO_2-GeO_2 glass particles also depends on the oxy-hydrogen flame temperature originating with the glass particle formation mechanism in the aforementioned flame [199]. Figure 4.91 shows a contour-line map for GeO_2 concentration in the SiO_2-GeO_2 system versus flame and substrate temperatures [134,20]. The numerals represent the GeO_2 concentration in molar percent. This figure shows that the GeO_2 concentration depends not only on the temperature of the substrate but also on that of the flame. Under fixed flame temperatures of 1,200° and 1,300°C, as indicated by the dotted and broken lines, respectively, the GeO_2 concentration increases as the substrate surface temperature increases. On the other hand, at a high fixed flame temperature of 1,400°C, shown by the dot-dashed line, the GeO_2 concentration tends to decrease at substrate temperatures above 700°C. The GeO_2 concentration decrease in the high flame and substrate temperature region seen in this figure might be attributed to the fact that the GeO_2 product does not deposit easily at a high temperature but remains in a vapor state because of its high saturated vapor pressure in the flame.

Using the temperature dependence of the GeO_2 concentration in SiO_2-GeO_2 glass particles, the graded-index profile was formed as shown in Figure 4.92. A practical and precise profile-control technique was achieved by applying not only

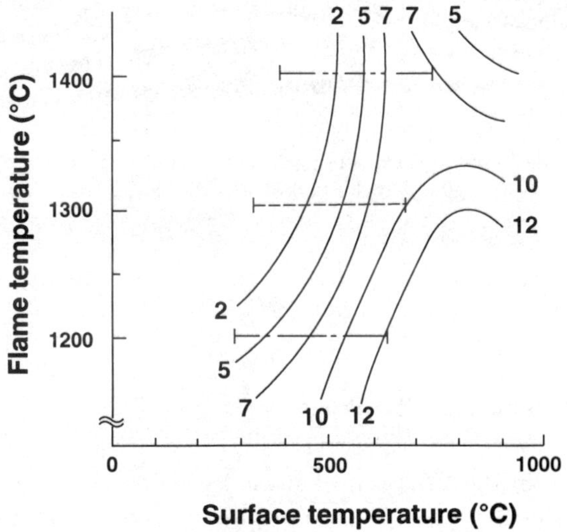

Figure 4.91 Contour-line map of GeO_2 concentration in the SiO_2-GeO_2 system versus flame and substrate temperatures [134,20].

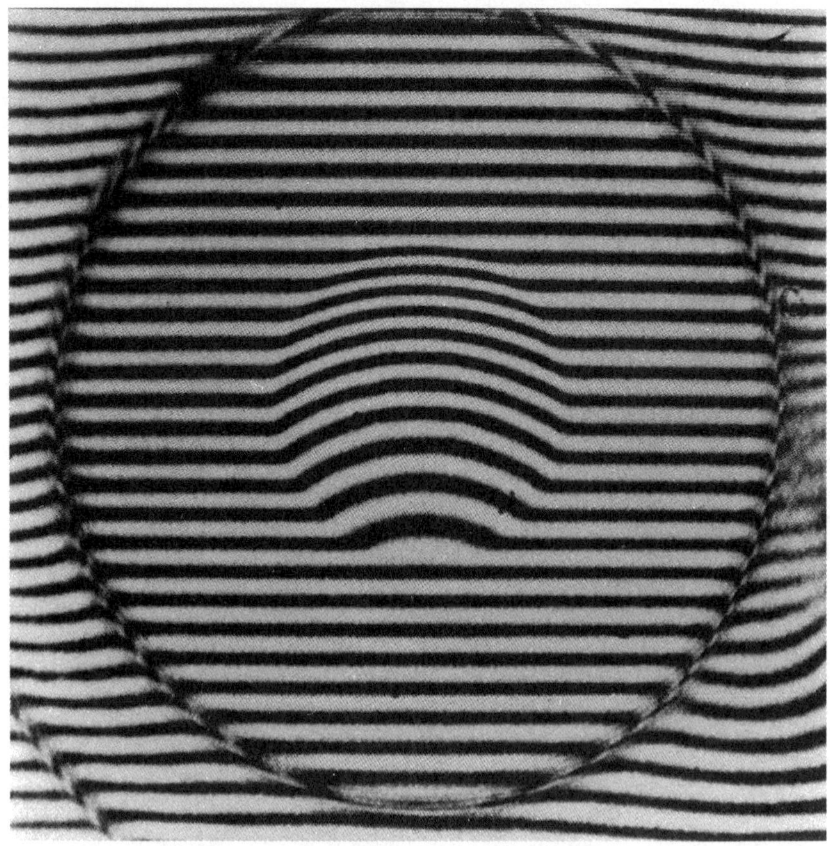

Figure 4.92 Graded-index profile formed by using the temperature dependence of the GeO_2 concentration in SiO_2-GeO_2 glass particles [200–202].

the temperature effect but also the mixing effect, and an index profile was formed with the desired value of $\alpha = 2.0$ [200–202].

4.4.5 Rare-Earth Doping in VAD Process

Rare-earth doping in the VAD process was already described in detail in Section 2.6. This involves not only the fabrication process but also the material structure of the rare-earth-doped high-silica glasses prepared by the VAD or soot process. In addition, the section described the strong interactions between the micro-material structure of rare-earth-doped high-silica glasses and amplification characteristics such as broadband and high gain.

The description of the doping method in Section 2.6 and of the VAD process in Section 4.4.4 should make it relatively easy to understand the technique of rare-earth doping in the VAD process. Therefore, we will not repeat the description here.

4.4.6 OVD Process

The OVD process is shown schematically in Figure 4.93. A vapor mixture of chloride raw materials react in an oxy-hydrogen flame to produce fine glass particles of the desired composition. The flame is directed toward a ceramic target rod, which is rotating and traversing in a lathe. The glass particles stick to the rod-forming porous layers. The composition of the glass particles is controlled layer by layer in order to form a radial refractive-index profile or the core/cladding structure. This process may be viewed as an intermediate approach between MCVD and VAD, because, in terms of the layer-by-layer preparation in the radial direction, OVD is similar to MCVD and, in terms of using flame hydrolysis to prepare the porous preform, OVD is closer to VAD. In fact, these three high-silica preform fabrication processes each have particular notable features and the common feature of their soot process or preparation direction.

When the soot deposition is completed, the porous preform is slipped off the ceramic target rod and then sintered to a transparent bubblefree preform in a furnace at about 1,500°C, which is almost the same as in the VAD process. In addition, the rare-earth doping process in the OVD process is the same as that in the VAD process and so requires no further description here.

4.4.7 High-Silica Fiber Drawing

High-silica fibers are made simply by drawing transparent preforms. One of the most important objects is to maintain the mechanical strength of the silica glass. Figure 4.94 shows a schematic diagram of the drawing apparatus, which is composed of a preform feed assembly, a graphite resistance furnace, a fiber diameter detector, a primary plastic coating assembly, a buffer plastic coating assembly, a coating diameter detector, a capstan, a dancer, and a take-up drum. The preform carrier is used to feed the preform into the furnace at a constant speed. The furnace is heated above 2,000°C to soften the end of the high-silica preform. The heater is usually made of high-purity carbon. The plastic coating assembly and the curing assembly are used to form the primary coating layer to prevent surface flaws from being formed as a result of the fiber surface touching the capstan and the take-up drum. The take-up speed of the fiber is controlled by a signal from the outer diameter measurement apparatus so as to maintain the outer diameter at a constant value. Figure 4.95 shows a small-scale fiber drawing apparatus.

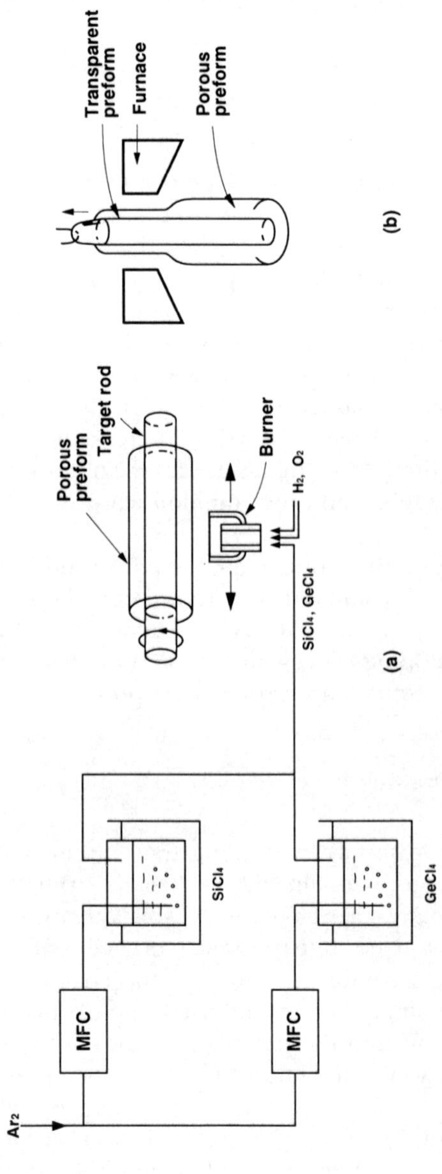

Figure 4.93 Schematic diagram of the simplest OVD preform fabrication apparatus: (a) porous preform preparation and (b) consolidation.

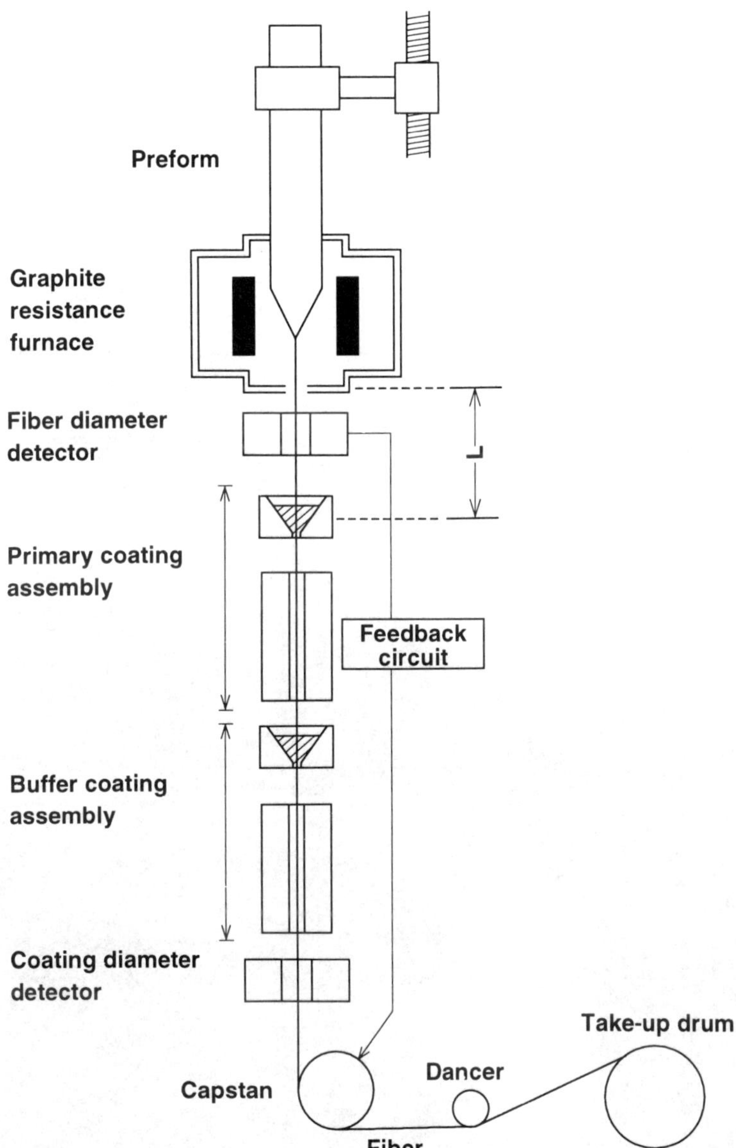

Figure 4.94 Schematic diagram of the drawing apparatus.

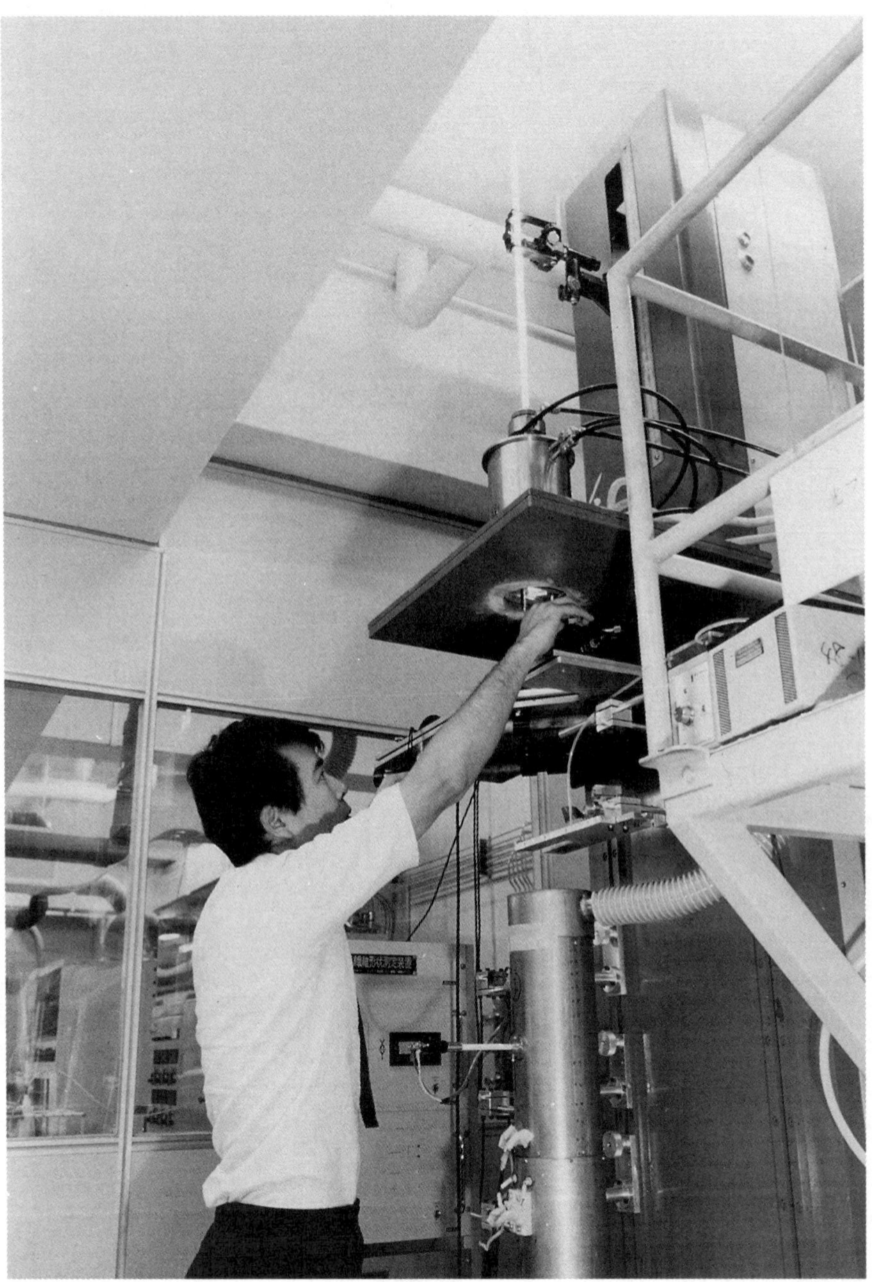

Figure 4.95 Small-scale fiber drawing apparatus.

There are two major origins of fiber diameter fluctuation: short-term fluctuation caused by temperature fluctuations in the furnace and long-term fluctuation caused by fluctuations in the outer diameter of the preform [20]. The short-term fluctuation, whose periodic length is less than about 100 cm, can be reduced to less than ±0.5 μm by reducing the temperature fluctuation to less than ±0.2°C. In order to realize precise temperature control, the inert gas flow rate in the furnace and the ratio of preform diameter to heater inner diameter must be optimized. The long-term fluctuation, whose periodic length is longer than 100 cm, can be reduced using a feedback control to the take-up speed through the signal from the outer diameter measurement apparatus.

4.5 MULTICOMPONENT GLASS FIBER FABRICATION PROCESS

Here we briefly describe the multicomponent glass fiber fabrication process and rare-earth doping to make it easier to understand fiber performance. The major differences between high-silica fiber and multicomponent glass fiber lie in their raw material purification and fiber drawing processes.

Many kinds of multicomponent oxide glass system have been investigated as transmission fiber material and as amplifier and laser host material. Among them, a typical glass composition for transmission fiber is soda-lime-silica because of its stable glassy state, relatively low dispersion, and low Rayleigh scattering loss [203]. In this glass system, SiO_2, GeO_2, Na_2O, CaO, Li_2O, and MgO are used as glass components. On the other hand, for many years, laser glasses have been manufactured for use in high-powered, Q-switched lasers. These multicomponent glass structures are mainly based on silicate and phosphate glasses [204,205]. A typical silicate host glass composition would be SiO_2 60 mol%, Li_2O 27.5 mol%, CaO 10 mol%, and Al_2O_3 2.5 mol%, with a rare-earth concentration of up to 8 mol% [204]. A typical phosphate host glass composition would be P_2O_5 60 mol%, K_2O 26.7 mol%, BaO 8.3 mol%, and Al_2O_3 5 mol%, with a rare-earth concentration up to 8 mol% [204]. Historically, the first glass laser was successfully operated using a neodymium-doped "barium crown" glass that consisted of 59 wt% SiO_2, 25 wt% BaO, 15% K_2O, and 1 wt% Sb_2O_3 [206].

When preparing multicomponent oxide glass fiber, the processes used for transmission fiber and active rare-earth-doped fiber are basically the same. Rare-earth-doped multicomponent oxide glass fibers are prepared by mixing rare-earth oxides into multicomponent oxides during the melting process followed by fiber drawing.

Multicomponent oxide glass fibers are commonly prepared in a double crucible from a high-purity multicomponent glass block and this clearly contrasts with the preform preparation method for high-silica fibers. The main process used for preparing multicomponent oxide glass fibers is as follows. First, high-purity component oxides such as SiO_2, GeO_2, Na_2O, CaO, Li_2O, and MgO are prepared.

The mixture of component oxides is then melted and refined to form multicomponent glass. Finally the fiber is drawn from the as-grown multicomponent glass block after it is cut and polished. The following starting materials are commonly used to make the high-purity component oxides: ultrahigh purity SiO_2 prepared by the flame hydrolysis of SiH_4, distilled $Ge(OC_4H_9)_4$ for GeO_2, $NaNO_3$, $LiNO_3$, $Ca(NO_3)_2$, and $MgSO_4$ [207]. These nitrates and sulfate are purified by an ion exchange process or a solvent extraction process. The starting materials are mixed and condensed into slurry. The slurry is then transferred to a fused-silica crucible by a plastic pump and completely dried in a resistance furnace. The glasses are melted and refined at around 1,400°C for 1 hr in a clean, dry N_2-O_2 gas flow. Multicomponent glass blocks made by this process are cut into several pieces, polished with abrasives, cleaned with a supersonic-wave cleaner in HNO_3-H_2SO_4 solution, and rinsed in distilled water. Individually prepared core and clad glasses are placed in a platinum double crucible and heated to 750°C to draw into fibers.

Figure 4.96 is a schematic diagram of a multicomponent glass fiber drawing system [20,207]. The fiber drawing is carried out under a cylindrical fused-silica cover in clean, dry N_2 gas. After drawing, the glass fibers are coated with silicone, as with high-silica fibers. Figure 4.97 is a photograph of the fiber drawing apparatus for multicomponent glass fiber.

One interesting problem that arises during the multicomponent glass fiber drawing process is bubble formation in the double crucible, which is not observed in the preform drawing method. The mechanism is the result of the difference between the alkali-ion concentrations of the core and cladding glasses, which forms a battery with platinum. Figure 4.98(a) shows the formation of a battery as a result of the different alkali-ion concentration in multicomponent glass and (b) bubble formation in a double crucible [20,208]. At the electrode immersed in the glass with the higher ion concentration, O^{2-} becomes O^2 and electrons as shown in the following equations [209]. At the second electrode immersed in the glass with the lower ion concentration, oxygen gas in the atmosphere is taken up in the molten glass and becomes O^{2-}:

$$\text{Electrode I: } 2O^{2-} \rightarrow O_2 + 4e^- \text{(bubble formation)} \qquad (4.80)$$

$$\text{Electrode II: } 4e^- + O_2 \rightarrow 2O^{2-} \qquad (4.81)$$

These bubbles are formed on the inner crucible surface when the alkali-ion concentration of the core glass is higher than that of the cladding glass. Some of the bubbles are introduced into the core-cladding interface of the prepared fiber and hence cause a large scattering loss. Therefore, it is preferable to have the same alkali-ion concentration in the core and cladding glasses and to draw the fiber in an oxygenfree atmosphere.

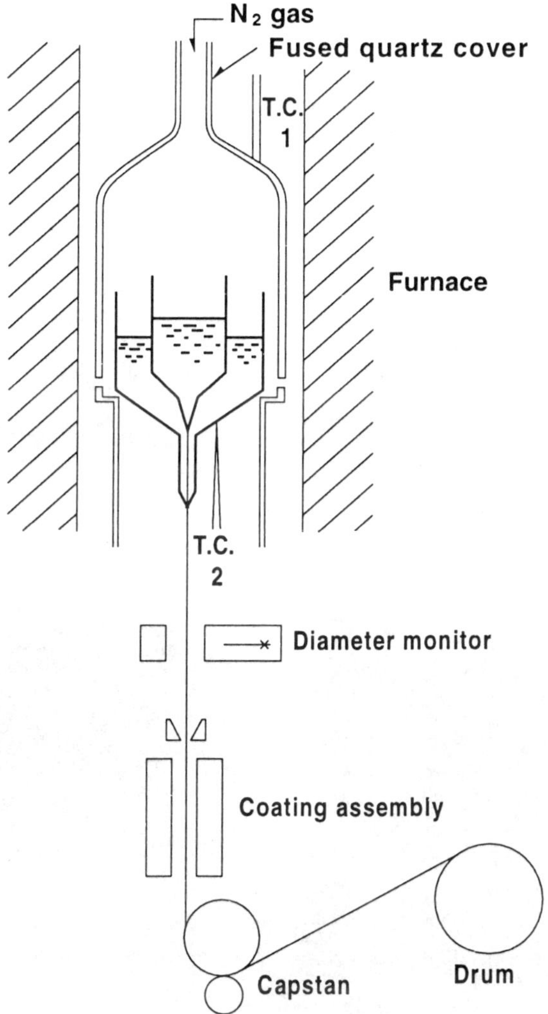

Figure 4.96 Schematic diagram of a multicomponent glass fiber-drawing system [20,207].

4.6 FLUORIDE FIBER FABRICATION PROCESS

Since heavy-metal fluoride glasses were first suggested as potential materials for ultra-low-loss optical fibers with transmission losses lower than those of silica fibers [210–213], there have been many studies on glass composition and characteristics and great efforts have been made to fabricate fibers and reduce their transmission losses. Among the fluoride glass systems developed to date, ZrF_4/HfF_4-based glasses,

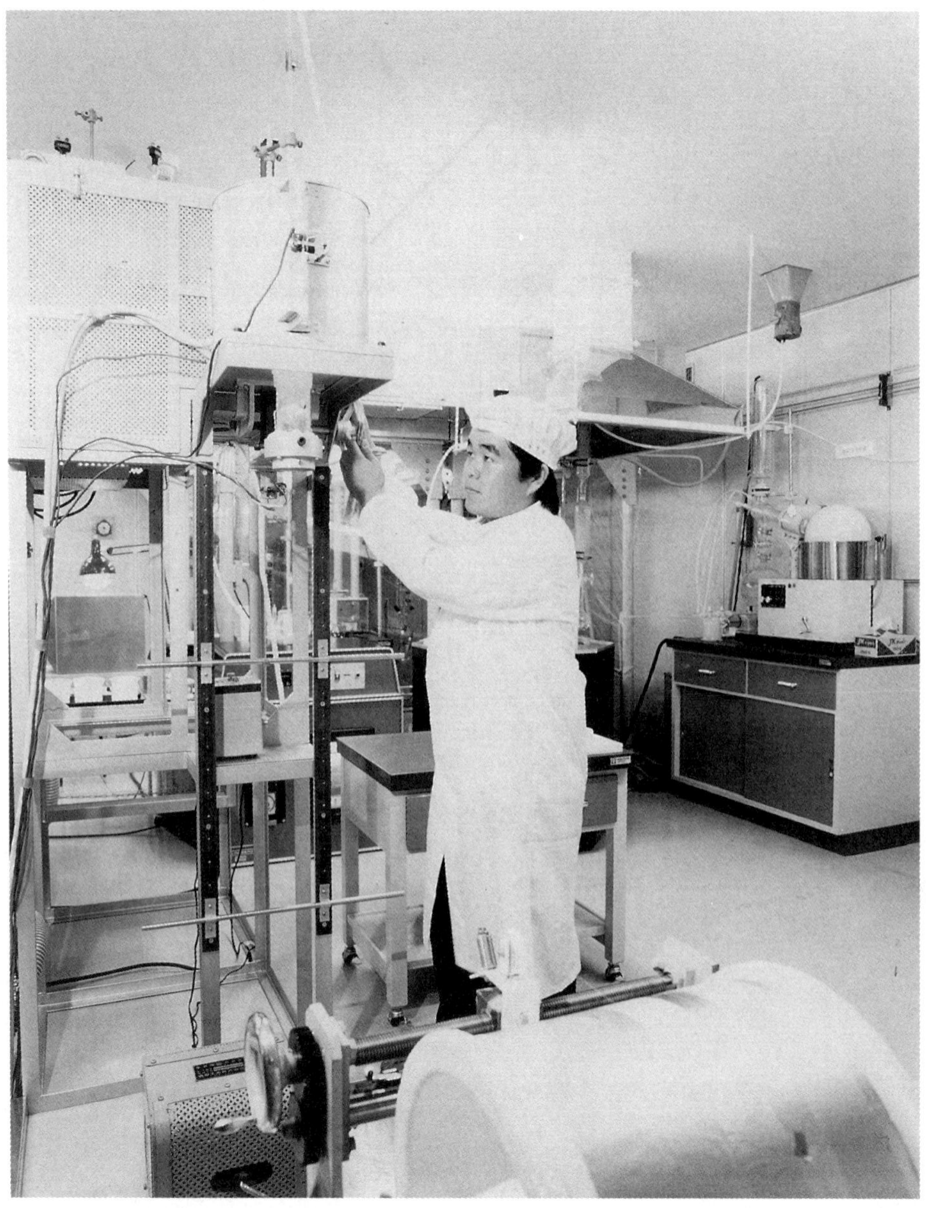

Figure 4.97 Photograph of fiber drawing apparatus for multicomponent glass fibers.

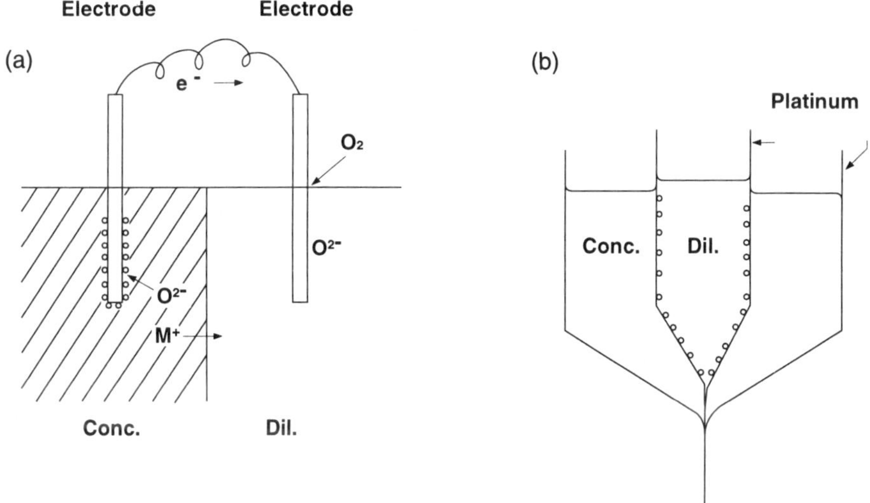

Figure 4.98 (a) Battery formation as a result of the different alkali-ion concentrations in multicomponent glass and (b) bubble formation in a double crucible [20,208].

AlF$_3$-based glasses, and InF$_3$/GaF$_3$-based glasses are regarded as the materials that offer the greatest potential as optical fibers because of their fairly large thermal stability and durability in addition to their high transparency. The first ZrF$_4$-based glass was fabricated in 1975 [214]. This was a ZrF$_4$-BaF$_2$-NaF system with a fairly good glass-forming ability. In 1980, ZrF$_4$-BaF$_2$-GdF$_3$ glass was first drawn into an unclad fiber with a loss of 480 dB/km [215]. The following AlF$_3$-based glasses were reported both before and after ZrF$_4$-based glasses: AlF$_3$-PbF$_2$-SrF$_2$-MgF$_2$ in 1949 [216], AlF$_3$-PbF$_2$-CdF$_2$ in 1955 [217], and AlF$_3$-CaF$_2$-BaF$_2$ in 1979 [218]. However, because of their poor glass-forming ability, fiber fabrication was delayed and AlF$_3$-CaF$_2$-BaF$_2$-YF$_3$ glass was drawn into an unclad fiber with a loss of 8,000 dB/km only in 1981 [25]. InF$_3$-based glass was reported in 1980 [219], and InF$_3$-ZnF$_2$-BaF$_2$-YbF$_3$-ThF$_4$ glass was drawn into fiber with a loss of 9,000 dB/km in 1988 [220]. Now, as a result of the development of both glass compositions with better glass-forming ability and fiber fabrication techniques that reduce extrinsic loss, the losses of multimode fibers have been reduced to about 1 dB/km for ZrF$_4$-based fluoride fiber [10,11], 6 dB/km for AlF$_3$-based fluoride fiber [12], and 73 dB/km for InF$_3$-based fiber [13].

These heavy-metal fluoride glasses with a low-phonon frequency have high radiative efficiencies compared with those of silica-based glasses and the potential to reveal new fluorescence bands that do not appear in silica-based glasses because the lower the highest stretching vibration frequency of the glass matrix, the lower the transition probability of nonradiative decay of active ions in the glass becomes

[221]. In particular, stable ZrF_4-based glasses, in which favorable numbers of rare-earth-metal ions below 8 mol% can be doped as a fluoride and where the fundamental phonon vibration is located at a low frequency of about 500 cm^{-1}, become good rare-earth-doped hosts for fiber amplifiers.

In Section 4.6.2, we describe the fabrication techniques commonly used for fluoride optical fibers and focus particularly on the fabrication of ZrF_4-based single-mode fluoride fiber. In Section 4.6.4, we describe fluoride glass synthesis using the CVD method because, although it has yet to produce low-loss fluoride glass fiber, this method has good potential.

4.6.1 Glass Compositions

Glass stability basically determines fiber quality and properties because the potential for reducing extrinsic scattering loss becomes higher as the stability against crystallization increases. Table 4.13 [222,223] and Figure 4.99 [225] show critical cooling rates R_c for fluoride glasses, which is a measure of glass-forming ability. Glass with a lower R_c is more stable during the melt cooling process. As shown in Table 4.13 and Figure 4.99, ZBLAN and ZBLYAN systems have the most favorable glass compositions for fiber fabrication, although their R_c values are about 100 times that of silica glass.

In order to fabricate an optical fiber, the refractive-index of the glass must be changed without any reduction in the glass stability. With ZrF_4-based glasses, the preferred techniques for obtaining an appropriate refractive-index difference (Δn) between a core and cladding are to dope AlF_3 and substitute HfF_4 for ZrF_4 to lower the index and to substitute LiF for NaF and PbF_2 for BaF_2 to raise the index. Using these techniques, a very high Δn of above 4% can be achieved in

Table 4.13

Critical Cooling Rate R_c for Fluoride Glasses

Composition *(mol%)*	R_c *(K/min)*
$63ZrF_4$-$33BaF_2$-$4GdF_3$	350
$60ZrF_4$-$32BaF_2$-$4GdF_3$-$4AlF_3$	70
$57ZrF_4$-$34BaF_2$-$4LaF_3$-$5AlF_3$	54.6–90[a]
$52ZrF_4$-$20BaF_2$-$5LaF_3$-$3AlF_3$-$20LiF$	26
$47.5ZrF_4$-$23.5BaF_2$-$2.5LaF_3$-$2YF_3$-$4.5AlF_3$-$20NaF$	1.1
$51ZrF_4$-$20BaF_2$-$4.5LaF_3$-$4.5AlF_3$-$20NaF$	0.7
$42AlF_3$-$24CaF_2$-$24BaF_2$-$10YF_3$	699–735[a]
$34AlF_3$-$25.5CaF_2$-$8.5BaF_2$-$8.5SrF_2$-$8.5MgF_2$-$15YF_3$	17[b]
$30.2AlF_3$-$20.2CaF_2$-$10.6BaF_2$-$13.2SrF_2$-$3.5MgF_2$-$8.3YF_3$-$10.2ZrF_4$-$3.8NaF$	16[c]
$Sb_{23.1}Ge_{15.4}Se_{61.5}$	20[d]
SiO_2	0.004[d]

From: [222,223]. a, [224]; b, [226]; c, [227]; d, [103].

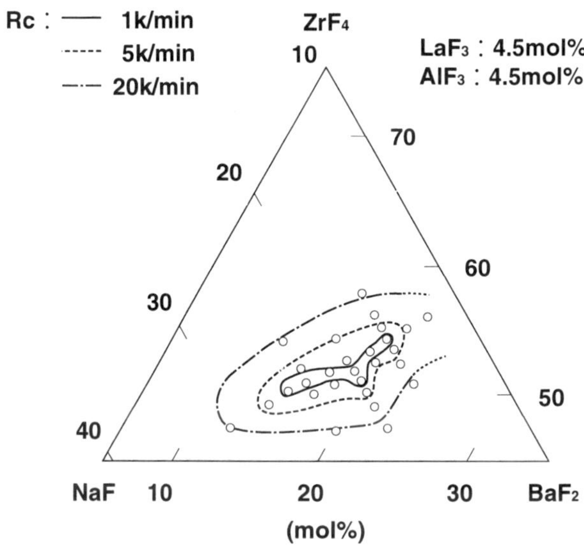

Figure 4.99 Compositional dependence of critical cooling rate in the ZrF_4-BaF_2-PbF_2-LaF_3-YF_3-AlF_3-NaF system [225].

ZrF_4/HfF_4-based fibers for optical amplifiers. However, the glass stability for high-Δn fiber decreases as the Δn increases. Figure 4.100 shows the change in Δn between ZrF_4-BaF_2-PbF_2-LaF_3-YF_3-AlF_3-LiF core and HfF_4-BaF_2-LaF_3-YF_3-AlF_3-LiF-NaF cladding glasses and the time t_g changes of the core glass as a function of PbF_2 concentration [104]. The t_g is the time required to reach a crystallized fraction of 10^{-6}, for which the material is considered to be a glass, at a flow point and is calculated from the Avrami and Arrhenius equations with measured parameters [222]. The value of t_g at a flow point represents the stability against crystallization during a reheating process; stable glass has a large t_g. The Δn increases rapidly with increasing PbF_2 concentration, but the t_g decreases rapidly with increasing PbF_2 content and falls to a particularly low level at a PbF_2 content of more than 15 mol%.

Some commonly used fiber compositions are listed in Table 4.14 [223,226]. As shown in the table, since fluoride glasses contain a few mol% of rare-earth fluorides, rare-earth elements can be homogeneously doped as radiative ions by substituting them for the rare-earth fluorides of glass components such as LaF_3 and YF_3.

The properties of typical fluoride glasses for optical fibers are listed in Table 4.15 [25,26,230–232].

4.6.2 Fabrication Process

In this section, we describe fluoride fiber fabrication techniques based on the preparation of glass preforms by a melting process. In this preform process, which

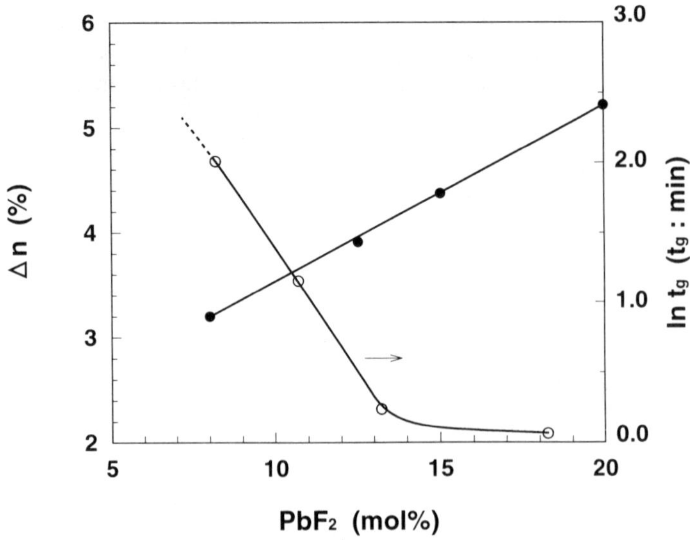

Figure 4.100 Change in the refractive index difference (Δn) between ZrF_4-BaF_2-PbF_2-LaF_3-YF_3-AlF_3-LiF core and HfF_4-BaF_2-LaF_3-YF_3-AlF_3-LiF-NaF cladding glasses and the time t_g changes of the core glass as a function of PbF_2 concentration [104].

is the most successful method for fabricating low-loss fluoride fibers, glass-melts are cast into preforms that are finally drawn into optical fibers.

A typical fabrication process is shown in Figure 4.101. The process consists of raw material purification, glass melting, preform fabrication, and fiber drawing.

4.6.2.1 Purification of Raw Materials

Ultrapure fluoride starting materials are required for the preparation of low-loss optical fiber, in which impurities must be kept at the sub-ppm level. Two types of purification process have been developed for fluoride materials. The first is a dry purification process consisting of sublimation [233,234] or chemical vapor purification [235], and the second is a wet process involving reprecipitation [234] or solvent extraction [236]. In this section, we describe sublimation and solvent extraction as successful purification techniques for use in practical fiber fabrication.

Sublimation

The most volatile components, such as ZrF_4, HfF_4, and AlF_3, are purified under a low pressure of 1- to 3-mm Hg at a high temperature using the sublimation apparatus as shown in Fig 4.102 [233]. This sublimation is effective in separating volatile

Table 4.14
Fluoride Fiber Compositions

Glass Composition (mol%)		Δn
Core	*Cladding*	*(%)*
$61ZrF_4$-$32BaF_2$-$3.9GdF_3$-$3.1AlF_3$	$59.5ZrF_4$-$31.2BaF_2$-$3.8GdF_3$-$5.5AlF_3$	0.33[a]
$53ZrF_4$-$20BaF_2$-$4LaF_3$-$3AlF_3$-$20NaF$	$39.9ZrF_4$-$13.3HfF_4$-$18BaF_2$-$4LaF_3$-$3AlF_3$-$22NaF$	0.40[b]
$51.5ZrF_4$-$19.5BaF_3$-$5.3LaF_3$-$3.2AlF_3$-$18NaF$-$2.5PbF_2$	$53ZrF_4$-$20BaF_2$-$4LaF_3$-$3AlF_3$-$20NaF$	0.47[c]
$49ZrF_4$-$25BaF_2$-$3.5LaF_3$-$2YF_3$-$2.5AlF_3$-$18LiF$	$47.5ZrF_4$-$23.5BaF_2$-$2.5LaF_3$-$2YF_3$-$4.5AlF_3$-$20LiF$	0.38
$49ZrF_4$-$25BaF_2$-$3.5LaF_3$-$2YF_3$-$2.5AlF_3$-$18NaF$	$47.5ZrF_4$-$23.5BaF_2$-$2.5LaF_3$-$2YF_3$-$4.5AlF_3$-$20NaF$	0.40
$49ZrF_4$-$25BaF_2$-$3.5LaF_3$-$2YF_3$-$2.5AlF_3$-$18NaF$	$23.7ZrF_4$-$23.8HfF_4$-$23.5BaF_2$-$2.5LaF_3$-$2YF_3$-$4.5AlF_3$-$20NaF$	0.80
$49ZrF_4$-$25BaF_2$-$3.5LaF_3$-$2YF_3$-$2.5AlF_3$-$18LiF$	$47.5ZrF_4$-$23.5BaF_2$-$2.5LaF_3$-$2YF_3$-$4.5AlF_3$-$20NaF$	0.95
$49ZrF_4$-$25BaF_2$-$3.5LaF_3$-$2YF_3$-$2.5AlF_3$-$18LiF$	$23.7ZrF_4$-$23.8HfF_4$-$23.5BaF_2$-$2.5LaF_3$-$2YF_3$-$4.5AlF_3$-$20NaF$	1.36
$30.2AlF_3$-$18CaF_2$-$9.3BaF_2$-$11.7SrF_2$-$3.5MgF_2$-$8.3YF_3$-$10.2ZrF_4$-$3.8NaF$-$5PbF_2$	$30.2AlF_3$-$19.2CaF_2$-$9.9BaF_2$-$12.4SrF_2$-$3.5MgF_2$-$8.3YF_3$-$10.2ZrF_4$-$2.5NaF$-$3.8PbF_2$	0.54[d]
$30InF_3$-$20ZnF_2$-$30BaF_2$-$10YbF_3$-$10ThF_4$	$30InF_3$-$20ZnF_2$-$25BaF_2$-$10YbF_3$-$10ThF_4$-$5NaF$	0.63[e]

From: [223,226]. a, [228]; b, [27]; c, [229]; d, [26]; e, [13].

components from transition-metal fluorides and other less-volatile impurities, where sub-ppm impurity concentrations are obtained. A higher degree of purification can be achieved by sublimation after converting impure metal fluorides to nonvolatile stable oxides [237]. Nonvolatile components such as BaF_2 can also be purified by sublimating out transition-metal fluorides from the raw materials. However, this sublimation process for the nonvolatile components is not suitable for reducing less-volatile impurities such as oxides. The sublimation temperatures are determined from their vapor pressure curves as shown in Figure 4.103 [234].

Solvent Extraction

Solvent extraction, which is an aqueous purification method, is effective in purifying nonvolatile fluoride materials such as BaF_2, LaF_3, YF_3, NaF, and LiF. Figure 4.104 shows solvent extraction using $Ba(CH_3COO)_2$ and diethylammonium diethyldithio-carbamate (DDDC). Fe^{3+} impurity species form stable complexes of $Fe(DDDC)_3$

Table 4.15
Physical Properties of Heavy-Metal Fluoride Glasses

Glass	ZBLA[a]	ZBLAN[b]	ACBYA[c]	ACBSMZYN[b]
Density (g/cm^3)	4.54	4.33	4.0	3.85
Microhardness (kg/mm^2)	250	225	360	315
Young's modulus (GPa)	55	53.8	63.8[d]	65.0
Shear modulus (GPa)	21.5	20.5	24.6[d]	24.8
Bulk modulus (GPa)	(41.5)	47.7	(52.3)[d]	57.0
Poisson's ratio	0.28	0.31	0.30[d]	0.31
Fracture toughness (MPa m$^{1/2}$)	0.31	–	0.38[d]	–
Thermal expansion coefficient (10^{-7}/K)	168	172	165	152
Glass transition temperature (°C)	320	257	430	392
Refractive index n_D	1.519	1.499	1.436[e]	1.439
Nonlinear refractive index n_2 (10^{-13} esu)	0.96	0.85	–	0.51
Abbe value	–	75.95	–	93.95
$(dn/dT)_{absolute}$ (10^{-6}/K)	–10	–13.45	–	–9.42

a, [230]; b, [26]; c, [25]; d, [231]; e, [232].
ZBLA: $57ZrF_4$-$34BaF_2$-$5LaF_3$-$4AlF_3$(mol%); ZBLAN: $53ZrF_4$-$20BaF_2$-$4LaF_3$-$3AlF_3$-$20NaF$; ACBY: $40AlF_3$-$22CaF_2$-$22BaF_2$-$16YF_3$; ACBSMYZN: $30.2AlF_3$-$20.2CaF_2$-$10.6BaF_2$-$13.2SrF_2$-$3.5MgF_2$-$8.3YF_3$-$10.2ZrF_4$-$3.8NaF$.

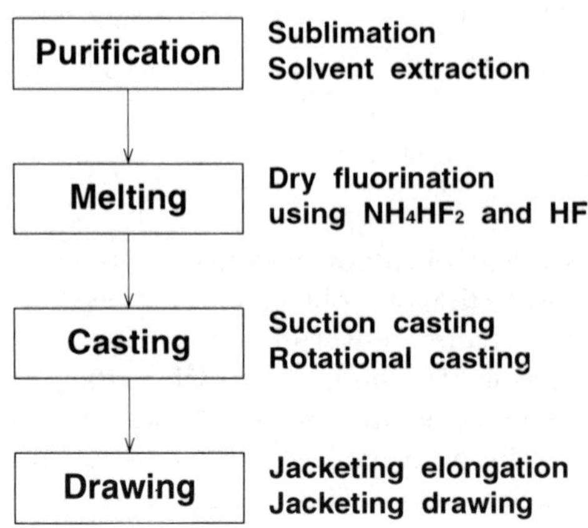

Figure 4.101 Typical fluoride fiber fabrication process.

and are then separated in an adjacent organic phase. By this method using DDDC with high distribution coefficients, we can achieve sub-ppb impurity concentrations of transition elements [238]. $Ba(CH_3COO)_2$ purified by solvent extraction is converted to BaF_2 via the reactions

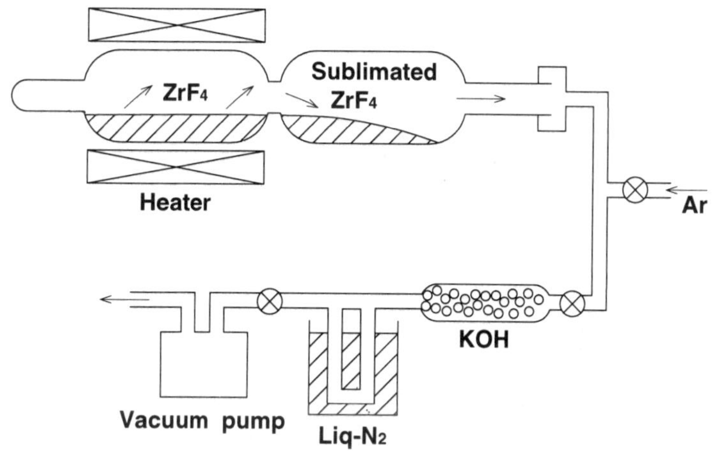

Figure 4.102 Purification of ZrF_4 by sublimation [233].

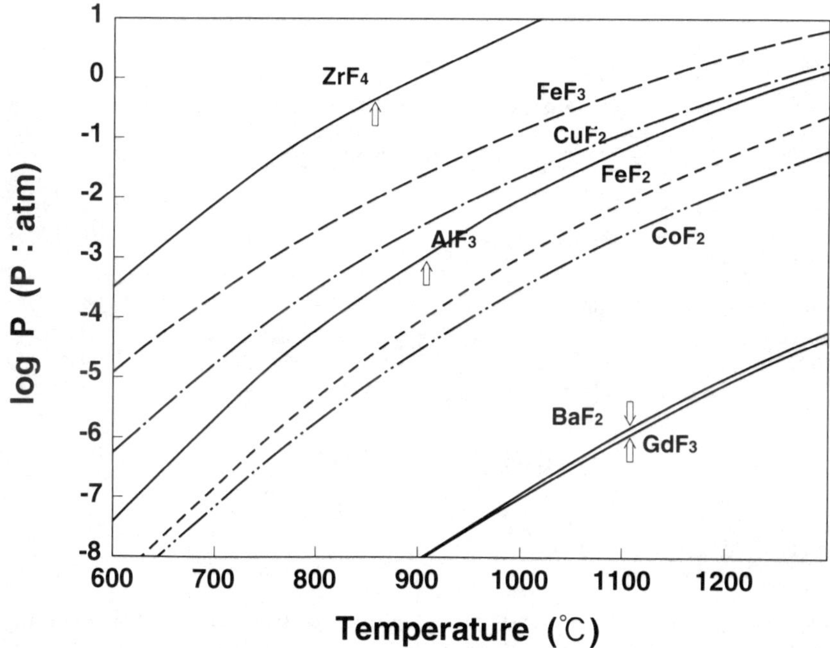

Figure 4.103 Vapor pressure curves for fluorides [234].

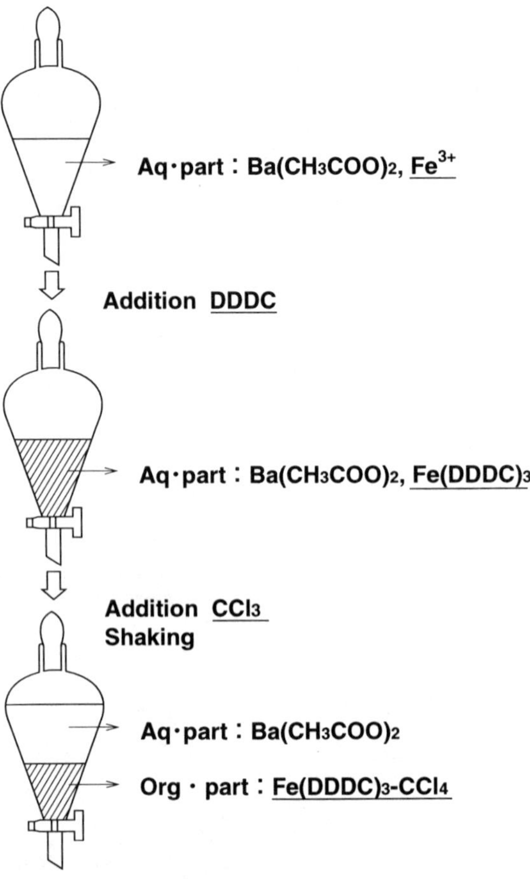

Figure 4.104 Solvent extraction process.

$$Ba(CH_3COO)_2 + 3HFaq \rightarrow BaHF_3 + 2CH_3COOH \qquad (4.82)$$

$$BaHF_3 \rightarrow BaF_2 + HF \qquad (4.83)$$

4.6.2.2 Glass Melting

The key requirements when using glass melting to attempt to fabricate low-loss fiber are: (1) to avoid contamination of the starting materials and the glass-melt by such impurities as H_2O, transition metals, and organic compounds and (2) to eliminate any resultant H_2O, OH^-, and oxide in the staring materials and reduced zirconium ions in the glass-melt.

To meet these requirements, the glasses are processed in glove boxes and fluorinating agents are used when the glass is melted. The storage of the raw materials, the weighing and batching of the fluoride powders, and the melting and casting of the glasses are all undertaken in a dry, clean inert-gas atmosphere to avoid contact with the environment [239,240]. Figure 4.105 shows typical apparatus consisting of a melting furnace, two glove boxes, and an exhaust box. The insides of the furnace and boxes are purged with dry gas during glass preparation. N_2 or Ar gas vaporized from liquid phase is used as the dry gas. This dry, clean atmosphere processing is effective in reducing OH^- absorption losses in fibers. NH_4HF_2 (ammonium bifluoride) is used as a fluorinating agent and is effective in eliminating the residual oxides present in the starting materials and OH^- adsorbed at their surface. Oxygen gas can be introduced to suppress the formation of reduced Zr.

Thermodynamic calculations have shown that the fluorination process using NH_4HF_2 corresponds to a wet HF atmosphere, where the hydrated system is stable [241], indicating that it is difficult to achieve complete deoxygenation using NH_4HF_2. In order to suppress oxide formation in glasses, dry fluorination is desirable. The use of HF gas can produce such a dry atmosphere. Fluorination with HF gas, however, is not sufficient because it is less active than NH_4HF_2. One of the best ways to overcome this is to use both HF and NH_4HF_2 [241].

The presence of reduced Zr, which is indicated when the glass is gray or contains a black phase, causes optical absorption loss in the short-wavelength region and the loss extends to the 2-μm wavelength region [242]. The reduced Zr is probably produced by one of two mechanisms. One is the elimination of fluorine by the reaction [243]

$$ZrF_4 \rightarrow ZrF_2 + F_2 \qquad (4.84)$$

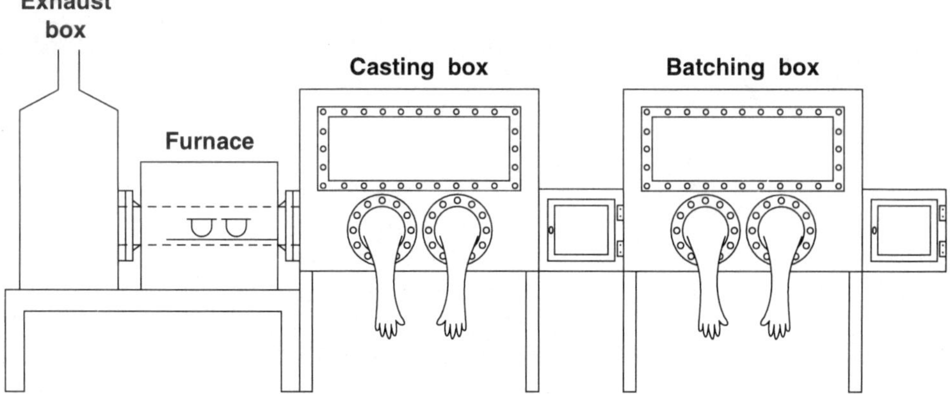

Figure 4.105 Typical apparatus for melting fluoride glass.

and the other is the elimination of oxygen that often occurs during the dehydration process when using NH_4HF_2 [244]. Therefore, NH_4HF_2 can reduce the oxide content but produces a lot of reduced Zr as a result of the reduction reactions. Photoluminescence observations and loss measurements have shown that the reduced Zr can be successfully eliminated by introducing O_2 gas into the glass-melting atmosphere and the absorption loss in the short-wavelength region is reduced as shown in Figure 4.106 [242]. In practice, fluoride glasses for optical fibers are ordinarily prepared under an O_2 gas flow to oxidize the reduced Zr.

Reactive atmosphere processing (RAP) using an active gas such as CF_4, NF_3, or SF_6 effectively reduces OH^- and oxides in glasses [245,246]. In the melting process using RAP, carbon crucibles must be used because noble metal crucibles react with these gases. These reagents are good fluorinating agents because of their high activity, but the use of carbon crucibles often causes contamination of the carbon particles in the glass.

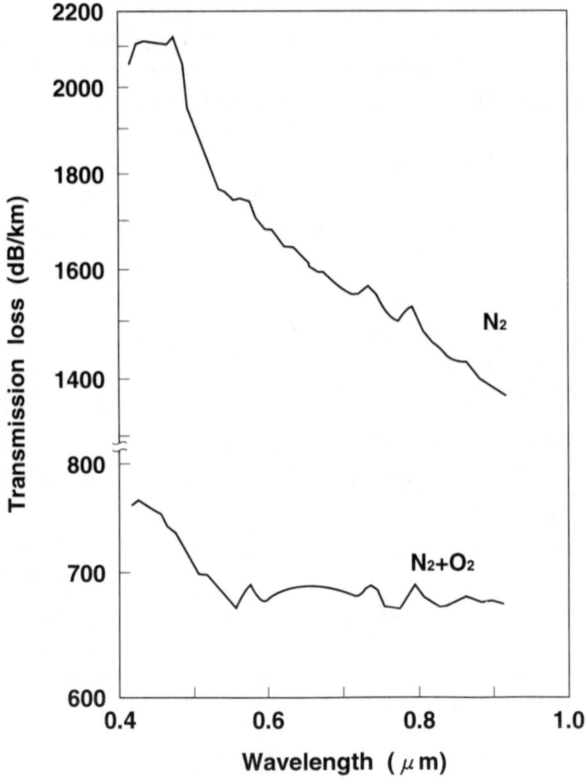

Figure 4.106 Transmission loss spectra of ZrF_4-based fluoride fibers prepared in a glass-melting atmosphere of N_2 and $N_2 + O_2$ [242].

A typical melting process with ZrF_4-based glass using noble metal crucibles and NH_4HF_2 as a fluorinating agent begins when batches of starting materials with small amounts of NH_4HF_2 are placed in gold or platinum crucibles in a weighing glove box. The crucibles containing the batch are transferred into another melting glove box in order to exclude the fine powders generated during the weighing and batching processes from the melting environment. The crucibles are placed in the melting furnace located in the second glove box and held at 300° to 400°C. At this stage, the starting materials are fluorinated by NH_4HF_2 (and HF gas). The furnace temperature is then raised to 800° to 900°C to fuse and melt the materials and held for a period to homogenize the melt. At this stage, the remaining NH_4HF_2 is vaporized out of the crucibles. Then the melt is cooled near the liquidus temperature (650° to 700°C) in order to enhance the cooling rate during the final casting and quenching by reducing the heat of the melt. Oxygen is introduced at this stage.

4.6.2.3 Preform Fabrication

Fluoride glass fiber preforms are prepared by a glass-melting process using crucibles followed by a mold-casting process that enables the melt to be rapidly quenched. The casting process using a metal mold makes use of the very low viscosity of fluoride glass-melts that correspond to its fairly low glass-forming tendency and is necessary to form high-quality fluoride glass for optical fiber. Build-in casting [247], modified build-in casting [248], suction casting [249], rotational casting [250], and extrusion methods [251] are established preform fabrication techniques for preparing multimode fibers with low losses of 1 to 10 dB/km. With all these methods except for extrusion, preforms with a waveguide structure are directly formed inside molds. With the extrusion method, two glass disks for the core and cladding are prepared by melt casting.

Build-in Casting

This was the first technique for preparing a fiber preform with a waveguide structure made of fluoride glass in both the core and cladding. Figure 4.107 shows the build-in casting process [228]. The cladding glass-melt is cast into a cylindrical mold preheated near the glass-transition temperature, and the mold is then immediately inverted. The melt in the central part of the mold flows out, and the solidified skin of glass formed on the inside walls is left behind as a cladding glass tube. The core glass-melt is then cast into the cladding tube to form the core part, and the assembly is cooled and annealed. Although preforms fabricated by this build-in casting method typically have core diameters that taper along its length, they have a smooth core-cladding interface because of the surface tension at the liquid-liquid interface. Good optical quality, therefore, is possible in the fiber.

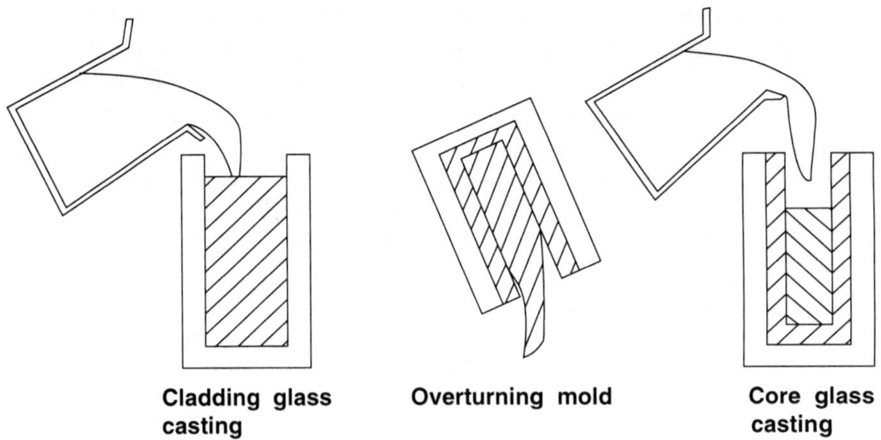

Figure 4.107 Build-in casting process [228].

Modified Build-in Casting

This technique enables large preforms to be made and reduces the taper in the core diameter found with the ordinary build-in casting method [248]. A schematic diagram of the modified build-in casting process is shown in Figure 4.108. The cladding glass-melt, whose volume is precisely controlled, is cast into a preheated

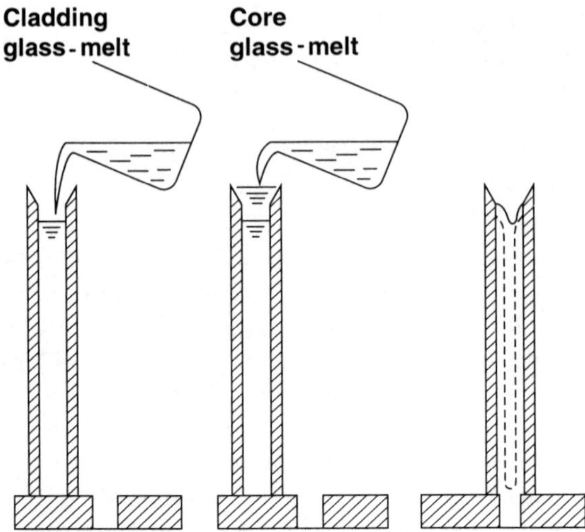

Figure 4.108 Modified build-in casting process.

mold with a specially designed bottom plate. Before the central part of the cladding melt solidifies, the core glass-melt is poured onto the cladding melt. The mold is then shifted laterally to a position where there is a hole in the bottom plate. The central part of the molten glass flows out into the hole, and the stream of core glass creates a core-cladding structure.

Suction Casting

This casting method, which is shown in Figure 4.109, involves a rapid quenching of the core and cladding melt and allows the preparation of a preform with a fairly large cladding/core diameter ratio, although the preform also has a tapered core [249]. With suction casting, a specially designed cylindrical mold is used that has a reservoir at the bottom. The cladding glass-melt is poured into the mold. Then, the core glass-melt is poured onto the cladding glass before the cladding glass has completely solidified. Significant volume constriction occurs during cooling. When this cladding glass volume constriction occurs in the reservoir, a cylindrical cladding tube is formed in the mold, which produces a suction effect on the core melt. The cladding/core diameter ratio and preform length can be controlled by the appropriate selection of the reservoir volume and mold diameter.

Rotational Casting

Rotational casting, which is based on an old established method used to form glass tubes, makes it easy to prepare a preform with excellent radial and longitudinal

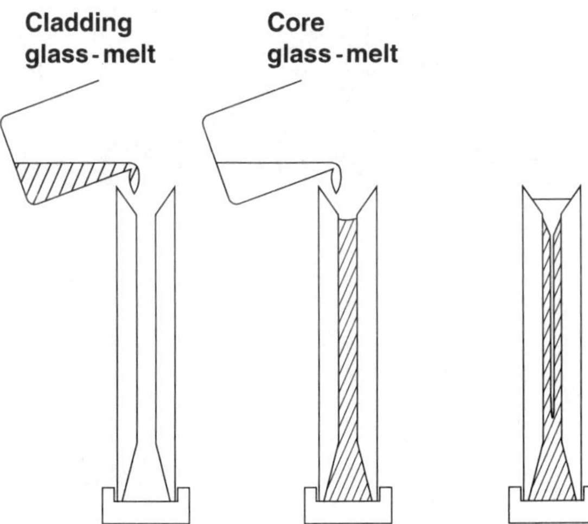

Figure 4.109 Suction casting process.

uniformity. The technique is illustrated in Figure 4.110 [250]. The cladding glass-melt is poured into a preheated mold, and then the mold is swung to a horizontal position and rotated at speeds of more than 3,000 rpm until the melt cools to form a glass tube. The obtained tube is highly concentric, and the inner diameter can be precisely controlled by the initial volume of injected glass. The core melt is then cast into this tube, and the assembly is cooled and annealed.

Extrusion

This method is advantageous for use with glasses that exhibit a strong tendency to crystallize and vaporize because preforms can be formed at a higher viscosity (10^8 to 10^9 poises) and hence at a lower temperature than with the previously mentioned methods. Core and cladding glass disks formed by melt casting are both polished without water. As shown in Figure 4.111 [12], the polished glass disks are placed inside a cylinder and heated to the deformation temperature in a dry atmosphere after which they are extruded under a pressure of 50 bars using a punch.

These fabrication techniques cannot be directly applied to single-mode fiber preparation because the core diameter is not small enough for single-mode waveguides. Jacketing methods are necessary to prepare a small and uniform core structure. Rotational casting is ordinarily used to form jacketing tubes for fabricating single-mode fluoride glass fibers.

4.6.2.4 Fiber Drawing

Multimode fibers can be drawn from preforms prepared by the aforementioned methods using conventional fiber-drawing apparatus, and the process is similar to that for jacketing drawing. The two fabrication techniques for drawing fluoride

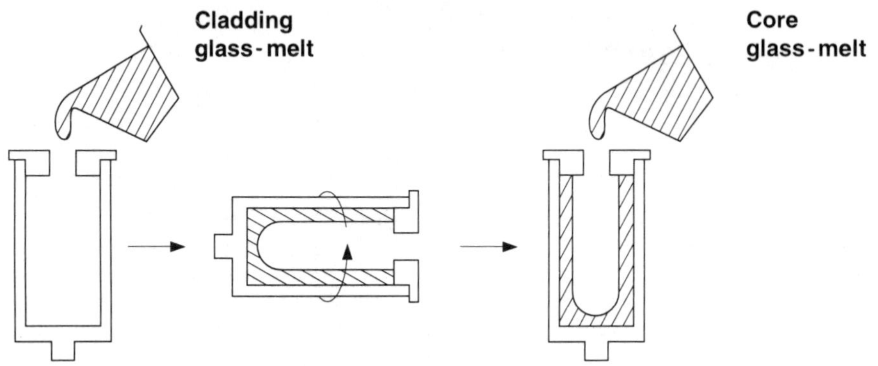

Figure 4.110 Rotational-casting process.

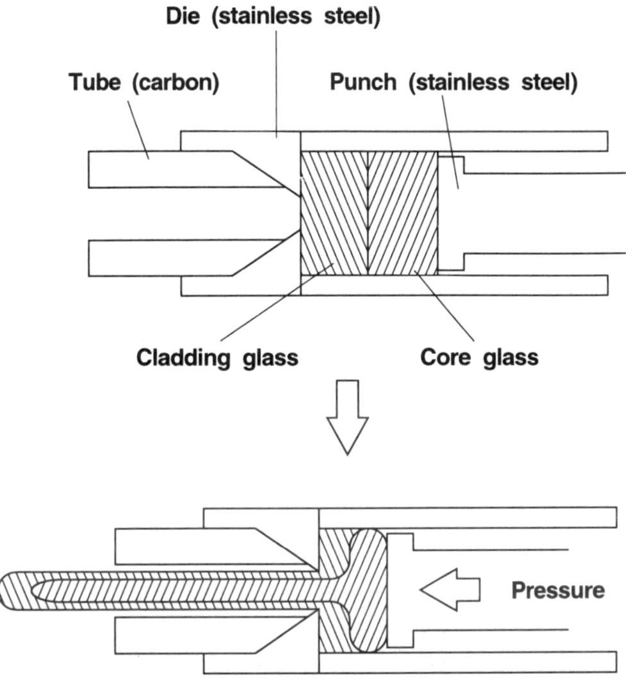

Figure 4.111 Extrusion process [12].

single-mode optical fibers are jacketing drawing and double-crucible drawing. The tendency of fluoride glass to crystallize easily makes it very difficult to fabricate fibers with the double-crucible technique. Some attempts have been made to apply this technique to fluoride glass fiber drawing [252], but no data on optical properties have been reported. Jacketing drawing is currently the only successful technique for making low-loss fluoride single-mode optical fibers.

Jacketing drawing consists of the elongation of a jacketed preform from which a fiber is drawn with another jacketing tube. Although the basic techniques needed for jacketing drawing, such as diameter control, capstan driving, and plastic coating, are based on standard methods developed for oxide glass fibers, some important differences exist. Elongation and drawing take place at a much lower temperature of 300° to 400°C using a electric furnace. The temperature of the elongation and drawing processes must be precisely controlled because the viscosity of fluoride glasses has a strong temperature dependence and the working range between the deformation and crystallization temperatures is very narrow. Moisture, which induces devitrification through hydration and dehydration reactions at the glass surface, must be eliminated at all stages of the process. If these conditions are not realized, bulk and surface crystallization will occur during elongation and fiber drawing and cause high fiber loss and poor mechanical strength. A dry inert atmosphere or a reactive fluorinating atmosphere such as NF_3 is effective in reducing moisture and, thus, minimizing surface crystallization [115,253]. The surface treatment of preforms by polishing and chemical etching, followed by reactive fluorination, is also effective in eliminating H_2O and OH^- and mechanical imperfections at the surface [253,254]. Solutions of $ZrOCl_2$-HCl [255] and NH_4NO_3-HNO_3 [254] are the preferable etchants.

Typical procedures for surface treatment and atmosphere control are described in subsequent sections [254]. Surfaces are etched by immersing the preforms in 1M NH_4NO_3 in 1 liter of 1N HNO_3 solution at 20°C for about 2 min, followed by heating in an HF-Ar atmosphere at 200°C for 2 hr to eliminate H_2O and OH^- absorbed at the surface. As shown in Figure 4.112, the surface-treated preforms are drawn into 125 ± 1-μm diameter fibers using a furnace purged with dry Ar and in a dry gas environment at 10 m/min and coated in-line with commercially available ultraviolet-cured acrylate with a thickness of 50 μm. At this stage, the drawing temperature is optimized using an improved furnace with an extended heat zone.

Figure 4.113 shows single-mode fiber fabrication using the tapered-elongation and jacketing-drawing method [256]. Preforms and jacketing tubes are prepared by the suction-casting and rotational-casting methods, respectively. The obtained straight preform with a tapered core is elongated together with the jacketing tube into a secondary preform with a smaller straight core. This is achieved by programming the elongation speed as a function of time. The secondary preform is reshaped into a cylinder and drawn together with another jacketing tube into a fiber coated with UV-curable acrylate.

With the jacketing method, it is essential that there is smooth contact between the preform and the jacketing tube during elongation and fiber drawing in order to obtain low-loss single-mode fiber. This can be achieved using a jacketing tube with its inner surface as cast and a preform with a treated surface by evacuating the space between the preform and the jacketing tube before elongation and fiber drawing and then filling it with dry gas.

This tapered-elongation and jacketing-drawing method has good potential for use in fabricating low-loss high-Δn fibers because the use of suction casting for preform fabrication in this method makes it possible to quench the core and cladding melts rapidly into glasses. This rapid quenching is essential if we are to produce high-quality preforms with few scattering centers from fluoride glasses for high-Δn fibers, which have a tendency to crystallize easily. Indeed, the high-Δn fiber with the lowest reported loss was realized using this jacketing method [104].

4.6.3 Characteristics

4.6.3.1 Transmission Loss

To date, minimum transmission loss values of about 1 dB/km at 2.6 μm, 10 dB/km at 2.5 μm, and 50 dB/km at 1.2 μm have been achieved for multimode

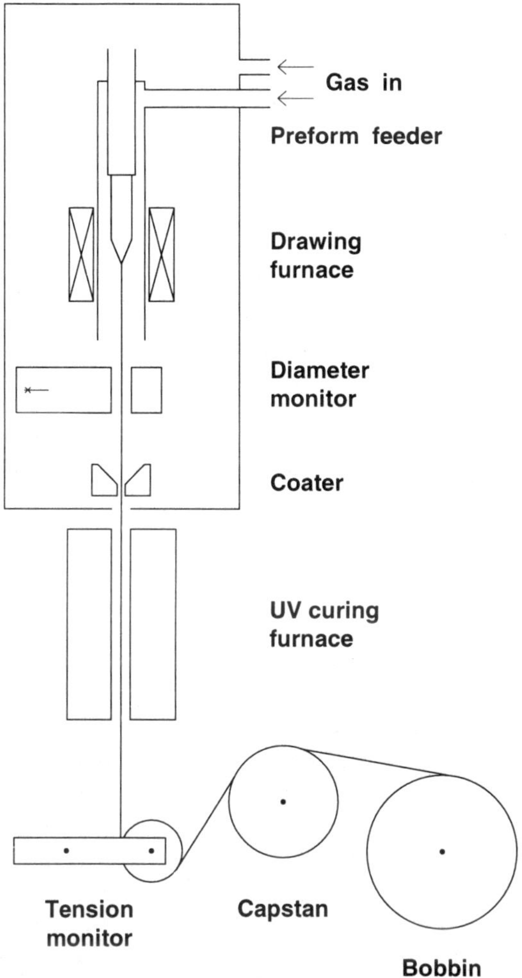

Figure 4.112 Fiber-drawing apparatus.

fluoride fibers, single-mode fibers, and high-Δn rare-earth-doped fibers, respectively. These fibers with low transmission losses are fabricated using ZrF_4-based glasses and, in terms of uniform transmission loss characteristics, are less than 200m in length. Single-mode fibers, which are reheated during the two processes of elongation and fiber drawing, exhibit high loss. High-Δn single-mode fibers with lower thermal stability have the highest loss. This is related to the stability of the glass against crystallization. However, these loss values have sufficient potential for short-distance applications.

The transmission loss spectrum of multimode ZrF_4-based fiber is shown in Figure 4.114 [9]. This fiber is drawn from a preform obtained by using the suction-

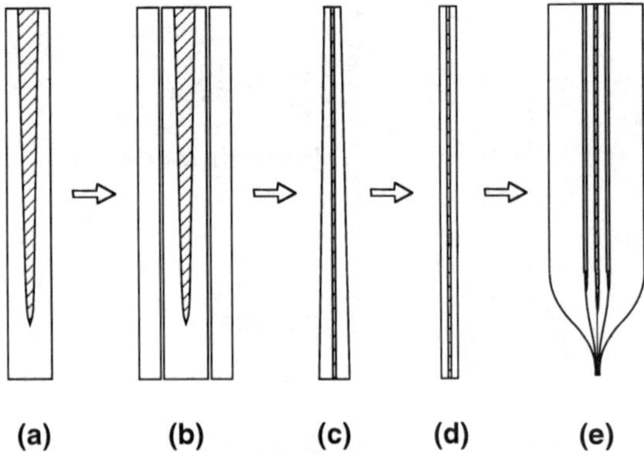

Figure 4.113 Process for single-mode fiber fabrication by the tapered-elongation and jacketing-drawing method: (a) preform prepared by suction casting, (b) setup for elongation, (c) secondary preform obtained by tapered-elongation, (d) secondary cylindrical preform reshaped by polishing, and (e) jacketing drawing.

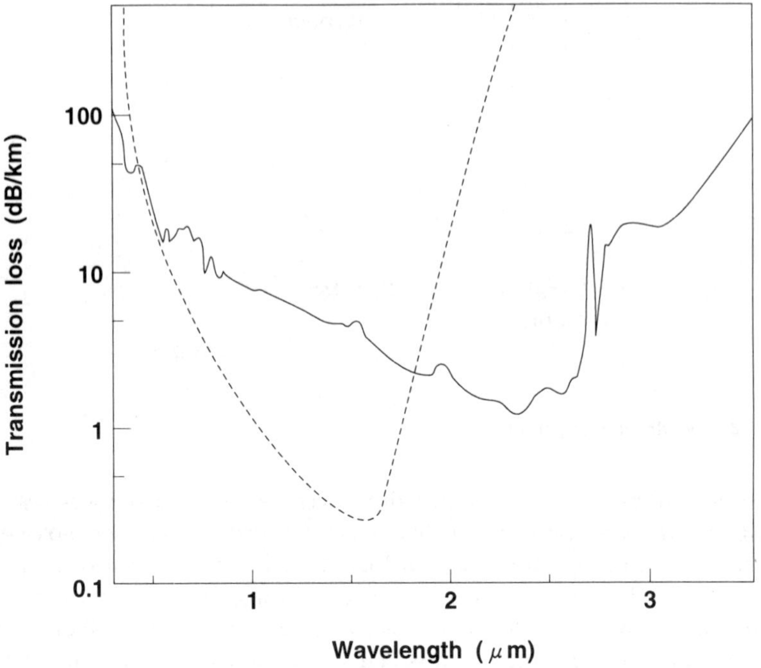

Figure 4.114 Transmission loss spectrum of multimode ZrF$_4$-based fiber [9]. The broken line is the spectrum of OH-free silica fiber [192].

casting method. As shown in Figure 4.114, there is a broad, transparent window, in which the loss is below 10 dB/km, between 0.8 and 2.7 μm, and the minimum loss is 1.2 dB/km at 2.30 μm. The loss in the 0.3- to 0.4-μm wavelength region is lower than that of silica-based fibers with a low OH^- content [257]. The absorption peaks below 2.6 μm originate from transition-metal and rare-earth-metal element impurities. The absorption peaks at wavelengths of 0.34 and 0.39 μm are due to Fe^{3+}; and the peaks at 0.42, 0.64, and 0.67 μm are due to Cr^{3+}. The hump at 0.51 μm and the peaks at 0.57, 0.73, 0.79, 0.86, and 2.49 μm are due to Nd^{3+} and the peaks at 1.45 and 1.54 μm are due to Pr^{3+}. The peak at 1.95 μm is due to Pr^{3+} and Ho^{3+}. The absorption peaks at longer wavelengths are attributed to molecular impurities. The peak at 2.69 μm and the hump at 2.78 μm are assigned to the combination modes of CO_2 fundamental vibrations, and the peak at 2.89 μm to the OH^- fundamental stretching vibration.

The estimated loss factors are Rayleigh and wavelength-independent scattering

$$0.67\lambda^{-4} + 0.9 \ (dB/km) \tag{4.85}$$

weak absorption tail (WAT)

$$2.0 \times 10^{-3} \ \exp(3.0/\lambda) \, (dB/km) \tag{4.86}$$

and metal impurity absorption; impurity concentrations are summarized in Table 4.16. Wavelength λ is expressed in micrometers.

The analyzed loss factors are also shown in Figure 4.115 [9]. The Rayleigh scattering, which is smaller than that in pure silica core fibers [6], is dominant in

Table 4.16
Impurity Concentrations in a ZrF_4-Based
Fluoride Fiber

Impurity	*Concentration (ppb)*
Cr^{3+}	20
Fe^{3+}	250
Co^{2+}	3
Ni^{2+}	6
Cu^{2+}	11
Pr^{3+}	10
Nd^{3+}	35
Ho^{3+}	5
OH^-	7
CO_2	120

From: [9].

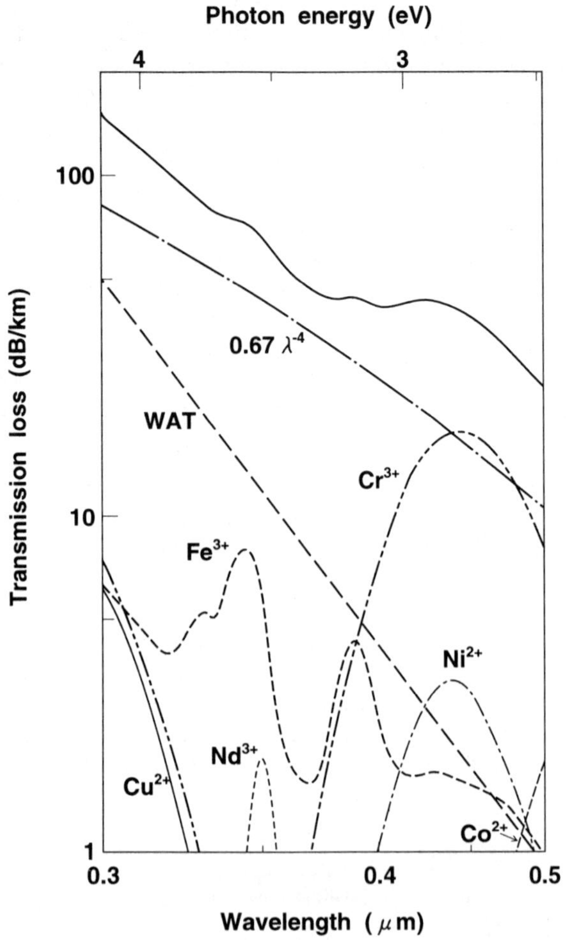

Figure 4.115 Transmission loss factors of ZrF_4-based fluoride fiber. Wavelength is expressed in microns [9].

the near-UV wavelength region. The weak absorption tail is the second major loss factor in this wavelength region and is lower than the Rayleigh scattering loss by more than one order of magnitude in the visible wavelength region. Metal-impurity absorption is a minor factor in the near-UV wavelength region, and the transmission loss has a transparent window near 0.37 μm. This loss analysis indicates that the transmission loss of fluoride fibers originates intrinsically from the Rayleigh scattering in the visible and near-IR wavelength regions.

The estimated extrinsic scattering loss is due to wavelength-independent scattering. Mie scattering with a λ^{-2} dependence is negligible. Mie scattering is mainly due to defects with small relative refractive indexes such as fluoride crystallites [70]. The negligibly small Mie scattering loss shows that there was hardly any nucleation or fluoride crystal growth. The wavelength dependence of the weak absorption tail, induced by electronic defects such as reduced Zr, is similar to that in optical materials [48]. However, the tail is smaller than that in pure silica glasses, estimated as the lower boundaries of experimental absorption data [3]. This indicates that the density of electronic defects in fluoride glass is intrinsically lower than that in pure silica glass. The metal-impurity content shown in Table 4.16 almost agrees with the sums of the analyzed content of the metal impurities in the raw materials.

Typical transmission loss spectra for single-mode ZrF_4-based fluoride fibers fabricated using the jacketing-drawing method are shown in Figures 4.116 and 4.117.

Figure 4.116 shows a typical transmission loss spectrum for single-mode fibers operated in a 2.5-μm transmission band [256]. The core diameter, cladding diameter, and Δn of the fiber are 11.3 μm, 123 μm, and 0.61%, respectively. A loss at 2.5 μm is 9.3 dB/km. The peak observed at 2.35 μm is due to the cutoff of the first higher order mode (LP_{11} mode), and the cutoff wavelength (λ_c) can be determined as 2.43 μm from the cutoff peak. This λ_c is found to be coincident with λ_c calculated from the observed values of the core diameter, Δn, and the refractive index of the core glass. This clear cutoff peak indicates that a single-mode waveguide structure is successfully formed by the jacketing method.

Figure 4.117 shows a typical transmission loss spectrum for Pr^{3+}-doped high-Δn single-mode fiber [104]. The core diameter, cladding diameter, Δn, λ_c, and Pr^{3+} concentration of the fiber were 1.6 μm, 120 μm, 3.7%, 0.88 μm, and 500 ppm,

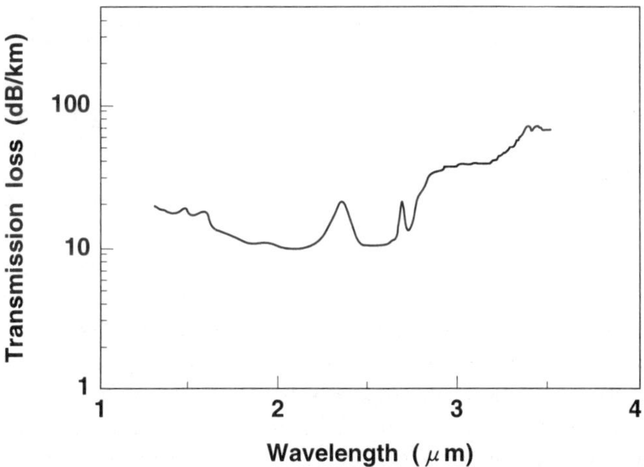

Figure 4.116 Transmission loss spectrum for single-mode fibers with a Δn of 0.3% and a λ_c of 2.43 μm [256].

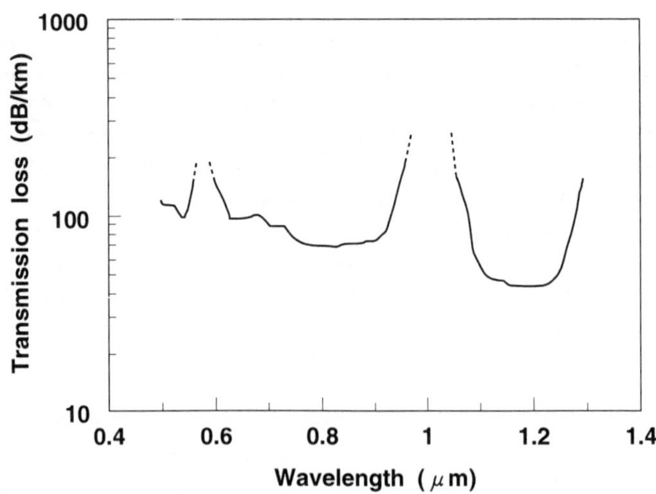

Figure 4.117 Transmission loss spectrum for Pr^{3+}-doped single-mode ZrF_4-based fluoride fiber with a Δn of 3.7%, a λ_c of 0.95 μm, and a Pr^{3+} concentration of 500 ppm [104].

respectively. The large peaks around 0.6 and 1 μm are caused by Pr^{3+}. A loss of 44 dB/km, which is the lowest value for high-Δn fluoride fibers, is achieved at 1.23 μm.

The transmission loss spectrum showed almost the same values and the OH absorption peak at 2.8 μm did not change when a Teflon FEP clad fluoride fiber was placed in air at room temperature for 387 days [258]. This suggests that the transmission losses of fluoride fibers does not vary during room-temperature operation.

4.6.3.2 Fiber Strength

In order to realize high fiber strength, it is essential to suppress surface crystallization during fiber drawing because both crystallization and mechanical defects at the fluoride fiber surface and interface formed by jacketing drawing generate large flaws that cause fiber fracture. Therefore, surface treatment and atmosphere control are key techniques for raising fiber strength. Specifically, drawing fibers in an atmosphere containing NF_3 results in significant strength improvement. The strengths of ZrF_4-based fluoride glass fibers measured under ambient conditions cover a wide range far below the theoretical value, which is estimated as 5.5 GPa [259]. The average values are 150 to 550 MPa for tensile tests and 0.5 to 1.6 GPa for bending tests. These strength values depend on the test type, the gauge length, and the loading rate. Bending test strengths, for example, are more than twice as high as tensile test strengths [260]. The higher values are obtained by minimizing H_2O contamination during preform fabrication and fiber drawing in addition to reducing surface imperfections. A mean bending strength of 1.4 GPa was obtained for ZBLA fibers, which was drawn in a N_2-NF_3 reactive atmosphere [253]. The preforms were prepared by a reactive atmosphere process using NF_3 and were chemically etched in a $ZrOCl_2$-HCl solution before fiber drawing. Moreover, a yield of 1.6 GPa was achieved after the drawn fibers were etched a second time.

A typical Weibull plot of the tensile strength of ZBLYAN fiber is shown in Figure 4.118 [261]. A mean tensile strength is 550 MPa. This characteristic was obtained by surface etching followed by an HF-Ar gas treatment and by fiber drawing using a dry gas. The steep distribution indicates uniformity in flaw size, which shows that these surface flaws have the same origin, which is related to surface crystallization. Indeed, fracture surface observation using SEM reveals that all the breaks originate from surface flaws. The results obtained to date indicate that fluoride fiber strength can be further increased by removing the surface flaws.

The strength of ZrF_4-based fluoride glass fibers is influenced by the environment and aging. It has been shown that the mean bending strength of ZBLALi fibers increases to 540 MPa from 460 MPa measured at an ambient temperature when the tests are carried out at liquid nitrogen temperature but that the ambient strength of ZBLA fibers decreases rapidly to half its initial value after 24 hr in water at room temperature [262]. In a humid atmosphere, mechanical strength decreases gradually. As described in Section 4.9.2, it is confirmed that ZrF_4-based fluoride fibers with the tensile strength of 500 MPa can be used over 25 years as components in fiber amplifiers without fracturing.

4.6.4 Vapor Phase Deposition

The vapor phase deposition technique is generally classified into two types: chemical and physical vapor deposition. The former is a fabrication technique that includes

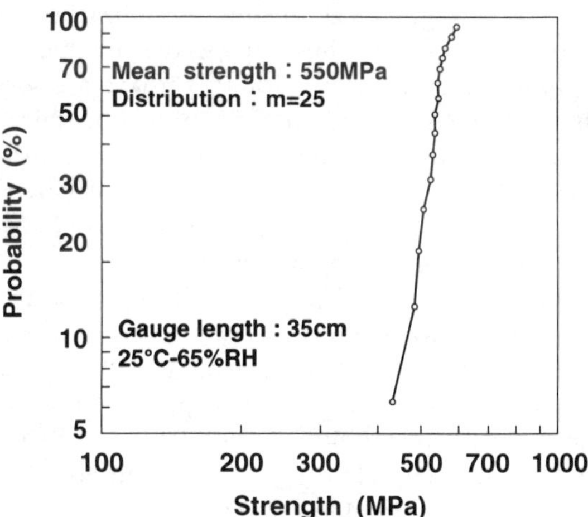

Figure 4.118 Weibull plot of the tensile strength of ZrF_4-based fluoride fiber [261].

a glass-forming chemical reaction; and the latter includes only physical phenomena such as sputtering, vaporization, and consolidation. Both techniques have been used for preparing fluoride glasses.

With CVD, BeF_2 glass [263] and ZBLAN glass [264] have been prepared as materials for ultra-low-loss optical fibers. The success of these trials led to the expectation that highly pure and homogeneous glasses could be prepared using a vapor phase reaction, but low-loss fibers have not yet been fabricated. However, CVD can be used to prepare fluoride glasses containing rare-earth ions homogeneously because a homogeneous vapor phase is quenched to form a solid in which the constituents are homogeneously dispersed. Therefore, it can be expected that this technique will be applicable to the preparation of rare-earth-doped fluoride glass film. Since it may be difficult to prepare a bulk glass by CVD, this is not useful for fiber fabrication.

Physical vapor deposition has also been used for rare-earth-doped fluoride glasses. With this technique, the compositions of the synthesized glass is determined by equilibrium vapor pressures and substrate temperature. This limits the glass systems that can be prepared to, for example, PbF_2-ZnF_2-GaF_3 glass. This technique has already been used to prepare rare-earth-doped glass film; rare-earth-doped planar waveguides are now under development [265].

4.7 CHALCOGENIDE FIBER FABRICATION PROCESS

Chalcogenide glasses are classified into three systems: sulfide glasses, selenide glasses, and telluride glasses. They have a large glass-forming area, fairly high stability against moisture, a high refractive index above 2, and a long cutoff wavelength compared with other nonoxide glass fiber materials. Moreover, these chalcogenide glasses offer the great advantage that their slow crystallization rate makes it easy to fabricate long homogeneous fiber. These properties are highly beneficial for an active ion host for optical amplification. However, there are some problems to be resolved before they can be used for amplification. It is fairly difficult to dope radiative ions such as rare-earth elements homogeneously in ordinary chalcogenide glasses. For lanthanum glasses such as La-Ga-S and La-Ga-Ge-Se, rare-earth ions can be doped by substituting them for La, but it is not easy to fabricate a low-loss fiber because they have a poor glass-forming ability.

In terms of chalcogenide glass fibers for IR transmission, As-S core-cladding fibers were first fabricated in the early 1960s by Kapany and Simms and exhibited a relatively high transmission loss of 20 dB/m at 5.5 μm [30]. No significant progress was made, however, in the development of chalcogenide glass fiber until the 1980s. Since the indication of the possibility of ultra-low loss [210–213], there have been many studies on the fabrication of unclad fibers [24,266–268], plastic clad fibers [269,270], and multimode and single-mode fibers with glass cladding [14,16,271–276].

As a result of the development of both fiber compositions with higher thermal stability against crystallization and fiber fabrication techniques that reduce extrinsic loss, the losses of chalcogenide fibers have now been reduced to less than 200 dB/km for multimode type [14,16,31,272,277] and to 400 dB/km for single-mode type [276].

In this section, we describe the techniques for fabricating chalcogenide optical fibers.

4.7.1 Glass Compositions

Compositional change, which causes a large scattering loss in fibers, occurs easily in chalcogenide glass during the reheating process because of the vaporization of volatile chalcogen elements above the deformation temperature. This characteristic, in addition to the bulk crystallization due to the intrinsic stability of the glass, limits the glass compositions available for fiber drawing. Table 4.17 lists typical glass systems drawn into fibers [14,273,275,277,278]. In this section, the glass compositions for fiber drawing is described for some ordinary chalcogenide systems, and a glass system proposed for fiber amplification is also shown.

4.7.1.1 Ge-P-S Glass System

Ge-P-S glasses have short cutoff wavelengths, relatively low refractive indexes near 2.0, high softening temperatures, and low toxicity. Figure 4.119 shows the composi-

Table 4.17
Chalcogenide Glass Fiber Compositions

Systems	Glass Composition (at %)		Δn (%)
	Core	Cladding	
Sulfide glasses	$Ge_{20}S_{80}$[a]	–	
	$Ge_{23}P_8S_{69}$[b]	–	
	$As_{40}S_{60}$	$As_{38}S_{62}$	1.3[c]
	$As_{38}S_{62}$	$AS_{37.4}S_{62.6}$	0.4[c]
Selenide glasses	$Ge_{20}S_{80}$[d]	–	
	$Ge_{28}Sb_{12}Se_{60}$	$Ge_{32}Sb_8Se_{60}$	NA = 0.6[e]
	$Ge_{10}Sb_{26}Se_{64}$[d]	–	
	$As_{40}Se_{60}$[a]	–	
	$As_{35}Ge_5Se_{57}$[a]	–	
	$As_{15}Ge_{30}Se_{55}$[f]	–	
Telluride glasses	$Ge_{25}As_{20}Se_{25}Te_{30}$	$Ge_{20}As_{30}Se_{30}Te_{20}$	2.7[g]

a, [14]; b, [266]; c, [275]; d, [277]; e, [278]; f, [24]; g, [273].

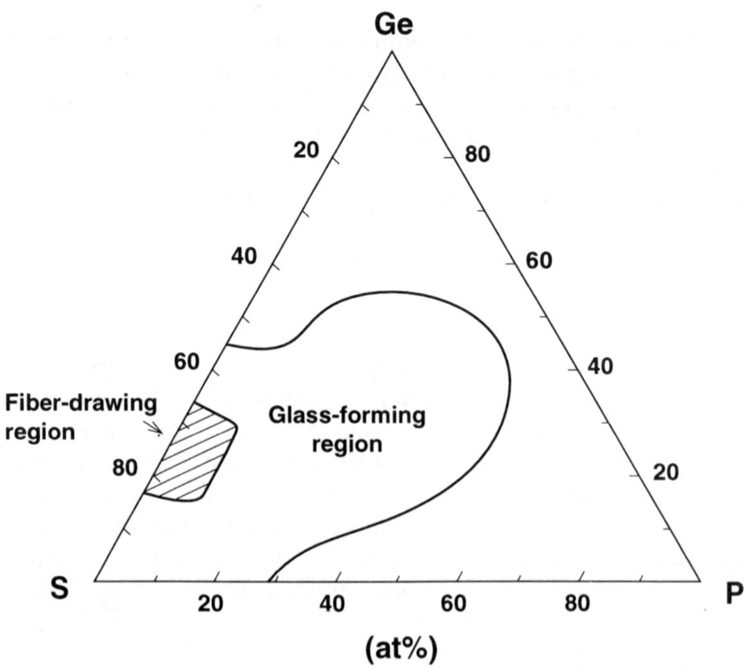

Figure 4.119 Fiber-drawing region in the Ge-P-S system with the preform method [266].

tions that can be drawn into fiber (fiber-drawing region) and the glass-forming compositions for the Ge-P-S system [266]. The former composition range is narrower than the latter because of crystallization in the Ge-rich region and the sublimation of sulfur in the S-rich region at a temperature above the deformation temperature. The appropriate glass composition in the Ge-S system for drawing fibers is limited to $Ge_{1-x}S_x$, where x ranges from 0.75 to 0.86. The addition of phosphorous to the germanium sulfide glass increases the glass-forming ability and optical homogeneity if the amount of phosphorous is less than 10 mol% [266].

4.7.1.2 As-S Glass System

As-S glasses have a strong glass-forming tendency and the highest stability against crystallization in two-component systems. Figure 4.120 shows the compositions for fiber drawing using the preform and the crucible methods [14]. The appropriate glass composition for drawing into unclad fibers using the preform method is limited to a narrow range in the glass-forming region. Glasses in the As-rich region have a tendency to crystallize during the fiber-drawing process. For glass in the S-rich region, the sublimation of sulfur, which causes compositional change and

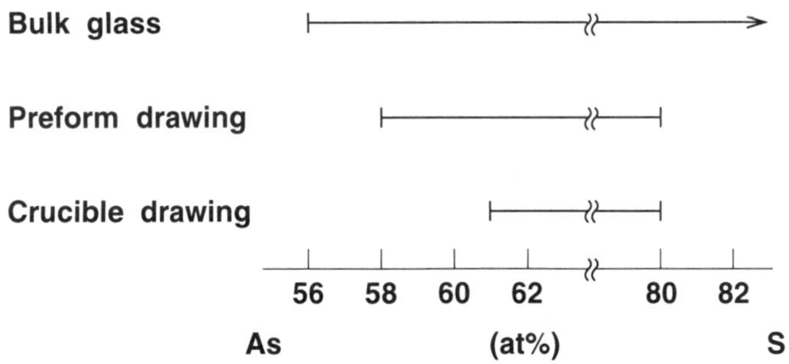

Figure 4.120 Fiber-drawing regions in the As-S system with the preform and crucible methods [14].

surface irregularity, occurs during drawing. These inhomogeneities increase the transmission loss of the drawn fiber. The composition region is very small with the ordinary crucible method because the glass is held at a higher temperature for a longer time and so requires greater stability.

In As-S fibers, a refractive index difference (Δn) is achieved between the core and cladding using glasses with a high and low As content as the core and cladding, respectively. If $As_{38}S_{62}$ and $As_{35}S_{65}$ are used as the core and the cladding, respectively, a Δn of 2.3% is obtained [14]. Small Δn values below 0.4% can be calculated based on the following equation for As_x-S_{100-x} glass [276]:

$$n_{1.3\mu m}(x) = 0.0134x + 1.9144 \qquad (40 < x < 39.3) \qquad (4.87)$$

4.7.1.3 As-Ge-Se Glass System

As-Ge-Se glasses have a high refractive index of about 2.6 and a long cutoff wavelength beyond 10 μm. Figure 4.121 shows the fiber-drawing composition region for the As-Ge-Se system [14]. Glass whose composition falls outside this region exhibits a tendency toward crystallization during preform drawing and cannot be drawn into homogeneous low-loss fiber. The appropriate glass compositions for drawing unclad fibers are also limited to the narrow part of the glass-forming region.

4.7.1.4 Lanthanum Glass System

La-Ga-S, La-Ga-Ge-S, and La-Ga-Ge-Se systems form glasses, but their glass-forming tendency is poor compared with ordinary chalcogenide glasses [32,279,280]. Glass-forming compositions for a La-Ga-Ge-S system are shown in Figure 4.122 [279]. There have been few reports on fiber drawing, so the appropriate composition has not been confirmed [281].

Some fiber compositions and their properties are listed in Table 4.18 [223].

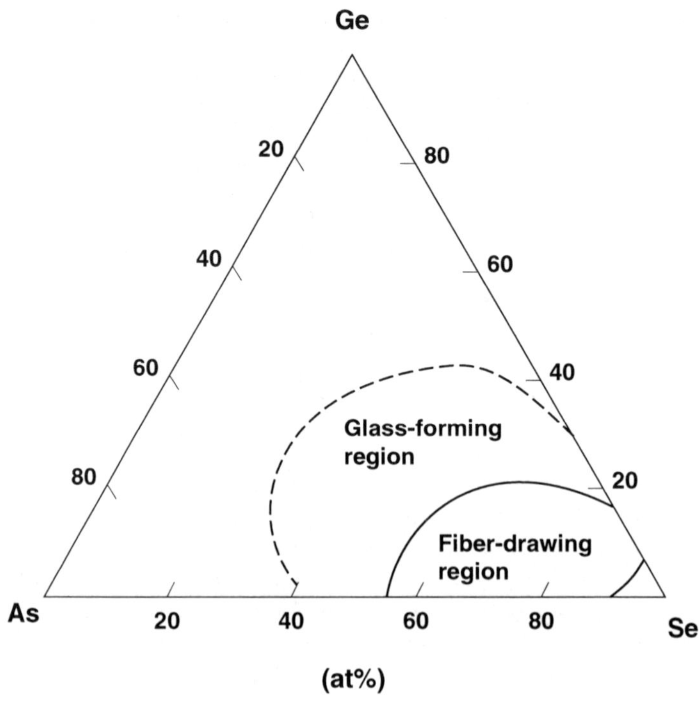

Figure 4.121 Fiber-drawing regions in the As-Ge-Se system with the preform method [14].

4.7.2 Fabrication Process

In this section, we outline fabrication techniques for chalcogenide fibers based on the preparation of bulk glasses by melting. This glass-melting process is the most successful method for fabricating low-loss chalcogenide fibers, where synthesized homogeneous glasses are converted by preform drawing or crucible drawing into optical fibers.

A typical fabrication sequence is shown in Figure 4.123. The process involves the purification of raw materials that are then weighed and sealed in a silica ampoule, glass melting using a rocking furnace, melt cooling to form a glass rod and a glass tube, and finally fiber drawing.

4.7.2.1 Purification of Raw Materials

The contamination of raw materials and glasses by such impurities as OH^-, H_2O, oxide, and organic compounds must be eliminated in order to reduce the extrinsic loss in fibers.

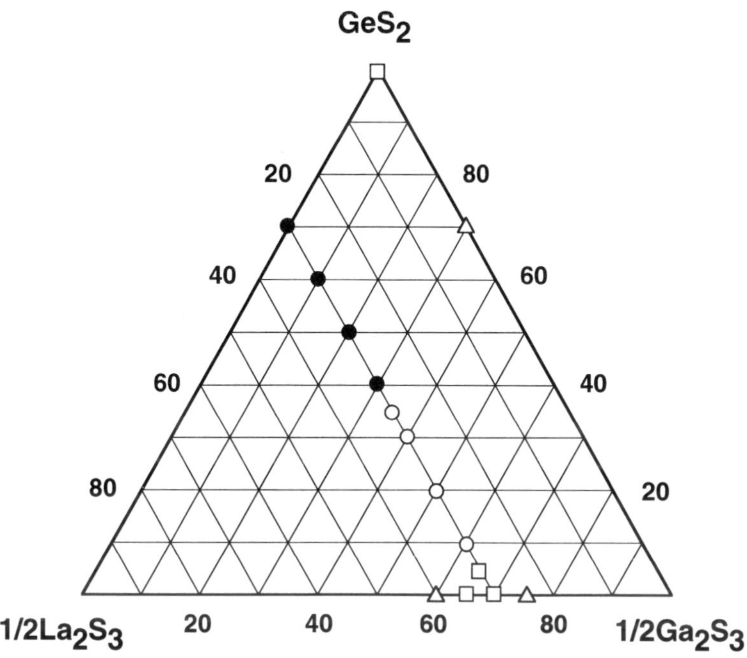

Figure 4.122 Glass-forming compositions for the La-Ga-Ge-S system [279]. Open circles, vitrified by quenching in air; open squares, vitrified by quenching in ice water; open triangles, partially vitrified; solid circles, crystallized.

Table 4.18

Physical Properties of Chalcogenide Glasses

Glass (at %)	Transmission Range$^+$ (μm)	n_D	T_g (°C)	T_x (°C)	α (°C^{-1})	ρ (g/cm^3)	H_k (kg/mm^2)
Ge$_{25}$S$_{75}$	0.51–7.2	2.113	260	500	25×10^{-6}	2.5	130
As$_{40}$S$_{60}$	0.62–9.2	2.41 (3 μm)	203	–	25×10^{-6}	–	190
As$_{40}$Se$_{60}$	0.87–11.9	2.65 (3 μm)	190	–	21×10^{-6}	4.6	156

n_D, refractive index at 0.589 μm; T_g, glass transition temperature; T_x, crystallization temperature; α, thermal expansion coefficient; ρ, density; H_k, Knoop hardness; $^+$, absorption coefficient less than 0.5 cm^{-1}. *From:* [223].

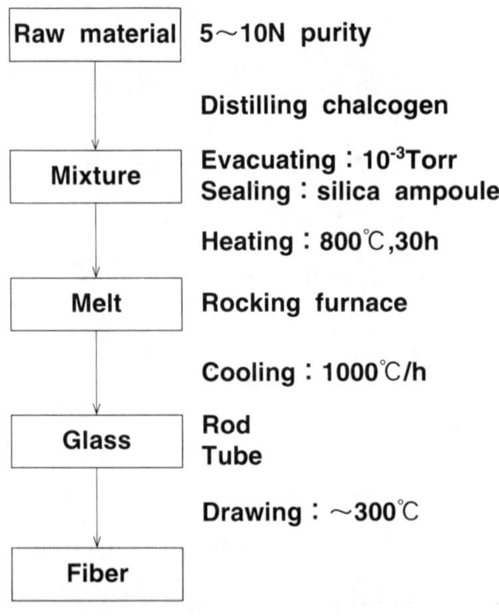

Figure 4.123 Chalcogenide fiber fabrication process.

In the weighing and batching process, elements with a high purity of 5N to 10N, which are available commercially, are used as starting materials. Chalcogen elements are purified by distillation using a dry inert gas, a dry inert gas + H_2 gas, or a dry inert gas + a reactive gas such as S_2Cl_2 or Se_2Cl_2. Figure 4.124 shows the purification of sulfur [223]. H_2 gas is used to reduce oxide impurities [282]. In the distillation using S_2Cl_2 gas, H_2O contained in the raw materials can be effectively eliminated by the chemical reaction [266]

$$2S_2Cl_2 + 4H_2O \rightarrow 2SO_2 + 4\,HCl + 2H_2S \tag{4.88}$$

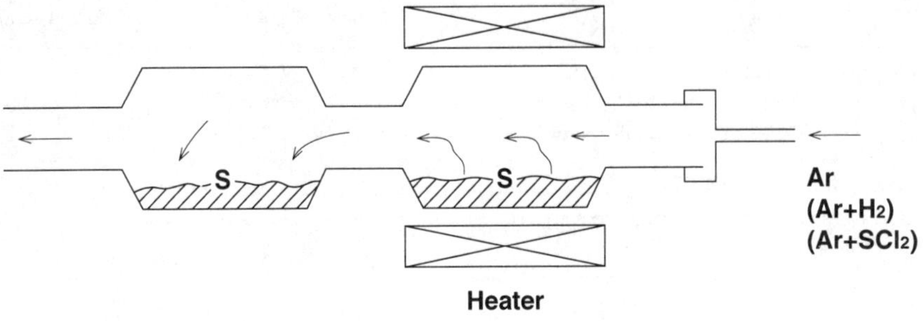

Figure 4.124 Sulfide distillation [223].

Metal elements are purified by removing surface oxide impurities. Arsenic is treated at 500°C in an Ar gas flow to sublimate out surface oxide impurities. Germanium is treated with acid or heated in an H_2 gas flow to clean the surface.

In the chemical vapor deposition process, gas-phase chlorides are used as starting materials. Figure 4.125 shows the preparation of Ge-Se particles by CVD using $GeCl_4$, $SeCl_2$, and H_2 gases [283]. The mixed gases are transported into the silica glass tube by Ar gas and heated at 800°C by an electric furnace. Fine Ge-Se particles are then deposited on the inner surface of the silica tube located outside the furnace. The fine particles contain crystalline GeSe and $GeSe_2$ formed by the following reactions

$$GeCl_4 + SeCl_2 + 3H_2 \rightarrow GeSe + 6HCl \qquad (4.89)$$

$$GeCl_4 + 2SeCl_2 + 4H_2 \rightarrow GeSe_2 + 8HCl \qquad (4.90)$$

After deposition, the tube is evacuated at a vacuum of about 10^{-3} Torr and sealed at the neck regions shown as A and B in Figure 4.125. Bulk glass is prepared by heating the sealed silica tube as described in the following section. It has been reported that distilling deposited Ge-Se fine particles in a CO or NH_3 gas atmosphere at 900°C is effective in reducing oxygen impurities.

4.7.2.2 Glass Melting

The following two points are important when melting glass for low-loss fiber fabrication:

1. Starting materials must be protected from impurities such as H_2O, transition metals, and organic compounds;
2. The glass melt must be homogenized without compositional change.

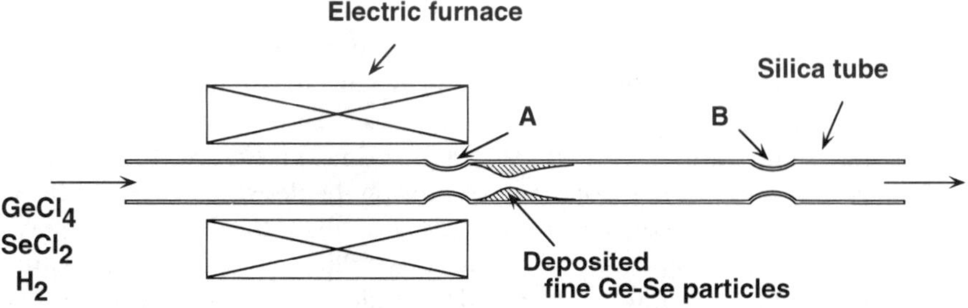

Figure 4.125 Synthesis of fine Ge-Se glass particles by chemical vapor deposition [283]. A and B are neck regions used for sealing and separating.

To achieve these conditions, an ampoule made from silica glass with a low OH content, a dry glove box, and a rocking furnace are used in the glass-melting process. Melting via an ampoule suppresses the loss of chalcogen elements with high vapor pressure that would lead to compositional change. Melting via a rocking furnace makes it possible to mix elements with extremely large atomic weight differences, homogeneously.

The silica ampoule is treated with acid to clean the surface and then evacuated at a high temperature to remove H_2O absorbed on the inside surface. The purified elements are weighed and put into the ampoule in a dry glove box, and then the ampoule is evacuated and sealed by fusing the inlet. The sealed ampoules are heated at 700° to 900°C for 20 to 100 hr in the rocking furnace to mix the constituents.

4.7.2.3 Melt Quenching

In the melt-quenching process, it is necessary to cool the melt rapidly in order to suppress both crystallization and the separation of its constituents and to avoid any cracking of the glass as a result of the thermal stress in the quenching process. Cracks cause fractures and make it difficult to clean the glass surface. Surface treatment is essential to obtaining low-loss fiber. The ampoules are generally removed from the rocking furnace to quench the melt naturally in air, held near the glass transition temperature in an annealing furnace, and then cooled gradually to room temperature.

Figure 4.126 shows the melt-quenching processes for forming a glass rod and a glass tube [31]. A glass rod is prepared by standing or rotating the ampoule vertically during cooling, and a glass tube is formed by rotating the ampoule horizontally. The inner diameter of the tube can be controlled by the volume of the glass melt in the ampoule. The obtained glass rods are used as the core or cladding glass for fibers. The glass tubes are used for cladding.

4.7.2.4 Fiber Drawing

The techniques needed for chalcogenide fiber drawing are based on standard methods developed for oxide glass fibers. There are, however, certain important differences, which are similar to the technique used for fluoride fibers. Fibers are drawn at low temperatures of 200° to 500°C using an electric furnace. Precise temperature and pressure control is needed during the drawing process because the vapor pressure of the glass constituents increases rapidly at the drawing temperature. Moisture and oxygen, which easily oxidize and crystallize chalcogenide glass, must be eliminated from the environment. H_2O, OH^-, oxide, and mechanical imperfections must also be removed from the glass surface. In general, a dry inert atmosphere is used to reduce the effects of moisture and oxidization, and the glass surface is cleaned by polishing, followed by dehydration in a vacuum.

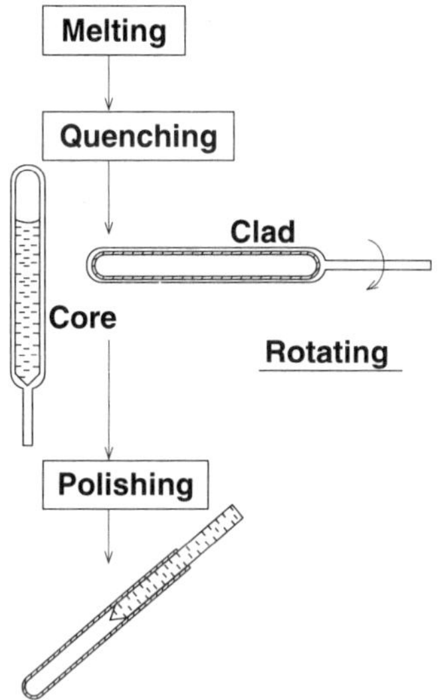

Figure 4.126 Melt-quenching processes for forming a glass rod and a glass tube [31].

There are two kinds of fiber-drawing techniques for fabricating chalcogenide fibers. One is conventional preform drawing and the other is crucible drawing developed especially for chalcogenide glasses.

The fabrication techniques necessary for fabricating low-loss fiber have not yet been well developed for chalcogenide glass preforms with a core and cladding waveguide structure. Therefore, unclad preform drawing, Teflon FEP jacketing drawing, or rod-in-tube drawing is generally employed using conventional preform-drawing apparatus. After surface treatment, a glass rod, a glass rod coated with Teflon FEP tube, or a core glass rod inserted into a cladding glass jacketing tube is heated per zone in an dry inert atmosphere and drawn into a fiber.

With regard to crucible drawing, two methods have been developed: one using a double crucible and the other a single crucible. The details are given in the following subsections.

Double-Crucible Method [14,274]

Figure 4.127 shows a silica glass double-crucible assembly [274]. Although its design is very similar to the conventional platinum double crucible used for producing silicate fibers, it has a rather long nozzle that is useful for cooling the glass-melt and additional pressure controllers that can individually pressurize the core and the cladding crucibles. These are required due to the rather narrow working range of chalcogenide glass.

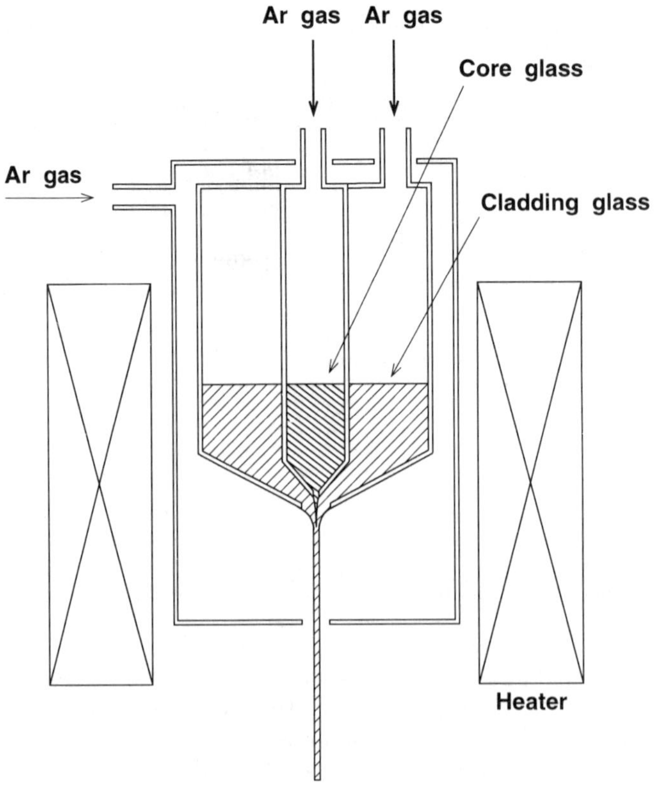

Figure 4.127 Double-crucible method [274].

Small glass blocks are cut from glass rods and their surfaces are cleaned. The core glass blocks are placed in the inner core section of the crucible and the cladding glass blocks in the outer cladding section. The crucible is then installed in the electric furnace. A dry inert atmosphere is maintained inside both the crucible and the furnace. The temperature of the assembly is increased above the deformation temperature. After the glasses have softened and adhered to the inner surface of the crucible, the insides of the core and cladding crucibles are pressurized with Ar gas. This allows drawing to be undertaken at a lower temperature and reduces the vaporization of the glass constituents. The core and cladding glasses descend from the nozzle and form a small drawing cone beneath it. The drawing can be carried out satisfactorily if there is a stable symmetrical cone of flowing glass beneath the nozzle. The core-to-cladding diameter ratio can be varied by changing the nozzle diameter ratio of the inner and outer crucibles and by changing their internal pressure. Fibers with a diameter of 100 to 200 μm can be drawn by this method.

Single-Crucible Method [31,273]

Figure 4.128 shows a schematic diagram of a single-crucible drawing [31]. After surface treatment, a core rod with a cladding tube is placed in a silica crucible. A dry inert atmosphere is maintained inside both the crucible and the furnace. The area around the nozzle is partially heated to the deformation temperature of the glass. After the bottom end of the cladding tube adheres uniformly to the inner surface of the crucible, the inside of the crucible is pressurized to 0.2 MPa with Ar gas and the space between the core rod and the cladding tube is evacuated to 1.3 Pa. The fiber drawn from the nozzle is coated with a UV curable acrylate polymer below the diameter monitor. Fibers with a diameter of 250 to 1,000 μm can be drawn by this method.

There are fewer inhomogeneities, inducing extrinsic loss, with the single-crucible method than with the double-crucible method because with the former method the glasses are partially heated for a short time and the core and cladding adhere in a vacuum, whereas with the latter method the glasses are heated throughout the drawing process and the core and cladding adhere under pressure. By contrast, it is easy to fabricate a single-mode fiber with a small core diameter using the double-crucible method, but an extra process such as elongation is required in order to achieve a small core with the single-crucible method.

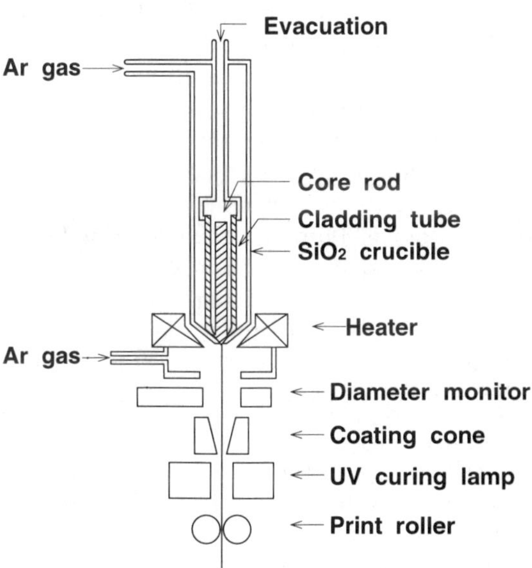

Figure 4.128 Single-crucible drawing apparatus [31].

4.7.3 Characteristics

4.7.3.1 Transmission Loss

The transmission losses of chalcogenide fibers depend on the glass systems and fabrication methods used. Distillation methods and drawing methods in particular affect the loss characteristics. In general, the minimum losses obtained for sulfide, selenide, and telluride fibers increase in that order; and those obtained by unclad preform drawing, crucible drawing, and rod-in-tube drawing increase in that order. These results reflect the thermal stability of the glass systems and the potential of the drawing method for extrinsic loss reduction. This section describes the typical loss characteristics of chalcogenide fibers.

Transmission Loss Spectra

Ge-S fiber. Figure 4.129(a) shows the transmission loss spectra of Ge-S unclad fibers [14]. The solid line indicates the loss curve for a fiber drawn from $Ge_{20}S_{80}(1)$ glass, which was prepared using S distilled in an Ar gas atmosphere; while the broken line represents the loss curve for a fiber drawn from $Ge_{20}S_{80}(II)$ glass, which was prepared using S distilled in an $Ar + H_2$ gas atmosphere. In both spectra, the peaks at 2.84 and 4.00 μm originate from OH and SH fundamental stretching vibrations, respectively [15]. The minimum losses are 500 dB/km at 2.40 μm for the $Ge_{20}S_{80}(1)$ fiber and 148 dB/km at 1.68 μm for the $Ge_{20}S_{80}(II)$ fiber. In spite of its higher loss in the long-wavelength region, which is due to the high hydrogen impurity content, the $Ge_{20}S_{80}(II)$ fiber shows lower loss in the short-wavelength region.

Figure 4.129(b) shows measured transmission loss versus photon energy for $Ge_{20}S_{80}(1)$ and $Ge_{20}S_{80}(II)$ unclad fibers and bulk glasses and the loss factors [14]. In the short-wavelength region, each loss consists of two exponential parts. The steep sloping line corresponds to the Urbach tail. The gradual sloping line, which exhibits a higher loss value at 0.633 μm than the scattering loss, is due to the WAT. In the model presented by Wood and Tauc, a WAT is induced by additional band-gap states such as dangling bonds and its magnitude is associated with the total concentration of the states [47]. Therefore, the small WAT obtained in the $Ge_{20}S_{80}(II)$ fiber with a high hydrogen impurity content suggests that hydrogen impurities reduce the number of gap states in the same way as in amorphous silicon [14].

As-S Fiber. Figure 4.130 shows the transmission loss spectrum of an $As_{40}S_{60}$ unclad fiber [14]. A minimum loss of 35 dB/km has been obtained at a short wavelength of 2.44 μm.

The transmission loss spectra of As-S glass core-cladding fibers, which were prepared by the double-crucible method, are shown in Figure 4.131 [276]. Figure 4.131(a) shows the loss curve for multimode fiber with a numerical aperture of

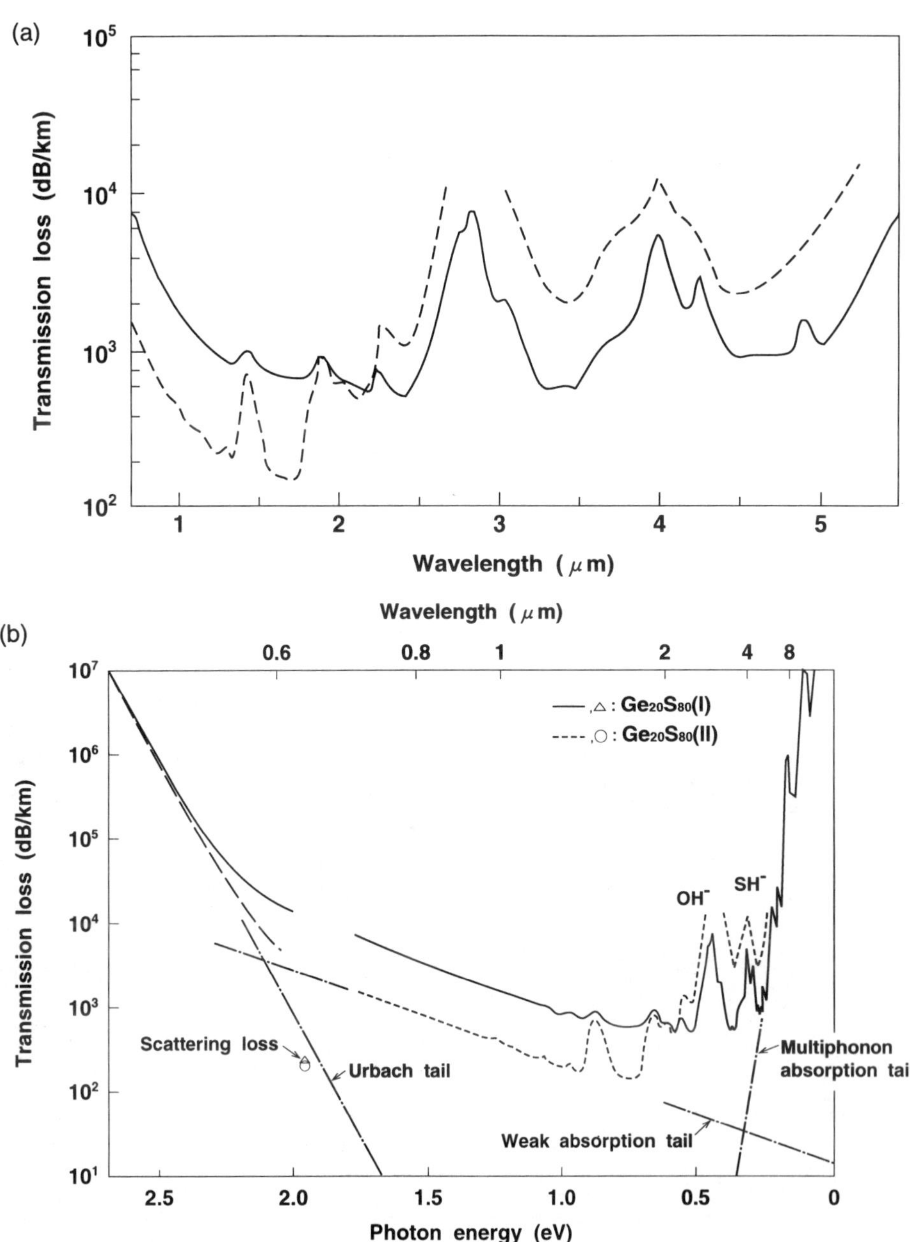

Figure 4.129 Transmission loss characteristics for unclad fibers and bulk glasses with $Ge_{20}S_{80}(I)$ (solid line) and $Ge_{20}S_{80}(II)$ (broken line), 200-μm diameter for both [14]: (a) transmission loss spectra and (b) loss factors; measured scattering losses at 0.633 μm for $Ge_{20}S_{80}(1)$ and $Ge_{20}S_{80}(II)$ bulk glasses are plotted as a triangle and a circle, respectively.

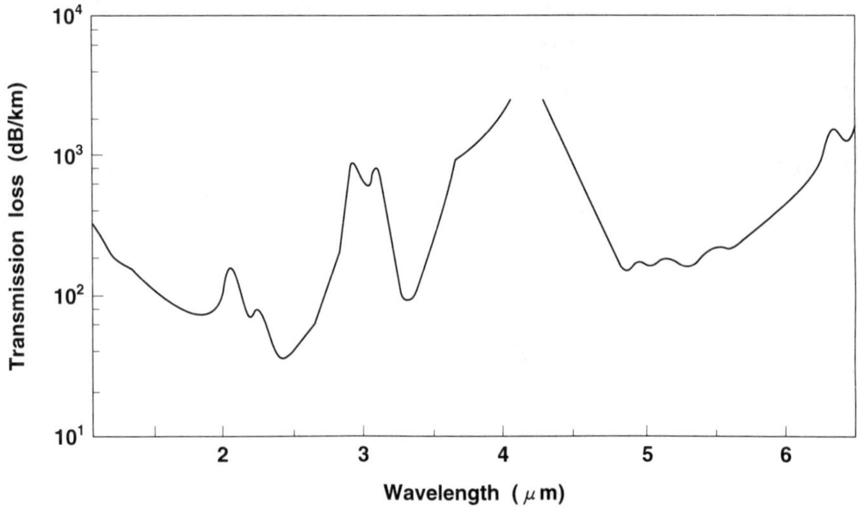

Figure 4.130 Transmission loss spectrum for a 200-μm diameter unclad $As_{40}S_{60}$ glass fiber [14].

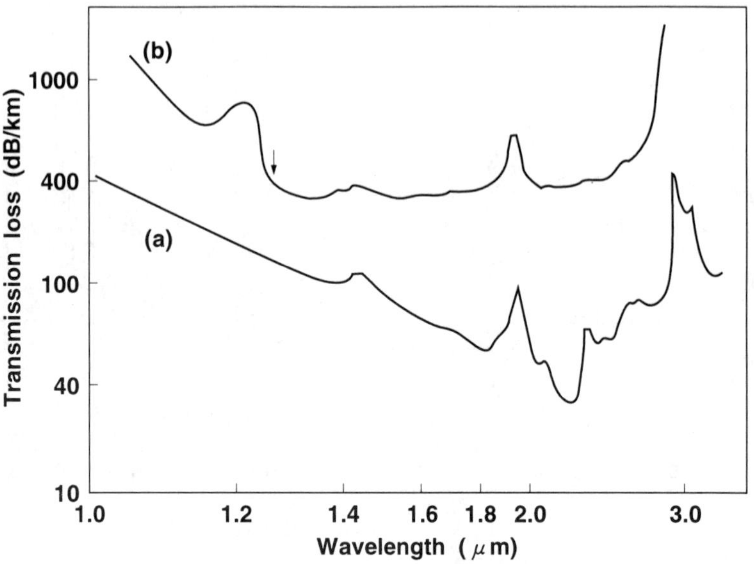

Figure 4.131 Transmission loss spectra for As-S glass core-cladding fibers [276]: (a) multimode fiber and (b) single-mode fiber.

about 0.49; and Figure 4.131(b) shows the loss curve for a single-mode As-S fiber with a Δn of 0.3%, which consists of an $As_{40}S_{60}$ glass core and $As_{39.5}S_{60.5}$ glass cladding. The peak observed around 1.2 μm is due to the cutoff of the first higher order mode and the cutoff wavelength can be estimated as about 1.25 μm from the cutoff peak. The losses of the single-mode fiber are almost constant in the 1.3- to 1.8-μm range and exhibit high values. The minimum losses are 23 $\pm$ 8 dB/km at 2.3 μm for the multimode fiber and about 400 dB/km at 1.3 to 1.8 μm for the single-mode fiber. This higher loss characteristic in the single-mode fiber is considered to be the result of using the double-crucible drawing process in which imperfections such as bubbles and microcrystallites arise at the interface between the core and cladding and cause higher scattering losses in a single-mode fiber with a smaller core diameter.

As-Ge-Se Fiber. Figure 4.132 shows the transmission loss spectra of As-Ge-Se fibers [14]. The solid line indicates the loss curve for an unclad fiber drawn from an $As_{38}Ge_5Se_{57}$ glass rod. The broken line represents the loss curve of a Teflon FEP clad fiber using the same $As_{38}Ge_5Se_{57}$ glass as that used for the unclad fiber. The transmission loss characteristics of this fiber in the short-wavelength region are the same as those of the unclad fiber. The loss in the wavelength region longer than 5 μm is higher than that for the unclad fiber due to the absorption of the Teflon FEP cladding. The minimum loss for the unclad fiber is 182 dB/km at 2.12 μm.

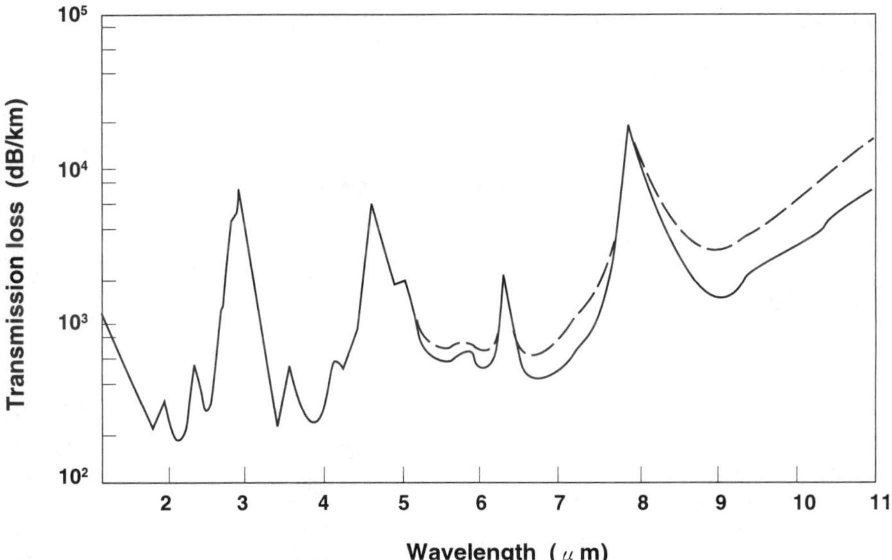

Figure 4.132 Transmission loss spectra for a 200-μm-diameter unclad $As_{38}Ge_5Se_{57}$ glass fiber (solid line) and an $As_{38}Ge_5Se_{57}$ glass core-Teflon FEP clad fiber with a 200-μm-diameter core and 220-μm-diameter cladding (broken line) [14].

Analyzed loss factors for $Ge_{20}S_{80}(1)$, $Ge_{20}S_{80}(II)$, $As_{40}S_{60}$, and $As_{38}Ge_5Se_{57}$ unclad fibers are listed in Tables 4.19 and 4.20 [14].

Ge-As-Se-Te Fiber. Figure 4.133 shows the transmission loss spectra for Ge-As-Se-Te fibers obtained by using the single-crucible method and the rod-in-tube method [273]. The higher losses for the rod-in-tube method result from the scattering due to such defects as bubbles at the interface between the core and cladding, suggesting that it is not easy to form a smooth interface between the core and cladding using the rod-in-tube method.

Rare-Earth-Doped Fiber. Rare-earth doping has been attempted for an As-S fiber [284]. To date, however, rare-earth-doped fiber has showed high losses, suggesting that

Table 4.19
Loss Characteristics of Chalcogenide Glass Fibers

Fiber Material (at %)	Urbach Tail (dB/km)	Multiphonon Absorption Tail (dB/km)	Weak Absorption Tail (dB/km)
$Ge_{20}S_{80}$	$1.0 \times 10^{-8} \exp(17/\lambda)$	$1.1 \times 10^{8} \exp(-59/\lambda)$	(I) $66 \exp(3.2/\lambda)$ (II) $16 \exp(3.2/\lambda)$
$As_{40}S_{60}$	$1.6 \times 10^{-11} \exp(23/\lambda)$	$1.8 \times 10^{9} \exp(-95/\lambda)$	$5.9 \exp(4.4/\lambda)$
$As_{38}Ge_5Se_{57}$	$9.3 \times 10^{-8} \exp(25/\lambda)$	$4.7 \times 10^{8} \exp(-124/\lambda)$	$5.7 \exp(6.4/\lambda)$

λ, in microns. *From*: [14].

Table 4.20
Impurity Absorption Bands

		Wavelength (μm)		
	Cause	Ge-S	As-S	As-Ge-Se
OH	Fundamental	2.84	2.91	2.92
	Overtone	1.43	1.44	1.45
	Combination	2.25, 1.88	2.29, 1.92	2.32, 1.92
SH	Fundamental	4.00	4.03	4.57
	Overtone	2.02	2.05	2.32
	Combination	3.78, 3.08	3.69, 3.11 2.54	4.15, 3.55
H_2O		6.32, 2.76	6.32, 2.88	6.32, 2.76
C		4.94	4.94	4.94
Oxide		–	–	7.90

From: [14].

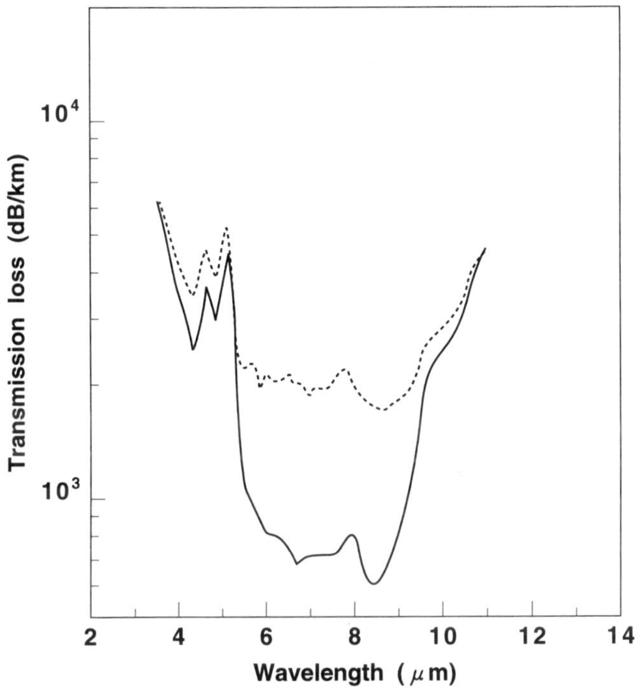

Figure 4.133 Transmission loss spectra for Ge-As-Se-Te fibers obtained when using the single-crucible method (solid line) and the rod-in-tube method (broken line) [273].

rare-earth doping degrades the glass stability, making it necessary to optimize the host glass composition.

For lanthanum glass fibers, which offer good potential for rare-earth doping, an unclad La-Ga-S fiber with a loss of 20 dB/m has been obtained by preform drawing [281]. The preform rod was prepared by casting remelted glass into a preheated vitreous carbon crucible. The extremely high fiber loss may reflect the poor stability of La-Ga-S glass.

Effect of Environment and Aging on Transmission Loss

The transmission loss of chalcogenide glass fibers is strongly affected by the environment and aging. A considerable loss increase occurs when the Teflon FEP clad fibers are placed in air at room temperature for a long period or in a high temperature and humid environment [31,285]. On the other hand, glass clad fibers show little loss increase in the same high temperature and humid environment. Since the two absorption peaks at 2.8 and 3.0 μm, which are due to H_2O molecules and the

OH group associated with the As-S glass network, respectively, grow in a humid environment and each OH absorption loss has a $1/T$ dependence, it is suggested that moisture in the atmosphere diffuses to the core through the Teflon FEP cladding or glass cladding and changes into OH groups via a chemical reaction [31]. The smaller loss increase for glass clad fiber indicates that the diffusion constant of H_2O for chalcogenide glasses is much smaller than that for Teflon FEP but not negligible.

Transmission losses at short wavelengths increase slightly when As-S and As-Ge-Se unclad fibers are exposed to light but do not change for Ge-S unclad fiber [14]. Measured optically induced losses are 2 dB/km at 2.44 μm for an unclad $As_{40}S_{60}$ fiber and 69 dB/km at 2.12 μm for an unclad $As_{38}Ge_5Se_{57}$ fiber at room temperature and under 0.3-mW/cm^2 light from a halogen lamp, and the losses increase with decreasing wavelength [223]. The same mid-gap absorption is reported to occur at temperatures lower than 15K in $As_{40}S_{60}$, $As_{40}Se_{60}$, and Se bulk glasses when exposed to light; and it is indicated that the absorption originates from valence alternation pairs [286]. These results suggest that arsenic glass fibers are practically unsuitable for application to fiber amplifiers.

4.7.3.2 Strength

The theoretical strength of chalcogenide glasses, which is estimated as 1.6 to 1.9 GPa [287], is essentially low because of their weak atomic binding energy compared to that of silica glasses and fluoride glasses. The average strength of chalcogenide glass fibers measured in tensile tests under ambient conditions is 100 to 170 MPa [31,272]. In bending tests higher values of 300 to 600 MPa are obtained, as shown in Table 4.21 [31]. Although the obtained strengths depend on the type of test and the glass system, they are far below the theoretical values. Volatilization during fiber drawing, which leads to roughness on the fiber surface, may be the main cause of the low strength.

Table 4.21
Strength of Chalcogenide Glass Fibers

Fiber Core / Cladding	Fiber Diameter (μm) Core / Cladding / Buffer	Before Treatment Tensile / Bending	After Treatment Tensile / Bending
As-S / Teflon	260 / 300 / –	110 / –	66 / –
As-S-Se / As-S	205 / 265 / 403	101 / 621	80 / 580
Ge-As-Se / Teflon	220 / 240 / –	116 / –	112 / –
Ge-As-Se-Te / Ge-As-Se	260 / 325 / 430	97 / 354	69 / 280
Ge-Se-Te / Ge-As-Se-Te	260 / 335 / 430	111 / 345	87 / 359

Treatment: 80°C, 85% RH, 24 hr. Strength: in MPa. *From*: [31].

In a humid atmosphere, the mechanical strength of chalcogenide fibers decreases [31]. Table 4.21 shows results for fibers exposed to a temperature of 80°C at 85%RH for 24 hr. The results indicate that the fibers must be protected from moisture, as this causes a loss increase, if they are to be practically employed.

4.8 CRYSTALLINE FIBER FABRICATION PROCESS

When we use crystal materials for active fiber, the most difficult issue is how to prepare crystal fibers with high quality. From the late 1970s to the early 1980s, polycrystalline fibers were made by a hot extrusion technique. The most widely studied polycrystalline fibers are KRS-5 [210] and AgCl [288] fibers. In addition, single-crystalline fibers were also made by the *edge-defined film-fed growth* (EFG) method with AgBr [33,289] and CsI [290]. However, the purpose of all these investigations was to develop an ultra-low-loss fiber better than high-silica glass fibers, not to develop active fibers for lasers or amplifiers.

Many materials have been investigated with a view to preparing crystal fibers for lasers or amplifiers. The most widely studied materials include Nd:YAG, Nd:GGG, Nd:LiNbO$_3$, Ti:Al$_2$O$_3$, and Cr:Al$_2$O$_3$. The early work on active single-crystal fibers was conducted in 1975 by C. A. Burrus and J. Stone [37,291–293]. They prepared Nd:YAG single-crystal fibers using a CO$_2$ laser beam as a heat source and succeeded in constructing a miniature LD-pumped fiber laser [37,291–293]. In 1982, M. Fejer, R. L. Byer, R. Feigelson, and W. Kway developed a stable method for fabricating single-crystal fibers that they then prepared with various materials such as Nd:YAG, LiNbO$_3$, and Al$_2$O$_3$ with a view to such applications as passive, active, and nonlinear devices [294–298]. Collaborating with them, M. J. F. Digonnet, C. J. Gaeta, and H. J. Shaw attempted to construct Nd:YAG single-crystal fiber lasers operating at lower pump power levels [299,300]. In addition to such experimental demonstrations, they carried out a theoretical analysis of optical fiber amplifiers and lasers in 1985 [301] and, eventually, in 1990, established a fundamental theory for the gain in three- and four-level fiber amplifiers and lasers [302]. In 1991, a 1.3-μm crystal fiber amplifier was demonstrated that employed Nd:YAG fiber [303].

Figure 4.134 shows a schematic diagram of single-crystal fiber fabrication using the *laser-heated pedestal growth* (LHPG) method [294–296,304–306]. In this noncrucible method, first a CO$_2$ laser is used to melt the end of a rod of feed material with a 360-degree axially symmetric irradiation. An oriented seed crystal is dipped into the molten zone thus formed, and then the fiber is grown by pulling the seed crystal away from the melt while fresh feed material is simultaneously fed into the molten zone. Typical fiber growth rates ranged from 1 to 3 mm/min for LiNbO$_3$ fibers and 10 to 40 mm/min for Al$_2$O$_3$ fibers. The diameter reduction ratio was about 1/3 to 1/6. The fibers had typical diameters of 30 to 170 μm and lengths of 100 to 300 mm. The diameter of each fiber was maintained to within a few tenths of a percent over tens of millimeter lengths by computer control of the fiber growth process.

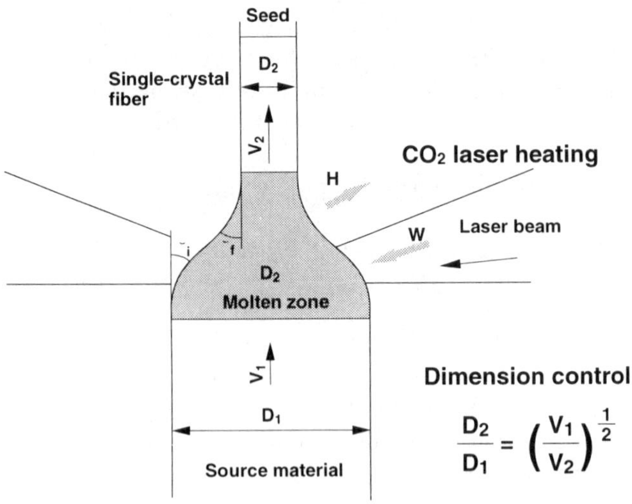

Figure 4.134 Schematic diagram of single-crystal fiber fabrication using the LHPG method [294–296,304–306].

Figure 4.135 shows a block diagram of single-crystal fiber fabrication using the LHPG method [294–296,304–306]. Typically, a 15-W polarized waveguide CO_2 laser serves as the heat source for single-crystal fiber growth. The output power is stabilized to within 1% fluctuation. An electro-optic power control system is used to adjust the laser power incident on the molten zone. The EO modulator consists of a ZnSe quarter waveplate, a CdTe electro-optic crystal, and a ZnSe polarizer/analyzer. The dynamic range is greater than 100:1. Provisions for modulating the incident laser power are incorporated into the control circuitry to allow a fiber to be grown with controlled diameter variations. The CO_2 laser beam emerging from the polarizer is combined with a HeNe beam on a dichroic mirror. The coincident HeNe beam facilitates CO_2 alignment down the remainder of the optical train. After passing through a ZnSe-focusing telescope and some beam-steering optics, the CO_2 beam enters the atmosphere-controlled growth chamber. Within the growth chamber a well-designed optical system focuses the CO_2 laser beam onto the fiber with a 360-degree axially symmetric distribution, as shown in Figure 4.136. The part of this optical system that focuses the CO_2 laser beam onto the fiber is called a "refraxicon," which consists of an inner cone surrounded by a large coaxial cone [304,305]. In order to achieve good optical performance it is essential that the refraxicon's two cones can be accurately aligned. A gold coating on the copper optical surfaces enhances reflectivity and protects the copper substrate. The refraxicon and parabolic mirror provide near-diffraction limited 1/2 focusing, yielding a minimum spot size of 30 μm. This tight focus is important for the stable growth of small-diameter fibers. The focus spot size can be controlled by modifying the

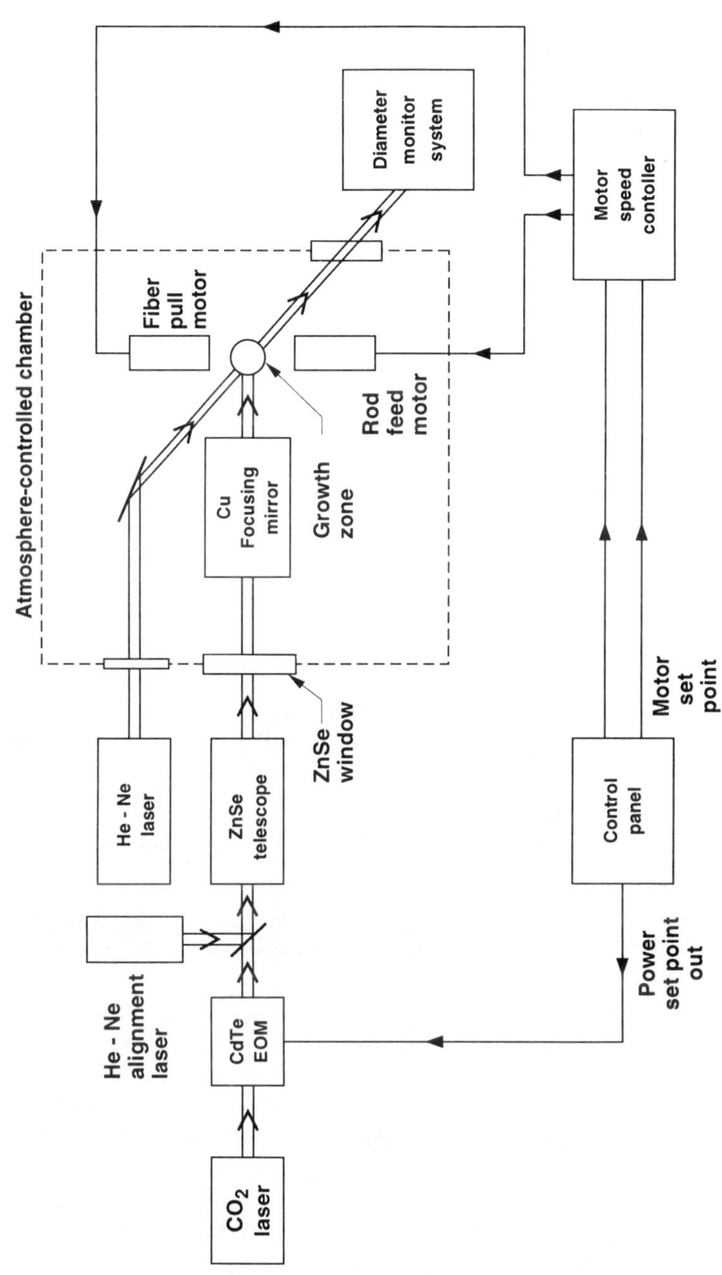

Figure 4.135 Block diagram of single-crystal fiber fabrication using the LHPG method [294–296,304–306].

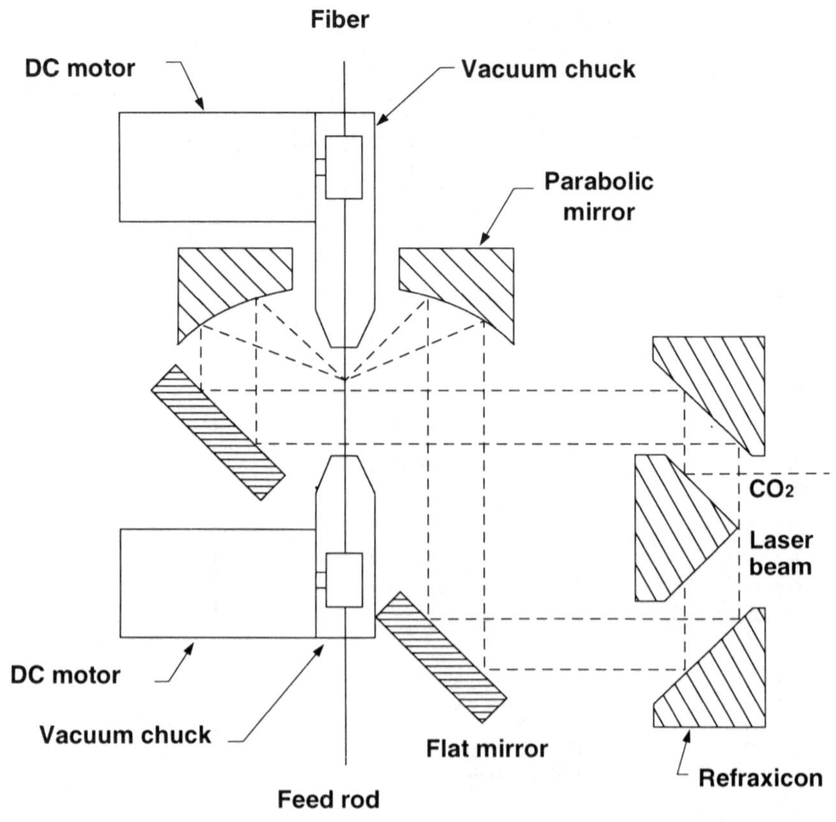

Figure 4.136 Cross-sectional diagram of the focusing optics and fiber pulling mechanism [296,304,305].

input beam divergence with the focusing telescope. Motorized x-y stages on the fiber and source rod translation devices permit adjustment of the fiber position with respect to the fixed laser focus spot.

This single-crystal fiber growth apparatus can be used to prepare fiber from various materials such as polycrystalline, sintered, or pressed powder materials as well as single-crystal. Nd:YAG single-crystal fibers with diameters of 10, 16, and 40 μm were prepared [303]. Figure 4.137 indicates the experimental and theoretical gain characteristics of Nd:YAG single-crystal fibers for the fiber amplifier application [303]. As seen in Figure 4.137, gains of up to 6.5 dB at 1.32 μm were observed using unclad Nd:YAG single-crystal fibers 10 to 40 μm in diameter and 1-cm long and a laser diode pumping power of 83 mW at a wavelength of 0.81 μm [303]. The results shown in the figure also indicate that the extrapolation of experimental and theoretical gain characteristics predicts gains in excess of 30 dB for fiber diameters of 5 μm.

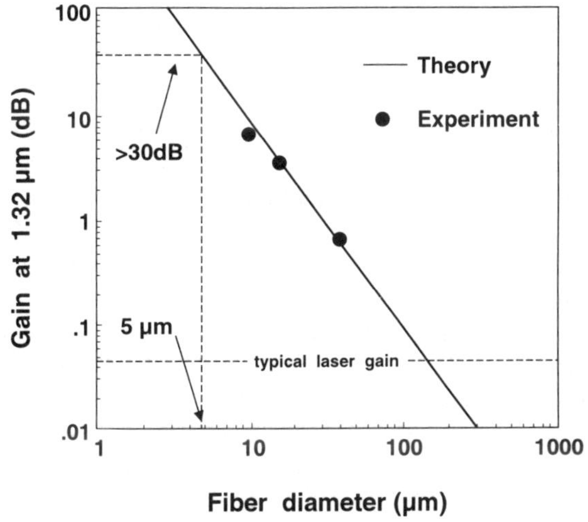

Figure 4.137 Experimental and theoretical gain characteristics of Nd:YAG single-crystal fibers [303].

To date, there have been many investigations into various methods for types of single-crystal fiber growth and their applications, in particular, to nonlinear optics. However, these are beyond the scope of this book and will be described elsewhere.

4.9 RELIABILITY OF AMPLIFIER HOST FIBERS

4.9.1 Reliability Requirements

Optical fiber amplifiers will be installed in various kinds of transmission networks and will operate in the system as an in-line device. Therefore, the increasing interest in OFA reliability springs from the high density of traffic supplied via fiber optics over a cable cross section, which means that a breakdown assumes major proportions in terms of economic loss to the customer and the communication supplier. The reliability of a fiber amplifier depends on various factors including the reliability of the host fiber, pump laser, and other optical components and the reliability of the module, itself. Of these factors, this section focuses on fiber reliability; the reliability of other components will be described in the following chapter. With regard to host fibers, practical applications require that a fiber never breaks down even if coiled around a compact spool in an amplifier module under any given service condition, and there should also be no increase in transmission loss leading to a degradation in amplification performance. The potential for optical property degradation depends on such parameters as temperature, humidity, and γ-ray irradiation. In order to demonstrate the long-term capability of optical fiber amplifiers to perform their functions reliably, they must pass a comprehensive reliability test program that attempts to duplicate the stresses that fibers are expected to experience during 20 years of operation.

The qualification test requirements for optical fiber amplifiers are currently being established and are addressed in Bellcore Technical Advisory TW-NWT-001312 "Generic Requirements for Optical Fiber Amplifier Performance and Reliability Assurance" [307]. Since the standardization of their reliability requirements is now being discussed [308], it is desirable that the components of optical fiber amplifiers should pass Bellcore Technical Advisory TA-NWT-001221 "Generic Requirements for Passive Fiber Optic Component Reliability Assurance Practices" [309]. TA-NWT-001221 is the latest of several *Technical Advisories* (TA) and *Technical References* (TR) dealing with *passive fiber optic* (PFO) components intended for *fiber-in-the-loop* (FITL) applications [310–315]. These TAs describe practices that, in Bellcore's opinion, can augment the component manufacturers' efforts to provide reliable products. TA-NWT-001221 contains test procedures and test conditions designed to provide an estimate of PFO component reliability [316]. The test procedures are based on industry-accepted standard test procedures (principally TlA/EIA); the test conditions are based on the environments that the components are expected to encounter in the field. In addition to the environmental stress testing, which comprises a large portion of the document, the issues of qualification and requalification, lot-to-lot controls, feedback and corrective action, storage and handling, and documentation control are also addressed. A qualification program has been proposed that will ensure PFO components meet the appropriate reliability requirements. This program is designed to identify any weakness in the manufacturing process and wear out mechanisms. Test samples are to be produced by normal fabrication techniques; that is, they should not be subjected to special handling or screening. Each test sample should be composed of components from different lots in order to obtain a valid picture of the process failure rate rather than that of a particular lot. The test procedures included in TA-NWT-001221 and their names are listed in Table 4.22.

Another key issue is lifetime estimation by which to demonstrate long-term reliability. The theory of lifetime estimation has been established for silica fibers used as a transmission medium.

4.9.2 Catastrophic Failure of Optical Fibers Caused by Chemical and Mechanical Stresses

Of possible accidents related to the reliability of optical fiber, host fiber breakage is worse than the loss increase caused by H_2 diffusion or γ-ray irradiation. This breakage occurs catastrophically because glass is inherently brittle due to its lack of inelastic flow. The fracture mechanism and the theory of lifetime estimation have been thoroughly investigated based on fracture mechanics and glass surface

Table 4.22
Test Procedures Included in TA-NWT-001221 and Their Names

Industry Standard	Test Description
EIA/TIA-455-2A	Mechanical shock (Impact)
EIA/TIA-455-11A	Variable frequency vibration
EIA/TIA-455-71	Thermal shock
EIA/TIA-455-4A	High-temperature storage (Dry heat)
EIA/TIA-455-5A	High-temperature storage (Damp heat)
EIA/TIA-455-4A	Low-temperature storage
EIA/TIA-455-3A	Temperature cycling
IEC 68-2-38	Temperature-humidity cycling
EIA/TIA-455-16A	Salt atmosphere exposure
EIA/TIA-455-12A	Water immersion
ASTM B827-92	Airborne contaminants

From: [309].

chemistry. A broad overview of issues related to the reliability of host fibers is provided in this section.

4.9.2.1 Fracture Mechanism and Theory of Lifetime Estimation

The first attempt to describe the rupture of materials was by Griffith [317]. His analysis suggests that a flaw in an elastic solid becomes unstable when the rate of release of stored energy equals the rate of increase in energy as a result of creating new surfaces according to thermodynamics. The fracture strength (σ_f) is thereby related to the flaw size (a_c) by

$$\sigma_f = (E\gamma/Ya_c)^{1/2} \tag{4.91}$$

where E denotes Young's modulus, γ is surface energy, and Y is a geometric constant for the specimen configuration. The Griffith model is shown in Figure 4.138. This model proposes an inverse relationship between failure stress and the square root of the flaw size. This relationship remains the fundamental law governing the strength of optical fibers [318].

An alternative approach was established by Irwin [319], who analyzed the stresses close to the crack tip rather than throughout the body. Such an analysis, also known as fracture mechanics, shows that the stress at the crack tip exceeds the average stress experienced by body. Irwin defined a parameter called the stress intensity factor (K_I) and showed that fracture occurs when a critical value of the stress intensity factor (K_{IC}) is reached, leading to an equation of type

$$\sigma_f = K_{IC}/(Ya_c)^{1/2} \tag{4.92}$$

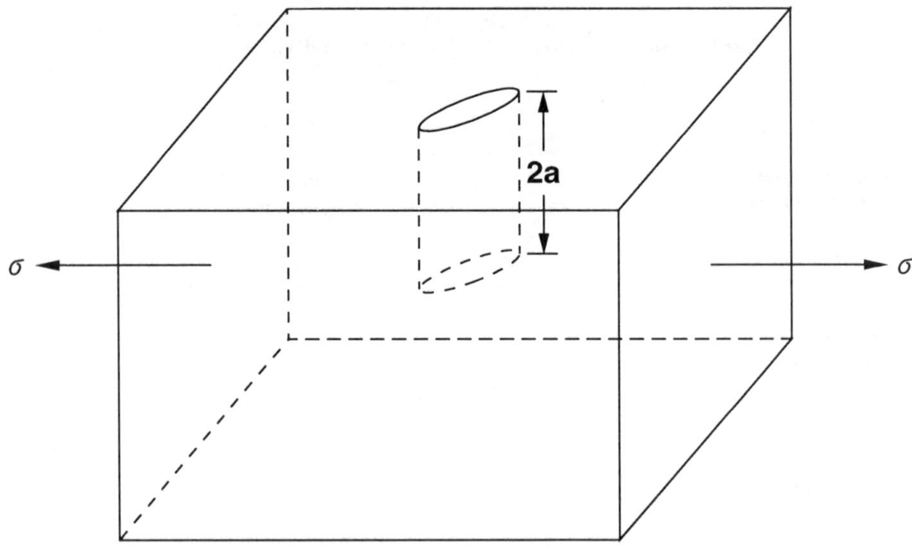

Figure 4.138 Griffith model for brittle fracture in glass initiating at a pre-existing flaw [318].

The critical stress intensity factor K_{IC} is a function of glass composition, the applied stress, and the crack size and geometry. The equations show that the failure stress decreases with increasing crack size [318].

This theory is extended to the lifetime estimation of optical fibers [319–323]. With regard to optical fibers in a noninert or active (ambient or hostile) environment, such as one where humidity, water, or a chemical species are present, an applied stress will cause crack growth to occur. This is called stress corrosion. The velocity is

$$da/dt = AK_I^{\,n}(t) \tag{4.93}$$

Here, A is a material proportionality parameter, while the dimension exponent n is the stress corrosion factor or n-value. Both A and n depend on the environmental conditions. Strength is distributed along the length of practical fibers. This can be expressed by a Weibull survival distribution.

$$P(S, L) = \exp\{-(S/S_0)^m (L/L_0)\} \tag{4.94}$$

Here, $P(S, L)$ is the cumulative survival probability of the fiber length L up to the inert strength S, and S_0 is the strength corresponding to a survival probability of e^{-1}, measured at a gauge length L_0. These equations are solved for a fiber crack in an active environment in order to derive a lifetime equation [323]. In principle, the time-to-failure can be estimated by integrating (4.94) under the installed stress conditions or stress history [323]. The static and dynamic fatigue parameters n_s and n_d are determined by measuring the time-to-failure of a fixed gauge length of fiber under a range of stress levels or stress rates, respectively, as illustrated in Figure 4.139. The use of a fixed gauge and the statistical nature of flaw distribution requires a scaling factor in estimating the time-to-failure of an installed fiber link. Lifetime can be calculated for various fiber stress histories. Examples are the constant tension in a buried cable or in a bend within a splice housing, or the variable tension due to temperature cycles, wind, or fiber payout from a bobbin or reel. However, fiber subjected to a constant applied stress σ_a, which does not vary with time, is the most common situation for which lifetime calculations are made. For a static fatigue test for the gauge length L_0, the lifetime t_f is described as

$$t_f(L, F, \sigma_f) = t_f(1)\{L_0/L \cdot F\}^{1/b}\sigma_a^{-n_s} \tag{4.95}$$

$$b = m/(n_s - 2) \tag{4.96}$$

where F is the allowable failure probability and m is the Weibull distribution parameter. For a dynamic fatigue test, the lifetime is described as

$$t_f(L, F, \sigma_f) = \sigma_f(1)^{n_d+1}/(n_d + 1)\{L_0/L \cdot F\}^{(n_d+1)/m_d}\sigma_a^{-n_d} \tag{4.97}$$

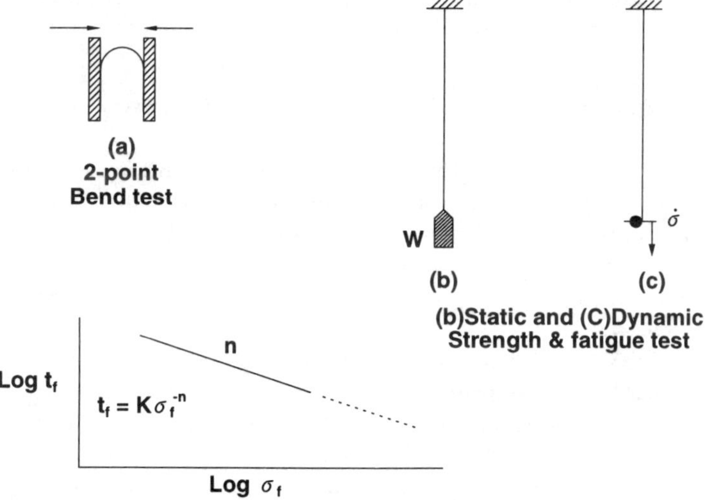

Figure 4.139 Measurement of strength and fatigue parameters [323]. For static fatigue (n_s) the failure time t_f is measured for several loads W. For dynamic failure (n_d) the variable is stress or strain rate.

$$\sigma_f(1)^{n_d+1} = (n_d + 1)t_f(1) \tag{4.98}$$

Here, σ_a is applied stress, $\sigma_f(1)$ is the intercept of the failure stress versus the applied stress rate plot, and F is the allowable failure probability.

Several researchers have found characteristic transitions in static fatigue experiments for fibers in water that lead to a shorter lifetime than predicted from these equations [321,322]. These transitions are explained by a combination of the conventional stress corrosion theory with the aging of fibers in water under zero stress [324–326]. Zero-stress aging is not taken into account in (4.95) and (4.97). The lifetime determined by zero-stress aging was estimated by fitting the results to an equation as [324]

$$\sigma_f = \sigma_{f0}/(1 + t/\tau)^p \tag{4.99}$$

Here, σ_{f0} is the measured strength at time $t = 0$. The lifetime can be calculated as the time at which σ_f is equal to the applied stress. With the static fatigue test, the lifetime can be estimated as follows. An equation has been formulated that combines zero-stress aging and stress corrosion [327]:

$$d(\sigma/\sigma_i)/dt = x/\tau(\sigma/\sigma_i)^{(x-1)/x} - 1/\{B\sigma_2^{n-2}(n-2)\}\sigma_a^n(\sigma/\sigma_i)^{3-n} \tag{4.100}$$

where σ_i is initial strength, B is a constant, and $x = -(n+1)/(n-2)p$. The differential equation was solved numerically; the relationship between the lifetime and the applied stress was derived from (4.100).

4.9.2.2 Mechanical Strength and Lifetime Estimation of Silica Fibers

A great deal of work has been done on the strength of silica-based optical fibers, and a great deal of progress has been made. Very high quality fiber can now be made that has high strength and reasonable uniformity. Several coating materials have been investigated and polymer, inorganic, and metal coatings have been developed. These coatings can protect the fiber from damage and are satisfactory optically [328,329].

Figure 4.140 shows the lifetime in years predicted from dynamic fatigue experiments for a silica fiber with a nominal glass diameter of 125 μm, coated with UV-cured epoxy acrylate polymer to an overall diameter of 250 μm under applied stresses between 100 and 350 MPa. In Figure 4.141, the predicted lifetime is presented in terms of bending radius of between 2 and 10 mm applied to the fiber based on static fatigue experiment results. These figures reveal that there is a very large variation in the lifetime depending on the environment and applied stress level [323]. However, this prediction clearly demonstrates that long-term reliability can be guaranteed for silica-based fibers even if the fiber is coiled around a small

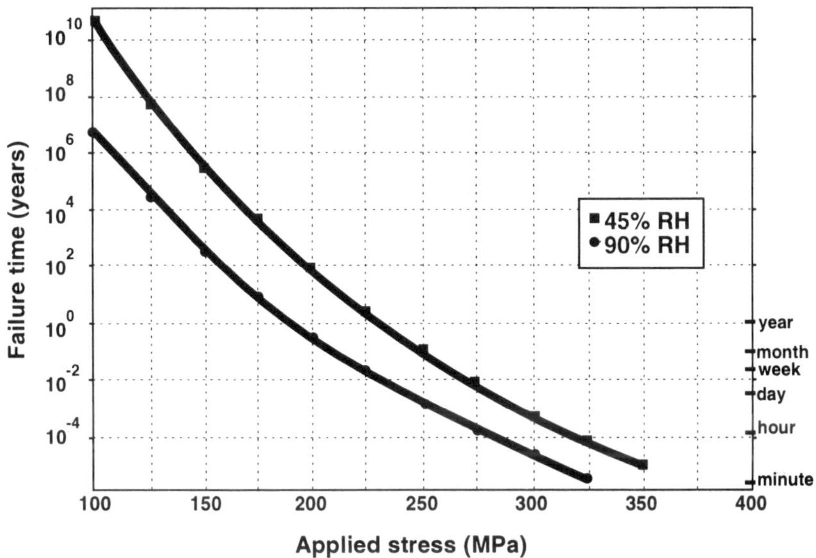

Figure 4.140 Lifetime prediction from dynamic fatigue experiments [328].

bobbin to facilitate the fabrication of a compact amplifier module. With regard to the erbium-doped silica-based fibers used in optical fiber amplifiers, carbon-coated fiber has been demonstrated to have high-fatigue resistivity [330]. Carbon coating has also been shown to effectively prevent any increase in loss caused by hydrogen diffusion as described in Section 4.9.3, and a carbon coating technique has already been developed in the field of transmission media [331–334]. Figure 4.142 shows the dynamic fatigue characteristics of carbon-coated and non-carbon-coated erbium-doped fibers [330]. The stress corrosion factor n of carbon-coated fiber is greater than 150, much higher than the value of 24 for non-carbon-coated fibers. Figure 4.143 shows the relationship between bending radius and failure probability estimated from the results in Figure 4.142. In this figure, the definition of failure probability is that the fiber will break once in a 100-m span over 25 years. Since erbium-doped fibers generally have a high Δn to improve their gain characteristics, they are relatively resistant to an increase in loss caused by bending, for example, about 0.01 dB/100m. Carbon-coated fiber, however, even if kept bent for 25 years at a radius that would cause problems in terms of bending loss, would only break at a probability of less than 10^{-5} [230]. This means that silica-based fiber is reliable enough for practical use in optical amplifiers.

4.9.2.3 Mechanical Strength and Lifetime Estimation of Fluoride Fibers

In the prediction, it is assumed that the applied stress is 36 MPa, which corresponds to that applied by a 10-cm-diameter fiber bobbin, and that the allowable failure

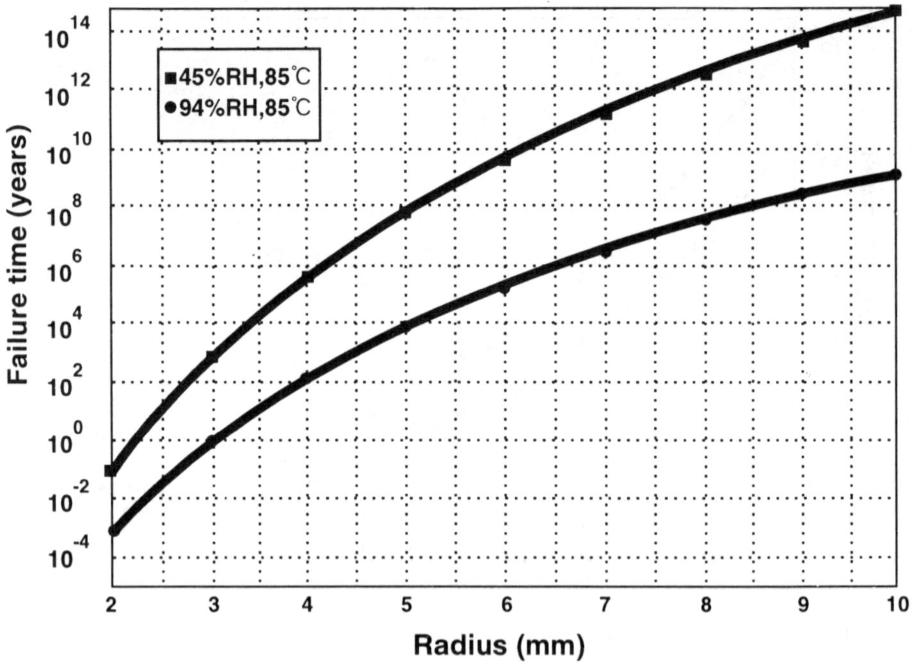

Figure 4.141 Lifetime prediction from static fatigue experiments [328].

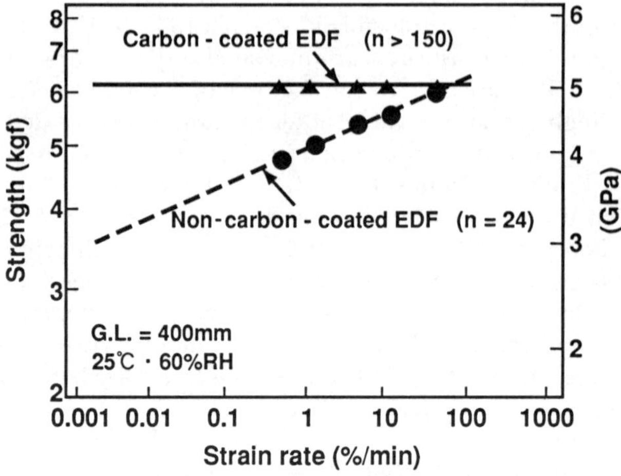

Figure 4.142 Dynamic fatigue characteristics of carbon-coated and non-carbon-coated Er-doped fibers [330].

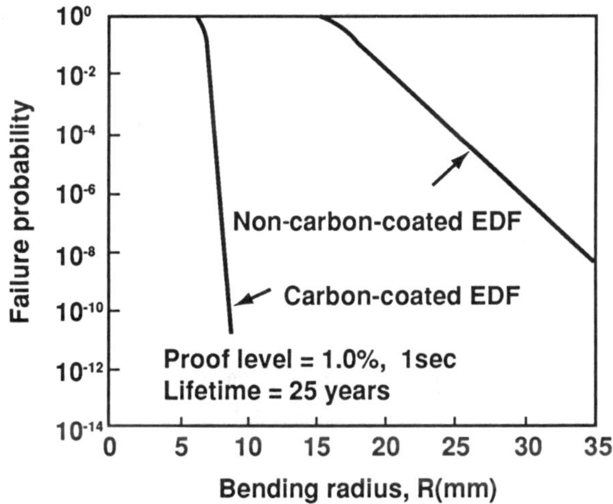

Figure 4.143 Predicted failure probability for carbon-coated and non-carbon-coated Er-doped fibers [330].

probability $F = 10^{-6}$. The lifetime was predicted for fibers with strengths of 350 and 500 MPa using the Weibull distribution factor m, the stress corrosion factor n, and the temperature dependence of the strength degradation measured for both fibers.

In general, fluoride glasses are known to be susceptible to moisture damage. Therefore, both stress corrosion and zero-stress aging must occur simultaneously in the presence of environmental moisture. The effect of stress corrosion can be estimated from (4.99) by measuring the stress corrosion factor. In a dry atmosphere, this factor was larger than 100. Since the stress corrosion factor decreased to 22 at 80°C and 70%RH with increasing humidity, H_2O diffuses to a crack tip, thus reducing the activation energy of chemical bond dissociation and the crack growth rate may increase in the high-humidity region. According to the stress corrosion mechanism, this effect may be dominant in the high-stress region. On the other hand, the strength degradation caused by zero-stress aging may be dominant in the low-stress region. Figure 4.144 shows the strength degradation that results from zero-stress aging. With fluoride glass fibers, this is mainly caused by H_2O, which forms a hydrate or an oxide on the fiber surface, and these crystals may be the origin of fractures. The mechanism of zero-stress aging for fluoride fibers is different from that for silica fibers. For fluoride fibers, strength degradation is caused by crystal growth on the fiber surface. These crystals, formed during fiber drawing, are grown during zero-stress aging after fiber fabrication. Therefore, zero-stress aging results may be strongly dependent on the initial strength.

Figure 4.145 shows the humidity dependence of the predicted lifetime for fluoride fibers with several different initial strengths at 80°C. Stress corrosion is the dominant mechanism for fluoride fibers in a dry atmosphere. The effect of zero-

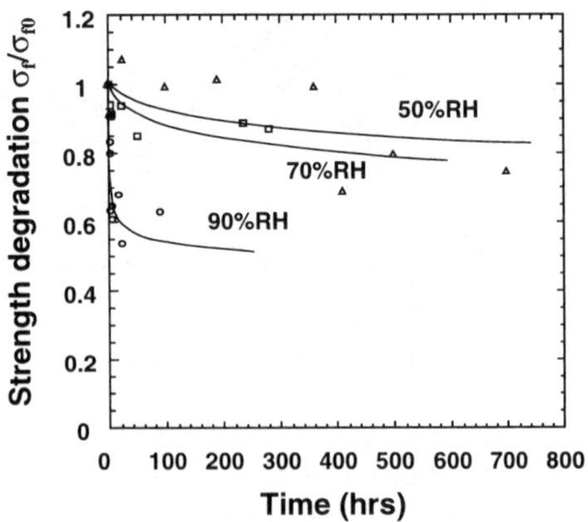

Figure 4.144 Strength degradation resulting from zero-stress aging [261].

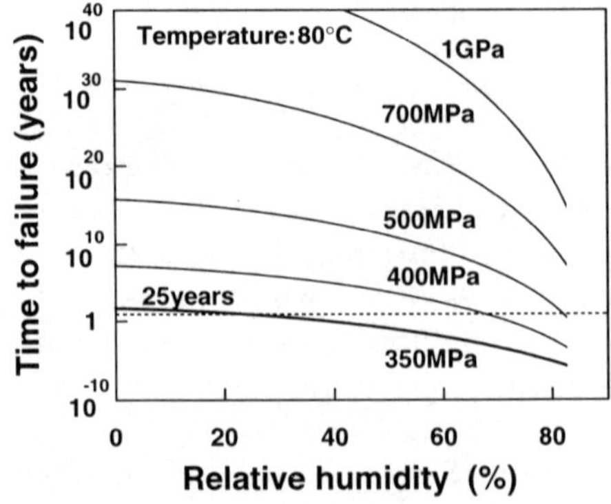

Figure 4.145 Humidity dependence of the predicted lifetime for fluoride fibers with several different initial strengths at 80°C [261].

stress aging on lifetime increased with increasing humidity, and the lifetime in Figure 4.145 was determined by zero-stress aging at humidities of over 50%RH. In the high-humidity region, the lifetime decreases considerably and the 350-MPa fiber no longer has a practical lifetime. However, the results indicate that a sufficient lifetime of over 25 years can be guaranteed for fluoride fibers with an initial strength of 500 MPa at 80°C and 80%RH.

Figure 4.146 shows the relationship between lifetime and applied stress for fiber with a strength of 350 MPa at 80°C and 50%RH. In this figure, the allowable failure probability was assumed to be 0.5. The results indicate that the predicted lifetime is about 65 years at an applied stress of 36 MPa. When the allowable failure probability used for the prediction is taken into account, this predicted lifetime is

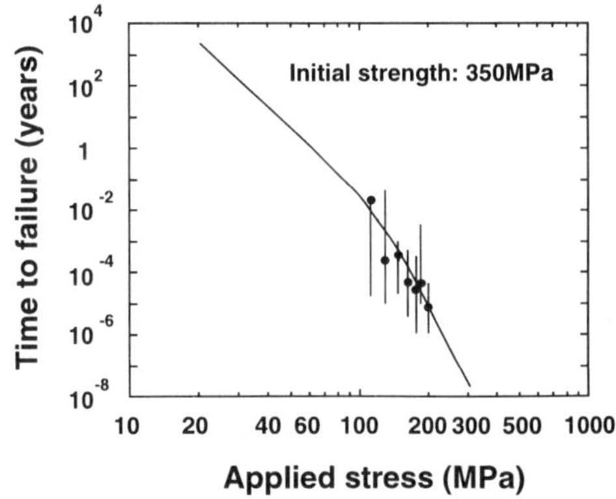

Figure 4.146 Relationship between lifetime and applied stress for fiber with a strength of 350 MPa at 80°C and 50%RH.

comparable to that obtained with the dynamic fatigue test. This result confirms that the dynamic fatigue test can provide a reasonable lifetime prediction for fluoride fibers.

Therefore, it has been confirmed that fluoride fibers with a strength of 500 MPa have a long-term reliability of over 25 years in optical fiber amplifiers in a practical environment such as 80°C and 80%RH [261].

4.9.2.4 Coating

As described in the previous sections, glass can suffer from zero-stress aging or experience fatigue as the result of stress under operational conditions. A good coating system helps to minimize the probability of fiber breakage by delaying or suppressing the adverse effects of service conditions. The coating should be applied within a clean environment after the fiber exits the heat source to reduce the risk of contaminants becoming attached to the surface of the glass. The coating must be applied concentrically and be of sufficient thickness without damaging the pristine surface of the fiber and must be solidified very rapidly before the fiber reaches a sheave or a capstan. Fast solidification generally excludes solvent-bearing coatings as being too slow. Liquid prepolymers capable of crosslinking by thermal or UV activation or thermoplastic materials that can be readily solidified by cooling are preferred.

With regard to coating materials for silica fibers, organic materials are most commonly used. Silicone rubber systems based on poly(dimethyl siloxane) or poly (methyl-phenyl siloxane) are examples of thermally cured systems that have low T_g's($-50°$ to $-125°$C) and low moduli (10^6 to 10^7 dyn/cm^2). This makes them extremely effective buffer coatings. The rather poor abrasion resistance of silicone-coated fibers, however, necessitates the use of an abrasion-resistant jacket. Typically, nylon is extruded over the silicone and produces a dual-coated structure. Epoxy or urethane acrylate materials are the most common UV-curable systems and can be used either as a hard-jacket or as a soft buffer coating. Low moduli and low T_g's are obtained in these systems by reducing the crosslink densities and increasing the chain flexibility. With silica fiber, since the resistivity to moisture is relatively high in comparison with multicomponent silicate fibers and fluoride fibers, it seems that the coating materials may have little effect on reliability when the fiber is used for amplifiers.

With regard to fluoride fibers, several kinds of material such as polymer, glass overcladding, and hermetic coating have been coated on both fibers and bulk glasses. Table 4.23 shows the mechanical properties of fluoride fibers with different coating materials. With regard to polymer coatings, UV-curable epoxy acrylate yields a higher strength than that of Teflon FEP coating. This result correlates with the adhesion to the glass surface and the permeability of moisture. Glass overcladding is useful for moisture protection, but high-strength fiber has not yet been fabricated because of the crystallization of the glass interface. Hermetic coating is the best for moisture protection, but it seems to be difficult to coat such materials on the fiber because of the low drawing temperature of fluoride fibers. Consequently, at present, the use of UV-curable epoxy acrylate is considered to be a realistic solution.

4.9.3 Hydrogen-Induced Loss Increase

Although hydrogen-induced losses are not usually a problem for standard single-mode fibers, they are of potential concern for the highly doped fibers used in Er amplifiers [335,336]. For instance, past work has shown that Ge/P-doped multimode fibers [337] and Al-doped single-mode fibers [338] can react rapidly with H_2, resulting in the formation of OH at around 1.4 μm. When a fiber is used as an

Table 4.23

Mechanical Properties of Fluoride Fibers With Various Coating Materials

	Materials	Substrate	Fiber Strength	H_2O Protection
Polymer coating	Teflon-FEP	Fiber	300–400 MPa	Poor
	UV-curable epoxy-acrylate	Fiber	500–700 MPa	Fair
Overcladding	Phosphate glass	Fiber	430 MPa	Fair
	Chalcogenide glass	Fiber	100–200 MPa	Fair
Hermetic coating	Carbon, Oxide, Nitride, Fluoride, etc.	Bulk glass	—	Excellent

optical amplifier host, loss increases at either the pump or the signal wavelength will reduce the gain of the amplifier and adversely affect its performance [339]. Figure 4.147 shows the predicted long-term loss increases at 1.55 μm for a 20-m-long erbium fiber doped with 18% GeO_2 and 1% Al_2O_3; the assumed H_2 pressure is 0.01 atm [340]. This estimation was carried out on the basis of data obtained from loss increase experiments performed at 92°, 123°, 153°, 180°, and 225°C with a H_2 pressure of 0.012 atm. Other experiments have shown that hermetic coating techniques such as carbon coating effectively prevent any loss increase in erbium-doped fiber at 1.55 μm [340,341].

4.9.4 Loss Increase due to γ-Ray Irradiation

The exposure of any solid materials to energetic ions, neutral particles, electrons, or photons with energies exceeding the optical bend gap results in the transfer of sufficient energy to the core and/or valence electrons to excite some of them into the conduction band. Most of these initially free electrons quickly thermalize, bind to hole states, and recombine by becoming separately trapped at isolated sites where they become "defect" centers. These centers are paramagnetic if the electronic spin is unpaired. Such paramagnetic centers are perforce located in the optical

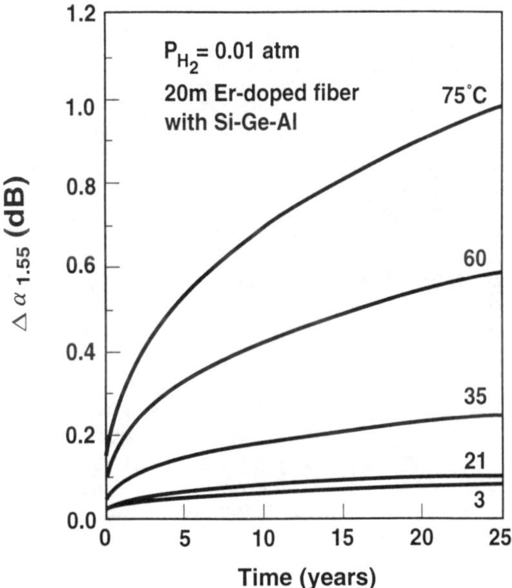

Figure 4.147 Predicted long-term loss increases at 1.55 μm for a 20-m-long Er-doped fiber codoped with 18% GeO_2 and 1% Al_2O_3 [340].

band gap and can thus absorb light by the excitation of the defect electron into the conduction band or lower-lying excited states [342].

The loss increase due to γ-ray irradiation was investigated to estimate environmental effects on EDFA performance in a similar way to the investigation of the loss increase due to H_2 diffusion [330,339–343]. Figure 4.148 shows the relationship between gain degradation and total dose of γ-ray irradiation [343]. The γ-ray exposure tests for erbium-doped fibers indicated that the loss increase and gain degradation at 1.55 μm are both negligible under realistic conditions for as long as 25 years [343].

In general, γ-ray irradiation results in the coloration of fluoride glasses in the UV to visible region. Figure 4.149 shows the effect of γ-ray radiation on the optical absorption spectra of a ZBLALi glass (ZrF_4-BaF_2-LaF_3-AlF_3-LiF) [344]. A resolved peak at 0.4 μm is evident in the irradiated fluoride glasses, and the damage does not extend to the IR region. Other research on the effects of radiation on the optical spectra of ZBGA (ZrF_4-BaF_2-GaF_3-AlF_3) and ZBGAP (ZrF_4-BaF_2-GaF_3-AlF_3-PbF_2) revealed that the induced losses due to 1-Mrad irradiation were 1,600 dB/km at 2.55 μm and 1,150 dB/km at 3.65 μm for ZBGA fiber and 2,300 dB/km at 2.55 μm and 1,050 dB/km at 3.65 μm for ZBGAP fiber [345]. Since it has been reported that the induced transmission losses for doped silica fibers irradiated by 1 Mrad are 2,000 to 30,000 dB/km at 1.5 μm [346], the induced loss for fluoride fibers is less than that for doped silica fibers. Based on these results

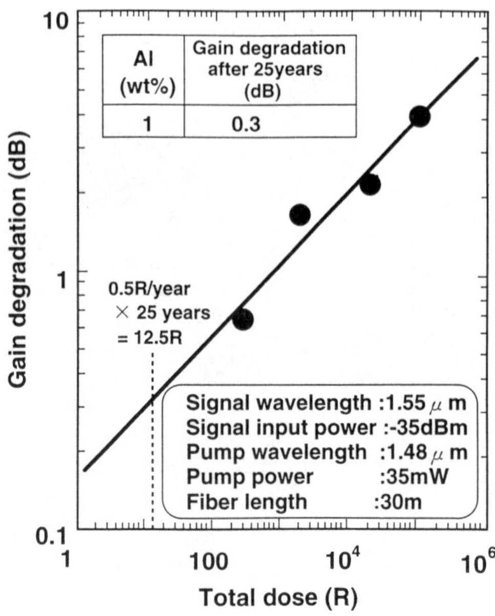

Figure 4.148 Relationship between gain degradation and total dose of γ-ray irradiation [343].

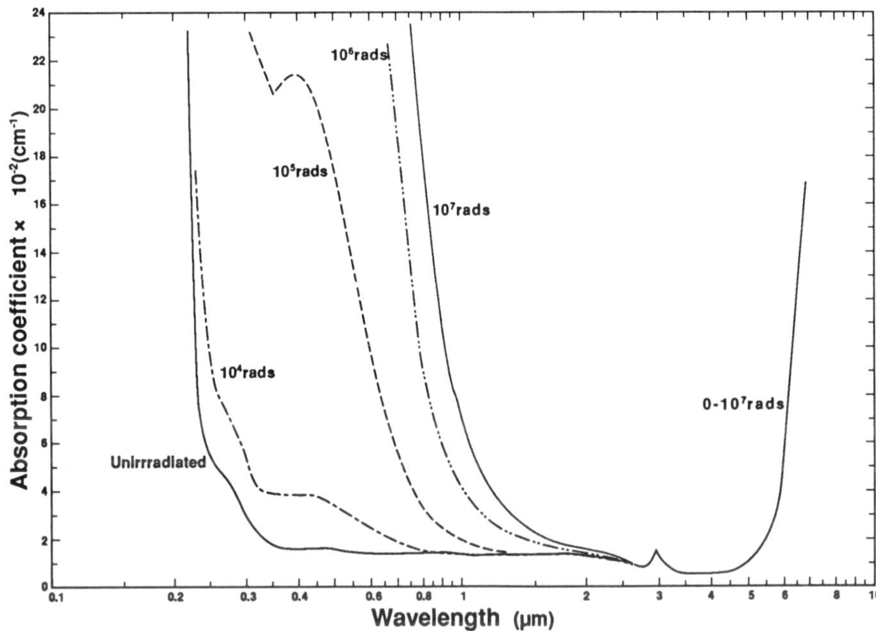

Figure 4.149 Effect of γ-ray radiation on the optical absorption spectra of a ZBLALi glass (ZrF_4-BaF_2-LaF_3-AlF_3-LiF) [344].

of the γ-ray exposure tests, the loss increase in fluoride fibers can be expected to be negligible under realistic conditions.

References

[1] Sir Isaac Newton, *Opticks, Book Three*, New York: Dover, 1952, p. 404.

[2] Gloge, D., "Weakly Guiding Fibers," *Appl. Opt.*, Vol. 10, 1971, pp. 2252–2258.

[3] Pinnow, D. A., T. C. Rich, F. W. Ostermayer, Jr., and M. DiDomenico, Jr., "Fundamental Optical Attenuation Limits in the Liquid and Glassy State with Application to Fiber Optical Waveguide Materials," *Appl. Phys. Lett.*, Vol. 22, 1973, pp. 527–529.

[4] Izawa, T., N. Shibata, and A. Takeda, "Optical Attenuation in Pure and Doped Fused Silica in the IR Wavelength Region," *Appl. Phys. Lett.*, Vol. 31, 1977, pp. 33–35.

[5] Miya, T., Y. Terunuma, T. Hosaka, and T. Miyashita, "Ultimate Low-Loss Single-Mode Fiber at 1.55 μm," *Electron. Lett.*, Vol. 15, 1979, pp. 106–108.

[6] Yokota, H., H. Kanamori, Y. Ishiguro, G. Tanaka, S. Tanaka, H. Takada, M. Watanabe, S. Suzuki, K. Yano, M. Hoshikawa, and H. Shimba, "Ultra-Low-Loss Pure-Silica-Core Single-Mode Fiber and Transmission Experiment," *Proc. for OFC'86*, Atlanta, GA, Postdeadline Papers, Feb. 24–26, 1986, pp. 11–18.

[7] Beales, K. J., C. R. Day, W. J. Duncan, and G. R. Newns, "Low Loss Compound Glass Optical Fiber," *Electron. Lett.*, Vol. 13, 1977, pp. 755–756.

[8] Takahashi, S., S. Shibata, and M. Yasu, "Low Loss and Low OH Content Soda-Lime-Silica Glass Fiber," *Electron. Lett.*, Vol. 14, 1978, pp. 151–152.

[9] Kanamori, T., K. Oikawa, Y. Terunuma, K. Kobayashi, and S. Takahashi, "Transmission Loss Characteristics of a Fluoride Optical Fiber in the Near-Ultraviolet to Mid-Infrared Wavelength Region," *Japan J. Appl. Phys.*, Vol. 28, 1989, pp. L1979–L1981.

[10] Kanamori, T., and S. Sakaguchi, "Preparation of Elevated NA Fluoride Optical Fibers," *Japan J. Appl. Phys.*, Vol. 25, 1986, pp. L468–L470.

[11] Carter, S. F., M. W. Moore, D. Szebesta, J. R. Williams, D. Ranson, and P. W. France, "Low Loss Fluoride Fiber by Reduced Pressure Casting," *Electron. Lett.*, Vol. 26, 1990, pp. 2115–2117.

[12] Itoh, K., K. Miura, I. Masuda, M. Iwakura, and T. Yamashita, "Low-Loss Fluorozirco-Aluminate Glass Fiber," *Proc. 7th Int. Symp. Halide Glasses*, Lorne, Victoria, Australia, Mar. 17–21, 1991, pp. 2.7–2.12.

[13] Soufiane, A., Y. Messaddeq, and M. Poulain, "Optimization of Thorium-Free Indium Fluoride Glass Compositions and Application to Optical Fibers," *Proc. 9th Int. Symp. Non-Oxide Glasses*, Hangzhou, China, May 1994, pp. 459–464.

[14] Kanamori, T., Y. Terunuma, S. Takahashi, and T. Miyashita, "Chalcogenide Glass Fibers for Mid-Infrared Transmission," *IEEE J. Lightwave Tech.*, Vol. LT2, 1984, pp. 607–613.

[15] Kanamori, T., Y. Terunuma, S. Takahashi, and T. Miyashita, "Transmission Loss Characteristics of $As_{40}S_{60}$ and $As_{38}Ge_5Se_{57}$ Glass Unclad Fibers," *J. Non-Crystalline. Solids*, Vol. 69, 1985, pp. 231–242.

[16] Vasilev, A. V., G. G. Devyatykh, E. M. Dianov, A. N. Guryanov, A. Y. Laptev, V. G. Plotnichenko, Y. N. Pyrkov, G. E. Snopatin, I. V. Skripachev, M. F. Churbanov, and V. G. Shipunov, "2-Layer Chalcogenide Glass Fiber Waveguides with Optical Losses Lower Than 30 dB/km," *Kvantovaya Elektron.*, Vol. 20, 1993, pp. 109–110, (Russian).

[17] Mimura, Y., Y. Okamura, Y. Komazawa, and C. Ota, "Growth of Fiber Crystals for Infrared Optical Waveguides," *Japan J. Appl. Phys.*, Vol. 19, 1980, pp. L269–L272.

[18] Takahashi, K., N. Yoshida, and M. Yokota, "Optical Properties of AgBr Polycrystalline Fibers," *Extended Abstracts 45th Autumn Meeting of the Japan Society of Applied Physics*, 1984, p. 49.

[19] Kachi, S., M. Kimura, H. Kikuchi, and K. Shiroyama, "Influence of Atmosphere Gas for Polycrystal-line Infrared Fibers," *Extended Abstracts 45th Autumn Meeting of the Japan Society of Applied Physics*, 1984, p. 50.

[20] Izawa, T., and S. Sudo, *Optical Fibers: Materials and Fabrication*, Dordrecht: D. Reidel, 1987.

[21] Zachariasen, W. H., "The Atomic Arrangement in Glass," *J. Amer. Chem. Soc.*, Vol. 54, 1932, pp. 3841–3851.

[22] Takahashi, H., I. Sugimoto, and T. Sato, "Germanium Oxide Glasses Optical Fiber Prepared by VAD Method," *Electron. Lett.*, Vol. 18, 1982, pp. 398–399.

[23] Lines, M. E., J. B. MacChesney, K. B. Lyons, A. J. Bruce, A. E. Miller, and K. Nassau, "Calcium Aluminate Glasses as Potential Ultralow-Loss Optical Materials at 1.5–1.9 μm," *J. Non-Crystalline Solids*, Vol. 107, 1989, pp. 251–260.

[24] Boniort, J. Y., C. Brehm, P. H. Dopont, D. Guignot, and C. LeSergent, "Infrared Glass Optical Fibers for 4 to 10 Micron Bands," *Proc. 6th European Conf. Optics Communication*, Sept. 16–19, 1980, pp. 61–64.

[25] Kanamori, T., K. Oikawa, S. Shibata, and T. Manabe, "BaF_2-CaF_2-YF_3-AlF$_3$ Glass Systems for Infrared Transmission," *Japan J. Appl. Phys.*, Vol. 20, 1981, pp. L326-L328.

[26] Izumitani, T., T. Yamashita, M. Tokida, K. Miura, and H. Tajima, "New Fluoroaluminate Glasses and Their Crystallization Tendencies and Physical, Chemical Properties," *Mater. Sci. Forum*, Vol. 19–20, 1987, pp. 19–26.

[27] Tokiwa, H., Y. Mimura, O. Shinbori, and T. Nakai, "A Core-Clad Composition for Crystallization-Free Fluoride Fibers," *IEEE J. Lightwave Tech.*, Vol. 3, 1985, pp. 569–573.

[28] Nishii, J., Y. Kaite, and T. Yamagishi, "Preparation and Properties of ZnF_2-InF_3-GaF_3-PbF_2-LaF_3 Glasses," *Phys. Chem. Glasses*, Vol. 30, 1989, pp. 55–58.

[29] Bartholomew, R. F., "Praseodymium-Doped Cadmium Mixed Halide Glasses for 1.3 Micron Amplification," *Proc. 9th Int. Symp. on Non-Oxide Glasses*, Hangzhou, China, May 1994, pp. 366–371.

[30] Kapany, N. S., and R. J. Simms, "Recent Developments of Infrared Fiber Optics," *Infrared Phys.*, Vol. 5, 1965, pp. 69–80.

[31] Nishii, J., S. Morimoto, I. Inagawa, R. Iizuka, T. Yamashita, and T. Yamagishi, "Recent Advances and Trends in Chalcogenide Glass Fiber Technology," *Proc. 7th Int. Symp. Halide Glasses*, Lorne, Victoria, Australia, Mar. 17–21, 1991, pp. 9.3–9.12.

[32] Loireau-Lozac'h, A. M., M. Guittard, and J. Flahaut, "Verres Formes les Sulfures L_2S_3 des Terres Rares avec le Sulfure de Gallium Ga_2S_3," *Mater. Res. Bull.*, Vol. 11, 1976, pp. 1489–1496.

[33] Bridges, T. J., J. S. Hasiak, and A. R. Strand, "Single-Crystal AgBr Infrared Optical Fibers," *Opt. Lett.*, Vol. 5, 1980, pp. 85–88.

[34] Tangonan, G., A. G. Pastor, and R. C. Pastor, "Single Crystal KCl Fibers for 10.6 μm Integrated Optics," *Appl. Opt.*, Vol. 12, 1980, pp. 1110–1111.

[35] Okamura, Y., Y. Mimura, Y. Komazawa, and C. Ota, "CsI Crystalline Fiber for Infrared Transmission," *Japan J. Appl. Phys.*, Vol. 19, 1980, pp. L649–L651.

[36] LaBelle, Jr., H. E., and A. I. Mlavsky, "Growth of Controlled Profile Crystals from the Melt: Part I Sapphire Filaments," *Mater. Res. Bull.*, Vol. 6, 1971, pp. 571–579.

[37] Burrus, C. A., and L. Stone, "Single-Crystal Fiber Optical Devices: a Nd: YAG Fiber Laser," *Appl. Phys. Lett.*, Vol. 26, 1975, pp. 318–320.

[38] Mitra, S. S., and B. Bendow, *Optical Properties of Highly Transparent Solids*, New York: Plenum Press, 1975.

[39] Osanai, H., T. Shinoda, T. Moriyama, A. Araki, M. Horiguchi, T. Izawa, and H. Takata, "Effects of Dopants on the Transmission Loss of Low-OH-Content Optical Fibers," *Electron. Lett.*, Vol. 12, 1976, pp. 549–550.

[40] J. D. Dow and D. Redfield, "Toward a Unified Theory of Urbach's Rule and Experimantal Absorption Edges," *Phys. Rev. B*, Vol. 5, 1972, pp. 549–610.

[41] B. Bendow, "Transparency of Bulk Halide Glasses," in *Fluoride Glass Fiber Optics*, I. D. Aggarwal and G. Lu (eds.), New York: Academic Press, 1991, pp. 127–129.

[42] T. Tomiki, "Optical Constants and Exciton States in KCl Single Crystal II—The Spectra and Reflectivity and Absorption Constants," *J. Phys. Soc. Japan*, Vol. 23, 1967, pp. 1280–1296.

[43] Martienssen, W., "Uber Die Excitonenbanden der Alkalihalogenid-kristalle," *J. Phys. & Chem. Solids*, Vol. 2, 1957, p. 257.

[44] Nagasawa, M., and S. Shinoya, "Urbach's Rule Exhibited in SnO_2," *Solid State Comm.*, Vol. 7, 1969, pp. 1731–1733.

[45] Haupt, U., "Uber Temperaturabhangigkeit und Form der Langwellingsten Exitonenbande in KI—Kristallen," *Z. Phys.*, Vol. 157, 1959, pp. 232–246.

[46] Miyata, T., and T. Tomiki, "The Urbach Tail and Reflection Spectra of NaCl Single Crystal," *J. Phys. Soc. Japan*, Vol. 22, 1967, pp. 209–218.

[47] Wood, D. L., and J. Tauc, "Weak Absorption Tail in Amorphous Semiconductors," *Phys. Rev. B*, Vol. 5, 1972, pp. 3144–3151.

[48] Mori, H., and T. Izawa, "A New Loss Mechanism in Ultra-Low-Loss Optical Fiber Materials," *J. Appl. Phys.*, Vol. 51, 1980, pp. 2270–2271.

[49] Theye, M. L., *Physics of Structurally Disordered Solids*, S. Mitra (ed.), New York and London: Plenum Press, 1979.

[50] Paul, W., G. A. N. Connell, and R. J. Temkin, "Properties of Tetrahedrally Bonded Amorphous Semiconductors," *Adv. in Phys.*, Vol. 22, 1973, pp. 531–580.

[51] Vasko, A., *The Physics of Selenium and Tellurium*, W. C. Cooper (ed.), New York: Pergamon Press, 1969, p. 241.

[52] Tauc, J., F. J. DiSalvo, G. E. Peterson, and D. L. Wood, *Amorphous Magnetism*, H. O. Hooper and A. M. deGraaf (eds.), New York: Pergamon Press, 1973, p. 1991.

[53] Kanamori, T., Y. Terunuma, and T. Miyashita, "Preparation of Chalcogenide Optical Fiber," *Rev. ECL*, Vol. 32, 1984, pp. 469–477.

[54] Hass, M., and B. Bendow, "Residual Absorption in Infrared Materials," *Appl. Opt*, Vol. 16, 1977, pp. 2882–2890.

[55] Boyer, L. L., J. A. Harrington, M. Hass, and H. B. Rosenstock, "Multiphonon Absorption in Ionic Crystals," *Phys. Rev. B*, Vol. 11, 1975, pp. 1665–1680.

[56] Sparks, M., "Infrared Absorption by Higher-Order-Dipole-Moment Mechanism," *Phys. Rev. B*, Vol. 10, 1974, pp. 2581–2589.

[57] Bendow, B., H. G. Lipson, and S. P. Yukon, "Resial Lattice Absorption in Semiconducting Crystals: Frequency and Temperature Dependence," *Appl. Opt.*, Vol. 16, 1977, pp. 2909–2913.

[58] Bendow, B., "Fundamental Optical Phenomena in Infrared Window Materials," *Ann. Rev. Mater. Sci.*, Vol. 7, 1977, pp. 23–53.

[59] Bendow, B., *Solid State Physics*, Vol. 33, H. Ehrenreich, F. Seitz, and D. Turnbull (eds.), New York: Academic Press, 1978.

[60] Deutsch, T. F., "Absorption Coefficient of Infrared Laser Window Materials," *J. Phys. & Chem. Solids*, Vol. 34, 1973, pp. 2091–2104.

[61] Strom, U., J. R. Hendrickson, R. J. Wagner, and P. C. Tayler, "Disorder-Induced Far Infrared Absorption in Amorphous Materials," *Solid State Comm.*, Vol. 15, 1974, pp. 1871–1875.

[62] Pidgeon, C. R., G. D. Holah, F. Al-Berkdar, P. C. Taylor, and U. S. Strom, "Application of Submillimeter Waveguide Laser to the Study of Absorption in Elemental Amorphous Solids," *Infrared Phys.*, Vol. 18, 1987, pp. 923–927.

[63] Schroeder, J., R. Mohr, P. B. Macedo, and C. J. Montrose, "Rayleigh and Brillouin Scattering in $K_2O\text{-}SiO_2$ Glasses," *J. Amer. Ceram. Soc.*, Vol. 56, 1973, pp. 510–514.

[64] Lines, M. E., "Scattering Losses in Optical Fiber Materials. I A New Parameterization," *J. Appl. Phys.*, Vol. 55, 1984, pp. 4052–4057.

[65] Lines, M. E., "Scattering Loss in Optical Fiber Materials. II Numerical Estimate," *J. Appl. Phys.*, Vol. 55, 1984, pp. 4058–4063.

[66] Chu, B., *Laser Light Scattering*, New York: Academic Press, 1974.

[67] Wemple, S. H., and M. DiDomenico, "Behavior of the Electronic Dielectric Constant in Covalent and Ionic Materials," *Phy. Rev. B*, Vol. 3, 1971, pp. 1338–1351.

[68] Wemple, S. H., J. Gabbe, and G. D. Boyd, "Refractive-Index Behavior of Ternary Chalcopyride Semiconductors," *J. Appl. Phys.*, Vol. 46, 1975, pp. 3597–3605.

[69] Wemple, S. H., "Material Dispersion in Optical Fibers," *Appl. Opt.*, Vol. 18, 1979, pp. 31–35.

[70] Hattori, H., S. Sakaguchi, T. Kanamori, and Y. Terunuma, "Scattering Characteristics of Crystallies in Fluoride Optical Fibers," *Appl. Opt.*, Vol. 26, 1987, pp. 2683–2687.

[71] Hart, R. W., and E. W. Montroll, "On the Scattering of Plane Waves by Soft Obstacles. 1. Spherical Obstacles," *J. Appl. Phys.*, Vol. 22, 1951, pp. 376–386.

[72] Van de Hulst, H. C., "Rigorous scattering theory for spheres of arbitrary size (Mie theory)," *Light Scattering by Small Particles*, New York: Dover, 1981, pp. 114–130.

[73] Van de Hulst, H. C., "Particles small compared to the wavelength," *Light Scattering by Small Particles*, New York: Dover, 1981, pp. 63–84.

[74] Buck, J. A., *Fundamentals of Optical Fibers*, New York: Wiley-Interscience, 1995.

[75] Banford, C. R., "The Application of the Ligand Field Theory to Coloured Glasses," *Phys. Chem. Glasses*, Vol. 3, 1962, pp. 189–202.

[76] Smith, H. L., and A. J. Cohen, "Absorption Specra of Cations in Alkali-Silicate Glasses of High Ultra-Violet Transmission," *Phys. Chem. Glasses*, Vol. 4, 1963, pp. 173–187.

[77] Schultz, P. C., "Optical Absorption of the Transition Elements in Vitreous Silica," *J. Amer. Ceram. Soc.*, Vol. 57, 1974, pp. 309–313.

[78] Ohishi, Y., S. Mitachi, T. Kanamori, and T. Manabe, "Optical Absorption of 3d Transition Metal and Rare Earth Elements in Zircomium Fluoride Glasses," *Phys. Chem. Glasses*, Vol. 24, 1983, pp. 77–82.

[79] Kaiser, P., A. R. Tynes, H. W. Astle, A. D. Pearson, W. G. Frence, R. E. Yaeger, and A. H. Cherin, "Spectral Losses of Unclad Vitreous Silica and Soda-Lime-Silicate Fiber," *J. Opt. Soc. Amer.*, Vol. 63, 1973, pp. 1141–1148.

[80] Mitachi, S., and T. Miyashita, "Preparation of Low-Loss Fluoride Glass Fiber," *Electron. Lett.*, Vol. 18, 1982, pp. 170–171.

[81] Szebesta, D., S. T, Davey, J. R. Williams, and M. W. Moore, "OH Absorption in the Low Loss Windows of ZBLAN(P) Glass Fibre," *Proc. 8th Int. Symp. Halide Glasses*, Perros-Guirec, France, Sept. 22–24, 1992, pp. 26–32.

[82] Patek, K., *Glass Lasers*, London: Butterworth, 1970.

[83] Snitzer, E., "Glass Lasers," *Appl. Opt.*, Vol. 5, 1966, pp. 1487–1499.

[84] Poole, S. B., D. N. Payne, and M. E. Fermann, "Fabrication of Low-Loss Optical Fibers Containing Rare-Earth Ions," *Electron. Lett.*, Vol. 21, 1985, pp. 737–738.

[85] Fujiura, K., T. Kanamori, Y. Ohishi, Y. Terunuma, K. Nakagawa, S. Sudo, and K. Sugii, "Fabrication of Low-Loss and High-Δn Single-Mode Fluoride Fiber for 1.3 μm Praseodymium-Doped Fiber Amplifiers," *Appl. Phys. Lett.*, Vol. 67, 1995, pp. 3063–3065.

[86] Doremus, R. H., "Molecular Structure," *Glass Science*, New York: John Wiley & Sons Inc., 1973, pp. 23–43.

[87] Sun, K. H., "Fundamental Condition of Glass Formation," *J. Amer. Ceram. Soc.*, Vol. 30, 1947, pp. 277–281.

[88] Rawson, H., *Properties and Applications of Glass, Glass Science and Technology 3*, Amsterdam: Elsevier, 1980, pp. 26–31.

[89] Simpson, J. R., and J. B. MacChesney, "Optical Fibers with Alminum Oxide-Doped Silicate Core Composition," *Electron. Lett.*, Vol. 19, 1983, pp. 261–262.

[90] Ainslie, B. J., S. P. Craig, S. T. Davey, and B. Wakefield, "The Fabrication, Assesment and Optical Properties of High-Concentration Nd^{3+} and Eu^{3+}-Doped Silica-Based Fibers," *Mater. Lett.*, Vol. 6, 1988, pp. 139–144.

[91] Riebling, E. F.,"Structure of Molten Oxides. II. A Density Study of Binary Germataes Containing Li_2O, Na_2O, and Rb_2O," *J. Chem. Phys.*, Vol. 39, 1963, p. 3022.

[92] Nagel, S. R., J. B. MacChesney, and K. L. Walker, "An Overview of the Modified Chemical Vapor Deposition (MCVD) Processes and Performance," *IEEE J. Quant. Electron.*, Vol. QE18, 1982, pp. 459–476.

[93] Ainslie, B. J., K. J. Beales, C. R. Day, and J. D. Rush, "The Design and Fabrication of Monomode Optical Fiber," *IEEE J. Ouant. Electron.*, Vol. QE18, 1982, pp. 514–523.

[94] Apling, A., *Electronics and Structural Properties of Amorphous Semiconductors*, P. Comber, and J. Mort (eds.), New York: Academic Press, 1973, p. 234.

[95] Tsuchihashi, S., and Y. Kawamoto, "Properties and Structure of Sulfide Glasses. III Structure of Glasses in the Arsenic-Sulfur System," *Yogyo Kyoukai-Shi*, Vol. 77, 1969, pp. 35–39.

[96] Apling, A., and A. Leadbetter, *Amorphous Liquid Semiconductors*, J. Stuke and W. Brenig (eds.), London: Taylor and Francis, 1974, p. 457.

[97] Kreidl, N. J., "Glass-Forming System," in *Glass: Sience and Technology, Vol. 1*, D. R. Uhlmann and N. J. Kreidl (eds.), New York: Academic Press, 1983, p. 241.

[98] Simmons, J. H., C. J. Simmons, and R. Ochoa, "Fluoride Glass Structure," in *Fluoride Glass Fiber Optics*, I. D. Aggarwal and G. Lu (eds.), New York: Academic Press, Inc., 1991, pp. 37–84.

[99] Yasui, I., and H. Inoue, "Structure of Heavy Metal Fluoride Glasses," *Proc. 4th Int. Symp. Hailde Glasses*, Montery, Jan. 26–29, 1987, pp. 86–92.

[100] Poulain, M., *Fluoride Glasses*, A. E. Comyns (ed.), New York: Wiley, 1989.

[101] France, P. W., and M. C. Brierley, "Fluoride fibre laser amplifiers," *Optical Fiber Lasers & Amplifiers*, P. W. France (ed.), London: Blackie, 1991, pp. 183–211.

[102] Aggarwal, I. D., and G. Lu, *Fluoride Glass Fiber Optics*, Boston, Academic Press Inc., 1991, pp. 4–5.

[103] Barandiaran, J. M., and J. Colmenero, "Continuous Cooling Approximation for the Formation of a Glass," *J. Non-Cryst. Solids*, Vol. 46, 1981, pp. 277–287.

[104] Kanamori, T., Y. Terunuma, K. Fujiura, Y. Ohishi, and S. Sudo, "Fabrication of Low-Loss, High-Δn, Pr^{3+}-Doped Fluoride Single-Mode Fibers for 1.3 μm Optical Amplifiers," in *Extended Abstracts of 9th International Symposium on Non-Oxide Glasses*, Hangzhou, China, May 24–28, 1994, 25-PM-B-2, pp. 74–79.

[105] Hammel, J. J., *Advances in Nucleation and Crystallization in Glasses*, L. L. Hench and S. W. Freiman (eds.), The American Ceramic Society, Ohio, 1971, pp. 1–9.

[106] Avrami, M., *Kinetics of Phase Change*, Vol. 7, 1939, pp. 1103–1112.

[107] Johnson, W. A., and R. F. Mehl, "Reaction Kinetics in Processes of Nucleation and Growth," *Tran. AIME*, Vol. 135, 1939, pp. 416–442.

[108] MacMillan, P. W., *Glass-Ceramics*, New York: Academic Press, 1964, p. 229.

[109] Lu, G., P. Hart, and I. Aggarwal, "Calculation of Scattering Loss as a Function of Cooling Rate for Fluorozirconate Glass," *Phys. Chem. Glasses*, Vol. 31, 1990, pp. 245–255.

[110] Drehman, A. J., "Crystallite Formation in Fluoride Glasses," *Mater. Sci. Forum*, Vol. 19–20, 1987, pp. 483–490.

[111] Shineider, H. W., and A. Staudt, "Influence of Various Techiniques on Optical Losses in Fluoride Glass Fibers," in *Proc. 6th Int. Symp. Halide Glasses*, Oct. 1–5, 1989, (Clausthal-Zellerfeld, Germany), pp. 57–70.

[112] Hopgood, A. A., and G. Rosman, "Effect of Thermal History on Crystal Size Distributions and Scattering in Fluroide Glass Fibers," in *Proc. 7th Int. Symp. Halide Glasses*, Mar. 17–21, 1991, (Lorne, Australia), pp. 3.13–3.18.

[113] Besse, L. E., G. Lu, D. C. Tran, and G. H. Sigel, Jr., "A Combined DSC/Optical Microscopy Study of Crystallization in Fluorozirconate Glasses upon Cooling From the Melt, " *Mater. Sci. Forum*, Vol. 5, 1985, pp. 219–228.

[114] Sakaguchi, S., "Evaluation of Extrinsic Dcattering Loss due to Crystallization in Fluoride Optical Fibers," *J. Lightwave Tech.*, Vol. 11(2), 1993, pp. 187–191.

[115] Sakaguchi, S., Y. Terunuma, Y. Ohishi, and T. Kanamori, "Fluoride Fibre Drawing with Improved Tensile Strength," *J. Mater. Sci. Lett.*, Vol. 6, 1987, pp. 1063–1065.

[116] MacFarlane, D. R., "Continuous Cooling (CT) Diagrams and Critical Cooling Rates: A Direct Method of Calculation Using the Concept of Additivity," *J. Non-Cryst. Solids*, Vol. 53, 1982, pp. 61–72.

[117] Carter, S. F., and J. M. Parker, "Computer Similated Casting of Fluoride Glass Preforms and Crystal Growth," *Glass Tech.*, Vol. 31, 1990, pp. 245–255.

[118] Hyde, J. F., "Method of Making a Transparent Article of Silica," U. S. Patent, 2.272.342 (Filed Aug. 27, 1934. Patented Feb. 10, 1942).

[119] Nordberg, M. E., "Glass Having an Expansion Lower than that of Silica," U. S. Patent, 2.326.059 (Filed Apr. 22, 1939. Patented Aug. 3, 1943).

[120] Keck, D. B., and P. C. Schultz, "Method of Producing Optical Waveguide Fibers," U. S. Patent, 3.711.262 (Filed May 11, 1970).

[121] Kapany, F. P., D. B. Keck, and R. D. Maurer, "Radiation Losses in Glass Optical Waveguides," *Technical Digest of Conference on Trunk Telecommunications by Guided Waves*, 29 Sept.–2 Oct., 1970, pp. 148–153.

[122] Kapany, F. P., D. B. Keck, and R. D. Maurer, "Radiation Losses in Glass Optical Waveguides," *Appl. Phys. Lett.*, Vol. 17, 1970, pp. 423–425.

[123] MacChesney, J. B., P. B. O'Connor, F. V. DiMarcello, J. R. Simpson, and P. D. Lazay, "Preparation of Low Loss Optical Fibers Using Simultaneous Vapor Phase Deposition and Fusion," *Proceedings of the International Congress on Glass*, Vol. 6, 1974, pp. 40–45.

[124] French, W. G., J. B. MacChesney, P. B. O'Connor, and G. W. Tasker, "Optical Waveguides with Very Low Losses," *Bell Syst. Tech. J.*, Vol. 53 , May–June 1974, pp. 951–954.

[125] MacChesney, J. B., and P. B. O'Connor, "Optical Fiber Fabrication and Resulting Production," U. S. Patent, 4.217.027 (Filed Aug. 29, 1977. Patented Aug. 12, 1980).

[126] Keck, D. B., and P. C. Schultz, "Method of Forming Optical Waveguide Fibers," U. S. Patent, 3.737.292 (Filed Jan. 3, 1972. Patented June 5, 1973).

[127] Schultz, P. C., "Fabrication of Optical Waveguide by the Outside Vapor Deposition Process," *Proc. IEEE*, Vol. 68. 1980, pp. 1187–1190.

[128] Izawa, T., S. Kobayashi, S. Sudo, and F. Hanawa, "Continuous Fabrication of High Silica Fiber Preform," *Technical Digest of IOOC'77 (Tokyo, Japan)*, July 1977, Paper C1–1.

[129] Izawa, T., T. Miyashita, and F. Hanawa, "Continuous Optical Fiber Preform Fabrication Method," U. S. Patent, 4.062.665 (Filed Apr. 5, 1977. Patented Dec. 13, 1977).

[130] Izawa, T., S. Sudo, and F. Hanawa, "Continuous Fabrication Process for High Silica Fiber Preforms," *Trans. IECE Japan*, Vol. E62, 1979, pp. 779–785.

[131] Schultz, P. C., "Vapor Phase Materials and Processes for Glass Optical Waveguides" in *Fiber Optics*, B. Bendow and S. S. Mitra (eds.), New York and London: Plenum Press, 1979.

[132] *CRC Handbook of Chemistry and Physics*, 52nd ed., Cleveland: CRC Press, 1973, D171–D176.

[133] Samsonov, G. V., *The Oxide Handbook*, New York: IFI/Plenum, 1973, p. 178.

[134] Sudo, S., "Studies on the Vapor-Phase Axial Deposition Method for Optical Fiber Fabrication," Ph.D. Thesis, University of Tokyo, Tokyo, 1982, p. 164.

[135] Kobayashi, S., H. Nakagome, N. Shimizu, H. Tsuchiya, and T. Izawa, "Low Loss Optical Glass Fiber with Al_2O_3-SiO_2 Core," *Electron. Lett.*, Vol. 10, Oct. 1975, pp. 410–411.

[136] Audsley, A., and R. K. Bayliss, "The Induced Plasma Torch as a High-Temperature Chemical Reactor, I. Oxidation of Silicon Tetrachloride," *J. Appl. Chem.*, Vol. 19, 1969, pp. 33–38.

[137] Nassau, K., T. C. Rich, and J. W. Schiever, "Low Loss Fused Silica Made by the Plasma Torch," *Appl. Opt.*, Vol. 13, 1974, pp. 744–745.

[138] Nassau, K., and J. W. Schiever, "Plasma Torch Preparation of High Purity, Low OH Content Fused Silica," *Am. Ceram. Soc. Bull.*, Vol. 54, 1975, pp. 1004–1011.

[139] Nassau, K., J. W. Shiever, and T. Krause, "Preparation and Properties of Fused Silica Containing Alumina," *J. Am. Ceram. Soc.*, Vol. 58, 1975, p. 461.

[140] Kobayashi, S., S. Sudo, T. Miyashita, and T. Izawa, "Preparation of Silica Glass Using a CO_2 Laser," *Appl. Opt.*, Vol. 14, 1975, pp. 2817–2818,.

[141] Kobayashi, S., S. Sudo, T. Miyashita, and T. Izawa, "Preparation of SiO_2-Al_2O_3 Glass Using a CO_2 Laser," presented at the *Japan Soc. Appl. Phys. Conf.*, Oct. 1975.

[142] Kobayashi, S., S. Sudo, T. Miyashita, and T. Izawa, "Preparation of Low-Loss Fibers Using a CO_2 Laser," presented at the *IECE Japan Nat. Conf.*, Paper 834, Mar. 1976.

[143] French, W. G., A. D. Pearson, G. W. Tasker, and J. B. MacChesney, "Low-Loss Fused Silica Optical Waveguide with Borosilicate Cladding," *Appl. Phys. Lett.*, Vol. 23, 1973, pp. 338–339.

[144] MacChesney, J. B., R. E. Jaeger, D. A. Pinnow, F. W. Ostermayer, T. C. Rich, and L. G. Van Uitert, "Low-Loss Fused Silica Optical Waveguide with Borosilicate Cladding," *Appl. Phys. Lett.*, Vol. 23, 1973, pp. 340–341.

[145] MacChesney, J. B., P. B. O'Connor, J. T. Simpson, and F. V. DiMarcello, "Multimode Optical Waveguide Having a Vapor Deposition Core of Germania Doped Bolosilicate Glass," *Am. Ceram. Soc. Bull.*, Vol. 52, 1973, p. 704.

[146] Kawachi, M., M. Horiguchi, A. Kawana and T. Miyashita, "OH-ion Distribution Profiles in Rod Preform of High Silica Optical Waveguides," *Electron. Lett.*, Vol. 13, 1977, pp. 247–249.

[147] Horiguchi, M., and M. Kawachi, "Measurement Technique of OH–ion Distribution Profile in Rod Preform of Silica Based Optical Fiber Waveguides," *Appl. Opt.*, Vol. 17, 1978, pp. 2570–2574.

[148] Walker, K. L., F. T. Geyling, and S. R. Nagel, "Thermophoretic Deposition of Small Particles in the Modified Chemical Vapor Deposition (MCVD) Process," *J. Amer. Ceramic Soc.*, Vol. 63, 1980, pp. 552–558.

[149] Miya, T., "Studies on Loss Reduction and Dispersion Minimization of Single Mode Optical Fibers for Long-Wavelength Region," Ph.D. Thesis, Tohoku Univ., Sendai, 1983.

[150] Yariv, A., *Optical Electronics*, 3rd ed., New York: Holt, Rinehart and Winston, 1985, p. 130.

[151] Shimazaki, E., and N. Kichizo, "Dampfdruckmessungen an Halogeniden der Seltenen Erden," *Z. Anorg. Allg. Chem.*, Vol. 314, 1962, pp. 21–34.

[152] Spedding, F. H., and A. H. Daane, *The Rare Earths*, New York: Wiley, 1961, p. 98.

[153] Sicre, J. E., J. T. Dubois, K. J. Eisentraut, and R. E. Sievers, "Volatile Lanthanide Chelates: II. Vapor Pressures, Heats of Vaporization, and Heats of Sublimation," *J. Amer. Chem. Soc.*, Vol. 91, 1969, pp. 3476–3481.

[154] Simpson, J., "Fabrication of Rare-Earth Doped Glass Fibers," SPIE, Vol. 1171, 1989, pp. 2–7.

[155] Simpson, J. R., "Rare Earth Doped Fiber Fabrication: Techniques and Physical Properties," in *Rare Earth Doped Fiber Lasers and Amplifiers*, M. J. F. Digonnet (ed.), New York: Marcel Dekker, 1993, pp. 1–18.

[156] Craig-Ryan, S. P., and B. J. Ainslie, "Glass Structure and Fabrication Techniques," in *Optical Fiber Lasers and Amplifiers*, P. W. France (ed.), Glasgow: Blackie, 1991, pp. 50–78.

[157] Poole, S. B., D. N. Payne, R. J. Mears, M. E. Fermann, and R. I. Laming, "Fabrication and Characterization of Low–Los Optical Fibers Contaning Rare–Earth Ions," *J. Lightwave Tech.*, Vol. LT-4, 1986, pp. 870–876.

[158] Ainslie, B. J., S. P. Craig, and S. T. Davey, "The Fabrication and Optical Properties of Nd^{3+} in Silica-Based Optical Fibres," *Mat. Lett.*, Vol. 5, 1988, pp. 143–146.

[159] Morse, T. F., L. Reinhart, A. Kilian, W. Risen, Jr., and J. W. Cipolla, Jr., "Aerosol Doping Technique for MCVD and OVD," SPIE, Vol. 1171, 1989, pp. 72–79.

[160] Townsend, J. E., S. B. Poole, and D. N. Payne, "Solution Doping Technique for Fabrication of Rare-Earth Doped Optical Fibers," *Electron. Lett.*, Vol. 23, 1987, pp. 329–331.

[161] Cognolato, L., B. Sordo, E. Modone, A. Gnazzo, and G. Cocito, "Aluminum/Erbium Active Fibre Manufactured by a Non-Aqueous Solution Doping Method," SPIE, Vol. 1171, 1989, pp. 202–208.

[162] Craig-Ryan, S. P., B. J. Ainslie, and C. A. Millar, "Fabrication of Long Lengths of Low Excess Loss Erbium-Doped Optical Fibre," *Electron. Lett.* Vol. 26, Feb. 1990, pp. 185–186.

[163] Stone, J., and C. A. Burrus, "Neodymium-Doped Silica Lasers in End-Pumped Fiber Geometry," *Appl. Phys. Lett.*, Vol. 23, 1973, pp. 388–389.

[164] Izawa, T., F. Hanawa, S. Kobayashi, S. Shibata, S. Sudo, and T. Miyashita, "Optical Fibers Fabricated by the Vapor-Phase Verneuil Method," *IECE Jpn. Nat. Conf.*, Tech. Dig., 1977, p. 792, (Japanese).

[165] Verneuil, A., "Peproduction Artificielle Du rubis Par Fusion," *Nature (Paris)*, Vol. 32, 1904, p. 77.

[166] Sudo, S., F. Hanawa, M. Kawachi, T. Edahiro, and M. Nakahara, "Method of Fabricating MultiMode Optical Fiber Preforms," U. S. Patent. 4.367.085 (Filed Oct. 1980).

[167] Sudo, S., F. Hanawa, M. Kawachi, and T. Edahiro, "Torch for Synthesizing Fine Glass Particles," Japanese Patent 1.121.387, 1982.

[168] Sudo, S., M. Kawachi, and F. Hanawa, "Torch for Synthesizing Fine Glass Particles," Japanese Patent 1.103.644, 1981.

[169] Kawachi, M., S. Tomaru, S. Sudo, and T. Edahiro, "Fabrication Method of Single-Mode Optical Fiber Prefrom," U.S. Patent 4.345.928 (Filed Sept. 19, 1980).

[170] Sudo, S., M. Kawachi, and F. Hanawa, "Vapor-Phase Axial Deposition Method for Optical Fiber Preform Fabrication IV," Private Communication Paper in ECL, No. 14176, 1979, (Japanese).

[171] French, W. G., and L. J. Pase, "Chemical Kinetics of the Modified Chemical Vapor Deposition Process," *IOOC'77*, Tokyo, Tech. Dig., 1977, Paper C1–1.

[172] French, W. G., L. J. Pace, and V. A. Foertmeyer, "Chemical Kinetics of the Reactions of $SiCl_4$, $SiBr_4$, $GeCl_4$, $POCl_3$, and BCl_3 with Oxygen," *J. Phys. Chem.*, Vol. 82, 1978, pp. 2191–2194.

[173] Powers, D. R., "Kinetics of SiCl4 Oxidation," *J. Amer. Ceram. Soc.*, Vol. 61, 1978, pp. 295–297.

[174] Hastie, J. W., "Molecular Basis of Flame Inhibition," *J. Res. Natl. Bur. Standards—A. Phys. Chem.*, Vol. 77A, 1973, pp. 733–754.

[175] Day, M. J., G. Dixon-Lewis, and K. Thompson, "Flame Structure and Flame Reaction Kinetics VI," *Proc. Roy. Soc. London Ser. A.*, Vol. 330, 1972, pp. 199–218.

[176] Frenkel, J., "Viscous Flow of Crystalline Bodies under the Action of Surface Tension," *J. Phys. (USSR)*, Vol. 9, 1945, pp. 385–391.

[177] Sherer, G. W., "Sintering of Low Density Glass I," *J. Amer. Ceram. Soc.*, Vol. 61, 1977, pp. 236–239.

[178] Sudo, S., T. Edahiro, and M. Kawachi, "Sintering Process of Porous Preforms Made by a VAD Method for Optical Fiber Fabrication," *Trans. IECE Japan*, Vol. E63, 1980, pp. 731–737.

[179] Perkins, W. G., and D. R. Begeal, "Diffusion and Permeation of He, Ne, Ar, Kr and D_2 Through Silicon Oxide Thin Film," *J. Chem. Phys.*, Vol. 54, 1971, pp. 1683–1694.

[180] Petzold, A., "Versuch zur Klassifizierung Von Komponenten Nach Ihrem Einfluß Auf Die Oberfachenspannung Vcon Silikatschmelzen," *Silikattechnik 5.* Jg. Helf 1, 1954, pp. 11–12.

[181] Kech, D. B., R. D. Maurer, and P. C. Schultz, "On the Ultimate Lower Loss Limit of Attenuation in Glass Optical Waveguide," *Appl. Phys. Lett.*, Vol. 221, 1973, pp. 307–309.

[182] Hair, H. L., "Hydroxyl Groups on Silica Surface," *J. Non-Cryst. Solids,* Vol. 19, 1975, pp. 299–309.

[183] Peri, J. B., "Infrared Study of OH and NH_3 Groups on the Surface of a Dry Silica Aerogel," *J. Phys. Chem.,* Vol. 70, 1966, pp. 2937–2945.

[184] Little, L. H. (Hasegawa et al., transl.), "Adhesion and Infrared Absorption Spectrau," *Kagaku Doginsha,* 1975.

[185] Sudo, S., M. Kawachi, T. Izawa, T. Edahiro, T. Shioda, and H. Gotoh, "Low OH Content Optical Fiber Fabricated by Vapor-Phase Axial Deposition Method," *Electron. Lett.,* Vol. 17, 1978, pp. 534–535.

[186] Edahiro, T., M. Kawachi, S. Sudo, and H. Takata, "OH-ion Reduction in VAD Fiber," *Electron. Lett.,* Vol. 15, 1979, pp. 482–483.

[187] Edahiro, T., M. Kawachi, S. Sudo, and N. Inagaki, "OH-ion Reduction in the Optical Fibers Fabricated by the Vapor-Phase Axial Deposition Method," *Trans. IECE Japan,* Vol. E63, 1980, pp. 574–580.

[188] Shackelford, J. F., and J. S. Masaryk, "The Thermodynamics of Water and Hydrogen Solubility in Fused Silica," *J. Non.-Cryst. Solids,* Vol. 211, 1976, pp. 55–56.

[189] Sudo, S., M. Kawachi, T. Edahiro, and N. Inagaki, "Dehydration and Consolidation Techniques in the Vapor-Phase Axial Deposition Method," *ECL Tech. J.,* Vol. 29, 1980, pp. 1719–1729.

[190] Pauling, L., *General Chemistry,* New York: Dover, 1970, p. 182.

[191] Uchida, N., N. Uesugi, Y. Murakami, M. Nakahara, T. Tanifuji, and N. Inagaki, "Infrared Loss Increase in Silica Optical Fiber due to Chemical Reduction of Hydrogen," *Tech. Digest of ECOC'83,* 1983, Postdeadline Paper.

[192] Hanawa, F., S. Sudo, M. Kawachi, and M. Nakahara, "Fabrication of Completely OH-Free VAD Fiber," *Electron. Lett.,* Vol. 16, 1981, pp. 699–700.

[193] Sudo, S., S. Kobayashi, T. Izawa, and Y. Masuda, "Fabrication of Graded-Index Fiber Preforms by the Vapor-Phase Veruneuil Method," *Nat. Conf. IECE Japan, Tech. Digest,* Mar. 1977, Paper 795.

[194] Kawachi, M., S. Sudo, N. Shibata, and T. Edahiro, "Deposition Properties of SiO_2-GeO_2 Particles in the Fame Hydrolysis Reaction for Optical Fiber Fabrication," *Japan J. Appl. Phys.,* Vol. 19, 1980, pp. 169–171.

[195] Borreli, N., "The Infrared Spectra of SiO_2-GeO_2 Glass," *Phys. Chem. Glasses,* Vol. 10, 1969, pp. 43–45.

[196] Tenny, A. S., and J. Wong, "Vibrational Spectra of Vapor-Deposited Binary Borosilicate Glasses," *J. Chem. Phys.,* Vol. 56, 1972, pp. 5516–5523.

[197] Smith, C. F., Jr., R. A. Condrate, Sr., and W. E. Votava, "The Difference Infrared Spectra of Titanium-Containing Vitreous Silica," *Appl. Spectrosc.,* Vol. 29, 1975, pp. 79–81.

[198] ASTM cards.

[199] Kawachi, M., S. Sudo, and T. Edahiro, "Threshold Gas Flow Rate of Halide Material for the Formation of Oxide Particles in the VAD Process for Optical Fiber Fabrication," *Japan J. Appl. Phys.,* Vol. 20, 1981, pp. 709–712.

[200] Sudo, S., M. Kawachi, T. Edahiro, and K. Chida, "21.2 km Graded-Index VAD Fiber with Low-Loss and Wide-Bandwidth," *Electron. Lett.,* Vol. 16, 1980, pp. 151–153.

[201] Sudo, S., M. Kawachi, H. Suda, M. Nakahara, and T. Edahiro, "Refractive-Index Profile Control Techniques in the Vapor-Phase Axial Deposition Method," *Trans. IECE Japan,* Vol. E64, 1981, pp. 536–543.

[202] Nakahara, M., S. Sudo, N. Inagaki, K. Yoshida, S. Shibuya, K. Kokura, and T. Kuroha, "Ultra Wide Bandwidth VAD Fiber," *Electron. Lett.,* Vol. 16, 1980, pp. 301–302.

[203] Shibata, S., and S. Takahashi, "Effect of Some Manufacturing Conditions on the Optical Loss of Compound Glass Fibers" *J. Non-Cryst. Solids,* Vol. 23, 1977, p. 111.

[204] Rapp, C. F., "Laser Glasses: 17.1 Bulk Glasses," *CRC Handbook of Laser Science and Technology, Supplement 2: Optical Material,* M. J. Weber (ed.), CRC Press, 1995, pp. 619–634.

[205] Hall, D. W., and M. J. Weber, "Glass Lasers," *CRC Handbook of Laser Science and Technology, Supplement 1: Lasers,* M. J. Weber (ed.), Boca Raton, FL: CRC Press, 1991, pp. 137–158.

[206] Snitzer, E., "Optical Maser Action of Nd^{+3} in a Barium Crown Glass," *Phys. Rev. Lett.*, Vol. 7, 1961, pp. 444–446.

[207] Takahashi, S., S. Shibata, and M. Yasu, "Preparation of Low Loss Multi-Component Glass Fiber," *Rev. ECL*, Vol. 27, 1979, pp. 123–130.

[208] Takahashi, S., "Studies on the Qualification of Glass Fibers for Optical Communications," Ph.D. Thesis, Waseda University, Tokyo, 1979.

[209] Cowan, J. H., W. M. Buehl, and J. R. Hutchins, "III: An Electrochemical Theory for Oxygen Reboil," *J. Amer. Ceram. Soc.*, Vol. 49, 1966, pp. 559–565.

[210] Pinnow, D. A., A. L. Gentile, A. G. Standlee, A. J. Timper, and L. M. Hobrock, "Polycrystalline Fiber Optical Waveguide for Infrared Transmission," *Appl. Phys. Lett.*, Vol. 33, 1978, pp. 28–29.

[211] Van Uitert, L. G., and S. H. Wemple, "$ZnCl_2$ Glass: A Potential Ultralow-Loss Optical Fiber Material," *Appl. Phys. Lett.*, Vol. 33, 1978, pp. 57–59.

[212] Goodman, C. H. L., "Devices and Materials for 4 μm Band Fiber Optical Communication," *IEEE J. Solid-State Circuits*, Jt. issue, Vol. 2, 1978, pp. 129–137.

[213] Shibata, S., M. Horiguchi, K. Jinguji, S. Mitachi, K. Kanamori, and T. Manabe, "Prediction of Loss Minima in the Infra-Red Optical Fibers," *Electron. Lett.*, Vol. 17, 1981, pp. 775–777.

[214] Poulain, M., M. Poulain, J. Lucas, and P. Brun, "Verres Fluores au Tetrafluorure de Zirconium Proprietes Optiques d'un Verre Dope au Nd^{3+}," *Mater. Res. Bull.*, Vol. 10, 1975, pp. 243–246.

[215] Mitachi, S., and T. Manabe, "Fluoride Glass Fiber for Infrared Transmission," *Japan J. Appl. Phys.*, Vol. 19, 1980, pp. L313–L314.

[216] Sun, K. H., "Fluoride glass," U.S. Patent 2.466.509, 1949.

[217] Imaoka, M., "Fluoride Glass," Seisan Kenkyuu, Vol. 7, 1955, pp. 14–19, (Japanese).

[218] Videau, J. J., J. Portier, and B. Piriou, "Sur de Nouveaux Verres Aluminofluorés," *Rev. Chim. Min.*, Vol. 16, 1979, pp. 393–399.

[219] Miranday, J. P., C. Jacoboni, and R. De Pape, "Glasses Containing Fluorine, Their Preparation and Application," U.S. Patent 4.328.318, 1980.

[220] Christensen, P. S., P. Le Gall, G. Fonteneau, J. Lucas, J. Y. Boniort, J. Leboucq, D. Tregoat, and H. Poignant, "Attempts on Fibering Indium Based Fluoride Glass," *Mater. Sci. Forum*, Vols. 32–33, 1988, pp. 467–470.

[221] Layne, C. B., W. H. Lowdermilk, and M. J. Weber, "Multiphonon Relaxation of Rare-Earth Ions in Oxide Glasses," *Phys. Rev. B*, Vol. 16, 1977, pp. 10–20.

[222] Kanamori, T., and S. Takahashi, "Crystallization Characteristics of ZrF_4-Based Fluoride Glasses in Cooling and Reheating Processes," *Japan J. Appl. Phys.*, Vol. 24, 1985, pp. L758–L760.

[223] Kanamori, T., "Study on Infrared Transmitting Glass Fibers," Ph.D. Thesis, Nagoya University, Nagoya, 1988, (Japanese).

[224] Mitachi, S., and P. A. Tick, "Oxygen Effects on Fluoride Glass Stability," *Mat. Sci. Forum*, Vols. 32–33, 1988, pp. 197–202.

[225] Kanamori, T., "Recent Advances in Fluoride Glass Fiber Optics in Japan," *Mat. Sci. Forum*, Vol. 19–20, 1987, pp. 363–373.

[226] Hu, H., F. Lin, Y. Yuan, and J. Feng, "Crystallization of Fluoroaluminate Glasses," *Proc. 5th. Int. Symp. Halide Glasses*, Shizuoka, Japan, May 29–June 2, 1988, pp. 370–374.

[227] Miura, K., I. Masuda, M. Tokida, and T. Yamashita, "Chlorine Doped Fluorozirco-Aluminate Glass," *Mat. Sci. Forum*, Vols. 32–33, 1988, pp. 367–372.

[228] Mitachi, S., T. Miyashita, and T. Manabe, "Preparation of Fluoride Optical Fibers for Transmission in the Mid-Infrared," *Phys. Chem. Glasses*, Vol. 23, 1982, pp. 196–201.

[229] France, P. W., S. F. Carter, M. W. Moore, and C. R. Day, "Progress in Fluoride Fibers for Optical Communications," *Br. Telecom Tech. J.*, Vol. 5, 1987, pp. 28–44.

[230] Aggarwal, I. D., and G. Lu (eds.), *Fluoride Glass Fiber Optics*, San Diego: Academic Press, 1991, p. 30.

[231] Tesar, A., R. C. Bradt, and C. G. Pantano: "Effects of NaF upon the Mechanical Properties of Fluorozirconate Glasses," *Proc. 2nd Int. Symp. on Halide Glasses*, Troy, NY, Aug. 2–5, 1983, Paper No. 49.

[232] Jinguji, K., M. Horiguchi, S. Shibata, T. Kanamori, S. Mitachi, and T. Manabe, "Material Dispersion in Fluoride Glasses," *Electron. Lett.*, Vol. 18, 1982, pp. 164–165.

[233] Mitachi, S., Y. Terunuma, Y. Ohishi, and S. Takahashi, "Reduction of Impurities in Fluoride Glass Optical Fibers," *Japan J. Appl. Phys.*, Vol. 22, 1983, pp. L537–L538.

[234] Mitachi, S., Y. Terunuma, Y. Ohishi, and S. Takahashi, "Reduction of Impurities in Fluoride Glass Fibers," *IEEE J. Lightwave Tech.*, Vol. LT-2, 1984, pp. 587–592.

[235] Fujiura, K., Y. Ohishi, S. Sakaguchi, and T. Terunuma, " Synthesis of High-Purity ZrF_4 by Chemical Vapor Deposition," *Proc. 5th. Int. Symp. Halide Glasses*, Shizuoka, Japan, May 29–June 2, 1988, pp. 174–179.

[236] Kobayashi, K., "Preparation of Highly Pure Lithium and Sodium Fluorides Using Solvent Extraction," *Bull. Chem. Soc. Japan*, Vol. 61, 1988, pp. 2965–2966.

[237] Tatsuno, T., "Synthesis of iron-impurity free ZrF_4," *Mat. Sci. Forum*, Vols. 19–20, 1987, pp. 181–186.

[238] Kobayashi, K., "Preparation of High-Purity Barium Hydrogenfluoride for Fluoride-Optical Fiber," *Nippon Kagaku Kaishi*, 1995, pp. 478–482, (Japanese).

[239] Ohishi, Y., S. Mitachi, and S. Takahashi, "Influence of Humidity During the Preparation Process on Transmission Loss for ZrF_4-Based Optical Fibers," *Mat. Res. Bull.*, Vol. 19, 1984, pp. 673–679.

[240] Mitachi, S., Y. Ohishi, and S. Takahashi, "Fabrication of Low OH and Low Loss Fluoride Optical Fiber," *Japan J. Appl. Phys.*, Vol. 23, 1984, pp. L726–L727.

[241] Takahashi, S., T. Kanamori, Y. Ohishi, K. Fujiura, and Y. Terunuma, "Reduction of Oxygen Impurity in ZrF_4-Based Fluoride Glass," *Mat. Sci. Forum*, Vols. 32–33, 1988, pp. 87–92.

[242] Sakaguchi, S., T. Kanamori, Y. Terunuma, Y. Ohishi, and S. Takahashi, "Photoluminescence at 0.5 μm Wavelength in Fluoride Optical Fibers," *Japan J. Appl. Phys.*, Vol. 25, 1986, pp. L481–L483.

[243] Robinson, M., R. C. Pastor, R. R. Turk, D. P. Devor, M. Braunstein, and R. Braunstein, "Infrared-Transparent Glasses Derived from the Fluorides of Zirconium, Thorium and Barium," *Mat. Res. Bull.*, Vol. 15, 1980, pp. 735–742.

[244] Ohishi, Y., S. Sakaguchi, and S. Takahashi, "Photoluminescence and Absorption of a Zirconium Fluoride Glass and ZrF_4," *J. Amer. Ceram. Soc.*, Vol. 70, 1987, pp. C81–C83.

[245] Robinson, M., "Preparation and Purification of Fluoride Glass Starting Materials," *Mat. Sci. Forum*, Vol. 5, 1985, pp. 19–34.

[246] Noda, Y., T. Nakai, N. Norimatsu, Y. Mimura, O. Shinbori, and H. Tokiwa, "Reactive Atmosphere Processing of Fluoride Glasses," *Mat. Sci. Forum*, Vols. 32–33, 1988, pp. 1–8.

[247] Mitachi, S., T. Miyashita, and T. Kanamori, "Fluoride-Glass-Cladded Optical Fibers for Mid-Infra-Red Ray Transmission," *Electron. Lett.*, Vol. 17, 1981, pp. 591–592.

[248] Sakaguchi, S., and S. Takahashi, "Low-Loss Fluoride Optical Fibers for Mid Infrared Optical Communication," *IEEE J. Lightwave Tech.*, Vol. LT-5, 1987, pp. 1219–1228.

[249] Ohishi, Y., "Transmission Loss Characteristics of Fluoride Glass Single-Mode Fiber," *Electron. Lett.*, Vol. 22, 1986, pp. 1034–1035.

[250] Tran, D. C., C. F. Fisher, and G. H. Sigel, "Fluoride Glass Preforms Prepared by a Rotational Casting Process," *Electron. Lett.*, Vol. 18, 1982, pp. 657–658.

[251] Miura, K., I. Masuda, K. Itoh, and T. Yamashita, "Fluorozirco-Aluminate Glass Fiber," *Proc. 6th Int. Symp. on Halide Glasses*, Clausthal-Zellerfeld, Oct. 1989, pp. 407–412.

[252] Tokiwa, H., Y. Mimura, T. Nakai, and O. Shinbori, "Fabrication of Long Single-Mode and Multimode Fluoride Glass Fibers by the Double-Crucible Technique," *Electron. Lett.*, Vol. 21, 1985, pp. 1131–1132.

[253] Schneider, H. W., "Strength in Fluoride Glass Fibers," *Mat. Sci. Forum*, Vols. 32–33, 1988, pp. 561–570.

[254] Sakaguchi, S., Y. Hibino, Y. Ohishi, and S. Takahashi, "Surface Flaws in ZrF_4-Based Fluoride Optical Fibers with Improved Tensile Strength," *J. Mat. Sci. Lett.*, Vol. 6, 1987, pp. 1440–1442.

[255] Schneider, H. W., A. Schoberth, A. Staudt, and C. Gerndt, "Fluoride Glass Etching Method for Preparation of Infra-Red Fibers with Improved Tensile Strength," *Electron. Lett.*, Vol. 22, 1986, pp. 949–950.

[256] Kanamori, T., Y. Terunuma, K. Fujiura, K. Oikawa, and S. Takahashi, "Fabrication of Fluoride Single-Mode Optical Fibers," *NTT R&D*, Vol. 39, 1990, pp. 1353–1362, (Japanese).

[257] Hibino, Y., and H. Hanafusa, "Consolidation-Atmosphere Influence on Drawing-Induced Defects in Pure Silica Optical Fibers," *IEEE J. Lightwave Tech.*, Vol. 6, 1984, pp. 172–178.

[258] Mitachi, S., Y. Ohishi, Y. Terunuma, and S. Takahashi. "Preparation of Fluoride Optical Fibers," *ECL Tech. J.*, Vol. 32, 1983, pp. 2732–2736, (Japanese).

[259] Lau, J., A. M. Nakata, and J. D. Mackenzie, "Influence of Drawing Temperature of the Strength of Fluoride Glass Fibers," *J. Non-Cryst. Solids*, Vol. 70, 1985, pp. 233–242.

[260] Klein, P. H., P. C. Pureza, W. I. Roberts, and I. D. Aggarwal, "Strengthening of ZBLAN Glass Fibers by Preform Treatment with Active Fluorine," *Mat. Sci. Forum*, Vols. 32–33, 1988, pp. 571–576.

[261] Fujiura, K., K. Hoshino, T. Kanamori, Y. Nishida, Y. Ohishi, and S. Sudo, "Reliability of Fluoride Fibers for Use in Fiber Amplifiers," *Tech. Digest of Optical Amplifiers and Their Applications*, Vol. 18, Davos, Switzerland, June 15–17, 1995, ThE4–1, pp. 104–107.

[262] Sanghera, J. S., C.-J. Chu, and J. D. Mackenzie, "Temperature Related Volumetric Changes in Fluorozirconate Glasses and Melts," *Mat. Sci. Forum*, Vols. 19–20, 1987, pp. 685–692.

[263] Sarhangi, A., and D. A. Thompson, "Chemical Vapor Deposition of Fluoride Glasses," *Proc. Int. Symp. Halide Glasses*, Monterey, CA, July 26–29, 1987, pp. 274–282.

[264] Fujiura, K., Y. Nishida, H. Sato, S. Sugawara, K. Kobayashi, Y. Terunuma, and S. Takahashi, "Plasma-Enhanced Chemical Vapor Deposition of ZrF_4-Based Fluoride Glasses," *J. Non-Cryst. Solids*, Vol. 161, 1993, pp. 14–17.

[265] Jacoboni, C., "Vapor Phase Deposition of Rare-Earth Doped PZG Glasses," *Proc. Int. Symp. Non-Oxide Glasses*, Hangzhou, China, May 24–28, 1994, pp. 302–307.

[266] Shibata, S., Y. Terunuma, and T. Manabe, "Ge-P-S Chalcogenide Glass Fibers," *Japan J. Appl. Phys.*, Vol. 19, 1980, pp. L603–L605.

[267] Miyashita, T., and Y. Terunuma, "Optical Transmission Loss of As-S Glass Fiber in 1.0-5.5 μm Wavelength Region," *Japan. J. Appl. Phys.*, Vol. 21, 1982, pp. L75–L76.

[268] Dianov, E. M., V. G. Plotnichenko, G. G. Devyatykh, M. F. Churbanov, and I. V. Scripachev, "Middle-Infrared Chalcogenide Glass Fibers with Losses Lower Than 100 dB km^{-1}," *Infrared Phys.*, Vol. 29, 1995, pp. 303–307.

[269] Shibata, S., Y. Terunuma, and T. Manabe, "Sulfide Glass Fibers for Infrared Transmission," *Mat. Res. Bull.*, Vol. 16, 1981, pp. 703–714.

[270] Brehm, C., M. Cornebois, C. Le Sergent, and J. P. Parant, "Plastic-Clad Chalcogenide Glass Fibers," *J. Non-Cryst. Solids*, Vol. 47, 1982, pp. 251–254.

[271] Dianov, E. M., "Materials for Infrared Low Loss Fibers," *Adv. in IR Fibers II*, Vol. 320, Los Angeles, CA: SPIE, Jan. 1982, Paper 320–04.

[272] Devyatykh, G. G., M. F. Churbanov, I. V. Scripachev, E. M. Dianov, V. G. Plotnichenko, and Yu. N. Pirkov, "Low-Loss Core-Clad Chalcogenide Glass Optical Fibrs," *Proc. 8th Int. Symp. on Halide Glasses*, Perros-Guirec, France, Sept. 1992, pp. 535–537.

[273] Nishii, J., T. Yamashita, and T. Yamagishi, "Chalcogenide Glass Fiber with a Core-Cladding Structure," *Appl. Optics.*, Vol. 28, 1989, pp. 5122–5127.

[274] Kanamori, T., M. Asobe, T. Nishi, K. Oikawa, and S. Sudo, "Preparation of Chalcogenide Glass Single Mode Fibers for Optical Switches," *Extended Abstracts of the 54th Autumn Meeting of the Japan Society of Applied Physics*, 1993, p. 1036.

[275] Asobe, M., H. Itoh, T. Miyazawa, and T. Kanamori, "Efficient and Ultra All-Optical Switching Using High Δn, Small Core Chalcogenide Glass Fiber," *Electron. Lett.*, Vol. 29, 1993, pp. 1966–1967.

[276] Devyatykh, G. G., E. M. Dianov, V. G. Plotnichenko, I. V. Scripachev, G. E. Snopatin, and M. F. Churbanov, "Single-Mode As-S Chalcogenide Glasses Fiber Waveguide," *Kvantovaya Elektron.*, Vol. 22, 1995, pp. 287–288, (Russian).

[277] Katsuyama, T., K. Ishida, S. Satoh, and H. Matsumura, "Low Loss Ge-Se Chalcogenide Glass Optical Fibers," *Appl. Phys. Lett.*, Vol. 45, 1984, pp. 925–927.

[278] Klocek, P., M. Roth, and R. D. Rock, "Chalcogenide Glass Optical Fibers and Image Bundles: Properties and Applications," *Opt. Engrg.*, Vol. 26, 1987, pp. 88–95.

[279] Kadono, K., H. Higuchi, M. Takahashi, Y. Kawamoto, and H. Tanaka, "Upconversion Luminescence of Ga_2S_3-Based Sulfide Glasses Containing Er^{3+} Ions," *Proc. 9th Int. Symp. on Non-Oxide Glasses*, Hangzhou, China, May 1994, pp. 500–504.

[280] Lucazeau, G., S. Barnier, and A. M. Loireau-Lozac'h, "Spectres Vibrationnels, Transitions Electroniques et Structures a Courtes Distances dans les Verres: Sulfures de Terres Rares-Sulfure de Gallium," *Mat. Res. Bull.*, Vol. 12, 1977, pp. 437–448.

[281] Medeiros Neto, J. A., E. R. Taylor, J. Wang, B. N. Samson, D. W. Hewak, R. I. Laming, D. N. Payne, E. Tarbox, P. D. Maton, G. M. Roba, B. Kinsman, and R. Hanney, "The Application of Ga:La:S-Based Glass for Optical Amplification at 1.3 μm," *Proc. 9th Int. Symp. on Non-Oxide Glasses*, Hangzhou, China, May 1994, pp. 465–470.

[282] Kettlewell, B. R., B. E. Kinsman, A. R. Wilson, A. M. Pitt, J. A. Savage, and P. J. Webber, "An Assessment of the Technique of Hydrogen Distillation as a Production Process for Ge-As-Se Glasses," *J. Mat. Sci.*, Vol. 12, 1977, pp. 451–458.

[283] Katsuyama, T., S. Satoh, and H. Matsumura, "Fabrication of High-Purity Chalcogenide Glasses by Chemical Vapor Deposition," *J. Appl. Phys.*, Vol. 59, 1986, pp. 1446–1449.

[284] Ohishi, Y., A. Mori, T. Kanamori, K. Fujiura, and S. Sudo, "Fabrication of Praseodymium-Doped Arsenic Sulfide Chalcogenide Fiber for 1.3-μm Fiber Amplifiers," *Appl. Phys. Lett.*, Vol. 65, 1994, pp. 13–15.

[285] Saito, M., "Optical Loss Increase in an As-S Glass Infrared Fiber due to Water Diffusion," *Appl. Optics.*, Vol. 26, 1987, pp. 202–203.

[286] Bishop, S. G., U. Strom, and P. C. Taylor, "Optically Induced Localized Paramagnetic States in Chalcogenide Glasses," *Phys. Rev. Lett.*, Vol. 34, 1975, pp. 1346–1350.

[287] Heo, J., J. S. Sanghera, and J. D. Mackenzie, "Chalcohalide Glasses for Infrared Fiber Optics," *Opt. Engrg.*, Vol. 30, 1991, pp. 470–479.

[288] Garfunkel, J. S., R. A. Skogman, and R. A. Walterson, "Infrared Transmitting Fibers of Polycrystalline Silver Halides," *IEEE/OSA*, Laser Engineering and Application, Optical Communication 8.1, 1979.

[289] Peterson, G. E., "Optical Waveguide Material: the Present and the Future," *Topical Meeting on Optical Fiber Communication*, Washington D.C., Tech. Dig., 1979, Paper TuA4.

[290] Okumura, Y., Y. Mimura, Y. Komazawa, and C. Ota, "CsI Crystalline Fiber for Infrared Transmission," *Japan. J. Appl. Phys.*, Vol. 19, 1980, pp. L649–L652.

[291] Stone, J., C. A. Burrus, and A. G. Dentai, "Nd:YAG Single-Crystal Fiber Laser: Room-Temperature cw Operation Using a Single LED as an End Pump," *Appl. Phys. Lett.*, Vol. 29, 1976, pp. 37–39.

[292] Burrus, C. A., J. Stone, and A. G. Dentai, "Room-Temperature 1.3 μm CW Operation of a Glass-Clad Nd:YAG Single-Crystal Fiber Laser End Pumped with a Single LED," *Electron. Lett.*, Vol. 12, 1976, pp. 600–602.

[293] Stone, J., and C. A. Burrus, "Self-Contained LED-Pumped Single-Crystal Nd:YAG Fiber Laser," *Fiber and Intergrated Optics*, Vol. 2, 1979, pp. 19–46.

[294] Fejer, M., R. L. Byer, R. Feigelson, and W. Kway, "Growth and Characterization of Single Crystal Refractory Oxide Fibers," *Advances in Infrared Fibers II*, Vol. 320, Los Angeles, CA: SPIE, Jan. 1982, pp. 50–55.

[295] Fejer, M. M., J. L. Nightingale, G. A. Magel, and R. L. Byer, "Laser Assisted Growth of Optical Quality Single Crystal Fibers," *Processing of Guided Wave Optoelectronic Materials*, Vol. 460, Los Angeles, CA: SPIE, Jan. 1984, pp. 26–32.

[296] Fejer, M. M., J. L. Nightingale, G. M. Magel, and R. L. Byer, "Laser-Heated Miniature Pedestal Growth Apparatus for Single-Crystal Optical Fibers," *Rev. Sci. Instrum.*, Vol. 55, 1984, pp. 1791–1796.

[297] Feigelson, R. S., W. L. Kway, and R. K. Route, "Single-Crystal Fibers by the Laser-Heated Pedestal Growth Method," *Opt. Engrg.*, Vol. 24, 1985, pp. 1102–1107.

[298] Nightingale, J. L., and R. L. Byer, "A Guided Wave Monolithic Resonator Ruby Fiber Laser," *Opt. Comm.*, Vol. 56, 1985, pp. 41–45.

[299] Digonnet, M. J. F., C. J. Gaeta, and H. J. Shaw, "1.064- and 1.32-μm Nd:YAG Single Crystal Fiber Lasers," *IEEE J. Lightwave Tech.*, Vol. LT-4, 1986, pp. 454–460.

[300] Digonnet, M. J. F., C. J. Gaeta, D. O'Meara, and H. J. Shaw, "Clad Nd:YAG Fibers for Laser Applications," *IEEE J. Lightwave Tech.*, Vol. LT-5, 1987, pp. 642–646.

[301] Digonnet, M. J. F., and C. J. Gaeta, "Theoretical Analysis of Optical Fiber Laser Amplifiers and Oscillators," *Appl. Optics*, Vol. 24, 1985, pp. 333–342.

[302] Digonnet, M. J. F., "Closed-Form Expressions for the Gain in Three- and Four-Level Laser Fibers," *IEEE J. Quantum Electron.*, Vol. 26, 1990, pp. 1788–1796.

[303] Davis, G. M., I. Yokohama, S. Sudo, and K. Kubodera, "1.3-μm Nd:YAG Crystal Fiber Amplifiers," *IEEE Photon. Tech. Lett.*, Vol. 3, 1991, pp. 459–461.

[304] Nightingale, J. L., "The Growth and Optical Applications of Single-Crystal Fibers," Ph.D. Thesis, Stanford University, Stanford, CA, 1985.

[305] Fejer, M. M., "Single-Crystal Fibers: Growth Dynamics and Nonlinear Optical Interactions," Ph.D. Thesis, Stanford University, Stanford, CA, 1987.

[306] Magel, G. A., "Optical Second Harmonic Generation in Lithium Niobate Fibers," Ph.D. Thesis, Stanford University, Stanford, CA, 1990.

[307] TA-NWT-001312, "Generic Requirements for Optical Fiber Amplifier Performance and Reliability," *Bellcore Technical Advisory*, Issue 2, 1993.

[308] Vita, P. D., "Status and Perspectives in Optical Amplifiers Standardisation," *Tech. Digest of OAA'95*, Vol. 18, Davos, Switzland, June 1995, Paper ThA2-1, pp. 3–6.

[309] TA-NWT-001221, "Generic Requirements for Passive Fiber Optic Component Reliability Assurance Practices," *Bellcore Technical Advisory*, Issue 1, 1992.

[310] TA-NWT-000020, "Generic Requirements for Optical Fiber and Optical Fiber Cable," *Bellcore Technical Advisory*, Issue 4, 1989.

[311] TA-NWT-000039, "Quality Program Analysis (QPA)," *Bellcore Technical Advisory*, Issue 3, 1989.

[312] TA-NWT-000326, "Generic Requirements for Optical Fiber Connectors and Connectorized Jumper Cables," *Bellcore Technical Advisory*, Issue 2 and Bulletin, 1991.

[313] TA-NWT-000357, "Component Reliability Assurance Requirements for Telecommunications Equipment," *Bellcore Technical Advisory*, Issue 1, 1987, and TW-NWT-001217, Issue 1, 1991.

[314] TA-NWT-000909, "Generic Requirements and Objectives for Fiber in the Loop," *Bellcore Technical Advisory*, Issue 1, 1990.

[315] TA-NWT-001209, "Generic Requirements for Fiber Optic Branching Devices," *Bellcore Technical Advisory*, Issue 1, 1991.

[316] Dugan, M. P., "Initiating Qualification Program Requirements for Passive Fiber Components," *SPIE*, Vol. 1791, 1992, pp. 202–216.

[317] Griffith, A. A., "The Phenomena of Rupture and Flaw in Solids," *Philos. Trans. Roy. Soc.*, Vol. 211, 1920, p. 163.

[318] Saifi, M. A., "Long-Term Fiber Reliability," *SPIE*, Vol. 1366, 1990, pp. 58–70.

[319] Irwin, G. R., "Fracture," *Encyclopedia of Physics*, Vol. 6, 1958, pp. 551–590.

[320] Evans, A. G., and S. M. Wiederhorn, "Proof Testing of Ceramic Materials—An Analytical Basis for Failure Prediction," *Int. J. Fract.*, Vol. 10, 1974, pp. 379–392.

[321] Chandon, C., and D. Kalish, "Temperature Dependence of Static Fatigue of Optical Fibers," *J. Amer. Ceram. Soc.*, Vol. 65, 1982, pp. 171–173.

[322] Krause, J., and C. Shute, "Temperature Dependence of the Transition in Static Fatigue of Fused Silica Optical Fiber," *Adv. Ceram. Mat.*, Vol. 3, 1988, pp. 118–121.

[323] Yuce, H. H., and F. P. Kapron, "Use of Fatigue Measuremants for Fiber Lifetime Prediction," *SPIE*, Vol. 1366, 1990, pp. 144–156.

[324] France, P. W., W. J. Duncan, D. J. Smith, and K. J. Beales, "Strength and Fatigue of Multicomponent Optical Glass Fibers," *J. Mat. Sci.*, Vol. 18, 1983, pp. 785–792.

[325] Matthewson, M. W., and C. R. Kurkjian, "Environmental Effects and Static Fatigue of Silica Fiber," *J. Amer. Ceram. Soc.*, Vol. 71, 1988, pp. 177–183.

[326] Griffioen, W., W. Ahn, A. T. De Boer, and G. Segers, "Stress-Induced and Stress-Free Aging of Optical Fibers in Water," *Proc. Int. Wire Cable Symp.*, Vol. 40, 1991, pp. 673–678.

[327] Haslov, P., K. B. Jensen, and N. H. Skovgaard, "Degradation of Stressed Optical Fibers in Water: New Worst-Case Lifetime Estimation Model," *J. Amer. Ceram. Soc.*, Vol. 77, 1994, pp. 1531–1536.

[328] Kurkjian, C. R., and D. Inniss, "Understanding Mechanical Properties of Lightguides: A Commentary," *Opt. Engrg.*, Vol 30, 1991, pp. 681–689.

[329] Lopez, A. R., "The Effects of Coating Moisture Permeability on the Mechanical Reliability," *SPIE*, Vol. 2290, 1994, pp. 42–51.

[330] Arai, S., A. Oyobe, A. Umeda, and K. Kokura, "Long-Term Reliability of Carbon-Coated Erbium-Doped Amplifer Fiber," *Tech. Digest OEC'94*, 1994, Paper 13B2–4, pp. 43–43.

[331] Bubnov, M. M., E. M. Dianov, A. M. Prokhorov, S. L. Semjonov, A. G. Schebunyaev, and C. R. Kurkjian, "High-Strength Carbon-Coated Optical Fibre," *Sov. Lightwave Comm.*, Vol. 2, 1992, pp. 245–250.

[332] Yonezawa, N., H. Tada, and Y. Katsuyama, "Strength Improvement and Fusion Splicing for Carbon-Coated Optical Fiber," *J. Lightwave Tech.*, Vol. 9, 1991, pp. 417–421.

[333] Oohashi, K., T. Shimomichi, S. Arai, and N. Satoh, "A High-Strength Carbon-Coated Optical Fiber Manufactured by Controlling Hydrogen Radicals in a CVD Reaction," *Proc. ECOC'91*, 1991, Paper MoB1–3, pp. 33–36.

[334] Bubnov, M. M., and S. L. Semijonov, "Strength of Carbon and Dual Hermetically Coated Fibers at Ambient and High (>400°C) Temperatures," *Proc. EUROPTO*, Vol. 1973, 1993, pp. 244–249.

[335] Pitt, N. J., and A. Marshall, "Long-Term Loss Stability of Single-Mode Optical Fibers Exposed to Hydrogen," *Electron. Lett.*, Vol. 20, 1984, pp. 512–514.

[336] Tomita, A., and P. J. Lemaire, "Hydrogen-Induced Loss Increases in Germanium-Doped Single-Mode Optical Fibres: Long-Term Predictions," *Electron. Lett.*, Vol. 21, 1985, pp. 71–72.

[337] Noguchi, K., N. Shibata, N. Uesugi, and Y. Negishi, "Loss Increase for Optical Fibers Exposed to Hydrogen Atmosphere," *J. Lightwave Tech.*, Vol. 3, 1985, pp. 236–243.

[338] Lemaire, P. J., and A. Tomita, "Behavior of Single-Mode MCVD Fibers Exposed to Hydrogen," *Proc. 10th ECOC*, 1984, pp. 306–307.

[339] Lemaire, P. J., H. A. Watson, D. J. DiGiovanni, and K. L. Walker, "Prediction of Long-Term Hydrogen-Induced Loss Increases in Er-Doped Amplifier Fibers," *IEEE Photon. Tech. Lett.*, Vol. 5, 1993, pp. 214–217.

[340] Lemaire, P. J., D. P. Monroe, and H. A. Watson, "Hydrogen-Induced-Loss Increases in Erbium-Doped Amplifier Fiber," *Tech. Digest of OFC'94*, Vol. 4, Feb. 1994, pp. 301–302.

[341] Lemaire, P. J., H. A. Watson, D. J. DiGiovanni, and K. L. Walker, "Hydrogen-Induced-Loss Increases in Hermitic and Nonhermetic Erbium-Doped Amplifier Fiber," *Tech. Digest of OFC/IOOC'93*, Vol. 4, Feb. 1993, pp. 53–54.

[342] Griscom, D. L., and E. J. Friebele, "Effects of high energy radiation on halide glasses," *Fluoride Glass Fiber Optics*, I. D. Aggarwal and G. Lu (eds.), San Diego, CA: Academic Press, 1991, pp. 307–350.

[343] Fukuda, C., Y. Chigusa, T. Kashiwada, M. Onishi, and H. Kanamori, "Effect of γ-Ray Irradiation on Erbium-Doped Fibers," *Tech. Digest of OEC'94*, Vol. 4, July 1994, pp. 44–45.

[344] Levin, K. H., D. C. Tran, R. J. Ginther, and G. H. Sigel, Jr., "Optical Properties of Fiber and Bulk Zirconium Fluoride Glass," *Glass Tech.*, Vol. 24, 1983, pp. 143–145.

[345] Ohishi, y., S. Mitachi, S. Talahashi, and T. Miyashita, "γ-Ray Irradiation Effect on Transmission Loss for ZrF_4-Based Optical Fibers," *Electron. Lett.*, Vol 19, 1983, pp. 830–831.

[346] Freibele, E. J., G. H. Sigel, Jr., and M. E. Gingerich, "Radiation-Induced Optical Absorption Spectra of Fiber Optic Waveguide in the 0.4–1.7 μm Region," in *Fiber Optics*, B. Bendow and S. S. Mitra (eds.), New York: Plenum Press, 1979, pp. 355–367.

Chapter 5

Amplification Characteristics of a Fiber Amplifier: Components, Design, and Amplification Characteristics of a Fiber Amplifier Module

M. Yamada
M. Shimizu

5.1 INTRODUCTION

This chapter describes the practical characteristics of fiber amplifiers such as gain spectrum, saturation output power, noise spectrum, and distortion. A knowledge of these characteristics allows us to understand the operation and limitations of fiber amplifiers. On the other hand, these features are also very important when constructing transmission systems that employ fiber amplifiers as in-line repeaters, booster amplifiers, and preamplifiers because they are used to analyze the maximum repeater spacing and the maximum number of available amplifiers in terms of data format, the *signal-to-noise* (S/N) ratio of an optical transmitter, receiver sensitivity, and transmission loss.

The characteristics can be basically predicted from steady-state rate equations and differential equations for the power change rates of the pump, signal, and *amplified spontaneous emission* (ASE) light propagated along a fiber in the guided mode. Rate equations, which are derived from the spectroscopic data presented in Chapter 4, are used to estimate populations in the energy levels in rare-earth ions under any pump or signal power conditions. From these calculated populations we can derive a gain coefficient for a signal light and the absorption coefficient for a pump light by considering the absorption and stimulated emission cross sections. The gain and the absorption coefficients vary with the longitudinal position

---

of the fiber, just as the pump energy that determines populations in an energy level is absorbed along the fiber length. The actual signal gain and noise powers are given by the integral of the gain coefficient over the whole fiber length.

Theoretical simulation methods are very important when designing a fiber amplifier, but a technical knowledge of the amplifier components is also important. This chapter aims to provide an overview of some of the most important components and to describe the overall performance of Nd-, Er-, Pr-, and Tm-doped fiber amplifiers.

Section 5.2 outlines optical components related to fiber amplifiers. They include filters, wavelength-selective couplers, isolators, and various pump sources.

Section 5.3 describes theoretical and experimental performance of individual fiber amplifiers.

5.2 FIBER AMPLIFIER RELATED DEVICES

A fiber amplifier is an optically pumped amplifier with an end pump scheme. The amplifier is made of a strand of fiber doped with rare-earth ions, a pump source, a multiplexing device for the pump and signal lights, and various optical components. Although various types of fiber amplifier have been reported, they can be classified into three categories in terms of pumping scheme. Figure 5.1 shows the three basic amplifiers with a single-pass configuration corresponding to unidirectional and bidirectional pumping techniques. With the single-pass configuration, the incident signal light grows with a one-way amplification. This configuration includes one or two pump lasers with a fiber output (also called fiber pigtail), one or two multiplexing devices that combine the pump and signal light, one strand of rare-earth-ion-doped fiber, and two polarization-independent optical isolators. All the devices except for the rare-earth-ion-doped fiber have a pigtail made of either a standard or a dispersion-shifted single-mode silica fiber. Therefore, a fusion splicing technique is commonly used to connect these devices. The rare-earth-ion-doped fibers are sometimes made from nonsilica glass. They usually have smaller cores and higher *numerical apertures* (NA) than the pigtail fiber of other devices, even when they are made from silica glass. Therefore, splicing is an important technique for assembling practical fiber amplifier modules.

In a double-pass (or reflective) fiber amplifier, a reflective device (for example, mirror and fiber grating) is placed at the output end to reflect the signal [1,2]. Figure 5.2 is a diagram of the double-pass configuration. An optical circulator is used in place of the optical isolator.

The following subsections describe major devices and technologies for fiber amplifiers.

5.2.1 Fiber Grating

Fiber devices incorporating a grating structure are unique devices for use as wavelength-selective reflection mirrors, optical notch filters, and optical taps. To date,

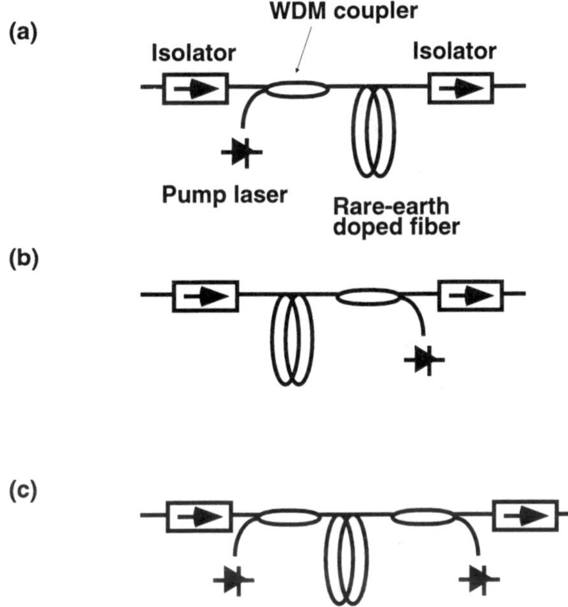

Figure 5.1 Schematic diagram of single-pass-type fiber amplifier: (a) forward pump scheme, (b) back-ward pump scheme, and (c) bidirectional pump scheme.

two types of fiber-grating structure have been reported. One is a fiber device, as shown in Figure 5.3, that consists of a fine-relief Bragg grating on the surface of a side-polished single-mode fiber [3–5], and the other is a Bragg grating written with UV light [6,7]. If we compare these two Bragg grating fiber devices, the UV-written grating has a great advantage because of its low insertion loss, ease of fabrication, and high reflectivity at the Bragg wavelength.

The major method for fabricating a UV-written grating is the side writing technique [7–11]. The side writing techniques that have been reported are shown in Figure 5.4. The technique, in which the fiber is exposed to a UV interference pattern from the side, enables us to fabricate a fiber Bragg grating with any desired Bragg resonance wavelength.

The mechanism of the UV-induced refractive index change is not completely understood. Several processes have been proposed based on the theory that Ge-Si bonds are broken by irradiation of the UV light [12–14]. This results in the disappearance of the 244-nm absorption band, the creation of negatively charged color centers and their absorption in the UV region, and an increase in the density of the glass. These phenomena cause the refractive index to change. Three techniques are effective in enhancing the photosensitivity: the highly germanium-doped technique, the flame deoxidization technique, and the hydrogen loading technique

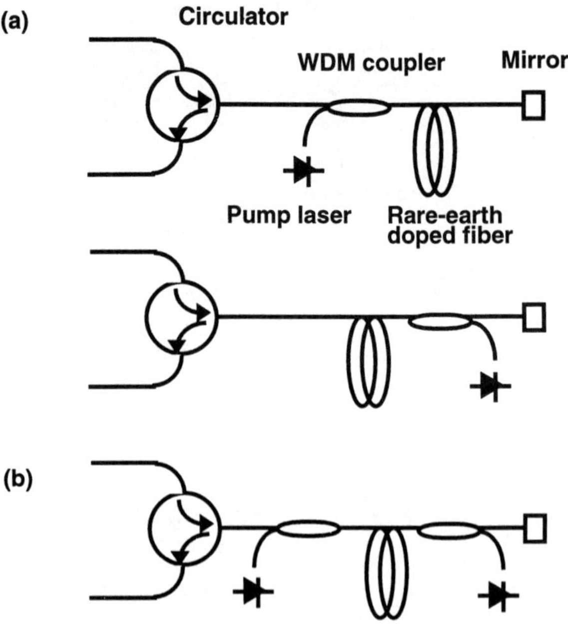

Figure 5.2 Schematic diagram of double-pass-type fiber amplifier: (a) single-directional pump scheme and (b) bidirectional pump scheme.

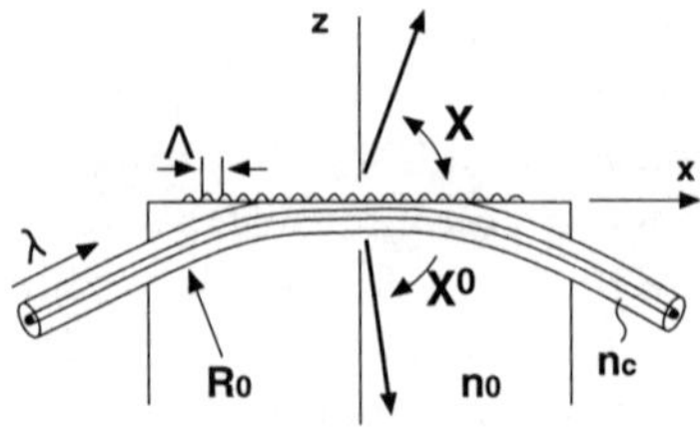

Figure 5.3 Schematic configuration of fine-relief-type fiber grating [3]. R_0, radius of curved fiber; n_c, refractive index of the cladding.

[15,16]. The UV-induced refractive index change exceeds 10^{-2} for ordinary commercial germanium-doped silica fiber after hydrogen loading for a week under the pressure of 200 atm at room temperature.

Figure 5.5 shows an example of a transmission spectrum of a fiber grating.

5.2.2 Thermally Expanded Core Fiber

The high NA structure of rare-earth-doped fiber enables us to construct a fiber amplifier with highly efficient gain characteristics. However, a high NA structure

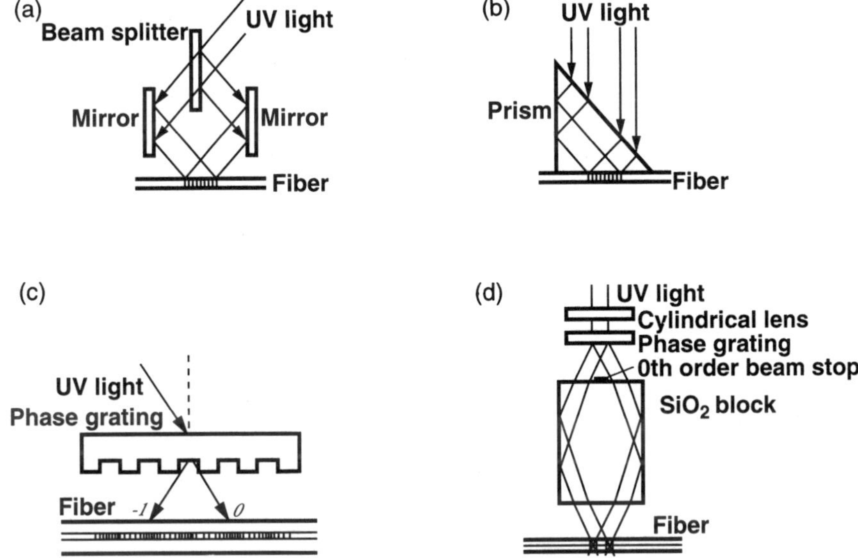

Figure 5.4 Fabrication method of UV-written grating: (a) holographic method [8], (b) prism interferometer method [9], (c) phase mask method [10], and (d) phase mask method (two beam type) [11].

reduces the mode-field diameter and results in an increase in the splicing loss between an active fiber and a transmission fiber. The use of single-mode optical fiber tapers, whose mode-field diameter is locally increased, is a powerful way to reduce the splicing loss caused by the mode-field mismatch [17].

The refractive index distribution of the fiber taper changes gradually along the fiber axis, and the spot size of the propagating mode also changes. Possible methods for realizing this type of mode conversion are to taper the fiber diameter with a constant core/cladding ratio [18,19] or to diffuse dopant material in the fiber by heat treatment [*thermally expanded core* (TEC) process] [20,21]. With the former method, the normalized frequency changes at the tapered position. This means the transmission losses of the fiber would be very sensitive to mechanical perturbations.

Figure 5.6 shows the schematic configuration of the TEC fiber [22]. Due to the diffusion of dopant materials, the core diameter gradually increases, and the mode field diameter of the propagated light is also expanded. In comparison with the taper method, the normalized frequency for the TEC fiber does not change along the fiber. In addition, the outer diameter of the TEC fiber does not change. Figure 5.7 shows the calculated transition losses of Gaussian-tapered TEC fibers as a function of the taper length [22]. It has been confirmed that transition losses are negligible for a core expansion ratio of 2 with a tapered length of more than

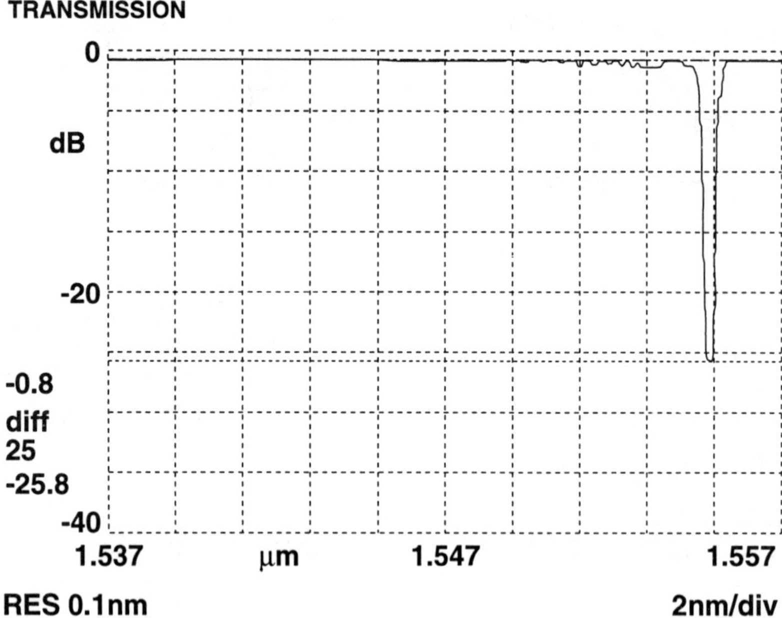

Figure 5.5 Typical transmission spectrum of a UV-written grating.

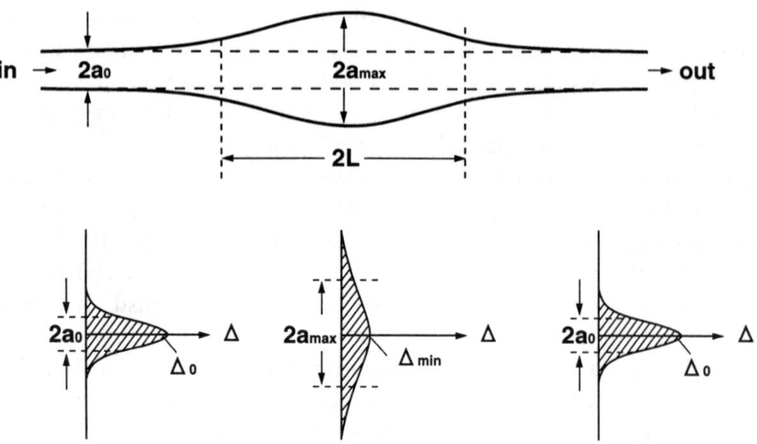

Figure 5.6 Schematic diagram of the Gaussian-tapered TEC fiber [22]. D_0, D_{min}, relative refractive-index difference of the fiber before and after TEC treatment; a_0, a_{max}, core radius of the fiber before and after TEC treatment; L, taper length of the TEC fiber.

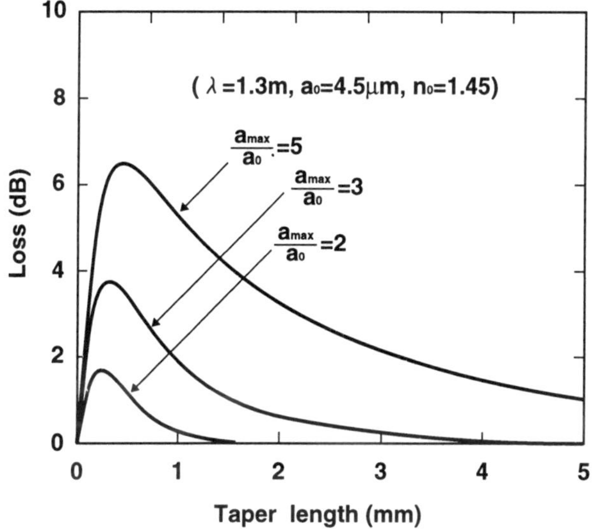

Figure 5.7 Calculàted transmission losses of Gaussian-tapered TEC fibers [22].

2 mm. For a core expansion ratio of 3, a tapered length of more than 5 mm enables us to achieve zero transition loss.

Of the several fabrication procedures that have been reported, localized heat treatment using flames from a microburner is particularly useful [17,20]. In the method, a fiber whose resin coating layer has been removed is heated directly by propane/oxygen flames produced by a microburner. The fiber tension, the position of the burner in relation to the fiber, and pressure at which the fuel gases are supplied are determined appropriately so that the fiber is heated without deforming the cladding. The heating temperature is up to 1,700°C. The mode-field diameter of standard single-mode fiber was broadened to 20 μm with a 5- to 10-min heat treatment.

5.2.3 Filters

Many types of optical filter are used in fiber amplifier modules and optical communication systems. A narrow bandpass filter is used to eliminate ASE components from an amplified signal and is indispensable for constructing preamplifiers with a high S/N ratio. A high-pass filter is useful for eliminating residual pump power from an amplifier signal. Dielectric thin-film multilayer filters are most commonly used to fabricate these devices [23].

5.2.3.1 Dielectric Thin-Film Multilayer Filter

First, we consider a parallel-sided, isotropic film with a refractive index of n_1 between media with indexes of n_0 and n_2 (substrate) [24]. This is a single-layer coated filter. Figure 5.8 shows a ray diagram of light incident to the plane at an angle F_0. The amplitude reflectance and transmittance are obtained from a summation of the multiply reflected beams. The result is

$$R = \frac{r_2 + r_1 e^{-2i\delta_1}}{1 + r_1 r_2 e^{-2i\delta_1}} \tag{5.1}$$

$$T = \frac{t_1 t_2 e^{-2i\delta_1}}{1 + r_1 r_2 e^{-2i\delta_1}} \tag{5.2}$$

$$2\delta_1 = \frac{4\pi}{\lambda} n_1 d \cos \Phi_1 \tag{5.3}$$

where R is the reflectance; T is the transmittance; r_1, r_2, t_1, t_2 are the Fresnel coefficients at the n_0/n_1 and n_1/n_2 interfaces; d_1 is the phase thickness of the

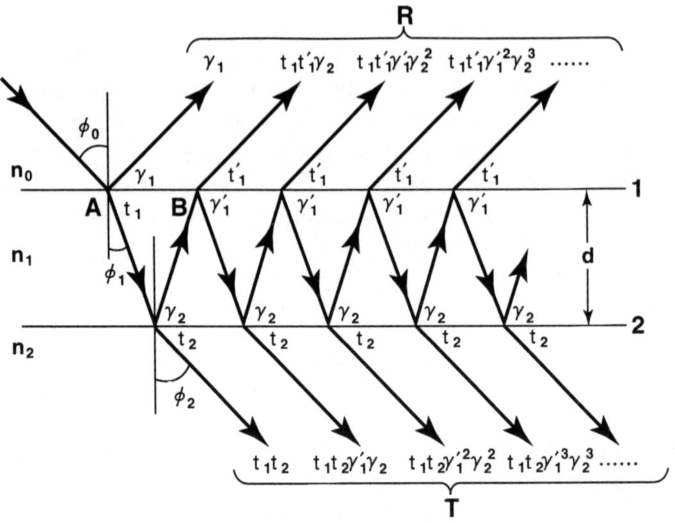

Figure 5.8 Ray diagram of a parallel-sided, isotropic film with a refractive index of n_1 between media with indexes of n_0 and n_2 [24].

film; and λ is the wavelength in a vacuum. It is clear that the reflectance and transmittance are affected by the incident angle.

In the special case of normal incidence and transparent media, the normal reflectance exhibits oscillatory variation as shown in Figure 5.9.

For $n_2 > n_1 > n_0$, the reflectance of the single-layer coated filter is lower than that of a noncoated filter. The optical thickness for the minimum reflectance is

$$n_1 d = \frac{(2m + 1)\lambda}{4} \tag{5.4}$$

where m is an integer.

Two kinds of $\lambda/4$ film with different refractive indexes are used to realize a multilayer filter. Since the low reflection of a $\lambda/4$ film results from the phase agreement between the beams reflected at the two surfaces, the same characteristic can be maintained in a stack composed of alternately high (H) and low (L) index $\lambda/4$ films.

A typical narrow bandpass filter is the Fabry–Perot interference filter, which consists of two partially reflecting films separated by a dielectric $l/2$ spacer layer. Here, high-reflecting HL all-dielectric stacks could be used. Figure 5.10 shows the normalized transmission curves of an all-dielectric multilayer-type Fabry–Perot filter [25]. The solid line is the spectral response for a *single half-wave* (SHW) type filter. The layer structure of this filter is

$$\text{air}/\underline{HLHL}\ \ \underline{HH}\ \ \underline{LHLH}/\text{substrate} \tag{5.5}$$
$$\text{mirror}\ \ \text{spacer}\ \ \text{mirror}$$

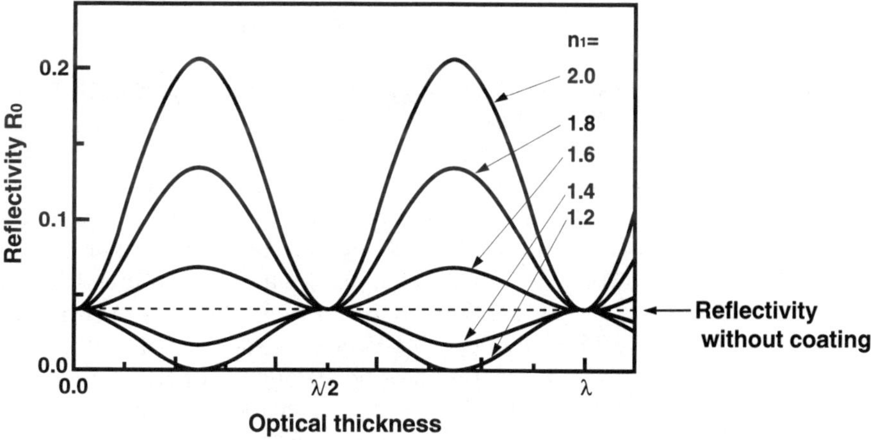

Figure 5.9 Reflectance characteristics of the single-layer coated filter.

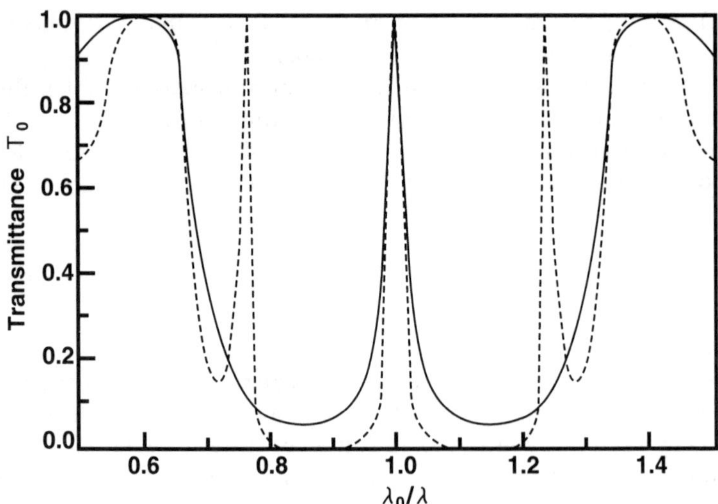

Figure 5.10 Normalized transmission curves of an all-dielectric multilayer-type Fabry–Perot filter (solid line: SHW type filter; dashed line: DHW type filter).

For narrower pass bandwidths, a *double half-wave* (DHW) filter, which consists of two SHW filters stacked together, is a very useful structure. The DHW structure is

$$\text{air}/\underline{HLHLHHLHLH}\ L\ \underline{HLHLHHLHLH}/\text{substrate} \tag{5.6}$$

$$\quad\ \ \text{SHW–filter}\qquad\qquad\ \text{SHW–filter}$$

Other functional filters such as the short wavelength pass filter are designed and manufactured using the HL stack.

Figure 5.11 shows an example of a bandpass filter module with a single-mode fiber pigtail [26]. *Graded index* (GRIN) rod lenses are used to collimate the beam from a single-mode fiber. The filter substrate is placed in the gap between the lenses to produce a low-loss device. To eliminate the reflected power from the filter surface, the lenses and fiber endfaces are antireflection coated. In addition, the filter substrate is tilted so that reflected light is not collected by the input fiber.

To realize a tunable bandpass filter, a bandpass filter can be tuned if it is designed to rotate slightly between the fiber collimators and thus vary the effective optical thickness. However, this type of tunable filter cannot change a center wavelength of the pass band without deterioration of polarization properties. As the tilt angle of the filter increases, the center wavelengths of the two polarization modes separate from each other. This phenomenon results in an increase of a *polarization-dependent loss* (PDL) and *polarization mode dispersion* (PMD). To overcome this problem, a bandpass filter has been developed whose pass wavelength is linearly chirped

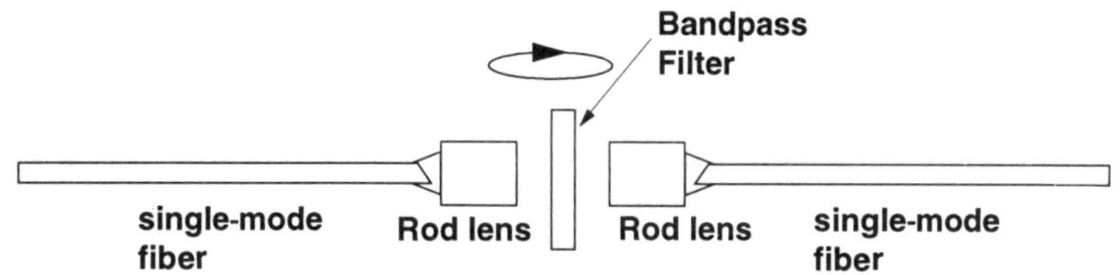

Figure 5.11 Schematic configuration of tunable bandpass filter.

in the filter plane. By using a tunable filter with a linearly sliding mechanism and a chirped bandpass filter, the PDL can be kept at less than 0.1 dB for the whole tunable wavelength region.

5.2.4 Isolators and Circulators

5.2.4.1 Principle of Optical Isolator and Materials

The basic principle behind optical isolators is shown in Figure 5.12. An isolator consists of a polarizer, a 45-degree Faraday rotator, and an analyzer [27,28]. The rotator is positioned between the polarizer and the analyzer. The azimuthal angle between the polarizer and the analyzer is set at 45 degrees. A linearly polarized forward light, whose polarization angle is parallel to the azimuthal angle of the polarizer, can pass through the polarizer without loss. After transmission through the Faraday rotator, the polarization plane of the light rotates +45 degrees and can

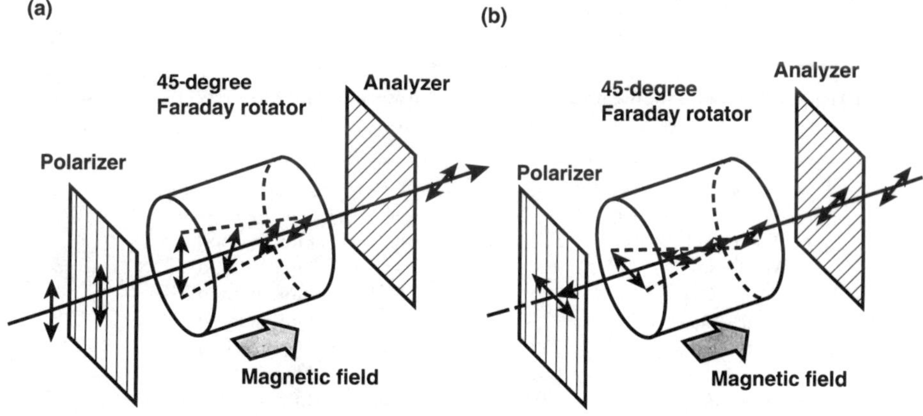

Figure 5.12 Schematic configuration of an optical isolator: (a) forward propagation and (b) backward propagation.

pass through the analyzer without loss. On the other hand, backward propagating light is totally blocked regardless of its polarization.

A Faraday rotator comprises Faraday rotation materials and a magnet to apply a magnetic field. Faraday effects, which are one type of magneto-optical effects, occur in gases, liquids, and solids. In general, solids exhibit the strongest effects. Solids are further classified into diamagnetic, paramagnetic, ferromagnetic, antiferromagnetic, and ferrimagnetic materials. Diamagnetic materials, which include glasses, crystalline materials, and some liquids, have a linear magneto-optic response to an applied field and are stable with respect to temperature. However, the Faraday effects of these materials are too small for them to be used to construct an optical isolator for use in telecommunication systems. Paramagnetic, ferromagnetic, antiferromagnetic, and ferrimagnetic materials generally have much stronger Faraday effects and are suitable for constructing compact isolator devices.

Yttrium-iron-garnet (YIG) is a nonmetallic ferrimagnetic crystal [29–31]. YIG and related crystals exhibit the most suitable performance with regard to fabricating 1.2- to 1.7-μm isolators because they have large Faraday effects and low transmission losses. By replacing specific ions in YIG with other ions, it is possible to substantially modify both its magnetic and magneto-optic characteristics. To enhance the magneto-optic properties, paramagnetic rare-earth ions, including Tb, Ce, Pr, and Nd, may be either partially or entirely substituted for the Y ions [32]. YIG and related materials perform extremely well in the 1.2- to 1.7-μm wavelength region; however, it is very difficult to use them at wavelengths of less than 1.1 μm because Fe ions have strong absorption in this wavelength region. Table 5.1 lists room-temperature saturation Faraday rotation and absorption data for YIG and related materials.

Good levels of performance at around 1.0 μm are provided by CdMnTe and related materials, which are paramagnetic crystals [33]. The Faraday effect in these materials is strongly dependent on both their composition and operating wave-

Table 5.1

Room-Temperature Saturation Faraday Rotation and Absorption Data for Selected Iron Garnets at $\lambda = 1,300$ nm

Material	$\theta_F'\,(°/cm)$	$\alpha\,(cm^{-1})$	Growth Technique
$Y_3Fe_5O_{12}$	210	0.3	Flux method
$Gd_3Fe_5O_{12}$	60	1.0	Flux method
$Tb_3Fe_5O_{12}$	320		Flux method
$Dy_3Fe_5O_{12}$	175		Flux method
$Tm_3Fe_5O_{12}$	110		Flux method
$Pr_3Fe_5O_{12}$	−1,060	70	Flux method
$Nd_3Fe_5O_{12}$	−690	< 50	LPE
$Y_{1.7}Bi_{1.3}Fe_5O_{12}$	−2,100		LPE
$Gd_{2.0}Bi_{1.0}Fe_5O_{12}$	−2,100	< 10	Flux method
$Y_{2.0}Ce_{1.0}Fe_5O_{12}$	-1.2×10^4	250	Sputtering

length. Recently CdHgMnTe was applied to optical isolators at a wavelength of 0.98 μm. Figure 5.13 shows the composition dependence of the absorption band tail wavelength and the Verdet constant. As Hg substitution increases, the absorption band tail shifts to a longer wavelength, and the Verdet constant at 0.98 μm increases [34]. Table 5.2 shows the characteristics of 0.98-μm optical isolators that employ a CdMnHgTe Faraday rotator.

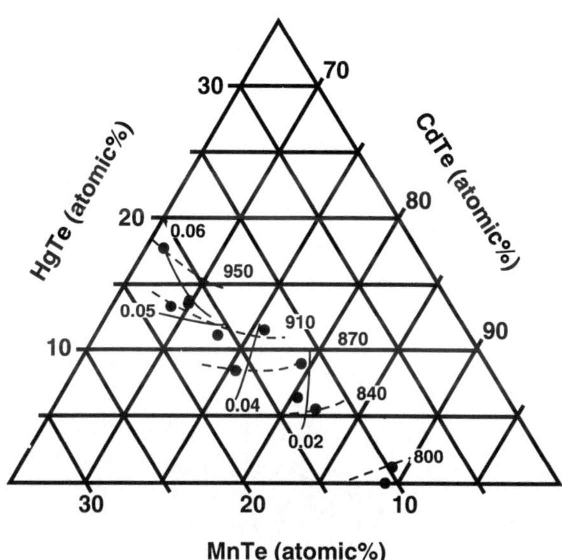

Figure 5.13 Composition dependence of the absorption band tail and Verdet constant on HgCdMnTe crystal [34] [solid line: Verdet constant at 980 nm (deg/cm/Oe); dashed line: absorption band tail (nm)].

Table 5.2

Characteristics of 0.98-μm Optical Isolators Using the $Cd_{1-x-y}Mn_xHg_yTe$ Faraday Rotator

Terms	
Wavelength	980 nm
Insertion loss	0.97 dB
Isolation	30.2 dB
Max. beam diameter	1.3 mm
Dimensions	ϕ4 × 4.5 mm
Faraday material	$Cd_{0.70}Mn_{0.17}Hg_{0.13}Te$
Loss of FR	0.63 db
Extinction ratio of FR	38.0 dB
Thickness of FR	3.0 mm

5.2.4.2 Polarization-Independent Isolator

A polarization-independent-type optical isolator consists of spatial walk-off polarizers, a Faraday rotator, and single-mode fiber pigtails with collimating optics. The Faraday rotator is inserted between the two polarizers.

To date, two types of isolator have been reported. One consists of a spatial walk-off polarizer with a flat plate shape [35], and the other has a polarizer that is wedge shaped [36].

Figure 5.14 shows a polarization-independent isolator based on the flat-plate spatial walk-off polarizer. The polarizer is made from birefringent crystal, and its optical axis is slightly tilted at an oblique angle to the light polarization direction. The combination of a 45-degree Faraday rotator and a polarization compensator, which is made from a quarter plate or a birefringent plate, acts as a 90-degree polarization rotator for forward propagating light. For backward propagating light, the polarization axis of the incident light does not rotate because the polarization

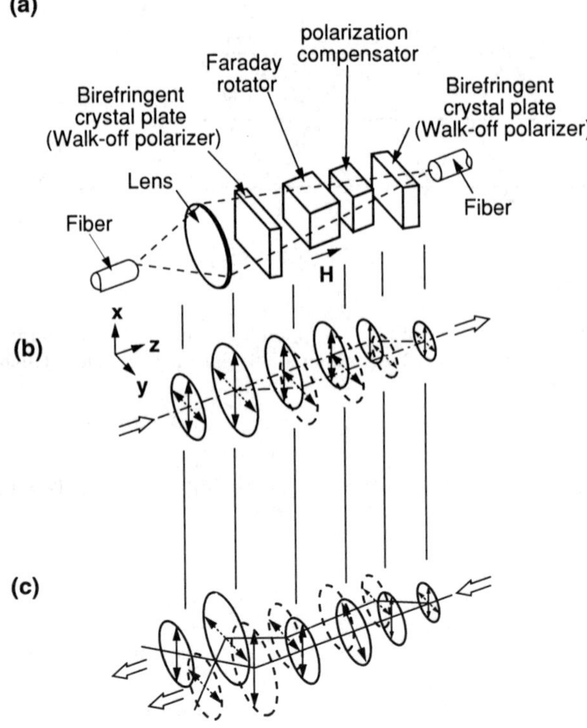

Figure 5.14 Schematic configuration of polarization-independent isolator with flat-shape birefringent crystals [35]: (a) configuration of the isolator, (b) and ray diagram of forward propagation light and (c) backward propagation light.

compensator cancels out the rotation of the polarization axis caused by the Faraday rotator.

When the incident light is forward propagating, it is separated into two orthogonally polarized elements with a small angle of separation between them. The polarization angle of the separated lights is rotated by the Faraday rotator/polarization compensator combination. By passing through the unit, the polarization angle of the two lights is rotated through 90 degrees. The two light rays are then refracted in the second spatial walk-off polarizer and are combined again. These light rays are coupled to the output fiber pigtail with low coupling loss.

For backward operation, two polarized lights pass through the Faraday rotator and are refracted in the second spatial walk-off polarizer. However, these refracted lights cannot be combined, and so these backward lights cannot be coupled to the fiber pigtail with low loss.

Figure 5.15 shows a polarization-independent isolator based on the wedge-shaped spatial walk-off polarizer. The operating concept for this isolator is almost the same as that of the aforementioned isolator. The incident light from the input fiber is converted to parallel rays by a lens and then separated into two orthogonal polarized elements by the first spatial walk-off polarizer. After this process, as each element passes through the Faraday rotator, it is refracted by the second spatial walk-off polarizer. The right deflection by the left wedge is canceled out by that of the right wedge for the two polarized light rays. The light rays are shifted parallel to each other after passing the right wedge and focused by a lens into the output fiber with low loss. For backward operation, the two orthogonal light rays are not

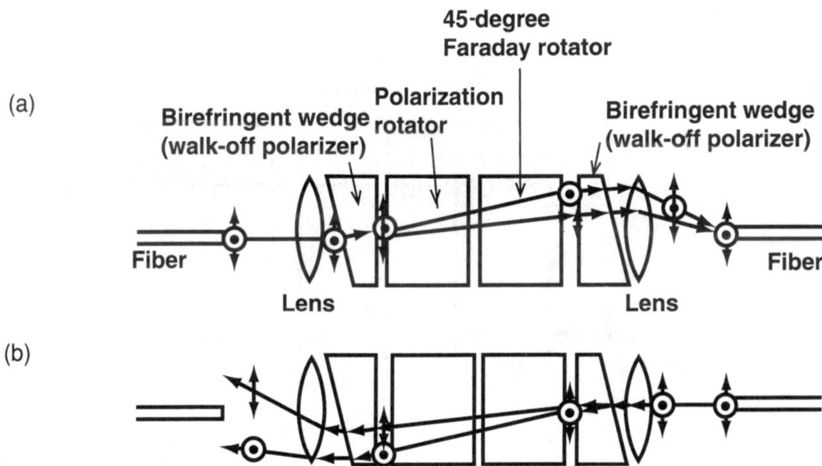

Figure 5.15 Schematic configuration of polarization-independent isolator with wedge-shaped birefringent crystals [36]: ray diagram of (a) forward propagation light and (b) backward propagation light.

parallel after passing the left wedge, which means they cannot be focused on the fiber.

A polarization-independent isolator with a YIG Faraday rotator is already commercially available in wavelengths of 1.3 or 1.5 μm.

5.2.4.3 Optical Circulator

Several types of circulator have been reported with polarization-independent features [37–41]. Figure 5.16 shows the schematic configuration of a circulator that consists of a polarization beam-splitter cube [37]. The Faraday rotator and polarization compensator combination functions in the same way as that of the isolator. When an unpolarized light enters port 1, the p-polarized component is transmitted and the s-polarized component is reflected 90 degrees from the incident beam by beam-splitting cube 3. After passing through the combined rotators 1 and 2, the p and s components are changed to s and p components, respectively. Therefore, the two components join together at the hypotenuse of cube 3 and exit through port 2 as unpolarized light. On the other hand, when unpolarized light enters port 2, the p and s components are divided by beam-splitting cube 4 and pass without any polarization change through the combined rotators. Then the p and s components join together and exit through port 3. Similarly, unpolarized light that enters port 3 exits through port 4. Consequently this structure operates as a polarization-independent four-port circulator.

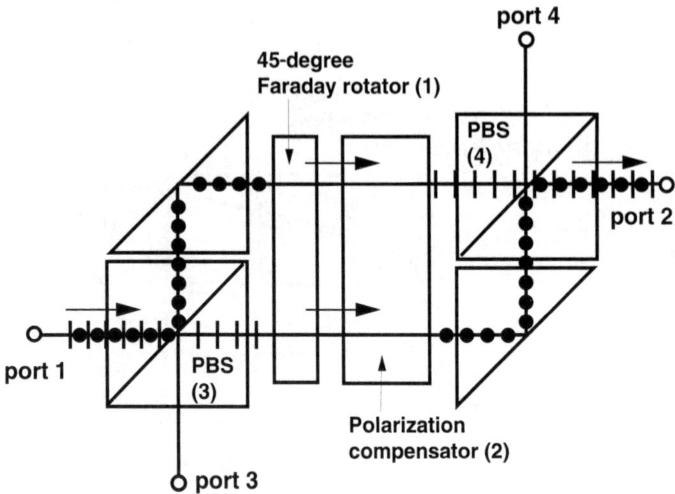

Figure 5.16 Schematic configuration of polarization-independent optical circulator [37]. PBS, polarization beam splitter.

5.2.5 Pump/Signal Multiplexing Devices

5.2.5.1 Fiber Couplers

The single-mode directional coupler is a very important device. Directional couplers have been made by polishing or fusion/tapering single-mode fibers to achieve close proximity of the fiber core. In the ideal directional coupler shown in Figure 5.17, two lossless waveguides are mirror images of each other. Light launched into port 1 is guided along the waveguides. In the coupling region, two waveguides are sufficiently close to produce an overlap of their transverse fields. Then, the eigenmodes are solutions of the wave function for the composite waveguides and are referred to symmetric and antisymmetric eigenmodes [42]. Because these two eigenmodes have different propagation constants, they beat together and power is transferred between the two waveguides. Based on the coupled mode theory, the power transfer follows the $\cos^2$ and $\sin^2$ form

$$P_1\,(z = L) = P_1\,(z = 0)\cos^2(\kappa L)$$

$$P_1\,(z = L) = P_2\,(z = 0)\sin^2(\kappa L)$$

where $P_1(z = 0)$ is the launched power, z is the length coordinate, and k is the coupling coefficient, which are calculated with the waveguide geometry, refractive indexes, and losses. Since the two waveguides have exactly the same propagation constants with very small losses, the power transferred between the waveguides varies from zero to unity. The wavelength dependence of k is an important feature because the mode-field diameters for longer wavelengths are larger than those for shorter wavelengths. We can manufacture a coupler with a high coupling ratio at one wavelength and a low coupling ratio at another. This dichroic directional coupler is very useful in the field of fiber amplifier technology.

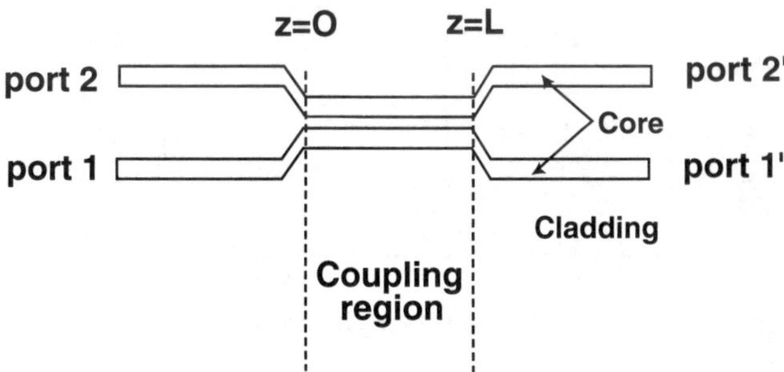

Figure 5.17 Basic structure of fused-taper-type fiber coupler.

Fused-Tapered-Type Fiber Coupler

In the fusion/tapering approach, the coupler has the biconical taper structure shown in Figure 5.18 [43]. Within each fiber, there is a long tapered section, then a uniform section of length L where they are fused together, and then another taper back to the original cross-sectional configuration of the individual fibers. Since the tapers are very gradual, injected optical power from either left port is not reflected to either left port. This means that this coupler has very low excess loss [44,45].

To fabricate the coupler, the resin coating of two fibers is removed along a short section of a few centimeters. The fibers are then twisted together, heated, and drawn while the throughput and the coupled powers are monitored. Figure 5.19 shows the relationship between the drawn length and the coupling ratio during the drawing process for a fused tapered coupler that consists of two fibers with exactly the same structure [46]. The solid and dashed lines indicate the coupling ratios at wavelengths of 1.31 and 1.54 μm, respectively. We can obtain a 1.31/1.54-μm WDM coupler for a drawn length of x.

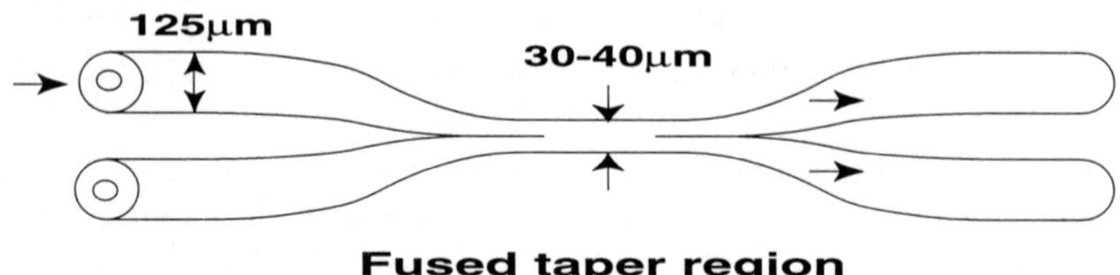

Figure 5.18 Structure and optical characteristics of fused-tapered-type fiber coupler [43].

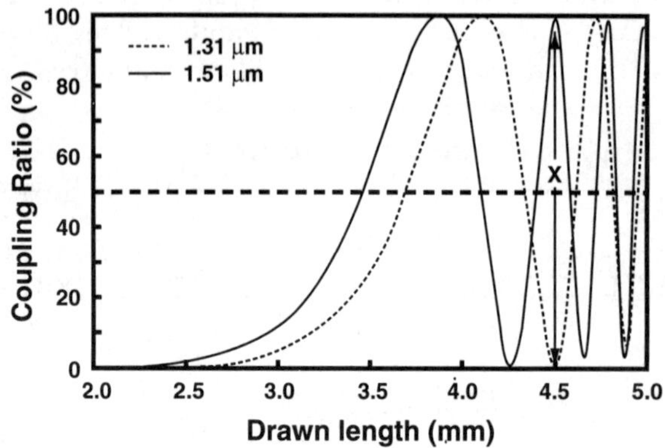

Figure 5.19 Relationship between coupling ratio and the drawn length [46]. X marks the point at which 1.31/1.54-μm WDM performance is achieved.

In order to manufacture a coupler with specific WDM characteristics, light is launched into one of the ports of the coupler while it is being made, and the output from the two ports on the opposite side is monitored to determine the coupling ratio. The drawing process is stopped when the coupling ratio reaches the required value. This method has been used to manufacture 1.48/1.55-μm, 0.98/1.55-μm, and 1.02/1.30-μm WDM fiber couplers.

Polished-Type Fiber Coupler

Other fiber-type WDM couplers include the polished-type fiber coupler and the D-fiber coupler [47–49]. Figure 5.20 shows schematic configurations for both types. The polished fiber coupler is made by burying an optical fiber in a curved groove in a glass block and then polishing the block to remove a semicircular potion of the cladding. Coupling is achieved by placing two half-coupler blocks back to back. A small quantity of refractive index matching oil (or adhesive) is introduced at the interface to realize good optical contact. The D-fiber coupler has almost the same configuration except for the cross-sectional shape of the fiber. With the D-fiber, almost half the cladding is removed. The coupler can be made by combining the flat sides of the D-fibers. A suitable amount of liquid is used between the D-fibers in order to realize good optical contact. The distance between the two cores is very carefully designed so as to realize the required WDM characteristics.

5.2.5.2 Bulk-Type Devices

Figure 5.21 shows some examples of bulk-type WDM devices. These devices generally consist of a fiber collimator and wavelength-selective optics such as a prism, a thin-film filter, and a grating. The fiber collimator is important in terms of constructing these devices with a low insertion loss [50]. The typical collimator structure is shown in Figure 5.22. It consists of a spherical lens, a glass spacer, and an optical fiber fixed in a ferrule. A ferrule with antireflection coating or angled polishing can be used to reduce reflection at the endface of the fiber. Typical optical coupling characteristics between collimators are shown in Figure 5.23. The coupling losses at the distances of 8 to 35 mm were 0.7 to 1.0 dB and 0.6 to 0.8 dB at wavelengths of 1.3 and 1.5 μm, respectively. The coupling loss can be reduced by replacing the spherical lens with an aspherical lens.

Figure 5.24 shows an example of a WDM coupler used for the polarization multiplexing of pump light [51]. For this module, the pump lights from two LDs are combined with a polarization-multiplexing technique to realize a redundant configuration. Polarized-multiplexed pump light is combined with a signal light by using a thin-film multilayer filter. Bulk-type WDM devices have an advantage compared with fused tapered couplers in that a nonsinusoidal spectral dependence on coupling ratio can be achieved. Bulk-type WDM devices are considered to be more

(a)

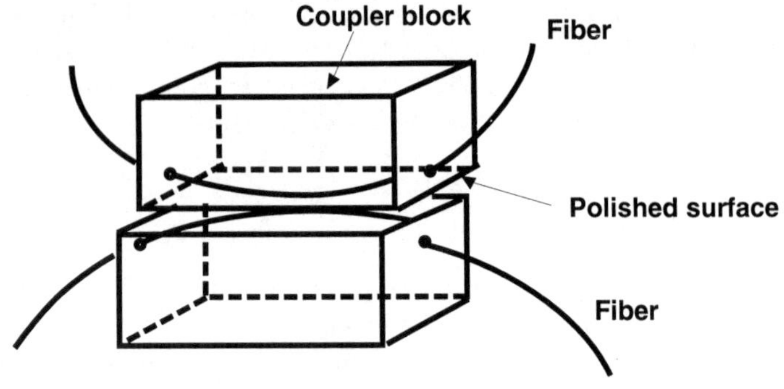

(b)

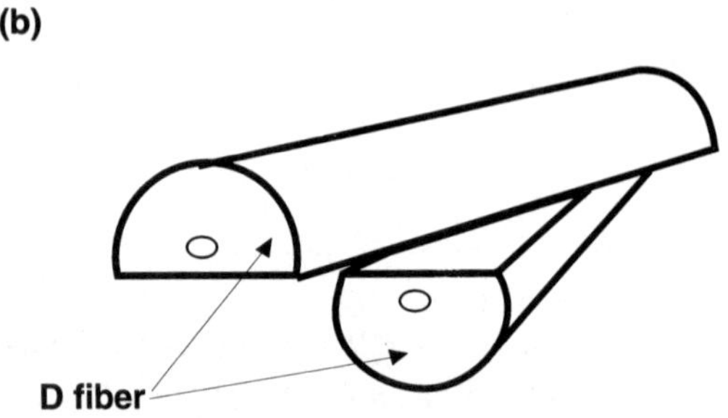

Figure 5.20 Schematic configuration of fiber-type WDM couplers [78–84]: (a) polished-type fiber coupler and (b) D-fiber-type coupler.

suitable than fiber couplers as 1.48/1.55-μm WDM devices because of their non-sinusoidal spectral response and flat pass band characteristics [52].

5.2.6 Splicing Devices and Technology

5.2.6.1 Connectors

Connectors are demountable fiber connections, whereas splices are permanent joints. They are used where it is necessary or convenient to disconnect and reconnect fibers easily, for example, at the input and output ports of optical amplifier modules.

(a)

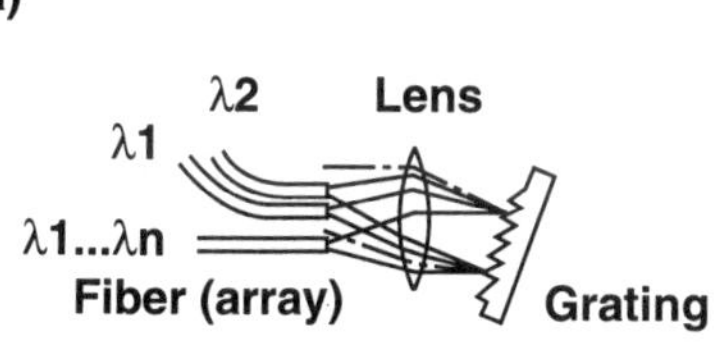

(b)

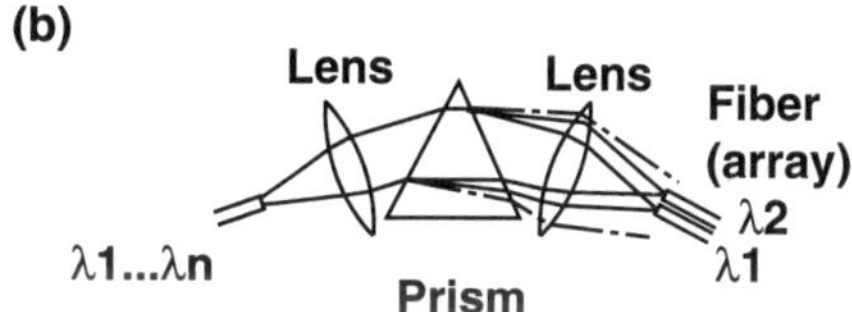

(c)

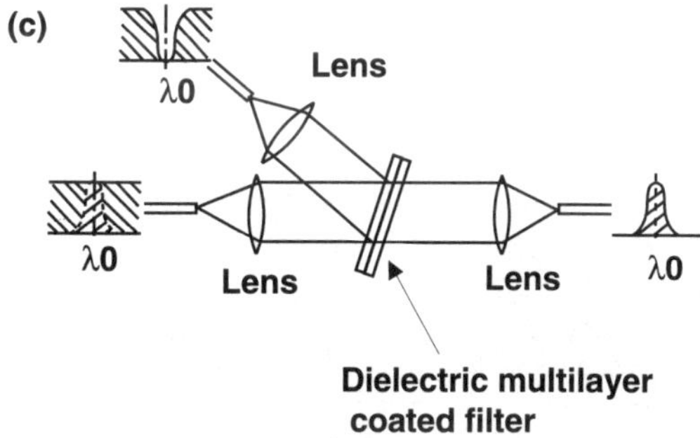

Figure 5.21 Some examples of bulk-type WDM devices: (a) grating-type device, (b) prism-type device, and (c) dielectric multilayer filter-type device.

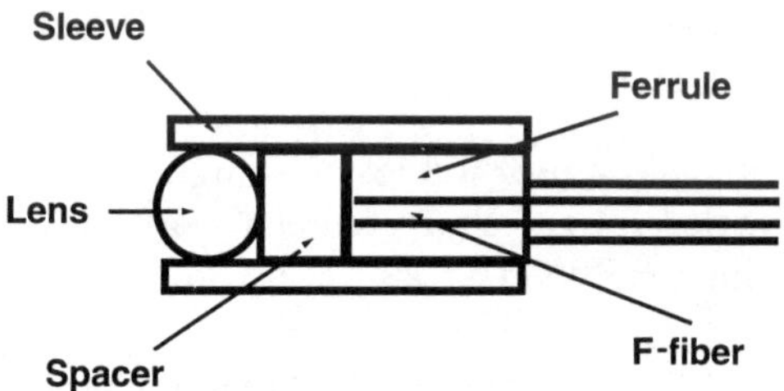

Figure 5.22 Structure of fiber collimator [50].

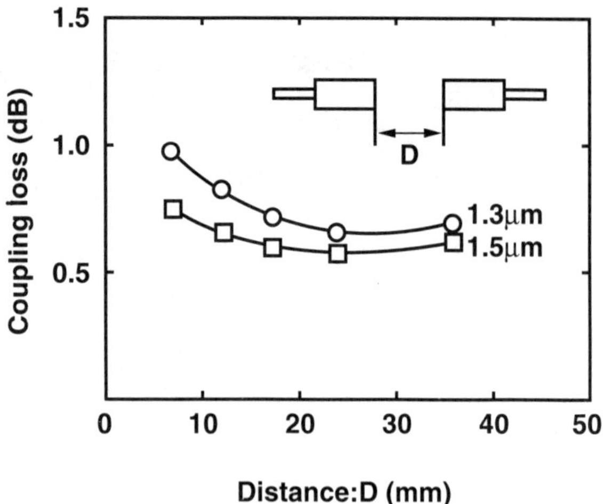

Figure 5.23 Coupling characteristics of fiber collimator [50].

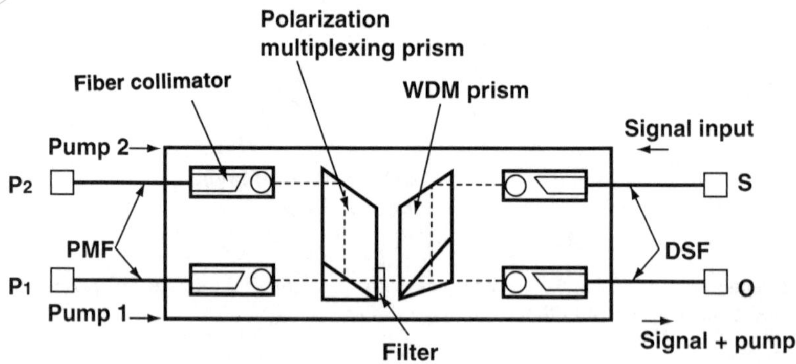

Figure 5.24 Example of a bulk-type WDM coupler.

Connectors must exhibit a low insertion loss and high return loss with good repeatability and reliability. Several types of connector are currently used with silica fibers in the telecommunication field and are briefly described in the following paragraphs.

The ST-type connector features a bayonet connection with keying to ensure consistent connector orientation within the bulkhead [53]. It is spring loaded to ensure that the faces of the two connectors in the bulkhead remain in contact at a constant pressure, thus eliminating the need to manufacture to a precise length. Inside the bulkhead the connectors are aligned within a ferrule sleeve.

The FC/PC connector, which is shown in Figure 5.25, is an improved FC connector [54,55]. The outer collar screws onto the bulkhead with little mechanical coupling to the inner fiber and all ceramic ferrule. The ferrule and fiber are keyed to improve repeatability. A small radius at the end of the ferrule allows point contact between connectors (physical contact).

The SC is made primarily of plastic, occupies very little space, and can be packed very densely into a panel [56]. Figure 5.26 shows a schematic configuration of the SC-type connector, which is a bricklike module. It is keyed, with an internal spring that allows the unit to snap positively in and out of place. The spring also maintains contact at a constant pressure between the connectors in the bulkhead.

The MU connector is a miniaturized SC connector [57] and features a plastic molded push-pull coupling connector with a rectangular housing and precision zirconia ceramic ferrule. The convex polished endface of the ferrule allows physical contact. The average connection losses of these connectors are less than about 0.3 dB, and the return losses are more than 40 dB.

5.2.6.2 Fusion Splicing

The arc fusion-splicing technique is the most commonly used permanent jointing procedure for silica fibers [58,59]. In this technique, localized heating applied at the interface between two butted, prealigned fiber ends causes them to soften and fuse together. A fusion splice can be made by heating the fiber ends in an electric arc. To minimize the splice loss, flat and perpendicular endfaces have to be pre-

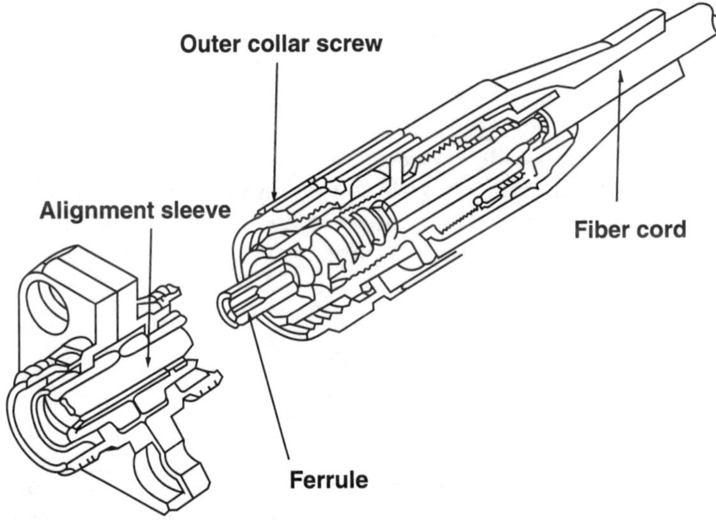

Figure 5.25 Schematic configuration of FC connector [54].

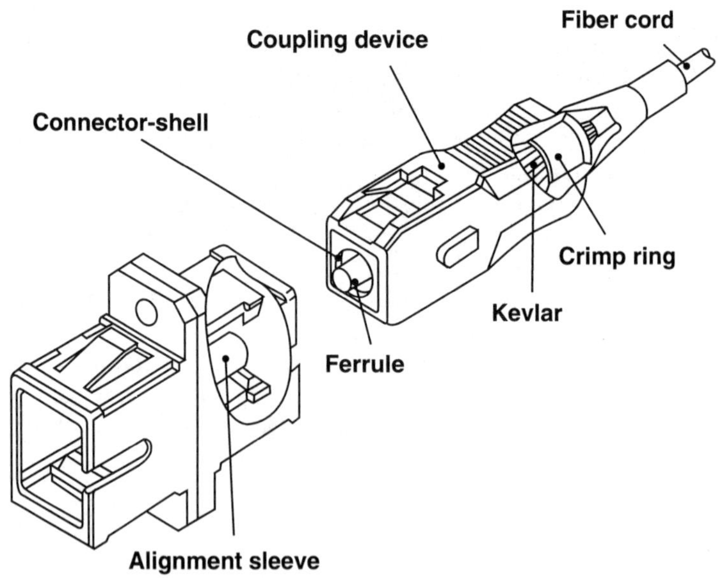

Figure 5.26 Schematic configuration of SC connector [56].

pared, and these must be aligned with minimum transverse core offset and minimum tilt angle. Figure 5.27 shows the main steps in the fusion-splicing procedure.

For precise alignment, the core can be directly monitored using a normal microscope combined with a high-resolution TV camera [60]. In a basic study of splicing using this direct monitoring system, it was found that if the detection accuracy is within 0.25 μm, an average splice loss of 0.2 dB can be guaranteed with 99% reliability. An automatic arc fusion-splicing machine that incorporates the direct monitoring system is widely used.

By contrast, arc fusion splicing is not commonly used for fluoride fibers. Major obstacles are the low melting temperature of about 300°C, the rapid decrease in fiber viscosity with increasing temperature, and crystallization during the heat treatment.

5.2.7 Pump Source

Pump sources, which supply energy to rare-earth-doped fiber, are major components of fiber amplifiers. The required pump wavelengths depend on the rare-earth ion. Typical reported pump sources are listed in Table 5.3. This section describes those in common use.

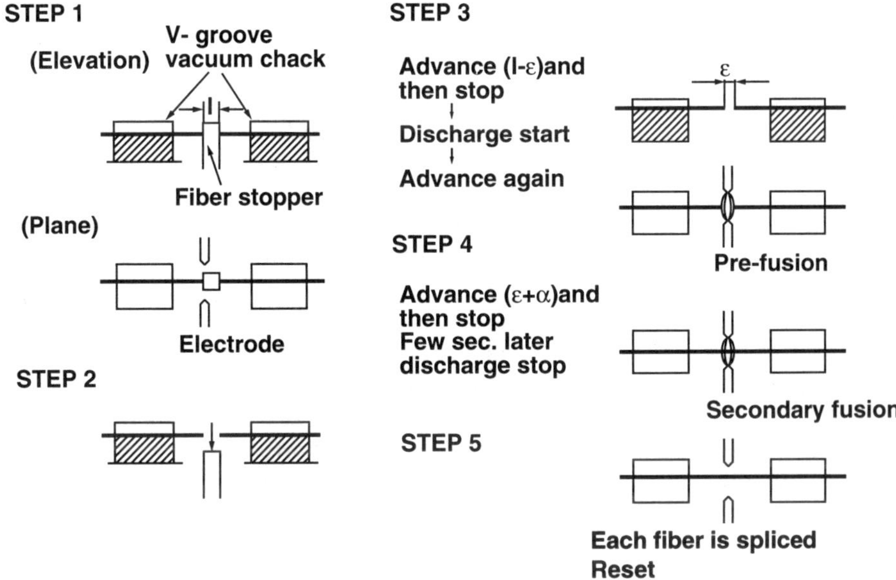

Figure 5.27 Fusion-splicing procedure.

Table 5.3
Pump Source Selection for Various Fiber Amplifiers

Active Ion	Classifications of Laser Mechanism	Pump Wavelength	Structure
Er	Semiconductor laser	1,480 nm	Single stripe LD (Bulk or MQW)
		980 nm	Single stripe LD (Strained QW)
			MOPA structure (Strained QW)
		807 nm	Single stripe LD (Bulk or MQW)
		660 nm	Single stripe LD (Bulk or MQW)
	Fiber Raman laser	1,480 nm	Cascaded Fiber Raman Laser
	Solid-state laser	1,064 nm	LD-pumped Nd:YAG
		1,047 nm	LD-pumped Nd:YLF
		980 nm	
Pr	Semiconductor laser	1,017 nm	Single stripe LD
			MOPA structure
	Solid-state laser	1,047 nm	LD-pumped Nd:YLF
	Fiber Raman laser	1,025 nm	Cascaded Fiber Raman Laser
Nd	Semiconductor laser	807 nm	Single stripe LD (Bulk or MQW)
Tm	Semiconductor laser	1,200 nm	Single stripe LD (Bulk or MQW)
		780 nm	Single stripe LD (Bulk or MQW)
	Solid-state laser	1,064 nm	LD-pumped Nd:YAG

5.2.7.1 Semiconductor Laser Diode

Principle of Semiconductor Laser Diode

Practical semiconductor LDs are based on the double heterojunction structure proposed by Woodall and first implemented by Hayashi and Panish in 1970 [61].

Figure 5.28 shows the energy diagram of the double heterojunction structure. The extrinsic materials of the *p*-region and *n*-region are shown with the same bandgap. The active layer in the middle has a smaller bandgap than either. As the electrons in the *n*-cladding layer move leftward in the conduction band, they enter the active layer and are prevented from going further because of the existence of a potential barrier. The same mechanism exists for the holes. As a result, the electrons and holes are strongly confined in the active region. When a forward bias is applied to the device, the injection of electrons and holes is confined to the active region and combined radiatively. The emission frequency is $f = Eg'/h$, where Eg' is the energy band gap of the active region.

Another advantage of the double heterojunction structure is that the refractive index of the active region materials is slightly greater than that of the extrinsic materials. This leads to the confinement of both the injected current and the light.

In order to realize pump lasers for fiber amplifiers, the band gap energy of the active layer must coincide with the absorption bands of active ions. In addition, the lattice parameters for both the substrate and the epitaxial layers must be almost equal to obtain good-quality epitaxial layers. Figure 5.29 shows the relationship

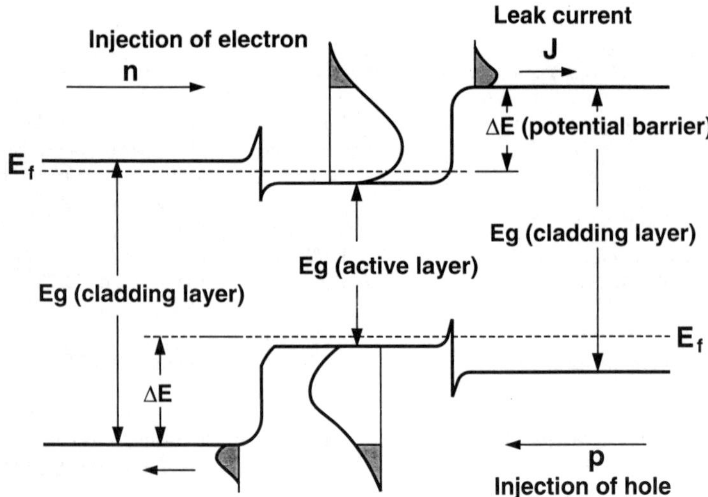

Figure 5.28 Energy diagram of the double heterojunction LD.

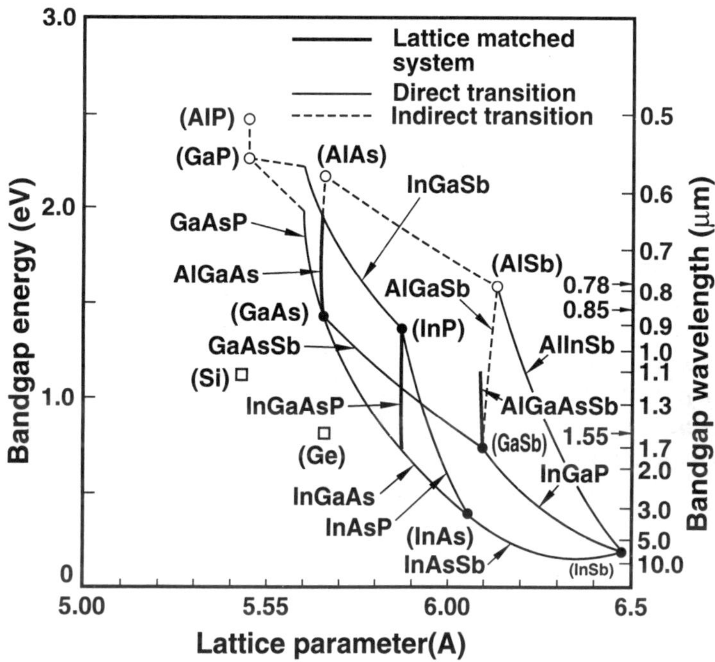

Figure 5.29 Relationship between bandgap and lattice parameters for a several-compound semiconductor [62].

between bandgap energies and lattice parameters for several compound semiconductor materials [62]. The circles indicate binary semiconductors such as GaAs. The hatched area indicates a compound semiconductor based on InGaAsP alloy. The solid and dashed lines indicate a three-element compound semiconductor with direct and indirect transitions, respectively.

For a lattice-matched system, GaAs and InP substrates are commonly used. AlGaAs, AlGaInP, and InGaAsP have all been used as the lattice-matched active layer of a GaAs substrate. A semiconductor laser with a AlGaAs active layer on a GaAs substrate has a CW lasing wavelength region from 0.75 to 0.9 μm at *room temperature* (RT). InGaAsP and InGaAlP are used to achieve a lower lasing wavelength. Lasers with InGaAlP on a GaAs substrate have a CW lasing wavelength region from 0.625 to 0.700 μm at RT. The lowest limit for the lasing wavelength is about 0.65 μm for the InGaAsP active layer.

With the InP substrate system, a compound semiconductor based on InGaAsP is used to obtain a CW lasing wavelength from 1.1 to 1.65 μm at RT. Below 1.1 μm, InGaAsP is still a direct-transition-type semiconductor, but a small difference between the bandgap energies of InP and InGaAsP results in insufficient carrier confinement and a large threshold current. Therefore, no semiconductor

LDs with a lasing wavelength from 0.9 to 1.1 μm have yet been developed for a lattice-matched system. In contrast, an InGaAs strained quantum well active layer on a GaAs substrate, which is a lattice-mismatched system, has an emission wavelength from 900 to 1,100 nm.

Module Configuration

Several techniques have been reported for realizing efficient coupling between a LD chip and a single-mode fiber. One is direct coupling with tapered hemispherical lenses, that is, drawn-tapered fiber endfaces with molten lenses on their tips. Others involve coupling with a suitable lens. Figure 5.30 shows the schematic configuration of an efficient technique for coupling a *laser diode* (LD) and a fiber. To reduce the optical feedback effect, a polarization-dependent isolator can be inserted between the LD and the fiber.

Figure 5.31 shows a 14-pin dual-in-line-type LD module with lensed fiber coupling. An LD chip is mounted on the thermoelectric cooler.

Conventional Pump Laser Diodes—1,480-nm Laser Diodes

The usual structure of these devices is a buried heterostructure Fabry–Perot laser based on an InGaAsP/InP lattice-matched system. For this lattice-matched system, the maximum output power from a LD is limited by the temperature increase in the heterojunctions. A design for achieving high-differential quantum efficiency with a long-cavity structure enables us to manufacture high-power 1,480-nm LDs.

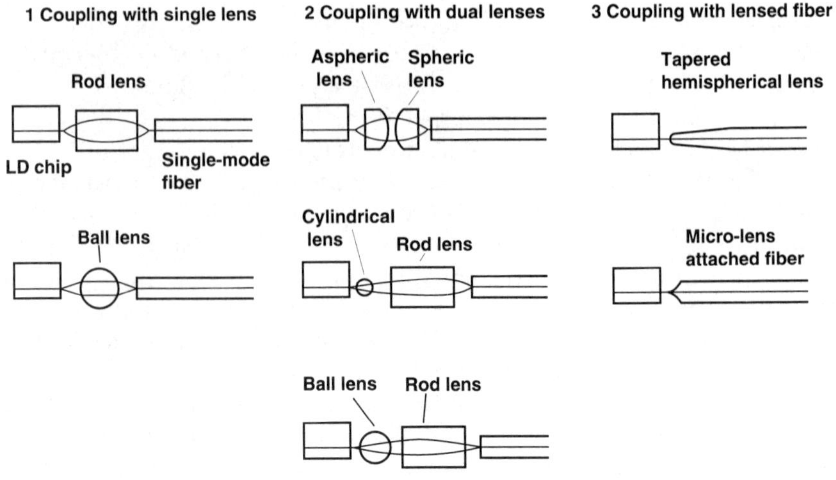

Figure 5.30 Example of efficient coupling between a LD chip and a single-mode fiber.

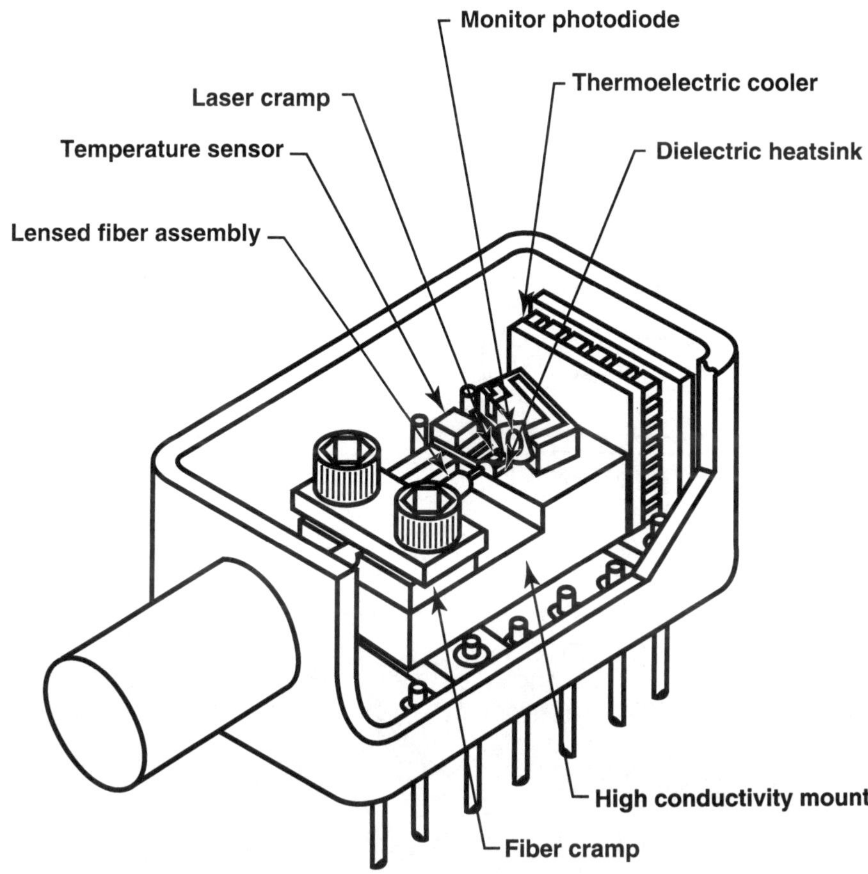

Figure 5.31 Schematic configuration of a 14-pin dual-in-line laser diode module.

The multiple quantum well active layer structure has become an indispensable technique with which to realize high-differential quantum efficiency. Several types of LD have already been reported [63–66].

Figure 5.32 shows a schematic diagram of a GaInAsP/InP GRIN-SCH strained MQW laser [66]. This laser was grown by low-pressure *metal-organic chemical vapor deposition* (MOCVD). The active layer consists of five comparatively strained GaInAsP quantum wells separated by GaInAsP barriers. The amount of strain in the well is about 1%. Low reflectivity (8%) at the front facet and high reflectivity (95%) at the rear facet are used to achieve higher output power from the front facet.

Figure 5.33 shows the light output versus injection current characteristics of GaInAsP/InP GRIN-SCH strained MQW lasers [66]. The maximum output power is 360 mW at an injection current of 1.3A. The reliability of the 1,480-nm laser is

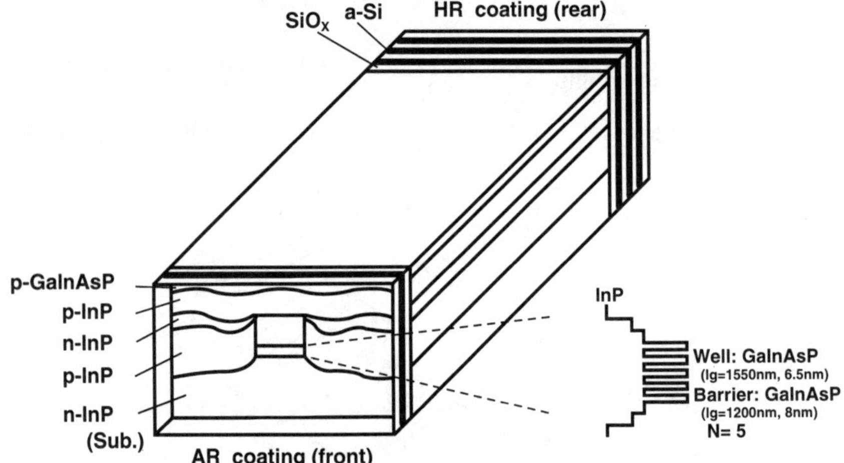

Figure 5.32 Crystal structure of 1,480-nm GaInAsP/InP GRIN-SCH strained MQW LD [66].

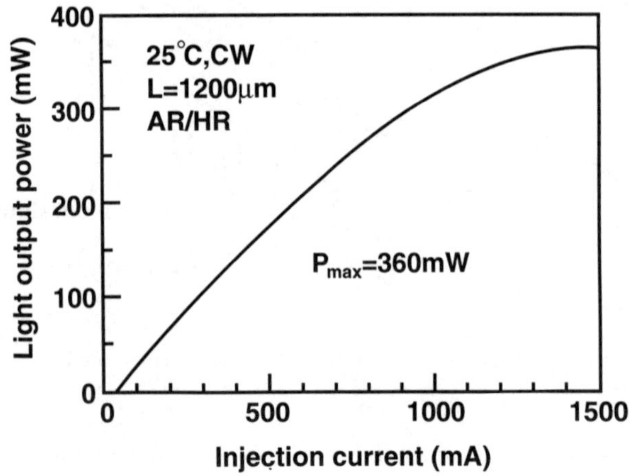

Figure 5.33 Example of I-L curve of 1,480-nm LD [66].

similar to that of 1,300-nm lasers. On the basics of an aging test, a predicted lifetime is more than 10^6 hr at room temperature.

Conventional Pump Laser Diodes—980/1,017-nm Laser Diodes

In these wavelength regions, a InGaAs strained quantum well layer is used for the active region [67,68]. The expression "strained quantum well" means that a very thin InGaAs active layer is inserted between cladding layers with different lattice parameters. These structures consist of very thin lattice-mismatched layers where the lattice mismatch is accompanied by a uniform elastic strain that reduces the cubic symmetry of the semiconductors and modifies the electronic and optical properties of the quantum well. To date, InGaAs/GaInAsP/GaInP and InGaAs/GaAs/AlGaAs structures have been reported. In both cases, the InGaAs active layer is compressed biaxially.

Figure 5.34 shows a schematic diagram and the crystal structure of the first reported 980-nm LD used as a pump source for an erbium-doped fiber amplifier [67]. The crystal has a *graded-index separate-confinement-heterostructure strained quantum well* (GRIN-SCH-SQW) structure with an 11-nm-thick $In_{0.2}Ga_{0.8}As$ well. The cladding layer is AlGaAs with an Al content of 0.6. The structure was grown by low-pressure MOVPE in a vertical configuration. A ridge structure was fabricated using electron-cyclotron-reactive ion etching. The ridge waveguide width was 3 μm. The reflectivity of the front and back facets were 10% and 90%, respectively. The cavity length was 600 μm.

Figure 5.35 shows the light output versus injection current characteristics of the 980-nm laser [67]. The threshold current is 9 mA and the slope efficiency is 0.77 W/A. The maximum output power from the chip is 85 mW.

The crystal structure of a recently reported high-power 980/1,017-nm SCH laser is shown in Figure 5.36 [69]. Layers with a relatively low refractive index were inserted between the cladding and guiding layers. Their aluminum content was larger than that of the cladding layers. These low-refractivity layers expand the vertical spot size and increase the vertical confinement factor per well. Figure 5.37 shows a typical far-field pattern and light output power versus injection current characteristics [69]. The full width at half-maximum of the vertical and horizontal divergence angles are 18 and 9 at an optical output power of 200 mW. The far-field patterns exhibit a stable fundamental transverse mode even at an optical

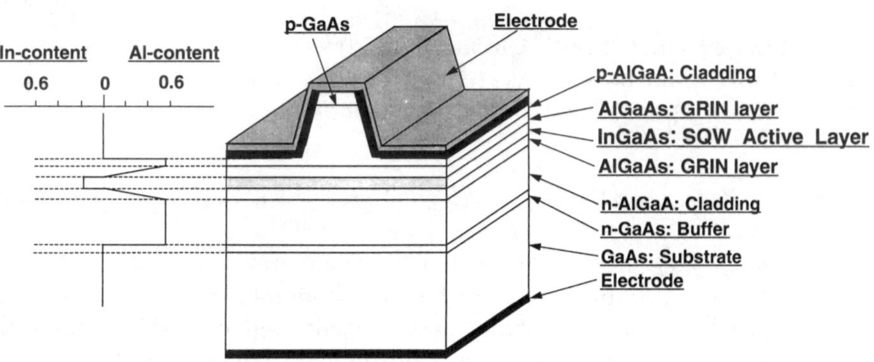

Figure 5.34 Crystal structure of 980-nm InGaAs/GaAs strained QW LD [67].

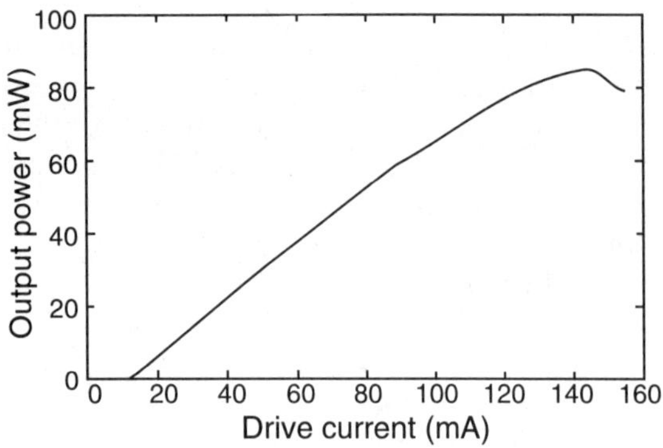

Figure 5.35 I-L curve of 980-nm LD [67].

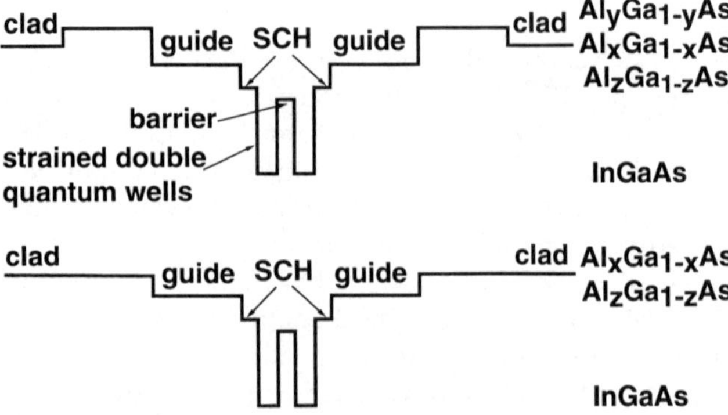

Figure 5.36 980-nm LD with guided layers and low-refractive-index layers [69].

output of more than 400 mW. The maximum free-space lasing output power was about 550 mW at an injection current of 750 mA. It was possible to couple an optical power of up to 250 mW into a single-mode fiber.

Complex-Type Pump Laser Diodes—Wavelength-Stabilized 980/1,017-nm Laser Diodes

As will be mentioned in more detail later, the lasing wavelength of a 980-nm laser should be tuned within the erbium 980-nm absorption band, which is typically 10- to 15-nm wide. Wavelength instability caused by reflection, mutual injection-locking among multiple lasers, and aging may shift the laser wavelength, thus degrading amplifier performance [70]. In particular, a low reflectivity front-facet coating, which is used to increase the laser output power, increases their sensitivity to outside

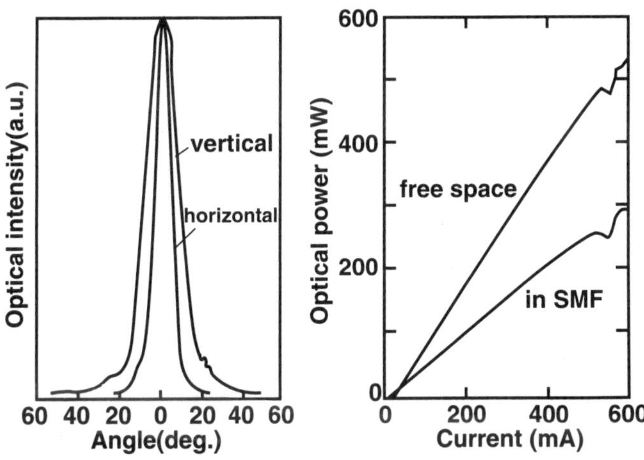

Figure 5.37 Far-field pattern and I-L curve of 980-nm LD with guided layers and low-refractive-index layers [69].

reflection. One solution is to fabricate a laser module incorporating an isolator. Another way is LDs with a wavelength-selective external cavity structure.

Figure 5.38 shows two types of LD that have this external cavity structure to stabilize the lasing wavelength. The combination of a narrow bandpass filter and a mirror acts as an external cavity LD [71]. In the case of an wavelength-stabilized LD with a fiber Bragg grating, the typical reflectivity of the fiber grating is a few percent [72].

Figure 5.39 shows the light output versus injection current characteristics of a 1,017-nm laser with an external fiber Bragg grating [73]. It was possible to couple

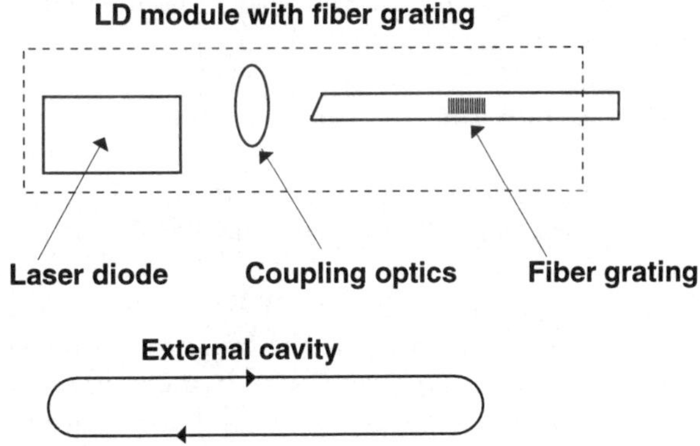

Figure 5.38 Schematic configuration of a wavelength-stabilized 980-nm LD with a fiber grating.

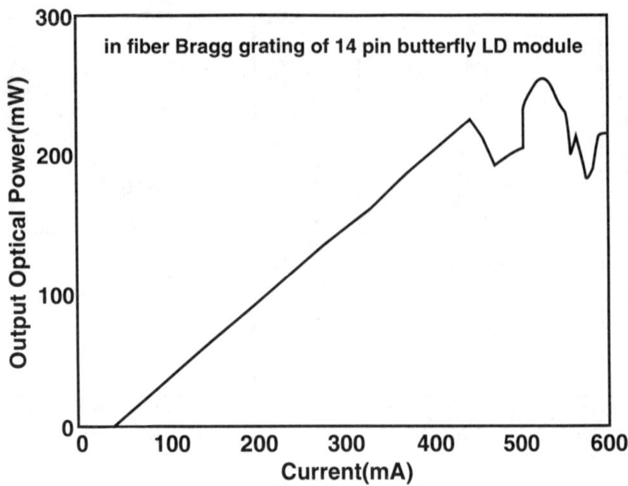

Figure 5.39 I-L curve of an wavelength-stabilized 1,017-nm LD [73].

an optical power of up to 200 mW into a single-mode fiber. The lasing spectra of this LD are shown in Figure 5.40. By comparison with a standard LD module, the lasing wavelength of the LD with the fiber grating is well stabilized at the Bragg wavelength.

980/1,017-nm MOPA Laser Module

A structure in which a flared power amplifier is monolithically integrated with master DBR-LD is very useful for realizing higher power semiconductor LDs. Figure 5.41 shows this *master oscillator power amplifier* (MOPA) structure [74]. The MOPA consists of a *distributed Bragg reflector* (DBR) master oscillator coupled to a flared power amplifier. The active layer is $In_{0.25}Ga_{0.5}As$ surrounded by AlGaAs cladding and a guiding layer. The gratings for the DBR master oscillator are second order and defined holographically. The flared amplifier region has a linearly tapered shape with an angle slightly larger than the free diffraction of the injected DBR master oscillator. The output aperture of the amplifier is 250 μm and antireflection coated. A MOPA without a DBR structure, which is called a tapered-gain-region laser, has also been reported [75].

The basic collimating optics from a MOPA laser chip are shown in Figure 5.42 [76]. A convex lens is used to collimate the fast vertical divergence of the laser (dashed lines) while bringing the beam to a waist in the horizontal plane. A plano-convex cylindrical lens collimates the beam in the horizontal plane (solid lines). The proper choice of focal lengths for these lenses yields a collimated beam with a near-unity aspect ratio and little wavefront distortion.

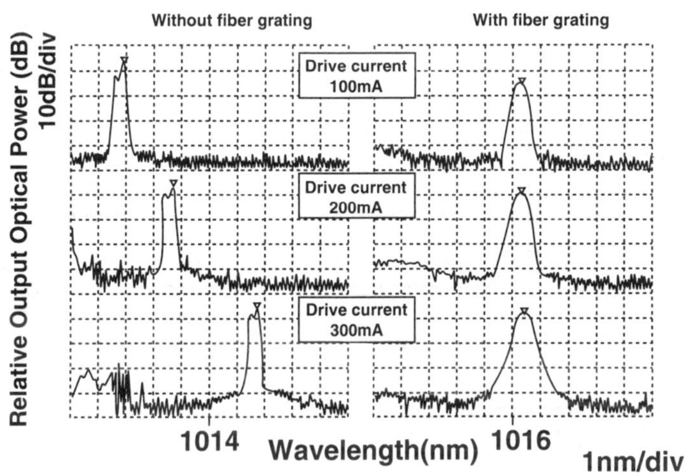

Figure 5.40 Lasing spectra of a wavelength-stabilized 1,017-nm LD [73].

Figure 5.43 shows typical I-L characteristics for the output power from a MOPA laser. When the drive current for the master DBR region is zero, the output power from the amplifier in the ASE mode radiates with large beam divergence. For a DBR drive current of 150 mA, the maximum output power in a single-diffraction limited output beam is 2-W CW at an input of 3.5A to the power amplifier.

Figure 5.44 shows typical I-L characteristics for a fiber-coupled MOPA laser module [77]. The maximum output power is about 600 mW from a single-mode fiber pigtail with a core diameter of about 6 μm.

5.2.7.2 Fiber Lasers

Fiber lasers are classified into two types: rare-earth-ion-doped fiber lasers with a dual cladding structure and fiber Raman lasers.

Rare-Earth-Ion-Doped Fiber Lasers

Rare-earth-ion-doped fiber lasers with single-mode fiber geometry and semiconductor LD pumping are attractive devices for constructing efficient lasers in the IR wavelength region. A good overlap between a pump and a laser light results in high

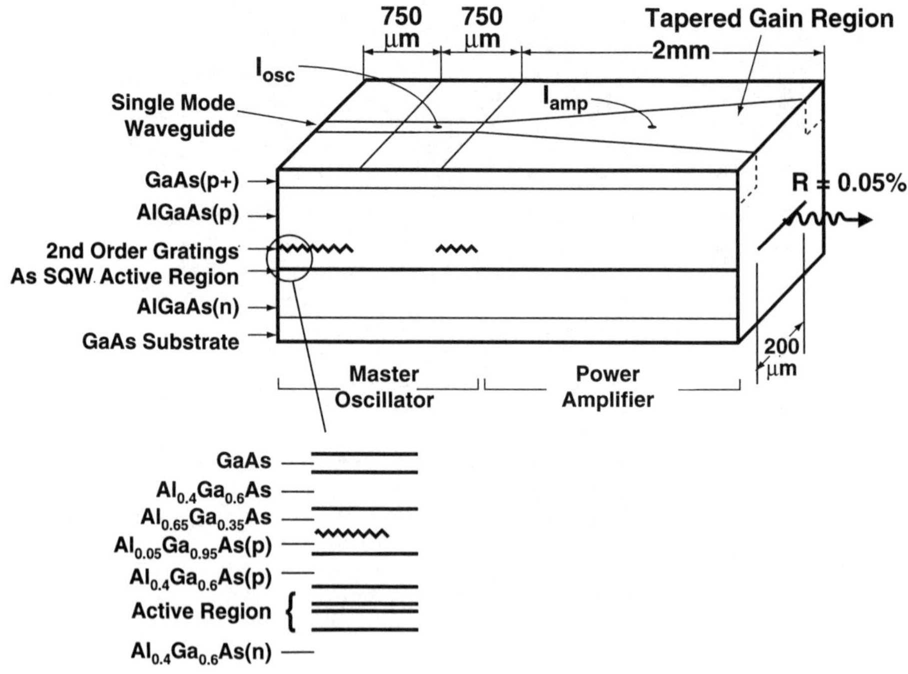

Figure 5.41 Structure of a MOPA laser [74].

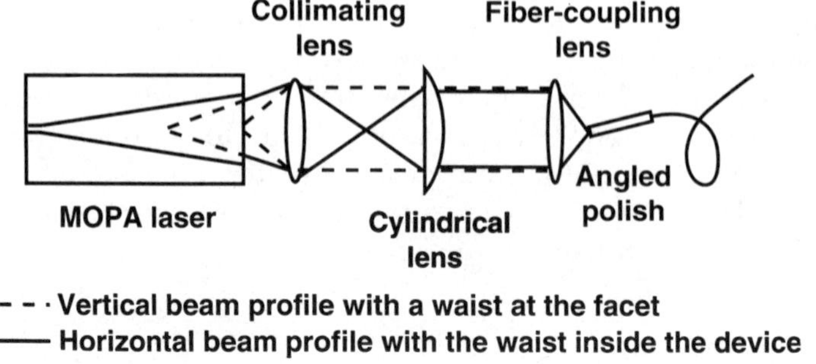

Figure 5.42 Example of collimating optics for a MOPA laser [75].

efficiency because both lights simultaneously propagate through a core. Increases in laser output power require that a high pump power is coupled into the core. As LDs with an output power of more than 1W have a wide-emitter width of over 100 μm, efficient coupling into the core has become a serious barrier to the construction of high-power fiber lasers. A dual cladding structure has been developed to overcome this problem.

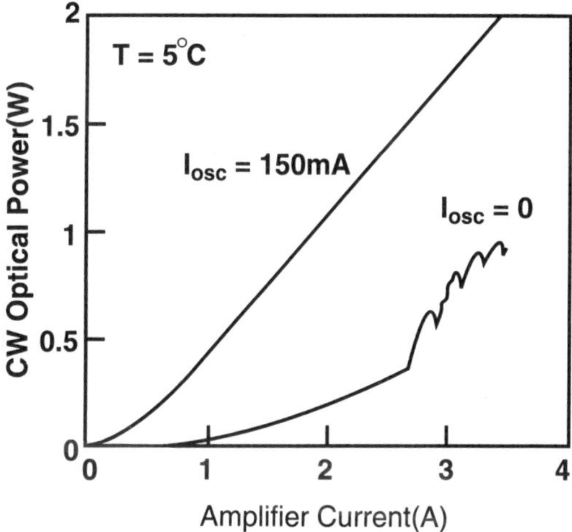

Figure 5.43 I-L curve of a 980-nm MOPA laser [74].

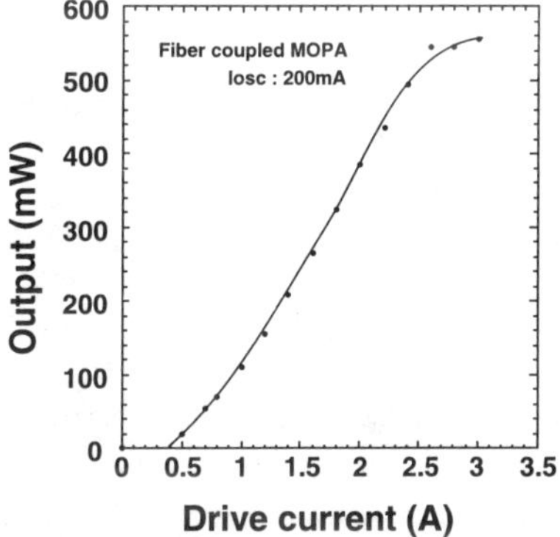

Figure 5.44 I-L curve of a fiber-coupled MOPA laser [77].

Figure 5.45 shows typical cross sections of a proposed dual-cladding fiber laser [78]. The key features of the dual-cladding fiber lasers are that they separate the propagating regions for the pump and lasing lights based on a coaxial structure and provide good coupling of the pump light into the outer cladding and good overlapping between the pump and lasing lights in the core. The overlapping of the two lights is illustrated schematically in Figure 5.46. The pump light from the wide-emitter LD is easily coupled into the inner cladding and propagates in a zig-zag manner. The structural parameters of the core and inner cladding satisfy the conditions for single-mode operation at the lasing wavelength. The refractive index difference between the inner and outer claddings is very large. A fiber laser with

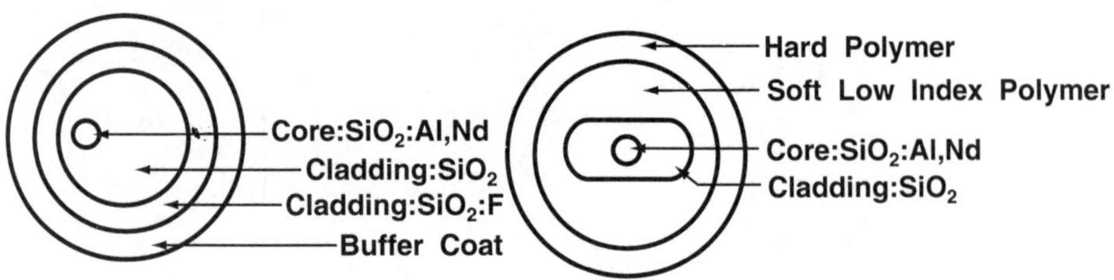

Figure 5.45 Cross sections of dual-cladding fiber lasers [78].

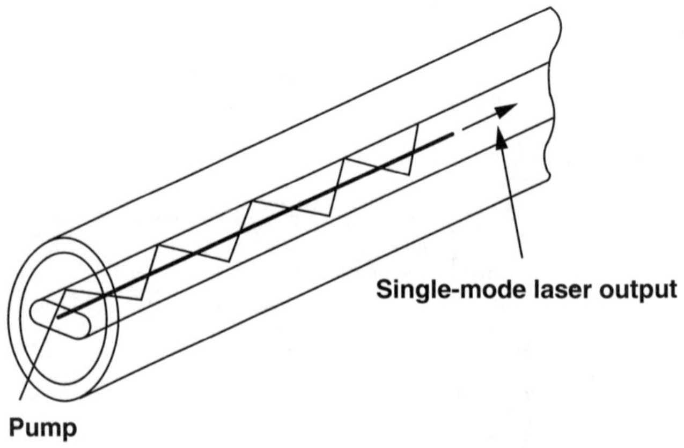

Figure 5.46 Overlapping of a pump and a lasing light in the dual-cladding fiber laser.

a dual cladding structure was first studied for a Nd-doped fiber laser and has also been applied to an Yb-doped fiber laser.

Figure 5.47 shows a refractive-index difference profile of a Yb-doped dual-cladding fiber laser, which uses silica and GeO_2-doped silica for the outer and the inner cladding [79]. The cavity mirrors are fiber Bragg gratings at a wavelength of 1.042 μm. The pump laser is a 0.975-μm semiconductor LD. Figure 5.48 shows the lasing characteristics of the laser [79]. A coupled pump power of 800 mW results in a lasing output power of about 500 mW.

Recently, a lasing output power of over 6W was reported from an Yb-doped fiber laser.

Fiber Raman Laser

The fiber Raman laser is a fiber-type wavelength converter that operates based on stimulated Raman scattering. Stimulated Raman scattering is a coherent light-

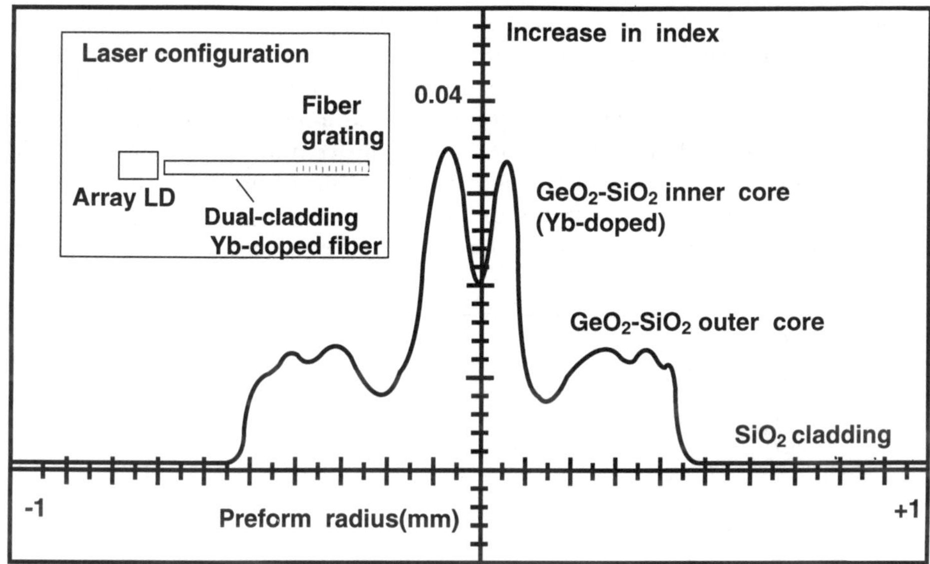

Figure 5.47 Refractive-index profile and laser cavity configuration of a Yb-doped dual-cladding fiber laser [79].

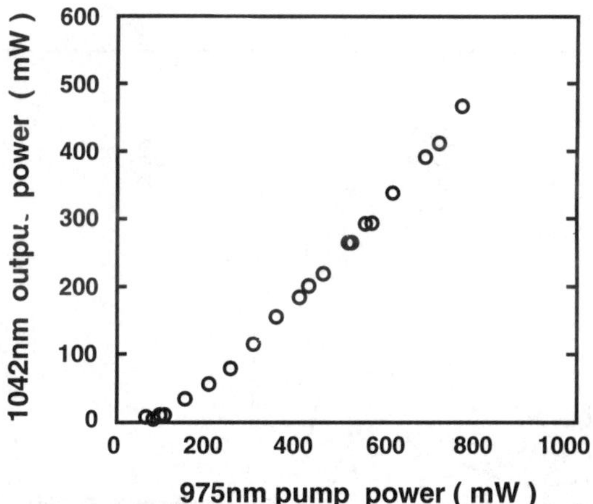

Figure 5.48 Lasing characteristics of a Yb-doped dual-cladding fiber laser [79].

scattering process by vibrational modes (optical phonons). Figure 5.49 shows a Raman gain spectrum for silica fibers [80]. Due to their glassy nature, fused silica fibers exhibit a broad gain spectrum over a 100-cm^{-1} bandwidth. Among the several Raman active vibration modes, the 440-cm^{-1} Raman mode exhibits the maximum Raman gain in silica fibers. At this peak, the gain coefficient for silica fibers is 9.4×10^{-14} mW^{-1} at a pump wavelength of 1,064 nm. Figure 5.50 shows Raman gain spectra for several glassy materials [81]. These oxide glasses are commonly used as dopants for fiber fabrication. The incorporation of these dopants has been

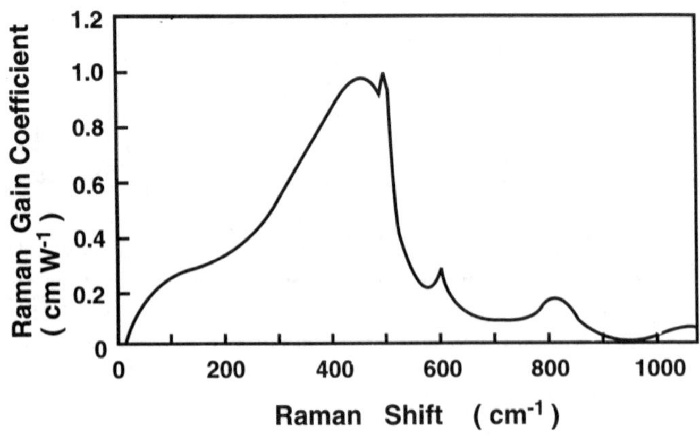

Figure 5.49 Raman gain spectra for silica fibers [80].

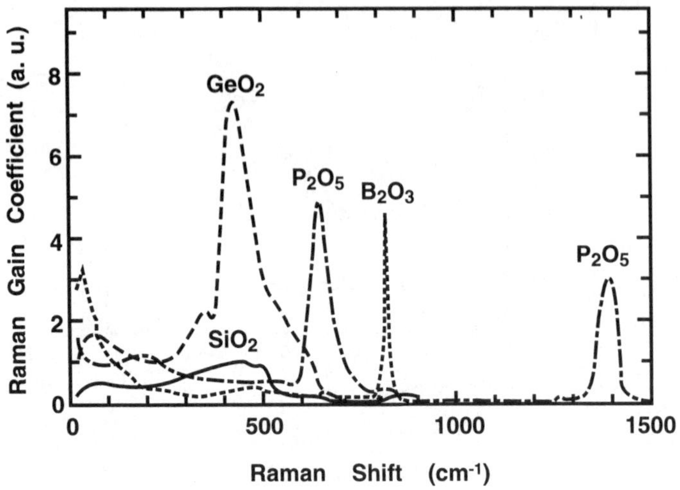

Figure 5.50 Raman gain spectra for glassy materials [81].

shown experimentally to modify the Raman gain coefficient for silica-based optical fibers. In particular, the Raman gain coefficient around 440 cm^{-1} can be increased by doping GeO_2 into silica. In order to achieve a consistently high Raman gain and to manufacture low-loss fibers, a silica-based single-mode fiber with a high GeO_2 content is a suitable gain medium [82].

Figure 5.51 shows the stimulated Raman scattering spectrum in a GeO_2-doped silica fiber pumped by a Q-switched Nd:YAG laser at a pump wavelength of 1,064 nm [83]. S_i represents the ith-order Stokes peak. The fifth-order stimulated Raman scattering enables a 1,064-nm incident light from the Nd:YAG laser to be successfully converted into a light around 1,450 nm. To construct highly efficient fiber Raman lasers, a cascaded laser cavity with a fiber Bragg grating is a very attractive and powerful configuration.

Figure 5.52 shows the cascaded fiber Raman laser configuration with an emission wavelength of 1,480 nm [49]. To increase the conversion efficiency, five nest-type laser cavities, whose emission wavelengths agree with the ith stimulated Raman scattering peak, are integrated with highly GeO_2-doped silica fiber. When a pump light is injected into the fiber Raman laser, it passes through the fiber Bragg grating without loss. The 1,117-nm pump light from the Yb-doped dual-cladding fiber laser

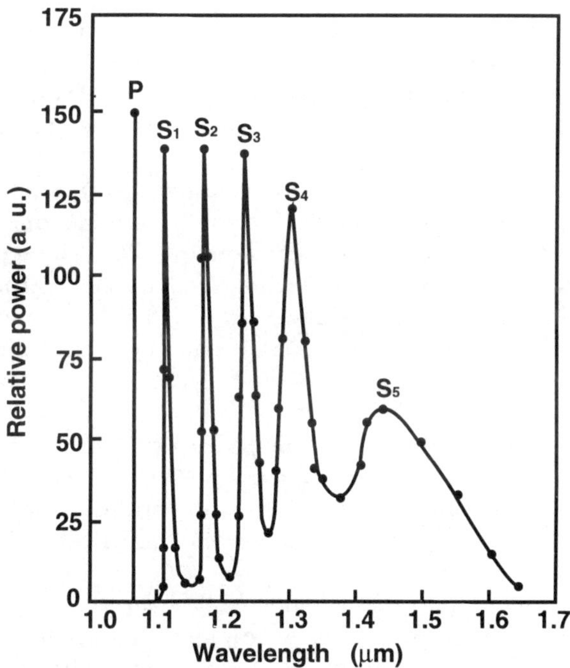

Figure 5.51 Stimulated Raman scattering spectrum in a GeO_2-doped silica fiber [83].

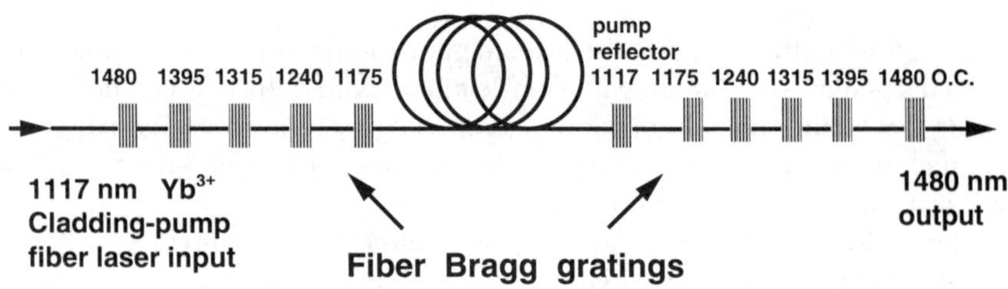

Figure 5.52 Schematic configuration of a cascaded fiber Raman laser [84].

is reflected at the 1.117-μm fiber grating. The central cavity with 1,175-nm gratings begins the lasing emission because the grating wavelength coincides with the first-order Stokes peak in the silica fiber. Then, this 1,175-nm laser emission becomes a pump source for the adjacent cavity with 1,240-nm gratings. With this cascaded process, the pump light from the Yb-doped fiber laser is successfully converted to 1,480-nm light with a conversion efficiency of 48%.

5.2.7.3 LD-Pumped Solid-State Lasers

Nd:YAG and Nd:YLF lasers are commonly used as pump sources for fiber amplifiers because these lasing wavelengths are in the absorption band of the ytterbium-sensitized erbium system and the praseodymium system. In addition, they achieve high quantum efficiency for lasing and have good thermal and mechanical properties.

Neodymium-doped yttrium aluminum garnet ($\mathrm{Nd:Y_3Al_5O_{12}, Nd:YAG}$) possesses a combination of properties uniquely favorable for laser operation. The YAG structure is stable from the lowest temperatures up to its melting point. Pure YAG is a colorless, optically isotropic crystal resulting from the cubic structure of garnets. The lasing transition at 1,064 nm is based on the stimulated transition from the $^4F_{3/2}$ level to the $^4I_{11/2}$ level. The upper laser level, $^4F_{3/2}$, has a radiative quantum efficiency of more than 99.5% and a fluorescence lifetime of 230 μs. The branching ratio from the $^4F_{3/2}$ level to the $^4I_{11/2}$ level is 60%.

Neodymium-doped lithium yttrium fluoride ($\mathrm{Nd:LiYF_4}$, Nd:YLF) has two lasing wavelengths at around 1,050 nm because YLF is a birefringent crystal. The lasing lines at 1,047 nm (extraordinary) and 1,053 nm (ordinary) are easily separated with an intracavity polarizer. The fluorescence lifetime in Nd:YLF is 480 μs, which is twice as long as in Nd:YAG.

The cavity configurations of LD-pumped solid-state lasers are classified into two types: end-pumping and side-pumping configurations. To achieve a high-conversion efficiency from electric to optic power, the end-pumping technique is highly advantageous for a laser output power of about 1W. Figure 5.53 shows the schematic configuration of an end-pumping type Nd:YAG laser [85]. Either a multi-emitter

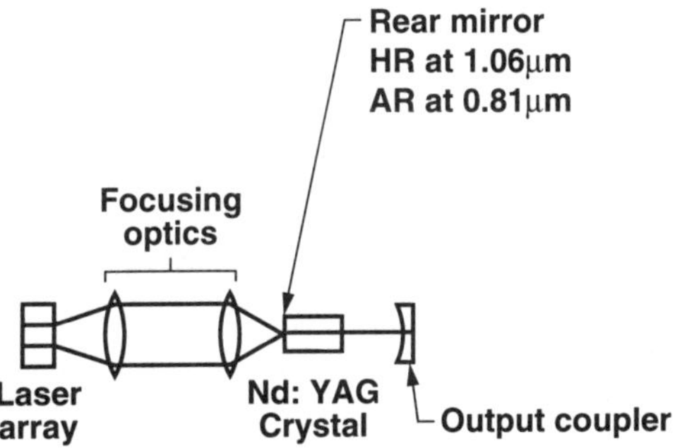

Figure 5.53 Schematic configuration of an LD-pumped Nd:YAG laser [85]

or a wide-emitter AlGaAs LD is usually used as the pump source. The lasing wavelength is precisely tuned to the absorption band of the Nd:YAG or the Nd:YLF. Figure 5.54 shows the lasing characteristics of the first Nd:YAG laser to be demonstrated with the end-pumping technique. The output power is about 80 mW for a pump power of 1W. The transversal mode of the output beam is TEM_{00}.

Currently, the typical output powers of wide-emitter LDs are 1W, 2W, and 4W for 100-, 200-, and 500-μm width-emitters, respectively. A TEM_{00} output power of 1W is therefore available for an LD-pumped Nd:YLF and Nd:YLF employing the end-pumping technique. Good beam quality results in high coupling efficiency between the laser and a single-mode fiber pigtail.

5.3 AMPLIFICATION CHARACTERISTICS OF FIBER AMPLIFIER MODULES

This section has three main parts. Section 5.3.1 describes *erbium* (Er^{3+})-doped fiber amplifiers for use in the 1.5-μm band. Section 5.3.2 focuses on *neodymium* (Nd^{3+})-doped fiber amplifiers and *praseodymium* (Pr^{3+})-doped fiber amplifiers for 1.3-μm-band applications. Section 5.3.3 presents *thulium* (Tm^{3+})-doped fluoride fiber amplifiers for 1.4-μm- and 1.65-μm-band use. All aspects of the four rare-earth-doped fiber amplifiers are covered to provide the reader with a comprehensive insight into their basic amplification principles, and device performance is reviewed.

5.3.1 1.5-μm-Band Fiber Amplifier

Since the first demonstration of 1.5-μm signal amplification with *Er^{3+}-doped fiber* (EDF) by Southampton University in 1987 [86] and the corroborative evidence of its advantages provided by the first successful large-capacity, long-haul optical transmission experiment with an *Er^{3+}-doped fiber amplifier* (EDFA) in 1989 [87], great efforts have been made throughout the world to improve amplification performance

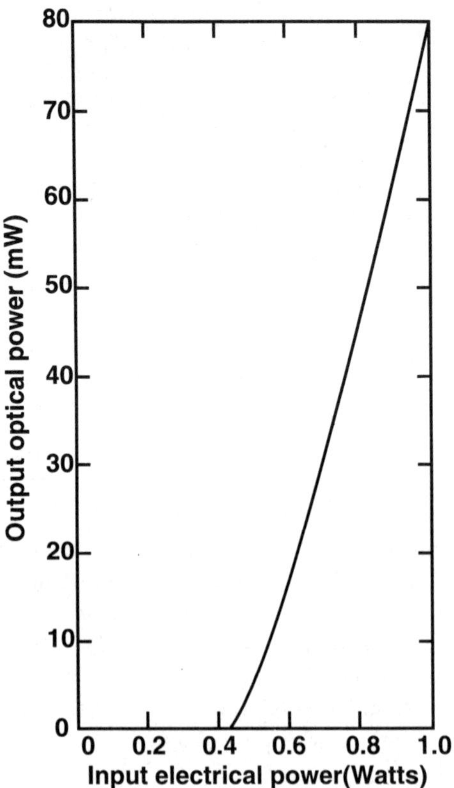

Figure 5.54 Lasing characteristics of an LD-pumped Nd:YAG laser [85].

and to develop 1.5-μm-band optical fiber amplifiers for practical use by means of theoretical and experimental investigations of the amplification characteristics. Practical 980- and 1,480-nm LDs and their modules were fabricated on the basis of earlier proposals of 980-nm [88,89] and 1,480-nm [90] pumping as efficient ways of exciting EDF. The first EDFA module was demonstrated in 1990 [91–93]. Today, several types of EDFA appear in the catalogs of optoelectronics component manufacturers, and various commercial 1.5-μm-band analog and digital transmission systems have been constructed with EDFAs.

In this section, we will first outline the amplification principle and the theoretical model of the EDFA. Then we will describe basic amplification characteristics such as the pump wavelength dependence of signal gain characteristics, the gain spectrum, saturation characteristics, noise characteristics, and the temperature dependence of the EDF. We will also mention optimization of the EDF structure that enables us to achieve efficient amplification. Furthermore, after describing an EDFA module with a basic amplifier configuration such as unidirectional or bidirectional pumping, we will say something about the advanced EDFA. Here, we will focus on EDFAs with cascade or double-pass configurations that improve the gain characteristics, broadband EDFAs with uniform gain and low noise for application to WDM transmission systems and all-optical self-routed wavelength-addressable networks, and high-output power EDFAs for long-haul repeaterless optical transmission and distribution applications.

5.3.1.1 Amplification Principle and Calculation Model of EDFA

Figure 5.55 shows the energy level diagram of the Er^{3+} ion and the associated transition [94]. Each energy level is split into several sublevels in a host glass environment by the Stark effect; that is, an electromagnetic field induced by atoms adjacent to an Er^{3+} ion affects its energy levels. For example, the $^4I_{13/2}$ and $^4I_{15/2}$ levels are split into seven and eight sublevels, respectively. Each level broadens because of site-to-site variations in the field caused by the amorphous nature of the glass.

In the diagram, the stimulated emission for the amplification of the signal photons in the 1.5-μm band is from the $^4I_{13/2}$ level to the $^4I_{15/2}$ level. There are several pump wavelength bands including 514 nm ($^4I_{15/2}{\rightarrow}^2H_{11/2}$ transition), 532 nm ($^4I_{15/2}{\rightarrow}^4S_{3/2}$ transition), 660 nm ($^4I_{15/2}{\rightarrow}^4F_{9/2}$ transition), 800 nm ($^4I_{15/2}{\rightarrow}^4I_{9/2}$ transition), 980 nm ($^4I_{15/2}{\rightarrow}^4I_{11/2}$ transition), and 1,480 nm

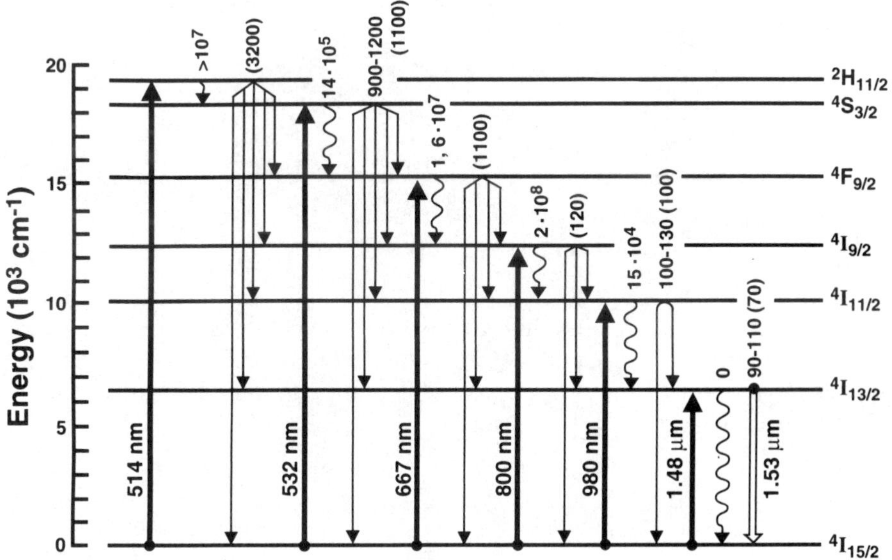

Figure 5.55 Energy level diagram of the Er^{3+} ion and the associated transition [94].

($^4I_{15/2} \rightarrow$ the upper portion of the $^4I_{13/2}$ transition). The excited Er^{3+} ions are essentially transferred to the $^4I_{13/2}$ level via fast nonradiative decay with one or several vibrational phonons of the surrounding crystal lattice [95,96], and the signal photons are amplified with a three-level amplification system without 1,480-nm pumping. Figure 5.55 also shows the range of radiative and nonradiative rates in reciprocal seconds, measured on each level in various glasses including silicate-, germanate-, fluorophosphate-, and ZrF_4-based fluoride glasses. The figure shows that, except for ZrF_4-based fluoride glasses, the relaxation process between the $^2H_{11/2}$ and $^4I_{13/2}$ levels is strongly dominated by fast nonradiative decay while the $^4I_{13/2} \rightarrow {}^4I_{15/2}$ transition is 100% radiative with a fluorescence lifetime of the order of 10 ms. With 1,480-nm pumping, which uses the upper portion of the $^4I_{13/2}$ level, the Er^{3+} ions are decayed by the intraband nonradiative transition and used for 1.5-μm-band amplification, where they operate as a quasi-three-level system. Of these, the 980- and 1,480-nm pumping band are mainly used for practical EDFAs because these bands offer good pump efficiency and high-output power characteristics. This section provides differential equations describing the evolution at the fiber coordinate z of the pump power $P_P(z)$ and signal power $P_S(z)$, for 980- and 1,480-nm pumped EDFAs.

EDFA amplification models have been reported in [94,97–113], and a unified description based on these models can be found in [114,115]. In this section we cite the descriptions given in [114,115]. For 980-nm pumping ($^4I_{15/2} \rightarrow {}^4I_{11/2}$ transition), rate equations on the ion population density of $N_1(^4I_{15/2}$ ground level), $N_2(^4I_{13/2}$ metastable level), and $N_3(^4I_{11/2}$ pump level) are given by

$$\frac{dN_1}{dt} = -W_{13}N_1 + W_{31}N_3 + W_{21}N_2 - W_{12}N_1 + \frac{N_2}{\tau_2} \tag{5.7}$$

$$\frac{dN_2}{dt} = A_{32}^{\text{non-rad}}N_3 - W_{21}N_2 + W_{12}N_1 - \frac{N_2}{\tau_2} \tag{5.8}$$

$$\frac{dN_3}{dt} = -A_{32}^{\text{non-rad}}N_3 + W_{13}N_1 - W_{31}N_3 \tag{5.9}$$

where τ_2 is the fluorescence lifetime of the $^4I_{13/2}$ level, W_{ij} is the transition rate from the ith to the jth levels, and $A_{32}^{\text{non-rad}}$ is the fast nonradioactive transition rate from the $^4I_{11/2}$ level to the $^4I_{13/2}$ level. The pumping rate W_{13} and stimulated emission rates W_{31} and $W_{21,12}$ are given by

$$W_{13,31} = \frac{\sigma_{a,e}^P(\nu_P)}{h\nu_P} \cdot P_P(\nu_P, z) \cdot \Psi_P(\nu_P, r, \theta) \tag{5.10}$$

$$W_{21,12} = \int \frac{\sigma_{e,a}^S(\nu)}{h\nu_S} \cdot (P_P^+(\nu, z) + P_S^-(\nu, z)) \cdot \Psi_S(\nu, r, \theta)\, d\nu \tag{5.11}$$

In (5.10) and (5.11), $\sigma_{a,e}^{S}$ and $\sigma_{a,e}^{P}$ are the absorption and emission cross sections for the signal and pump lights, respectively; ν_S and ν_P are the frequencies of the signal and pump lights, respectively; and h is Plank's constant. $P_P(\nu_P)$ are pump powers with the frequency ν_P, and $P_S^{\pm}(\nu, z)$ are forward (+) and backward (−) propagating signals and the spontaneous emission power in the 1,550-nm band, with the frequency ν. $\Psi_P(r, \theta)$, $\Psi_S(r, \theta)$ are the normalized transverse mode envelopes of the pump and signal, respectively; that is, $\iint \Psi_{S,P}(r, \theta)\, dr\, d\theta = 1$, where (r, θ) represent the cylindrical transverse coordinates (z is the fiber longitudinal coordinate). The signal power $P_S(\nu_S, z)$ with the frequency ν_S is described as $P_S(\nu, z)\delta(\nu - \nu_S)$, where $\delta(\nu - \nu_S)$ is the delta function.

The total population density $\rho(r, \theta)$ satisfies $\rho(r, \theta) = N_1 + N_2$, where N_3 is zero since it is assumed that the $^4I_{11/2}$ level relaxes instantaneously to the metastable $^4I_{13/2}$ level. Ion population densities N_1 and N_2 in a steady state for the $^4I_{15/2}$ and $^4I_{13/2}$ levels are given by

$$N_1 = \rho \frac{1 + W_{21}\tau_2}{1 + W_{13}\tau_2 + (W_{12} + W_{21})\tau_2} \tag{5.12}$$

$$N_2 = \rho \frac{W_{13}\tau_2 + W_{12}\tau_2}{1 + W_{13}\tau_2 + (W_{12} + W_{21})\tau_2} \tag{5.13}$$

On the other hand, for 1,480-nm pumping ($^4I_{15/2} \to$ the upper portion of the $^4I_{13/2}$ transition), the rates $W_{13,31} = 0$ and the stimulated emission rates $W_{12,21}$ in (5.11) must be modified. The modified $W_{12,21}^{M}$, which includes the pumping rates $W_{12,21}^{Pump}$ for the 1,480-nm pumping, is given by

$$W_{12,21}^{M} = W_{12,21}^{Pump} + W_{12,21} \tag{5.14}$$

where

$$W_{12,21}^{Pump} = \frac{\sigma_{a,e}^{P}(\nu_P)}{h\nu_P} \cdot P_P(\nu_P, z) \cdot \Psi_P(\nu_P, r, \theta) \tag{5.15}$$

The steady-state solutions of (5.7) to (5.9) for ion population densities N_1 and N_2 take the form

$$N_1 = \rho \frac{1 + W_{21}^{Pump}\tau_2 + W_{21}\tau_2}{1 + (W_{12}^{Pump} + W_{21}^{Pump})\tau_2 + (W_{12} + W_{21})\tau_2} \tag{5.16}$$

$$N_2 = \rho \frac{W_{12}^{Pump}\tau_2 + W_{12}\tau_2}{1 + (W_{12}^{Pump} + W_{21}^{Pump})\tau_2 + (W_{12} + W_{21})\tau_2} \tag{5.17}$$

The signal emission coefficient $\gamma_e(\nu_S, r, \theta)$ and signal absorption coefficient $\gamma_a(\nu_S, r, \theta)$ at signal frequency ν_S and the pump absorption coefficient $\gamma_P(\nu_P, r, \theta)$ at pump frequency ν_P, for both 980- and 1,480-nm pumping, can be expressed with transverse coordinates (r, θ) by (5.12), (5.13), (5.16), and (5.17) as

$$\gamma_e(\nu_S, r, \theta) = \sigma_e^S(\nu_S) N_2(r, \theta)$$

$$= \rho \sigma_e^S(\nu_S) \frac{R^* \tau_2 + W_{12}\tau_2}{1 + R^*\tau_2 \left[1 + \dfrac{1}{S(\nu_P)}\right] + (W_{12} + W_{21})\tau_2} \tag{5.18}$$

$$\gamma_a(\nu_S, r, \theta) = \sigma_a^S(\nu_S) N_1(r, \theta)$$

$$= \rho \sigma_a^S(\nu_S) \frac{1 + \dfrac{R^*\tau_2}{S(\nu_P)} + W_{21}\tau_2}{1 + R^*\tau_2 \left[1 + \dfrac{1}{S(\nu_P)}\right] + (W_{12} + W_{21})\tau_2} \tag{5.19}$$

$$\gamma_P(\nu_P, r, \theta) = \sigma_a^P(\nu_P) N_1(r, \theta)$$

$$= \rho \sigma_a^P(\nu_S) \frac{1 + \dfrac{R^*\tau_2}{S(\nu_P)} + W_{21}\tau_2}{1 + R^*\tau_2 \left[1 + \dfrac{1}{S(\nu_P)}\right] + (W_{12} + W_{21})\tau_2} \tag{5.20}$$

where

$$R^* = \begin{bmatrix} W_{13} = \dfrac{\sigma_a^P(\nu_P)}{h\nu_P} \cdot P_P(\nu_P, z) \cdot \Psi_P(\nu_P, r, \theta) & \text{(980-nm pumping)} \\[2em] W_{12}^{\text{Pump}} = \dfrac{\sigma_a^P(\nu_P)}{h\nu_P} \cdot P_P(\nu_P, z) \cdot \Psi_P(\nu_P, r, \theta) & \text{(1,480-nm pumping)} \end{bmatrix}$$

$$\tag{5.21}$$

$$S(\nu_P) = \frac{\sigma_a^P(\nu_P)}{\sigma_e^P(\nu_P)} \tag{5.22}$$

The signal emission coefficient, signal absorption coefficient, and pump absorption coefficient become identical using (5.18) to (5.20), for both 980- and 1,480-nm pumping. When $\sigma_e^P(\nu_P)$ is much less than $\sigma_a^P(\nu_P)$ for 980-nm pumping, the term proportional to $\sigma(\nu_P)^{-1}$ vanishes.

The total signal emission coefficient $\gamma_e(z, \nu)$, signal absorption coefficient $\gamma_a(z, \nu)$, and pump absorption coefficient $\gamma_P(z, \nu_P)$, are obtained by integrating $\gamma_e(\nu, r, \theta)$, $\gamma_a(\nu_S, r, \theta)$, and $\gamma_P(\nu_P, r, \theta)$ over the transverse plate (r, θ) weighted by the normalized signal and pump distribution $\Psi_{P,S}(r, \theta)$

$$\gamma_e(z, \nu) = \int\int \gamma_e(z, \nu, r, \theta) \cdot \Psi_S(\nu, r, \theta) r \, dr \, d\theta \tag{5.23}$$

$$\gamma_a(z, \nu) = \int\int \gamma_a(z, \nu, r, \theta) \cdot \Psi_S(\nu, r, \theta) r \, dr \, d\theta \tag{5.24}$$

$$\gamma_P(z, \nu_P) = \int\int \gamma_P(z, \nu, r, \theta) \cdot \Psi_P(\nu_P, r, \theta) r \, dr \, d\theta \tag{5.25}$$

Furthermore, (5.23) to (5.25), which are expressed as integral functions, can be simplified by assuming a uniform Er^{3+} density distribution $\rho(r, \theta) = \rho_0$ for $r \leq a$ and $\rho(r, \theta) = 0$ for $r > a$, where a is the fiber core radius, and by using a Gaussian approximation of $\Psi_{P,S}(r, \theta)$, that is, $\Psi_{P,S}(r, \theta) = [\exp(-r^2/\omega_{S,P}^2)]/\pi\omega_{S,P}^2$, where $\omega_{S,P}$ is the mode power radius for the signal and pump lights. They can then be rewritten as

$$\gamma_e(z, \nu) = \rho_0\sigma_e^S(\nu)\Gamma_S\frac{U + V\left[1 + \dfrac{1}{s(\nu)}\right]}{1 + U\left[1 + \dfrac{1}{s(\nu_P)}\right] + V} \tag{5.26}$$

$$\gamma_a(z, \nu) = \rho_0\sigma_a^S(\nu)\Gamma_S\frac{1 + V[1 + s(\nu)] + \dfrac{U}{s(\nu_P)}}{1 + U\left[1 + \dfrac{1}{s(\nu_P)}\right] + V} \tag{5.27}$$

$$\gamma_P(z, \nu_P) = \rho_0\sigma_a^P(\nu_P)\Gamma_P\frac{1 + V[1 + s(\nu)] + \dfrac{U}{s(\nu_P)}}{1 + U\left[1 + \dfrac{1}{s(\nu_P)}\right] + V} \tag{5.28}$$

In (5.26) to (5.28), $s(\nu) = \sigma_a^S(\nu)/\sigma_e^S(\nu)$ is the ratio of the absorption cross section to the emission cross section for the pump light; $\Gamma_{S,P} = 1 - \exp(-a^2/\omega_{S,P}^2)$ is a filling factor; and $U = P_P(z)/P_P^{th}$, and $V = P_S(z)/P^{sat}$, where P_P^{th} is the pump power threshold and $P^{sat}(\nu_S)$ are gain saturation powers, given as

$$P_P^{th} = \frac{h\nu_P \pi \omega_P^2}{\sigma_a^P(\nu_P)\tau_2} \tag{5.29}$$

$$P^{sat}(\nu_S) = \frac{h\nu_S \pi \omega_S^2}{[\sigma_a^S(\nu_S) + \sigma_e^S(\nu_S)]\tau_2} \tag{5.30}$$

From the expression of the total signal emission coefficient, the signal absorption coefficient and the pump absorption coefficient, given by (5.26) to (5.28), the differential equations describing the envelope at the fiber coordinate z for signal power $P_S(z, \nu)$ and pump power $P_P(z)$ have to be solved as

$$\frac{dP_S^{\pm}(z, \nu)}{dz} = \pm(\gamma_e(z, \nu) - \gamma_a(z, \nu))P_S^{\pm}(z, \nu) \pm P_0\gamma_e(z, \nu) \tag{5.31}$$

$$\frac{dP_P(z)}{dz} = -\gamma_P(z, \nu_P)P_P(z) \tag{5.32}$$

where $P_0 = 2h\nu_S\Delta\nu$ is the power of two photons per unit frequency in bandwidth and corresponds to spontaneous emission in the two polarization modes of the fiber.

The amplification characteristics will be obtained by solving differential equations (5.31) and (5.32) with the absorption and emission cross sections for the signal and pump lights $\sigma_{a,e}^{S,P}$, lifetime τ_2, and the mode power radius for the signal and pump light $\omega_{S,P}$. The absorption and emission cross sections and the lifetime are important basic EDF parameters. There have been several reports about these parameters for various glasses, indicating that they are dependent properties. This is because they can be modified by using a different host glass with slightly different electromagnetic strength at Er^{3+} ions and by a change in the total number of sites in the glass [116–123]. As an example, typical absorption and emission cross-section spectra between the $^4I_{13/2}$ and the $^4I_{15/2}$ levels are shown in Figure 5.56(a–e) for Ge/Er-doped silica glass [117], Al/Er-doped silica glass [117], Ge/Al/Er-doped silica glass [117], Al/P/Er-doped silica glass [116], and ZrF_4-based fluoride glass [116], respectively, and absorption cross sections from the $^4I_{15/2}$ to $^4I_{11/2}$ transition and the lifetimes of the $^4I_{13/2}$ level for these glasses are given in Table 5.4 [116,117]. By contrast, the mode power radius is a basic parameter for optical fiber. This parameter is expressed with core radius a and relative refractive-index difference $\Delta n = 1 - n_2/n_1$ where n_1 and n_2 are the refractive indexes of the core and cladding of the fiber, respectively, as [124]

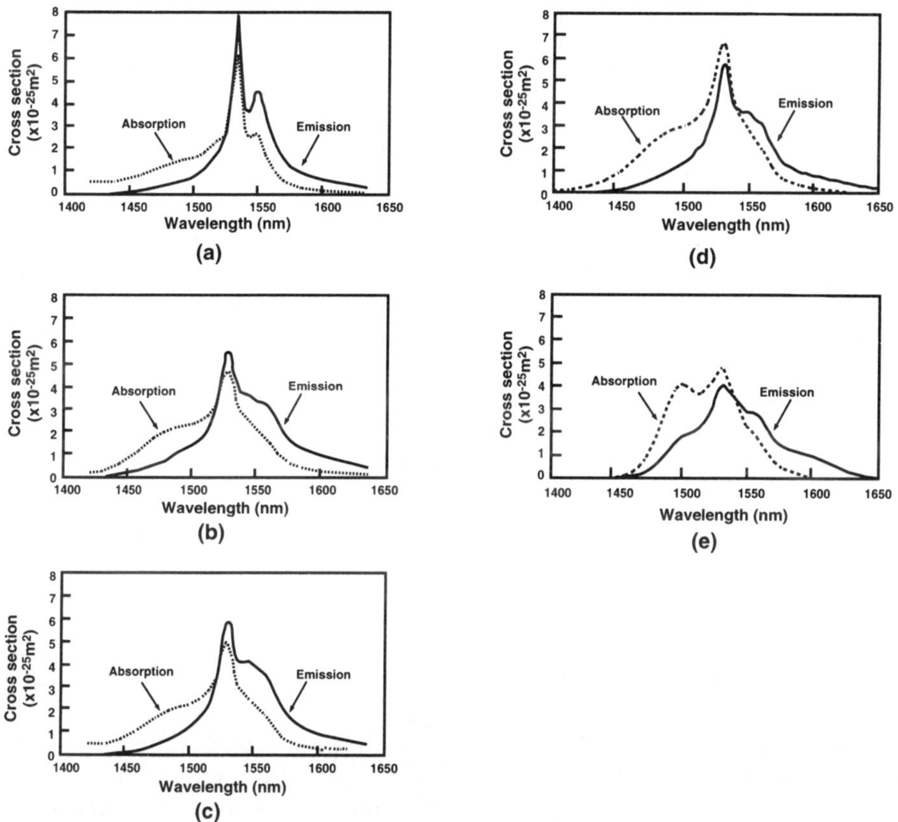

Figure 5.56 Typical absorption and emission cross-section spectra between the $^4I_{13/2}$ and the $^4I_{15/2}$ levels: (a) Ge/Er-doped silica glass [117]; (b) Al/Er-doped silica glass [117]; (c) Ge/Al/Er-doped silica glass [117]; (d) Al/P/Er-doped silica glass [116]; and (e) ZrF$_4$-based fluoride glass [116].

$$\frac{\omega_{S,P}}{a} = 0.65 + \frac{1.619}{V_{S,P}^{3/2}} + \frac{2.879}{V^6}$$

$$V_{S,P} = n_1 \frac{2\pi}{\lambda_{S,P}} a\sqrt{2\Delta n}$$

$$(5.33)$$

where λ_S and λ_P are the signal and pump wavelengths, respectively. The mode power radius can also be measured by the far-field scanning technique and analyzed using the Petermann II definition of the mode power radius [125] at pump and signal wavelengths.

Table 5.4
Absorption Cross Sections from the $^4I_{15/2}$ to $^4I_{11/2}$ Transition and the Lifetimes of the $^4I_{13/2}$ Level

	Absorption Cross Sections from the $^4I_{15/2}$ to $^4I_{11/2}$ Transition $(\times\ 10^{-25}\ m^2)$	Lifetimes of the $^4I_{13/2}$ Level (ms)	Cite
Ge/Er-doped silica glass	2.52 ± 0.03	12.1 ± 0.3	[117]
Al/Er-doped silica glass	1.9 ± 0.3	10.2 ± 0.3	[117]
Ge/Al/Er-doped silica glass	1.7 ± 0.3	10.2 ± 0.3	[117]
Al/P/Er-doped silica glass	3.12	10.8	[116]
ZrF$_4$-based fluoride glass	2.15	9.4	[116]

5.3.1.2 Basic Amplification Characteristics of EDFA

The basic amplification characteristics, such as the pump wavelength dependence of the signal gain characteristics, gain spectrum, saturation characteristics, noise characteristics, and temperature dependence, will be described in this subsection.

Pump Wavelength Dependence of Signal Gain

As explained in Subsection 5.3.1.1, there are several pump wavelength bands including the 514-, 532-, 667-, 800-, 980-, and 1,480-nm bands. Research results have been published on pump bands of 514.5 and 528 nm with an Ar laser [88,89,94,97]; 532 nm with a frequency doubled Nd:YAG laser [126,127]; 665 nm with InGaAlP/GaAs semiconductor LDs and a dye laser [86,102,112,128,129]; 800 to 830 nm with a dye laser, Ti-sapphire laser, and AlGaAs/GaAs LDs [129–134]; 970 to 990 nm with a dye laser, Ti-sapphire laser, and InGaAs/GaAs strained quantum well LDs [88,135–140]; and 1,460 to 1,500 nm with a color center laser and InGaAsP/InP LDs [87,141–145]. The amplification performance of EDFs pumped by each pump band have been compared in terms of gain coefficient, rate of pump *excited state absorption* (ESA) cross section $\sigma_{ESA}(\nu_P)$ and pump absorption cross section $\sigma_a^P(\nu_P)$, power conversion efficiency, and noise characteristics. First, we will explain each of these parameters.

The gain coefficient is defined as the ratio of the signal gain to the pump power launched into the Er^{3+}-doped fiber [88–90]. Figure 5.57 shows the typical relationship between the signal gain and the launched pump power [143]. As the pump power increases, the gain increases steeply and then levels off. The last regime of the signal gain corresponds to the case where almost all the Er^{3+} ions are excited to the $^4I_{13/2}$ level along the whole fiber length. The maximum gain coefficient is obtained as the slope of the tangent to the gain curve that intersects the origin, as shown by the dashed line in this figure. Moreover, because the gain coefficient

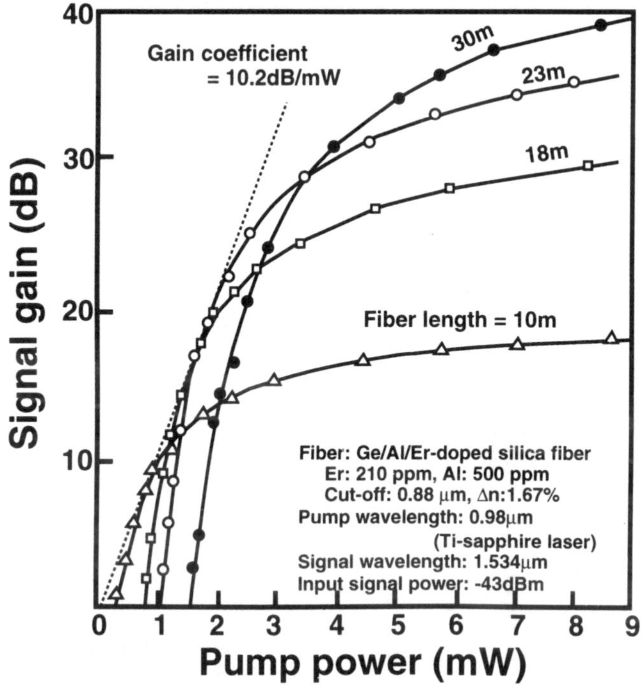

Figure 5.57 Typical relationship between signal gain and launched pump power [143].

depends not only on the fiber length as shown in Figure 5.57 but also on the input signal power as shown in Figure 5.58 [144], the value is usually estimated under a small signal condition with optimized fiber length to obtain the maximum value. To date, the gain coefficient has been improved by optimizing the EDF structure and the highest gain coefficient yet achieved is 11.0 dB/mW for 980-nm pumping [146,147]. In Subsection 5.3.1.3, we will describe EDF design optimization.

The pump ESA must be known in order to achieve efficient pumping because, as a result of pump ESA, the pump light at ν_P is not only absorbed from the ground level $^4I_{15/2}$ of the Er^{3+} ion but also from the excited level $^4I_{13/2}$ due to the excitation of the upper level whose energy gap with $^4I_{13/2}$ happens to match closely the pump photon energy $h\nu_P$. This process results in an excess loss for the pump light and the wasteful consumption of pump power. Because the pump ESA cross section $\sigma_{ESA}(\nu_P)$ is important with respect to the pump absorption cross section $\sigma_a^P(\nu_P)$, the parameter $\delta = \sigma_{ESA}(\nu_P)/\sigma_a^P(\nu_P)$ is usually used to express the influence of the ESA process on each pump band. ESA has been observed to occur in Er^{3+}-doped fiber in several wavelength bands [102,116,129,148–158]. It has been found that the pump ESA bands are 514 nm ($^4I_{13/2}{\rightarrow}^2H_{9/2}$ transition) [148–151], 630 nm

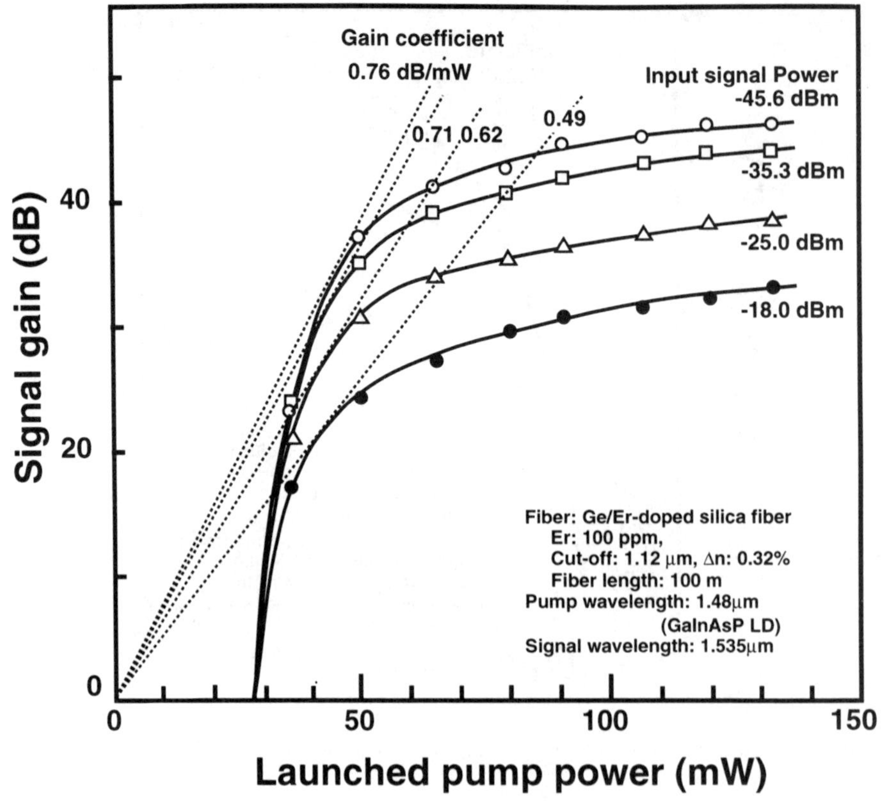

Figure 5.58 Signal gain against pump power for different signal power [144].

($^4I_{13/2} \rightarrow {}^4F_{5/2}$ transition) [102,129,148–151], 715 nm ($^4I_{13/2} \rightarrow {}^4F_{7/2}$ transition) [116,129,148–151], 790 nm ($^4I_{13/2} \rightarrow {}^2H_{11/2}$ transition) [116,129,148–154], 850 nm ($^4I_{13/2} \rightarrow {}^4S_{3/2}$ transition) [116,129,148–154], 1,140 nm ($^4I_{13/2} \rightarrow {}^4F_{9/2}$ transition) [155], 1,680 nm ($^4I_{13/2} \rightarrow {}^4I_{9/2}$ transition) [156], and 980 nm ($^4I_{11/2} \rightarrow {}^4F_{7/2}$ transition) [157,158]. Therefore, the 514-, 660-, and 800-nm pump bands are affected by the pump ESA process. Although the 980-nm pump band closely matches the $^4I_{11/2} \rightarrow {}^4F_{7/2}$ transition, the pump ESA process does not occur with 980-nm pumping because the Er^{3+} ions excited to the $^4I_{11/2}$ level are rapidly dumped by nonradiative decay for silica-based EDF without fluoride-based EDF [158]. The influence of the pump ESA process on fluoride-based EDF used for 980-nm pumping will be described in Subsection 5.3.1.5, ''Broadband EDFA.'' Figure 5.59(a,b) compares the δ spectrum with the $\sigma_a^P(\nu_P)$ spectrum for Ge/Er-doped fiber in the 800- to 850-nm band and the 640- to 680-nm band, respectively [129]. The minimum δ value of 0.65 is obtained at a pump wavelength of 827 nm for 800-nm pumping

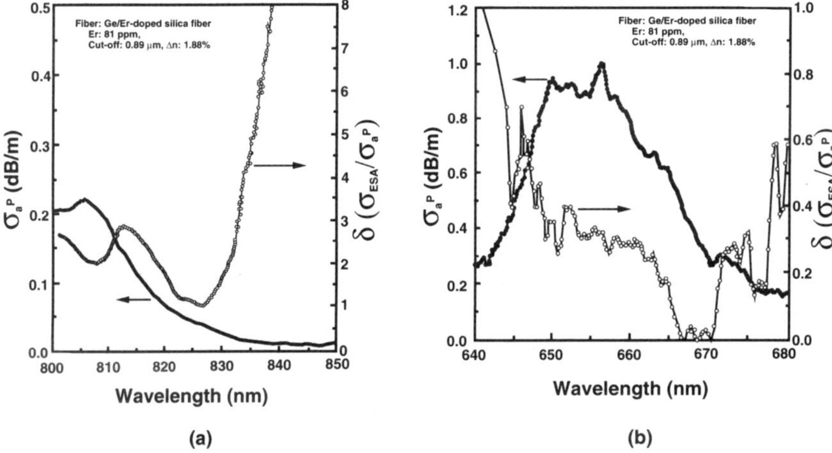

Figure 5.59 $\delta(\sigma_{ESA}/\sigma_a^P)$ spectrum compared with the σ_a^P spectrum for Ge/Er-doped fiber [138]: (a) 800- to 850-nm pump band and (b) 640- to 680-nm pump band.

[116,129,152–154], and the optimum pump wavelength for 660-nm pumping is from 663 to 671 nm, where the δ value is below 0.2 [129,148,150]. By contrast, it has been reported that the δ value for 660-nm pumping is about 0.15 for Al/Er-doped fiber [150]. In addition, the degradation in pump efficiency can be improved by changing the host glass. It is known that fluorophosphate glass suppresses the ESA process for 800-nm pumping [116,152,154].

Power conversion efficiency (PCE) is a useful parameter with regard to applying the EDFA as a booster (power) amplifier because it is related to the saturation characteristics of the EDFA. This value is usually used during module design to estimate the output signal power obtained from an EDFA module with the available pump power in the module and is defined as [115]

$$\text{PCE}(\%) = \frac{P_S^{OUT} - P_S^{IN}}{P_P^{IN}} \times 100 \tag{5.34}$$

where, P_S^{OUT}, P_S^{IN}, and P_P^{IN} are output signal power, input signal power, and launched pump power into an EDF, and the maximum value is given for signal wavelength λ_S and pump wavelength of λ_P by

$$\text{PCE}^{max}(\%) = \frac{\lambda_P}{\lambda_S} \times 100 \tag{5.35}$$

Therefore, since the PCE^{max} value increases as the pump wavelength approaches the signal wavelength, the highest PCE^{max} of 95.5% is obtained for 1,480-nm pumping. The saturation characteristics and actual PCE value will be described in Subsection 5.3.1.2, "Saturation Characteristics."

The *noise figure* (NF) represents the S/N degradation between the input and output of the EDF and indicates the amplification quality [159–161]. This value must be estimated for system application because there must be a certain level of S/N at the receiver in order for the signal to be detected. This value depends on the pump band, and it is well known that the lowest NF of 3.1 dB that is close to the quantum limit is achieved with 980-nm pumping [162–166]. The noise characteristics will be described in detail in Subsection 5.3.1.2, "Noise Characteristics."

Now, we return to a comparison of the pump bands, using the above parameters. Table 5.5 shows the highest gain coefficient (dB/mW) reported to date, δ, PCE^{max} (%), and NF (dB), for EDFs pumped by each pump band. This table reveals the superiority of the 980- and 1,480-nm pump bands that have a high gain coefficient, high PCE, and are free from the ESA process. Hence, 980-nm-band InGaAs/GaAs strained quantum well LDs and a 1,480-nm-band InGaAsP/InP LDs are usually used as pump light sources when fabricating practical EDFA modules. The former is suitable for constructing a low-noise amplifier because of its low-noise pumping, and the latter is suitable for use in a booster amplifier with a high PCE. Figure 5.60 compares 980- and 1,480-nm pumping by plotting the relationship between the EDF length and signal gain for both at the same pump power [138]. The optimum fiber length at which the signal gain is maximum for 980-nm pumping is about half that for 1,480-nm pumping, while the maximum signal gain is double. These results also reveal that 980-nm pumping is more effective than 1,480-nm pumping and show that a high-gain EDFA with a shorter EDF can be constructed by employing 980-nm pumping. The pump wavelength dependence of small signal gain in the 980- and 1,480-nm pump bands is shown in Figure 5.61 [170]. The pump bandwidth, defined at 3-dB gain suppression, is 6 nm (detuning tolerance of ±3 nm) for 980-nm pumping and 20 nm (detuning tolerance of ±10 nm) for

Table 5.5

Highest Gain Coefficient, δ, PCE^{max} (%), and NF (dB) for EDFs Pumped by Each Pump Band

Pump Wavelength (nm)	Gain Coefficient (dB/mW)	Cite	$\delta(=\sigma_{ESA}/\sigma_a^P)$	Cite	PCE^{max} (%)	Noise Figure (dB)	Cite
514	~0.6	[97]	0.5	[150]	33	No report	
532	2	[127]	–		34.3	No report	
664	3.8	[128,129]	~0.15	[129,150]	42.8	No report	
827	1.3	[129,132]	0.65	[129]	53.4	3.8	[129,168]
980	11	[147]	–		63.2	3.1	[163–166]
1,480	6.3	[167]	–		95.5	4.2	[163–165,169]

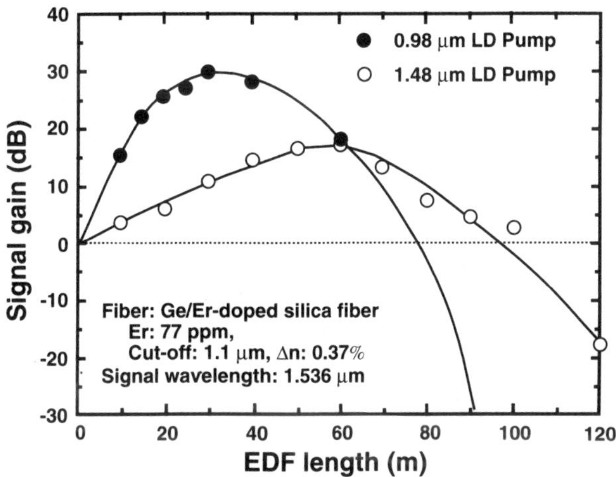

Figure 5.60 Relationship between the EDF length and signal gain at 20-mW pump power [138].

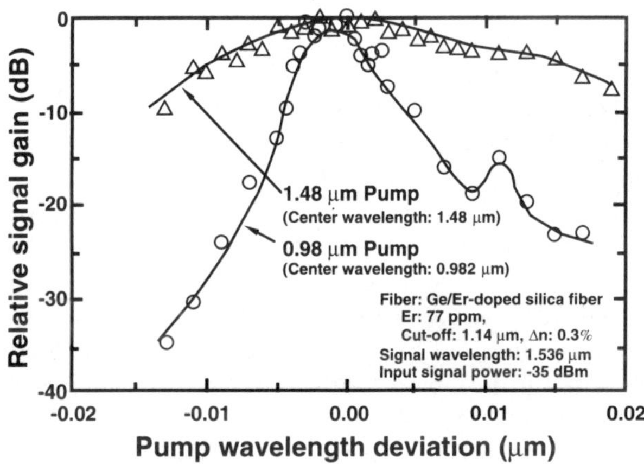

Figure 5.61 Pump wavelength dependence of small signal gain in the 980- and 1,480-nm pump bands [170].

1,480-nm pumping. It is considered that the difference between the two pump wavelengths is due to the fact that the cross section of the 980-nm-band pump absorption spectrum is narrow whereas that of the 1,480-nm band is broad. These results show that control of the lasing wavelength of the 980-nm-band LD is indispensable when constructing stable EDFAs. In addition, in both cases the pump wavelength dependence weakens when the pump power and input signal power are increased for booster applications, as shown in Figure 5.62(a) for 980-nm pumping [104] and Figure 5.62(b) for 1,480-nm pumping [171].

Gain Spectrum

The gain spectrum of the EDFA is important with regard to actual applications to a 1.5-μm transmission system because the amplification bandwidth is the key parameter in determining the transmission band. This property has therefore been investigated for EDFs with various glasses with a view to broadening the EDFA bandwidth. It is well known that the codoping of Al and/or P into the core of Er^{3+}-doped silica fiber is effective in broadening the gain spectrum [119–121,172–174]. (Attempts have also been made to use lanthanum and nitrogen as the codopant for EDF [175-177]). Figure 5.63(a,b) shows typical gain spectra of Ge/Er-doped silica fiber [164,165] and Al/P/Er-doped silica fiber [120], respectively, for various pump powers. The improvement in the bandwidth with Al and P codoping is apparent by comparing the two spectra. The spectrum of the Ge/Er-doped silica fiber has a peaked profile with two peaks—one at 1.536 μm and another at 1.552 μm. In contrast, the gain spectrum of the Al/P/Er-doped silica fiber has a wide amplification band in the 1.545- to 1.560-μm wavelength region, although there is a protruding gain peak around 1.530 μm. Recently, it was also found that extremely high levels of Al codoping flattens the amplification band in the 1.540- to 1.560-μm wavelength region [178,179]. Furthermore, changing the fiber host material from silica to ZrF_4-based fluoride glass and fluorophospate glass is also promising for broadening and flattening the bandwidth. Fluoride-based EDF in particular provides a flat amplification band in the 1.530- to 1.560-μm wavelength region [180–185]. The improvement in the gain spectrum with extremely high Al-codoped silica fiber, fluoride-based fiber, and fluorophospate glass fiber will be described in Subsection 5.3.1.5, "Broadband EDFA," along with other approaches for improving the amplification bandwidth characteristics. This relates to the application of EDFAs to a WDM transmission system.

Moreover, the EDFA gain spectrum can be shifted to the long wavelength region using a long EDF. Figure 5.64 shows the gain spectra of Al/Ge/Er-doped silica fiber [186]. By increasing the EDF length, a gain spectrum can be realized in the 1.570- to 1.620-μm wavelength region. The decrease in signal gain in the wavelength region above 1.620 μm is due to the ESA process ($^4I_{13/2} \rightarrow {}^4I_{9/2}$ transition), and the upper limit of the gain spectrum is decided by this ESA.

In addition, the gain spectrum is generally measured by scanning the signal wavelength with a single signal source. These gain spectra mentioned measured by this method. However, the spectrum varied when the input signal power was changed. For example, Figure 5.65 shows the gain spectra of fluoride-based EDF for various launched input signal powers [187]. As shown in this figure, the spectrum becomes flat when the input signal power is high. The input signal power level is

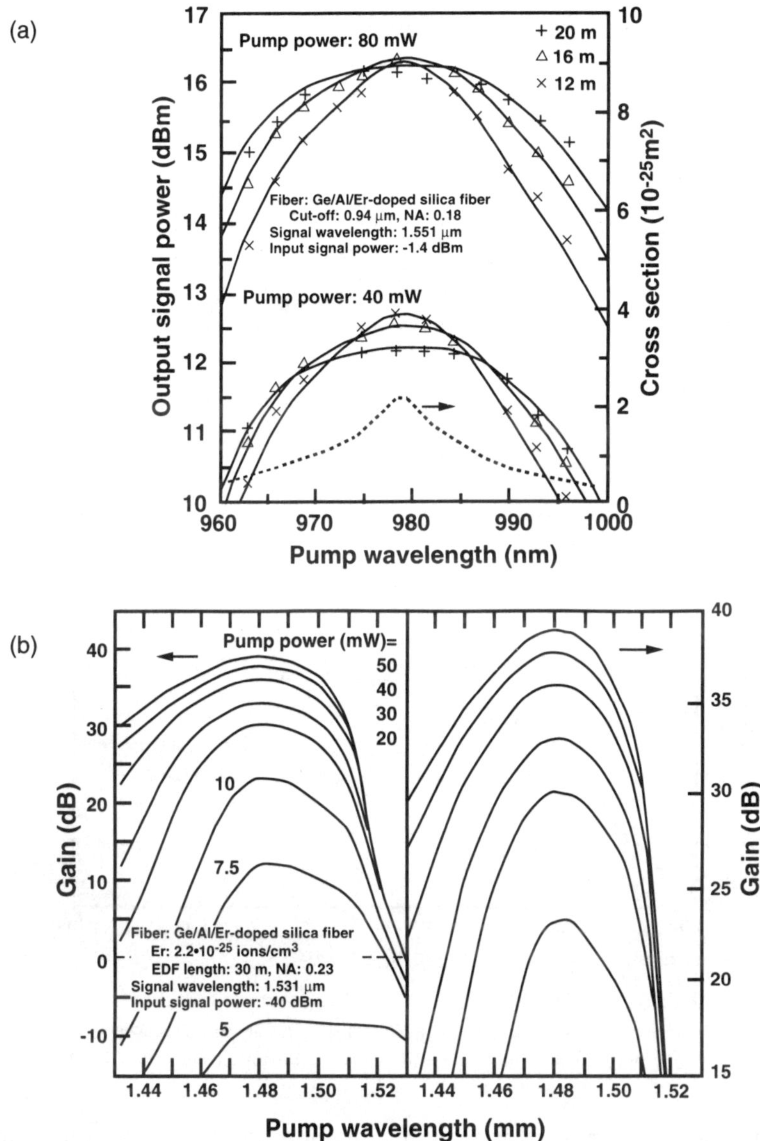

Figure 5.62 Pump wavelength dependence of signal gain: (a) output signal power as a function of pump wavelength for three different EDF lengths in the 0.98-μm band [104]; (b) signal gain as a function of pump wavelength for different launched pump powers in the 1.48-μm band (left and right figures represent the same curves, except for the last one, plotted with different vertical scales) [171].

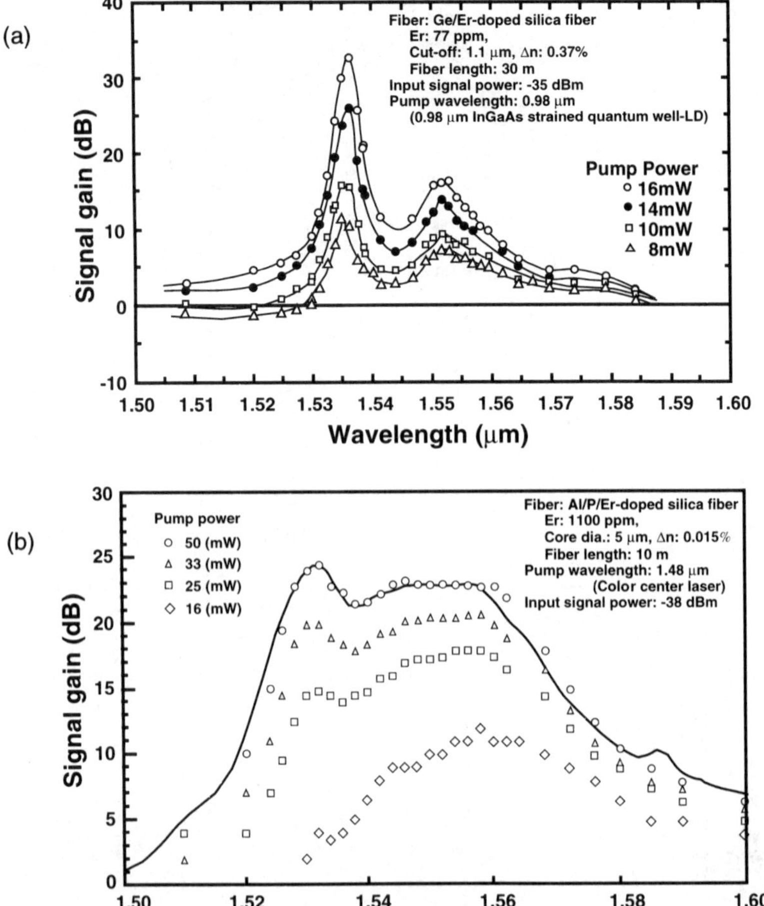

Figure 5.63 Typical gain spectra of Er^{3+}-doped fiber for various pump powers. Gain spectra of (a) Ge/Er-doped silica fiber [164] and (b) Al/P/Er-doped silica fiber [120].

an important parameter for this scanning method, and the spectrum is usually measured under small signal conditions.

Saturation Characteristics

This section describes the saturation characteristics of the EDFA. The signal output power obtained at the output end of an EDFA is a particularly interesting value because it is related to the transmission and repetition distances of long-haul trans-

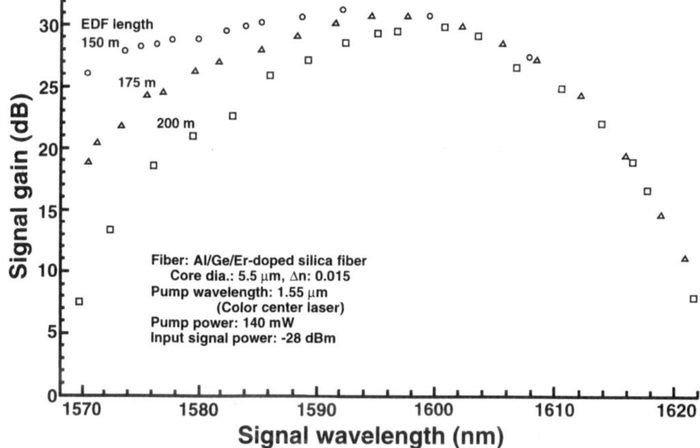

Figure 5.64 Gain spectra of Al/Ge/Er-doped silica fiber, obtained for long EDF [186].

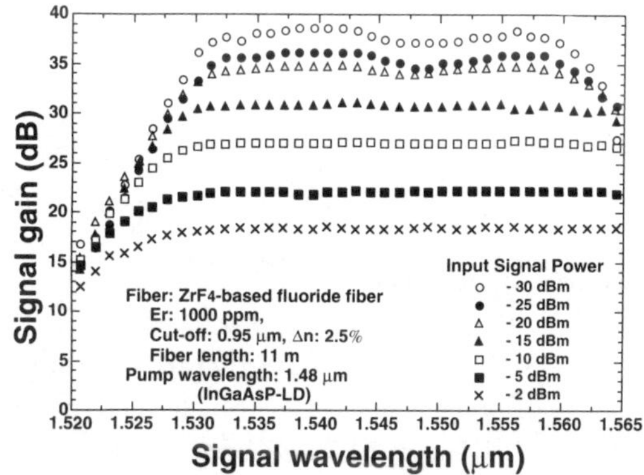

Figure 5.65 Gain spectra of fluoride-based EDF for various input signal powers [187].

mission systems and to increasing the number of output ports for distribution applications. In Figure 5.66, for example, the signal gain is plotted as a function of the signal output power for various launched pump powers [145]. The saturation output power $P_{sat}^{OUT}(\nu_S)$, which is defined as the output power where the signal gain is 3 dB below the unsaturated value, is usually used as a parameter to indicate the saturation characteristics of an optical amplifier. As Figure 5.66 shows, the saturation output power increases with increasing launched pump power. In this figure, the $P_{sat}^{OUT}(\nu_S)$ is 2.5, 7.8, 10.3, and 11.3 dBm for launched pump powers of 11.3, 24.5, 39.0, and 53.6 mW, respectively. The increase in $P_{sat}^{OUT}(\nu_S)$ can be explained by the calculation model in Subsection 5.3.1.1. From (5.26) and (5.27), the simplified signal gain coefficient $g_S(\nu_S) = \gamma_e(\nu_S) - \gamma_a(\nu_S)$ is expressed as [114]

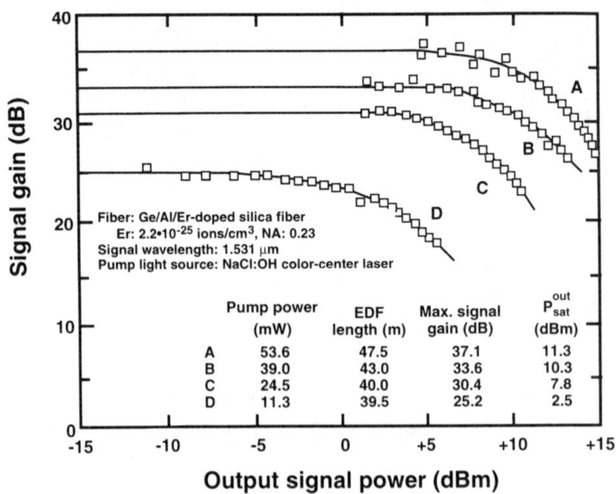

	Pump power (mW)	EDF length (m)	Max. signal gain (dB)	P_{sat}^{out} (dBm)
A	53.6	47.5	37.1	11.3
B	39.0	43.0	33.6	10.3
C	24.5	40.0	30.4	7.8
D	11.3	39.5	25.2	2.5

Figure 5.66 Signal gain versus the output signal power for various launched pump powers [145].

$$g_S(\nu_S) \approx \rho_0 \sigma_e^S(\nu_S) \frac{U - s(\nu_S)}{1 + U} \frac{1}{1 + \dfrac{P_S(\nu_S)}{P^{sat}(\nu_S)[1 + U]}} \tag{5.36}$$

Hence, $P_{sat}^{OUT}(\nu_S)$, where $g_S(\nu_S)$ becomes half, is given as

$$P_{sat}^{OUT}(\nu_S) \approx P^{sat}(\nu_S)[1 + U] \tag{5.37}$$

which shows that the saturation output power $P_{sat}^{OUT}(\nu_S)$ increases with increasing launched pump power. So the key to achieving a high power amplifier is to launch a high pump power into the EDF. Subsection 5.3.1.5, "Ultra-High-Power EDFA," describes recent approaches for increasing signal output power based on this concept.

The saturation characteristics depend on the pumping configuration and EDF length. Figure 5.67 shows the power conversion efficiency as a function of EDF length, for forward- and backward-pumping configurations [188]. The PCE is another important parameter indicating the saturation characteristics, as defined in (5.34). The most important conclusion we can draw from this figure is that backward pumping yields the highest PCE for any EDF length, and there is a certain EDF length at which the PCE reaches maximum. So backward pumping is employed with the optimum EDF length when constructing a booster amplifier module in order to achieve a high-output power.

Furthermore, as explained in Subsection 5.3.12, "Pump Wavelength Dependence of Signal Gain," the saturation characteristics also depend on the pump

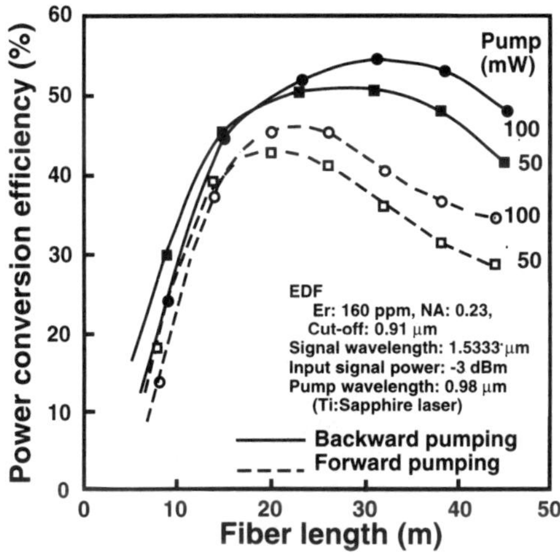

Figure 5.67 Power conversion efficiency as a function of EDF length for forward- and backward-pumping configurations [188].

band, and 1,480-nm pumping is the most effective among the pump bands because PCE^{max} is 95.5%. The actual PCE value has already been experimentally investigated for practical pump bands of 980 and 1,480 nm. The results reveal that the actual PCE is about 80% [189] for 1,480-nm pumping and the highest measured PCE for 980-nm pumping is 56% [190].

Noise Characteristics

The noise characteristics are of crucial importance for applications in optical communication systems since they represent a major parameter that determines the overall system performance such as maximum transmission distance and bit rate. Noise characteristics have therefore been studied in detail both theoretically and experimentally [98,105,163–166,169,172,191–198]. It has been confirmed that the noise characteristics depend strongly on the pump band, and the performance with 980-nm pumping is better than that with 1,480-nm pumping [105,162–166,191]. In this section, we first present a theoretical analysis of the noise in an EDFA and then describe the several noise characteristics reported to date.

With an optical amplifier that uses a stimulated emission, noise is generated when the optical signal is detected with ASE at an *avalanche photo diode* (APD) or p-i-n photodiode. The noise at the output terminal of an optical amplifier consists of the following five components: signal shot noise N_S, spontaneous emission shot noise (sp. shot noise) N_{SP}, beat noise between the signal light and the spontaneous emission light (signal-sp. beat noise) $N_{S\text{-}SP}$, beat noise between the spontaneous emission components (sp.-sp. beat noise) $N_{SP\text{-}SP}$, and excess noise N_{ex}. The noise power spectral density (A^2/Hz) in a unit resistance for each noise component is [160]

$$N_S = 2e^2 G n_0 \tag{5.38}$$

$$N_{SP} = 2e^2 (G - 1) n_{SP} m_t \Delta f_1 \tag{5.39}$$

$$N_{S-SP} = 4e^2 G(G - 1) n_{SP} n_0 \tag{5.40}$$

$$N_{SP-SP} = 2e^2 (G - 1)^2 n_{SP} m_t \Delta f_2 \tag{5.41}$$

$$N_{ex} = 2e^2 G^2 n_{ex} \tag{5.42}$$

The total noise N_{total} generated by the optical amplifier is

$$N_{total} = N_S + N_{SP} + N_{S-SP} + N_{SP-SP} + N_{ex} \tag{5.43}$$

where e is the electron change, G is the signal gain, n_0 is the photon density of the input signal light, n_{SP} is the population inversion parameter, Δf_1 and Δf_2 are the effective noise bandwidths of the sp. shot noise and the sp.-sp. beat noise, and n_{ex} is the excess noise parameter of the incident signal. Further, m_t is the transverse mode number of the spontaneous emission and equals 2 in the case of an optical fiber amplifier. From (5.38) to (5.42), it is considered that the signal-sp. beat noise and the signal shot noise increase in proportion to the signal output power Gn_0, while the sp. shot noise and the sp.-sp. beat noise are constant. The signal-sp. beat noise and sp.-sp. beat noise are larger than the signal shot noise and the sp. shot noise, respectively, by an amount almost corresponding to the signal gain G. The main noise components are the signal-sp. beat noise and sp.-sp. beat noise. Moreover, because the signal shot noise is contained in the optical signal itself and the excess noise N_{ex} can be ignored if the signal light incident on the optical amplifier is completely coherent so that the excess noise parameter n_{ex} becomes 0, the noise components inherent to the optical amplifier are the sp. shot noise, the signal-sp. beat noise, and the sp.-sp. beat noise, which are determined by the parameters Δf_1, Δf_2, and n_{SP}. Furthermore, the most important of these noise components is the signal-sp. beat noise because it cannot be eliminated, whereas the others can be suppressed by using an optical band pass filter. The NF that represents the SNR degradation from the input and output of the EDF is given by as follows if only the signal-sp. beat noise is considered [199]

$$NF = 2n_{sp} \tag{5.44}$$

This expression of the NF has a one-to-one correspondence with the population inversion parameter n_{sp}. Since the minimum value of n_{sp} is 1, the theoretical limit of the noise figure in the optical amplifier is $F = 3$ (dB).

The effective noise bandwidths Δf_1 and Δf_2 are quantities proportional to the integral of the amplifying bandwidth and the overlapping integral, respectively, of an optical amplifier. The population inversion parameter n_{sp} indicates the increment in the spontaneous emission caused by the imperfection of the population inversion of the Er^{3+}-doped optical fiber. When the pumping condition does not vary regardless of the length of the EDF, the population inversion parameter is given as $n_{sp} = N_2 / (N_2 - N_1)$, where N_2 and N_1 are the ion population densities of the $^4I_{13/2}$ metastable level and the $^4I_{15/2}$ ground level, respectively. In an actual fiber, however, the pump power decreases along the fiber length and the population inversion parameter n_{sp} for EDF is derived by solving a differential equation describing the signal photon density n_S. The differential equation for n_S is expressed based on (5.31) in Subsection 5.3.1.1 as

$$\frac{dn_S(z, \nu)}{dz} = (\gamma_e - \gamma_a) n_S \pm \gamma_e \tag{5.45}$$

From (5.45), the signal photon density n_S and population inversion parameter n_{SP} are obtained as

$$n_S = n_0 G + \int_0^L \frac{\gamma_e}{G} \, dz \tag{5.46}$$

$$n_{SP} = \frac{G \int_0^L \frac{\gamma_e}{G} \, dz}{G - 1} \tag{5.47}$$

for

$$G = \mathrm{Exp}\left[\int_0^L (\gamma_e - \gamma_a) \, dz \right] \tag{5.48}$$

where L and G are the EDF length and signal gain, respectively. The first term on the right-hand side of (5.46) indicates the amplification term of the photon density n_0 of the incident light, while the second term indicates the spontaneous emission caused by the amplification process. The population inversion parameter n_{SP}, which is expressed in (5.47), is obtained from the second term on the right-hand side of (5.46) [164]. In addition, by assuming that a small signal condition and the pump power does not change with EDF length, the population inversion parameter n_{SP} can be expressed as

$$n_{SP} \approx \frac{P_P(0)}{P_P(0) \cdot \left[1 - \dfrac{s(\nu_S)}{s(\nu_P)}\right]}$$

$$= \frac{1}{1 - \dfrac{s(\nu_S)}{s(\nu_P)}} \quad (P_P(0) \gg P_P^{th}) \tag{5.49}$$

where $P_P(0)$ is the pump power launched into the EDF.

After presenting a theoretical analysis of the noise in an EDFA, we will describe several important noise characteristics that have been reported to date. Figure 5.68 shows the experimental setup used to measure the aforementioned noise components and to compare the noise characteristics of EDFAs pumped at 980 and 1,480 nm [163,164]. The noise power $P_m(\omega)$ is measured using an RF spectrum analyzer with an InGaAs-APD [160]. The total beat noise $\langle i_n^2 \rangle_{beat} = N_{S\text{-}SP} + N_{SP\text{-}SP}$ and the total shot noise $\langle i_n^2 \rangle_{shot} = N_S + N_{SP}$ are estimated from the noise power $P_m(\omega)$ as

$$\langle i_n^2 \rangle_{beat} = \frac{P_m(\omega) - P_t(\omega)}{R_L B_0 G_e(\omega) \eta^2 M^2} - \frac{2e \langle i_{pho} \rangle M^x}{\eta} \tag{5.50}$$

$$\langle i_n^2 \rangle_{shot} = 2e \langle i_{pho} \rangle \tag{5.51}$$

where $P_t(\omega)$ is the sum of the thermal noise power of the measurement equipment and the APD dark current shot noise, R_L is the load resistance, B_0 is the resolution frequency bandwidth of the RF spectrum analyzer, $G_e(\omega)$ is the overall frequency response of the APD and RF amplifier, η is the product of the quantum efficiency of the APD and the coupling efficiency of the amplified signal output to the APD surface, M is the avalanche multiplication factor of the APD, and x is the excess noise component of the APD. $\langle i_{pho} \rangle$ is the detector photocurrent when the total output power from the EDF is detected using an APD with a quantum efficiency of unity and $M = 1$ and expressed by $\langle i_{pho} \rangle = \langle i_{APD} \rangle / \eta M$ where $\langle i_{APD} \rangle$ is the measured APD photocurrent. The signal-sp. beat noise is evaluated by subtracting the sp.-sp. beat noise from the total beat noise, and the signal shot noise is also evaluated in a similar way.

The measured results for the total beat noise, the total shot noise, and each noise component are shown in Figure 5.69 as a function of the output signal power

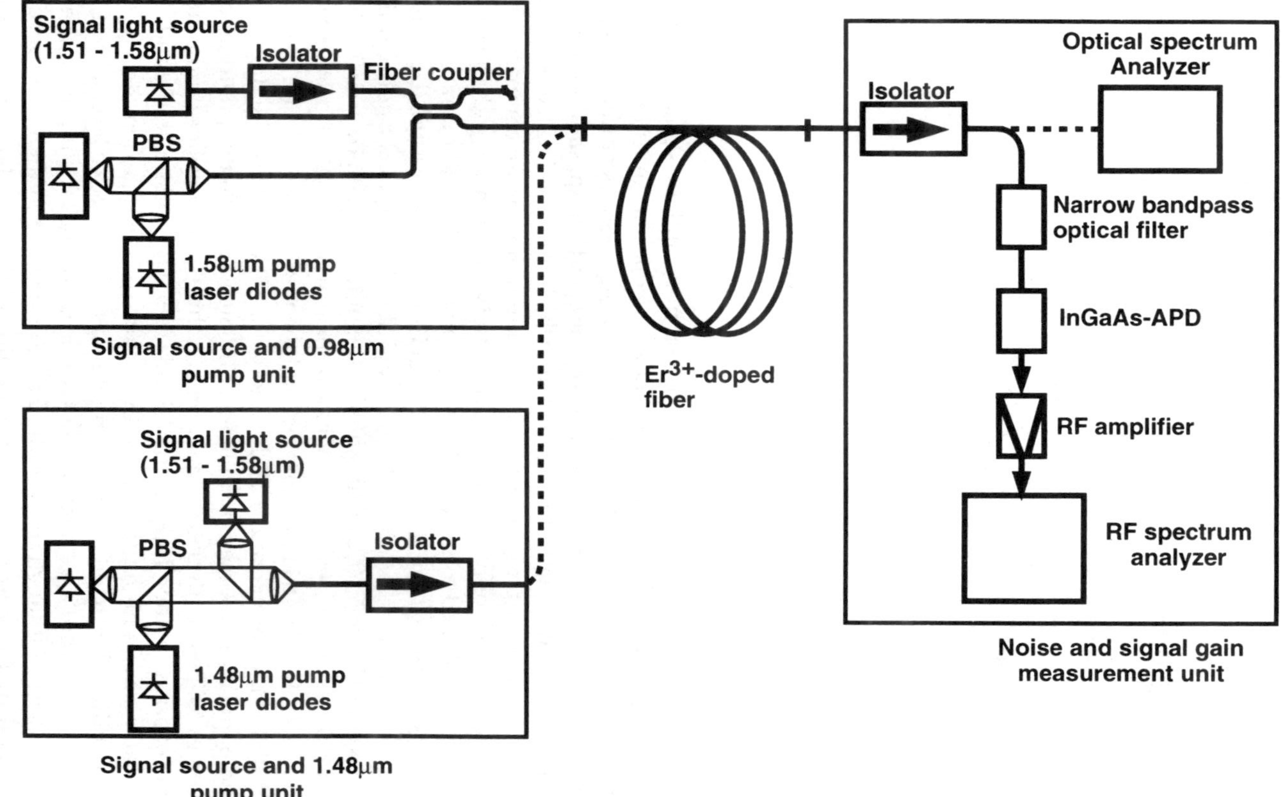

Figure 5.68 Experimental setup for measuring noise power [163,164].

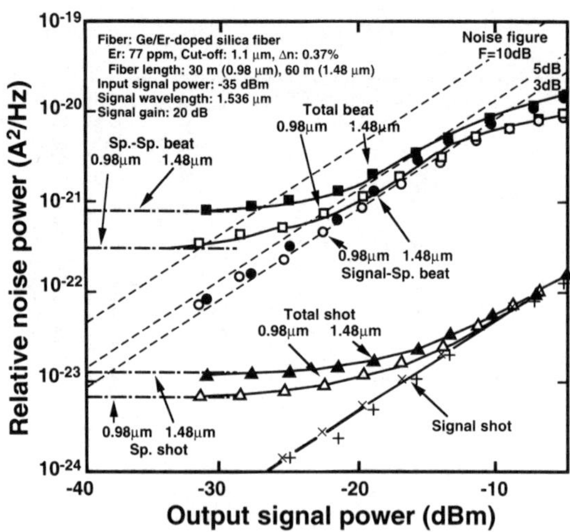

Figure 5.69 Experimental relationship between the relative noise power and the output signal power [163,164].

[163,164]. It has been confirmed that the s.-sp. beat noise and the signal shot noise increase in proportion to the signal output power, while the sp. shot noise and sp.-sp. beat noise are constant [160]. Moreover, the signal-sp. beat noise and the sp.-sp. beat noise are larger than the signal shot noise and the sp. shot noise by an amount almost corresponding to the signal gain, which is in good agreement with (5.38) to (5.42). By contrast, when 980- and 1,480-nm pumping are compared, the sp.-sp. beat noise, sp. shot noise, and signal-sp. beat noise are all larger with the latter. This difference is attributed to the fact that the population inversion parameter for 1,480 nm is larger than that for 980-nm pumping. Figure 5.70 shows the theoretical noise figure levels at NF = 3, 4, 5, and 10 dB and the signal-sp. beat noise, which is already shown in Figure 5.69, as a function of the signal output power [163,164]. The noise figure estimated from the signal-sp. beat noise is 3.2 dB for 980-nm pumping and 4.1 dB for 1,480-nm pumping, and the population inversion parameter nsp is 1.05 for the former and 1.29 for the latter. The cause of the difference between the population inversion parameters for 980- and 1,480-nm pumping can be explained as follows. With 980-nm pumping, a three-level amplification system is used that consists of the $^4I_{15/2}$ ground level, the $^4I_{11/2}$ pump level, and the $^4I_{13/2}$ metastable level, and all the Er^{3+} ions at the $^4I_{15/2}$ level are excited to the $^4I_{11/2}$ level. On the other hand, 1,480-nm pumping is a quasi-three-level amplification system in which the $^4I_{13/2}$ metastable level is used both for the pump level and metastable level. Although most of the Er^{3+} ions excited to the upper portion of the $^4I_{13/2}$ level are transferred to the lower portion of the $^4I_{13/2}$ level by infraband relaxation, some of the Er^{3+} ions are transferred directly to the $^4I_{15/2}$ ground level

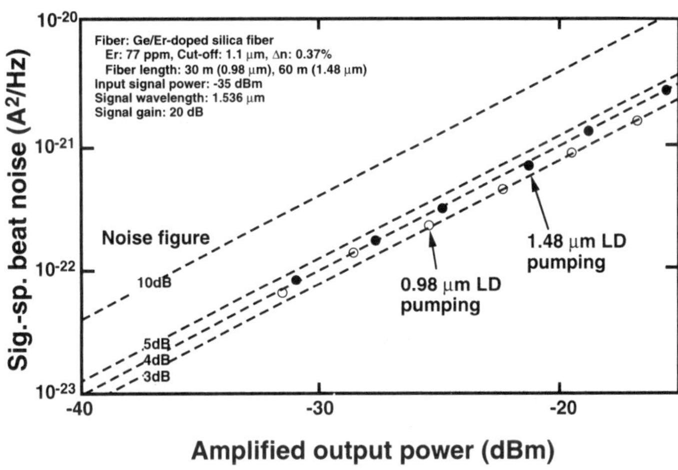

Figure 5.70 Theoretical noise figure levels at NF = 3, 4, 5, and 10 dB and signal-sp. beat noise [163,164].

by a stimulated transition from the upper portion of the $^4I_{13/2}$ level to the $^4I_{15/2}$ ground level. Hence, the population inversion is imperfect, and so parameter is larger with 1,480-nm-band pumping than with 980-nm-band pumping. The difference between the population inversion parameters is also explained theoretically with (5.49). That is, n_{sp} becomes 1 with 980-nm pumping because $s(\nu_P)$ $(= \sigma_a^P(\nu_P)/\sigma_e^P(\nu_P)) \approx \infty$, while n_{sp} for 1,480-nm pumping is limited to $1/[1 - \sigma_a^S(\nu_S)\sigma_e^P(\nu_P)/\sigma_e^S(\nu_S)\sigma_a^P(\nu_P))]$ because $\sigma_e^S(\nu_S) > 0$. For these reasons, a 980-nm pumped EDFA has a low noise figure that is close to the quantum limit, and the noise characteristics for 980-nm pumping are better than those for 1,480-nm pumping [192,193].

The use of an optical bandpass filter is an effective way to suppress noise. Figure 5.71 shows measured results for the noise characteristics versus the output signal power of a 980-nm pumped EDFA with and without the filter [164]. Although the signal-sp. beat noise and signal shot noise remain unchanged when the filter is installed, both the sp. shot noise and sp.-sp. beat noise decrease. For this reason, a bandpass filter is often attached to the output end of an EDFA to suppress the sp. shot noise and sp.-sp. beat noise when low noise characteristics are required, for example, when applied as a preamplifier or repeater.

The NF, which is estimated from the signal-sp. beat noise (or population inversion parameter n_{sp}), does not only depend on the pump band but also on the launched pump power, signal wavelength, and pumping configuration.

Figure 5.72 shows the theoretical value of the population inversion parameter n_{sp} and the signal gain of the EDF as a function of launched pump power [164]. The population inversion parameter decreases monotonically with increasing launched pump power, as predicted by the simple expression of n_{sp} in (5.48). This is due to

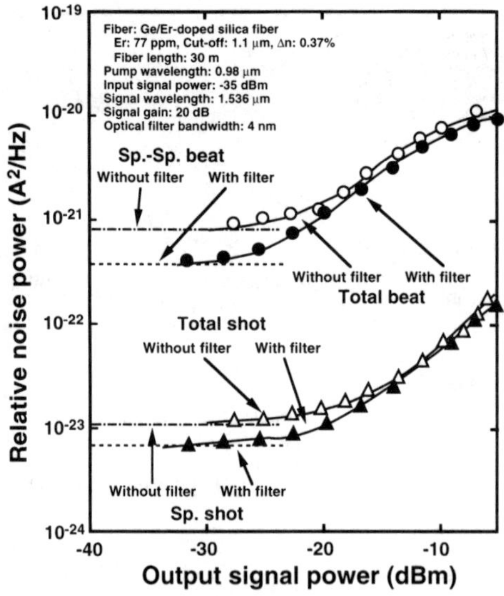

Figure 5.71 Noise characteristics versus output signal power of a 980-nm pumped EDFA with and without a filter [164].

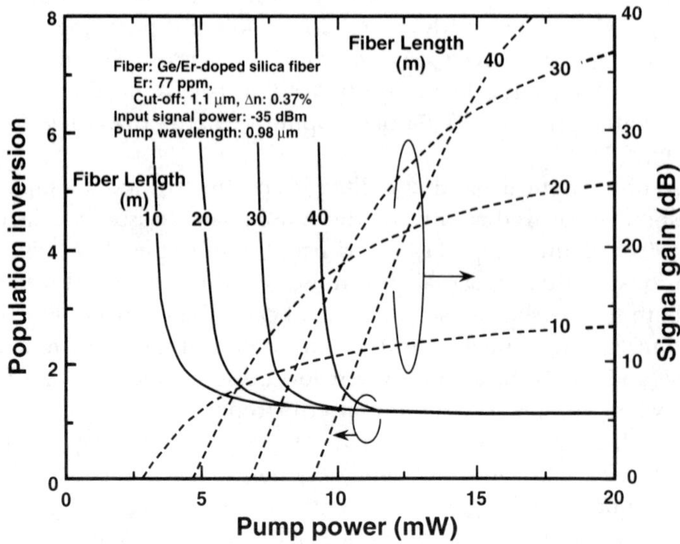

Figure 5.72 Theoretical value of the population inversion parameter n_{sp} and the signal gain of the EDF as a function of launched pump power [164].

the fact that the number of Er^{3+} ions excited from the $^4I_{15/2}$ ground level to the $^4I_{13/2}$ metastable level increase with increases in the launched pump power $P_P(0)$, and the population inversion parameter decreases monotonically. In addition, the noise figure $NF = 2n_{sp}$ decreases with increasing signal gain when the EDF length is fixed, as shown in Figure 5.73 [164].

The calculated relationship between the population inversion parameter and the signal wavelength for various pump powers (normalized to pump threshold) is shown in Figure 5.74 [105]. As shown in this figure, at values beyond 1,530 nm,

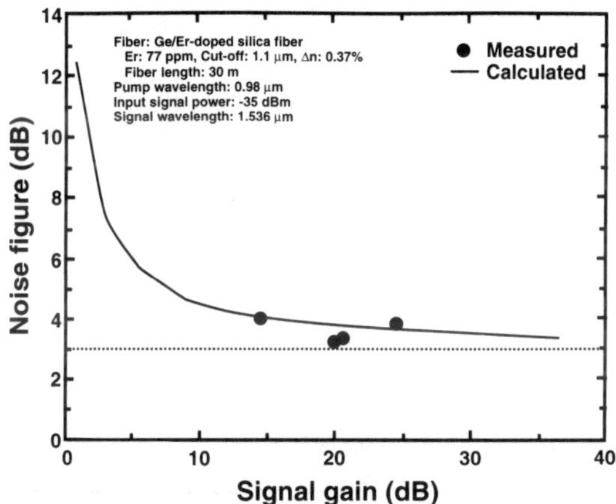

Figure 5.73 Noise figure as a function of signal gain [164].

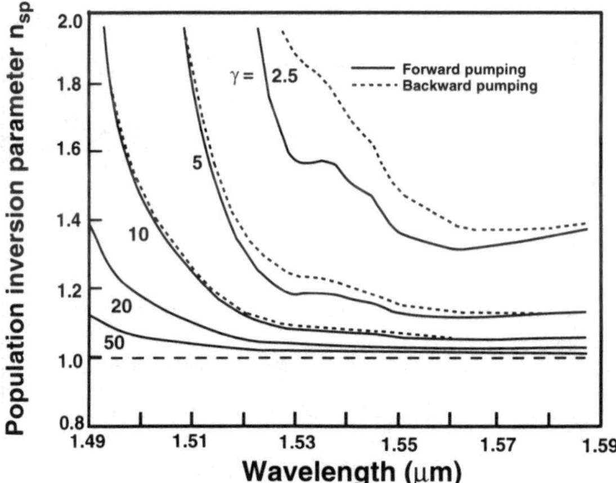

Figure 5.74 Calculated relationship between the population inversion parameter and signal wavelength for various pump powers (normalized to pump threshold) γ = 2.5 to 50 [105].

the population inversion parameter is almost independent of the signal wavelength. The slight signal dependence on the NF is due to the wavelength dependence of the absorption and emission cross sections of the signal and pump lights $\sigma_{a,e}^{S,P}$; and this slight dependence is expressed simply by the expression n_{sp} in (5.49), which includes $\sigma_{a,e}^{S,P}$. On the other hand, at less than 1,530 nm, the NF increases with decreasing signal wavelength because $s(\nu_S) = \sigma_a^S(\nu_S)/\sigma_e^S(\nu_S)$ in (5.49) increases with decreasing signal wavelength. The difference between forward and backward pumping is described as follows.

Figure 5.75 shows the calculated relationship between NF and fiber length for forward and backward pumping [98]. The noise performance for backward pumping can be significantly worse than that for forward pumping, since in the former, the population inversion at the input of the EDF is relatively low. If the EDF is divided into short segments, the NF of the Nth segment is expressed as [161]

$$NF = \frac{NF_1}{L_c} + \frac{NF_2}{L_c L_1 G_1} + \frac{NF_3}{L_c L_1 G_1 L_2 G_2} + \cdots + \frac{NF_N}{L_c L_1 G_1 L_2 G_2 \ldots L_{N-1} G_{N-1}} \tag{5.52}$$

where G_i and NF_i are the signal gain and NF of the ith segment. L_c and L_i are the coupling coefficient of the signal light into the EDF, and that between the ith and $(i+1)$th segment, respectively, and in this case, $L_c = 1$, $L_i = 1$. In this expression,

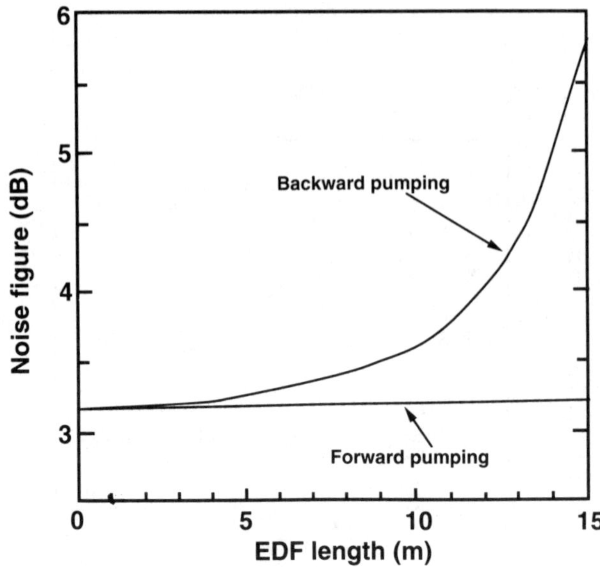

Figure 5.75 Calculated relationship between noise figure and fiber length for forward and backward pumping [98].

the first and *N*th segments are the input and output ends of the EDF, respectively. As shown by this expression, NF_1 for backward pumping is larger than that for forward pumping because the pump power of the former is smaller than that of the latter. Hence, the relationship between the total NF for forward pumping NF^F and backward pumping NF^B become $NF^B \geq NF^F$. On the basis of these results, forward pumping should be used with 980-nm pumping when an EDFA with low-noise characteristics is required.

Furthermore, the saturated noise characteristics of the EDFA are also important [197,198]. Figure 5.76 shows the saturation characteristics and NF for forward, backward, and bidirectional pumping [198]. A striking feature of these data is the reduction in the NF with increasing output signal power for forward and backward pumping at the onset of saturation, before the large increase in the NF with complete saturation. It is considered that this NF decrease arises from the suppression of backward-traveling ASE by the high forward-traveling signal light power. The backward-traveling ASE degrades the NF by reducing the population inversion at the low forward-traveling signal light power that leads to a higher signal-sp. beat noise. So, this dip is most pronounced for high-gain operation and disappears when the signal gain is very low because the ASE power never becomes large enough to saturate with low gain, and this effect is observed only under high-gain conditions.

Temperature Dependence

For practical applications in communication systems, it is very important to understand the temperature-dependent characteristics of EDFAs. The theoretical and

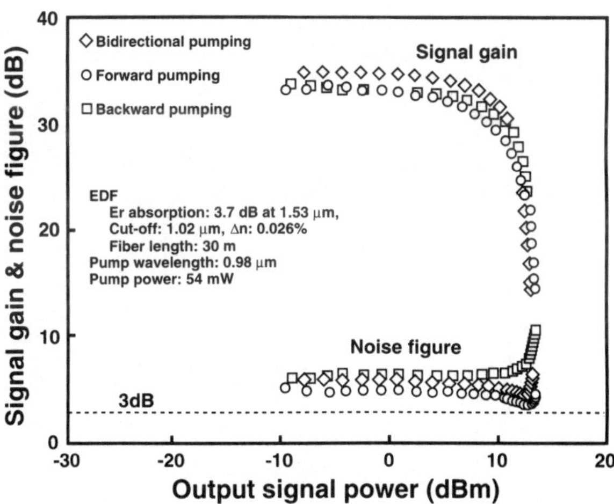

Figure 5.76 Signal gain and noise figure as a function of output signal power for forward, backward, and bidirectional pumping [198].

experimental temperature dependencies of the signal gain properties of EDF have been investigated for 980- and 1,480-nm pumping [200–204]. It has been confirmed that the temperature dependencies of the signal gain characteristics result from changes in the absorption and emission cross sections $\sigma_{a,e}^S$ and $\sigma_{a,e}^P$ of the signal and pump lights [201–204]. Figure 5.77 shows the deviations in the absorption and emission cross sections $\sigma_{a,e}^S$ and $\sigma_{a,e}^P$ of Ge/Er-doped fiber for a fiber temperature range of −40° to 80°C [201]. This figure focuses attention on the cross sections at the two signal gain peak wavelengths of 1.536 and 1.552 μm for Ge/Er-doped fiber. The changes in each cross section are caused by the temperature-dependent distribution of Er^{3+} ion populations within the metastable level $^4I_{13/2}$ and ground level $^4I_{15/2}$. It has been found that each cross section changes linearly with temperature.

The signal gain as a function of the fiber length for various temperatures is shown in Figure 5.78(a) for a 980-nm pump, 1.536-μm signal; (b) for a 1,480-nm pump, 1.536-μm signal; (c) for a 980-nm pump, 1.552-μm signal; and (d) for a 1,480-nm pump, 1.552-μm signal [201]. The temperature coefficient (in dB/°C) obtained from Figure 5.78(a–d), as a function of EDF length, is shown in Figure 5.79 [201]. The change in signal gain strongly depends on fiber length, pump wavelength, and signal wavelength [201,202]; and a 1,480-nm pumped EDFA is more sensitive to temperature than one pumped at 980 nm due to the fact that the temperature dependence with 1,480-nm pumping is larger than with 980-nm pumping at both signal wavelengths [201–204] because there is a finite value for the emission cross section σ_e^P at a pump wavelength of 1,480 nm, and this value is

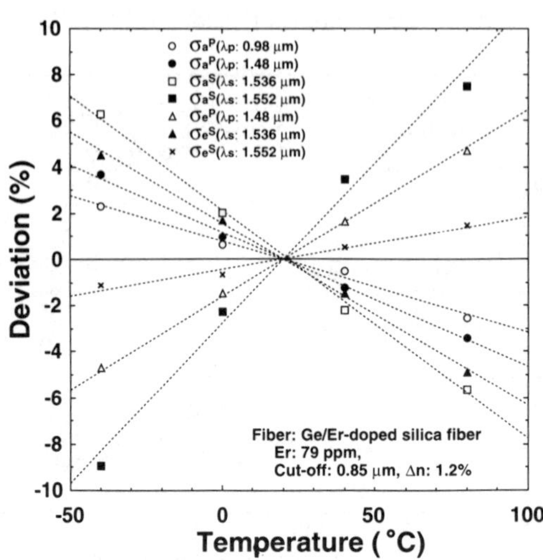

Figure 5.77 Deviations in the absorption and emission cross sections $\sigma_{a,e}^S$ and $\sigma_{a,e}^P$ of Ge/Er-doped fiber for a fiber temperature range of −40° to 80°C [201].

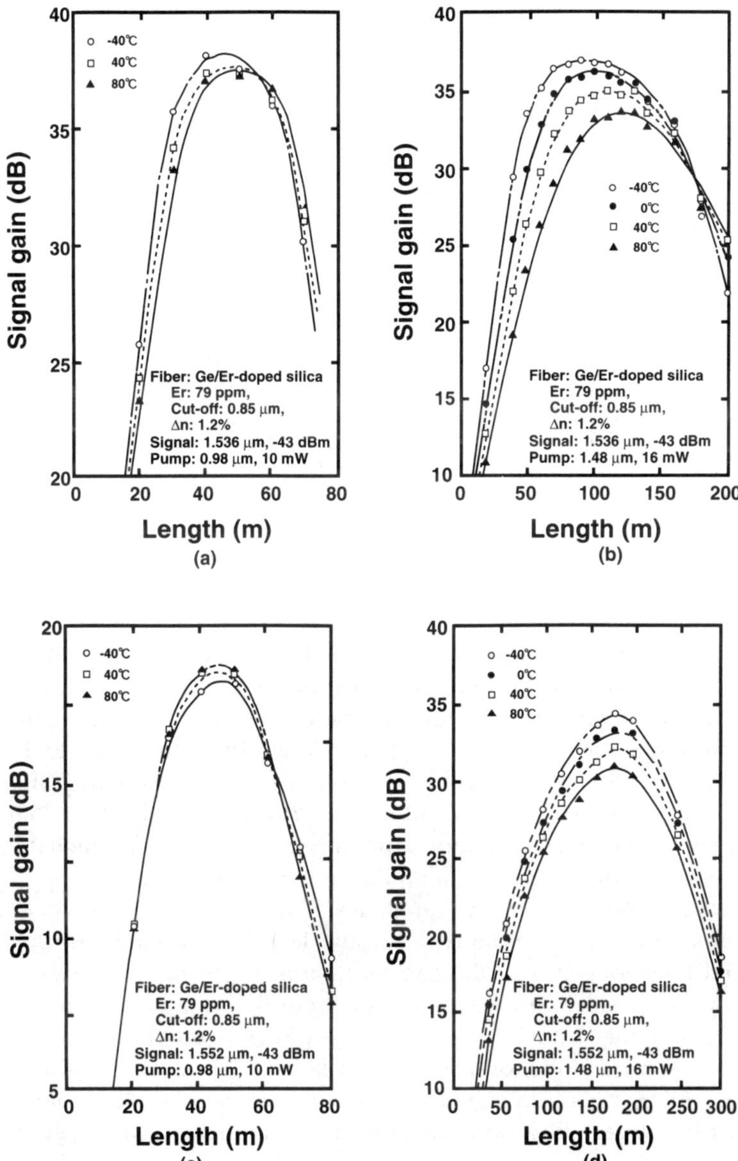

Figure 5.78 Measured relationship between EDF length and signal gain at various temperatures [162]. EDF, Ge/Er-doped silica fiber (Er: 79 ppm, Δn: 1.2%, Cut-off: 0.85 μm). (a) 980-nm pump, 1.536-μm signal; (b) 1,480-nm pump, 1.536-μm signal; (c) 980-nm pump, 1.552-μm signal; and (d) 1,480-nm pump, 1.552-μm signal.

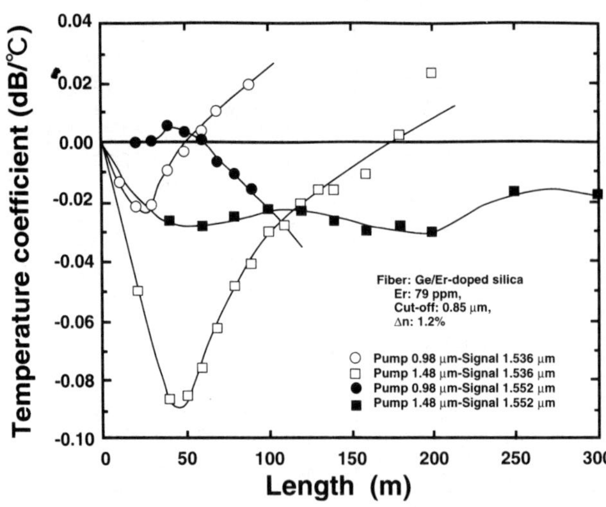

Figure 5.79 Temperature coefficient (in dB/°C) obtained from Figure 5.78(a–d) as a function of EDF length [201].

very sensitive to fiber temperature [201,202,204]. However, for a 1,480-nm pump, 1.536-μm signal, which is the most sensitive, the temperature gain coefficient is less than 0.1 dB/°C, which indicates that the use of *automatic gain control* (AGC) will easily overcome the problem of temperature sensitivity for actual EDFAs [204].

A unique solution to this problem is obtained by choosing an EDF length at which the temperature gain coefficient is zero [201,202]. As shown in Figure 5.79, it has been found that there are temperature-insensitive lengths at which the temperature gain coefficient is zero for amplification with a 980-nm pump, 1.536-μm signal; a 980-nm pump, 1.552-μm signal; and a 1,480-nm pump, 1.536-μm signal. The phenomenon of a temperature-insensitive length is explained by the fact that an Er-doped fiber longer than the gain optimum length can be divided into two sections. One is a positive gain region on the input side (for a copropagating pump and signal), and the other is a negative gain region on the output side. The temperature-insensitive phenomenon with a 980-nm pump, 1.536-μm signal and a 1,480-nm pump, 1.536-μm signal is explained as an example. When the fiber temperature decreases, the absorption and emission cross sections $\sigma_{a,e}^{s}$ of the signal increase as shown in Figure 5.77. The gain in the positive gain region then increases monotonically as the temperature decreases. For the negative gain region, the absorption loss increases with decreasing temperature, and this may outbalance the gain increase in the positive gain region.

Figure 5.80 shows actual temperature-insensitive gain characteristics reported in [201,202] of EDFAs tuned to their temperature-insensitive lengths for amplification with a 980-nm pump, 1.536-μm signal; a 980-nm pump, 1.552-μm signal; and

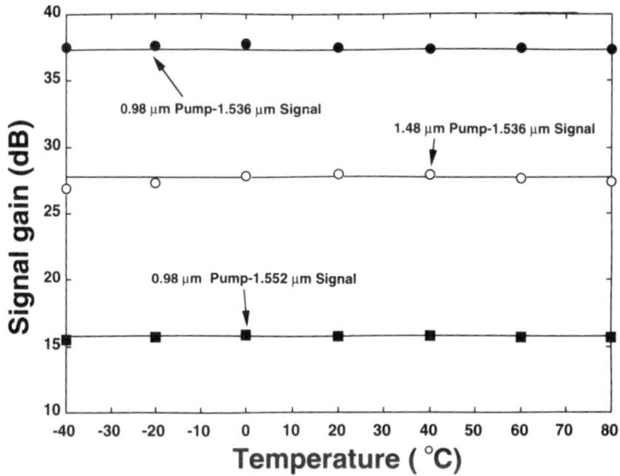

Figure 5.80 Temperature-insensitive gain characteristics of EDFAs tuned to their temperature-insensitive lengths for amplification with a 980-nm pump, 1.536-μm signal; a 980-nm pump, 1.552-μm signal; and a 1,480-nm pump, 1.536-μm signal [201,202].

a 1,480-nm pump, 1.536-μm signal. Signal gains have been stabilized using this unique operating method.

5.3.1.3 Er^{3+}-Doped Fiber Optimization

Two key factors must be taken into account when attempting to obtain highly efficient Er^{3+}-doped fiber. The first is fiber glass material optimization. A high doping concentration of Er^{3+} ions is desirable because it allows the EDF length to be shortened. However, the maximum doping concentration is limited due to the *cooperative up-conversion process* (CUP) with microclustering in the host glass matrix [119,205]. Figure 5.81 shows this process. When excited Er^{3+} ions in the $^4I_{13/2}$ metastable level are close to each other, an energy transfer can occur between them; that is, one ion is excited to an upper excited level $^4I_{9/2}$ and the other decays to the ground level $^4I_{15/2}$. The highly excited ion decays through a nonradiative process, and the excitation energy wastes away. Hence, too high an Er^{3+} concentration results in a degradation in pump efficiency [157,170,205–210]. Figure 5.82 shows the lifetime of the $^4I_{13/2}$ level of Er^{3+} ions as a function of Er^{3+} concentration [210]. The lifetime for Ge/Er-doped fiber (Al concentration of 0 wt%) is constant at 12.8 ms for Er^{3+} concentrations of less than 100 wt ppm but becomes shorter at higher concentrations due to the CUP [170,206–210]. On the other hand, the lifetime for Ge/Al/Er-doped fiber with an Al concentration of 0.5 wt% is constant at 10.3 ms up to nearly 1,000 wt ppm. The shorter lifetime of the Ge/Al/Er-doped fiber is ascribed to the transition probability increase because the environment is modified by the Al codoped fiber. Moreover, a portion of the Er^{3+} ions excited by

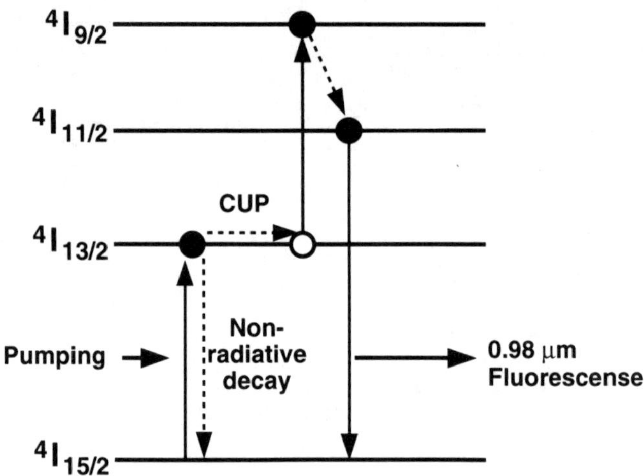

Figure 5.81 Cooperative up-conversion process and emission mechanism of 0.98 μm fluorescence

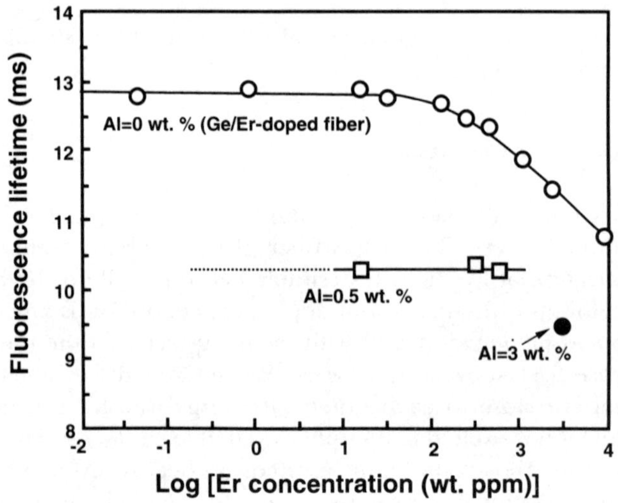

Figure 5.82 Fluorescence lifetime of the $^4I_{13/2}$ level of Er^{3+} ions as a function of Er^{3+} concentration [210].

the CUP is transferred radiatively from the $^4I_{9/2}$ level to the ground level $^4I_{15/2}$ with the 980-nm fluorescence, as shown in Figure 5.81 [170,207–210]. The intensity of the 980-nm fluorescence for EDF pumped at 1,480 nm as a function of Er^{3+} concentration is shown in Figure 5.83 [210]. The results, which show that the fluorescence intensity increases above 100 wt ppm for Ge/Er-doped fiber, suggest

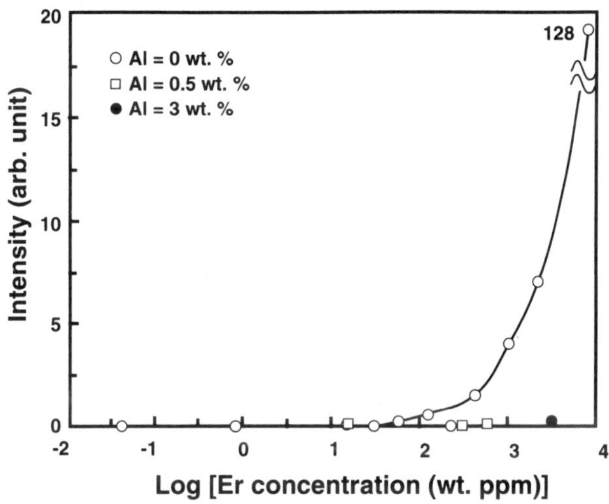

Figure 5.83 Intensity of the 980-nm fluorescence for EDF pumped at 1,480 nm as a function of Er^{3+} concentration [210].

that the CUP begins again. By contrast, no fluorescence is observed for the Ge/Al/Er-doped fiber. These results indicate that to avoid degradation of the pump efficiency, the maximum Er^{3+} concentration should be less than 100 wt ppm for Ge/Er-doped fiber while Al codoping is effective in increasing the allowable Er^{3+} concentration, and a concentration of up to 1,000 wt ppm is possible for Ge/Al/Er-doped fiber with an Al concentration of 0.5 wt%.

The introduction of glass network modifier such as Al_2O_3 to prevent the CUP and microclustering was investigated for silica glass hosts with Nd^{3+} and Er^{3+} ions [157,211]. These studies suggest that P codoping with P_2O_5 also is effective as a glass network modifier. The dissolution of the microclusters is attributed to the formation of solvation shells that surround the rare-earth ions. With Al codoping, a possible model for the solvation shell includes two aluminum atoms per rare-earth ion. An investigation of Nd^{3+}-doped silica glass [211] indicates that the maximum Er^{3+} concentration for Ge/Al/Er-doped fiber needed to prevent the CUP is about 1/10 of the Al concentration and 1/15 of the P concentration.

The dependence of gain characteristics on Er^{3+} concentration has been studied in detail experimentally for Ge/Er-doped fiber. The relationship between the maximum signal gain that is obtained when the fiber length is optimized and the Er^{3+} concentration for 980-nm pumping is shown in Figure 5.84 [170,206]. It has been confirmed that the maximum signal gain depends strongly on the Er^{3+} concentration.

The second important factor in obtaining highly efficient Er^{3+}-doped fiber is the design of the fiber structure. Thus far, two major techniques have been reported:

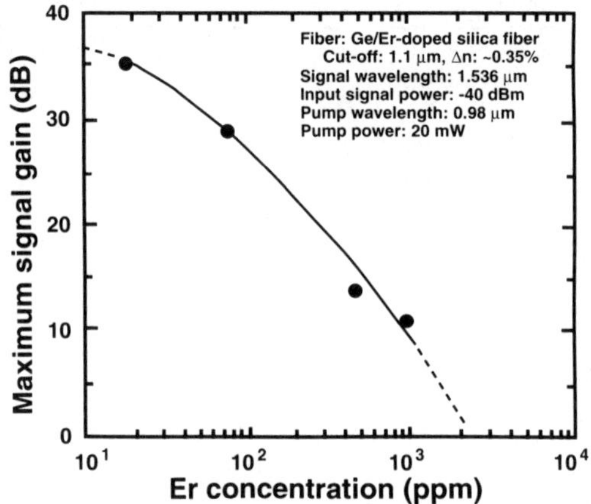

Figure 5.84 Relationship between the maximum signal gain that is obtained when the fiber length is optimized, and the Er^{3+} concentration for 980-nm pumping [170,206].

Er^{3+} confined structure [111,212–217] and the high NA structure [109,146,147, 167,190,214,218,219], as shown in Figure 5.85(a,b), respectively. Confining the Er^{3+} ions to the central core region enables the pump power to be used more efficiently [212]. Figure 5.86 shows the calculated gain coefficient versus confinement, that is, the ratio between the Er^{3+}-doped radius and the core radius, for the 980- and 1,480-nm pumping of fiber with three different compositions: Al/Er-doped,

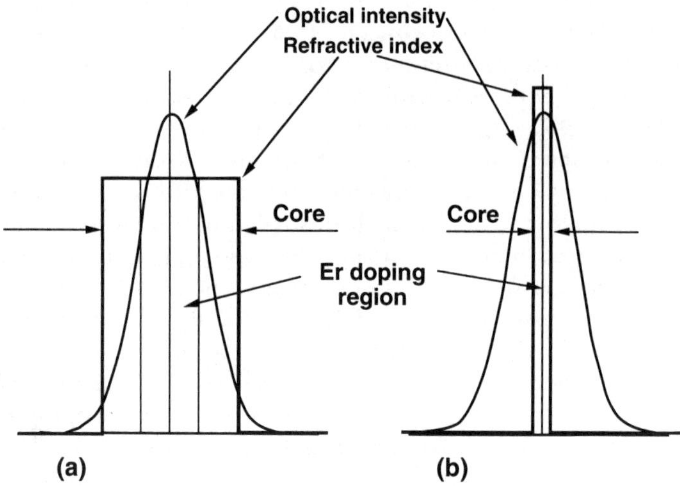

Figure 5.85 Design of the Er^{3+}-doped fiber structure: (a) Er^{3+} confined structure and (b) high NA structure.

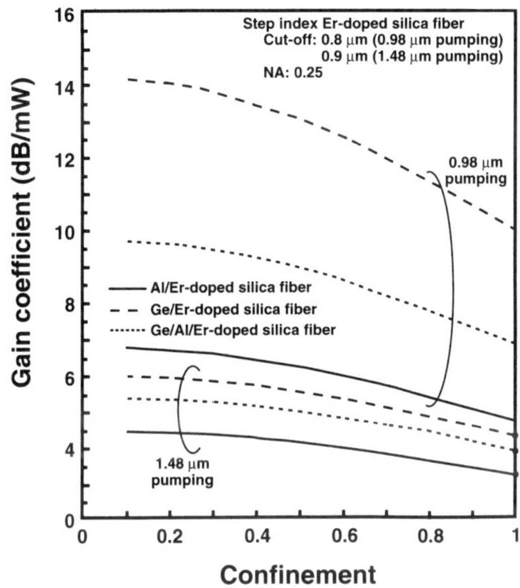

Figure 5.86 Calculated gain coefficient versus confinement, that is, the ratio between the Er^{3+}-doped radius and the core radius, for the 980- and 1,480-nm pumping of fiber with three different compositions: Al/Er-doped, Al/Ge/Er-doped, and Ge/Er-doped fiber [214].

Al/Ge/Er-doped, and Ge/Er-doped fiber [214]. The improvement in gain coefficient is between 40% and 45% in all cases, when the confinement is reduced from 1 to 0.1. A notable gain improvement has been confirmed experimentally. Figure 5.87 shows a typical measured signal gain as a function of Er^{3+}-doped fiber length for two confinement ratios [217]. The disadvantage of this confined structure is that it requires a long Er^{3+}-doped fiber that is dependent on the confinement ratio.

On the other hand, the very high NA structure can provide a high pump power density in an Er^{3+}-doped fiber, resulting in a high-gain coefficient [146,147]. Figure 5.88 shows the calculated gain coefficient versus the cutoff wavelength for the LP_{11} mode in step-index Al/Er-doped fiber for NAs of 0.1 to 0.4 in steps of 0.05 [219]. The optimum cutoff wavelength is independent of the NA and is 0.80 μm for 980-nm pumping and 0.90 μm for 1,480-nm pumping. Figure 5.89 shows the calculated relationship between the gain coefficient at the optimum cutoff wavelength and NA for Al/Er-doped and Ge/Er-doped fiber [219]. The gain coefficient with the optimum cutoff wavelength increases with increasing NA. Moreover, the high NA structure also has the advantage of improving the output characteristics of Er^{3+}-doped fiber. The calculated *quantum conversion efficiency* (QCE) as a function of cutoff wavelength of Er/Ge/Al/P-doped fiber for various NAs is shown in Figure 5.90(a) for 980-nm pumping and Figure 5.90(b) for 1,480-nm pumping [190]. QCE is defined as the number of photons the amplifier adds to

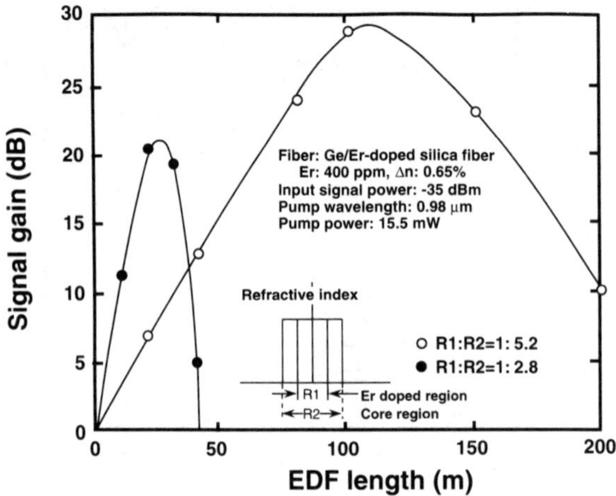

Figure 5.87 Measured signal gain as a function of Er^{3+}-doped fiber length for two confinement ratios [217].

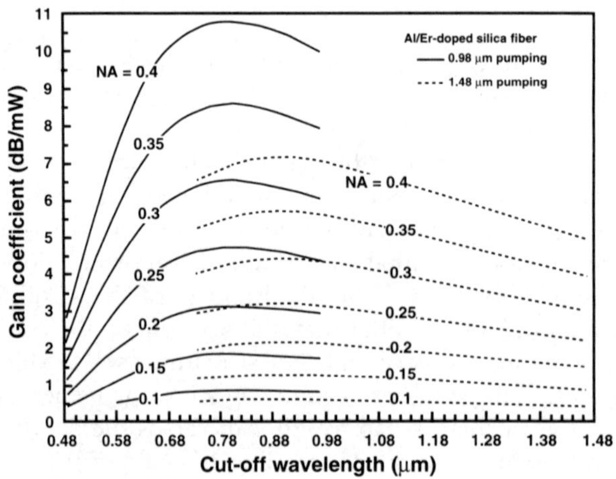

Figure 5.88 Calculated gain coefficient versus the cut-off wavelength for the LP_{11} mode in step-index Al/Er-doped fiber, for NAs of 0.1 to 0.4 in steps of 0.05 [219].

the signal divided by the number of launched pump photons and is related to the *power conversion efficiency* (PCE) as

$$QCE = \frac{\lambda_S}{\lambda_P} PCE \tag{5.53}$$

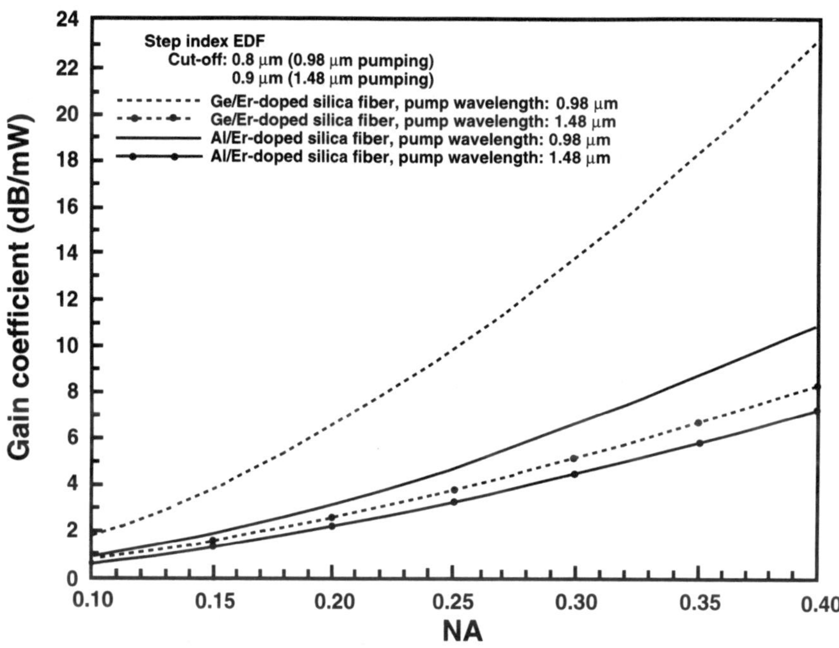

Figure 5.89 Calculated relationship between the gain coefficient at the optimum cut-off wavelength and NA for Al/Er-doped and Ge/Er-doped fibers [219].

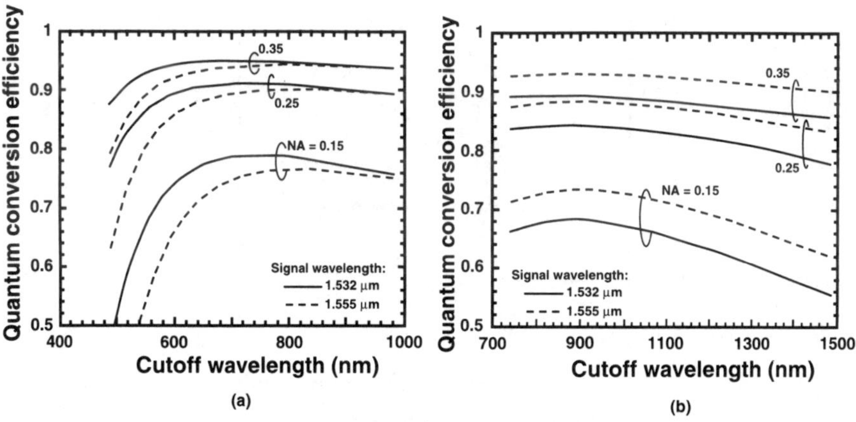

Figure 5.90 Calculated quantum conversion efficiency as a function of cut-off wavelength of Er/Ge/Al/P-doped fiber for various NAs [190]: (a) 980-nm pumping and (b) 1,480-nm pumping.

From the calculated results shown in Figure 5.90(a,b), it can be seen that the QCE increases with increasing NA.

The gain coefficient dependence on the relative refractive index difference Δn has been studied in detail experimentally for Ge/Er-doped fiber, as shown in Figure 5.91; and it has been confirmed that the gain coefficient increases with NA [146,147].

The optimization of the Er^{3+} concentration and the fiber structure has led to the development of highly efficient fiber. Figure 5.92 shows the gain characteris-

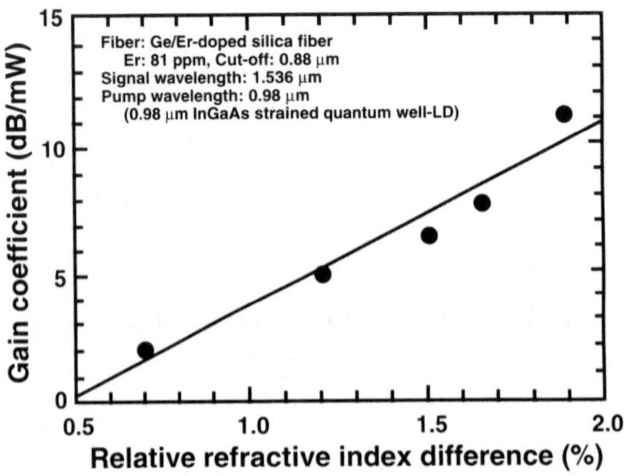

Figure 5.91 Gain coefficient versus relative refractive index difference Δn [146,147].

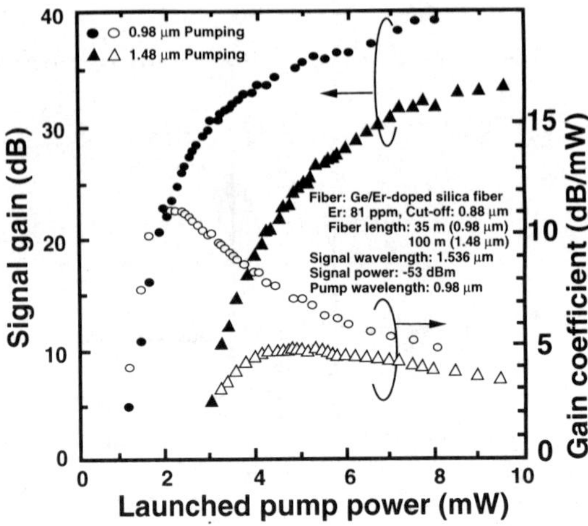

Figure 5.92 Gain characteristics of Ge/Er-doped fiber with an Er^{3+} concentration of 81 ppm and a high Δn of 1.88% [146,147].

tics of Ge/Er-doped fiber with an Er^{3+} concentration of 81 ppm and a high Δn of 1.88% [146,147]. Maximum gain coefficients of 11.0 and 5.0 dB/mW have been obtained for 0.98- and 1.48-μm pumping. In addition, this gain coefficient of 11.0 dB/mW is the highest value reported to date for an EDFA pumped in the 980-nm band. Incidentally, the highest gain coefficient reported to date for 1,480-nm pumping is 6.3 dB/mW [167].

5.3.1.4 EDFA Module

Figure 5.93 shows the three basic amplifier configurations corresponding to unidirectional (forward and backward) and bidirectional pumping. As explained in Subsection 5.4.1.2, forward and backward pumping are suitable for use in a preamplifier that requires low-noise characteristics and in a power amplifier with high-output power, respectively. Bidirectional pumping is used for applications that require both low-noise and high-output power characteristics. The main components of an EDFA are Er^{3+}-doped fiber, a pump light source, WDM couplers, and optical isolators. The function of the WDM coupler in the module is to multiplex or demultiplex the signal and pump lights. To prevent unexpected laser oscillation caused by the Fabry–Perot cavity mode, a polarization-independent optical isolator is inserted in both the input and output ends. For 980-nm pumping, the WDM coupler that is used to demultiplex the signal light and pump lights is sometimes removed because the optical isolator rejects the pump light that leaks to the outside of the EDFA module. A bandpass optical filter is often inserted at the output end of the EDF to suppress the sp. shot noise and sp.-sp. beat noise. Optical components with low insertion losses must be used for the signal light when constructing a high-performance EDFA because they reduce the signal gain G and output signal power P_S^{OUT} and increase the NF of the EDFA module, as expressed in

$$G(\text{dB}) = G^{\text{fiber}}(\text{dB}) - 2(L_{\text{iso}}(\text{dB}) + L_{\text{WDM}}(\text{dB}) + L_{\text{splice}}(\text{dB}) - L_{\text{BPF}}(\text{dB})) \tag{5.54}$$

$$P_S^{OUT}(\text{dBm}) = P_S^{\text{fiber-OUT}}(\text{dBm}) - (L_{\text{iso}}(\text{dB}) + L_{\text{WDM}}(\text{dB}) + L_{\text{BPF}}(\text{dB}) + L_{\text{splice}}(\text{dB})) \tag{5.55}$$

$$NF(\text{dB}) = NF_{\text{fiber}}(\text{dB}) + (L_{\text{iso}}(\text{dB}) + L_{\text{WDM}}(\text{dB}) + L_{\text{splice}}(\text{dB})) \tag{5.56}$$

where G^{fiber}, $P_S^{\text{fiber-OUT}}$, and NF_{fiber} are the signal gain, output signal power, and noise figure of the EDF, respectively. L_{iso}, L_{WDM}, and L_{BPF} are the insertion losses of the optical isolator, WDM coupler, and bandpass filter, respectively. L_{splice} is the coupling loss between a standard fiber and EDF. Furthermore, the insertion loss of the WDM coupler and the coupling loss between the standard fiber and EDF for the pump light reduce the pump power launched into the EDF, and this

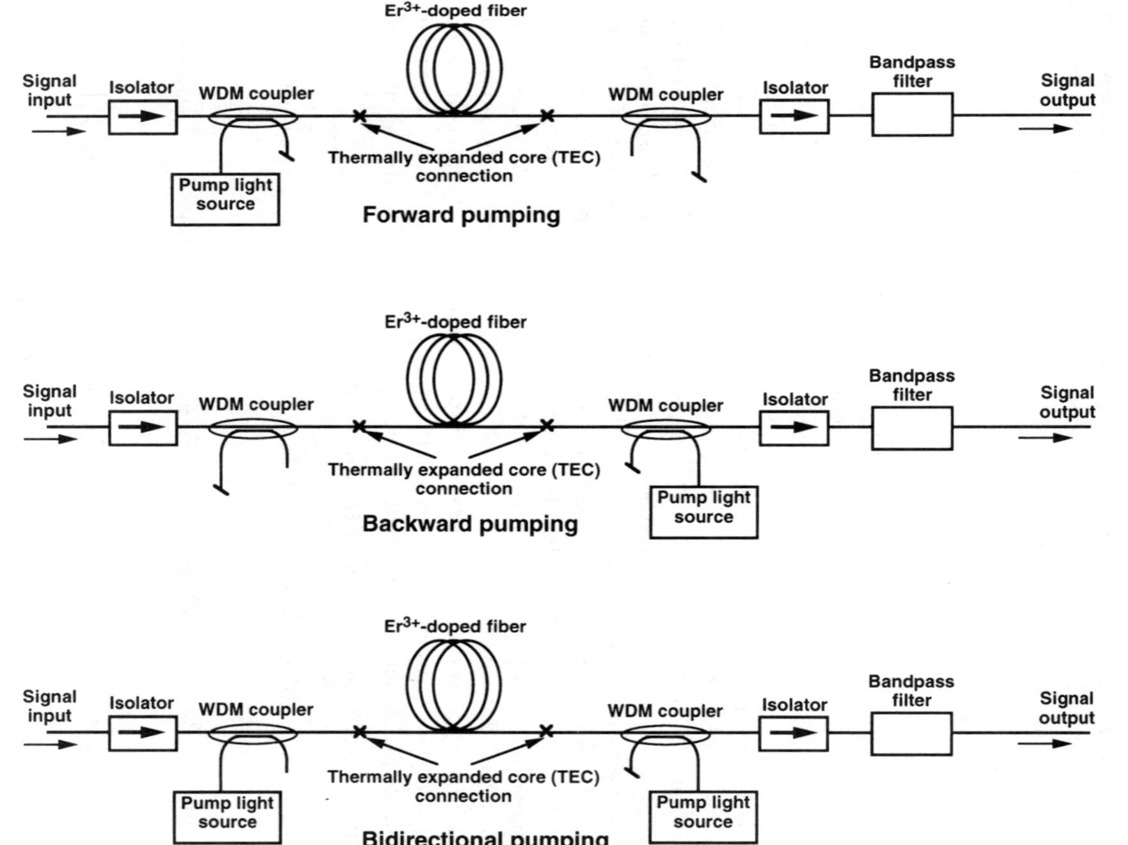

Figure 5.93 Three basic amplifier configurations corresponding to unidirectional (forward and backward) and bidirectional pumping.

reduction in launched pump power also induces a decrease in the signal gain G^{fiber} and output power $P_S^{\text{fiber-OUT}}$ and an increase in the NF_{fiber} of the EDF. So optical components with low insertion losses have been developed as described in Section 5.2, and low-loss coupling has been achieved between standard fiber and EDF. In fact, because optimized EDF usually has a small mode-field diameter with a high Δn compared to standard fiber, conventional fusion splicing usually yields a significant coupling loss. Today, the TEC connection technique [220,221] is used to realize low-loss splicing, and values of less than 0.1 dB have been achieved for both pump and signal lights. This technique will be explained in detail in Subsection 5.3.2.2, "PDFA Module."

Practical EDFA modules with the efficient pump bands of 980 and 1,480 nm have been demonstrated by the fabrication of an LD module with a fiber pigtail. The LD module is a very important component in terms of achieving stable pumping. Figure 5.94(a,b) shows a photograph and the gain characteristics of the first 980-nm LD-pumped EDFA module [91]. A forward pump configuration with a 980-nm-band InGaAsP-LD module was used in this EDFA module that was employed as an optical preamplifier. This EDFA module was only 36 cc (4 by 6 by 1.5 cm^3) in volume and provided a maximum signal gain of 33 dB with an electricity

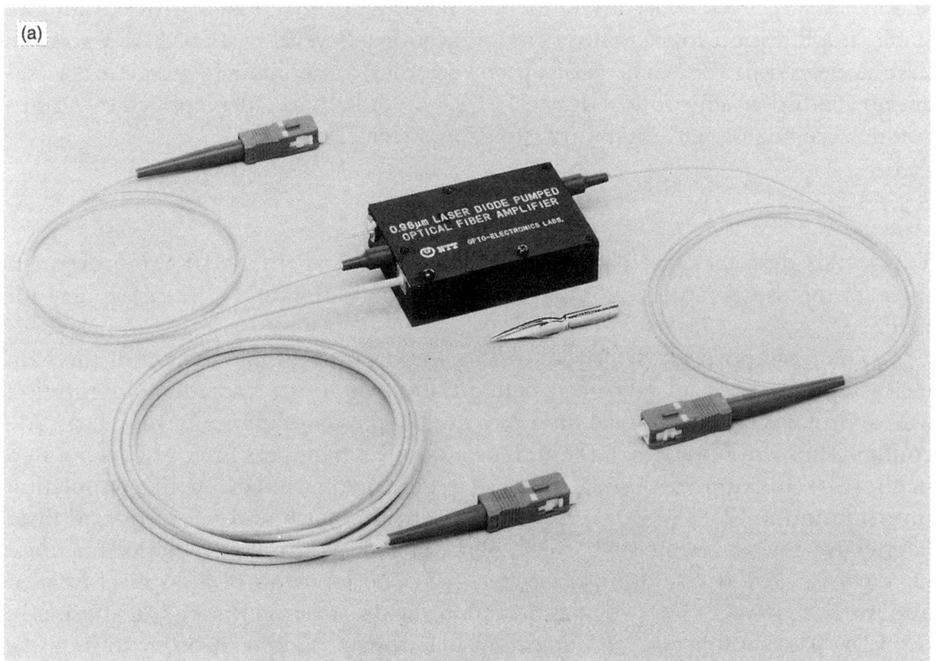

Figure 5.94 980-nm LD-pumped EDFA module [91]: (a) photograph and (b) gain characteristics.

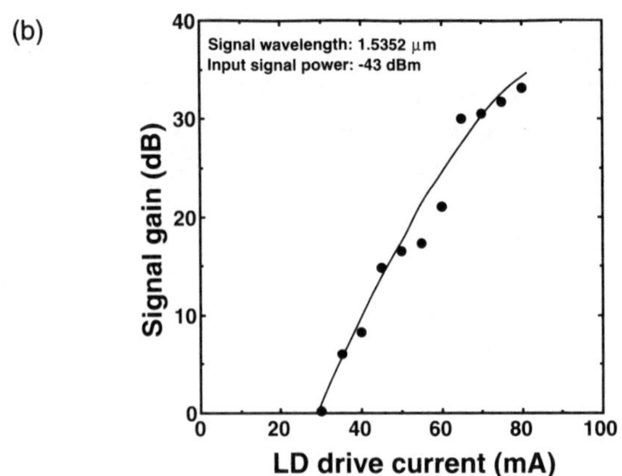

Figure 5.94 (continued).

consumption of only 175 mW, which corresponds to an LD drive current of 80 mA. Since this demonstration, the gain characteristics of EDFA modules have been improved by various laboratories and companies. *Automatic gain control* (AGC) and *automatic power control* (APC) functions have also been added in order to achieve stable amplification for actual system applications. Several types of EDFA module have appeared in the catalogs of optoelectronics component manufacturers and are now indispensable optical devices. EDFA modules have been applied to various commercial analog and digital transmission systems.

5.3.1.5 Advanced EDFA

This section describes and discusses a number of approaches to EDFA improvement through the use of a more advanced EDFA configuration and a new glass host for EDF.

The first approach dealt with in this section is the improvement of the gain characteristics using an advanced configuration such as a cascade configuration with an isolator and/or optical filter or a double-pass configuration. In the cascade configuration, an optical isolator and/or an optical bandpass filter is incorporated in the EDF to suppress ASE, which has a detrimental effect on the population inversion in the EDF. The insertion of an optical isolator and/or an optical filter is effective in improving such gain characteristics as the gain coefficient, noise characteristics, and saturation characteristics. The insertion of an optical isolator also works to prevent laser oscillation caused by the internal Rayleigh scattering in the EDF. The double-pass configuration has been proposed in order to improve pump efficiency. In this configuration, the signal passes through the EDF twice by employing an optical circulator and an endface reflective mirror. The gain coefficient will be twice that of the conventional single-pass configuration.

The second and third approaches are designed to achieve a wide amplification band and improve the output characteristics, respectively. Broadband EDFAs with uniform gain and low noise are now urgently required in order to apply EDFAs to WDM transmission systems. Several methods have been reported for silica-based EDF with a view to achieving these requirements. They include adding an extremely high Al concentration and optimizing fiber structure, adding a gain equalizer, and employing a hybrid configuration. Moreover, new glass hosts for EDF, typically fluoride-based EDF, have also been studied to obtain a drastic improvement in the gain spectrum. Approaches with silica-based EDF, fluoride-based EDF, and other multicomponent glass EDF will be described in Subsection 5.3.1.5, "Broadband EDFA." In addition, finding a way to increase the output signal power is an important problem with regard to long-haul repeaterless optical transmission, for head-end amplifiers for CATV distribution applications, and in terms of compensating for the insertion loss for $1 \times N$ splitters. Subsection 5.3.1.5, "Ultra-High-Power EDFA," describes various approaches that are based on the supply of a high pump power to the EDF by using high-power pump light sources and optimizing the EDF structure so that the pump power is effectively absorbed.

Improvement in Gain Characteristics With Advanced Configuration

Cascade configuration with midway isolator and/or optical filter. An ideal amplifier that employs the stimulated emission process amplifies only the signal light and not the spontaneous emission light, as shown in Figure 5.95. In this ideal amplifier, spontaneous emission light whose generation cannot be avoided when using the stimulated emission process would be eliminated immediately from the amplifying medium. However, in actual amplifiers, the spontaneous emission grows into ASE via the stimulated emission process, and this ASE induces a degradation in the amplification performance. This may consist of a decrease in the gain coefficient and saturation output power or an increase in the NF due to the reduction in

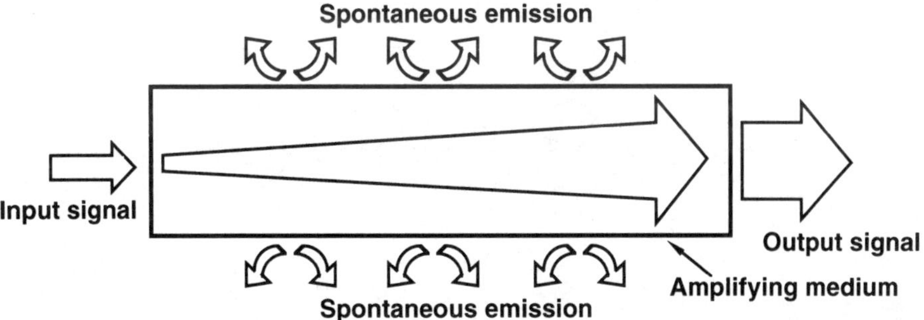

Figure 5.95 Ideal amplifier that employs the stimulated emission process.

the population inversion in the amplifying medium. Furthermore, with high-gain operation, the signal gain is limited by lasing and extra noise is added because of the multiple reflection, which causes internal Rayleigh scattering [222].

The incorporation of an optical isolator and/or an optical bandpass filter into the Er^{3+}-doped fiber is effective in suppressing ASE and thus moving the EDFA closer to an ideal amplifier [197,223–229]. The optical isolator is used to eliminate any back-propagating ASE and to prevent lasing and the addition of extra noise caused by the Rayleigh scattering. The optical bandpass filter is used to reduce both the forward- and back-propagating ASE, without the ASE around the signal wavelength. Figure 5.96(b) shows the basic cascade configuration that incorporates an optical isolator and/or an optical filter compared to a conventional configuration (a). The bypass coupling shown in (c) is used to guide the pump light pass through an isolator for 980-nm unidirectional pumping and through an optical filter for 980- and 1,480-nm unipumping if the filter cannot transmit the pump light. Although the effectiveness is enhanced by increasing the numbers of isolators and filters, the configuration then becomes complex. Almost all reports of theoretical and experi-

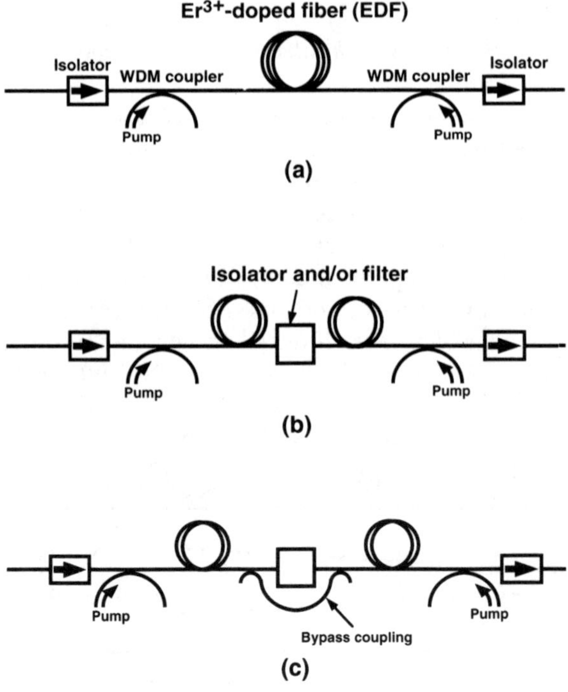

Figure 5.96 Basic cascade configuration that incorporates an optical isolator and/or an optical filter: (a) conventional configuration, (b) cascade configuration, and (c) cascade configuration with bypass coupling.

mental investigations related to this configuration employ one optical isolator and/ or one optical filter. In the following, several such results will be described.

Figures 5.97 and 5.98 show calculated signal gain and NF as a function of pump power for a conventional configuration and a cascade configuration with a bandpass filter, with an isolator, and with an optical bandpass filter and an isolator for 980- and 1,480-nm pumping, respectively [223]. It is confirmed that the greatest improvements in both the signal gain and the noise characteristics for both pump wavelengths are achieved by using the configuration either with an optical bandpass filter or an isolator. The insertion of an optical isolator provides a more marked improvement than the use of an optical bandpass filter. A NF close to the quantum limit of 3 dB is achieved simply by using an isolator.

An additional but nonetheless important factor is the relative position of the isolator or the filter in this configuration. The dependence of the calculated gain and NF on the relative position of the isolator and filter are shown in Figure 5.99 [228] and Figure 5.100 [229], respectively. The position is measured from the input end and is relative to the total EDF length. The optimum position is about 20%~30% for a configuration with an isolator and 40% for a configuration with a filter. The optimum position for a configuration with a filter and an isolator is possibly the same as that with just an isolator because of the drastic improvement in the gain

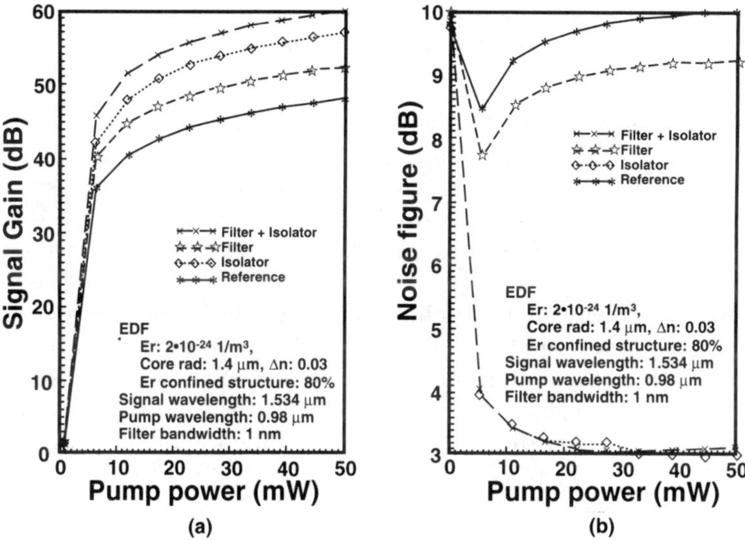

Figure 5.97 Calculated signal gain and noise figure as a function of pump power for a conventional configuration, and a cascade configuration with a bandpass filter with an isolator, an optical bandpass filter, and an isolator for 980-nm pumping [223]: (a) signal gain as a function of pump power and (b) noise figure as a function of pump power.

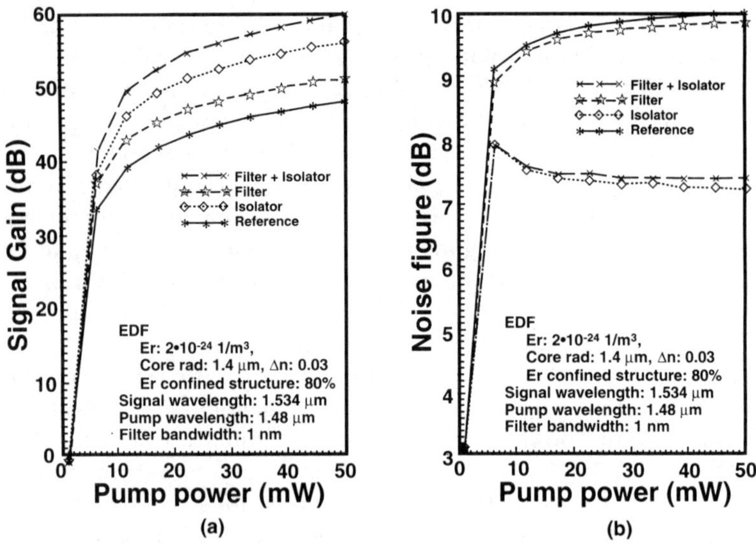

Figure 5.98 Calculated signal gain and noise figure as a function of pump power for a conventional configuration, and a cascade configuration with a bandpass filter, an isolator, and an optical bandpass filter, and an isolator for 1,480-nm pumping [223]: (a) signal gain as a function of pump power and (b) noise figure as a function of pump power.

characteristics resulting from the insertion of an isolator into an EDF. Furthermore, the incorporation of an isolator suppresses the influence of the multiple reflection of the Rayleigh scattering. This is confirmed by the fact that an EDFA with a cascade configuration incorporating an isolator has achieved a signal gain of 54 dB, which is above the upper limit due to the Rayleigh scattering, with a low NF of 3.1 dB [227].

There is another actual cascade configuration incorporating an isolator that has first and second amplification units, as shown in Figure 5.101 [230]. The advantage of this configuration is not only that it eliminates any back-propagating ASE but also that it enables an EDFA to be constructed that has both low-noise and high-power characteristics. The NF of this configuration NF_{cascade} is expressed from (5.52) by

$$NF_{\text{cascade}} = NF_{\text{first}} + \frac{NF_{\text{second}}}{G_{\text{first}} \cdot L} \tag{5.57}$$

where G_{first} is the signal gain of the first unit, L is the insertion loss of the optical isolator, and NF_{first} and NF_{second} are the noise figures of the first and second units, respectively. In this configuration, 0.98-μm forward pumping is used for the first

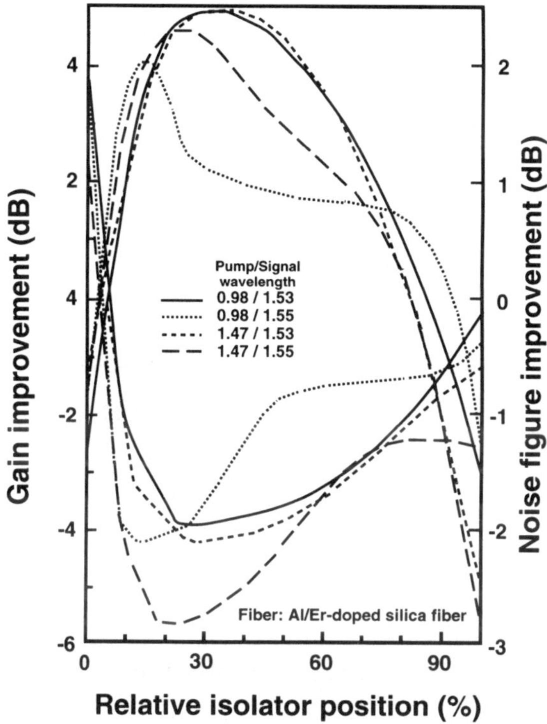

Figure 5.99 Dependence of the calculated signal gain and noise figure improvement on the relative position of the isolator [228].

unit to realize low-noise characteristics. By contrast, 1.48-μm backward pumping is used with the second unit to achieve an efficient power amplifier. (This is sometimes called a hybrid pump configuration because different pump wavelengths are used for the first and second amplification units.)

Double-pass configuration. The conventional amplifier configuration used for EDFAs employs one-way signal amplification. However, the signal can be amplified in both the forward and backward directions in an EDF under the same pump conditions, and its gain coefficient will then be twice that of the conventional configuration [231–234]. This amplification concept is based on an investigation of semiconductor laser amplifiers designed to obtain polarization independence [235]. The configuration based on this consideration is called the double-pass configuration, and the conventional amplifier configuration is the single-pass configuration.

Figure 5.102(a) shows the conventional (single-pass) A, and Figure 5.102(b,c) shows two double-pass configurations B,C [233]. The double-pass configurations use additional components such as an optical circulator and an reflective mirror.

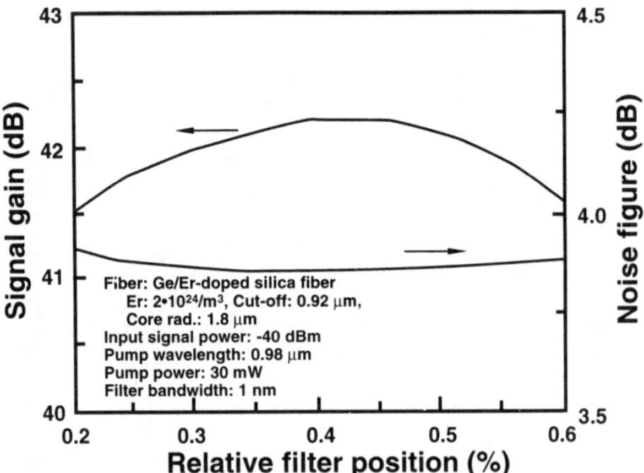

Figure 5.100 Dependence of the calculated signal gain and noise figure on the relative position of the filter [229].

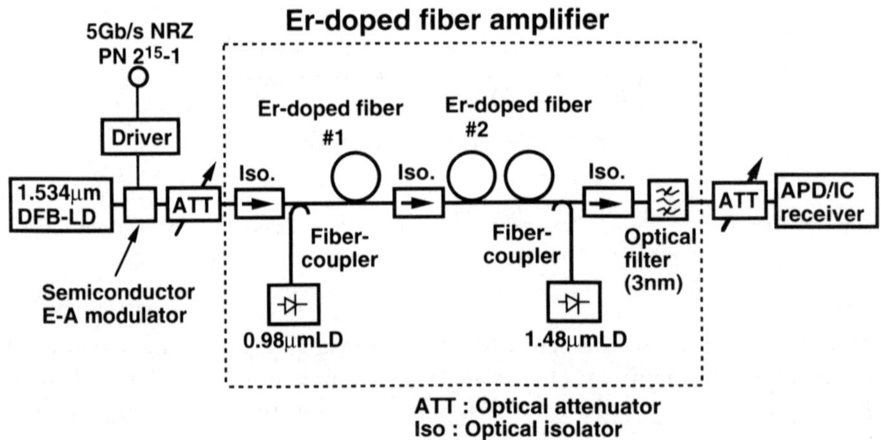

Figure 5.101 Cascade configuration with 0.98/1.48-μm hybrid pumping [230].

With configuration B, both the signal and pump lights are reflected from the endface reflective mirror. The signal light is amplified in both the forward and backward directions through the EDF. The double-pass configuration C also uses a bandpass filter to improve the noise characteristics by suppressing the ASE, and a WDM coupler is placed in front of the bandpass filter to lead the pump light to the mirror in order to avoid filtering it out. In these double-pass configurations, the reflected pump light is reused and the pump light is used efficiently to excite the EDF. So these configurations are sometimes also called reflective-pump configu-

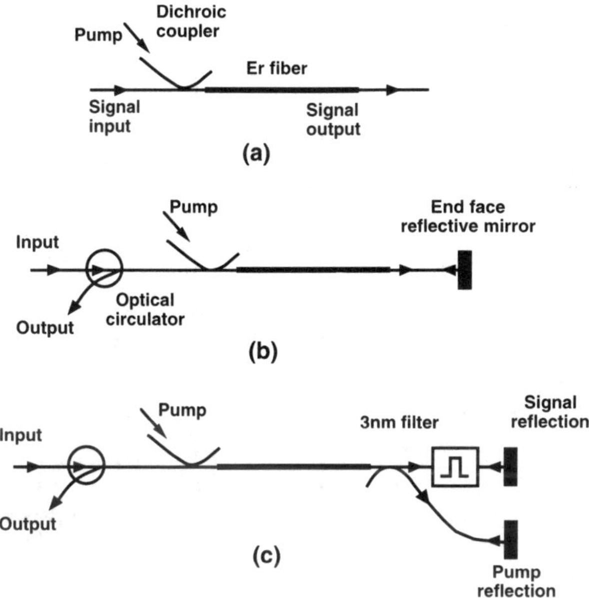

Figure 5.102 Double-pass configurations [233]: (a) conventional (single-pass) configuration, A; (b) double-pass configuration, B; and (c) double-pass configuration, C.

rations. Figure 5.103(a,b) shows the signal gain and NF as functions of pump power for all three configurations, respectively [233]. It is confirmed that both double-pass configurations B and C improve the gain coefficient and that the noise characteristics are improved with configuration C.

These double-pass configurations have been adopted for PDFAs because they are especially effective for a low-efficiency gain medium; the detail of the gain coefficient, which is twice that of the conventional PDFA configuration, will described in Subsection 5.3.2.2, "PDFA Module."

In addition, a configuration has been proposed that only uses reflected pump light [234], as illustrated Figure 5.104, and it has been confirmed that this configuration improves the gain coefficient.

Broadband EDFA

It is vitally important to realize an EDFA that has a wide amplification band with uniform gain and low-noise characteristics for various applications. Recently, these improvements are being demanded for WDM transmission systems and all-optical self-routed wavelength-addressable networks that are capable of increasing the transmission capacity by exploiting the vast bandwidth of optical fiber. If these improve-

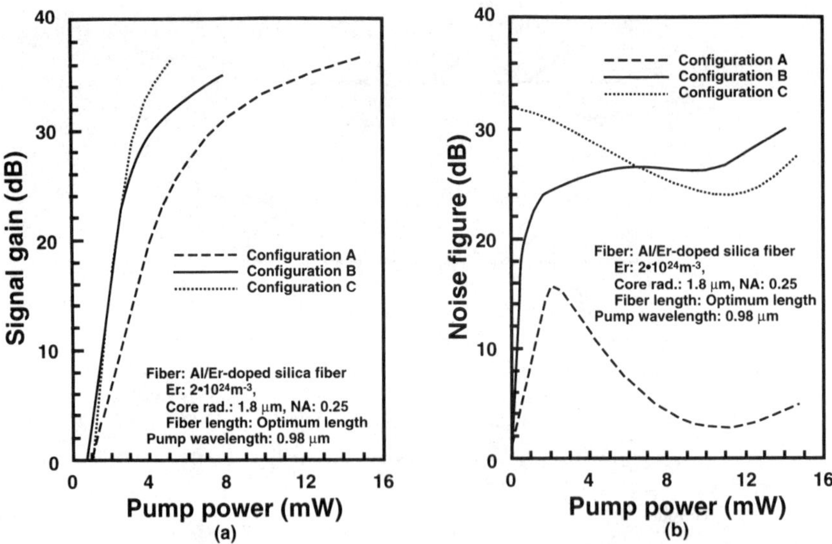

Figure 5.103 Signal gain and noise figure as a function of pump power for all three configurations [233]: (a) signal gain as a function of pump power and (b) noise figure as a function of pump power.

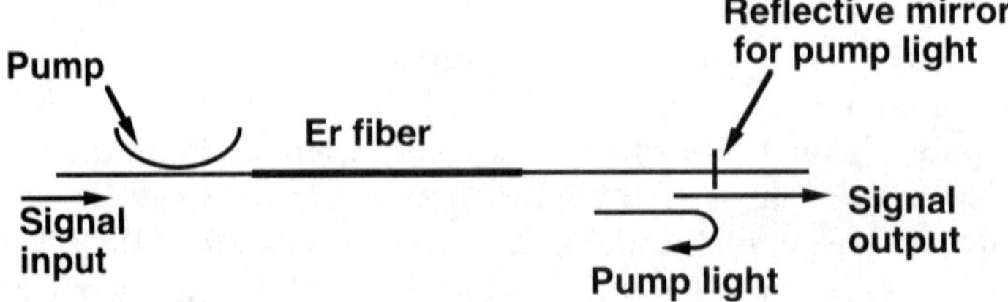

Figure 5.104 Configuration that only uses reflected pump light [234].

ments are achieved it will allow us to construct flexible networks that are transparent in terms of signal format and bit rate. The following approaches have been reported for realizing these requirements:

1. Improve the amplification bandwidth characteristics of Er^{3+}-doped silica fiber (silica-based EDF) by adding an extremely high concentration of Al or by optimizing the fiber structure;
2. Change the fiber host material from silica: (a) fluoride-based EDF and (b) other multicomponent glass EDFs;
3. Use a gain equalizer;

4. Hybrid configuration: (a) Al/Ge-codoped silica-based EDF + Al-codoped silica-based EDF, (b) Al-codoped silica-based EDF + Al/P-codoped silica-based EDF, (c) silica-based EDF + fluoride-based EDF, and (d) silica-based EDF + another multicomponent glass EDF.

In the following, we will describe various approaches, including those with silica-based EDF, fluoride-based EDF, and other multicomponent glass EDFs.

Approaches with silica-based EDF. One approach is extremely high-concentration Al codoping. It is well known that the codoping of Al into the core of Er^{3+}-doped silica fiber is effective in broadening the gain spectrum [119–121,172–174] as described in Subsection 5.3.1.2, "Gain Spectrum." The amplification characteristics of the Al-codoped silica-based EDFA have already been studied in detail, and practical EDFA modules have been fabricated. Furthermore, it was recently found that silica-based EDF with an extremely high aluminum concentration of as much as 3 wt% has a very flat gain spectrum in the 1,540- to 1,560-nm wavelength region [178]. Figure 5.105 shows the gain spectrum of EDF with Al concentrations of 0.5, 1.5, and 3 wt% [178]. The other fiber parameters of the three compared EDFs, such as Δn, cutoff wavelength, and Er^{3+} concentration, are almost the same. The gain variation in the 1.55-μm amplification band is suppressed as the Al concentration increases. The EDF with an Al concentration of 3 wt% in particular shows no reduction in gain near 1.54 μm, which can be found with the other two lower Al concentration EDFs, and a very flat signal gain spectrum with a gain fluctuation of less than 1 dB is achieved in the 1,540- to 1,560-nm signal region. The difference between the signal gains at 1,540 and at 1,550 nm is plotted as a function of Al

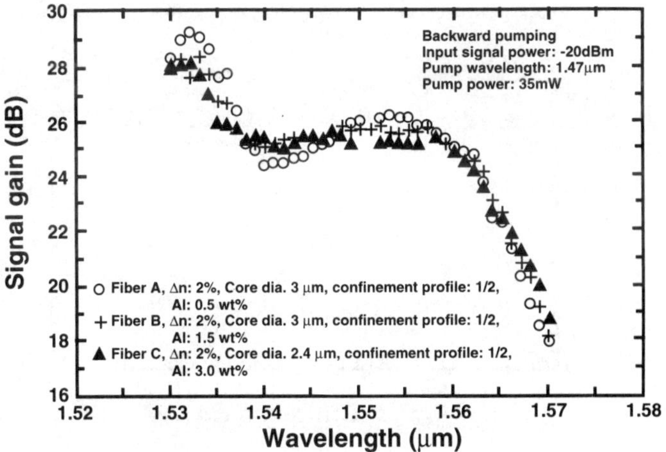

Figure 5.105 Gain spectrum of EDF with Al concentrations of 0.5, 1.5, and 3 wt% [178].

concentration in Figure 5.106 [178]. The wavelength dependence of the gain decreases clearly as a function of concentration.

A silica-based EDFA with a high Al concentration has been employed as a repeater in a WDM transmission experiment. An excellent gain uniformity of less than 0.2 dB was confirmed for WDM signals ranging from 1,545 to 1,557 nm [179].

The idea of employing a gain equalizer is a submissive approach because it is usually used in an electrical amplifier to improve amplification spectrum characteristics. Several equalizers have been reported for optical fiber amplifiers, and the methods use can be classified as (1) the insertion of a band-stop or notch filter into the EDF or (2) the addition of a passive filter to the output port of the EDFA.

The former filter is generally used to suppress the sharp gain peak around 1.53 μm. The insertion of a filter in the EDF also has the advantages of increasing the pump efficiency and improving the noise characteristics, which results from the suppression of ASE build-up, as achieved with the cascade configuration incorporating an isolator, which was described in Subsection 5.3.1.5, "Improvement in Gain Characteristics With Advanced Configuration." Several types of filter have been reported with notch characteristics around 1.53 μm. They include the mechanical deformation of the fiber to couple fundamental and leaky modes [236], the use of a grating filter [237–239], a twin-core fiber filter [240], a WDM coupler [241], and a tapered fiber filter [242]. As an example, Figure 5.107 shows the transmission spectra of a *tapered fiber filter* (TFF) for the signal wavelength and 980-nm pump wavelength regions [242]. The use of this filter completely suppresses the gain peak near 1.53 μm, as shown in Figure 5.108, and a wide flat gain spectrum is realized from 1,530 to 1,560 nm, which is the entire gain region of the EDFA.

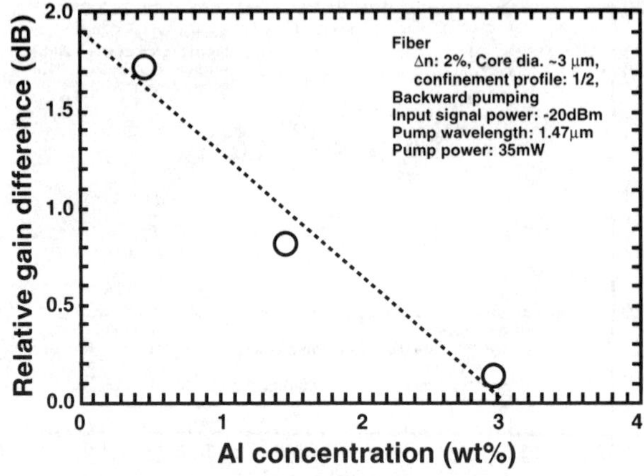

Figure 5.106 Difference between the signal gains at 1,540 and 1,550 nm is plotted as a function of Al concentration [178].

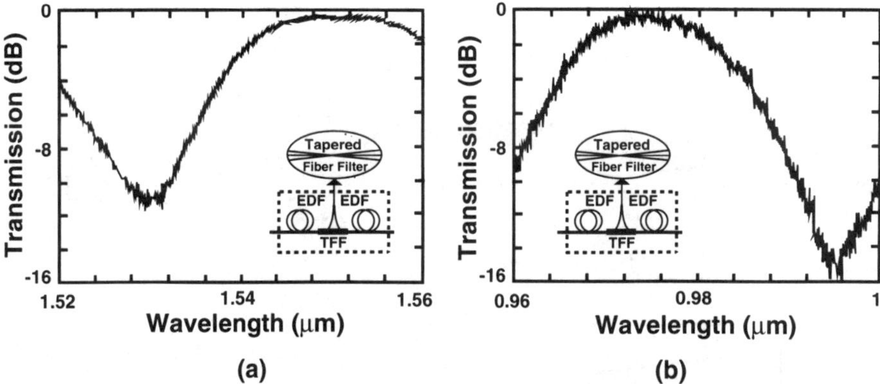

Figure 5.107 Transmission spectra of a tapered fiber filter for the signal wavelength and 980-nm pump wavelength regions [242]: (a) signal wavelength region and (b) 980-nm pump wavelength region.

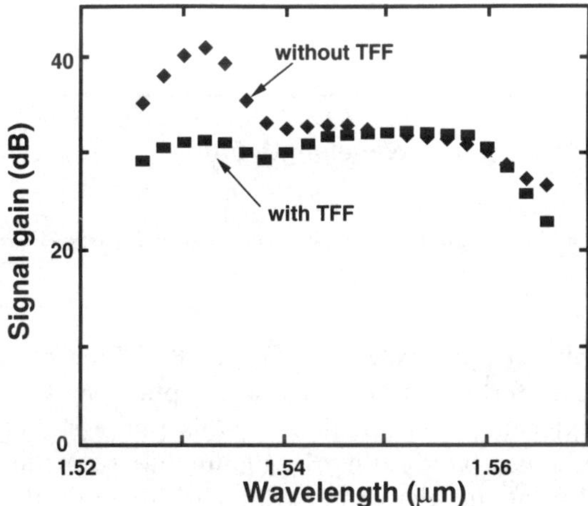

Figure 5.108 Gain spectrum with and without midway tapered fiber filter [242].

On the other hand, the passive filters that are added to the output port of the EDFA are mainly used to flatten the gain spectrum of 1,540- to 1,560-nm signals. A *Mach–Zehnder* (MZ) type *planar lightwave circuit* (PLC) filter [243,244], a split-beam Fourier filter [245], and an acousto-optic tunable filter [246] have already been used. These filters can flexibly change the transmission characteristics to correspond to a change in the gain spectrum of the EDFA. Figures 5.109 and 5.110 show the configuration and transmission spectrum of an MZ-type PLC filter and a

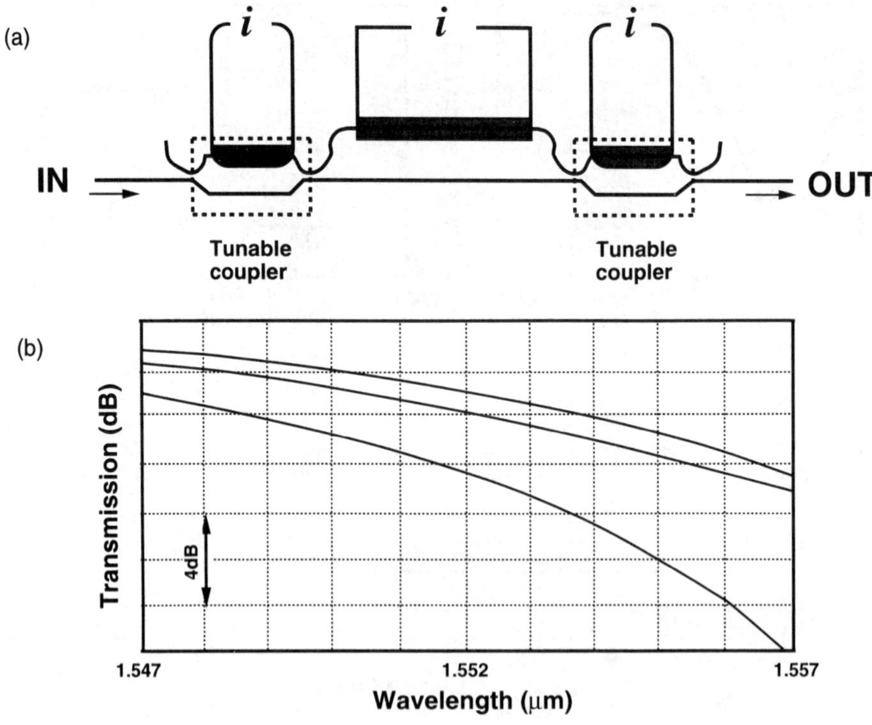

Figure 5.109 (a) Configuration and (b) transmission spectrum of an MZ type PLC filter [243,244].

split-beam Fourier filter, respectively. The MZ-type PLC filter has the configuration of a symmetric MZ interferometer equipped with a phase shifter. The transmission spectrum can be externally controlled using this phase shifter. Specifically, by applying this filter, a flat bandwidth of ~10 nm has been demonstrated for a 100-channel optical frequency-division-multiplexing system with six in-line EDFAs [244]. The split-beam Fourier filter also has MZ characteristics and comprises a flat plate of glass with one polished edge perpendicular to the plate surface that is inserted into a single-mode fiber beam expander. The flat plate can be moved into and out of the beam to change the transmission depth or rotated within the beam to change the transmission center wavelength. Figure 5.111 shows the improvement in flatness realized with the split-beam Fourier filter. After filtering, a flat gain region from 1,540 to 1,558 nm has been demonstrated with a gain fluctuation of less than 0.1 dB. In addition, the acousto-optic tunable filter can also be used to flatten the gain spectrum. The operation principle of the acousto-optic tunable filter is based on the fact that the transmittance of the filter at various signal wavelengths can be adjusted by tuning the frequency and power level of the

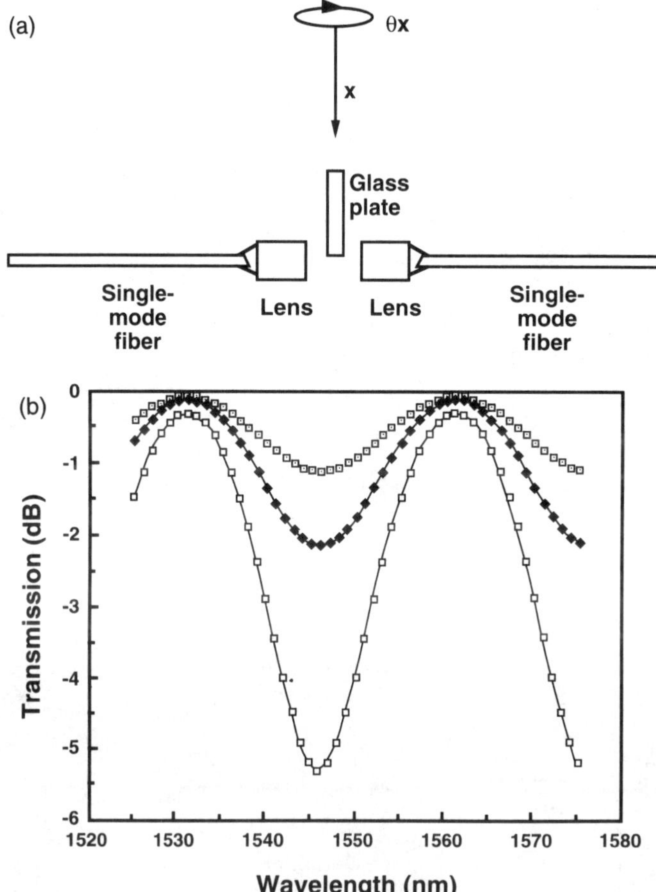

Figure 5.110 (a) Configuration and (b) transmission spectrum of a split-beam Fourier filter.

individual RF signals applied to it. The problem with this filter is that it requires relatively high RF power (that is, 1W for a 10-channel signal).

The concept has been proposed of an EDFA with a hybrid configuration. This is a two-stage fiber amplifier that has two doped-fiber compositions with different gain spectra and the first such hybrid configuration consisting of Al/Ge-codoped silica-based EDF and Al-codoped silica-based EDF has been reported [247]. A gain correction of 1 dB has been demonstrated for two signals 2.5 nm apart.

Recently, the gain flatness of the EDFA with a hybrid configuration has been improved for WDM signals using Al-codoped silica-based EDF and Al/P-codoped silica-based EDF [248,249]. Figure 5.112 shows the configuration of this hybrid-type EDFA, and Figure 5.113 shows its output spectrum. The amplification character-

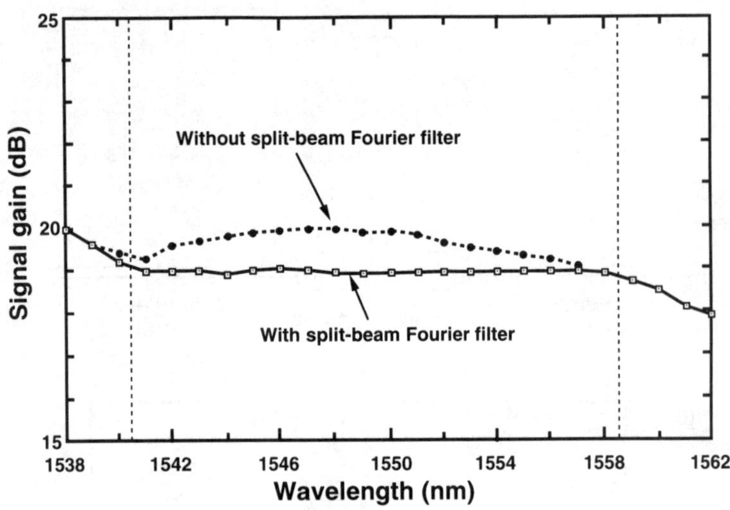

Figure 5.111 Improvement in flatness realized with the split-beam Fourier filter [245].

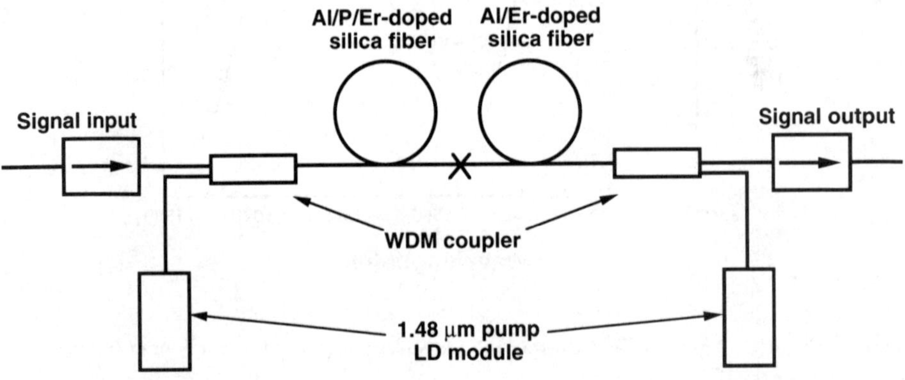

Figure 5.112 Configuration of hybrid EDFA constructed by using Al-codoped silica-based EDF and Al/P-codoped silica-based EDF [248,249].

istics are flat and have a gain nonuniformity of 0.3 dB for WDM signals in the 1,543- to 1,558-nm wavelength region. Figure 5.113 also shows the output spectrum of the Al-codoped silica-based EDF and Al/P-codoped silica-based EDF. The flat gain spectrum of the hybrid type EDFA compensates for their respective output spectra. The practical advantages of this hybrid EDFA founded on silica-based EDF have been confirmed from the results of a successful transmission experiment of over 509 km of 4 by 2.5 Gbps [249].

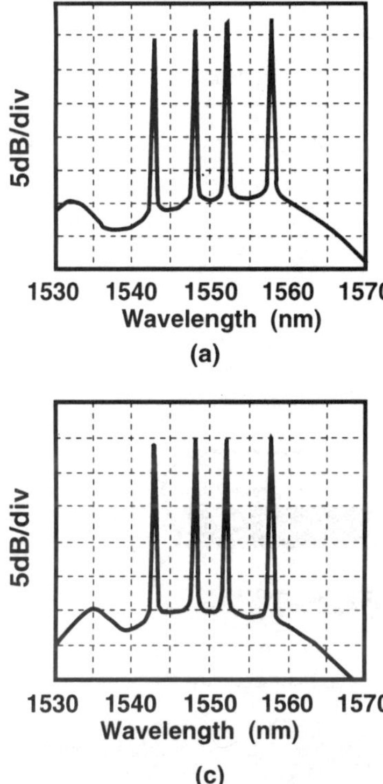

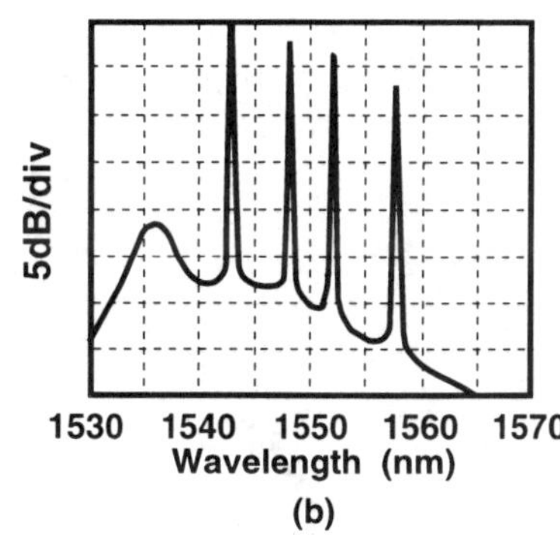

Figure 5.113 Output spectrum of the (a) Al-codoped silica-based EDF (22m), (b) Al/P-codoped silica-based EDF (9m), and (c) hybrid EDF (Al-codoped EDF: 16m; Al/P-codoped EDF: 9m) [248,249].

There have been two interesting reports on fiber structure design as it relates to WDM signals: one is on Er^{3+}-doped twin-core fiber [250,251] and the other is on ring-doped EDF [252]. They were proposed in order to equalize the two different wavelength signal powers of a two wavelength WDM system.

Figure 5.114 shows the Er^{3+}-doped twin-core fiber that can provide automatic spectral gain equalization as a result of spatial hole burning [250]. Both cores are Er^{3+}-doped. The amplifier is configured such that the signal and pump lights couple periodically between the two cores along the fiber length. One signal exhibits a certain periodic spatial intensity distribution and thus accesses a group of Er^{3+} ions, whereas a different signal wavelength accesses a different group of Er^{3+} ions. This practically decouples the gain at the two signal wavelengths; therefore, when one signal is larger than the other, spatial hole burning will preferentially reduce its gain resulting in spectral gain equalization.

The structure of the ring-doped EDF is shown schematically in Figure 5.115 [252]. The fiber has two Er^{3+}-doped regions, a center region and a concentric outer

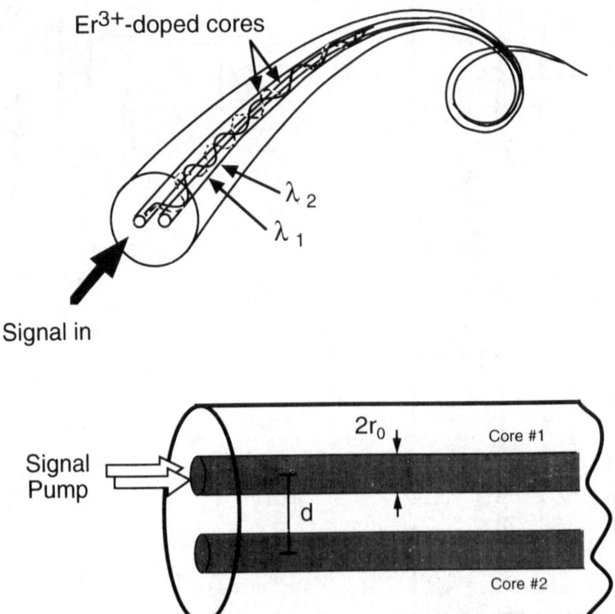

Figure 5.114 Er^{3+}-doped twin-core fiber geometry [250].

region, which are made of different glass material each having different emission and absorption spectra. The outer region is set where the power density of 980-nm pump wavelength light is sufficiently small and that of 1,480-nm pump wavelength light is not. This means that the Er^{3+} ions of the outer region are only pumped by 1,480-nm pump wavelength light. Pumping the fiber at both 980 and 1,480 nm results in two types of Er^{3+} ions becoming excited independently under the control of the two wavelength pump powers. This enables two different wavelength signal powers to be controlled independently by using the difference between the gain spectra of two Er^{3+}-doped regions.

Approaches with fluoride-based EDFA. Since the first observation of an ASE spectrum with Er^{3+}-doped ZrF$_4$-based fluoride single-mode fiber in 1998 [180], it has been confirmed that an EDFA that uses Er^{3+}-doped ZrF$_4$-based fluoride single-mode fiber (fluoride-based EDFA) displays a flatter ASE spectrum than a silica-based EDFA, without the need for any active or passive spectral filtering devices and with good gain performance and pump efficiency [181–185,253,254]. Figure 5.116 shows typical ASE spectra for fluoride-based and silica-based EDFAs [185]. The fluctuation in the ASE power of the fluoride-based EDFA in the 1,530- to 1,560-nm wavelength range is less than that of the silica-based EDFA where a large peak remains near 1,530 nm. In addition, there is a flat gain region where flat amplification is expected around the signal wavelength of 1,535 nm.

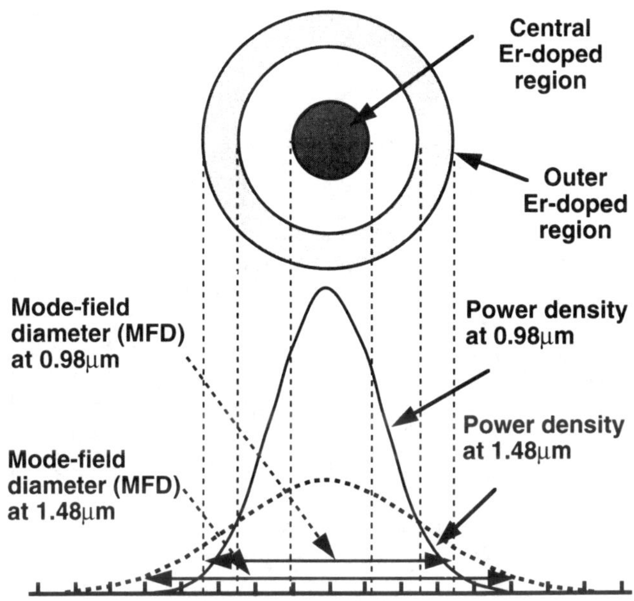

Figure 5.115 Structure of the ring-doped EDF [252].

The first part of this section describes the amplification characteristics of WDM signals with the fluoride-based EDFA and compares them with those of the silica-based EDFA. We will clarify its advantages with regard to flat gain and mention the noise characteristics. Next, we will describe a hybrid-type EDFA founded on the fluoride-based EDFA. This hybrid-type EDFA has been proposed to improve the noise characteristics of the fluoride-based EDFA while maintaining its flat gain characteristics.

The typical configuration of a fluoride-based EDFA is almost the same as that of a silica-based EDFA, without the splicing between the silica fiber and the Er^{3+}-doped fluoride fiber. The desire to fabricate a practical fluoride-based EDFA grew out of the difficulty regarding low-loss and low-reflection splicing presented by the silica-based EDFA [181,185]. A method has recently been developed to overcome this problem that combines tilted V-groove connection with an angled-polishing technique [185]. This method offers a splicing loss of less than 0.3 dB and a reflection of less than −60 dB and will be described in detail in Subsection 5.3.2.2, "PDFA Module." Practical fluoride-based EDFAs fabricated using this method have been demonstrated.

Figure 5.117 shows the amplification characteristics of a fluoride-based EDFA module for WDM signals [185] and, as a comparison, the typical amplification characteristics of a 980-nm pumped silica-based EDFA, which used Al-codoped Er^{3+}-doped silica fiber. The input signal is an 8-channel WDM signal allocated from

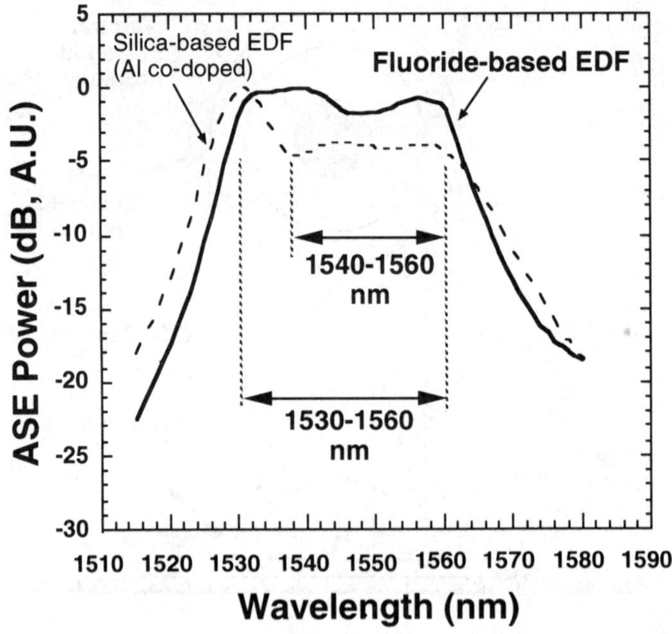

Figure 5.116 Typical ASE spectra for fluoride-based and silica-based EDFAs [185].

1,532 to 1,560 nm with a channel spacing of about 4 nm. It can be seen that the fluoride-based EDFA has flat amplification characteristics with a gain nonuniformity of less than 1.5 dB for the WDM signal in this wavelength region.

From the fact that the gain nonuniformity of the silica-based EDFA was about 4 dB due to a gain protrusion around 1,530 nm, it has been confirmed that the fluoride-based EDFA has a wide and flat amplification band. By contrast, the NF of the fluoride-based EDFA is higher than that of the 980-nm pumped silica-based EDFA because it cannot employ 980-nm pumping to achieve low-noise amplification due to the pump excited state absorption of the Er ions from the $^4I_{11/2}$ metastable level to the $^4F_{7/2}$ level in the fluoride glass. (Note that the noise characteristics of the fluoride-based EDFA are similar to those of the 1,480-nm pumped silica-based EDFA.)

Furthermore, the fluoride-based EDFA has a flat gain region around the signal wavelength of 1,535 nm, as shown in Figure 5.116. Figure 5.118 shows the amplification characteristics of a fluoride-based EDFA for 7-channel WDM signals allocated from 1,532 to 1,544 nm with channel spacing of 2 nm [185]. It is found that there is a flat gain region from 1,534 to 1,542 nm with a gain excursion of less than 0.2 dB for WDM signals, which is considered to be the fluoride-based EDFA's own amplification region.

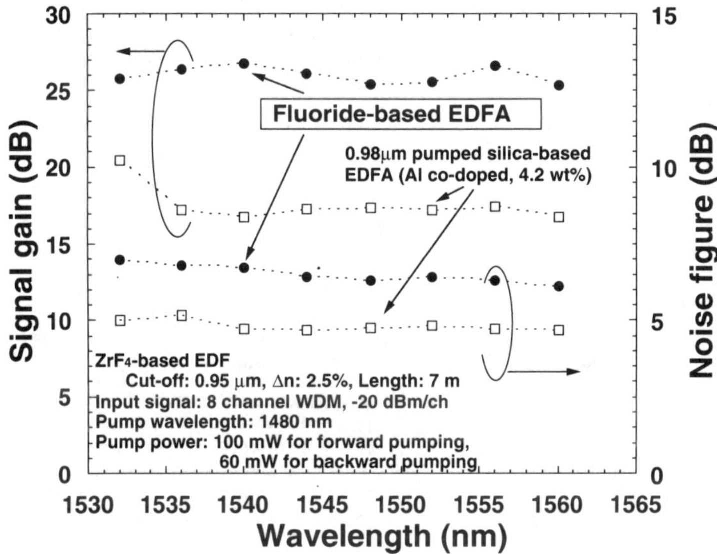

Figure 5.117 Amplification characteristics of a fluoride-based EDFA module for 8-channel WDM signals allocated from 1,532 to 1,560 nm with a channel spacing of 4 nm [185].

Several transmission experiments with fluoride-based EDFAs, such as 100-Gbps by 4-channel 100-km TDM-WDM transmission [255], 8- by 10-Gbps over 100-km WDM transmission [256], 10-Gbps by 10-channel over 600-km WDM transmission [257], 2.5-Gbps by 16-channel 440-km WDM transmission [258], have been demonstrated; and it has been confirmed that the flat gain characteristics of a fluoride-based EDFA have practical advantages in transmissions. As one example of a transmission experiment result, Figure 5.119 compares silica-based and fluoride-based EDFAs in four-amplifier chains [254].

In addition, the gain flatness of the fluoride-based EDFA has been improved with a gain equalizer whose configuration and transmittance spectrum are shown in Figure 5.120(a,b), respectively, and gain nonuniformity has been reduced to 0.25 dB for a 32-channel WDM signal from 1,535 to 1,560 nm [259].

A hybrid-type EDFA that is composed of a silica-based and a fluoride-based Er^{3+}-doped fiber amplifier in a cascade configuration has been proposed as a way of improving the noise characteristics of fluoride-based EDFAs [260].

The configuration of this hybrid-type EDFA with a 980-nm pumped *silica-based Er^{3+}-doped fiber amplifier* (S-EDFA) and a 1,480-nm pumped *fluoride-based Er^{3+}-doped fiber amplifier* (F-EDFA) is shown in Figure 5.121 [260]. Because the NF for a hybrid configuration with first and second amplification units is almost the same as the NF of the first unit, as explained in Subsection 5.4.1.5, the first unit is the S-EDFA

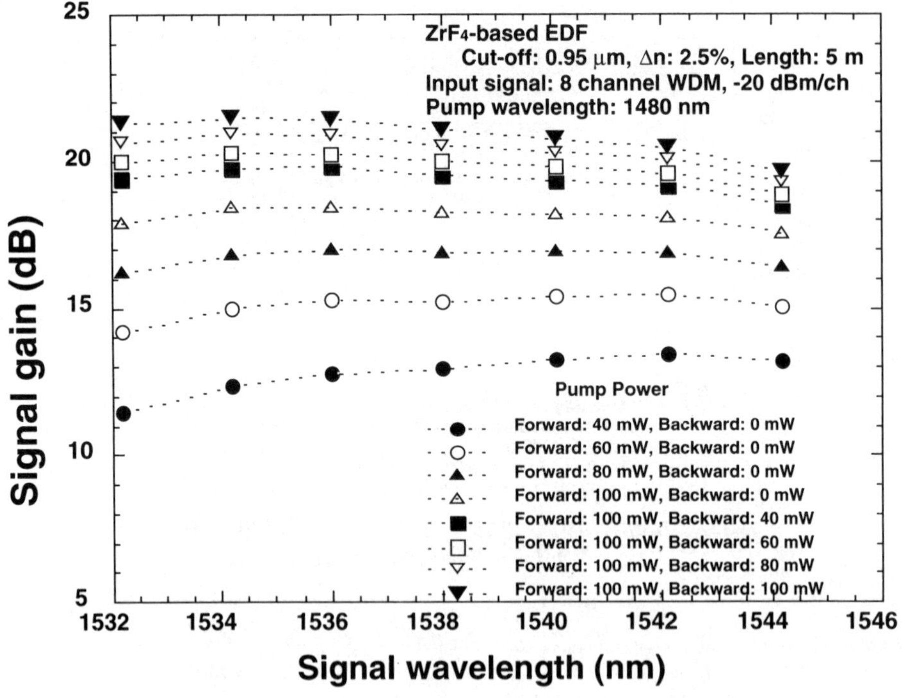

Figure 5.118 Amplification characteristics of a fluoride-based EDFA for 7-channel WDM signals allocated from 1,532 nm to 1,544 nm with a channel spacing of 2 nm [185].

that is forward pumped at 980 nm in order to realize low-noise characteristics. This configuration incorporates an optical isolator between the S-EDF and F-EDF to eliminate any back-propagation of either ASE or of the 1,480-nm pump light of the F-EDFA to the 980-nm pumped S-EDFA. Figure 5.122 shows a photograph of fabricated hybrid-type EDFA modules.

The signal gain and NF of the hybrid-type EDFA for an 8-channel WDM signal are shown in Figure 5.123, and the amplification characteristics of the silica-based and fluoride-based EDFA are also indicated [260]. This figure shows that by using the hybrid configuration, the NF of the fluoride-based EDFA is reduced and flattened to the same level as that of the S-EDFA, while retaining its flat gain characteristics. The gain nonuniformity and NF of the hybrid-type EDFA are 1.4 dB and 5 ± 0.2 dB, respectively. It is found that the hybrid-type EDFA offers a flat gain spectrum with flat and low-noise characteristics for WDM signals. Furthermore, this low-noise and gain-flattened optical amplifier has been applied in WDM transmission trials such as a 100-Gbps by 10-channel transmission [261], and system tests with this amplifier have confirmed its wide amplification characteristics with uniform gain and low noise.

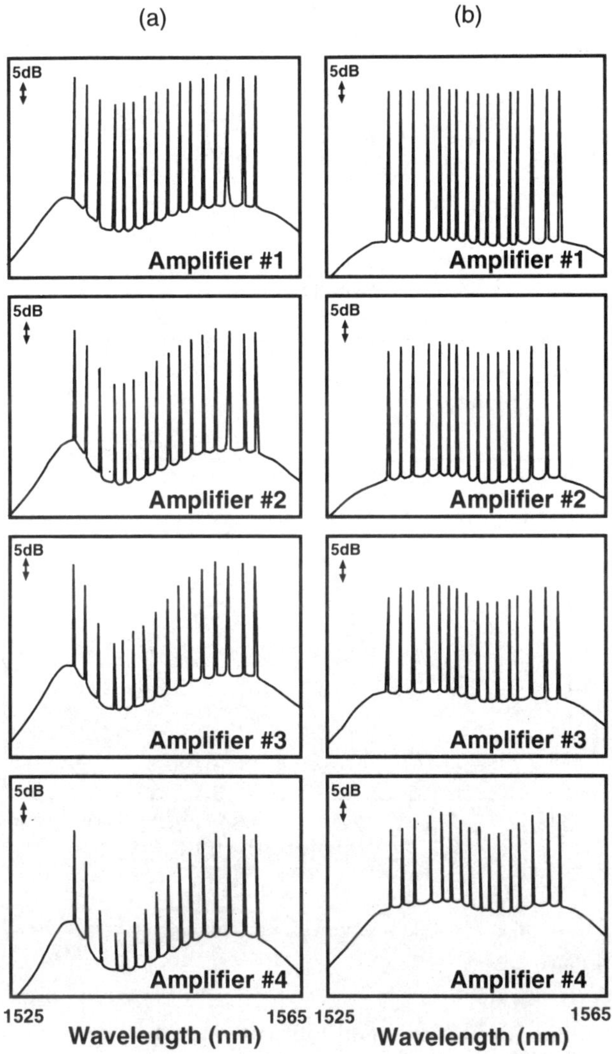

Figure 5.119 Experimental output spectra for (a) silica-based and (b) fluoride-based EDFAs in four-amplifier chains [254].

Table 5.6 reveals that the advantage of the fluoride-based EDFA is its flat gain spectrum in the 1,530- to 1,560-nm wavelength region. The advantages of the hybrid-type EDFA are its flat gain spectrum with low-noise characteristics for WDM signals.

Approaches with other multicomponent glass EDFs. To date, Er^{3+}-doped phosphate fiber [262], germania glass fiber [263], aluminosilicate glass fiber [264,265], and fluoro-

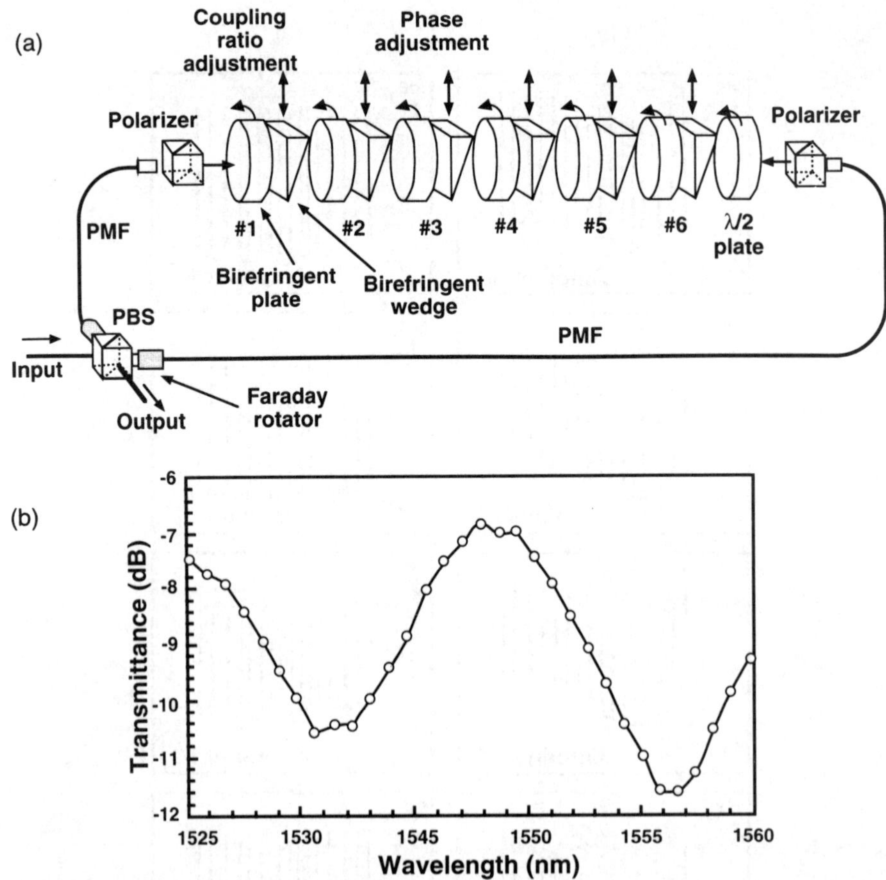

Figure 5.120 (a) Configuration and (b) transmittance spectrum of a gain equalizer for a fluoride-based EDFA [259]

phosphate glass fiber [266] have been fabricated; and some of their gain spectra have been investigated from the viewpoint of a wide amplification band, flat gain, and low noise for WDM transmission systems. From these investigations, it is found that Er^{3+}-doped fluorophosphate glass fiber has the potential to provide a flat signal gain with a broad bandwidth. The fiber can be used to achieve low-noise amplification with 980-nm pumping. Figure 5.124 shows the signal gain and noise figure of Er^{3+}-doped fluorophosphate glass fiber for an 8-channel WDM signal with 980-nm pumping [266]. The gain nonuniformity and NF of the hybrid-type EDFA are 1.1 dB and less than 6 dB, respectively. The NF is higher than the quantum limit of 3 dB because of the scattering loss of the fiber. So reducing the scattering loss is a very important issue for practical use.

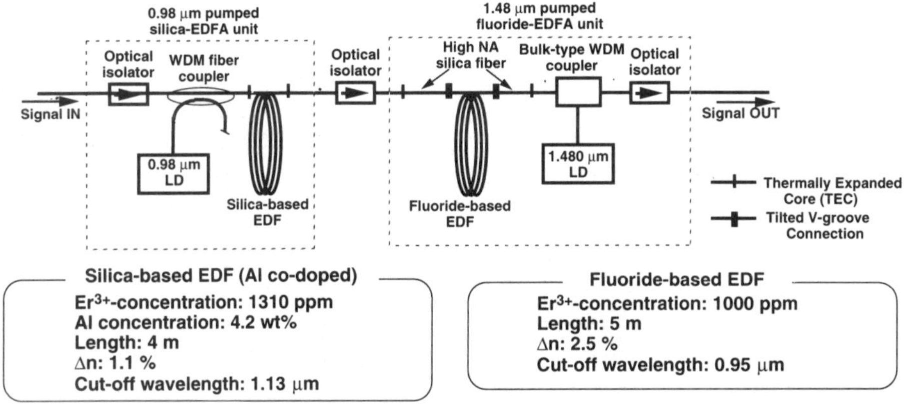

Figure 5.121 Configuration of a hybrid-type EDFA with a 980-nm pumped silica-based Er^{3+}-doped fiber amplifier and a 1,480-nm pumped fluoride-based Er^{3+}-doped fiber amplifier [260].

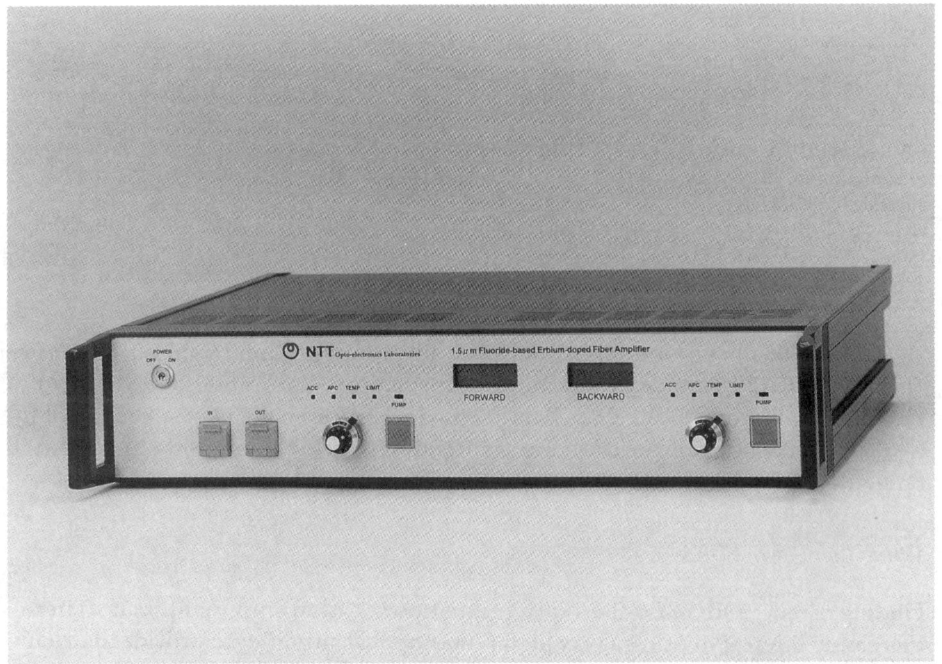

Figure 5.122 Photograph of fabricated hybrid-type EDFA modules [260].

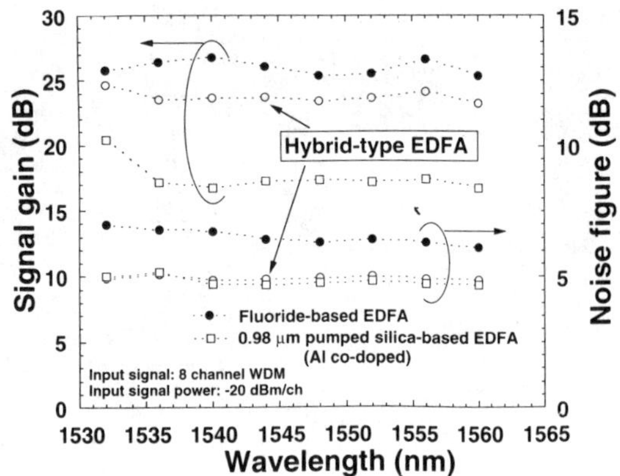

Figure 5.123 Signal amplification characteristics of a hybrid-type EDFA, a silica-based EDFA, and a fluoride-based EDFA for an 8-channel WDM signal [260].

Table 5.6
Comparison of Various EDFAs

	Amplification Band (nm)	Gain Nonuniformity (dB)	Noise Figure (dB)
Silica-based EDFA (980-nm pump, Al co-doped)	1,530–1,560	~4	5 ± 0.2
Fluoride-based EDFA	1,530–1,560	1.5	< 7
Hybrid-type EDFA (Silica- + Fluoride-based EDFA)	1,530–1,560	1.4	5 ± 0.2

There has been one report on a cascade configuration based on a new host glass-based EDF. This describes an EDF formed by connecting a S-EDF and an Er^{3+}-doped aluminosilicate glass fiber. Figure 5.125 shows the gain spectrum of this EDF [267]. A flat gain spectrum was achieved in the 1,543- to 1,558-nm bandwidth range.

Ultra-High-Power EDFA

Finding a way to increase the output signal power is an important issue. There is increasing interest in using a very high power optical amplifier to provide alternative system architectures in a portion of the optical network. Potential high-power EDFA applications include its use as a booster amplifier for long-haul repeaterless optical links as a head-end amplifier for CATV distribution architectures and as an amplifier for $1 \times N$ lossless splitters.

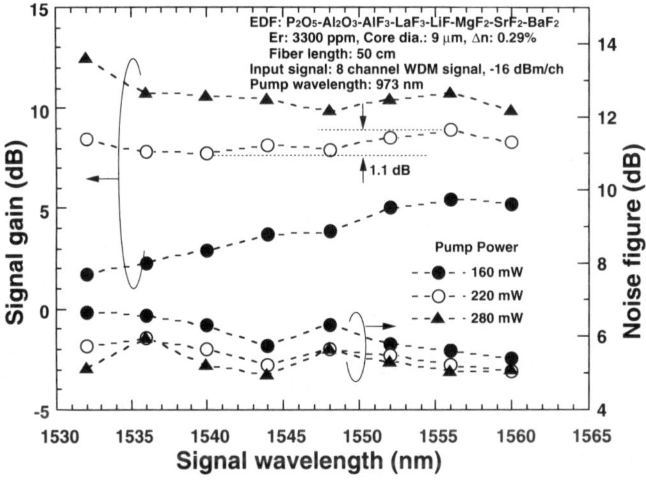

Figure 5.124 Signal gain and noise figure of Er[3+]-doped fluorophosphate glass fiber for an 8-channel WDM signal with 980-nm pumping [266].

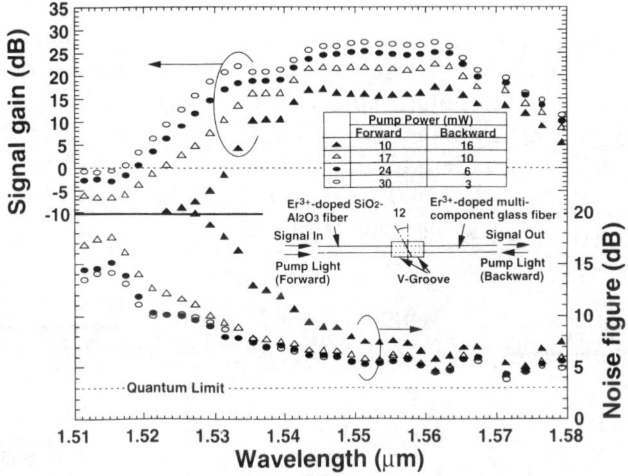

Figure 5.125 Gain spectrum of EDF formed by connecting a silica-based EDF and an Er[3+]-doped aluminosilicate glass fiber [267].

To realize a high-power EDFA requires high pump power to be supplied into the Er[3+]-doped fiber. There are several possible approaches including the use of high-power pump light sources and the optimization of the EDF structure in order to absorb pump power effectively.

The approaches using high-power pump light sources include effective pump source schemes such as polarization and wavelength multiplexing [268], newly designed 980-nm-band high-power pump sources such as the MOPA [269] and tapered-gain-region laser [270], and optimizing the conventional 980-nm LD module laser structure for high-power and efficient coupling to the fiber [271]. Figure 5.126 shows the pump source scheme with polarization and wavelength multiplexing [268]. By using these two pump sources and a bidirectional pump configuration, a maximum signal output power of about 30 dBm is expected with eight 1,480-nm LD modules (output power of ~200 mW pumping), taking into account the coupling efficiency of 1.5 dB between the LD module and Er^{3+}-doped fiber, and the PCE of 80%. A fiber-coupled MOPA laser with an output power of over 600 mW has been fabricated. A high-power EDFA pumped by one fiber-coupled MOPA laser was reported with an output signal power of 25 dBm [272]. The use of 980-nm MOPA laser pumping also offers advantages for the fabrication of low-noise amplifiers. A tapered-gain-region laser operating at 980 nm, which has a similar laser structure to MOPA, has also been fabricated for the purpose of supplying high pump power to EDF; and an EDFA with an output signal power of 28.5 dBm has been demonstrated with four tapered-gain-region lasers on an optical bench [270]. In contrast, the output power of a conventional 980-nm LD module laser, which of course has a fiber pigtail, has been increased to over 300 mW by optimizing the active layer structure for high-power and by achieving efficient coupling with fiber of over 65%. The predicted maximum output signal power of an EDFA is 26 dBm with four LD module lasers, polarization multiplexing, and a bidirectional pump configuration, by taking account of the coupling efficiency of 1.5 dB between the LD module and EDF, and the PCE of 50%. In addition, a Raman laser [273] and a Tm^{3+}-doped fiber laser [274], operating at 1,480 nm, have recently been reported as pump light sources with which to realize high-power EDFAs.

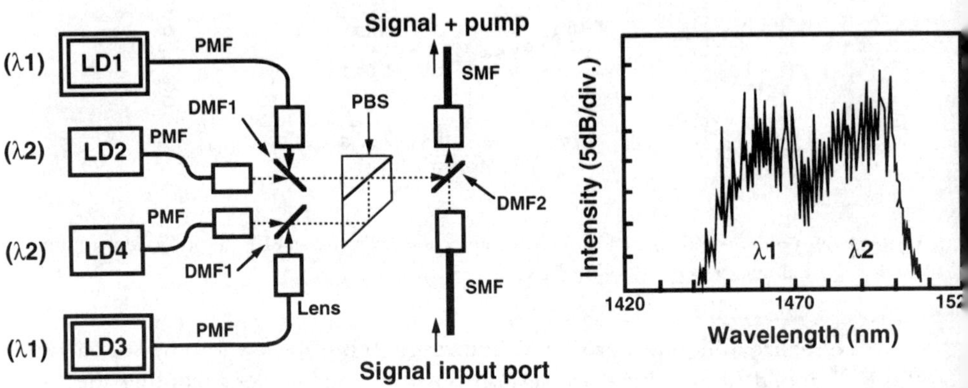

Figure 5.126 Pump source scheme with polarization and wavelength multiplexing [268].

There are two approaches to optimizing the fiber structure so that pump power is absorbed effectively; they consist of codoping Yb^{3+} ions into the Er^{3+}-doped fiber as a sensitizer [90,275–277] for applications to high-power Nd:YAG and YLF lasers [278–280] and a double cladding fiber structure [281,282] for expanding the aperture of the pump light. Figure 5.127(a,b) shows the energy level diagrams of Yb^{3+} and Er^{3+} ions and the absorption spectrum of Yb^{3+}/Er^{3+}-codoped fiber, respectively [277]. The advantage of this fiber is that the 980-nm pump band for Er^{3+}-doped fiber widens, and so a diode-pumped Nd:YLF operating at 1,047 nm and a diode-pumped Nd:YAG laser operating at 1,064 nm, which can supply a pump power of over 1W, can be applied due to the expiation of the Er^{3+} ions through pumping Yb^{3+}. A high Yb^{3+} ion concentration is required in this fiber, such as an $Yb^{3+}:Er^{3+}$ ratio of 30:1, in order to achieve an efficient energy transfer from Yb^{3+} ions to Er^{3+} ions. This fiber has been used to fabricate a commercially available high-power EDFA with an output signal power of over 27 dBm that is fabricated with two diode-pumped Nd:YAG lasers. Furthermore, ultra-high output signal power has recently been successfully demonstrated with a three-stage amplification configuration and by employing three Nd^{3+} cladding-pumped lasers with a

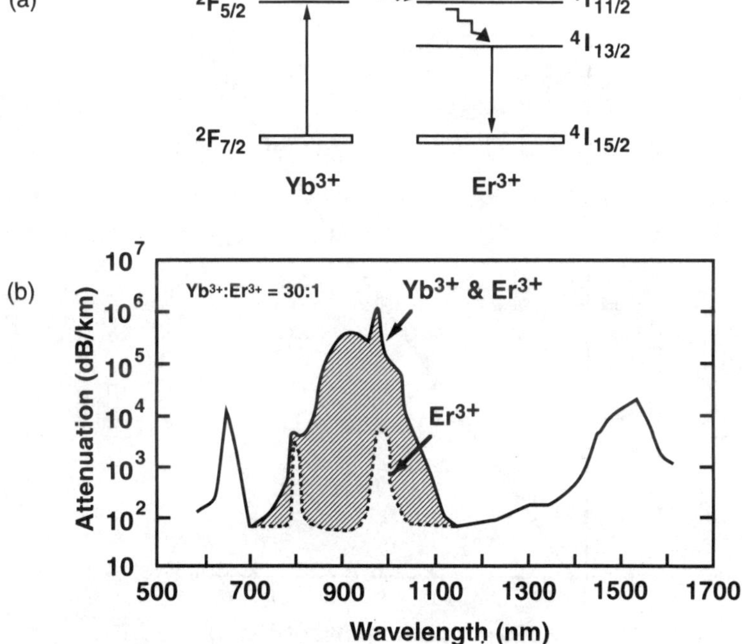

Figure 5.127 (a) Energy level diagrams of Yb^{3+} and Er^{3+} ions and (b) the absorption spectrum of Yb^{3+}/Er^{3+}-codoped fiber [277].

single-mode fiber-coupled output power of 5W and operating at 1,060 nm. Figure 5.128(a,b) shows the configuration of this EDFA and its output signal power characteristics [279]. This achieved the highest output signal power yet achieved of 4W. The noise characteristics of a Yb^{3+}/Er^{3+}-codoped fiber amplifier have also been studied, and it has been found that the NF of this amplifier is less than 4.2 dB, as shown in Figure 5.129 [280]. On the other hand, a double-cladding fiber structure is also effective. This fiber design is proposed with a view to achieving a high-power Nd^{3+}-doped fiber laser. Figure 5.130 shows a typical double-cladding fiber structure [281,282]. The core, which is doped with Er^{3+}-ions, is for signal amplification; and the first and second claddings, which surround the core region, provide a multimode

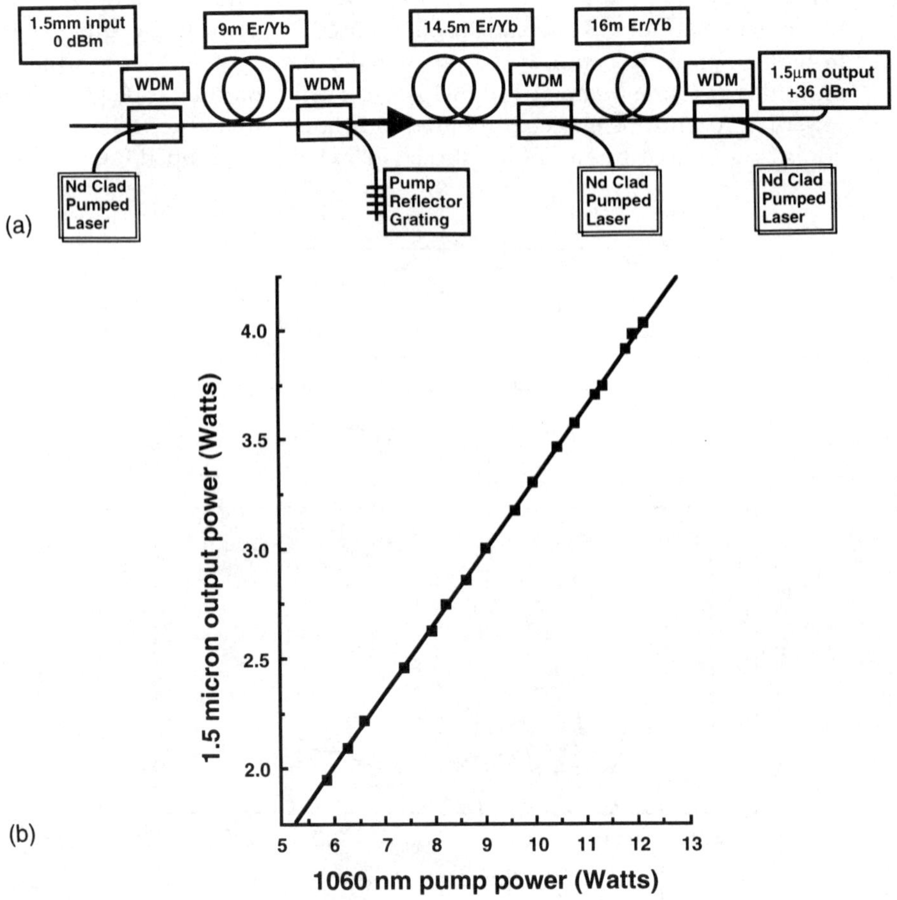

Figure 5.128 High-power three-stage Yb^{3+}/Er^{3+}-codoped fiber amplifier [279]: (a) configuration and (b) output signal power characteristics.

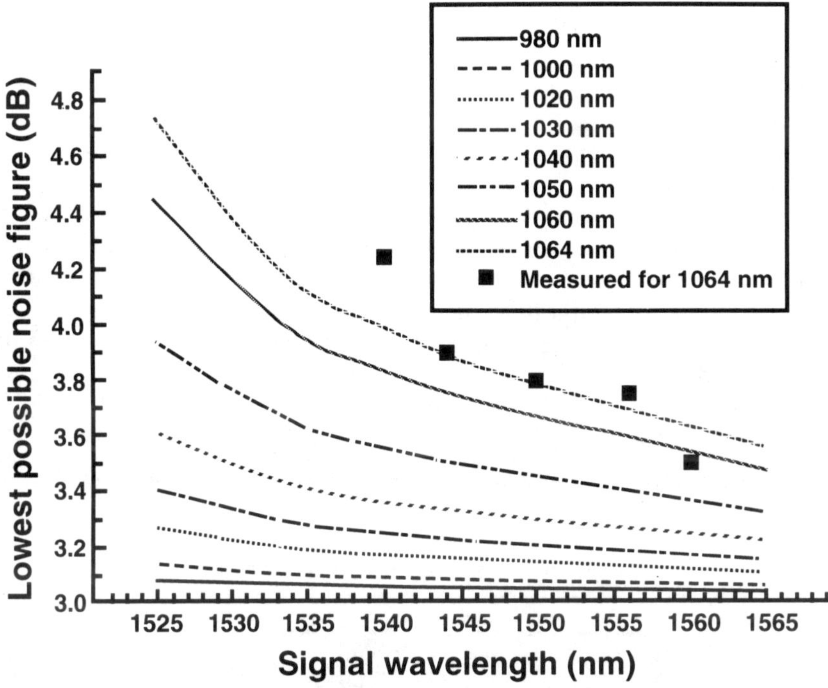

Figure 5.129 Noise characteristics of Yb^{3+}/Er^{3+}-codoped fiber amplifier [280].

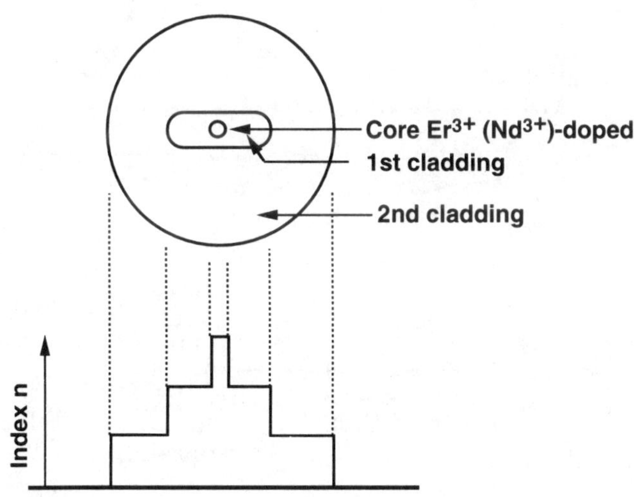

Figure 5.130 Typical double-cladding fiber structure [281,282].

waveguide for the pump light. The pump light propagates in the fiber and is absorbed in the core region. The main advantage of this fiber is that it allows the use of high-power multiple-stripe laser diodes because it provides a multimode waveguide for the pump light. As an example, Figure 5.131 shows the configuration of a recently reported EDFA with double-clad fiber [283]. The output signal power of the EDFA was 24 dBm.

In addition, we think that the combination of the codoping of Yb^{3+} ions into Er^{3+}-doped fiber and a double-cladding fiber structure is interesting from the viewpoint of increasing the output signal power of an EDFA, but to our knowledge, no report has yet been published. In the future, the EDFA output signal power may be increased by improving the described techniques and by combining several techniques.

5.3.2 1.3-μm-Band Fiber Amplifier

Almost all optical telecommunication systems throughout the world operate in the 1.3-μm band because it is the second optical window and the chromatic dispersion

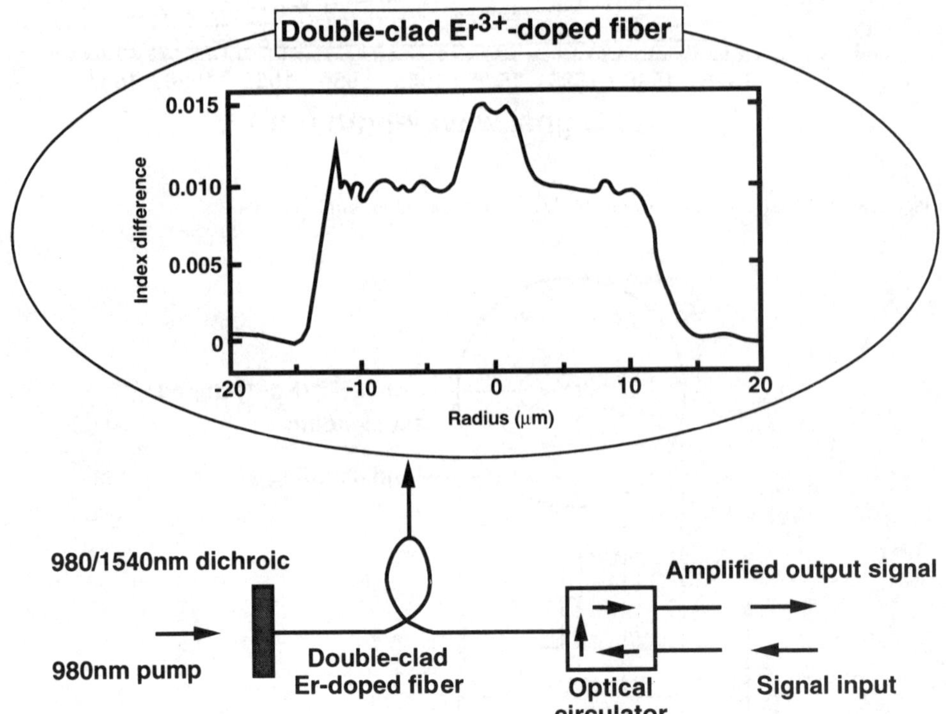

Figure 5.131 Configuration of EDFA with double-clad fiber [283].

of standard fiber is zero in this band. Therefore, there is great interest in the development of practical optical amplifiers for use at this wavelength. There have been two proposals to meet this need: the Nd^{3+}-doped fiber amplifier and the Pr^{3+}-doped fiber amplifier.

In 1988, it was predicted that amplification at 1.3 μm could be achieved using a *Nd^{3+}-doped ZrF_4-based fluoride fiber amplifier* (NDFA) [284]; in the same year, this was actually demonstrated by Brierly et al. [285]. ZrF_4-based fluoride glass is used as a Nd^{3+}-doped fiber host to suppress the signal ESA of Nd^{3+}-doped silica-based fiber and to shift its spectrum from the 1.32- to 1.34-μm range to a shorter wavelength region [284,285]. However, experimental work to assess the performance of *Nd^{3+}- doped ZrF_4-based fluoride fiber* (NDF) has shown that it is difficult to achieve efficient signal gain for the 1.3-μm transmission window because of the strong signal ESA located around 1.30 μm, which remains despite the use of ZrF_4-based fluoride glass as the host glass, and the competing ASE transition at 0.88 and 1.06 μm. These factors have meant that the highest signal gain ever reported is just 10 dB [286,287]. All these properties will be further clarified in Subsection 5.3.2.1, which describes the theoretical model and the basic characteristics of the NDFA. The same section presents research on material considerations with a view to reducing the strength of the signal ESA and shifting the ESA spectrum to a shorter wavelength. It also provides several amplifier configurations designed to suppress ASE with the competing transition.

The *Pr^{3+}-doped ZrF_4-based fluoride fiber amplifier* (PDFA) was proposed by Ohishi et al. in 1990 as a way to overcome the problems of the NDFA, and it offered a possible new way to develop fiber amplifiers for 1.3-μm use [288,289]. Durteste et al. and Carter et al. also reported optical amplification using *Pr^{3+}-doped ZrF_4-based fluoride fiber* (PDF) at almost the same time [290,291]. The PDFA is superior to the NDFA in that its gain spectrum covers the operational 1.29- to 1.33-μm wavelength band of the 1.3-μm-band optical telecommunication window, as shown in Figure 5.132. Although the PDFA is problematic in terms of its low-gain coefficient resulting from the low quantum efficiency of the stimulated transition for 1.3-μm-band amplification [288], the possibility of realizing a PDFA with a signal gain of over 30 dB and an output power of 17.8 dBm was demonstrated using a Ti:Sapphire pump laser [292]. Subsequent achievements include an increase in the gain coefficient that was realized using a high numerical aperture fiber and by decreasing the scattering loss of the PDF [293,294], and the experimental demonstration of a PDFA module pumped with high-power pump sources such as InGaAs-LD modules, a MOPA-LD, or a Nd:YLF laser [295–297]. Performance has been successively improved and the PDFA is now attracting attention as a promising device for use in 1.3-μm optical telecommunication systems [298–301]. In Subsection 5.3.2.2, we discuss the theoretical model and basic characteristics of the PDFA. Furthermore, we describe a recently fabricated PDFA module and outline the future prospects.

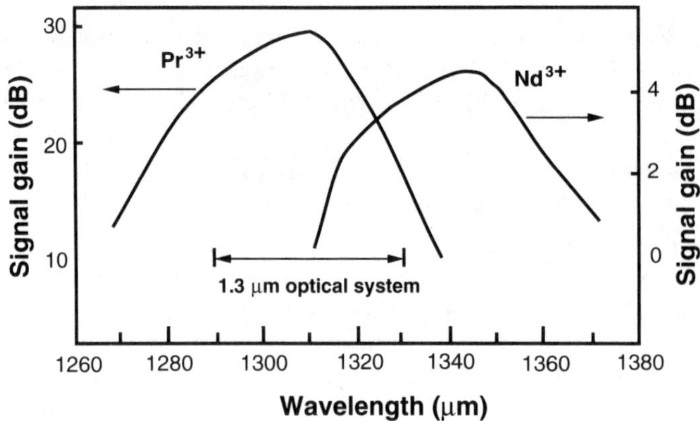

Figure 5.132 Gain spectra of Pr³⁺- and Nd³⁺-doped ZrF₄-based fluoride fiber amplifiers.

5.3.2.1 Neodymium-Doped Fiber Amplifier

In this section, first we describe the amplification principle and the theoretical model of the NDFA. Then the basic amplification characteristics such as the gain spectrum and the saturation and noise characteristics of the NDF will be explained, including the three major problems regarding the effective amplification of the 1.3-μm signal used in a 1.3-μm transmission system. The first problem is that the NDFA gain spectrum does not fully cover the operational wavelength of the 1.3-μm-band transmission system from 1.29 to 1.33 μm. The second is that the highest gain is limited by the signal saturation effect caused by the build up of strong ASE at $1,050$ nm, and the third is that the 1.3-μm amplification transition has a lower gain coefficient. These three problems are caused by the signal ESA and the fact that the transitions of the 0.88- and 1.05-μm bands are stronger than that of the 1.3-μm band. This section also outlines research on material considerations with a view to reducing the signal ESA strength and shifting its spectrum to a shorter wavelength and describes various amplifier configurations designed to suppress ASE at 880 and $1,050$ nm.

Amplification Principle and Calculation Model of NDFA

Amplification models of the NDFA have been reported in [302–306], and in this section we will use the description provided in [302].

The amplification model of the NDFA is described by the energy level diagram shown in Figure 5.133 [307]. The fast nonradiative transitions—${}^4I_{13/2}\rightarrow{}^4I_{11/2}$, ${}^4I_{13/2}\rightarrow{}^4I_{9/2}$, ${}^4I_{11/2}\rightarrow{}^4I_{9/2}$, ${}^4F_{5/2}\rightarrow{}^4F_{3/2}$ and ${}^4G_{7/2}\rightarrow{}^4I_{9/2}$—are assumed, which means that the lifetimes of τ_{32}, τ_{31}, τ_{21}, τ_{54}, and τ_{61} equal 0. The lifetimes related

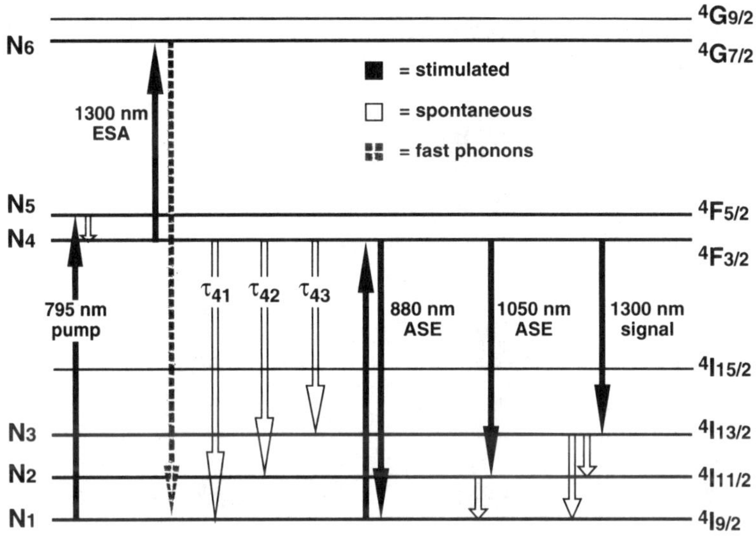

Figure 5.133 Energy level diagram of Nd^{3+} ions [307].

to the transitions—$^4F_{3/2}{\to}^4I_{13/2}$ (880-nm spontaneous emission), $^4F_{3/2}{\to}^4I_{11/2}$ (1,050-nm spontaneous emission), and $^4F_{3/2}{\to}^4I_{9/2}$ (1,300-nm spontaneous emission)—are indicated by τ_{43}, τ_{42}, and τ_{41}, respectively. Pump photon ground state absorption occurs from the $^4I_{9/2}$ level to the $^4F_{5/2}$ level, and stimulated emission for the signal photons of the 1.3-μm band occurs from the $^4F_{3/2}$ metastable level to the $^4I_{13/2}$ ground level. The signal photons are also absorbed by the signal ESA from the $^4F_{3/2}$ metastable level to the $^4G_{7/2}$ level.

Rate equations for the ion population density N_4 ($^4F_{3/2}$ metastable level) and N_1 ($^4I_{13/2}$ ground level), taking those transitions into account and assuming $N_6 = N_5 = N_3 = N_2 = 0$, are given by

$$\frac{dN_4}{dt} = N_1(W_{15} + W_{14}) - N_4\left(W_{43} + W_{46} + W_{41} + W_{42} + \frac{1}{\tau_4}\right)$$

$$\frac{dN_1}{dt} = N_4\left(W_{43} + W_{46} + W_{41} + W_{42} + \frac{1}{\tau_4}\right) - N_1(W_{15} + W_{14}) \qquad (5.58)$$

where

$$\frac{1}{\tau_4} = \frac{1}{\tau_{41}} + \frac{1}{\tau_{42}} + \frac{1}{\tau_{43}} \qquad (5.59)$$

and W_{ij} is the transition rate from the ith to the jth level. W_{15}, W_{14}, W_{43}, W_{46}, W_{41}, and W_{42} are given by

$$W_{15} = \frac{\sigma_{15}(\nu_P)}{h\nu_P} \cdot P_P(z) \cdot \Psi(r, \theta, \nu_P) \tag{5.60}$$

$$W_{43} = \frac{\sigma_{43}(\nu_S)}{h\nu_S} \cdot P_S(z) \cdot \Psi(r, \theta, \nu_S)$$
$$+ \int \frac{\sigma_{43}(\nu)}{h\nu} (P_{1300}^+(z, \nu) + P_{1300}^-(z, \nu))\Psi(r, \theta, \nu)\, d\nu \tag{5.61}$$

$$W_{46} = \frac{\sigma_{46}(\nu_S)}{h\nu_S} \cdot P_S(z) \cdot \Psi(r, \theta, \nu_S)$$
$$+ \int \frac{\sigma_{46}(\nu)}{h\nu} (P_{1300}^+(z, \nu) + P_{1300}^-(z, \nu))\Psi(r, \theta, \nu)\, d\nu \tag{5.62}$$

$$W_{41} = \int \frac{\sigma_{41}(\nu)}{h\nu} (P_{880}^+(z, \nu) + P_{880}^-(z, \nu))\Psi(r, \theta, \nu)\, d\nu \tag{5.63}$$

$$W_{42} = \int \frac{\sigma_{42}(\nu)}{h\nu} (P_{1050}^+(z, \nu) + P_{1050}^-(z, \nu))\Psi(r, \theta, \nu)\, d\nu \tag{5.64}$$

$$W_{14} = \int \frac{\sigma_{14}(\nu)}{h\nu} (P_{880}^+(z, \nu) + P_{880}^-(z, \nu))\Psi(r, \theta, \nu)\, d\nu \tag{5.65}$$

where σ_{ij} is the stimulated transition cross section from the ith to the jth level; ν_S and ν_P are the frequencies of the signal and pump lights, respectively; h is Plank's constant; $P_P(z)$ and $P_S(z)$ are pump and signal powers, respectively, where z represents the fiber longitudinal coordinate; and $\Psi(r, \theta, \nu)$ is the normalized transverse mode envelope of the light with a frequency of ν, where (r, θ) represent the cylindrical transverse coordinates; and $P_{880}^\pm(z)$, $P_{1050}^\pm(z)$, and $P_{1300}^\pm(z)$ are the spectrally dependent powers of forward (+) and backward (−) propagating ASE in the 880-, 1,050-, and 1,300-nm bands, respectively.

The total population density $\rho(r, \theta)$ satisfies $\rho(r, \theta) = N_1 + N_4$. Ion population densities N_1 and N_4 in a steady state for the $^4I_{9/2}$ ground level and the $^4F_{3/2}$ metastable level are given by

$$N_1 = \rho \frac{W_{43} + W_{46} + W_{41} + W_{42} + \dfrac{1}{\tau_4}}{W_{15} + W_{14} + W_{43} + W_{46} + W_{41} + W_{42} + \dfrac{1}{\tau_4}} \tag{5.66}$$

$$N_4 = \rho \frac{W_{15} + W_{14}}{W_{15} + W_{14} + W_{43} + W_{46} + W_{41} + W_{42} + \dfrac{1}{\tau_4}} \qquad (5.67)$$

The local population densities have been found as indicated. So the set of coupled differential equations describing the spatial development of the pump power and signal power $P_P^{\pm}(z)$, $P_S^{\pm}(z)$, and ASE power $P_{880}^{\pm}(z)$, $P_{1050}^{\pm}(z)$, $P_{1300}^{\pm}(z)$ for the 880-, 1,050-, and 1,300-nm bands along the fiber longitudinal coordinate z are

$$\frac{dP_P^{\pm}}{dz} = \pm g_P(z) \cdot P_P^{\pm}(z) \qquad (5.68)$$

$$\frac{dP_S^{\pm}}{dz} = \pm g_S(z) \cdot P_S^{\pm}(z) \qquad (5.69)$$

$$\frac{dP_{880,1050,1300}^{\pm}}{dz} = \pm g_{880,1050,S}(z) \cdot P_{880,1050,1300}^{\pm}(z) \pm 2h\nu \cdot a_{880,1050,1300}(z, \nu) \qquad (5.70)$$

where

$$g_P(z) = \sigma_{15} \iint N_1(r, \theta, z) \cdot \Psi(r, \theta, \nu) r \, dr \, d\theta \qquad (5.71)$$

$$g_S(z) = (\sigma_{43} - \sigma_{46}) \iint N_4(r, \theta, z) \cdot \Psi(r, \theta, \nu) r \, dr \, d\theta \qquad (5.72)$$

$$g_{880}(z) = \iint (N_4(r, \theta, z) \cdot \sigma_{41} - N_1(r, \theta, z) \cdot \sigma_{14}) \cdot \Psi(r, \theta, \nu) r \, dr \, d\theta \qquad (5.73)$$

$$g_{1050}(z) = \sigma_{42} \iint N_4(r, \theta, z) \cdot \Psi(r, \theta, \nu) r \, dr \, d\theta \qquad (5.74)$$

$$a_{880}(z, \nu) = \sigma_{41} \iint N_4(r, \theta, z) \cdot \Psi(r, \theta, \nu) r \, dr \, d\theta \qquad (5.75)$$

$$a_{1050}(z, \nu) = \sigma_{42} \iint N_4(r, \theta, z) \cdot \Psi(r, \theta, \nu) r \, dr \, d\theta \qquad (5.76)$$

$$a_{1300}(z, \nu) = \sigma_{43} \iint N_4(r, \theta, z) \cdot \Psi(r, \theta, \nu) r \, dr \, d\theta \qquad (5.77)$$

$a_{880,1050,1300}(z, \nu)$ in (5.75) to (5.77) represents the addition of new photons into the two bound orthogonal polarization modes due to spontaneous decay.

The amplification characteristics of NDF can be calculated by solving the differential equations (5.68) to (5.70) with the stimulated transition cross sections σ_{15}, σ_{14}, σ_{41}, σ_{42}, σ_{43}, σ_{46} and lifetime τ_4. Typical values for each fundamental parameter follow. Figure 5.134 shows the emission cross-section spectra of σ_{41}, σ_{42}, σ_{43} that are obtained by measuring the fluorescence spectrum from a short length of NDF pumped to full inversion and from the measured lifetime $\tau_4 = 720$ μs [307]. The emission and absorption cross sections $\sigma_{14,41}$ for the 880-nm band, which are derived for the corresponding absorption coefficient of the NDF, are also reported as 20.5×10^{-25} m^2 and 5.7×10^{-25} m^2, respectively [307]. Another important cross section is that of the signal ESA σ_{46}. This spectrum will be shown in Subsection 5.3.2.1, "Basic Amplification Characteristics of NDFA," along with the gain spectrum of NDF.

Basic Amplification Characteristics of NDFA

Since the prediction that 1.3-μm-band amplification could be achieved through the use of NDF [284] and the first report of signal gain in NDF [285], the NDF gain characteristics have been studied using multimode or single-mode fiber [286]. It has become clear that there are three major obstacles to the efficient amplification of signals in the 1.3-μm transmission system [285,286,308,309].

One of the problems is that the NDF gain spectrum does not fully cover the operational wavelength band, from 1.29 to 1.33 μm, of 1.3-μm-band optical

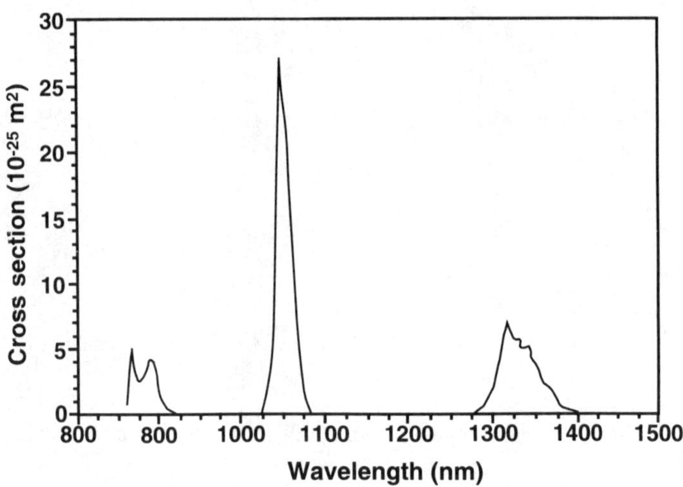

Figure 5.134 Emission cross-section spectra of σ_{41}, σ_{42}, and σ_{43} [307].

telecommunication systems. Figure 5.135 shows the gain spectrum, ESA, and emission cross section associated with the 1.3-μm-band amplification of NDF [308]. The 3-dB down spectral bandwidth is from 1,318 to 1,362 nm, although the emission cross section is observed from 1,280 to 1,370 nm. The disappearance of the signal gain below 1,310 nm is a result of signal ESA because the gain coefficient for a 1.3-μm signal is proportional to $\sigma_{43}(\nu) - \sigma_{46}(\nu)$, as shown in (5.72).

The other two problems are limiting the highest signal gain by the signal saturation effect because of the build up of strong ASE at 1,050 nm and the lower gain coefficient. Figure 5.136 shows the relationship between signal gain and fluorescence intensity at 1.318 and 1.048 μm as a function of pump power [309]. The transitions corresponding to the 0.88- and 1.05-μm bands are two to four times stronger than the gain transition of the 1.3-μm band. The 1.05-μm transition is particularly strong, and the 1,050-nm ASE builds up very quickly. Once the 1,050-nm ASE has built up (pump power of over 45 mW in this figure), the ASE clamps the population density of the metastable level $^4F_{3/2}$ and saturates the signal gain. Hence, the highest signal gain ever reported is about 6 dB with 0.8-μm pumping. (A signal gain of over 10 dB has been achieved by using 0.49-μm pumping, which is related to the $^4I_{9/2}$-$^4G_{9/2}$ transition [286].) Furthermore, the gain coefficient is 0.13 dB/mW as the result of the low branching ratio to $^4I_{13/2}$ from $^4F_{3/2}$ and hence the low value of $\sigma_{43} \times \tau_{43}$, even though Nd^{3+}-doped single-mode fluoride fiber with a high Δn of 3.7% was used to obtain a high-gain coefficient [309].

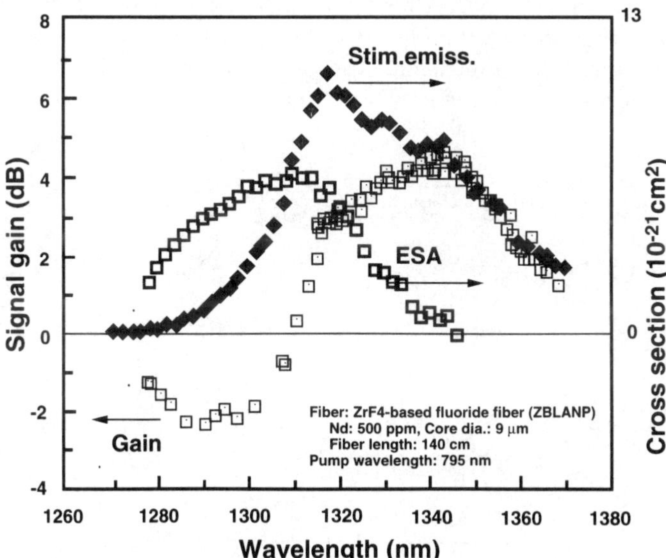

Figure 5.135 Gain spectrum, ESA, and emission cross section associated with the 1.3-μm-band amplification of Nd^{3+}-doped ZrF$_4$-based fluoride fiber [308].

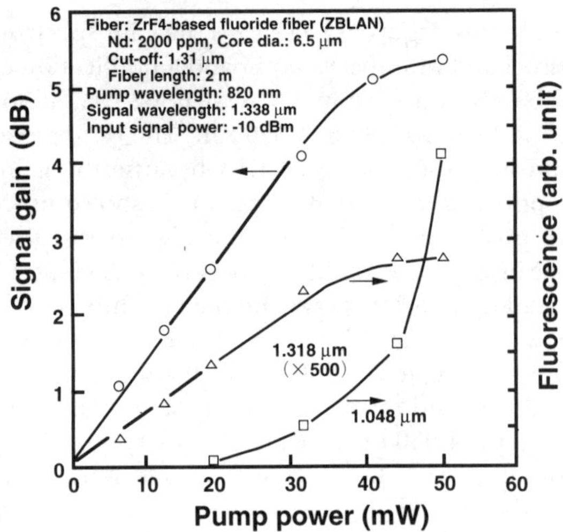

Figure 5.136 Relationship between signal gain and fluorescence intensity at 1.318 and 1.048 μm as a function of pump power [309].

Next, we describe the saturation and noise characteristics. Figure 5.137 shows typical saturation characteristics [303]. The highest saturation output power ever reported is 18.5 dBm. The NDFA has good saturation characteristics, despite the relatively poor gain performance. Moreover, by comparing the saturation characteristics for 1.337 and 1.320 μm, it has also been found that the signal saturation performance for a longer wavelength is better than that for a shorter wavelength. The saturation output power $P_{\text{sat}}^{\text{OUT}}(\nu_S)$ for 3-dB gain compression is given by [302]

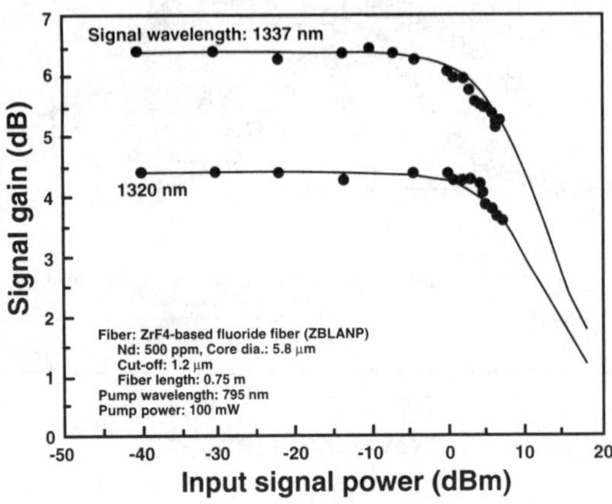

Figure 5.137 Typical saturation characteristics of Nd^{3+}-doped ZrF_4-based fluoride fiber [303].

$$P_{\text{sat}}^{\text{OUT}}(\nu_S) \approx \frac{A_{\text{eff}}h\nu_S}{(\sigma_{43}(\nu_S) + \sigma_{46}(\nu_S))\tau_4} \tag{5.78}$$

where A_{eff} is the effective core area. Therefore, it is believed that NDF with a low $\sigma_{43} \times \tau_4$ value results in good saturation characteristics. As expressed in (5.78), the saturation output power $P_{\text{sat}}^{\text{OUT}}(\nu_S)$ decreased with decreasing signal wavelength because of the increase in signal ESA. It may be expected that the photon conversion efficiency, which is an important parameter when an NDFA is used as a booster amplifier, will reach 100% [306].

Furthermore, as the NDFA is classified as a four-level system, its NF will be close to the value of 3.0 dB for an ideal optical amplifier using stimulated transition if there is no signal ESA ($^4F_{3/2}$-$^4G_{7/2}$ transition). The NF is expressed as (5.79) from (5.72)

$$NF \approx 2N_{\text{SP}} = 2\frac{\sigma_{43}(\nu)\displaystyle\iint N_4(r,\,\theta,\,z) \cdot \Psi(r,\,\theta,\,\nu)r\,dr\,d\theta}{g_S(\nu)}$$

$$= 2\frac{\sigma_{43}}{\sigma_{43} - \sigma_{46}} \tag{5.79}$$

where N_{sp} is a population inversion parameter. The measured noise characteristics reported in [305] are shown in Figure 5.138. Here, the NF increased with increasing signal wavelength from 2.5 to 3.8 dB for wavelengths below 1.345 μm. The increase

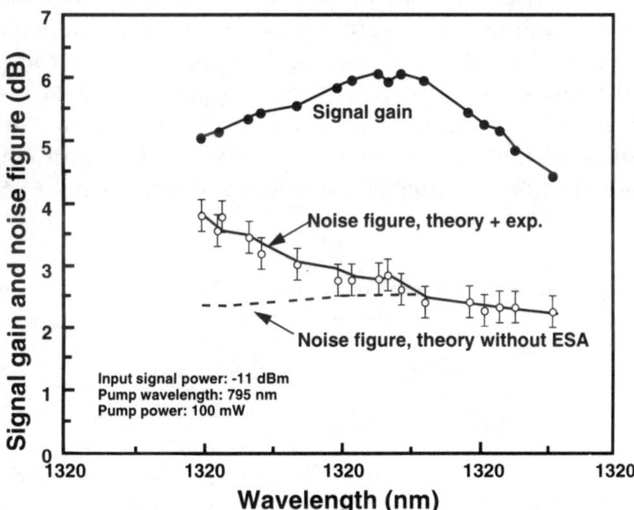

Figure 5.138 Noise characteristics of Nd^{3+}-doped ZrF$_4$-based fluoride fiber [305].

in the NF is due to the increase in signal ESA cross-section values, where signal ESA becomes significant.

As explained, although it has been predicted that the NDFAs will have good saturation and noise performance, NDF has a problem in that its signal gain and efficiency are influenced by signal ESA and ASE. In Subsection 5.3.2.1, "Various Approaches for Improving NDFA Amplification Characteristics," we will describe various approaches designed to overcome this problem.

Various Approaches for Improving NDFA Amplification Characteristics

Efficient Nd^{3+}-doped fiber that has a high gain coefficient and supplies a high signal gain is essential for realizing a practical NDFA. There have been two approaches: (1) to change to a host material that can shift the signal ESA spectrum to a shorter wavelength and reduce the ESA strength and (2) to employ a configuration that effectively suppresses ASE buildup. These approaches are explained in the following subsections.

New host glasses for Nd^{3+}-doped fiber. The performance of several potential Nd^{3+}-doped fiber glasses has been studied [310–313] using the figure of merit [304]

$$ m = \frac{4.34}{h\upsilon_\mathrm{p}}\left(1 - \frac{\sigma_{46}}{\sigma_{43}}\right)\frac{\sigma_{43}\tau_4}{A_\mathrm{eff}} \tag{5.80} $$

in units of decibels gain per unit of absorbed pump power. Table 5.7 summarizes the emission characteristics around 1.3 μm of commercially available phosphate, ZrF_4-based fluoride, and fluorophosphate glass [312]. In this table, λ_em and λ_ESA are the peak wavelengths of the emission and the signal ESA spectrum, respectively, and a modified figure of merit m' $[= (1 - S_\mathrm{ESA}/S_\mathrm{em})(\sigma_{43}\tau_4/A_\mathrm{eff})]$ is used instead of m, where S_ESA and S_em are the line strength of the ESA and the stimulated emission transition of the 1.3-μm band, respectively. It has been confirmed that fluorophosphate glass has a better m' value and a shorter signal ESA wavelength

Table 5.7

Emission Characteristics Around 1.3 μm of Commercially Available Phosphate, ZrF_4-Based Fluoride, and Fluorophosphate Glass

	$S_\mathrm{ESA}/E_\mathrm{em}$	$S_\mathrm{em}\tau$ $cm^2 ms$ $(\times 10^{-19})$	m' ms $(\times 10^{-13})$	λ_em (nm)	λ_ESA (nm)
LHG8 (phosphate)	0.675	63.0	25.1	1,324	1,311
ZBLA (fluoride)	0.492	61.7	38.4	1,319	1,316
FPCO (fluorophosphate)	0.406	84.4	61.5	1,323	1,305

From: [313].

than phosphate or fluoride glass. (In addition, it has also been reported that fluoroberyllate glass may have more potential than fluorophosphate glass [313].) Nd^{3+}-doped fluorophosphate glass single-mode fiber was fabricated using theoretical predictions based on those spectroscopic studies. As shown in Figure 5.139, a slightly improved gain spectrum has been demonstrated with a gain peak wavelength of 1.32 μm, which is shorter than that of NDF [312].

However, as yet there have been no reports of a new host glass where the gain spectrum covers the whole operational wavelength region of 1.29 to 1.33 μm or that can offer a high signal gain of, for example, over 20 dB.

Configuration for suppressing ASE buildup. There are three methods by which to suppress the ASE buildup: (1) use a cascade configuration that incorporates optical isolators and/or optical filters into the NDF [287,307,314], (2) optimize the Nd^{3+} ion distribution [315], and (3) dope the NDF cladding with ASE-absorbing ions such as Yb^{3+}.

Among these methods, the cascade configuration that incorporates optical filters has been well studied theoretically. Figure 5.140 shows the calculated improvement in signal gain versus the total NDF length as a function of the number of filters inserted into the NDF [307]. It is indicated that increasing the number of filters is effective for achieving high-gain operation. A dielectric multilayer coated filter or a grating notch filter has been used to limit the 1,050-nm ASE, and signal gains of 10 dB [287] and 9.2 dB [314] have been demonstrated. For example, Figure 5.141 (a,b) shows a cascade configuration with a dielectric multilayer coated filter and its gain characteristics, respectively [287].

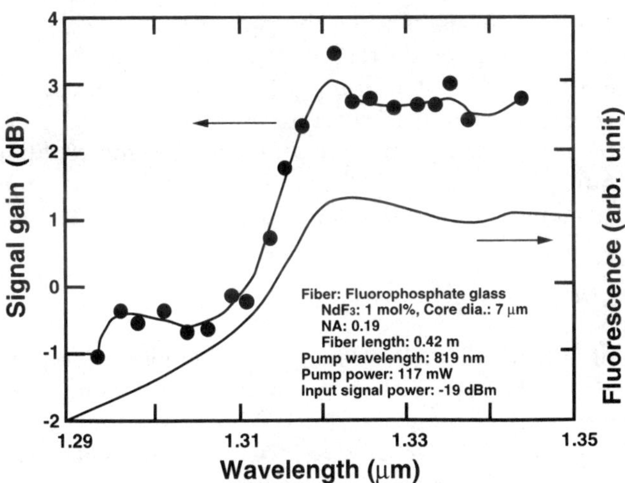

Figure 5.139 Gain spectrum of Nd^{3+}-doped fluorophosphate glass single-mode fiber [312].

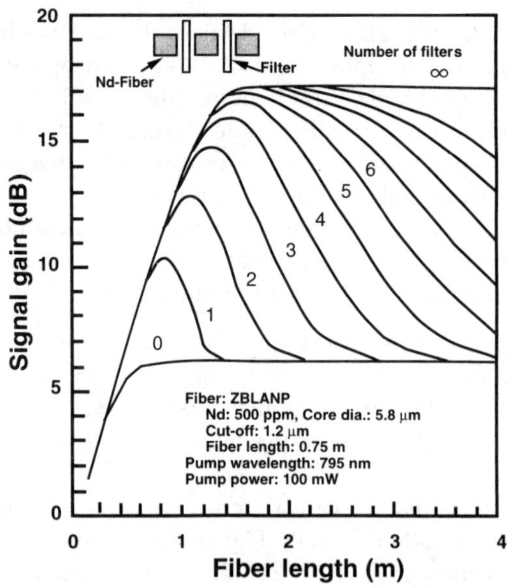

Figure 5.140 Calculated improvement in signal gain versus the total NDF length as a function of the number of filters inserted into the NDF [307].

Furthermore, as shown in Figure 5.142(a), an NDF with a unique Nd^{3+}-ion distribution has been proposed. In this fiber, the Nd^{3+} ions are located at a greater distance from the fiber axis, where the signal intensity is larger than that at 1,050-nm ASE [315]. Using this ion distribution, the gain coefficient of the 1.05-μm band, as expressed in (5.76), can be decreased, and the decrease in the 1.05-μm-band gain coefficient is larger than that of the 1.3-μm band. Figure 5.142(b) shows the expected improvement in the gain characteristics.

In addition, although it would seem that doping the cladding of the NDF with ASE absorbing ions such as Yb^{3+} is a comparatively easy approach, to our knowledge it has not yet been attempted.

5.3.2.2 Praseodymium-Doped Fiber Amplifier

The first part of Subsection 5.3.2.2 describes the amplification principle and the theoretical model of the PDFA. Then we discuss the optimum fiber structure for achieving efficient amplification and basic amplification characteristics such as the gain spectrum, saturation characteristics, and noise characteristics of PDF. Furthermore, after an explanation of the optimum pump wavelength and pump scheme considerations, we describe the construction technique, pump light source, and various configurations for PDFA modules designed to achieve practical

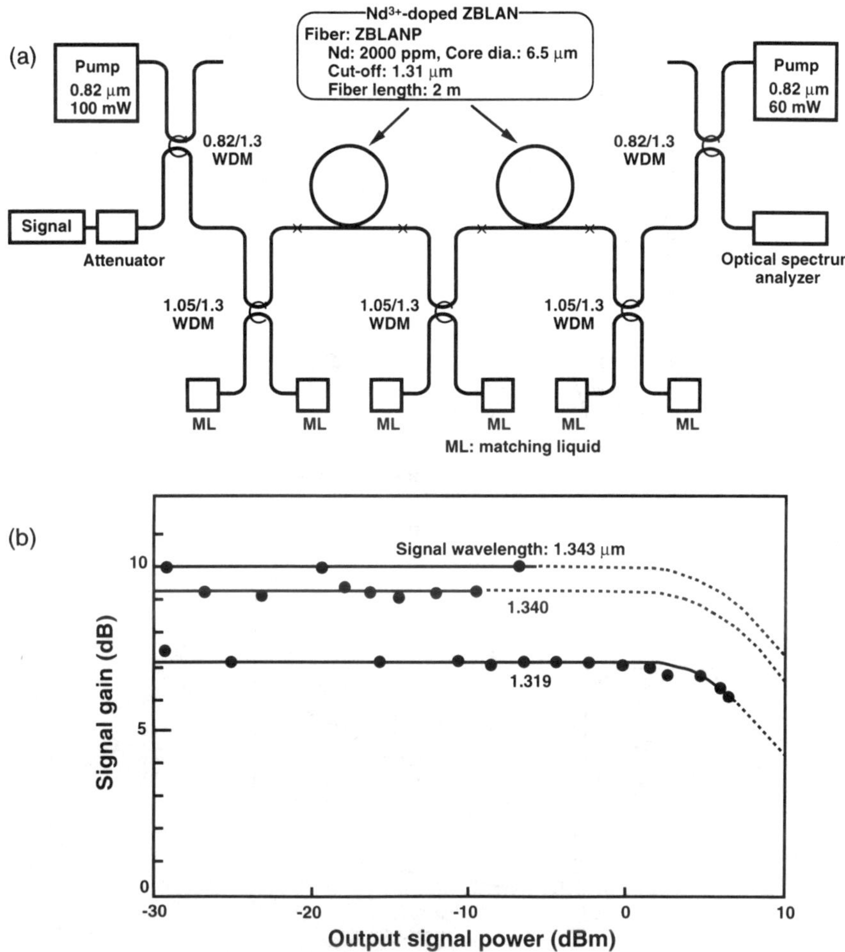

Figure 5.141 Cascade configuration with a dielectric multilayer coated filter [287]: (a) cascade configuration and (b) gain characteristics of NDFA with cascade configuration.

1.3-μm-band amplifiers. Finally, we outline the prospects for a PDFA with new host glasses for Pr^{3+} ions, whose phonon energy is as low as possible. As the new Pr^{3+} ion host glasses, we discuss mixed halide, InF_3-based fluoride, and chalcogenide glasses.

Amplification Principle and Calculation Model of PDFA

Amplification models of the PDFA have been described in [316–319]; in this section, we will use the description provided in [317]. The amplification model of the four-

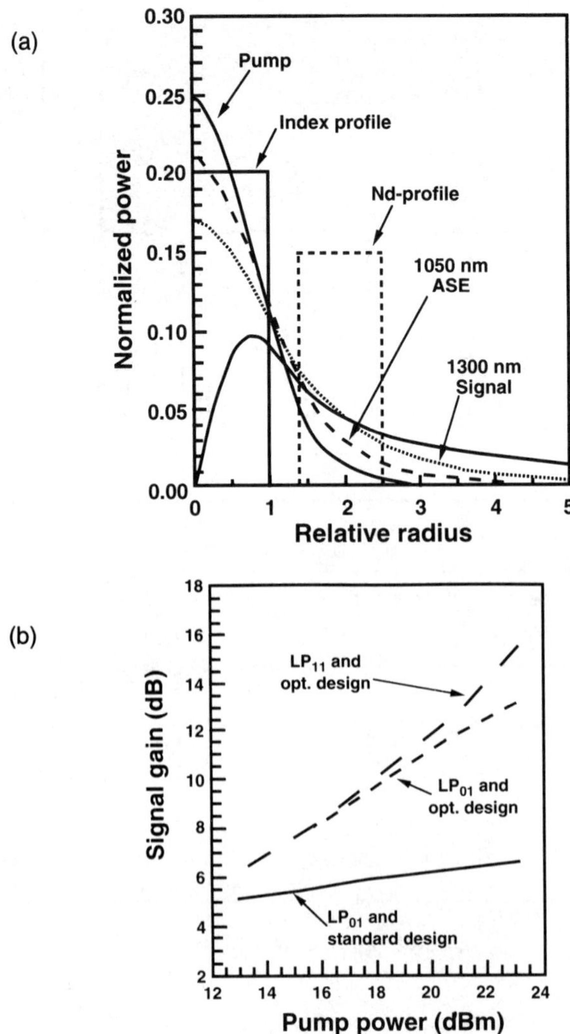

Figure 5.142 (a) Amplified spontaneous emission suppressing NDF which has a unique Nd^{3+}-ion distribution and (b) expected improvement in gain characteristics [315].

level $^1G_4$-$^3H_5$ transition of the PDFA is based on the simplified energy diagram shown in Figure 5.143. The spontaneous lifetimes of the $^1G_4$, $^1D_2$, and $^3P_0$ levels are indicated by τ_i for $i = 4, \ldots, 6$. The measured values for τ_4, τ_5, and τ_6 were 110, 350, and 58 μs, respectively, for PDF [316–318]. Pump photon *ground state absorption* (GSA) occurs between the $^3H_4$ level and the $^1G_4$ level. Pump photons are also absorbed by the pump ESA between the $^1G_4$ level and the $^3P_0$ level and

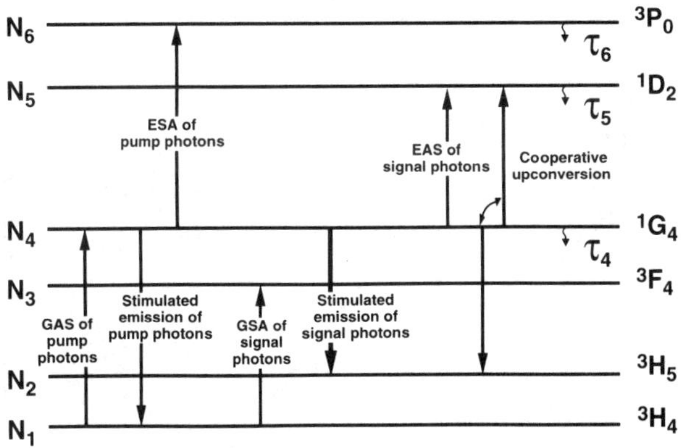

Figure 5.143 Simplified energy diagram of Pr^{3+} ions [317].

by the stimulated emission between the $^1G_4$ metastable level and the $^3H_4$ level. Signal photons are absorbed by the $^3H_4$-$^3F_4$ GSA as well as the $^1G_4$-$^1D_2$ ESA. In addition, the population of the $^1G_4$ metastable level can be reduced by the cooperative upconversion caused by the $(^1G_4$-$^1D_2) - (^1G_4$-$^3H_5)$ transition. Cooperative upconversion must be considered as one factor that degrades the gain characteristics because the energy difference between the $^1G_4$ and the $^1D_2$ levels matches that between the $^1G_4$ and $^3H_5$ levels. The $^1G_4$ level is populated by decay from the $^3P_0$ and $^1D_2$ levels. The branching ratios for the $^3P_0$-$^1G_4$ and $^1D_2$-$^1G_4$ transitions obtained by Judd–Ofelt analysis are $B_{64} = 2\%$ and $B_{54} = 9\%$ for PDF, respectively [316–318].

In this amplification mechanism, the excited Pr^{3+} ions of the $^1G_4$ level descend to the $^3F_4$ level very easily, due to multiphonon relaxation. Therefore, efficient amplification requires the suppression of the $^1G_4$-$^3F_4$ nonradiative transition, whose energy gap is 3,000 (1/cm), by choosing a host glass whose phonon energy is as low as possible. Figure 5.144 shows the relationship between the multiphonon relaxation rate and the energy gap for various host glasses. So, in view of the low multiphonon relaxation rate and the fact that a low-loss fiber fabrication technique has been developed, ZrF_4-based fluoride glass is employed for PDFAs.

Rate equations for the ion population density N_i, taking these transitions into account, are given by

$$\frac{dN_4}{dt} = W_{14}N_1 - \left(W_{46} + W_{45} + W_{42} + W_{41} + \frac{1}{\tau_4} + cN_4 \right) N_4 + \frac{B_{54}N_5}{\tau_5} + \frac{B_{64}N_6}{\tau_6}$$

$$(5.81)$$

$$\frac{dN_5}{dt} = \left(W_{46} + \frac{cN_4}{2} \right) N_4 - \frac{N_5}{\tau_5}$$

$$(5.82)$$

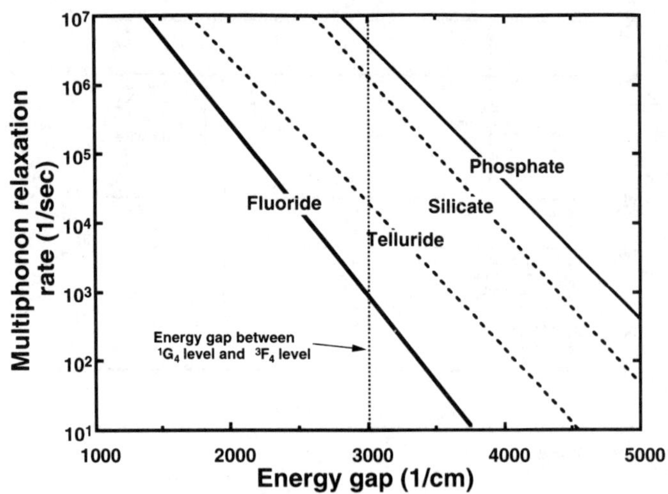

Figure 5.144 Relationship between the multiphonon relaxation rate and the energy gap for various host glasses.

$$\frac{dN_6}{dt} = W_{46}N_4 - \frac{N_6}{\tau_6} \tag{5.83}$$

where c is the cooperative upconversion coefficient and W_{ij} is the transition rate from the ith to the jth levels. By assuming a uniform Pr^{3+} density distribution $\rho(r, \theta) = \rho_0$ for $r \le a$ and $\rho(r, \theta) = 0$ for $r > a$, where a is the fiber core radius and (r, θ) represent the cylindrical transverse coordinates (z is the fiber longitudinal coordinate), and by using a Gaussian approximation of the normalized transverse mode envelopes $\psi_{S,P}(r, \theta)$ of the pump and signal, respectively, that is, $\psi_{P,S}(r, \theta) = [\exp(-r^2/\omega_{S,P}^2)]/\pi \omega_{S,P}^2$, where $\omega_{S,P}$ is the mode power radius for the signal and pump lights, W_{14}, W_{41}, W_{42}, W_{45}, and W_{46} are given by

$$W_{14,41} = \frac{\Gamma_P \sigma_{14,41}(\nu_P)}{Ah\nu_P} \cdot P_P(\nu_P, z) \tag{5.84}$$

$$W_{42,45,46} = \int \frac{\Gamma_S \sigma_{42,45,46}(\nu)}{Ah\nu_S} \cdot (P_S^+(\nu, z) + P_S^-(\nu, z)) \, d\nu \tag{5.85}$$

where $P_P(\nu_P)$ are pump powers with the frequency ν_P, $P_S^{\pm}(\nu, z)$ are forward (+) and backward (−) propagating signals and spontaneous emission power in the 1.3-μm band with the frequency ν, σ_{ij} is the stimulated transition cross section from the ith to the jth level, $\Gamma_{S,P} = 1 - \exp(-a^2/\omega_{S,P}^2)$ is a filling factor, $A = \pi a^2$ is the core area, and h is Planck's constant. Figure 5.145 shows the cross-section spectra of σ_{14}, σ_{41}, σ_{13}, σ_{45}, and σ_{42} [319]. σ_{42} is 0.6×10^{-22} cm^2 [317,318]. The value of the cooperative upconversion coefficient c will be given in Subsection 5.3.2.2, "Optimum Design of Pr^{3+}-Doped Fiber."

The total population density ρ satisfies $\rho = N_1 + N_4 + N_5 + N_6$, N_2 and N_3 are taken to be zero since it was assumed that the Pr^{3+} ions on the ^{3}H$_5$ and ^{3}F$_4$ levels relax instantaneously to the ^{3}H$_4$ ground level. Ion population densities N_1, N_4, N_5, and N_6 in a steady state for the ^{3}H$_4$, ^{1}G$_4$, ^{1}D$_2$ and ^{3}P$_0$ levels are given by

$$N_1 = \rho - \left[1 + \tau_4 \left(W_{35} + \frac{cN_3}{2} \right) \tau_5 W_{35} \right] N_3 \tag{5.86}$$

$$N_4 = \frac{-\beta + \sqrt{\beta^2 + 4\alpha\rho}}{2\alpha} \tag{5.87}$$

$$N_5 = \tau_5 \left(W_{45} + \frac{cN_4}{2} \right) N_4 \tag{5.88}$$

$$N_6 = \tau_6 W_{46} N_4 \tag{5.89}$$

where

$$a = \frac{c(\tau_5 + (2 - B_{54})/W_{14})}{2} \tag{5.90}$$

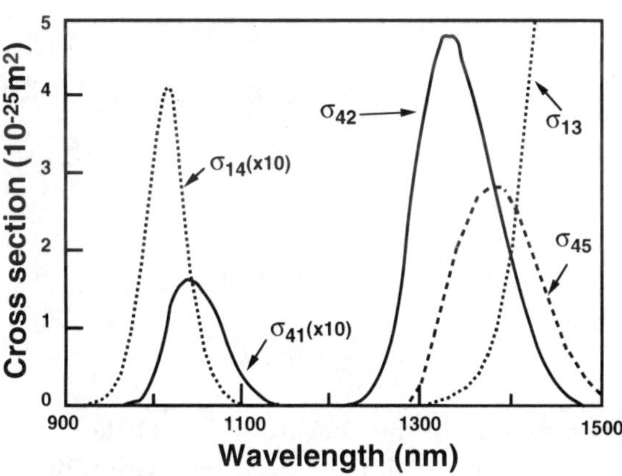

Figure 5.145 Cross-section spectra of σ_{14}, σ_{41}, σ_{13}, σ_{45}, and σ_{42} [319].

$$\beta = 1 + W_{45}\tau_5 + W_{46}\tau_6 + \frac{W_{46} + W_{45} + W_{42} + W_{41} - B_{54}W_{45} - B_{64}W_{46} + 1/\tau_4}{W_{14}}$$

$$(5.91)$$

The local population densities have been found as indicated previously. So the set of coupled differential equations describing the spatial development of pump power and signal power P_P, P_S, and ASE along the fiber are

$$\frac{dP_P(\nu_P,\ z)}{dz} =$$

$$[\gamma_e^P(\nu_P,\ z) - \gamma_P(\nu_P,\ z) - \gamma_{P-ESA}(\nu_P,\ z) - \alpha(\nu_P,\ z)] \cdot P_P(\nu_P,\ z) \qquad (5.92)$$

$$\frac{dP_S^{\pm}(\nu,\ z)}{dz} = \pm g_S(z,\ \nu)\,P_S^{\pm}(\nu,\ z) \pm P_0 \gamma_e(z,\ \nu) \qquad (5.93)$$

$$g_S(\nu,\ z) = \gamma_e^S(\nu,\ z) - \gamma_{S-ESA}(\nu,\ z) - \gamma_{GSA}(\nu,\ z) - \alpha(\nu,\ z) \qquad (5.94)$$

where $P_0 = 2h\nu_S\Delta\nu$ is the power of two photons per unit frequency in bandwidth and corresponds to spontaneous emission in the two polarization modes of the fiber and γ_e^P, γ_P, and γ_{P-ESA} are the stimulated emission coefficient, absorption coefficient, and ESA coefficient for the pump light, respectively. g_S, γ_e^S, γ_{S-ESA}, and g_{GSA} are the gain coefficient, stimulated emission coefficient, ESA coefficient, and GSA coefficient for the signal light, respectively. α is the scattering loss of the PDF. The coefficients, γ_e^P, γ_P, γ_{P-ESA}, γ_e^S, γ_{S-ESA} and γ_{GSA} are given by

$$\gamma_e^P = \sigma_{41}N_4 A \qquad (5.95)$$

$$\gamma_P = \sigma_{14}N_1 A \qquad (5.96)$$

$$\gamma_{P-ESA} = \sigma_{46}N_4 A \qquad (5.97)$$

$$\gamma_e^S = \sigma_{42}N_4 A \qquad (5.98)$$

$$\gamma_{GSA} = \sigma_{13}N_1 A \qquad (5.99)$$

$$\gamma_{S-ESA} = \sigma_{45}N_4 A \qquad (5.100)$$

where σ_{13} is the cross section of the GSA from the $^3H_4$ level to the $^3F_4$ level.

In addition, (5.86) to (5.89) for the local population density N_i are difficult to use for qualitative discussion, although these equations are useful for precisely calculating the amplification characteristics. By omitting the contribution of pump

ESA and cooperative upconversion, simplified expressions for N_1, N_4, and N_5 are given by [317,320,321]

$$N_1 = \rho\,\frac{(W_{41} + W_{42} + W_{45} + 1/\tau_4)}{W_{14} + W_{41} + W_{41} + W_{45} + 1/\tau_4} \tag{5.101}$$

$$N_4 = \rho\,\frac{W_{14}}{W_{14} + W_{41} + W_{41} + W_{45} + 1/\tau_4} \tag{5.102}$$

$$N_5 = \rho\,\frac{W_{14}W_{45}\tau_4}{W_{14} + W_{41} + W_{41} + W_{45} + 1/\tau_4} \tag{5.103}$$

Hence the gain coefficient g_S and population inversion parameter N_{sp} are expressed as

$$g_S = (\sigma_{42} - \sigma_{45})\Gamma_S N_4 - \sigma_{14}\Gamma_S N_1 - \alpha \tag{5.104}$$

$$N_{\mathrm{sp}} = \frac{\sigma_{42}\Gamma_S N_4}{g_S}$$

$$= \frac{1}{1 - \sigma_{45}/\sigma_{42} - \sigma_{41}N_1/\sigma_{42}N_4 - \alpha/\sigma_{42}\Gamma_S N_4} \tag{5.105}$$

Furthermore, the saturation output power for 3-dB gain compression, $P_{\mathrm{sat}}^{\mathrm{OUT}}$, which can be approximated by the signal power at which the number of ions at the $^1G_4$ level with no signal input becomes one-half, is given by

$$P_{\mathrm{sat}}^{\mathrm{OUT}} = \frac{Ah\nu_S}{(\sigma_{42} + \sigma_{45})\Gamma_S}\left[\frac{P_P(\sigma_{14} + \sigma_{41})}{Ah\nu_P} + \frac{1}{\tau_4}\right]$$

$$\approx \frac{Ah\nu_S}{(\sigma_{42} + \sigma_{45})\Gamma_S\tau_4} \tag{5.106}$$

These simplified expressions in (5.101) to (5.106) are helpful in terms of an intuitive understanding of the PDFA characteristics.

Optimum Design of Pr^{3+}-Doped Fiber

The main problem with the PDFA is the low-gain coefficient due to the low quantum efficiency of the stimulated $^1G_4$-$^3G_5$ transition for 1.3-μm-band amplification [288]. The excited Pr^{3+} ions of the $^1G_4$ level descend to the $^3F_4$ level very easily due to multiphonon relaxation. The quantum efficiency, which is defined as $\tau_{\mathrm{Measured}}/\tau_{\mathrm{Radiative}}$, is 3.4% (the measured lifetime τ_{Measured} and radiative lifetime $\tau_{\mathrm{Radiative}}$

are 110 μs and 3.24 ms for PDF). So the first reported PDF required a launched pump power of 180 mW to achieve a signal gain of 5.2 dB, and the gain coefficient was 0.05 dB/mW [288]. To solve this problem, the best approach is to change the host glass to one with a low phonon energy; several candidates such as chalcogenide glass, InF_4-based fluoride glass, and mixed halide glass have been reported as achieving efficient amplification. However, today, there is no practical fiber that uses these host glasses. The present state of Pr^{3+}-doped fiber fabrication with these new host glasses will be described in Subsection 5.3.2.2, "Prospects for PDFA."

Another practical approach is to optimize the parameters of PDF. Several investigations have shown that optimizing the Pr^{3+} concentration in the core, increasing the NA, and reducing the scattering loss of the fiber are the keys to increasing the gain coefficient.

The Pr^{3+} concentration dependence of the signal gain has been studied experimentally and theoretically [316]. This study obtained an upconversion coefficient c of 0.2×10^{-22} cm^3/s for a Pr^{3+} concentration of 500 ppm and 1.1×10^{-22} cm^3/s for a Pr^{3+} concentration of 1,000 ppm, so the calculated gains fit the measured gains. It is clear that there is no great improvement in the gain coefficient even if the Pr^{3+} concentration is reduced to less than 500 ppm. As an example, Figure 5.146 shows the internal gain characteristics of two PDF with Pr^{3+} concentrations of 500 and 1,000 ppm [316]. The fiber is forward pumped with a pump wavelength of 1,017 nm. The product of PDF length and Pr^{3+} concentration for both fibers is equal for a fixed amount of Pr^{3+} ions in each fiber, and the fiber lengths with Pr^{3+}

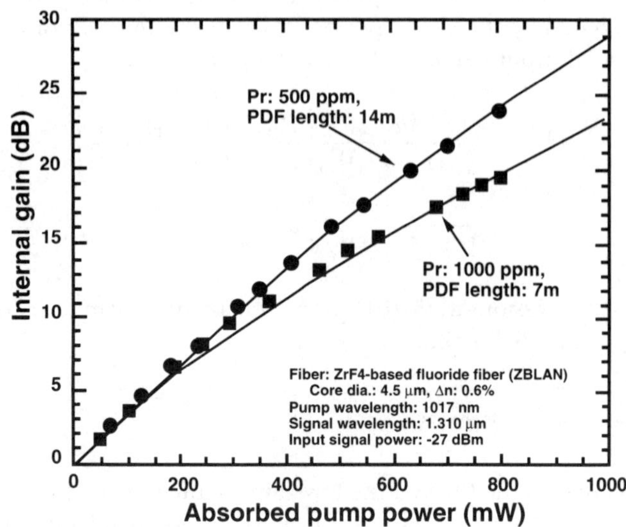

Figure 5.146 Internal gain characteristics of two Pr^{3+}-doped ZrF_4-based fluoride fibers with Pr^{3+} concentrations of 500 and 1,000 ppm [316].

concentrations of 500 and 1,000 ppm are 14m and 7m, respectively. The other parameters are the same for both PDFs. As shown in this figure, although the gains of both fibers start to increase with almost the same slope as the pump power increases, the signal gain of the 1,000-ppm fiber started to saturate earlier than that of the 500-ppm fiber. The signal gain of the 500-ppm fiber did not saturate noticeably unlike that of the 1,000-ppm fiber, which indicates that a concentration of 1,000 ppm is still too high to fabricate PDF with a high gain coefficient. In addition, the internal gain used in this figure is the sum of the signal gain and fiber loss (scattering loss + GSA from $^3H_4$ to $^3F_4$) at the signal wavelength, which was usually used in the early years of PDFA research to show the potential of this active fiber because there were no low-loss PDFs and no satisfactory technique for connecting PDF to silica fiber, in order to input the signal and pump power for measurement. The internal gain was measured as the ratio of the pumped-to-unpumped signal level at the output end of the PDF.

Other key PDF parameters with regard to increasing the gain coefficient are NA and scattering loss [288,293,294]. The calculated relationship between the gain coefficient and core radius of PDF for various relative reflective-index differences Δn between the core and cladding are shown in Figure 5.147 [288,289]. This

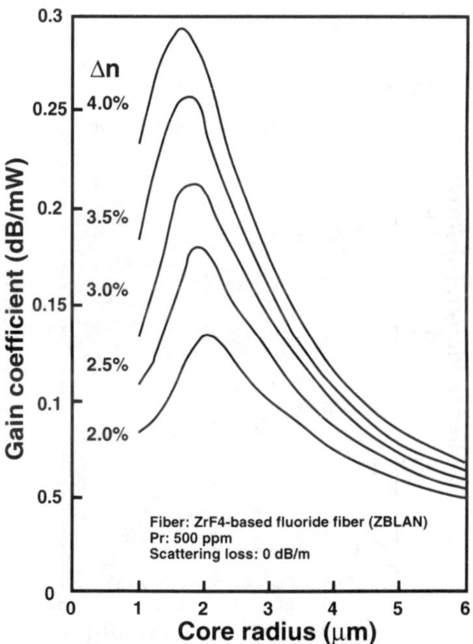

Figure 5.147 Calculated relationship between the gain coefficient and core radius of Pr^{3+}-doped ZrF$_4$-based fluoride fiber, for various Δn values [288].

calculation only considers a forward-pumped fiber with a step-index core profile and a uniform distribution of Pr^{3+}-ions in the core region with a pump wavelength of 1,017 nm. The scattering loss of the fiber is assumed to be zero. The gain coefficients increase with increasing Dn due to an improving overlap between the fiber core and the pump and signal power, as with Er^{3+}-doped fiber. The cutoff wavelengths of the PDF with the peaks of the gain coefficient in various Δn curves are all approximately 0.8 μm [293,322]. From this calculated value, it is expected that a gain coefficient of more than 0.2 dB/mW can be achieved when the Δn is larger than 3%. However, to fabricate an actual amplifier, a PDF with a high Δn must be connected to a silica fiber with low splicing loss. When this is taken into consideration, it is considered that the maximum Δn of the PDF is about 3.7% from the fact that the Δn of the highest NA silica fiber fabricated so far is 3.7%. The gain coefficient of a PDF with a Δn of 3.7% and a core radius of 1.8 μm is shown as a function of the scattering loss in Figure 5.148 [293]. To achieve highly efficient fiber of over 0.2 dB/mW requires a low scattering loss of less than 0.05 dB/m for a PDF with a high Δn of 3.7% [293]. Recently, by improving the fiber fabrication technique and optimizing the composition of ZrF_4-based fluoride glass to suppress crystallization, the scattering loss of PDF with a Δn of 3.7% has been reduced to 0.02 dB/m, as shown in Figure 5.149(a), and the highest gain coefficient of 0.21 dB/mW has been achieved with a Pr^{3+} concentration of 500 ppm, as shown in Figure 5.149(b) [293].

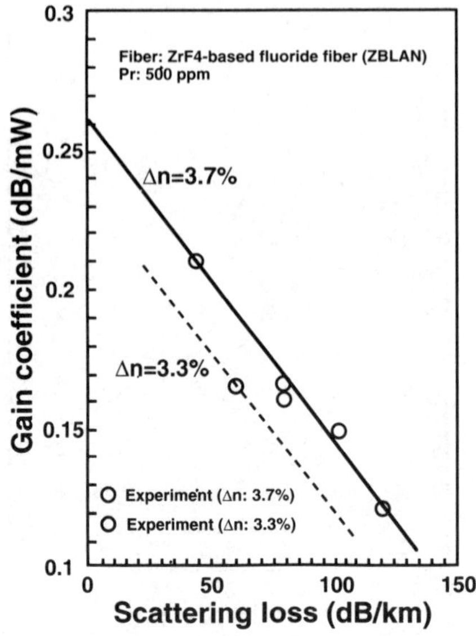

Figure 5.148 Gain coefficient of a Pr^{3+}-doped ZrF_4-based fluoride fiber as a function of the scattering loss [293].

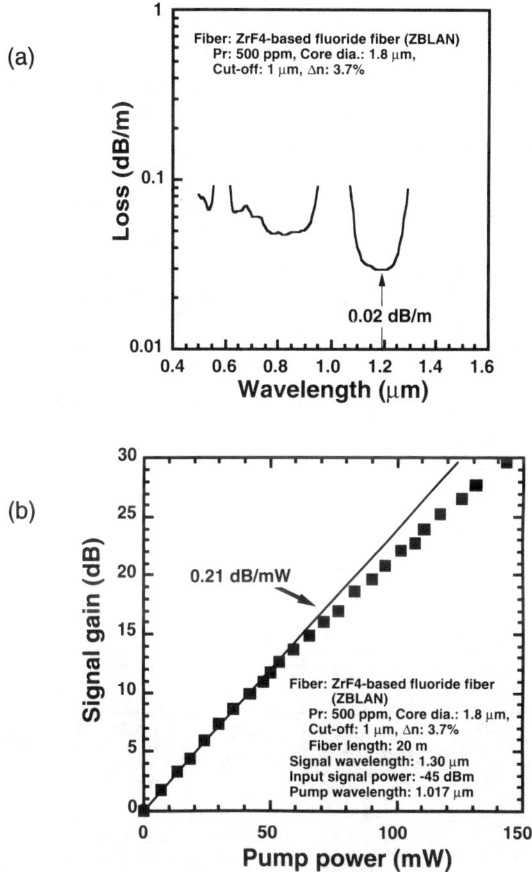

Figure 5.149 Characteristics of highly efficient Pr^{3+}-doped ZrF_4-based fluoride fiber [293]: (a) loss spectrum and (b) gain coefficient.

Furthermore, the gain coefficient is also improved by carefully selecting the pump and amplification configurations. The efficiency has been increased to 0.24 dB/mW from 0.21 mW by employing a bidirectional pump configuration. Moreover, the gain coefficient has been doubled by using a double-pass amplification configuration, and a coefficient of 0.4 dB/mW has been realized. These two configurations will be explained in detail in Subsection 5.3.2.2, "Optimum Pump Wavelength and Pump Scheme Considerations" and "PDFA Module."

Basic Amplification Characteristics of PDFA

The basic PDFA amplification characteristics such as gain spectrum, saturation characteristics, and noise characteristics will be described in this section.

Gain spectrum. The PDFA spectrum has no structure, unlike that of the EDFA. Figure 5.150(a,b) shows the typical gain spectra for various pump powers launched into the PDF and for various input signal powers, respectively [323]. The signal gain spectrum has a peak at around 1.30 μm and provides good coverage of the 1.29- to 1.33-μm operational wavelength band of the 1.3-μm-band optical telecommunication system, irrespective of pump condition and signal power. The 3-dB down spectral bandwidth increases with decreasing pump power and increasing input power.

Furthermore, Figure 5.151 shows the PDF length dependence of the signal gain spectrum [323]. The signal gain spectrum of the PDFA depends strongly on

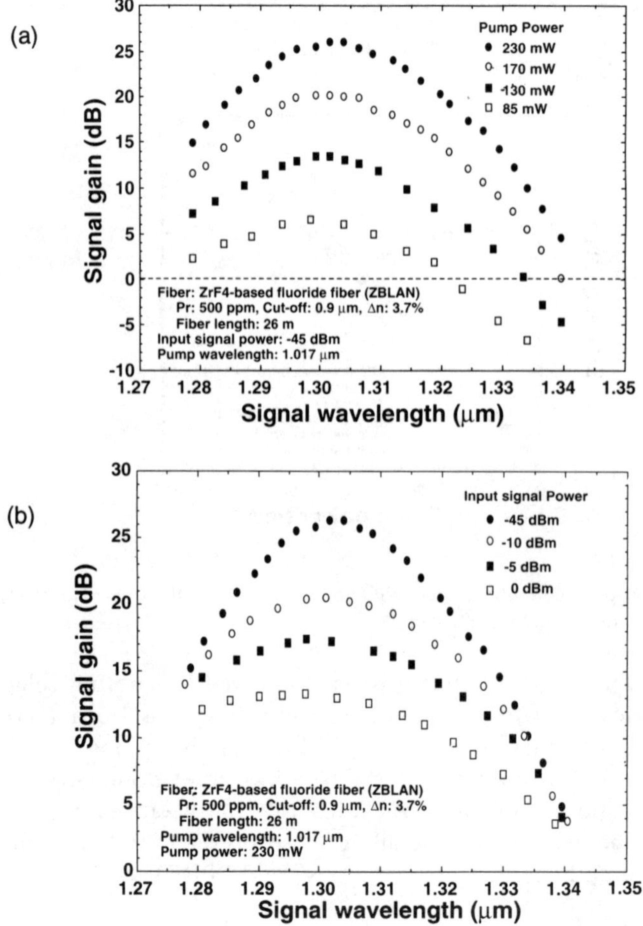

Figure 5.150 Typical gain spectra of PDFA for various (a) pump powers and (b) input signal powers [323].

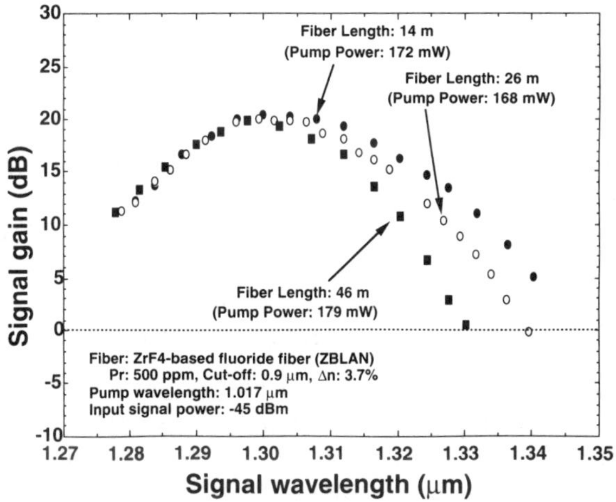

Figure 5.151 PDF length dependence of the signal gain spectrum [323].

the PDF length. The 3-dB down spectral bandwidth decreases with increasing fiber length because the signal gain at a signal wavelength of over 1.30 μm decreases with increasing PDF length due to the influence of the GSA. Therefore, it is important to optimize the PDF length to achieve a PDFA with a high gain and a wide amplification band. Figure 5.152 shows the relationship between the signal gain and the 3-dB down spectral bandwidth [323]. The signal gain increases with increasing PDF length until an optimum length is reached at which the PDF provides maximum signal gain. The signal gain then decreases slightly with further increases in PDF length due to the scattering loss of the PDF. By contrast, the spectral bandwidth decreases monotonously with increasing PDF length. Hence, it is concluded that

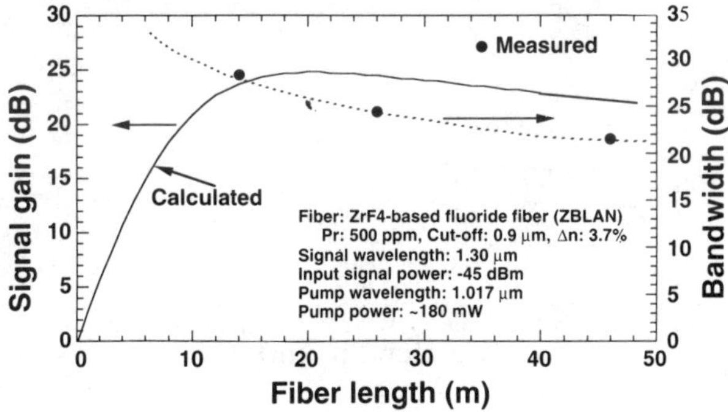

Figure 5.152 Relationship between the signal gain and the 3-dB down spectral bandwidth [323].

the Pr^{3+}-doped fiber used in an amplifier module must be tuned to its optimum length.

In addition, here we explain the relationship between signal gain, internal gain, and fiber loss using their respective spectra. Figure 5.153 shows the measured signal gain, internal gain, and fiber loss spectrum of the PDF [320,321]. The wavelength dependence of the fiber loss that consists of the scattering loss and the GSA from the $^3H_4$ level to the $^3F_4$ level is due to the wavelength dependence of the GSA as shown in Figure 5.145. Although the peak wavelength of the signal gain is around 1.30 mm, the peak of the internal gain spectrum is at 1.31 μm. The peak shift from 1.30 to 1.31 μm is due to the influence of the GSA spectrum.

Gain saturation. Figure 5.154 shows the typical gain saturation characteristics of PDF. The solid lines in this figure are calculated values [317]. The gain saturation power for 3-dB gain compression decreases with increasing signal wavelength. In this figure, the 3-dB down saturation powers are 12, 10, and 9.5 dBm at signal wavelengths of 1.29, 1.31, and 1.33 μm, respectively. This is due to signal ESA whose presence reduces the saturation power, as expressed in (5.106). However, the signal ESA cross section in the 1.29- to 1.33-μm signal region of the 1.3-μm-band system is more than 10 times smaller than the stimulated cross section. The effect of signal ESA on signal gain is not thought to be serious with regard to telecommunication signals.

Furthermore, PCE and slope efficiency are important parameters that characterize the saturation performance of PDF. It has been reported that the former parameter is about 36% for a pump wavelength of 1,010 nm and 21% for 1,047 nm [324]. (The maximum power conversion efficiency PCE^{max} is about 80%.) The latter

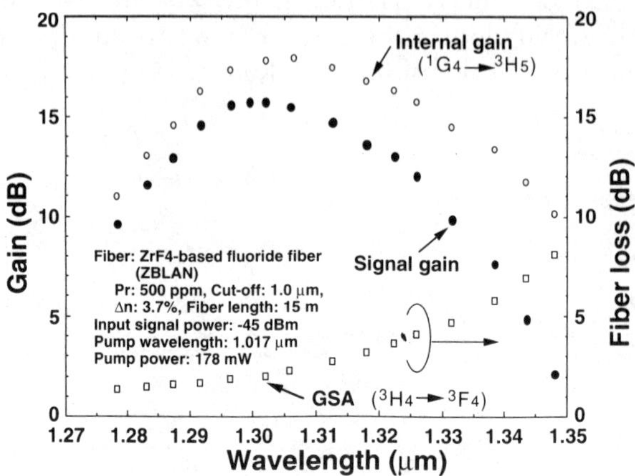

Figure 5.153 Measured signal gain, internal gain, and fiber loss spectrum of Pr^{3+}-doped ZrF_4-based fluoride fiber [320,321].

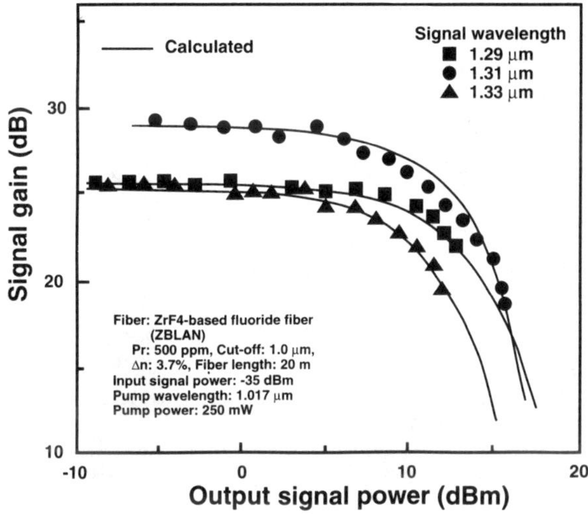

Figure 5.154 Typical gain saturation characteristics of Pr^{3+}-doped ZrF_4-based fiber [317].

parameter will be discussed in Subsection 5.3.2.2, "Optimum Pump Wavelength and Pump Scheme Considerations," where the slope efficiency will be used to evaluate various pump light sources.

Noise figure characteristics. If there is no signal ESA ($^1G_4$-$^1D_2$ transition) or GSA ($^3H_4$-$^3F_4$ transition) in the PDFA, the noise figure of this amplifier is close to the value of 3.0 dB for an ideal optical amplifier using stimulated transition, just as the PDFA is classified as a four-level system optical amplifier [317,320,321,324,325]. However, Pr^{3+}-doped fluoride fiber is severely influenced by the signal ESA and GSA, and therefore the noise characteristics worsen. From (5.105), the NF is given by [317,320,321]

$$NF = 2N_{SP}$$

$$\approx 3(dB) \qquad\qquad (\lambda_S < 1.31 \ \mu m)$$

$$10 \log\left[\frac{2}{1 - \sigma_{45}/\sigma_{42} - \sigma_{41}N_1/\sigma_{42}N_4}\right](dB) \qquad (\sigma_S > 1.31 \ \mu m) \quad (5.107)$$

In this equation, the scattering loss of the PDF is ignored.

Figure 5.155 shows the measured NF of PDF, as a function of signal wavelength [320,321]. The NF is 3.4 dB at wavelengths shorter than 1.31 μm. The 0.4-dB difference between the measured NF and the quantum limit value of 3.0 dB is thought to be due to the scattering loss of the PDF. On the other hand, at wavelengths longer than 1.31 μm, the NF worsens with increasing signal wavelength

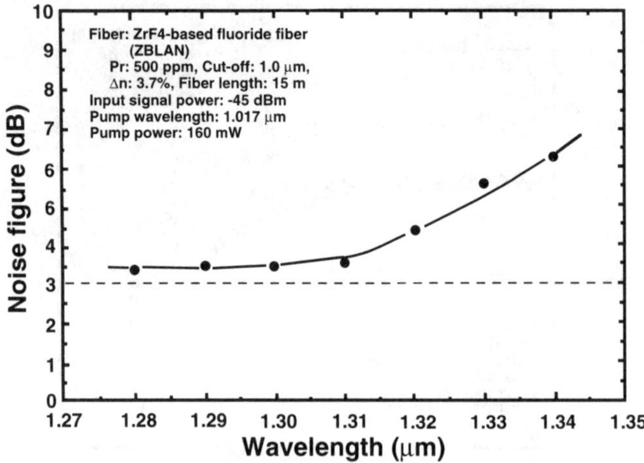

Figure 5.155 Noise figure of Pr^{3+}-doped ZrF_4-based fluoride fiber, as a function of signal wavelength [320,321].

due to the signal ESA cross section σ_{45} and the GSA cross section σ_{14} as expressed by the second and third terms of the denominator in (5.107).

Furthermore, Figure 5.156 shows the relationship between the NF and the signal output power [324]. These is no change in the NF for signal wavelengths below 1.31 μm because of the four-level amplification system of the PDFA. The slight change in the NF at signal wavelengths longer than 1.31 μm is due to the GSA, which is expressed theoretically as the GSA cross section in (5.107) by the third term of the denominator.

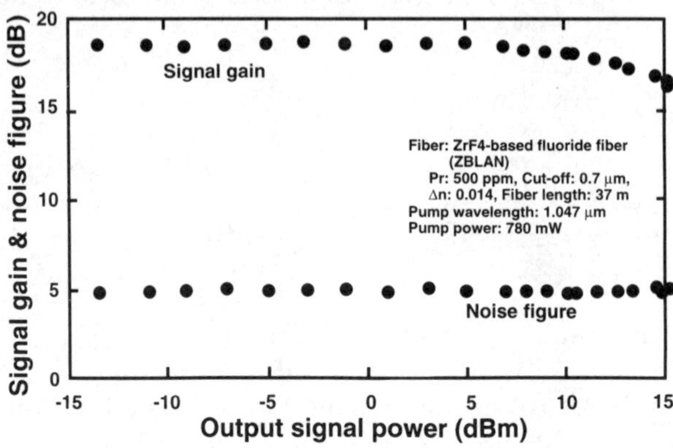

Figure 5.156 Relationship between the noise figure and the signal output power [324].

In conclusion, PDFs can amplify signals with a noise of around the quantum limit of 3 dB except in the signal wavelength region above 1.31 μm.

Temperature Dependence

The temperature-dependent characteristics of Pr^{3+}-doped fluoride fiber are important in terms of practical applications and have been investigated both theoretically and experimentally [320,321,326].

Figure 5.157(a,b) shows the temperature dependence of the signal gain and NF, respectively, of PDF for various signal wavelengths [320,321]. The signal wavelength is varied from 1.28 to 1.33 μm, and the fiber temperature range is from $-40°$ to 80°C. Regardless of the signal wavelength, the signal gain increases monotonically as the fiber temperature is reduced. By contrast, the NF increases with increasing fiber temperature at signal wavelengths higher than 1.31 μm. However, the change in NF is negligible at signal wavelengths below 1.31 μm. The temperature dependence of the saturation output power for 3-dB gain compression is shown in Figure 5.157(c) and reveals that the saturation output power increases with increasing fiber temperature [320,321].

In the simplified equations (5.101) to (5.107), the parameters that exhibit temperature dependence are the cross sections σ_{41}, σ_{45}, σ_{13} and the fluorescence lifetime τ_4. The changes in τ_4 and σ_{41} are of particular importance when considering the temperature dependence of a PDFA [320,321,326].

Figure 5.158 shows the temperature dependence of the lifetime τ_4 in a ZrF_4-based fluoride glass. The lifetime changes monotonically from 77 μs at 335K to 155 μs at 9K [326]. This temperature dependence is caused by a variation in the nonradiative relaxation probability due to multiphonon relaxation from the $^1G_4$ metastable level to the $^3F_4$ level [327]. It should be noted that with EDFAs, the temperature-dependent gain characteristics result only from the stimulated-emission cross section because the energy interval between the $^4I_{13/2}$ metastable level and the $^4I_{15/2}$ ground level is sufficiently large and multiphonon relaxation does not affect the metastable lifetime. This change in τ_4 affects the Pr^{3+} ion population density N_4 and N_1 as indicated in (5.101) and (5.102) and as the fiber temperature is reduced, N_4 increases while N_1 decreases.

Figure 5.159 shows the change in the $^1G_4$-$^3H_5$ emission band in ZrF_4-based fluoride glass in the 9- to 320-K temperature range [326]. A change in the emission means a change in the cross section s41. The emission intensity in the 1.28- to 1.33-μm wavelength region increases with decreasing temperature until 250K, although the emission peak wavelength shifts toward longer wavelengths and the emission intensity near 1.30 μm is noticeably reduced because of the red shift in the emission peak with decreasing temperature below 250K. This gain cross-section variation is due to the Boltzmann distribution change.

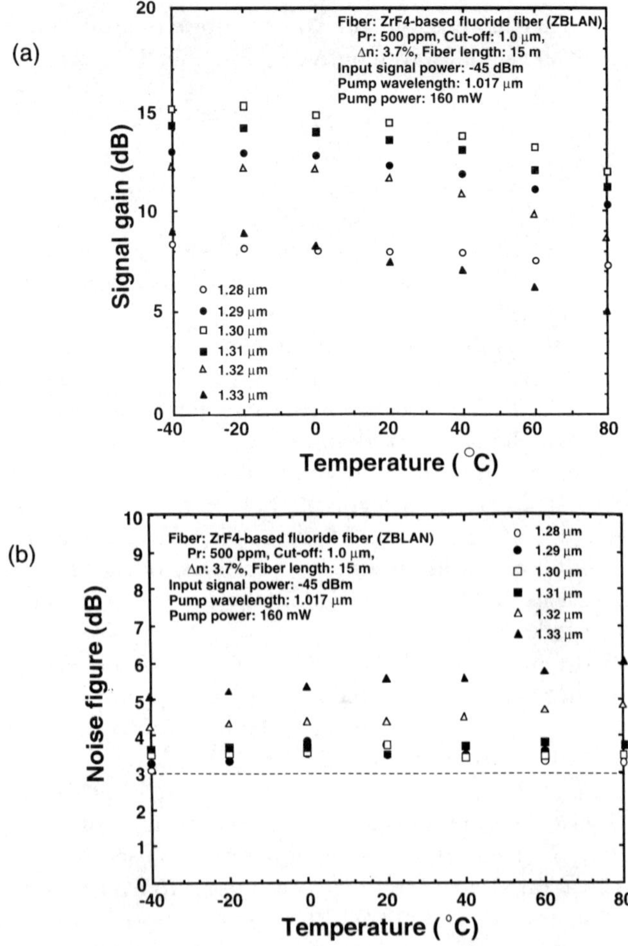

Figure 5.157 Temperature dependence of PDFA: (a) signal gain, (b) noise figure, and (c) saturation output power [320,321].

Taking account of the changes in τ_4 and σ_{41}, the temperature dependence of the gain characteristics in the fiber temperature region between $-40°$ and $80°$C, can be described as follows.

1. As the fiber temperature is reduced, the signal gain increases due to the increase in N_4, the decrease in N_1, and the increase in σ_{41}, as shown in (5.104).
2. At signal wavelengths of less than 1.31 μm, the NF is constant as shown in (5.107). But when the fiber temperature is increased at signal wavelengths above 1.31 μm, the NF increases because of the increase in the third term

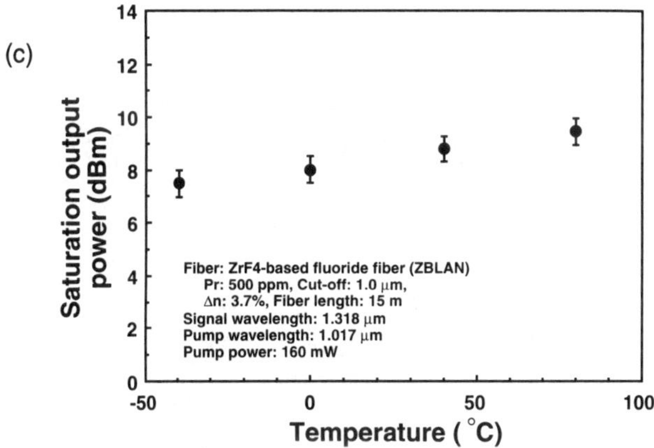

Figure 5.157 (continued).

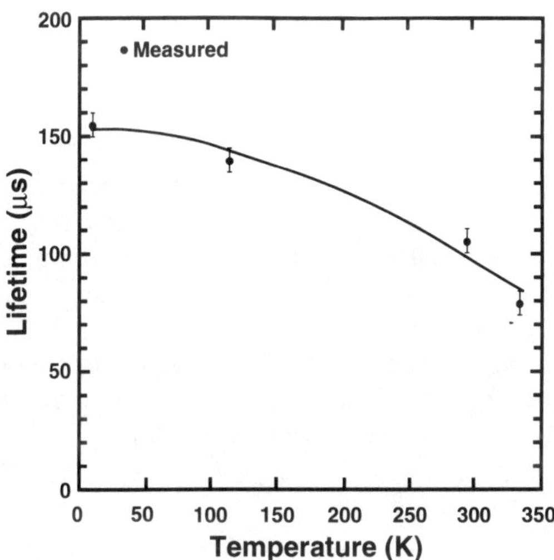

Figure 5.158 Temperature dependence of the lifetime τ_4 in a ZrF$_4$-based fluoride glass [326].

in the denominator in (5.104) due to the decrease in N_4, the increase in N_1, and the decrease in σ_{41}.

3. As the fiber temperature is increased, the lifetime τ_4 and σ_{41} both decrease and the saturation output power increases, as shown in (5.106).

As mentioned, the signal gain, NF, and saturation output power are temperature dependent. However, the signal gain and output power can be controlled at

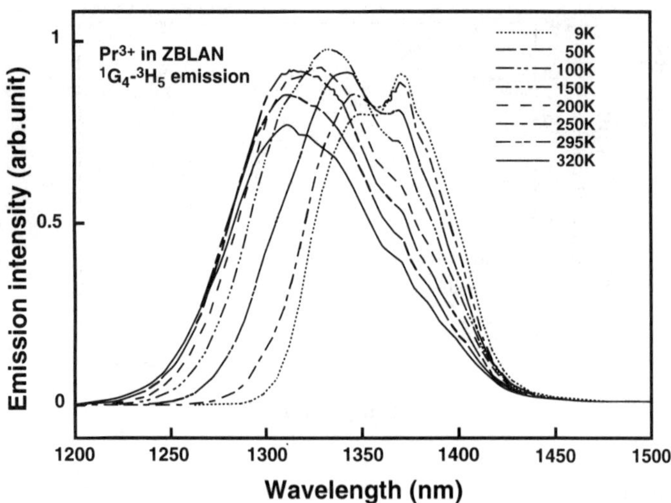

Figure 5.159 Change in the ${}^1G_4\text{-}{}^3H_5$ emission band in ZrF_4-based fluoride glass in the 9- to 320-K temperature range [326].

constant values using the AGC and APC techniques because the change in the signal gain and output power is sufficient to allow control in the $-40°$ to $80°C$ temperature range.

In addition, Figure 5.160 shows the measured and calculated temperature dependence of the small signal gain at several wavelengths from 1.29 to 1.33 μm, which includes the temperature region below $-40°C$ [326]. Signal gains do not increase monotonically below $-40°C$ because the decrease in the stimulated emission cross section σ_{41} with decreasing temperature, shown in Figure 5.159, degrades the signal gain at such low temperatures, although the quantum efficiency is improved.

Optimum Pump Wavelength and Pump Scheme Considerations

To achieve efficient pumping, it is useful to study the gain dependence on pump wavelength [300,301,317,328] and a pump scheme [317,318] with small and large input signals. In the following, we present calculated and measured characteristics and results that have been obtained with a view to realizing efficient PDFAs.

Figure 5.161 shows the measured pump wavelength dependence of the small signal gain [317]. There is a peak in the excitation spectrum at around 1,017 nm. However, the tolerance of the pump wavelength is wide, although the bandwidth for 3-dB gain compression is 45 nm from 988 to 1,033 nm. Figure 5.162 shows the calculated slope efficiency, which is normalized at a pump wavelength of

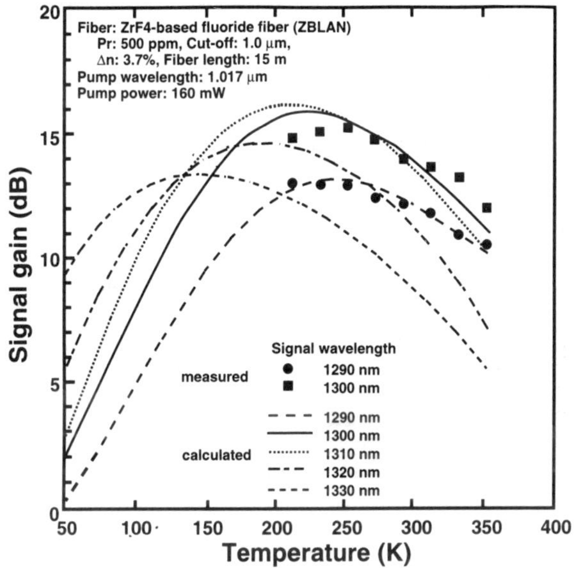

Figure 5.160 Measured and calculated temperature dependence of the small signal gain at several wavelengths from 1.29 to 1.33 μm which includes the temperature region below $-40°$C [326].

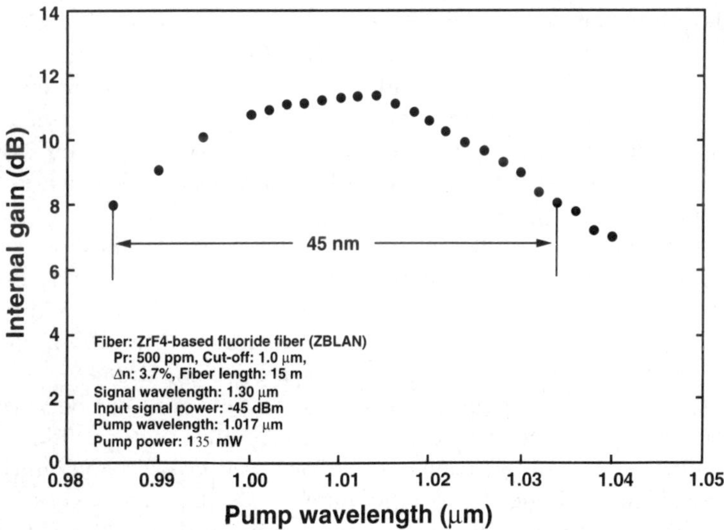

Figure 5.161 Measured pump wavelength dependence of the small signal gain [317].

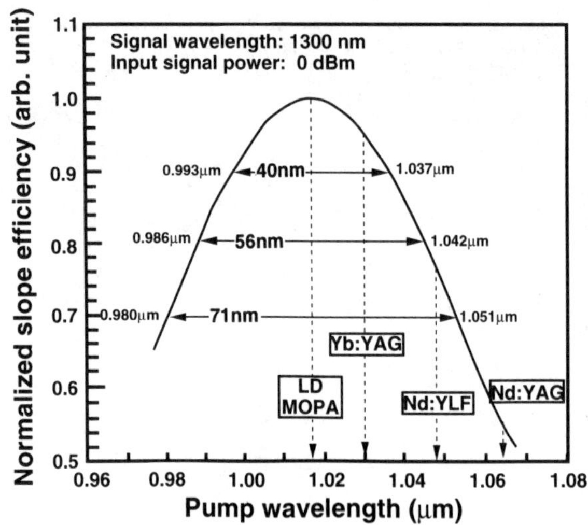

Figure 5.162 Calculated slope efficiency, which is normalized at a pump wavelength of 1,017 nm where the efficiency is maximum, as a function of the pump wavelength for an input signal of 0 dBm [301].

1,017 nm where the efficiency is maximum, as a function of the pump wavelength for an input signal of 0 dBm [301]. In this calculation, the PDF length is optimized for each pump wavelength. A normalized slope efficiency of more than 70% maintains a wide pump wavelength region from 980 to 1,051 nm. This wide pump wavelength region of the PDF implies that an Nd:YLF solid-state laser operating at 1,047 nm, a Yb:YAG solid-state laser, and a Yb^{3+}-doped fiber laser are available as pump light sources for a PDFA in addition to the semiconductor laser that can operate at an optimum pump wavelength of around 1,015 nm. The choice of pump light source for a practical PDFA module will be discussed in Subsection 5.3.2.2, "PDFA Module."

Next, we outline pump scheme considerations. Figure 5.163 shows the measured and calculated pump power dependence of internal gains at 1.30 μm for forward-pumped and bidirectionally pumped PDF [317,318]. When the PDF is bidirectionally pumped, the same pump power is launched into both ends. The gains of the bidirectionally pumped fiber are larger than those for a forward-pumped PDF. A differential gain coefficient of 0.21 dB/mW has been reported with a forward pump scheme while the value was 0.24 dB/mW for a bidirectionally pumped scheme. The gain coefficient can be increased by using the bidirectional pump scheme. Figure 5.164 compares the pump power dependence of the output signal power from bidirectionally pumped and forward-pumped PDF [317,318]. It is seen that the output signal powers for the former are roughly double those for the latter at pump powers of more than 100 mW. The bidirectional pump scheme

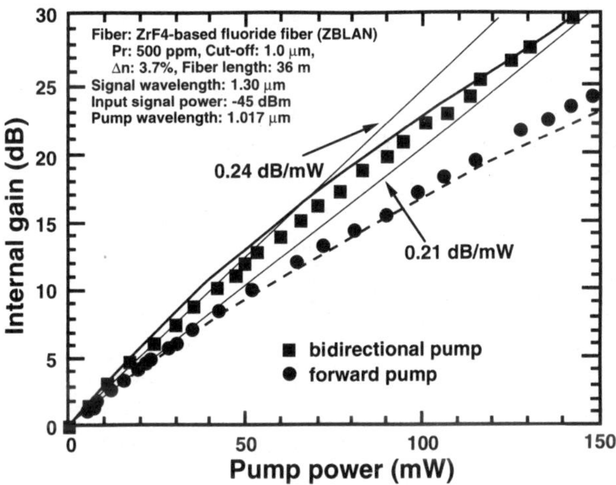

Figure 5.163 Measured and calculated pump power dependence of internal gains at 1.30 μm for forward-pumped and bidirectionally pumped PDF [317,318].

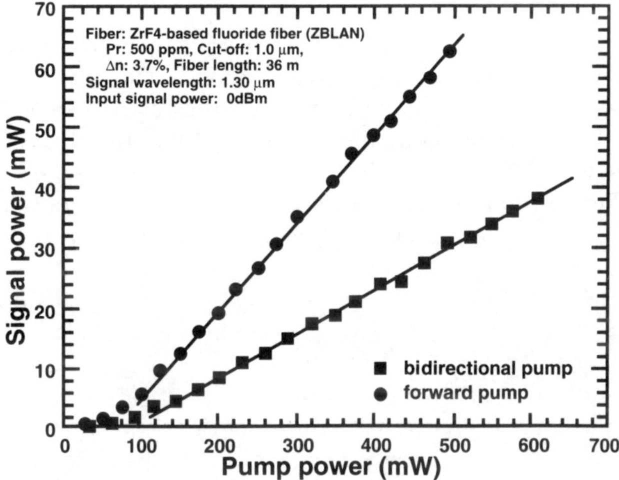

Figure 5.164 Pump power dependence of the output signal power from bidirectionally pumped and forward-pumped PDF [317,318].

is also more advantageous for obtaining a higher output signal power from a PDFA because the slope efficiency is more than doubled by using the bidirectional scheme. This difference between forward-pumped and bidirectionally pumped PDF can be explained as follows. Two kinds of process, pump ESA and cooperative upconversion, which depend on the square of the pump power intensity, reduce the population of the $^1G_4$ metastable level without contributing to the stimulated transition at 1.3 μm. Since pump ESA is a two-step excitation process, the population of the $^3P_0$ level is proportional to the square of the pump power. As a result, the $^1G_4$ population does not increase linearly with pump intensity due to the pump ESA. The $(^1G_4\text{-}^3H_5)\text{-}(^1G_4\text{-}^1D_2)$ cooperative upconversion also prevents the $^1G_4$ population from increasing in magnitude in proportion to pump power because the number of ions per unit time moving to other levels from the $^1G_4$ level is proportional to the square of the $^1G_4$ population, that is, the square of pump power intensity. Hence, there is a greater population reduction due to pump ESA and cooperative upconversion in a fiber segment where the pump power is strong than in a segment where it is weak. When the total pump power launched into Pr^{3+}-doped fibers is constant, the total number of Pr^{3+} ions excited at the $^1G_4$ metastable level in a bidirectional pump scheme is thought to be larger than that in a unidirectional (forward or backward) pump scheme due to both pump ESA and cooperative upconversion. In addition, the improvement in the population of the $^1G_4$ metastable level with a bidirectional pump scheme is mainly caused by the suppression of pump ESA because the Pr^{3+} concentration must usually be less than 500 ppm to achieve a high gain coefficient, as explained in Subsection 5.3.2.2, "Optimum Design of Pr^{3+}-Doped Fiber."

As described, a bidirectional pump scheme can suppress the degradation of the $^1G_4$ metastable level population caused by pump ESA and cooperative upconversion. Moreover, it has been shown that the gain coefficient increases from 0.21 to 0.24 dB/mW with a bidirectional pump scheme and that the signal output power with this scheme is roughly double that with a forward pump scheme. This shows that the bidirectional pump scheme is effective in extracting the full potential of PDF and is an appropriate scheme with which to construct an efficient PDFA module.

PDFA Module

The first attempt to construct a PDFA module was in 1991, where strained-quantum-well semiconductor LDs were used that operated at 1,017 nm [295]. After this attempt, an LD module with a fiber pigtail was fabricated [329] and the pump power of the LD module was increased by optimizing the laser structure [271,330] or by employing a new structure such as MOPA [269,331] to achieve a PDFA module with high gain and high power. Furthermore, a compact laser diode-pumped Nd:YLF laser with a fiber pigtail was also employed as a pump source with a low-noise and high-power laser [332]. Another key to the construction of the practical PDFAs was the development of splicing between PDF with a high Δn and silica fiber with a conventional Δn (~0.3%). This problem was overcome using the TEC connection technique [220,221,332] and the tilted V-groove connection technique, or angled-polish and active-alignment technique, and excellent connections with low loss and low reflection were realized [329,332,333]. Using these connection techniques and high power pump sources, high-performance PDFA modules have been fabricated with high signal gains of over 30 dB and signal output powers of over 18 dBm. This subsection will describe the connection technique and pump light sources for a PDFA. Then, we will report the performance of recently fabricated PDFA modules and mention various PDFA configurations designed to achieve high-gain and high output power. Finally, we will introduce the performance of recently fabricated commercially available PDFAs.

Technique for connecting Pr^{3+}-doped fluoride fiber and silica fiber. Figure 5.165 shows a schematic PDFA module configuration that is basically the same as that of an EDFA module, without the tilted V-groove connections and with high NA silica fiber. The high NA silica fiber has almost the same fiber parameters, including Δn, core diameter, and cutoff wavelength, as PDF, thus making it possible to achieve low-loss butt jointing between the two types of fiber. The tilted V-groove connection, which is illustrated in Figure 5.166(a), is used for connections between high NA silica fiber and PDF and has a high Δn to realize high-gain performance [329,332]. In this connection, both fibers are first fixed into the V-groove and their endfaces are polished. After polishing, the two fibers are carefully aligned and connected with UV-curable adhesive. The alignment accuracy for both fibers was better than 0.1 μm, providing a connection loss of less than 0.2 dB. A connection angle of 25 degrees is used to achieve a low reflection of less than -60 dBm from the relationship between the connection angle and the reflection, as shown in Figure 5.166(b). In addition, Figure 5.167 shows another connection technique, namely,

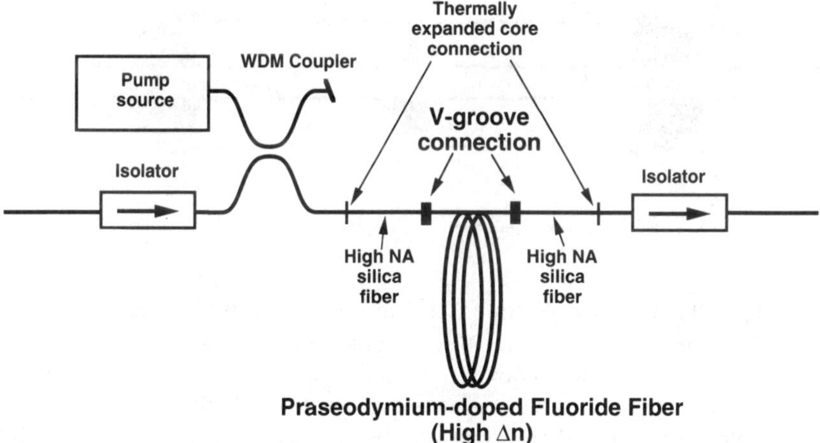

Figure 5.165 Schematic PDFA module configuration.

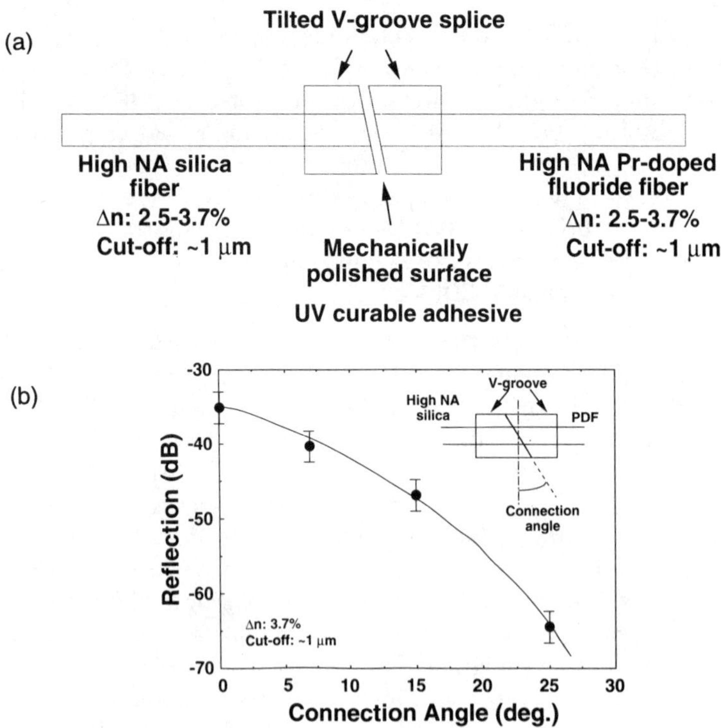

Figure 5.166 (a) Tilted V-groove connection technique and (b) relationship between the connection angle and the reflection [329,332].

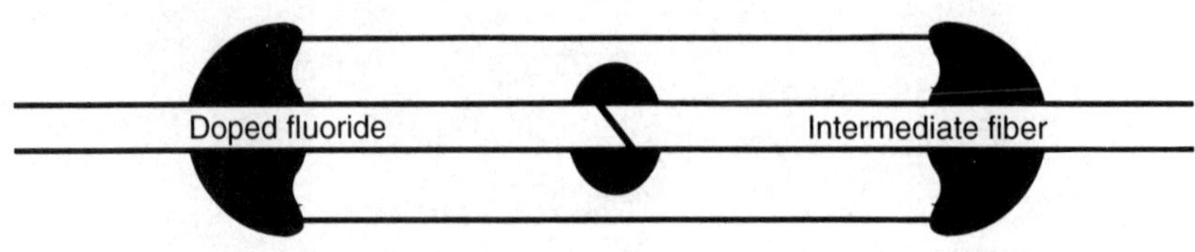

Figure 5.167 Angled-polish and active-alignment technique [333].

the angled-polish and active-alignment technique [333]. This connection procedure is almost the same as that used for tilted V-groove connection.

Furthermore, a low-loss connection can be obtained between the other end of the high NA silica fiber and a conventional single-mode fiber ($\Delta n \sim 0.3\%$) from a fiber coupler or an optical isolator using fusion splicing with the TEC technique [220,221,332]. Here, the connection between the fibers, made by the conventional fusion-splicing method, is directly heated by microburners using propane/oxygen to a temperature of about 1,700°C. Figure 5.168 shows the connection loss as a function of heating time [332]. After heating for several minutes, the connection loss decreases to less than 0.1 dB.

The use of these techniques has resulted in a low splicing loss of less than 0.3 dB and a reflection loss of less than −60 dB between Pr^{3+}-doped fluoride fiber with a high Δn and conventional silica fiber ($\Delta n \sim 0.3\%$) [332].

Pump light sources. As explained in Subsection 5.3.2.2, the LD operating at around 1,017 nm, the Nd:YLF solid-state laser operating at 1,047 nm, the Yb:YAG solid-state laser operating at 1,029 nm, and the Yb^{3+}-doped fiber laser operating from 1,010 to 1,030 nm are all candidates for the pump light source for PDF because of its wide pump wavelength region. Figure 5.169 shows the calculated output signal power of a PDFA as a function of launched pump power into the PDF for various pump wavelengths, which are related to the operating wavelength of each candidate pump light source [300]. This figure also includes the Nd:YAG solid-state laser that operates at 1,064 nm. In this calculation, the PDF length is optimized for each pump wavelength. The Δn of the PDF is 2.5%. To achieve an output signal power of over 16 dBm, the pump power must be 440 mW for the LD and the Yb^{3+}-doped fiber laser operating at around 1,017 nm. The corresponding pump powers are 470 mW for a Yb:YAG laser, 520 mW for an Nd:YLF laser, and 720 mW for an Nd:YAG laser.

Table 5.8 summarizes a comparison of several pump light sources for the PDFA module. The standard semiconductor laser is good in terms of both its lasing wavelength and device size, but it must provide high output power and reliability. In contrast, another semiconductor laser, the MOPA laser is very powerful and can exceed the required pump power at the optimum pump wavelength. A problem with this source may be reliability. On the other hand, an LD-pumped Nd:YLF laser can provide a high pump power, but this pump light source is larger than a standard

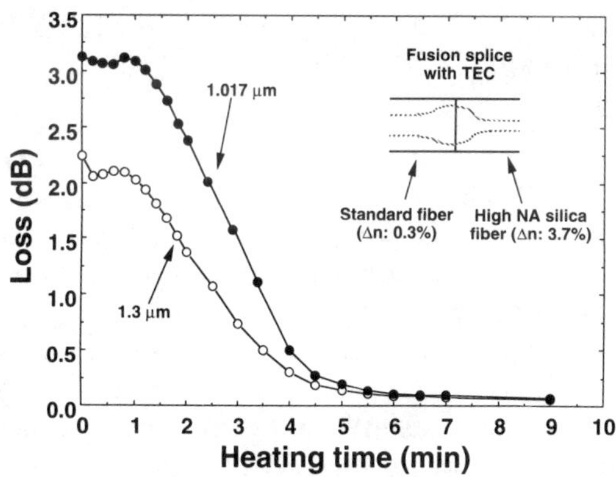

Figure 5.168 Connection loss as a function of heating time [332].

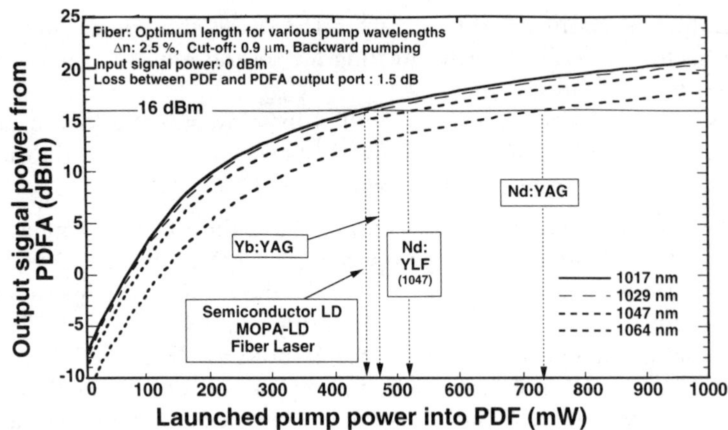

Figure 5.169 Calculated output signal power of a PDFA as a function of launched pump power for various pump wavelengths, which are related to the operating wavelength of each candidate pump light source [300].

LD module or a MOPA LD module. The next development targets for this laser are therefore to reduce its size to approximately that of a microchip solid-state laser and to move the operating wavelength closer to the optimum pump wavelength of 1017 nm by replacing the laser crystal with a Yb:YAG crystal. Furthermore, LD-pumped, Yb^{3+}-doped, and Nd^{3+}-Yb^{3+}-doped fiber lasers that are currently being developed are interesting because of their lasing wavelength.

Of these candidates, the LD module, MOPA-LD module, and Nd:YLF laser are the most advanced and are already commercially available. In addition, it is considered that the LD module is suitable for use in a preamplifier that requires a small size and a high signal gain and that the LD-pumped Nd:YLF laser is suitable for use in a booster amplifier with high output power. It may be possible to use the MOPA in the construction of both these amplifiers.

Various PDFA module configurations. In the first stage of PDFA module development, various configurations were investigated using the LD module [295,296,300, 329,332,334,335] or the LD-pumped Nd:YLF laser [298,299,333]. Figure 5.170(a–d) shows typical configurations, which comprise the single-pass [296,329,335], double-pass [334], cascade [332] with LD modules, and bidirectional pump with an Nd:YLF laser [298,299,333], respectively.

The single-pass configuration is the standard configuration for current PDFAs, and a signal gain of over 30 dB has been achieved using four pump LD modules with a pump power of about 100 mW and a polarization multiplexing module [296]. Recently, the pump power of the LD module has been increased to over 200 mW, and a PDFA module pumped by two LD modules has achieved a signal gain of over 30 dB with this configuration without a polarization multiplexing

Table 5.8
Comparison of Several Pump Light Sources for the PDFA Module

	Lasing Wavelength (μm)	Max. Output Power (into PDF) (mW)	(PDFA: 18 dBm) Required Power (mW)	Required Number	Size	Problem	Time Scale
Semiconductor-LD							
• Standard LD module*	1.017 (Optimum pump wavelength)	1~200	440	3~4	Small	High Power Reliability	Now in use
• MOPA-LD (Fiber-coupled)*		~600	440	1	Small	Reliability	Now in use
Solid-State Laser (LD Pump)							
• Nd:YLF Laser*	1.047	~800	520	1	Large	Size reduction (Micro chip solid laser structure)	Now in use
• Yb Laser	1.03	~700 (?)	470	1 (?)			Future
Fiber Laser (LD Pump)							
• Yb, Yb-Nd	1.00–1.15	470 (Pump 0.98 μm, 800 mW)	460 (1.017 μm)	1 (?)	Large (?)		Future

*indicates commercially available light source. *From*: [300].

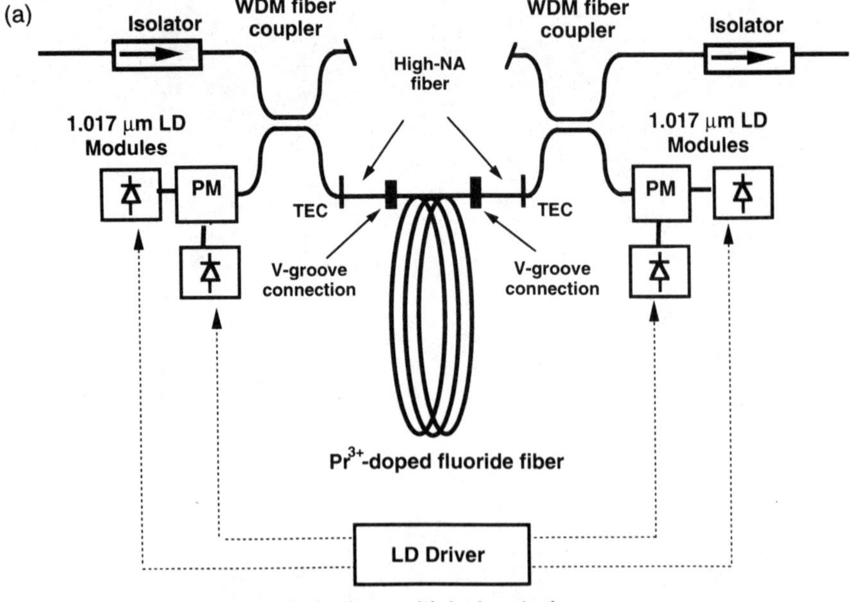

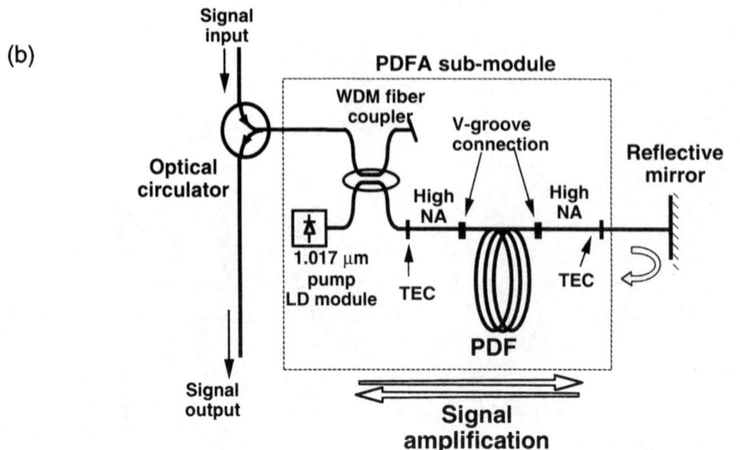

Figure 5.170 Typical PDFA module configurations: (a) single-pass configuration with four LD modules [296,329,335], (b) double-pass configuration with an LD module [334], (c) cascade with four LD modules [332], and (d) bidirectional pump configuration with a Nd:YLF laser [298,299,333].

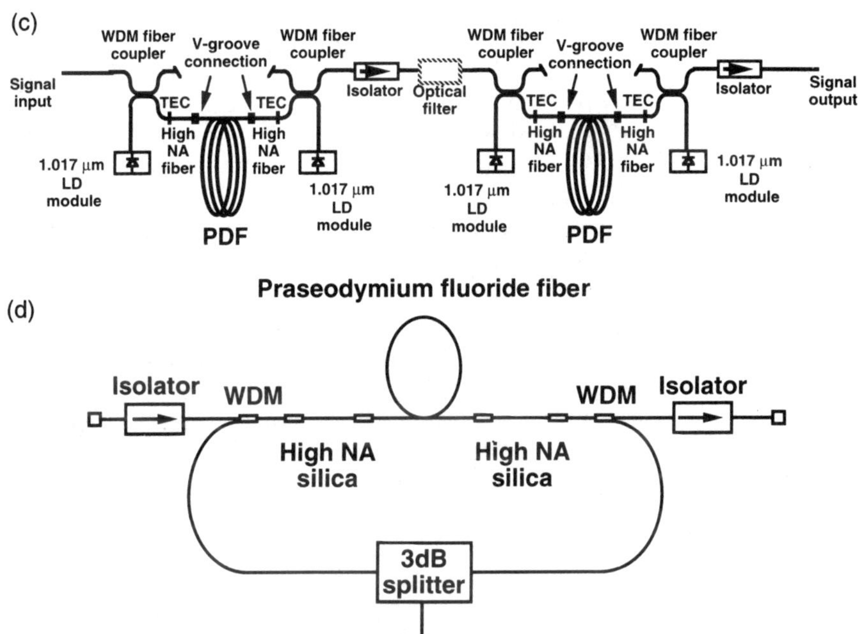

Figure 5.170 (continued).

module [301]. The double-pass configuration with an optical circulator and an endface reflective mirror is more practical for obtaining higher gain with lower pump power. The basic design idea is the same as that previously reported for the EDFA, as described in Subsection 5.3.1.5, "Improvement in Gain Characteristics With Advanced Configuration." In this configuration, both the signal and pump lights are reflected from the endface reflective mirror. The signal light is amplified in both the forward and backward directions through the PDF. Hence, its gain coefficient is twice that with the single-pass configuration. This configuration has been used to achieve the highest gain coefficient of 0.4 dB/mW, and a one-LD-pumped PDFA has been constructed with a 23-dB signal gain at a pump power of about 100 mW [334]. The cascade configuration, which incorporates an isolator and/or an optical filter in the PDF, yields stable amplification characteristics and a higher maximum gain by suppressing the ASE in the PDF. This configuration has also been proposed for the EDFA in order to realize high-gain and low-noise characteristics, as described in Subsection 5.3.1.5, "Improvement in Gain Characteristics With Advanced Configuration." The optical isolator is used to eliminate any ASE that back-propagates from the second unit to the first. An optical bandpass filter is used to reduce the ASE from the first unit to the second. Therefore, a significant improvement in signal gain characteristics can be achieved by using this cascade configuration to stop the build-up of the ASE. This configuration was used to construct the PDFA with a highest reported signal gain of 42 dB, which was realized by installing an isolator and an optical filter in the PDF [332]. Furthermore, the use of a bidirectional pump with a 3-dB splitter as a pump light laser is interesting with a view to realizing efficient pumping with one Nd:YLF laser. A PDFA module pumped by one Nd:YLF laser has been demonstrated with a signal gain of over 30 dB and an output signal power of 18 dBm [333].

However, these configurations may induce extra noise due to interaction between LD modules or by the optical light fed back to the LD module and Nd:YLF laser. When pump sources are injected with optical feedback light, or light from another source that operates at almost the same wavelength, the pump light becomes unstable and extra noise is induced in the radio frequency region [336]. In particular, this noise degrades the picture quality in a subcarrier multiplexed multichannel AM-VSB video signal transmission, which is an important application of the PDFA module [336,337]. This problem will be solved by attaching an optical isolator to the pump light source. In fact, recently, an optical isolator has been demonstrated with a low loss of less than 0.5 dB for the PDF pump light wavelength region [34]. However, a pump light source with an isolator has yet to be fabricated.

To avoid such problems, recently fabricated practical PDFA modules basically employ unidirectional pumping, although bidirectional pumping is attractive in terms of efficiency [300,301,338–340]. Figures 5.171 and 5.172 show photographs, configurations, and gain characteristics of recently constructed practical and commercially available Nd:YLF-pumped and MOPA LD-pumped PDFA modules, respectively. The PDFA module with two Nd:YLF lasers has a cascade configuration with an optical isolator [339]. The optical isolator prevents any interaction between the two Nd:YLF lasers, as shown in Figure 5.171. This module has a maximum signal gain of 40.6 dB at 1.30 μm, a maximum output signal power of 20.1 dB at an input signal power of 0 dBm, and a NF of 5 dB. The module with a MOPA LD, as shown in Figure 5.172, employs backward pumping [340]. This module has a maximum signal gain of 30.5 dB at 1.30 μm and a NF of 5.5 dB. The signal output power is 18.5 and 15 dBm for signal input powers of 10 and 0 dBm, respectively.

A practical PDFA module with two Nd:YLF lasers has been used for a 40-channel AM-VSB video signal transmission [300,339,341] and a 10-Gbps digital transmission system [300,339]. Table 5.9 shows the measured CNR and three types of distortion: CSO, *composite triple beat* (CTB), and *cross modulation* (XM), for a 40-channel AM-VSB video signal transmission with and without the PDFA [339]. The signal source for this system was a DFB laser operating at 1.302 μm that was directly modulated by 40-channel carriers ranging from 91.25 to 403.25 MHz with a modulation depth of 7.1%/ch. As shown in this table, the distortion values almost meet CATV trunk line specifications. The degradation in picture quality monitored on a television set, before and after PDFA amplification, was imperceptible in subjective tests. It has been found that the PDFA improves the loss budget by

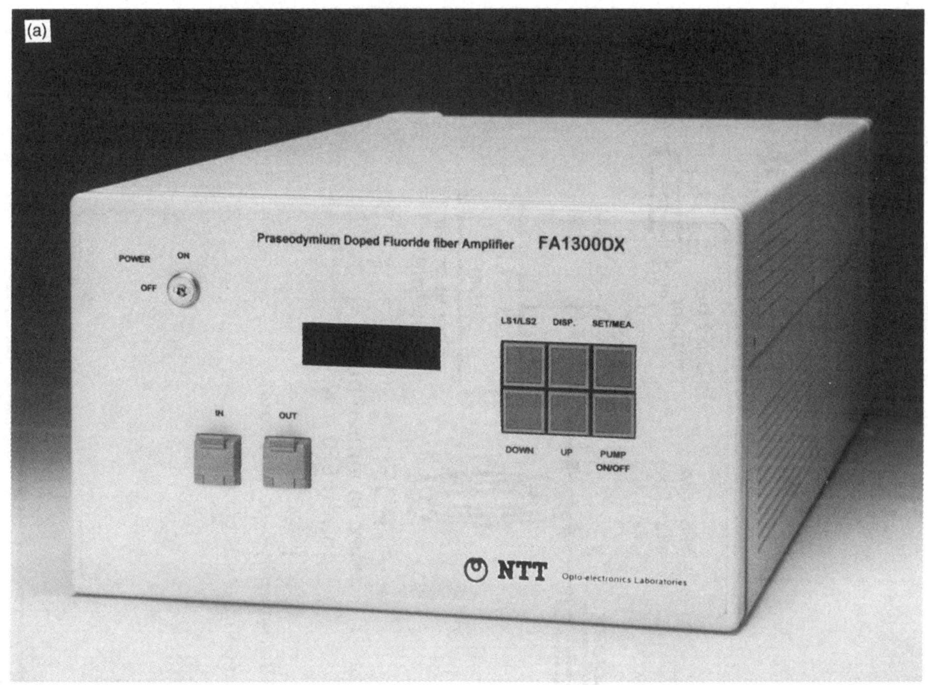

Figure 5.171 Nd:YLF-pumped PDFA module [339]: (a) photograph, (b) configuration, and (c) gain characteristics.

10 dB for 52-dB CNR and distortion levels satisfy the CATV trunk line specifications, as shown in Figure 5.173 [341]. Figure 5.174 shows the BER performance for a 10-Gbps digital signal transmission with the PDFA module used as a booster amplifier [339]. The transmitter was a strained MQW DFB laser operating at 1.315 μm. The laser was directly modulated using an NRZ pseudorandom signal with a pattern length of 2^{23-1}. An optical signal from the optical transmitter was fed to a 110.8-km standard single-mode fiber through a PDFA. The transmitted signal was received by an APD with an average received power of −25.4 dBm. In Figure 5.174, the white circles indicate the performance for a 2-m-long fiber transmission, and the black circles plot results for a 110.8-km-long fiber transmission. We obtained a total system power budget of 39.1 dB, and it is clear that a 10-Gbps transmission of more than 110 km was successfully achieved with the PDFA. From the above results, it has been confirmed that this PDFA module performs well for both analog and digital transmission.

Prospects for PDFA

A major problem with regard to improving the performance of PDFAs is to increase the quantum efficiency of 1.3-μm amplification. If the quantum efficiency can be

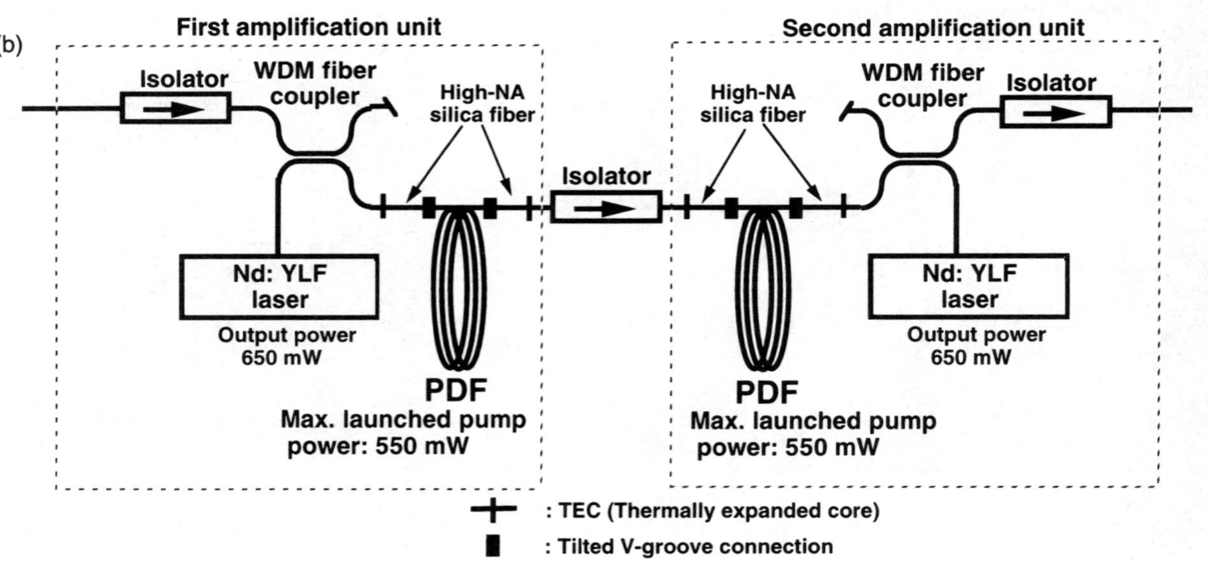

Figure 5.171 (continued).

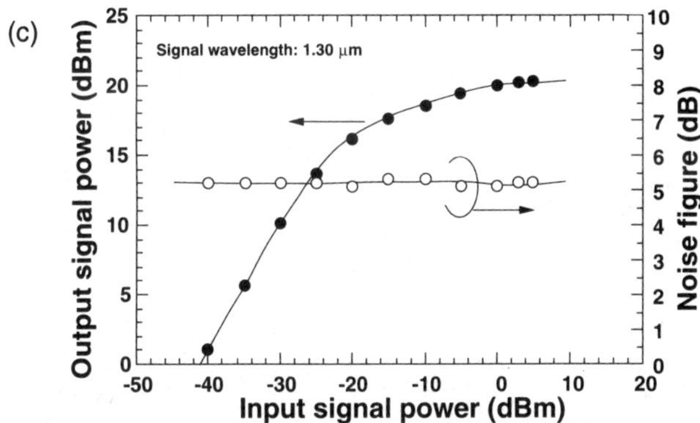

Figure 5.171 (continued).

significantly improved, the load that the pump light source has to bear will decrease. Multiphonon relaxation from the $^1G_4$ level, which is the initial level of the $^1G_4$-$^3H_5$ stimulated transition for 1.3-μm-band amplification to the lower lying $^3F_4$ level, reduces the gain coefficient. The most effective approach to increasing the gain coefficient is to use Pr^{3+} host materials whose phonon energy is as low as possible because the multiphonon relaxation rate is inversely related to the number of phonons required to bridge the energy gap between the $^1G_4$ and $^3F_4$ levels [288,317].

Table 5.10 compares the spectroscopic parameters of the $^1G_4$-$^3H_5$ transition of Pr^{3+} ions in several glasses [288,317,342–348]. The quantum efficiency has been investigated in InF_3-based fluoride glasses, mixed halide glasses and chalocogenide glasses. The $^1G_4$ lifetime in chlorine-doped ZBLAN and InF_3-based fluoride glasses increased to 150 to 180 μs. Therefore, the radiative quantum efficiency of a 1.3-μm emission transition given by $\eta \times \beta$ is roughly double that in *ZrF₄-based fluoride glass* (ZBLAN). On the other hand, the quantum efficiency in chalcogenide glasses is almost 20 times larger than that in ZBLAN. A noticeable improvement is expected in the quantum efficiency if chalcogenide glasses are used as Pr^{3+} ion hosts. Figure 5.175 shows the gain coefficient for *Pr^{3+}-doped InF₃-based fluoride fiber* (IBSPZ) and *chalcogenide glass fiber* (La-Ga-S) compared with that of ZBLAN [317]. A Δn of 3.7% was assumed in the calculations. With the InF_3-based fluoride fiber, gain coefficients of more than 0.4 dB/mW can be expected. Gain coefficients greater than 2 dB/mW are achievable using chalcogenide fiber. As the highest gain coefficient ever achieved using ZrF_4-based fluoride fibers is 0.24 dB/mW, the value can be increased by more than double and up to 10 times using these new hosts. So the development of these new hosts is the next major target of PDFA research.

Recently, a Pr^{3+}-doped InF_3-based single-mode fluoride fiber was successfully fabricated using IBSPZ, InF_3/GaF_3, and PbF_2/InF_3 glasses. Both the InF_3/GaF_3 and PbF_2/InF_3-based single-mode fibers with extra-high Δn values of 6% have produced particularly notable data such as high signal gain coefficients of over

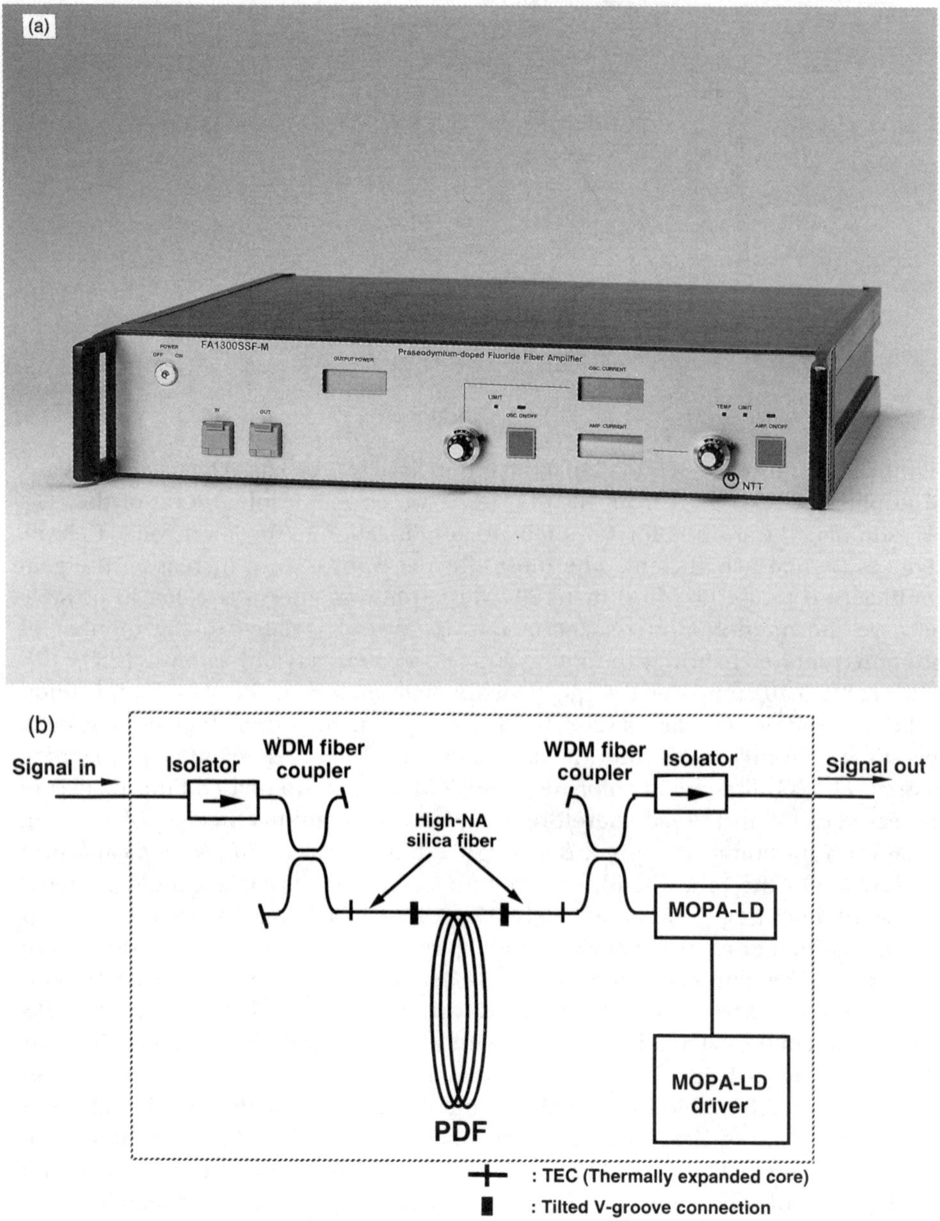

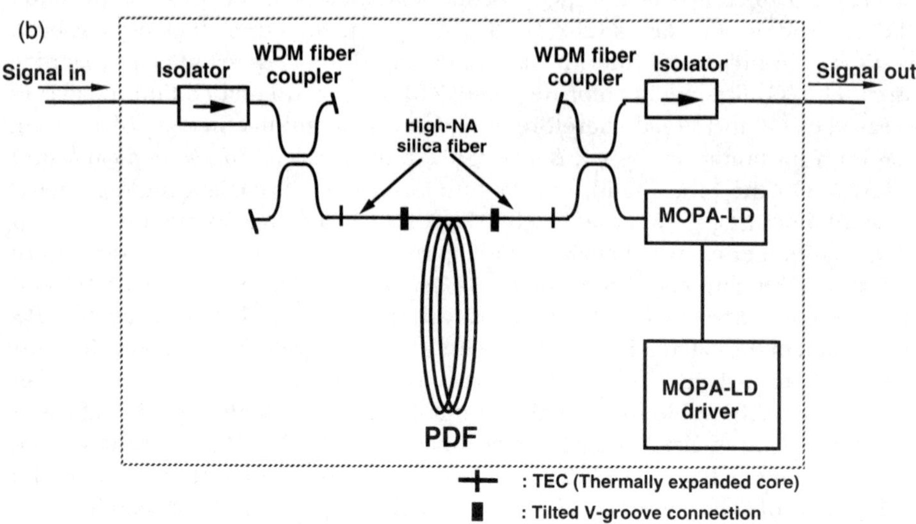

Figure 5.172 MOPA LD-pumped PDFA module [340]: (a) photograph, (b) configuration, and (c) gain characteristics.

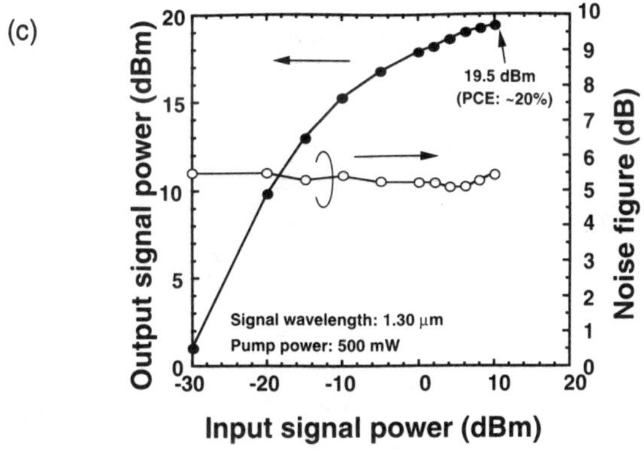

Figure 5.172 (continued).

Table 5.9

Measured Carrier-to-Noise Ratio and Three Types of Distortion for a 40-Channel AM-VSB Video
Signal Transmission With and Without a PDFA

| | *Carrier Frequency (MHz)* | | | |
| | *91.25* | *211.25* | *403.25* | |
	With PDFA	*With PDFA*	*Without PDFA*	*With PDFA*
CNR (dB)	51.6	51.3	55.9	52.1
CSO (dB)	71.7	69.2	63.5	66.5
CTB (dB)	70.2	68.4	69.9	67.8
XM (dB)	71.5	65.6	63.7	60.2

Optical modulation depth: 7.1%. *From:* [339].

0.3 dB/mW, although their scattering losses are not low [345–347]. However, these
fibers may have problems with regard to splicing them to silica fiber because of
their extra-high Δn. On the other hand, Pr^{3+}-doped AsS-based chalcogenide glass
fiber has been prepared that achieve fluorescence at 1.3 μm [348]. However, there
remains the problem of how to reduce the scattering loss.

5.3.3 1.4-μm- and 1.65-μm-Band Fiber Amplifiers

In addition to the practical 1.3-μm- and 1.5-μm-band fiber amplifiers described
in Subsections 5.3.1 and 5.3.2, which have already been developed, 1.4-μm- and
1.65-μm-band fiber amplifiers must be developed to make better use of the vast
bandwidth of optical fiber. Specifically, the 1.65-μm-band amplifier may be used
for an optical transmission line-monitoring system in the 1.65-μm band [349,350].

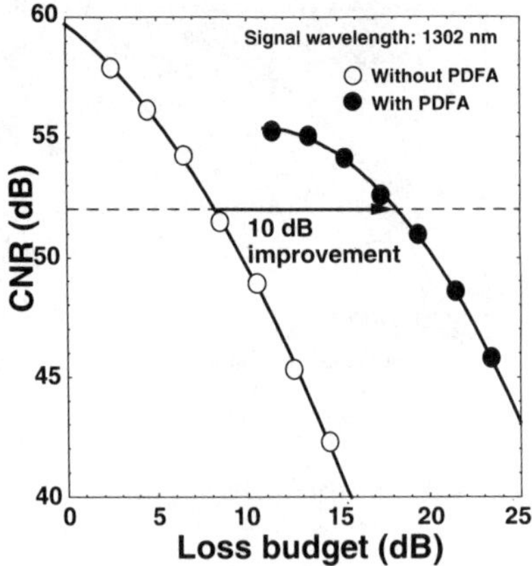

Figure 5.173 Carrier-to-noise ratio against loss budget [341].

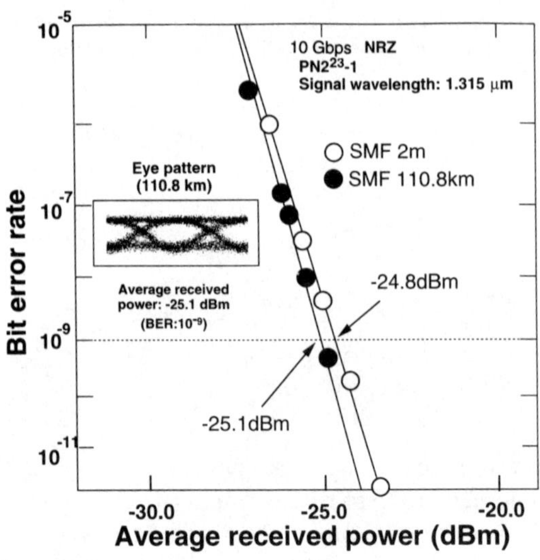

Figure 5.174 Bit error-rate performance for a 10-Gbps digital signal transmission with a PDFA module used as a booster amplifier [339].

A *Tm³⁺-doped fiber amplifier* (TDFA) has been investigated with a view to amplifying 1.4- and 1.65-μm signals. The efficient amplification of both bands has been demonstrated by employing ZrF_4-based fluoride glass as the Tm^{3+}-doped fiber host.

In this section, we describe the principle of the Tm^{3+}-doped ZrF_4-based fluoride fiber amplifier for 1.4-μm- and 1.65-μm-band amplification. Then we introduce the basic amplification characteristics of 1.4-μm- and 1.65-μm-band TDFAs.

Table 5.10
Spectroscopic Parameters of the $^1G_4$-$^3H_5$ Transition of Pr^{3+} Ions in Several Glasses

	τ_4 (μs)	σ_{42} (10^{-21} cm^2)	η (%)	B	$\eta \times B$	*Cite*
ZrF$_4$-based fluoride						
ZBLAN	110	3.48	3.4	0.64	2.18	[288,317,342]
Mixed halide						
ZBLAN-Cl	180	3.75	6.5	0.61	1.98	[317,342]
InF$_3$-based fluoride						
IZBSC	150	3.79	5.1	0.65	3.96	[317,342]
IBSPZ	180	3.90	6.2	0.65	3.32	[317,342]
InF$_3$/GaF$_3$	186	4.40	8.6	0.64	5.50	[345]
PbF$_2$/InF$_3$	172	3.89	6.1	0.67	4.08	[346,347]
Chalcogenide						
La-Ga-S	300	10.5	60	0.6	36	[343]
As-S	250					[348]
Ge-Ga-S	354	13.3	71.7	0.57	40.87	[348]

τ_4: lifetime
η: radiative quantum efficiency
B: branching ratio

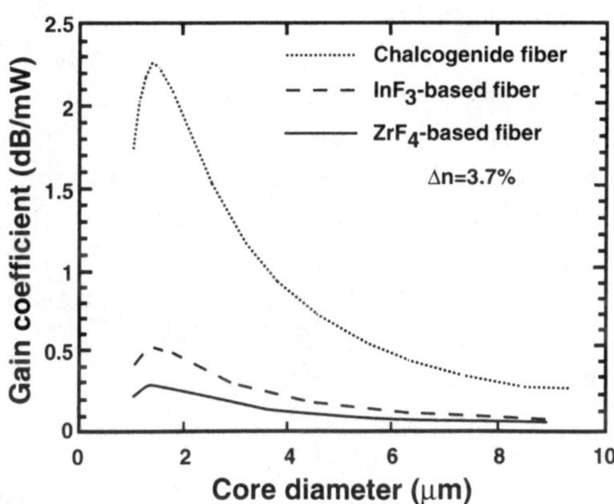

Figure 5.175 Gain coefficient for Pr^{3+}-doped InF$_3$-based fluoride fiber and chalcogenide glass fiber compared with that of Pr^{3+}-doped ZrF$_4$-based fluoride fiber [317].

5.3.3.1 1.4-μm- and 1.65-μm-Band TDFA Amplification Principle

Figure 5.176 shows the Tm^{3+} ion energy diagram. 1.4-μm-band amplification is based on the four-level transition from the $^3H_4$ level to the $^3F_4$ level, and 1.6-μm-band amplification is based on the three-level transition from the $^3F_4$ level to the $^3H_6$

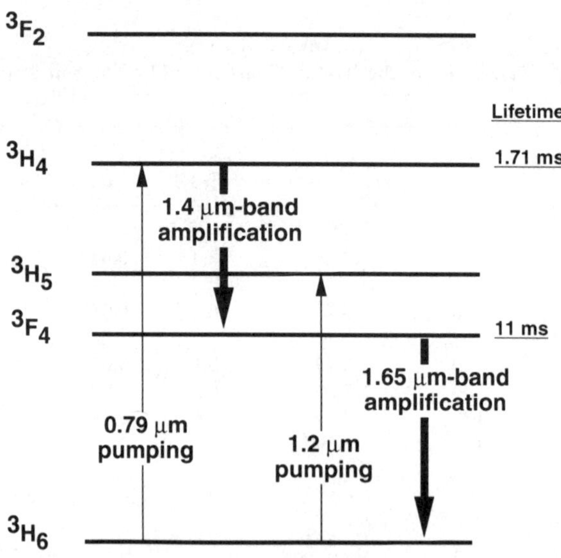

Figure 5.176 Tm^{3+} ion energy diagram.

level [351–353]. The ZrF_4-based fluoride glasses are used as fiber materials in order to obtain efficient amplification by suppressing the nonradiative transition from the $^3H_4$ upper level to the $^3H_5$ level caused by multiphonon relaxation for 1.4-μm-band amplification and by increasing the fluorescence lifetime of the 3F$_4$ upper level to ~10 ms from ~0.5 ms in silica glass for 1.6-μm-band amplification. However, there are problems that must be overcome in order to achieve practical TDFAs for both 1.4-μm- and 1.65-μm-band amplification. In the following, these problems and the approaches to overcoming them for both amplification bands will be described.

The problem with the 1.4-μm-band Tm^{3+}-doped fiber amplifier is that 1.4-μm-band amplification is generally limited by the fact that the lifetime of the $^3H_4$ upper level (1.7 ms) is shorter than that of the $^3F_4$ lower level (11 ms), which is a "self-terminating system." Therefore, it is difficult to form a population inversion between the $^3H_4$ upper level and the $^3F_4$ lower level. The lower level must be depopulated in order to achieve a high gain.

Three approaches have been proposed to solve the problem of depopulation. One is colasing with the $^3F_4$-$^3H_6$ transition, which corresponds to an emission of 1.8 μm [351,354,355]. In this approach, the population density of the $^3F_4$ lower level is reduced by the stimulated emission of the laser oscillation around 1.8 μm. The second and third approaches are a 1.06-μm-band upconversion pumping method [356] and the codoping of acceptor ions [357–361], as shown in Figure 5.177. As regards upconversion pumping using a Nd:YAG laser operated at

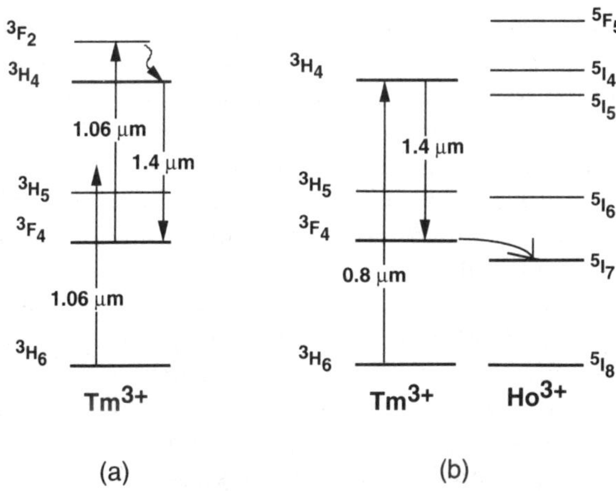

Figure 5.177 (a) 1.06-μm-band upconversion pumping method [356] and (b) the codoping acceptor ions [357].

1.06 μm, the ground state ^{3}H$_6$ ions are excited to the ^{3}H$_5$ level and decay to the lower ^{3}F$_4$ level. Then the Tm^{3+} ions excited in the lower ^{3}F$_4$ level are re-excited to the upper ^{3}H$_4$ level through the nonradiative transition from the ^{3}F$_2$ level. As a result the 3F$_4$ level is depopulated of Tm^{3+} ions that are excited to the ^{3}H$_4$ level. In this upconversion pumping method, the 1.4-μm-band signal is amplified like a three-level amplification system that uses the lower ^{3}F$_4$ level as the virtual ground state. With codoping, the ^{3}F$_4$ level is depopulated of excited Tm^{3+} ions as a result of the energy transfer to the acceptor ions, as observed from the ^{3}F$_4$ level of Tm^{3+} ions to the ^{5}I$_7$ level of Ho^{3+} ions. The Ho^{3+} ion is one of the most efficient acceptor ions because Ho^{3+} ions shorten the lifetime of the 3F$_4$ level of Tm^{3+} but do not shorten the lifetime of the ^{3}H$_4$ level [357]. It has been reported that, by adding 1 wt% of Ho^{3+} ions to a ZrF$_4$-based fluoride glass doped with 0.05 wt% of Tm^{3+}, the lifetime of the ^{3}F$_4$ lower level is greatly reduced from 11 to 1.2 ms while the lifetime of the ^{3}H$_4$ level is slightly reduced from 1.71 to 1.48 ms [357]. The Tb^{3+} ions are also used as acceptor ions, since the Tm^{3+} ions excited in the ^{3}F$_4$ level are transferred to the ^{7}F$_0$ level of the Tb^{3+} ions. It has been reported that the lifetime of the ^{3}H$_4$ upper and the ^{3}F$_4$ lower level of Tm^{3+} ions are 0.5 and 0.43 ms, respectively, with 0.1 wt% Tm^{3+} and 1 wt% Tb^{3+} dopant concentrations [358]. The advantage of this codoping method is that a high-power commercially available 0.8-μm-band LD can be used as a pump source. Of the above three approaches, the 1.06-μm-band upconversion pumping method and the codoping of Ho^{3+} ions have both already been applied to 1.4-μm-band TDFAs in order to obtain efficient 1.4-μm amplification.

By contrast, the problem with the 1.65-μm-band Tm^{3+}-doped fiber amplifier that is pumped in the 1.2-μm pump band (^{3}H$_6$ → ^{3}H$_5$ transition) is that

1.65-μm-band amplification is limited by the large ASE or laser oscillation around 1.8 μm [362,363]. Figure 5.178 shows the absorption and emission cross sections of the $^3F_4$-$^3H_6$ transition for Tm^{3+} ions in ZrF_4-based fluoride glass [364,365]. The Tm^{3+} ion has a wide emission cross section from 1.6 to 2.0 μm. The high-gain region is from 1.75 to 2.0 μm, where the emission cross section is larger than the absorption cross section. Therefore, 1.65-μm-band amplification is always accompanied by a large ASE and often by laser oscillation around 1.8 μm. This greatly suppresses the signal gain of the 1.6-μm-band signal.

The doping of Tb^{3+} ions in the cladding has been proposed as a way of overcoming these problems [364,365]. The energy level diagram and absorption spectrum of Tb^{3+} ions are shown in Figure 5.179(a,b), respectively. Tb^{3+} ions have a large absorption from 1.75 to 2.0 μm, which is related the $^7F_6$-$^7F_0$ transition, and a small absorption around 1.6 μm. Therefore, the large ASE and laser oscillation are suppressed and efficient amplification becomes possible by doping the cladding with Tb^{3+} ions. The cladding is doped to avoid any crossrelaxation between Tm^{3+} ions in the $^3F_4$ level and Tb^{3+} ions. Figure 5.180 shows the ASE spectrum of TDF with Tb^{3+}-doped cladding. The Tm^{3+}-ion and Tb^{3+}-ion concentrations are 0.2 wt% and 0.4 wt%, respectively [323]. The peak of the ASE from the Tm^{3+} and TDF shifted to a shorter wavelength than that from TDF. This result shows that the Tb^{3+} ions in the cladding effectively absorb the unwanted ASE around 1.8 μm.

In the following section, we describe the basic amplification characteristics of a 1.4-μm-band TDFA that employs the 1.06-μm-band upconversion pumping method or the codoping of Ho^{3+} ions and also those of a 1.65-μm-band TDFA with

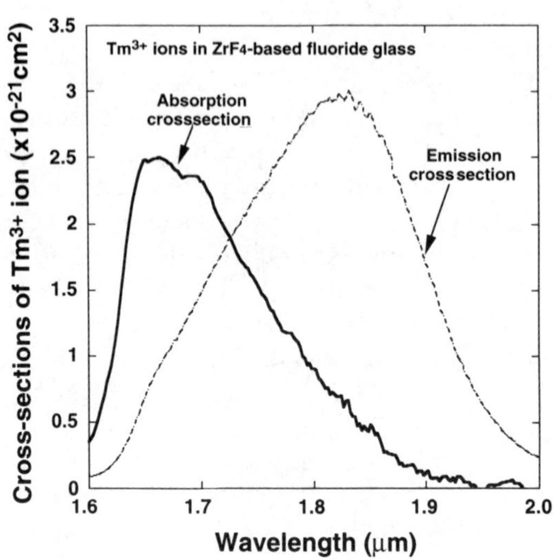

Figure 5.178 Absorption and emission cross sections of the $^3F_4$-$^3H_6$ transition for Tm^{3+} ions in ZrF_4-based fluoride glass [364,365].

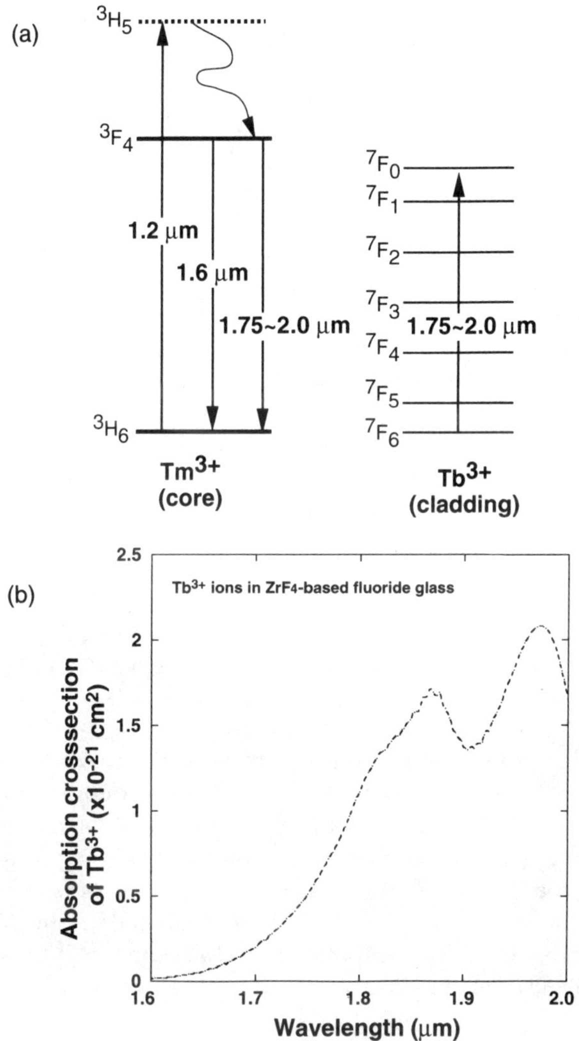

Figure 5.179 (a) Energy level diagram and (b) absorption spectrum of Tb^{3+} ions [364,365].

Tb^{3+}-doped cladding. Any reader interested in the calculation model of the TDFA can obtain the information by referring to [356] for 1.4-μm-band amplification, and [365] for 1.65-μm-band amplification.

5.3.3.2 Basic Amplification Characteristics of 1.4-μm-Band TDFA

In this section, we describe and compare the basic amplification characteristics of a 1.4-μm-band TDFA with either the 1.06-μm-band upconversion pumping method or the codoping of Ho^{3+} ions.

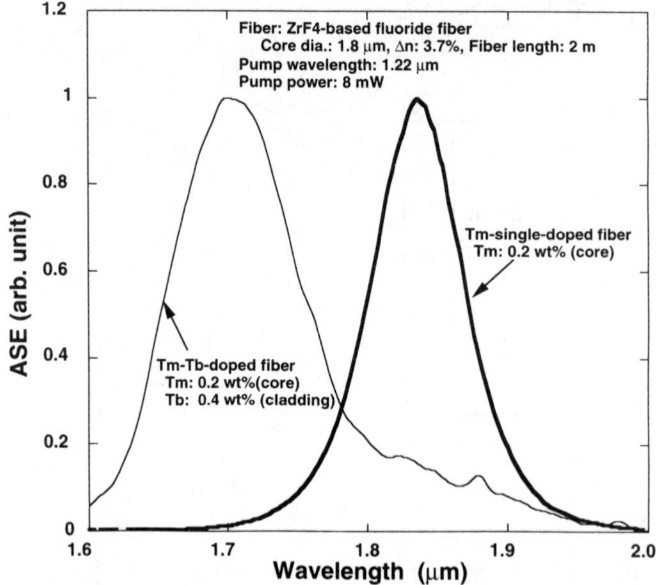

Figure 5.180 ASE spectrum of Tm^{3+}-doped ZrF_4-based fluoride fiber with Tb^{3+}-doped cladding [364].

Figure 5.181 shows the gain spectra of a TDFA with upconversion pumping with 2,000 ppm of Tm^{3+} in the core and a Tm^{3+}-Ho^{3+}-doped fluoride fiber amplifier with 500 ppm of Tm^{3+} and 10,000 ppm of Ho^{3+} in the core [366]. Both amplifiers have wide gain spectra that cover the entire 1.4-μm band. These spectra are much wider than the gain spectrum from the 0.79-μm pumped TDFA because the Tm^{3+} ions pumped in the lower $^3F_4$ level are effectively depopulated for the upconversion amplifier and the Tm^{3+}-Ho^{3+}-doped amplifier; therefore, the signal absorption from the $^3F_4$ lower level to the $^3H_4$ upper level for 1.4-μm-band amplification is much smaller than with the 0.79-μm pumped TDFA.

Figure 5.182(a,b) shows the noise spectra of 1.4-μm-band TDFAs with the 1.06-μm-band upconversion pumping method or codoping with Ho^{3+} ions [356,357]. The NF exhibits its minimum value at around 1.48 μm for both amplifiers. The increase in the NF for signal wavelengths below 1.5 μm is due to the signal reabsorption from the $^3F_4$ lower level to the $^3H_4$ upper level, while the increase for the region above 1.50 μm is due to the GSA from the $3H_6$ ground level to the $3F_4$ level. When the noise characteristics of the two amplifiers are compared, we find that the NF of the Tm^{3+}-Ho^{3+}-doped fluoride fiber amplifier is higher than that of the upconversion amplifier because of the signal reabsorption from the $^3F_4$ lower level to the $^3H_4$ upper level with the Tm^{3+}-Ho^{3+}-doped fluoride fiber amplifier. By contrast, with the upconversion amplifier, there is almost no signal reabsorption

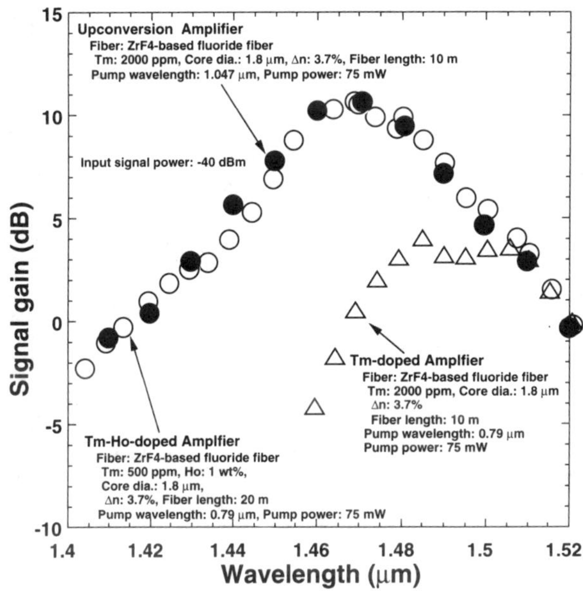

Figure 5.181 Gain spectra of a Tm^{3+}-doped fluoride fiber amplifier with upconversion pumping with 2,000 ppm of Tm^{3+} in the core, and a Tm^{3+}-Ho^{3+}-doped fluoride fiber amplifier with 500 ppm of Tm^{3+} and 10,000 ppm of Ho^{3+} in the core [366].

because almost all of the Tm^{3+} ions in the $^3F_4$ lower level are re-excited to the $^3H_4$ upper level.

Figure 5.183 shows the small signal gain characteristics of 1.4-μm-band TDFAs with the 1.06-μm-band upconversion pumping method or codoped with Ho^{3+} ions [366]. The gain efficiencies are almost the same for both amplifiers. The crucial difference is that the gain increases with increasing pump power for the upconversion amplifier, while the gain saturates for the Tm^{3+}-Ho^{3+}-doped amplifier with 500 ppm of Tm^{3+} and 10,000 ppm of Ho^{3+}. This saturation of the Tm^{3+}-Ho^{3+}-doped amplifier is due to the large ASE in the 0.8-μm band ($^3H_4$-$^3H_6$ transition). With the Tm^{3+}-Ho^{3+}-doped amplifier, direct pumping to the $^3H_4$ level by 0.79-μm pump light causes a population inversion between the $^3H_4$ level and the $^3H_6$ ground state, and the ASE grows in the 0.8-μm band. With 1.06-μm upconversion pumping, the excitation from the $^3H_6$ ground state to the $^3H_5$ level is very small, and almost all the ions remain in the $^3H_6$ ground state. It is hard to form a population inversion between the $^3H_4$ upper level and the $^3H_6$ ground state, so the problematic ASE does not occur in the 0.8-μm region.

The signal output power of the upconversion amplifier is much larger than that of the Tm^{3+}-Ho^{3+}-doped fiber amplifier. The output power of the upconversion amplifier is 15 dBm when the signal input power is 0 dBm and the pump power

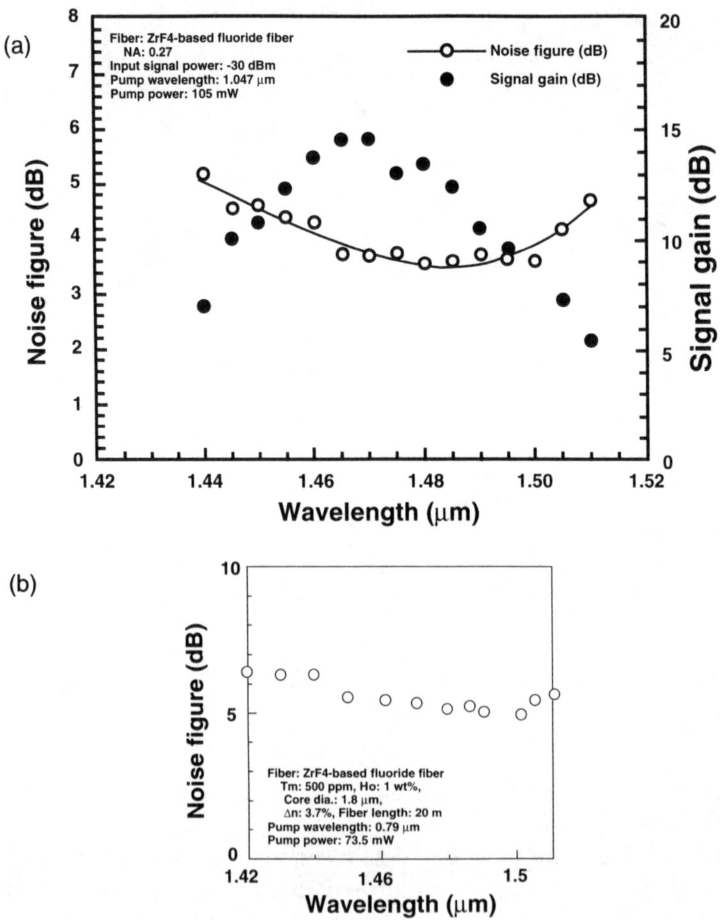

Figure 5.182 Noise spectra of 1.4 μm-band TDFA: (a) 1.06-μm band upconversion pumping method [356] and (b) codoping method with Ho^{3+} ions [357].

is 150 mW, while that of Tm^{3+}-Ho^{3+}-doped fiber amplifier is 4.5 dBm. With regard to the Tm^{3+}-Ho^{3+}-doped fiber amplifier, the large ASE in the 0.8-μm band consumes the pump energy; therefore, the output power is small [366]. Given these results, it is confirmed that the upconversion amplifier achieves a large gain, wide gain spectrum, large output signal, and low noise. The LD-pumped Nd:YAG laser operated at 1.06 μm makes it possible to construct practical 1.4-μm-band TDFA modules. On the other hand, the Tm^{3+}-Ho^{3+}-doped fiber amplifier, which has the advantage that a commercial high-power LD can be used as a pump source, has the disadvantage that the large ASE around 0.8 μm must be suppressed in order to construct an efficient 1.4-μm-band TDFA.

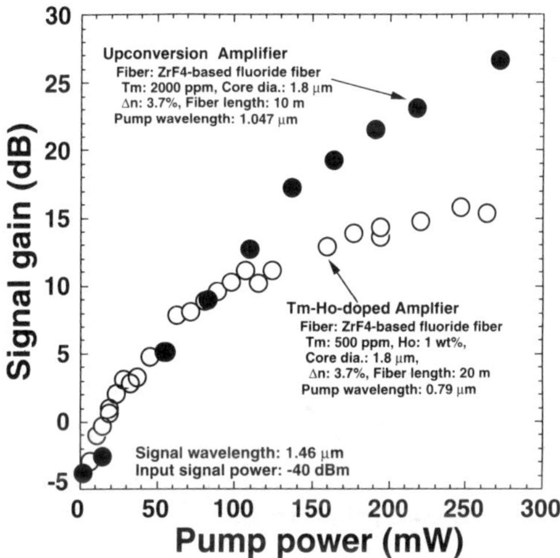

Figure 5.183 Small signal gain characteristics of 1.4-μm-band TDFAs with the 1.06-μm-band upconversion pumping method or codoped with Ho^{3+} ions [366].

5.3.3.3 Basic Amplification Characteristics of 1.65-μm-Band TDFA

In this section, we describe the gain characteristics of the 1.65-μm-band TDFA with Tb^{3+}-doped cladding.

Figure 5.184 shows the calculated signal gain and NF of the amplifier at a signal wavelength of 1.65 μm, when the Tb^{3+} concentration in the cladding is varied for ZrF$_3$-based fluoride fiber with a Tm^{3+} concentration of 0.2 wt% in the core [364–366]. In this calculation, the fiber lengths are optimized for each Tb^{3+} concentration. As the Tb^{3+} ion concentration increases, the gain increases up to a certain level and then decreases. This result confirms that efficient amplification is possible by doping Tb^{3+} ions in the cladding to suppress the ASE in the 1.75- to 2.0-μm region but that too much doping reduces the gain because of the signal loss caused by the Tb^{3+} ions. It is found that the optimum Tb^{3+} concentration for achieving the highest small signal gain is 0.4 wt%. Furthermore, the NF increases with increases in the Tb^{3+} ion concentration due to the signal loss caused by the Tb^{3+} ions.

By using the two-stage-type fiber amplifier configuration shown in Figure 5.185(a), the highest signal gain yet reported of 35 dB at a signal wavelength of 1.65 μm has been achieved with ZrF$_3$-based fluoride fiber with a Tm^{3+} concentration of 0.2 wt% in the core and a Tb^{3+} concentration of 0.4 wt%, as shown in Figure 5.185(b) [364–366]. However, the NF of this amplifier is still high at 8 dB, which is due to the signal loss caused by the Tb^{3+} ions in the cladding because the Tb^{3+} concentration of this amplifier is optimized only for the signal gain. So to realize

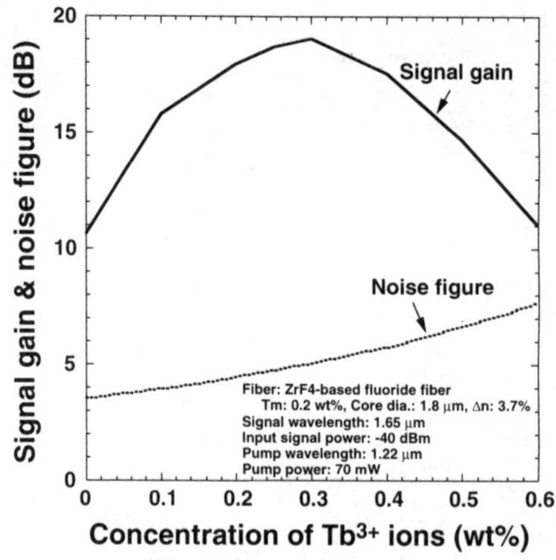

Figure 5.184 Calculated signal gain and noise figure of a 1.65-μm-band TDFA at a signal wavelength of 1.65 μm, when the Tb^{3+} concentration in the cladding is varied for ZrF_3-based fluoride fiber with a Tm^{3+} concentration of 0.2 wt% in the core [364–366].

a practical high-gain TDFA with low noise for a 1.65-μm-band signal, the Tb^{3+} concentration must be optimized for both the signal gain and noise characteristics.

References

[1] Nishi, S., K. Aida, and K. Nakagawa, "Highly Efficient Configuration of Erbium-Doped Fibre Amplifiers," *Proc. European Conf. on Optical Communication, ECOC'90*, Vol. 1, 1990, pp. 99–103.

[2] Lauridsen, V., R. Tadayoni, A. Bjaklev, J. H. Povlsen, and B. Pedersen, "Gain and Noise Performance of Fiber Amplifiers Operating in New Pump Configurations," *Electron. Lett.*, Vol. 27, 1991, pp. 327–328.

[3] Russell, P. St. J., and R. Ulrich, "Grating-Fiber Coupler Is a High Resolution Spectrometer," *Opt. Lett.*, Vol. 10, 1985, pp. 291–293.

[4] Sorin, W. V., and H. J. Shaw, "A Single-Mode Fiber Evanescent Grating Reflector," *IEEE J. Lightwave Tech.*, Vol. LT-3, 1985, pp. 1041–1043.

[5] Bennion, I., D. C. J. Reid, C. J. Rowe, and W. J. Stewart, "High-Reflectivity Monomode-Fibre Grating Filters," *Electron. Lett.*, Vol. 22, 1986, pp. 341–343.

[6] Hill, K. O., Y. Fujii, D. C. Johnson, and B. S. Kawasaki, "Photosensitivity in Optical Fiber Waveguides: Application to Reflection Filter Fabrication," *Appl. Phys. Lett.*, Vol. 32, 1978, pp. 647–649.

[7] Meltz, G., W. W. Monrey, and W. H. Glenn, "Formation of Bragg Gratings in Optical Fibers by a Transverse Holographic Method," *Opt. Lett.*, Vol. 14, 1989, pp. 823–825.

[8] Kashyap, R., J. R. Armitage, R. Wyatt, S. T. Davey, and D. L. Williams, "All-Fiber Narrowband Reflection Gratings at 1500 nm," *Electron. Lett.*, Vol. 26, 1990, pp. 730–732.

[9] Hill, K. O., B. Malo, F. Bilodeau, D. C. Johnson, and J. Albert, "Bragg Gratings Fabricated in Monomode Photosensitive Optical Fiber by UV Exposure Through a Phase Mask," *Appl. Phys. Lett.*, Vol. 62, 1993, pp. 1035–1037.

[10] Anderson, D. Z., V. Mizrahi, T. Erdogan, and A. E. White, *Proc. Optical Fiber Communication, OFC'93*, San Jose, 1993, Paper PD16.

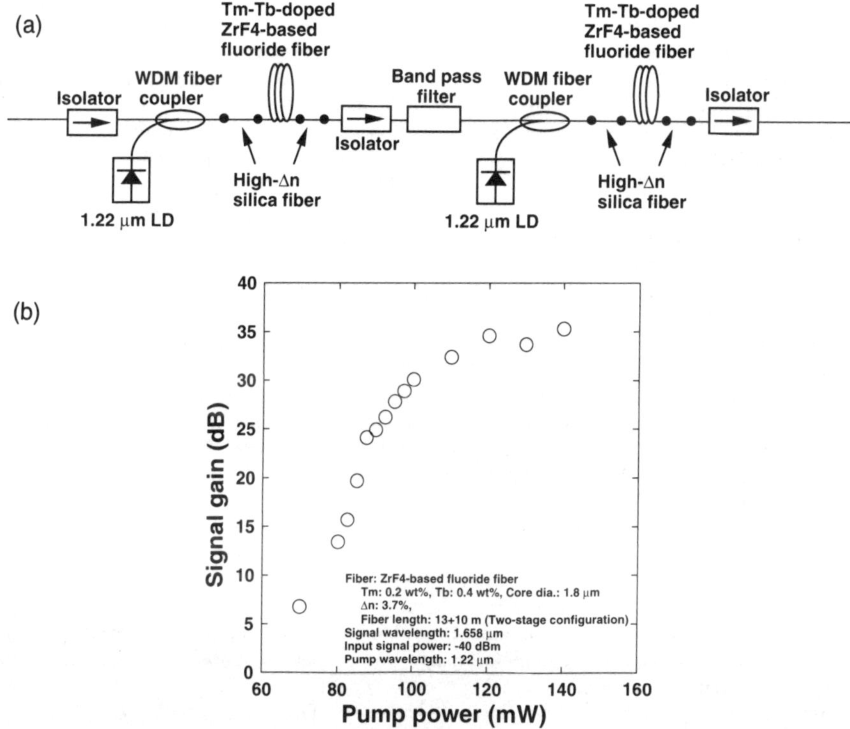

Figure 5.185 (a) Two-stage-type 1.65-μm-band TDFA configuration and (b) its gain characteristics [364–366].

[11] Armitage, J. R., "Fiber Bragg Reflectors Written at 262 nm Using a Frequency Quadrupled Diode-Pumped Nd^{3+}:YLF Laser," *Electron Lett.*, Vol. 29, 1993, pp. 1181–1183.

[12] Bernardin, J. P., and N. M. Lawandy, "Dynamics of the Formation of Bragg Gratings in Germanosilicate Optical Fibers," *Opt. Comm.*, Vol. 79, 1990, pp. 194–199.

[13] Hand, D. P., and P. St. J. Russel, "Photoinduced Refractive-Index Changes in Germanosilicate Fibers," *Opt. Lett.*, Vol. 15, 1990, pp. 102–104.

[14] Poyntz-Wright, L. J., and P. St. J. Russel, *Proc. Integrate Photonics Research*, 1990, Paper MJ3.

[15] Hill, K. O., F. Bilodeau, B. Malo, J. Albert, D. C. Johnson, Y. Hibino, M. Abe, and M. Kawachi, *Proc. Optical Fiber Communication, OFC'93*, San Jose, 1993, Paper PD15.

[16] Leamire, P. J., A. M. Vengsarkar, W. A. Reed, V. Mizrahi, and K. S. Kranz, *Proc. Optical Fiber Communication, OFC'94*, San Jose, 1994, Paper Tul-1.

[17] Hanafusa, H., M. Horiguchi, and J. Noda, "Thermally-Dffused Expanded Core Fibers for Low-loss and Inexpensive Photonic Components," *Electron. Lett.*, Vol. 27, 1991, pp. 1968–1969.

[18] Jedrzejewshi, K. P., F. Martinez, J. D. Minelly, C. D. Hussey, and F. P. Payne, "Tapered-Beam Expander for Single-Mode Optical-Fiber Gap Devices," *Electron Lett.*, Vol. 22, 1986, pp. 105–106.

[19] Amitay, N., H. M. Presby, F. V. Dimarcello, and K. T. Nelson, "Single-Mode Optical Fiber Tapers for Self-Aligned Beam Expansion," *Electron. Lett.*, Vol. 22, 1986, pp. 702–703.

[20] Harper, J. S., C. P. Botham, and S. Hornung, "Tapers in Single-Mode Optical Fiber by Controlled Core Diffusion," *Electron. Lett.*, Vol. 24, 1988, pp. 245–246.

[21] Kawakami, S., K. Shiraishi, and Y. Aizawa, "A Method To Realize Fiber Embedded Optical Devices," *Proc. 2nd Optoelectronics Conf., OEC'88*, Makuhari, 1988, pp. 172–173.

[22] Shiraishi, K., O. Hanaizumi, T. Sato, and S. Kawakami, "Vertical Integration Technology for Fiber-Optic Circuit," *Opto-electronics*, Vol. 10, 1995, pp. 55–74.

[23] Chopra, K. L., *Thin Film Phenomena*, New York: Robert E. Krieger Publishing Company, 1979.

[24] Heavens, O. S., *Optical Properties of Thin Solid Films*, New York: Dover Publications Inc., 1965.

[25] Smith, S. D., *J. Opt. Soc. Amer.*, Vol. 48, 1958, p. 43.

[26] Reeve, M., et al., *Brit. Telecom Tech.*, Vol. 7, 1989, p. 89.

[27] Rayleigh, L., "On the Rotation of Magnetic Rotation of Light in Disulfide of Carbon," *Phil. Trans. Roy. Soc. London*, Vol. 176, 1885, p. 343.

[28] Geusic, J. E., and H. E. D. Scovil, "A Unidirectional Traveling-Wave Optical Meser," *Bell Sys. Tech. J.*, Vol. 41, 1962, p. 1371.

[29] Dillon, J. F., Jr., "Magneto-Optical Properties of Magnetic Garnets," in *Physics of Magnetic Garnets*, A. Paoletti (ed.), New York: North-Holland, 1978.

[30] Scott, G. B., D. E. Lacklison, H. I. Ralph, and J. L. Page, "Magnetic Crcular Dichroism and Faraday Rotation Spectra of $Y_3Fe_5O_{12}$," *Phys. Rev. B*, Vol. 12, 1975, p. 2562.

[31] Crossley, W. A., R. W. Coopeer, J. L. Page, and R. P. Van Stapele, "Faraday Rotation in Rare-Earth Iron Garnets," *Phys. Rev.*, Vol. 181, 1969, p. 896.

[32] Booth, R. C., and E. A. D. White, "Magneto-Optic Properties of Rare Earth Iron Garnet Crystals in the Wavelength Range 1.1–1 7 μm and Their Use in the Device Fabrication," *J. Phys. D: Appl. Phys.*, Vol. 17, 1984, pp. 579–587.

[33] Lullendroff, N., and B. Hök, "Temperature Independent Faraday Rotation Near the Band Gap in $Cd_{1-x}Mn_xTe$," *Appl. Phys. Lett.*, Vol. 46, 1985, p. 1016.

[34] Onodera, K., M. Kimura, and T. Masumoto, "Optimum Composition of $Cd_{1-x-y}Mn_xHg_yTe$ Single Crystals for Fabrication of 0. 98 mm Compact Optical Isolators," *Tokin Tech. Rev.*, Vol. 21, 1994, pp. 33–36. (Japanese)

[35] Matsumoto, T., "Polarization-Independent Isolators for Fiber Optics," *IECE Japan*, Vol. J62-C, 1979, pp. 505–512. (Japanese)

[36] Shiraishi, M., and K. Asama, "Compact Optical Isolator for Fiber Using Birefringent Wedge," *Appl. Opt.*, Vol. 21, 1982, pp. 4296–4299.

[37] Iwamura, H., H. Iwasaki, K. Kubodera, Y. Torii, and J. Noda, "Simple Polarization-Independent Optical Circulator for Optical Transmission Systems," *Electron. Lett.*, Vol. 15, 1979, pp. 830–831.

[38] William, L. E., "A Polarization-Independent Optical Circulator for 1.3 μm," *IEEE J. Lightwave. Tech.*, Vol. LT-1, 1983, pp. 466–469.

[39] Yokohama, I., K. Okamoto, and J. Noda, "Polarization-Independent Optical Circulator Consisting of Two Fiber-Optic Polarizing Beam Splitter and Two YIG Spherical Lenses," *Electron. Lett.*, Vol. 22, 1986, pp. 370–372.

[40] Fujii, Y., "High-Isolation Polarization-Independent Quasi-Optical Circulator," *IEEE J. Lightwave Tech.*, Vol. LT-10, 1991, pp. 1226–1229.

[41] Koga, M., and T. Matsumoto, "Polarization-Insensitive High-Isolation Nonreciprocal Device for Optical Circulator Application," *Electron. Lett.*, Vol. 27, 1991, pp. 903–905.

[42] Yariv, A., *Quantum Electronics*, 3rd edition, New York: Wiley, 1989.

[43] Kawasaki, B. S., K. O. Hill, and R. G. Lamont, "Biconical-Taper Single-Mode Fiber Coupler," *Opt. Lett.*, Vol. 6, 1981, pp. 327–328.

[44] Johnson, D. C., and K. O. Hill, "Control of Wavelength Selectivity of Power Transfer in Fused Biconical Monomode Directional Coupler," *Appl. Opt.*, Vol. 25, 1986, pp. 3800–3803.

[45] Lamont, R. G., D. C. Johnson, and K. O. Hill, "Power Transfer in Fused Biconical-Taper Single-Mode Fiber Couplers: Dependence on External Refractive Index," *Appl. Opt.*, Vol. 24, 1985, pp. 327–332.

[46] Tekippe, V. J., *Fiber and Integrated Optics*, Vol. 9, 1990, pp. 97–.

[47] Bergh, R., G. Kotler, and H. J. Shaw, "Single Mode Fiber Optic Directional Coupler," *Opt. Lett.*, Vol. 6, 1981, pp. 327–328.

[48] Digonnet, M. J. F., and H. J. Shaw, "Analysis of a Tunable Single-Mode Optical Fiber Coupler," *IEEE J. Quantum Electron.*, Vol. QE-18, 1982, pp. 746–754.

[49] Dyott, R. B., V. A. Handrek, and J. Bello, "Polarization Holding Directional Coupler Using D-Fiber," *SPIE Technical Symposium*, East Arlington, VA, 1984.

[50] Kato, K., and I. Nishi, "Simple Assembly Technique for Single-Mode Optical Fiber Multi/Demultiplexers," *Electron. Comm. in Japan*, Vol. 74, 1991, pp. 40–48.

[51] Shikii, S., and Y. Tamura, "Effects of Polarization Multiplexing of High Power LDs in Fiber Raman Amplifier," *1989 Autumn National Conv. IEICE Jpn.*, Vol. C-278, 1989. (Japanese)

[52] Lord, A., I. J. Wilkinson, A. Ellis, D. Cleland, R. A. Garnham, and W. A. Stallard, "Comparison of WDM Coupler Technologies for Use in Erbium-Doped Fibre Amplifier Systems," *Electron. Lett.*, Vol. 26, 1990, pp. 900–901.

[53] Alameel, G. M., and A. W. Carlisle, "The Performance of the AT&T Single-Mode ST Connector and Its Latest Enhanced Features," *SPIE Components for Fiber Optic Applications and Coherent Lightwave Comm.*, Vol. 988, 1988, pp. 226–234.

[54] Suzuki, N., Y. Iwahara, M. Saruwatari, and K. Nawata, "Ceramic Capillary Connector for 1.3 μm Single-Mode Fibers," *Electron Lett.*, Vol. 15, 1979, pp. 809–811.

[55] Suzuki, N., M. Saruwatari, and M. Okuyama, "Low Insertion- and High Return-Loss Optical Connector with Spherically Convex-Polished End," *Electron. Lett.*, Vol. 22, 1986, pp. 110–112.

[56] Sugita, E., R. Nagase, K. Kanayama, and T. Shintaku, "SC-type Single-Mode Optical Fiber Connectors," *IEEE J. Lightwave Tech.*, Vol. 7, 1989, pp. 1689–1696.

[57] Nagase, R., E. Sugita, S. Iwano, K. Kanayama, and Y. Ando, "Miniature Optical Connector with Small Zirconia Ferrule," *IEEE Photon. Tech. Lett.*, Vol. 3, 1991, pp. 1045–1047.

[58] Murata, H., and N. Inagaki, "Low-Loss Single-Mode Fiber Development and Splicing Research in Japan," *IEEE J. Quatum Electron.*, Vol. QE-17, 1981, pp. 835–849.

[59] Kato, Y., S. Seikai, and M. Tateda, "Arc-Fusion Splicing of Single-Mode Fibers. 1: Optimum Splicing Conditions," *Appl. Opt.*, Vol. 21, 1982, pp. 1332–1336.

[60] Kawata, O., K. Hoshino, Y. Miyajima, M. Ohnishi, and K. Ishihara, "A Splicing and Inspection Technique for Single-Mode Fibers Using Direct Core Monitoring," *IEEE J. Lightwave Tech.*, Vol. LT-2, 1984, pp. 185–191.

[61] Hayashi, I., P. B. Panish, P. W. Foy, and S. Sumiki, "Junction Lasers Which Operate Continuously at Room Temperature," *Appl. Phys. Lett.*, Vol. 17, 1970, pp. 109–111.

[62] Iga, K. (ed.), *Semiconductorlaser*, Tokyo: Ohm, 1994.

[63] Oshiba, S., and Y. Tamura, "Recent Progress in High-Power GaInAsP Lasers," *IEEE J. Lightwave Tech.*, Vol. 8, 1990, pp. 1350–1356.

[64] Asano, H., S. Takano, M. Kawaradani, M. Kitamura, and I. Mito, "1.48-μm High-Power InGaAs/InGaAsP MQW LD's for Er-Doped Fiber Amplifier," *IEEE Phonon. Tech. Lett.*, Vol. 3, 1991, p. 415.

[65] Kamei, H., N. Tatoh, J. Shinkai, H. Hayashi, and M. Yoshimura, "Ultrahigh output power of 1.48-μm GaInAsP/GaInAsP strained-layer MQW laser diodes," *Proc. Opt. Fiber Comm.*, 1992, p. 45.

[66] Kasukawa, A., M. Ohkubo, T. Namegaya, T. Ijichi, Y. Ikegami, N. Tsukiji, S. Namiki, and Y. Shirasaki, "980 nm and 1480 nm High Power Laser Modules for Er-Doped Fiber Amplifiers," *Optoelectronics*, Vol. 9, 1994, pp. 219–230.

[67] Uehara, S., M. Horiguchi, T. Takeshita, M. Okayasu, M. Yamada, M. Shimizu, O. Kogure, and K. Oe, "0.98-μm InGaAs Strained Quantum Well Lasers for Erbium-Doped Fiber Amplifiers," *Proc. IOOC'89*, Kobe, 1989, Paper 20PDB-11.

[68] Laming, R. I., V. Shah, L. Curtis, R. S. Vodhanel, F. J. Favire, W. L. Barnes, J. D. Minelly, D. P. Bour, and E. J. Tarbox, "Highly Efficient 978 nm Diode Pumped Erbium-Doped Fibre Amplifier with 24 dB Gain," *Proc. IOOC'89*, Kobe, 1989, Paper 20PDA-4.

[69] Temmyo, J., and M. Sugo, "Design of High-Power Strained InGaAs/AlGaAs Quantum-Well Lasers with Vertical Divergence Angle of 18°," *Electron Lett.*, Vol. 31, 1995, pp. 642–644.

[70] Giles, C. R., T. Erdogan, and V. Mizrahi, "Reflection-Induced Changes in the Optical Spectra of 980-nm QW Lasers," *IEEE Photon. Tech. Lett.*, Vol. 6, 1994, pp. 903–906.

[71] Masuda, H., K. Aida, K. Nakagawa, K. Yoshino, M. Wada, and J. Temmyo, "Pump-Wavelength-Locked Erbium-Doped Fiber Amplifier Employing Novel External Cavity for 0.98-μm Laser Diode," *Electron. Lett.*, Vol. 28, 1992, pp. 1855–1857.

[72] Giles, C. R., T. Erdogan, and V. Mizrahi, "Simultaneous Wavelength-Stabilization of 980-nm Pump Lasers," *IEEE Photon. Tech. Lett.*, Vol. 6, 1994, pp. 907–909.

[73] Temmyo, J., M. Sugo, T. Nishiya, T. Tamamura, M. Yamada, F. Yamamoto, F. Bilodeau, and K. O. Hill, "A Wavelength-Stabilized 1.016 μm InGaAs Strained Quantum Well Laser Using a Fiber Bragg Grating and Its Application for a 1.3 μm Pr-Doped Optical Fiber Amplifier," *Proc. Annual Meeting of IEICE Japan*, Vol. B-11-13, 1996, pp. 759–760. (Japanese)

[74] Parke, R., D. F. Welch, A. Hardy, R. Lang, D. Mehuys, S. O'Brien, K. Dzurko, and D. Scifres, "2 W CW Diffraction Limited Operation of a Monolithically Integrated Master Oscillator Power Amplifier," *IEEE Photon. Tech. Lett.*, Vol. 5, 1993, pp. 297–300.

[75] Kintzer, E. S., J. N. Walpole, S. R. Chinn, C. A. Wang, and L. J. Missaggia, "High-Power, Strained-Layer Amplifier and Lasers with Tapered Gain Region," *IEEE Photon. Tech. Lett.*, Vol. 5, 1993, pp. 605–608.

[76] Livas, J. C., S. R. Chinn, E. S. Kintzer, J. N. Walpole, C. A. Wang, and L. J. Missaggia, "High-Power Erbium-Doped Fiber Amplifier with 975nm Tapered-Gain-Region Laser Pumps," *Electron. Lett.*, Vol. 30, 1994, pp. 1054–1055.

[77] Sanders, S., F. Shum, R. J. Lang, J. D. Ralston, D. G. Mehuys, R. G. Waarts, and D. F. Welch, "Fiber-Coupled M-MOPA Laser Diode Pumping a High-Power Erbium-Doped Fiber Amplifier," *Technical Digest of OFC'96*, San Jose, 1996, Paper TuG5.

[78] Snitzer, E., H. Po, F. Hakimi, R. Tummineli, and B. C. McCollum, *Technical Digest of OFS'88*, 1988, Paper PD5.

[79] Hanna, D. C., H. M. Pask, J. E. Townsend, J. L. Archambault, L. Reekie, and A. C. Tropper, "Yb-Doped Silica Cladding-Pumped Fiber Laser," *Proc. Advanced Solid State Lasers*, Vol. 20, 1994, pp. 86–90.

[80] Stolen, R. H., and E. P. Ippen, "Raman Gain in Glass Optical Waveguide," *Appl. Phys. Lett.*, Vol. 22, 1973, pp. 276–278.

[81] Galeener, F. L., J. C. Mikkelsen, Jr., R. H. Geils, and W. J. Mosby, "The Relative Raman Cross Sections of Vitreous SiO_2, GeO_2, B_2O_3, and P_2O_5," *Appl. Phys. Lett.*, Vol. 32, 1978, pp. 34–36.

[82] Nakashima, T., S. Sikai, M. Nakazawa, and Y. Negisi, "Theoretical Limit of Repeater Spacing in an Optical Transmission Line Utilizing Raman Amplification," *IEEE J. Lightwave Tech.*, Vol. LT-4, 1986, pp. 1267–1272.

[83] Auyeung, J., and A. Yariv, "Spontaneous and Stimulated Raman Scattering in Long Low Loss Fibers," *IEEE J. Quantum Electron.*, Vol. QE-14, 1978, pp. 347–352.

[84] Grubb, S. G., et al., "High-Power 1.48 μm Cascaded Raman Laser in Germanosilicate Fibers," *Proc. OAA*, Davos, 1995, pp. 197–199.

[85] Berger, J., D. F. Welch, D. R. Scifres, W. Streifer, and P. S. Cross, "High-Power, High Efficient Neodymium:yttrium Aluminum Garnet Laser End Pumped by a Laser Diode Array," *Appl. Phys. Lett.*, Vol. 51, 1987, pp. 1212–1214.

[86] Mears, R. J., L. Reekie, I. M. Jauncey, and D. N. Payne, "High-Gain Rare-Earth-Doped Fiber Amplifier Operating at 1.54 μm," *Proc. OFC/IOOC'87*, Reno, Nevada, 1987, Paper W12, p. 167; and "Low Noise Erbium-Doped Fiber Amplifier Operating at 1.54 μm," *Electron. Lett.*, Vol. 23, 1987, pp. 1026–1028.

[87] Hagimoto, K., K. Iwatsuki, M. Nakazawa, M. Saruwatari, K. Aida, M. Nakagawa, and M. Horiguchi, "A 212 km Non-repeated Transmission Experiment at 1.8 Gbit/s Using LD Pumped Er^{3+}-Doped Fiber Amplifiers in IM/Direct-Detection Repeater System," *Proc. OFC'89*, Huston, Texas, 1989, Paper PD15.

[88] Laming, R. I., M. C. Farries, P. R. Morkel, L. Reekie, D. N. Payne, P. L. Scrivener, F. Fontana, and A. Righetti, "Efficient Pump Wavelengths of Erbium-Doped Fiber Amplifier," *Electron. Lett.*, Vol. 25, 1989, pp. 12–14.

[89] Laming, R. I., L. Reekie, D. N. Payne, P. L. Scrivener, F. Fontana, and A. Righetti, "Optimal Pumping of Erbium-Doped Fiber Amplifiers," *Proc. ECOC'88*, part 2, Brighton, United Kingdom, 1988, pp. 25–28.

[90] Snitzer, E., H. Po, F. Hakimi, R. Tumminelli, and B. C. McCollum, "Erbium Fiber Laser Amplifier at 1.55 μm with Pump at 1.49 μm and Yb Sensitized Er Oscillator," *Proc. OFC/OFS'88*, New Orleans, Louisiana, 1988, Paper PD2.

[91] Shimizu, M., M. Horiguchi, M. Yamada, I. Nishi, J. Noda, T. Takeshita, M. Okayasu, S. Uehara, and E. Sugita, "Compact and Highly Efficient Fiber Amplifier Module Pumped by a 0.98 μm Laser Diode," *Proc. OFC'90*, San Francisco, CA, 1990, Paper PD17.

[92] Shimizu, M., M. Horiguchi, M. Yamada, M. Okayasu, T. Takeshita, I. Nishi, S. Uehara, J. Noda, and E. Sugita, "Highly Efficient Integrated Optical Fiber Amplifier Modules Pumped by a 0.98 μm Laser Diode," *Electron. Lett.*, Vol. 26, 1990, pp. 498–499.

[93] Shimizu, M., M. Horiguchi, M. Yamada, I. Nishi, J. Noda, T. Takeshita, M. Okayasu, S. Uehara, and E. Sugita, "Compact and Highly Efficient Fiber Amplifier Modules Pumped by a 0.98-μm Laser Diode," *IEEE J. Lightwave Tech.*, Vol. 9, 1991, pp. 291–296.

[94] Desurvire, E., C. R. Giles, and J. R. Simpson, "Gain Saturation Effects in High-Speed, Multichannel Erbium-Doped Fiber Amplifiers at λ = 1.53 μm," *IEEE J. Lightwave Tech.*, Vol. 7, 1989, pp. 2095–2104.

[95] Reisfeld, R., and C. K. Jorgensen, "Lasers and Excited States of Rare Earths," *Organic Chemistry Concepts*, Vol. 1, Berlin, Heidelberg, New York: Springer-Verlag, 1977.

[96] Layne, C. B., W. H. Lowdermilk, and M. J. Weber, "Nonradiative Relaxation of Rare-Earth Ions in Silicate Glass," *IEEE J. Quantum Electron.*, Vol. 11, 1975, pp. 798–799.

[97] Desurvire, E., J. R. Simpson, and P. C. Becker, "High-Gain Erbium-Doped Traveling-Wave Fiber Amplifier," *Opt. Lett.*, Vol. 12, 1987, pp. 888–890.

[98] Olshansky, R., "Noise Figure for Erbium-Doped Optical Fibre Amplifiers," *Electron. Lett.*, Vol. 24, 1988, pp. 1363–1365.

[99] Armitage, J. R., "Three-Level Fiber Laser Amplifier: a Theoretical Model," *Appl. Optics*, Vol. 27, 1988, pp. 4831–4836.

[100] Desurvire, E., and J. R. Simpson, "Amplification of Spontaneous Emission in Erbium-Doped Single-Mode Fibers," *IEEE J. Lightwave Tech.*, Vol. 7, 1989, pp. 835–845.

[101] Bjarklev, A., S. L. Hanse, and J. H. Povlsen, "Large-Signal Modeling of an Erbium-Doped Fiber Amplifier," *Proc. SPIE Conf. on Fiber Laser Sources and Amplifiers*, Vol. 1171, 1989, pp. 118–129.

[102] Morkel, P. R., and R. I. Laming, "Theoretical Modeling of Erbium-Doped Fiber Amplifiers with Excited-State Absorption," *Optics Lett.*, Vol. 14, 1989, pp. 1062–1064.

[103] Desurvire, E., "Analysis of Transient Gain Saturation and Recovery in Erbium-Doped Fiber Amplifiers," *IEEE Photonics Tech. Lett.*, Vol. 1, 1989, pp. 196–199.

[104] Desurvire, E., "Analysis of Erbium-Doped Fiber Amplifiers Pumped in the $^4I_{13/2}$-$^4I_{15/2}$ Band," *IEEE Photonics Tech. Lett.*, Vol. 1, 1989, pp. 293–296.

[105] Desurvire, E., "Analysis of Noise Figure Spectral Distribution in Erbium-Doped Fiber Amplifiers Pumped Near 980 and 1480 nm," *Appl. Optics*, Vol. 29, 1990, pp. 3118–3125.

[106] Marcerou, J. F., H. A. Fevrier, J. Ramos, J. C. Auge, and P. Bousselet, "General Theoretical Approach Describing the Complete Behavior of the Erbium-Doped Fiber Amplifier," *Proc. SPIE Conf. on Fiber Laser Sources and Amplfiers*, Vol. 1373, 1990, pp. 168–186.

[107] Peroni, M., and M. Tamburrini, "Gain in Erbium-Doped Fiber Amplifiers: a Simple Analytical Solution for the Rate Equations," *Optics Lett.*, Vol. 15, 1990, pp. 842–844.

[108] Saleh, A. A. M., R. M. Jopsori, J. D. Evankow, and J. Aspell, "Modeling of Gain in Erbium-Doped Fiber Amplifiers," *IEEE Photonics Tech. Lett.*, Vol. 2, 1990, pp. 714–717.

[109] Digonnet, M. J. F., "Closed-Form Expressions for the Gain in Three- and Four-Level Laser Fibres," *IEEE J. Quantum Electron.*, Vol. 26, 1990, pp. 1788–1796.

[110] Desurvire, E., "Study of the Complex Atomic Susceptibility of Erbium-Doped Fiber Amplifiers," *IEEE J. Lightwave Tech.*, Vol. 8, 1990, pp. 1517–1527.

[111] Desurvire, E., J. L. Zyskind, and C. R. Giles, "Design Optimization for Efficient Erbium-Doped Fiber Amplifiers," *IEEE J. Lightwave Tech.*, Vol. 8, 1990, pp. 1730–1741.

[112] Pedersen, B., K. Dybdal, C. D. Hansen, A. Bjarklev, J. H. Povlsen, H. Vendeltorp-Pommer, and C. C. Larsen, "Detailed Theoretical and Experimental Investigation of High-Gain Erbium-Doped Fiber Amplifier," *IEEE Photonics Tech. Lett.*, Vol. 2, 1990, pp. 863–865.

[113] Giles, C. R., and E. Desurvire, "Modeling Erbium-Doped Fiber Amplifiers," *IEEE J. Lightwave Tech.*, Vol. 9, 1991, pp. 271–283.

[114] Digonnet, M. J. F., *Rare Earth Doped Fiber Lasers and Amplifiers*, New York: Marcel Dekker, 1993.

[115] Desurvire, E., *Erbium-Doped Fiber Amplifiers*, New York: John Wiley & Sons, 1994.

[116] Miniscalco, W. J., "Erbium-Doped Glasses for Fiber Amplifiers at 1500 nm," *IEEE J. Lightwave Tech.*, Vol. 9, 1991, pp. 234–250.

[117] Barnes, W. L., R. I. Laming, E. J. Tarbox, and P. R. Morkel, "Absorption and Emission Cross Section of Er^{3+} Doped Silica Fibers," *IEEE J. Quantum Electron.*, Vol. 27, 1991, pp. 1004–1010.

[118] Sandoe, J. N., P. H. Sarkies, and S. Parke, "Variation of Er^{3+} Cross Section for Stimulated Emission with Glass Composition," *J. Phys. D: Appl. Phys.*, Vol. 5, 1991, pp. 1788–1799.

[119] Ainslie, B. J., "A Review of the Fabrication and Properties of Erbium-Doped Fibers for Optical Amplifiers," *IEEE J. Lightwave Tech.*, Vol. 9, 1991, pp. 220–227.

[120] Atkins, C. G., J. F. Massicott, J. R. Armitage, R. Wyatt, B. J. Ainslie, and S. P. Criaig-Ryan, "High-Gain Broad Spectral Bandwidth Erbium-Doped Fiber Pumped Near 1.5 μm," *Electron. Lett.*, Vol. 25, 1989, pp. 910–911.

[121] Dybdal, K., N. Bjerre, J. E. Pedersen, and C. C. Larsen, "Spectroscopic Properties of Er-Doped Silica Fibers and Preforms," *Proc. SPIE Conf. on Fiber Laser Sources and Amplfiers*, Vol. 1171, 1989, pp. 209–218.

[122] Zemon, S., G. Lambert, W. J. Miniscalco, L. J. Andrews, and B. T. Hall, "Characterization of Er^{3+} Doped Glasses by Fluorescence Line Narrowing," *Proc. SPIE Conf. on Fiber Laser Sources and Amphfiers*, Vol. 1171, 1989, pp. 229–236.

[123] Miniscalco, W. J., and R. S. Quimby, "General Procedure for the Analysis of Cross-Sections," *Optics Lett.*, Vol. 16, 1991, pp. 258–260.

[124] Marcuse, D., "Loss Analysis of Single-Mode Fiber Splices," *The Bell System Tech. J.*, Vol. 56, 1977, pp. 703–718.

[125] Petermann, K., "Constraints for Fundamental-Mode Spot Size for Broadband Dispersion-Compensated Single-Mode Fibers," *Electron. Lett.*, Vol. 19, 1983, pp. 712–714.

[126] Fafarries, M. C., R. I. Laming, P. R. Morkel, T. A. Birks, and D. N. Payne, "Efficient High-Gain Erbium-Doped Fiber Amplifier Pumped with a Frequency-Doubled Nd:YAG Laser," *Proc. OFC'89*, Houston, TX, 1989, Paper TuG5, pp. 23–24.

[127] Choy, M. M., C. Y. Chen, M. Anderejco, M. Saifi, and C. Lin, "A High-Gain, High-Output Saturation Power Erbium-Doped Fiber Amplifier Pumped at 532 nm," *IEEE Photonics Tech. Lett.*, Vol. 2, 1990, pp. 38–40.

[128] Horiguchi, M., M. Shimizu, M. Yamada, K. Yoshino, and H. Hanafusa, "Highly Efficient Er-Doped Fiber Amplifiers Pumped in 660 nm Band," *Electron. Lett.*, Vol. 27, 1991, pp. 2319–2320.

[129] Horiguchi, M., K. Yoshino, M. Shimizu, M. Yamada, and H. Hanafusa, "Erbium-Doped Optical Fiber Amplifiers Pumped in the 660- and 820-nm Bands," *IEEE J. Lightwave Tech.*, Vol. 12, 1994, pp. 810–820.

[130] Whitley, T. J., and T. G. Hodgkinson, "1.54 μm Er^{3+}-Doped Fiber Amplifier Optically Pumped at 807 nm," *Proc. ECOC'88*, part 1, Brighton, 1988, pp. 58–61.

[131] Whitley, T. J., "Laser Diode Pumped Operation of Er^{3+}-Doped Fiber Amplifier," *Electron. Lett.*, Vol. 24, 1988, pp. 1537–1539.

[132] Horiguchi, M., M. Shimizu, M. Yamada, K. Yoshino, and H. Hanafusa, "Highly Efficient Optical Fiber Amplifier Pumped by a 0.8 μm Band Laser Diode," *Electron. Lett.*, Vol. 26, 1990, pp. 1758–1759.

[133] Nakazawa, M., Y. Kimura, and K. Suzuki, "High Gain Erbium Fiber Amplifier Pumped by 800 nm Band," *Electron. Lett.*, Vol. 26, 1990, pp. 548–550.

[134] Suzuki, K., Y. Kimura, and M. Nakazawa, "High Gain Er^{3+}-Doped Fiber Amplifier Pumped by 800 nm GaAlAs Laser Diodes," *Electron. Lett.*, Vol. 26, 1990, pp. 948–949.

[135] Uehara, S., M. Horiguchi, T. Takeshita, M. Okayasu, M. Yamada, M. Shimizu, O. Kogure, and K. Oe, "0.98-μm InGaAs Strained Quantum Well Lasers for Erbium-Doped Fiber Optical Amplifier," *Proc. IOOC'89*, Kobe, Japan, 1989, Paper 20PDB-11.

[136] Laming, R. I., V. Shah, L. Curtis, R. S. Vodhanel, F. J. Favire, W. L. Barnes, J. D. Minelly, D. P. Bour, and E. J. Tarbox, "Highly Efficient 978 nm Diode Pumped Erbium-Doped Fiber Amplifier with 24 dB Gain," *Proc. IOOC'89*, Kobe, Japan, 1989, Paper 20PDA-4.

[137] Vodhanel, R. S., R. I. Laming, V. Shah, L. Curtis, D. P. Bour, W. L. Barnes, J. D. Minelly, E. J. Tarbox, and F. J. Favire, "Highly Efficient 978 nm Diode-Pumped Erbium-Doped Fiber Amplifier with 24 dB Gain," *Electron. Lett.*, Vol. 25, 1989, pp. 1386–1388.

[138] Yamada, M., M. Shimizu, T. Takeshita, M. Okayasu, M. Horiguchi, S. Uehara, and E. Sugita, "Er^{3+}-Doped Fiber Amplifier Pumped by 0.98 μm Laser Diodes," *IEEE Photonics Tech. Lett.*, Vol. 1, 1989, pp. 422–424.

[139] Miniscalco, W. J., B. A. Thompson, E. Eichen, and T. Wei, "Very High Gain Er^{3+} Fiber Amplifier Pumped at 980 nm," *Proc. OFC'90*, San Francisco, CA, 1990, Paper FA9, p. 192.

[140] Becker, P. C., A. Lidgard, J. R. Simpson, and N. A. Olsson, "Erbium-Doped Fiber Amplifier Pumped in the 950–1000 nm Region," *IEEE Photonics Tech. Lett.*, Vol. 2, 1990, pp. 35–37.

[141] Nakazawa, M., Y. Kimura, and K. Suzuki, "Efficient Er^{3+}-Doped Optical Amplifier Pumped by a 1.48 μm InGaAsP Laser Diode," *Electron. Lett.*, Vol. 25, 1989, pp. 12–14.

[142] Nakazawa, M., Y. Kimura, and K. Suzuki, "Efficient Er^{3+}-Doped Optical Fiber Amplifier Pumped by a 1.48 μm InGaAsP Laser Diode," *Appl. Phys. Lett.*, Vol. 54, 1989, pp. 295–297.

[143] Nakazawa, M., Y. Kimura, and K. Suzuki, "An Ultra-Efficient Erbium-Doped Fiber Amplifier of 10.2 dB/mW at 0.98 μm Pumping and 5.1 dB/mW at 1.48 μm Pumping," *Proc. OAA'90*, Monterey, CA, 1990, Paper Pdp1.

[144] Nakazawa, M., K. Suzuki, and Y. Kimura, "46.5 dB Gain in Er^{3+}-Doped Fiber Amplifier Pumped by 1.48 μm GaInAsP Laser Diodes," *Electron. Lett.*, Vol. 25, 1989, pp. 1656–1657.

[145] Desurvire, E., C. R. Giles, J. R. Simpson, and J. L. Zyskind, "Efficient Erbium-Doped Fiber Amplifier at a 1.53-μm Wavelength with a High Output Saturation Power," *Optics Lett.*, Vol. 14, 1989, pp. 1266–1268.

[146] Shimizu, M., M. Yamada, M. Horiguchi, and T. Takeshita, "0.98 μm Laser Diode Pumped Erbium-Doped Fiber Amplifiers with a Gain Coefficient of 7.6 dB/mW," *Proc. OAA'90*, Monterey, CA, 1990, Paper MB2, pp. 12–15.

[147] Shimizu, M., M. Yamada, M. Horiguchi, T. Takeshita, and M. Okayasu, "Erbium-Doped Fiber Amplifiers with an Extremely High Gain Coefficient of 11.0 dB/mW," *Electron. Lett.*, Vol. 26, 1990, pp. 1641–1642.

[148] Armitage, J. R., C. G. Atkins, R. Wyatt, B. J. Ainslie, and S. P. Craig, "Spectroscopic Studies of Er^{3+}-Doped Single-Mode Silica Fiber," *Proc. Topical Meeting on Tunable Solid State Lasers*, 1987, Paper WD3, pp. 193–196.

[149] Andrews, L. J., W. J. Miniscalco, and T. Wei, "Excited State Absorption of Rare Earths in Fiber Amplifiers," *Proc. 1st Int. School on Excited States of Transition Elements*, Poland, 1988, pp. 9–30.

[150] Laming, R. I., S. B. Poole, and E. J. Tarbox, "Pump Excited-State Absorption in Erbium-Doped Fiber," *Optics Lett.*, Vol. 13, 1988, pp. 1084–1086.

[151] Zemon, S., G. Lambert, W. J. Miniscalco, R. W. Davies, B. T. Hall, R. C. Folweiler, T. Wei, L. J. Andrews, and M. P. Singh, "Excited State Cross Sections for Er-Doped Glasses," *Proc. SPIE Conf. on Fiber Laser Sources and Amplfiers*, Vol. 1373, 1990, pp. 21–32.

[152] Pedersen, B., S. Zemon, and W. J. Miniscalco, "Erbium-Doped Fibers Pumped in 800 nm Band," *Electron. Lett.*, Vol. 27, 1991, pp. 1295–1297, 1991.

[153] Delavaque, E., T. Georges, and J. F. Bayon, "Pump Wavelength Optimisation of Erbium-Doped Fiber Amplifier in 800 nm Band," *Electron. Lett.*, Vol. 27, 1991, pp. 1421–1422.

[154] Pedersen, B., W. J. Miniscalco, and S. Zemon, "Evaluation of the 800-nm Pump Band for Erbium-Doped Fiber Amplifiers," *IEEE J. Lightwave Tech.*, Vol. 10, 1992, pp. 1041–1049.

[155] Farries, M. C., "Excited-State Absorption and Gain in Erbium-Fiber Amplifiers Between 1.05 and 1.35 μm," *IEEE Photonics Tech. Lett.*, Vol. 3, 1991, pp. 619–621.

[156] Blixt, P., J. Nilsson, J. Babonas, and B. Jaskorzynska, "Excited-State Absorption at 1.5 μm in Er^{3+}-Doped Fiber Amplifiers," *Proc. OAA'92*, Santa Fe, New Mexico, 1992, Paper WE2, pp. 63–66.

[157] Quimby, R. S., "Output Saturation in a 980-nm Pumped Erbium-Doped Fiber Amplifier," *Appl. Optics*, Vol. 30, 1991, pp. 2548–2552.

[158] Quimby, R. S., W. J. Miniscalco, and B. T. Thomson, "Excited State Absorption at 980 nm in Erbium-Doped Silica Glass," *Proc. OAA'92*, Santa Fe, New Mexico, 1992, Paper WE3, pp. 67–70.

[159] Yamamoto, Y., "Noise and Error Rate Performance of Semiconductor Laser Amplifiers in PCM-IM Optical Transmission Systems," *IEEE J. Quantum Electron.*, Vol. 16, 1980, pp. 1073–1081.

[160] Mukai, T., and Y. Yamamoto, "Noise in an AlGaAs Semiconductor Laser Amplifier," *IEEE J. Quantum Electron.*, Vol. 18, 1982, pp. 564–575.

[161] Mukai, T., Y. Yamamoto, and T. Kimura, "S/N and Error Rate Performance in AlGaAs Semiconductor Laser Preamplifier and Linear Repeater Systems," *IEEE J. Quantum Electron.*, Vol. 18, 1982, pp. 1560–1568.

[162] Laming, R. I., P. R. Morkel, D. N. Payne, and L. Reekie, "Noise in Erbium-Doped Fiber Amplifiers," *Proc. ECOC'89*, part 1, Brighton, United Kingdom, 1988, pp. 54–57.

[163] Yamada, M., M. Shimizu, M. Okayasu, T. Takeshita, M. Horiguchi, Y. Tachikawa, and E. Sugita, "Noise Characteristics of Er^{3+}-Doped Fiber Amplifiers Pumped by 0.98 and 1.48 μm Laser Diodes," *IEEE Photonics Tech. Lett.*, Vol. 2, 1990, pp. 205–207.

[164] Yamada, M., M. Shimizu, M. Horiguchi, M. Okayasu, and E. Sugita, "Noise Characteristics of Er^{3+}-Doped Fiber Amplifiers Pumped by Laser Diodes," *Electronics and Comm. in Japan*, part 2, Vol. 74, 1991, pp. 19–31.

[165] Yamada, M., M. Shimizu, M. Horiguchi, M. Okayasu, and E. Sugita, "Noise Characteristics of Er^{3+}-Doped Fiber Amplifiers Pumped by Laser Diodes," *Trans. Institute of Electron., Information and Communication Engineers* (IEICE), Japan, Vol. 74-C-I, 1991, pp. 126–136. (Japanese)

[166] Laming, R. I., and D. N. Payne, "Noise Characteristics of Erbium-Doped Fiber Amplifier Pumped at 980 nm," *IEEE Photonics Tech. Lett.*, Vol. 2, 1990, pp. 418–421.

[167] Kashiwada, T., M. Shigematsu, T. Kougo, H. Kanamori, and M. Nishimura, "Erbium-Doped Fiber Amplifier Pumped at 1.48 μm with Extremely High Efficiency," *IEEE Photonics Tech. Lett.*, Vol. 3, 1991, pp. 721–723.

[168] Kimura, Y., K. Suzuki, and M. Nakazawa, "Noise Figure Characteristics of Er^{3+}-Doped Fiber Amplifier Pumped in 0. 8 μm Band," *Electron. Lett.*, Vol. 27, 1991, pp. 146–148.

[169] Giles, C. R., E. Desurvire, J. L. Zyskind, and J. R. Simpson, "Noise Performance of Erbium-Doped Fiber Amplifier Pumped at 1.49 μm, and Application to Signal Preamplification at 1.8 Gbits/s," *IEEE Photonics Tech. Lett.*, Vol. 1, 1989, pp. 367–369.

[170] Shimizu, M., M. Yamada, and M. Horiguchi, "Fabrication and Characteristics of Erbium-Doped Silica Single-Mode Fibers," *Trans. Institute of Electron., Information and Communication Engineers* (IEICE), Japan, Vol. 74-C-I, 1991, pp. 545–551. (Japanese)

[171] Pedersen, B., J. Chirravuri, and M. J. Miniscalco, "Gain and Noise Penalty of Detuned 980-nm Pumping of Erbium-Doped Fiber Amplifiers," *IEEE Photonics Tech. Lett.*, Vol. 4, 1992, pp. 351–353.

[172] Fevrier, H., J. Auge, V. Parlier, P. Bousselet, A. Dursin, J. F. Marcerou, and B. Jacquier, "Erbium-Doped Fiber Amplifier with Outstanding Gain Characteristics," *Proc. ECOC'89*, part 1, Brighton, United Kingdom, 1989, Paper TUA5-2, pp. 66–69.

[173] Welter, R., R. I. Laming, R. S. Vodhanel, W. B. Sessa, M. W. Maeda, and R. E. Wagner, "Performance of Erbium-Doped Fiber Amplifier in a 16-Channel Coherent Broadcast Network Experiment," *Proc. CLEO'89*, Baltimore, Maryland, 1989, Paper PD22.

[174] Armitage, J. R., "Spectral Dependence of the Small-Signal Gain Around 1.5 μm in Erbium-Doped Silica Fiber Amplifiers," *IEEE J. Quantum Electron.*, Vol. 26, 1990, pp. 423–425.

[175] Lumholt, O., K. Dybdal, C. C. Larsen, S. Dahl-Petersen, K. Schusler, A. Bjarklev, J. H. Povlsen, T. Rasmussen, and K. Rottwitt, "Er-La Doped Fiber Amplifier, Pumped at 1.47 μm," *Proc. ECOC'91*, Paris, 1991, Paper TuPS-1, pp. 285–288.

[176] Nakazawa, M., and Y. Kimura, "Lanthanum Codoped Erbium Fiber Amplifier," *Electron. Lett.*, Vol. 27, 1991, pp. 1065–1067.

[177] Bogatyrjov, V. A., E. M. Djanov, K. M. Bolant, V. I. Karpov, R. R. Khrapko, A. S. Kurkov, and V. N. Protopopov, "Passive Selective Filter for Flattening the Erbium-Doped Fiber Amplifier Gain Spectrum Based on a Feature of the Silicon Oxynitride Fiber Absorption," *Electron. Lett.*, Vol. 31, 1995, pp. 61–62.

[178] Kashiwada, T., K. Nakazato, M. Ohnishi, H. Kanamori, and M. Nishimura, "Spectral Gain Behavior of Er-Doped Fiber with Extremely High Aluminum Concentration," *Proc. OAA'93*, Yokohama, Japan, 1993, Paper MA6.

[179] Yoshida, S., S. Kuwano, and K. Iwashita, "Gain-Flattened EDFA with High Al Concentration for Multistage Repeatered WDM Transmission Systems," *Electron. Lett.*, Vol. 31, 1995, pp. 1765–1767.

[180] Millar, C. A., M. C. Brierley, and P. W. France, "Optical Amplification in an Erbium-Doped Fluorozirconate Fiber Between 1480 nm and 1600 nm," *Proc. ECOC'88*, part 1, Brighton, United Kingdom, 1988, pp. 66–69.

[181] Marcerou, J. F., S. Artigaud, J. Hervo, and H. Fevrier, "Basic Comparison Between Fluoride- and Silica-Doped Fiber Amplifier in the 1550 nm Region," *Proc. ECOC'92*, Berlin, Germany, 1992, Paper Mo A2. 3, pp. 53–56.

[182] Ronarc'h, D., M. Guibert, H. Ibrahim, M. Monerie, H. Poignant, and A. Tromeur, "30 dB Optical Net Gain at 1.543 mm in Er^{3+}-Doped Fluoride Fiber Pumped Around 1.48 μm," *Electron. Lett.*, Vol. 27, 1991, pp. 908–909.

[183] Bayart, D., B. Clesca, L. Hamon, and J. L. Beylat, "Experimental Investigation of the Gain Flatness Characteristics for 1.55 μm Erbium-Doped Fluoride Fiber Amplifiers," *IEEE Photonics Tech. Lett.*, Vol. 6, 1994, pp. 613–615.

[184] Clesca, B., D. Ronarc'h, D. Bayart, Y. Sorel, L. Hamon, M. Guibert, J. L. Beylat, J. F. Kerdiles, and M. Semenkoff, "Gain Flatness Comparison Between Erbium-Doped Fluoride and Silica Fiber Amplifiers with Wavelength-Multiplexed Signals," *IEEE Photonics Tech. Lett.*, Vol. 6, 1994, pp. 509–512.

[185] Yamada, M., T. Kanamori, Y. Terunuma, K. Oikawa, M. Shimizu, K. Sagawa, and S. Sudo, "Fluoride-Based Erbium-Doped Fiber Amplifier with Inherently Flat Gain Spectrum," *IEEE Photonics Tech. Lett.*, Vol. 8, 1996, pp. 882–884.

[186] Massicott, J. F., J. R. Armitage, R. Wyatt, B. J. Ainslie, and S. P. Craig-Ryan, "High Gain, Broadband, 1.6 μm Er^{3+} Doped Silica Fiber Amplifier," *Electron. Lett.*, Vol. 26, 1990, pp. 1645–1646.

[187] Yamada, M., T. Kanamori, Y. Terunuma, K. Oikawa, M. Shimizu, Y. Ohishi, and S. Sudo, "Erbium-Doped Fluoride Fiber Amplifier with Inherently Flat Gain Spectrum," *Proc. Institute of Electron., Information and Communication Engineers* (IEICE), Japan, 1995, Paper c-221. (Japanese)

[188] Laming, R. I., J. E. Townsend, D. N. Payne, F. Meli, G. Grasso, and E. J. Tarbox, "High-Power Erbium-Doped-Fiber Amplifiers Operating in the Saturated Regime," *IEEE Photonics Tech. Lett.*, Vol. 3, 1991, pp. 253–255.

[189] Massicott, J. F., R. Wyatt, B. J. Ainslie, and S. P. Craig-Ryan, "Efficient, High Power, High Gain, Er^{3+} Doped Silica Fiber Amplifiers," *Electron. Lett.*, Vol. 26, 1990, pp. 1038–1039.

[190] Pedersen, B., M. L. Dakss, B. A. Thompson, W. J. Miniscalco, T. Wei, and L. J. Andrews, "Experimental and Theoretical Analysis of Efficient Erbium-Doped Fiber Power Amplifier," *IEEE Photonics Tech. Lett.*, Vol. 3, 1991, pp. 1085–1087.

[191] Desurvire, E., "Spectral Noise Figure of Er^{3+}-Doped Fiber Amplifiers," *IEEE Photonics Tech. Lett.*, Vol. 2, 1990, pp. 208–210.

[192] Payne, D. N., and R. I. Laming, "Fiber Optical Amplifiers," *Proc. OFC'90*, San Francisco, CA, 1990, Paper THF1.

[193] Armitage, J. R., "Spectral Dependence of the Small-Signal Gain Around 1.5-μm in Erbium-Doped Silica Fiber Amplifiers," *IEEE J. Quantum Electron.*, Vol. 26, 1990, pp. 423–425.

[194] Vendeltorp-Pommer, H., B. Pedersen, A. Bjarklev, and J. H. Povlsen, "Noise and Gain Performance for an Er^{3+} Doped Fibre Amplifier Pumped at 980 nm or 1480 nm," *Proc. Society of Photo-Optical Instrumentation Engineers Conf., OE/Fibers '90*, Fiber Laser Sources and Amplifiers ll, San Jose, CA, 1990, pp. 254–265.

[195] Dakss, M. L., B. Pedersen, G. R. Joyce, and W. J. Miniscalco, "Signal Output Power and Noise in Erbium-Doped Fiber Power Amplifiers," *Proc. SPIE Conf. on Fiber Laser Sources and Amplifiers*, Vol. 1581, 1991, pp. 236–250.

[196] Pedersen, B., J. Chinavuri, and W. J. Miniscalco, "Gain and Noise Properties of Small-Signal Erbium-Doped Fiber Amplifiers Pumped in the 980-nm Band," *IEEE Photonics Tech. Lett.*, Vol. 4, 1992, pp. 556–558.

[197] Marcerou, J. F., H. Fevrier, J. Hervo, and J. Auge, "Noise Characteristics of the EDFA in Gain Saturation Regimes," *Proc. OAA'91*, Snowmass Village, Co, Paper ThE1, 1991, pp. 162–165.

[198] Smart, R. G., J. L. Zyskind, J. W. Sulhoff, and D. J. DiGiovanni, "An Investigation of the Noise Figure and Conversion Efficiency of 0.98 μm Pumped Erbium-Doped Fiber Amplifier under Saturated Conditions," *IEEE Photonics Tech. Lett.*, Vol. 4, 1992, pp. 11261–1264.

[199] Saito, T., and T. Mukai, "1.5-μm GaInAsP Traveling-Wave Semiconductor Laser Amplifier," *IEEE J. Quantum Electron.*, Vol. 23, 1987, pp. 1010–1020.

[200] Kagi, N., A. Oyobe, and K. Nakamura, "Temperature Dependence of the Gain in an Erbium-Doped Fiber," *Proc. CLEO'90*, Anaheim, CA, 1990, Paper CPDP36.

[201] Yamada, M., M. Shimizu, M. Horiguchi, and M. Okayasu, "Temperature Dependence of Signal Gain in Er-Doped Optical Fiber Amplifiers," *IEEE J. Quantum Electron.*, Vol. 23, 1992, pp. 640–649.

[202] Yamada, M., M. Shimizu, M. Okayasu, and M. Horiguchi, "Temperature Insensitive Er-Doped Optical Fiber Amplifiers," *Electron. Lett.*, Vol. 26, 1990, pp. 1649–1650.

[203] Masuda, H., A. Takada, and K. Aida, "Temperature Characteristics of Erbium-Doped Single-Mode Fiber Amplifiers," *Tech. Digest of Technical Group, OCS'90-22*, Institute of Electron., Information and Communication Engineers (IEICE), Japan, 1990, pp. 17–23. (Japanese)

[204] Suyama, M., R. I. Laming, and D. N. Payne, "Temperature Dependent Gain and Noise Characteristics of a 1.480 nm-Pumped Fiber Amplifier," *Electron. Lett.*, Vol. 26, 1990, pp. 1756–1757.

[205] Ainslie, B. J., S. P. Craig-Ryan, S. T. Davey, J. R. Armitage, C. G. Atkins, and R. Wyatt, "Optical Analysis of Er^{3+} Doped Fiber for Efficient Lasers and Amplifiers," *Proc. IOOC'89*, Kobe, Japan, 1989, Paper 20A3-2.

[206] Shimizu, M., M. Yamada, M. Horiguchi, and E. Sugita, "Concentration Effect on Optical Amplification Characteristics of Er-Doped Silica Single-Mode Fibers," *IEEE Photonics Tech. Lett.*, Vol. 2, 1990, pp. 43–45.

[207] Kagi, N., A. Oyobe, and K. Nakamura, "Efficient Optical Amplifier Using a Low-Concentration Erbium-Doped Fiber," *IEEE Photonics Tech. Lett.*, Vol. 2, 1990, pp. 559–561.

[208] Blixt, P., J. Nilsson, T. Carlnas, and B. Jaskorzynska, "Amplification Reduction in Fiber Amplifiers Due to Up-Conversion at Er^{3+}-Concentrations Below 1000 ppm," *Proc. OAA'91*, Snowmass Village, CO, 1991, Paper WD3, pp. 52–55.

[209] Georges, T., E. Delevaque, M. Monerie, P. Lamouler, and J. F. Bayon, "Pair-Induced Quenching in Erbium-Doped Silicate Fibers," *Proc. OAA'92*, Santa Fe, NM, 1992, Paper WE4, pp. 71–74.

[210] Yamauchi, R., "Erbium Doped Fiber Amplifiers for 1.55 μm Band and Their Key Technologies," *Optoelectronics-Devices and Technologies*, Vol. 10, 1995, pp. 39–54.

[211] Arai, K., H. Namikawa, K. Kumata, and T. Honda, "Aluminum or Phosphorus Codoping Effects on the Fluorescence and Structural Properties of Neodymium-Doped Silica Glass," *J. Appl. Phys.*, Vol. 59, 1986, pp. 3430–3436.

[212] Ainslie, B. J., J. R. Armitage, S. P. Craig, and B. Wakefield, "Fabrication and Optimization of the Erbium Distribution in Silica Based Doped Fiber," *Proc. ECOC'88*, part 1, Brighton, United Kingdom, 1988, pp. 62–65.

[213] Pedersen, B., A. Bjarklev, O. Lumholt, and J. H. Povlsen, "Detailed Design Analysis of Erbium-Doped Fiber Amplifiers," *IEEE Photonics Tech. Lett.*, Vol. 3, 1991, pp. 548–550.

[214] Pedersen, B., M. L. Dakss, B. A. Thompson, W. J. Miniscalco, T. Wei, and L. J. Andrews, "Experimental and Theoretical Analysis of Efficient Erbium-Doped Fibre Power Amplifiers," *IEEE Photonics Tech. Lett.*, Vol. 3, 1991, pp. 548–550.

[215] Kagi, N., A. Oyobe, T. Morikawa, Y. Sasaki, and K. Nakamura, "Gain Characteristics of Er^{3+} Doped Fiber with Quasiconfined Structure," *Proc. OFC '90*, San Francisco, CA, 1990, Paper FA8, p. 200.

[216] Ohashi, M., and M. Tsubokawa, "Optimum Parameter Design of Er^{3+}-Doped Fiber for Optical Amplifiers," *Electron. Lett.*, Vol. 3, 1991, pp. 121–123.

[217] Shimizu, M., and T. Kitagawa, "Fabrication and Characterization of Er-Doped Single-Mode Fibers and Waveguides," *Proc. Meeting on Optical Fiber Amplifiers and Semiconductor Amplifiers*, Microoptics News Group of MICROOPTICS (The Optical Society of Japan), Tokyo, Japan, 1900, pp. 13–18. (Japanese)

[218] Pedersen, B., A. Bjarklev. and J. H. Povlsen, "Design of Erbium-Doped Fiber Amplifiers for 980 nm or 1480 nm Pumping," *Electron. Lett.*, Vol. 27, 1991, pp. 255–256.

[219] Pedersen, B., A. Bjarklev, J. H. Povlsen, K. Dybdal, and C. C. Larsen, "The Design of Erbium-Doped Fiber Amplifiers," *IEEE J. Lightwave Tech.*, Vol. 9, 1991, pp. 1105–1112.

[220] Kawakami, S., K. Shiraishi, and Y. Aizawa, "A Method To Realize Fiber-Embedded Optical Devices," *2nd Optoelectron. Conf. Tech. Dig.*, 1988, pp. 172–173.

[221] Hanafusa, H., M. Horiguchi, and J. Noda, "Thermally-Diffused Expanded Core Fibers for Low-Loss and Inexpensive Photonic Components," *Electron. Lett.*, Vol. 27, 1991, pp. 1968–1969.

[222] Hansen, S. L., K. Dybdal, and C. C. Larsen, "Upper Gain Limit in Er-Doped Fiber Amplifiers Due to Internal Rayleigh Backscattering," *Proc. OFC'92*, San Jose, CA, 1992, Paper TuL4, p. 68.

[223] Povlsen, J. H., A. Bjarklev, O. Lumholt. H. Vendeltorp-Pommer, and K. Rottwitt, "Optimizing Gain and Noise Performance of EDFAs with Insertion of a Filter or an Isolator," *Proc. SPIE Conf. on Fiber Laser Sources and Amplifiers*, Vol. 1581, 1991, pp. 107–113.

[224] Yamashita, S., and T. Okoshi, "Performance Improvement and Optimization of Fiber Amplifier with a Midway Isolator," *IEEE Photonics Tech. Lett.*, Vol. 4, 1992, pp. 1276–1278.

[225] Zervas, M. N., R. I. Laming, and D. N. Payne, "Efficient Erbium-Doped Fiber Amplifier with an Integral Isolator," *Proc. OAA'92*, Santa Fe, NM, 1992, Paper FB2, pp. 162–165.

[226] Lumholt, O., K. Schusler, A. Grunnet-Jepsen, A. Bjarklev, S. Dahl-Petersen, J. H. Povlsen, T. P. Rasmussen, K. Rottwitt, and C. C. Larsen, "Low Noise High Gain Preamplifier Using an Isolator Within the Erbium Doped Fibre," *Proc. ECOC'92*, Berlin, Germany, 1992, Paper MoA3.5, pp. 93–96.

[227] Laming, R. I., M. N. Zervas, and D. N. Payne, "54dB Gain Quantum-Noise-Limited Erbium-Doped Fiber Amplifier," *Proc. ECOC'92*, Berlin, Germany, 1992, Paper MoA3.4, pp. 89–92.

[228] Lumholt, O., J. H. Povlsen, K. Schusler, A. Bjarklev, S. Dahl-Petersen, T. P. Rasmussen, and K. Rottwitt, "Quantum Limited Noise Figure Operation of High Gain Erbium-Doped Fiber Amplifiers," *IEEE J. Lightwave Tech.*, Vol. 11, 1991, pp. 1344–1352.

[229] Yu, A., M. J. O'Mahony, and A. S. Siddiqui, "Analysis of Optical Gain Enhanced Erbium-Doped Fiber Amplifiers Using Optical Filters," *IEEE Photonics Tech. Lett.*, Vol. 5, 1993, pp. 773–775.

[230] Aoki, Y., T. Saito, K. Fukagawa, Y. Sunohara, S. Ishikawa, and S. Fujita, "Low Noise and High Saturation Output Power Erbium-Doped Fiber Amplifiers Pumped with 0.98 μm and 1.48 μm LDs for Long-Distance Optical Communication," *Proc. ECOC'90*, Amsterdam, Netherlands, 1990, Paper WeC9-2, pp. 585–588.

[231] Nishi, S., K. Aida, and K. Nakagawa, "Highly Efficient Configuration of Erbium-Doped Fibre Amplifier," *Proc. ECOC'90*, Amsterdam, Netherlands, 1990, pp. 99–102.

[232] Giles, C. R., J. Stone, L. W. Stulz, K. Walker, and C. A. Burns, "Gain Enhancement in Reflected-Pump Erbium-Doped Fiber Amplifiers," *Proc. OAA '91*, Snowmass, CO, 1991, Paper ThD2. pp. 148–151.

[233] Lauridsen, V., R. Tadayoni, A. Bjarklev, J. H. Povlsen, and B. Pedersen, "Gain and Noise Performance of Fibre Amplifiers Operating in New Pump Configurations," *Electron. Lett.*, Vol. 27, 1991, pp. 327–329.

[234] Farries, M. C., C. M. Ragdale, and D. C. Reid, "Fibre Bragg Filters for Pump Rejection and Recycling in Erbium-Doped Fibre Amplifiers," *Proc. OAA '91*, Snowmass, CO, 1991, Paper ThDl, pp. 144–147.

[235] Olsson, N. A., "Polarization-Independent Configuration Optical Amplifier," *Electron. Lett.*, Vol. 24, 1989, pp. 1075–1076.

[236] Tachibana, N., R. I. Laming, P. R. Morkel, and D. N. Payne, "Gain-Shaped Erbium-Doped Fiber Amplifier (EDFA) with Broad Spectral Bandwidth," *IEEE Photonics Tech. Lett.*, Vol. 3, pp. 118–120.

[237] Wilkinson, M., A. Bebbington, S. A. Cassidy, and P. Mckee, "D-Fiber for Erbium Gain Spectrum Flattening," *Electron. Lett.*, Vol. 28, 1992, pp. 131–132.

[238] Kashyap, R., R. Wyatt, and R. J. Campbell, "Wideband Gain Flattened Erbium Fiber Amplifier Using a Photosensitive Fiber Blazed Grating," *Electron. Lett.*, Vol. 29, 1993, pp. 154–156.

[239] Kashyap, R., R. Wyatt, and P. F. Mckee: "Wavelength Flattened Saturated Erbium Amplifier Using Multiple Side-Tap Bragg Gratings," *Electron. Lett.*, Vol. 29, 1993, pp. 1025–1026.

[240] Grasso, G., F. Fontana, A. Righetti, P. Scrivener, P. Turner, and P. Maton, "980-nm Diode-Pumped Er-Doped Fiber Optical Amplifiers with High Gain-Bandwidth Product," *Proc. OFC'91*, San Diego, CA, 1991, Paper FA3, p. 195.

[241] Pan, J. Y., M. A. Ali, A. F. Elrefaie, and R. E. Wagner, "Multiwavelength Fiber-Amplifier Cascades with Equalization Employing Mach-Zehnder Optical Filter," *IEEE Photonics Tech. Lett.*, Vol. 7, 1995, pp. 1501–1503.

[242] Belov, A. V., E. M. Dianov, V. I. Karpov, V. N. Protopopov, M. Y. Tsvetkov, A. N. Guryanov, V. F. Khopin, W. H. Choe, and S. J. Kim, "Gain Spectrum Flattening of Erbium-Doped Fiber Amplifier Using Tapered Fiber Filter," *Proc. OAA '95*, Davos, Switzerland, 1995, Paper ThDl. pp. 32–35.

[243] Inoue, K., T. Kominato, and H. Toba, "Tunable Gain Equalization Using a Mach-Zehnder Optical Filter in a Multistage Fiber Amplifier," *IEEE Photonics Tech. Lett.*, Vol. 3, 1991, pp. 718–720.

[244] Toba, H., K. Nakanishi, K. Oda, K. Inoue, and T. Kominato, "A 100-Channel Optical FDM In-Line Amplifier System Employing Tunable Gain Equalizers," *Proc. ECOC'92*, Berlin, Germany, 1992, Paper TuA4.2, pp. 113–116.

[245] Betts, R. A., S. J. Frisken, and D. Wong, "Split-Beam Fourier Filter and Its Application in a Gain-Flattened EDFA," *Proc. OFC'95*, San Diego, CA, 1995, Paper TuP4, pp. 80–81.

[246] Su, S., R. Olshansky, D. A. Smith, and J. E. Baran, "Flattening of Erbium-Doped Fiber Amplifier Gain Spectrum Using an Acoustooptic Tunable Filter," *Proc. ECOC'92*, 1992, Paper WeP2.3, pp. 477–480.

[247] Giles, C. R., and D. J. DiGiovanni, "Dynamic Gain Equalization in Two-Stage Fiber Amplifiers," *IEEE Photonics Tech. Lett.*, Vol. 2, 1990, pp. 866–868.

[248] Shigematsu, M., M. Kakui, and M. Nishimura, "120 Channel AM-VSB Signal Transmission by 2 Wavelength Multiplexing Through Gain Flattened Hybrid Erbium-Doped Fiber Amplifier," *Proc. OAA'95*, Davos, Switzerland, 1995, Paper ThB3, pp. 13–16.

[249] Kashiwada, T., M. Shigematsu, M. Kakui, M. Onishi, and M. Nishimura, "Gain-Flattened Optical-Fiber Amplifiers with a Hybrid Er-Doped-Fiber Configuration for WDM Transmission," *Proc. OFC'95*, San Diego, CA, 1995, Paper TuP1, pp. 77–78.

[250] Zervas, M. N., and R. I. Laming, "Twin-Core Fiber Erbium-Doped Channel Equalizer," *IEEE J. Lightwave Tech.*, Vol. 13, 1995, pp. 721–731.

[251] Laming, R. I., J. D. Minelly, L. Dong, and M. N. Zervas, "Twincore Erbium-Doped Fiber Amplifier with Passive Spectral Gain Equalization," *Electron. Lett.*, Vol. 29, 1993, pp. 509–510.

[252] Inagaki, S., N. Shukunami, and K. Okamura, "Wavelength Multiplexed Gain Characteristics of Ring Doped EDF," *Proc. OAA'96*, Monterey, CA, 1996, Paper FA4, pp. 85–88.

[253] Bayart, D., B. Kowalski, J. Hervo, M. L. Crom, S. Lempereur, J. J. Girard, and F. Chiquet, "Power Conversion Efficiency of Fluoride-Based EDFAs in Flat-Gain Operation," *Proc. OFC'96*, San Jose, CA, 1996, Paper WK8, pp. 157–158.

[254] Beylat, J. L., D. Bayart, B. Clesca, and B. M. Desthieux, "Flat-Gain Fiber Optical Amplifiers for WDM Networks," *Proc. ECOC'95*, Brussels, Belgium, 1995, Paper Th.L.7, pp. 1095–1102.

[255] Morioka, T., S. Kawanishi, H. Takara, O. Kamatani, M. Yamada, T. Kanamori, K. Uchiyama, and M. Saruwatari, "100 Gbit/s × 4 ch, 100 km Repeaterless TDM-WDM Transmission Using a Single Supercontinuum Source," *Proc. ECOC'95*, Brussels, Belgium, 1995, Paper Th.A.3.4, pp. 979–982.

[256] Clesca, B., S. Artigaud, L. Pierre, and J. P. Thiery, "8 × 10 Gbit/s Repeaterless Transmission over 150 km of Dispersion-Shifted Fiber Using Fluoride-Based Fiber-Amplifiers," *Proc. OAA'95*, Davos, Switzerland, 1995, Paper SaB5, pp. 220–223.

[257] Yoshida, S., S. Kuwano, M. Yamada, T. Kanamori, K. Iwashita, and N. Takachio, "100 km Repeater Spacing 10-Gb/s, 10-Channel Transmission Experiment with 1.55 μm Dispersion Shifted Fiber and Fluoride-Based Erbium-Doped Fiber Amplifier," *Proc. IOOC'95*, Hong Kong, 1995, Paper PD2-3.

[258] Clesca, B., D. Bayart, C. Coeurjolly, L. Berthelon, L. Hamon, and J. L. Beylat, "Over 25-nm, 16 Wavelength-Multiplexed Signal Transmission Through Four Fluoride-Based Fiber-Amplifier Cascade and 440 km Standard Fiber," *Proc. OFC'94*, San Jose, CA, 1994, Paper PD20.

[259] Fukutoku, M., M. Fukui, M. Yamada, K. Oda, and H. Toba, "25 nm Bandwidth Optical Gain Equalization for 32-Channel WDM Transmission with a Lattice Type Optical Circuit," *Proc. OAA'96*, Monterey, CA, 1996, Paper FA4, pp. 66–69.

[260] Yamada, M., H. Ono, T. Kanamori, T. Sakamoto, Y. Ohishi, and S. Sudo, "A Low-Noise and Gain-Flattened Amplifier Composed of a Silica-Based and a Fluoride-Based Er^{3+}-Doped Fiber Amplifier in a Cascade Configuration," *IEEE Photonics Tech. Lett.*, Vol. 8, 1996, pp. 620–622.

[261] Morioka, T., H. Takara, S. Kawanishi, O. Kamatani, K. Takiguchi, K. Uchiyama, M. Saruwatari, H. Takahashi, M. Yamada, T. Kanamori, and H. Ono, "100 Gbit/s × 10 Channel OTDM/WDM Transmission Using a Single Supercontinuum WDM Source," *Proc. OFC'96*, San Jose, CA, 1996, Paper PD21.

[262] Nishi, T., K. Nakagawa, Y. Ohishi, and S. Takahashi, "The Amplification Properties of a Highly Er^{3+}-Doped Phosphate Fiber," *Japan J. Appl. Phys.*, Vol. 31, 1992, pp. L177–L179.

[263] Saifi, M. A., M. J. Andrejco, W. I. Way, A. Von Lehman, A. Y. Yan, C. Lin, F. Bilodeau, and K. O. Hill, "Er^{3+}-Doped GeO_2-CaO-Al_2O_3 Silica Core Fiber Amplifier Pumped at 813 nm, "*Proc. OFC'91*, San Diego, CA, 1991, Paper FA6, p. 198.

[264] Yamada, M., M. Shimizu, M. Horiguchi, M. Okayasu, and E. Sugita, "Gain Characteristics of an Er^{3+}-Doped Multicomponent Glass Single-Mode Optical Fiber," *IEEE Photonics Tech. Lett.*, Vol. 2, 1990, pp. 656–658.

[265] Ohashi, M., and K. Shiraki, "Er^{3+}-Doped Multicomponent Glass Core Fiber Amplifier Pumped at 1.48 μm," *Electron. Lett.*, Vol. 27, 1991, pp. 2143–2145.

[266] Ono, H., K. Nakagawa, M. Yamada, and S. Sudo, "Gain Characteristics of Er^{3+}-Doped Fluorophosphate Glass Single-Mode Fiber for a WDM Signal," *Electron. Lett.*, Vol. 32, 1996, pp. 1586–1587.

[267] Yamada, M., M. Shimizu, Y. Ohishi, M. Horiguchi, S. Sudo, and A. Shimizu, "Flattening the Gain Spectrum of an Erbium-Doped Fiber Amplifier by Connecting an Er^{3+}-Doped SiO_2-Al_2O_3 Fiber and an Er3+-Doped Multicomponent Fiber," *Electron. Lett.*, Vol. 30, 1994, pp. 1762–1763.

[268] Shikii, S., and Y. Tamura, "240 mW Pump Module Using Semiconductor Laser," *Proc. Institute of Electron., Information and Communication Engineers* (IEICE), General Spring Conf., Japan, 1989, Paper c-623. (Japanese)

[269] Welch, D. F., R. Parke, D. Mehuys, A. Hardy, R. Lang, S. O'Brien, and S. Scifres, "1.1 W CW, Diffraction-Limited Operation of a Monolithically Integrated Flared-Amplifier Master Oscillator Power Amplifier," *Electron. Lett.*, Vol. 28, 1992, pp. 2011–2013.

[270] Livas, J. C., S. R. Chinn, E. S. Kintzer, J. N. Walpole, C. A. Wang, and L. J. Missaggia, "High Power Erbium-Doped Fiber Amplifier Pumped by Tapered-Gain-Region Lasers," *Proc. OAA'94*, Breckenridge, CO, 1994, Paper WC4, pp. 29–31.

[271] Sugo, M., J. Temmyo, T. Nishiya, E. Kuramochi, and T. Tamamura, "1.02-μm Pump Laser Diodes with High Power Above 300 mW into Single-Mode Fiber," *Proc. OAA'95*, Davos, Switzerland, 1995, Paper ThC3, pp. 28–30.

[272] Sanders, S., F. Shum, R. J. Lang, J. D. Ralston, D. G. Mehuys, R. G. Waarts, and D. F. Welch, "Fiber-Coupled M-MOPA Laser Diode Pumped a High-Power Erbium-Doped Fiber Amplifier," *Proc. OFC'96*, San Jose, CA, 1996, Paper TuG5, pp. 31–32.

[273] Grubb, S. G., T. Strasser, W. Y. Cheung, W. A. Reed, V. Mizrahi, T. Erdogan, P. J. Lemaire, A. M. Vengsarkar, and D. J. DiGiovanni, "High-Power 1.48 μm Cascaded Raman Laser in Germanosilicate Fibers," *Proc. OAA'95*, Davos, Switzerland, 1995, Paper SaA4, pp. 197–199.

[274] Bousselet, P., A. Marquant, F. Chiquet I. Riant, and J. L. Beylat, "1 W Output Power All-Fiber Laser Based Tm^{3+}-Doped Fluoride Fiber," *Proc. OAA'95*, Davos, Switzerland, 1995, Paper ThD4, pp. 44–47.

[275] Snitzer, E., and R. Woodcock, "Yb^{3+}-Er^{3+} Glass Laser," *Appl. Phys. Lett.*, Vol. 6, 1965, pp. 45–46.

[276] Hanna, D. C., R. M. Percival, I. R. Perry, R. G. Smart, and A. C. Tropper, "Efficient Operation of an Yb Sensitized Er Fiber Laser Pumped in the 0.8 μm Region," *Electron. Lett.*, Vol. 24, 1988, pp. 1068–1069.

[277] Barnes, W. L., S. B. Poole, J. E. Townsend, L. Reekie, D. J. Taylor, and D. N. Payne, "Yb^{3+}-Er^{3+} and Yb^{3+} Doped Fiber Lasers," *IEEE J. Lightwave Tech.*, Vol. 7, 1989, pp. 1461–1465.

[278] Grubb, S. G., W. Humer, R. S. Cannon, S. W. Vendetta, K. L. Sweeney, P. A. Leilabady, M. R. Keur, J. G. Kwasegroch, T. C. Munks, and D. W. Anthon, "24.6 dBm Output Er/Yb Codoped Optical Amplifier Pumped by Diode-Pumped Nd:YLF Laser," *Electron. Lett.*, Vol. 28, 1992, pp. 1275–1276.

[279] Grubb, S. G., D. J. DiGiovanni, J. R. Simpson, W. Y. Cheung, S. Sanders, D. F. Welch, and B. Rockney, "Ultrahigh Power Diode-Pumped 1. 5-μm Fiber Amplifiers," *Proc. OFC'96*, San Jose, CA, 1996, Paper TuG4, pp. 30–31.

[280] Wysocki, P. F., G. Nykolak, D. S. Shenk, and K. Eason, "Noise Figure Limitation in Ytterbium-Codoped Erbium-Doped Fiber Amplifiers Pumped at 1064 nm," *Proc. OFC'96*, San Jose, CA, 1996, Paper TuG5, pp. 31–32.

[281] Snitzer, E., H. Po, F. Hakimi, R. Tumminelli, and B. C. McCollum, "Double-Clad, Offset Core Nd Fiber Laser," *Proc. OFC/OFS'88*, New Orleans, Louisiana, 1988, Paper PD5.

[282] Po, H., E. Snitzer, R. Tumminelli, L. Zenteno, F. Hakimi, N. M. Cho, and T. Haw, "Double-Clad High Brightness Nd Fiber Laser Pumped by GaAlAs Phased Array," *Proc. OFC'89*, Houston, TX, 1989, Paper PD7.

[283] Minelly, J. D., Z. J. Chen, R. I. Laming, and J. E. Caplen, "Efficient Cladding Pumping of an Er^{3+} Fiber," *Proc. ECOC'95*, Brussels, Belgium, 1995, Paper Th.L.1.2, pp. 917–920.

[284] Miniscalco, W. J., L. J. Andrews, and B. A. Thompson, "1.3 μm Fluoride Fiber Laser," *Electron. Lett.*, Vol. 24, 1988, pp. 28–29.

[285] Brierley, M. C., and C. A. Millar, "Amplification and Lasing at 1350 nm in a Neodymium Doped Fluorozirconate Fibre," *Electron. Lett.*, Vol. 24, 1988, pp. 438–439.

[286] Miyajima, Y., T. Komukai, and T. Sugawa, "1. 31–1.36 μm Optical Amplification in a Nd^{3+}-Doped Fluorozirconate Fibre," *Electron. Lett.*, Vol. 26, 1990, pp. 194–195.

[287] Sugawa, T., Y. Miyajima, and T. Komukai, "10 dB Gain and High Saturation Power in a Nd^{3+}-Doped Fluorozirconate Fibre Amplifier," *Electron. Lett.*, Vol. 26, 1990, pp. 2042–2044.

[288] Ohishi, Y, T. Kanamori, T. Kitagawa, S. Takahashi, E. Snitzer, and G. H. Sigel, Jr., "Pr^{3+}-Doped Fluoride Fiber Amplifier Operating at 1.31 μm," *Proc. OFC'91*, San Diego, CA, 1991, Paper PDP2.

[289] Ohishi, Y., T. Kanamori, T. Kitagawa, S. Takahashi, E. Snitzer, and G. H. Sigel, Jr., "Pr^{3+}-Doped Fluoride Fiber Amplifier Operating at 1.31 μm," *Opt. Lett.*, Vol. 16, 1991, pp. 1747–1749.

[290] Durteste, Y., M. Monerie, J. Y. Allain, and H. Poignant, "Amplification and Lasing at 1.3 μm in Praseodymium-Doped Fluorozirconate Fibres," *Electron. Lett.*, Vol. 27, 1991, pp. 626–628.

[291] Carter, S. F., D. Szebesta, S. T. Davey, R. Wyatt, M. C. Brierley, and P. W. France, "Amplification at 1.3 μm in a Pr^{3+}-Doped Single-Mode Fluorozircinate Fibre," *Electron. Lett.*, Vol. 27, 1991, pp. 628–629.

[292] Ohishi, Y., T. Kanamori, T. Nishi, and S. Takahashi, "A High Gain, High Output Saturation Power Pr^{3+}-Doped Fluoride Fiber Amplifier Operating at 1.3 μm," *IEEE Photonics Tech. Lett.*, Vol. 3, 1991, pp. 715–717.

[293] Kanamori, T., Y. Terunuma, K. Fujiura, Y. Ohishi, and S. Sudo, "Fabrication of Low-Loss High-Δn, Pr^{3+}-Doped Fluoride Single-Mode Fibers for 1.3 μm Optical Amplifiers," *9th Int. Symp. on Non-oxide Glasses*, Hangzhou, 1994, pp. 74–79.

[294] Miyajima, Y., T. Sugawa, and Y. Fukasaku, "38.2 dB Amplification at 1.3 μm and Possibility of 0.98 μm Pumping in Pr^{3+}-Doped Fluoride Fiber," *Proc. OAA'91*, Snowmass Village, CO, 1991, Paper PDP1.

[295] Ohishi, Y., T. Kanamori, T. Temmyo, M. Wada, M. Yamada, M. Shimizu, K. Yoshino, H. Hanafusa, M. Horiguchi, and S. Takahashi, "Laser Diode Pumped Pr^{3+}-Doped and Pr^{3+}-Yb^{3+}-Codoped Fluoride Fibre Amplifier Operating at 1.3 mm," *Electron. Lett.*, Vol. 27, 1991, pp. 1995–1996.

[296] Shimizu, M., T. Kanamori, J. Temmyo, M. Wada, M. Yamada, Y. Terunuma, Y. Ohishi, and S. Sudo, "28.3 dB Gain 1.3 μm-Band Pr-Doped Fluoride Fiber Amplifier Module Pumped by 1.017 μm InGaAs-LD's," *IEEE Photonics Tech. Lett.*, Vol. 5, 1993, pp. 654–657.

[297] Whitley, T., R. Wyatt, D. Szebesta, S. Davey, and J. R. Williams, "Quarter Watt Output at 1.3 μm from a Praseodymium Doped Fluoride Fibre Amplifier Pumped with a Diode-Pumped Nd:YLF Laser," *Proc. OAA'92*, Santa Fe, NM, 1992, Paper PDP4.

[298] Whitley, T., R. Lobbett, R. Wyatt, and D. Szebesta, "5 Gbps Transmission over 100 km of Optical Fiber Using a Directly Modulated DFB Laser and an Engineered 1.3 Micron Pr^{3+}-Doped Fluoride Fiber Power Amplifier," *Proc. OAA'94*, Breckenridge, CO, 1994, Paper WB1.

[299] Whitley, T., "A Review of Recent System Demonstrations Incorporating 1.3-μm Praseodymium-Doped Fluoride Fiber Amplifiers," *IEEE J. Lightwave Tech.*, Vol. 13, 1995, pp. 744–760.

[300] Yamada, M., and M. Shimizu, "Progress Toward the Telecommunication Application of 1.3-μm Praseodymium-Doped Fiber Amplifiers," *Proc. OAA'94*, Breckenridge, CO, 1994, Paper WC1.

[301] Ohishi, Y., T. Kanamori, K. Fujiura, Y. Nishida, M. Yamada, and S. Sudo, "Recent Progress in 1.3-μm Fiber Amplifiers," *Proc. OFC'96*, San Jose, CA, 1996, Paper TuG1.

[302] Bjarklev, A., *Optical Fiber Amplifiers: Design and System Applications*, Boston and London: Artech House, Inc., 1993.

[303] Pedersen, J. E., and M. C. Brierley, "High Saturation Output Power from a Neodymium-Doped Fluoride Fiber Amplifier Operating in the 1300 nm Telecommunications Window," *Electron. Lett.*, Vol. 26, 1990, pp. 819–820.

[304] Dakss, M. L., and W. J. Miniscalco, "Fundamental Limits on Nd^{3+}-Doped Fiber Amplifier Performance at 1.3 μm," *IEEE Photonics Tech. Lett.*, Vol. 2, 1990, pp. 650–652.

[305] Pedersen, J. E., M. Brierley, and R. A. Lobbett, "Noise Characterization of a Neodymium-Doped Fluoride Fiber Amplifier and Its Performance in a 2.4 Gb/s System," *IEEE Photonics Tech. Lett.*, Vol. 2, 1990, pp. 750–752.

[306] Dakss, M. L., and W. J. Miniscalco, "A Large Signal Model and Signal/Noise Ratio Analysis for Nd^{3+}-Doped Fiber Amplifiers at 1.3 μm," *Proc. SPIE Conf. on Fiber Laser Sources and Amplfiers*, Vol. 1373, 1990, pp. 111–124.

[307] Øbro, M., B. Pedersen, A. Bjarklev, J. H. Povlsen, and J. E. Pedersen, "Highly Improved Fibre Ampilfier for Operation Around 1300 nm," *Electron. Lett.*, Vol. 27, 1991, pp. 470–472.

[308] Brierley, M. C., S. F. Carter, P. W. France, and J. E. Pedersen, "Amplification in the 1300nm Telecommunications Window in a Nd-Doped Fluoride Fibre," *Electron. Lett.*, Vol. 26, 1990, pp. 329–330.

[309] Miyajima, Y., T. Sugawa, and T. Komukai, "Efficient 1.3 μm Band Amplification in a Nd^{3+}-Doped Single-Mode Fluoride Fiber," *Electron. Lett.*, Vol. 26, 1990, pp. 1397–1398.

[310] Zemon, S. A., B. Pedersen, G. Lambert, W. J. Miniscalco, B. T. Hall, R. C. Folwefiler, B. A. Thompson, and L. J. Andrews:, "Excited-State-Absorption Cross Sections and Amplifier Modeling in the 1300-nm Region for Nd-Doped Glasses," *IEEE Photonics Tech. Lett.*, Vol. 4, 1992, pp. 224–227.

[311] Miniscalco, W. J., B. A. Thompson, M. L. Dakss, S. A. Zemon, and L. J. Andrews, "The Measurement and Analysis of Cross Sections for Rare-Earth-Doped Glasses," *Proc. SPIE Conf. on Fiber Laser Sources and Amplfiers*, Vol. 1581, 1991, pp. 80–90.

[312] Ishikawa, E., H. Aoki, T. Yamashita, and Y. Asahara, "Laser Emission and Amplification at 1.3 μm in Neodymium-Doped Fluorophosphate Fiber," *Electron. Lett.*, Vol. 28, 1992, pp. 1497–1499.

[313] Zemon, S., W. J. Miniscalco, and B. A. Thompson, "Nd^{3+}-Doped Fluoroberyllate Glasses for Fiber Amplifiers at 1300 nm," *Proc. OAA'92*, Santa Fe, NM, 1992, Paper WB3, pp. 12–15.

[314] Øbro, M., J. E. Pedersen, and M. C. Brierley, "Gain Enhancement in Nd^{3+} Doped ZBLAN Fibre Amplifier Using Mode Coupling Filter," *Electron. Lett.*, Vol. 28, 1992, pp. 99–100.

[315] Bjarklev, A., T. Rasmussen, H. Povlsen, O. Lumholt, K. Rottwitt, S. Dahl-Petersen, and C. C. Larsen, "9 dB Gain Improvement of 1300 nm Optical Amplifier by Amplified Spontaneous Emission Suppressing Fibre Design," *Electron. Lett.*, Vol. 27, 1991, pp. 1701–1702.

[316] Ohishi, Y., T. Kanamori, T. Nishi, S. Takahashi, and E. Snitzer, "Concentration Effect on Gain of Pr^{3+}-Doped Fluoride Fiber for 1.3 μm Amplification," *IEEE Photonics Tech. Lett.*, Vol. 4, 1992, pp. 1338–1341.

[317] Ohishi, Y., T. Kanamori, M. Shimizu, M. Yamada, Y. Terunuma, J. Temmyo, M. Wada, and S. Sudo, "Praseodymium-Doped Fiber Amplifier at 1.3 μm," *The Institute of Electron., Information and Communication Engineers (IEICE) Trans. Comm.*, Vol. E77-B, 1994, pp. 421–440.

[318] Ohishi, Y., T. Kanamori, Y. Terunuma, M. Shimizu, M. Yamada, and S. Sudo, "Investigation of Efficient Pump Scheme for Pr^{3+}-Doped Fluoride Fiber Amplifier," *IEEE Photon. Tech. Lett.*, Vol. 6, 1994, pp. 195–198.

[319] Pedersen, B., W. J. Miniscalco, and R. S. Quimby, "Optimization of Pr^{3+}:ZBLAN Fiber Amplifiers," *IEEE Photon. Tech. Lett.*, Vol. 4, 1992, pp. 446–448.

[320] Yamada, M., M. Shimizu, Y. Ohishi, T. Kanamori, and S. Sudo, "Gain Characteristics of Pr^{3+}-Doped Fluoride Fiber Amplifier," *Electronics and Comm. in Japan*, part 2, 1994, Vol. 77, pp. 75–87.

[321] Yamada, M., M. Shimizu, M. Horiguchi, M. Okayasu, and E. Sugita, "Gain Characteristics of Pr^{3+}-Doped Fluoride Fiber Amplifier," *Trans. Institute of Electron., Information and Communication Engineers* (IEICE), Japan, Vol. J77-C-I, 1994, pp. 462–469. (Japanese)

[322] Karasek, M., "Numerical Analysis of Pr3+-Doped Fluoride Fiber Amplifier," *IEEE Photon. Tech. Lett.*, Vol. 4, 1992, pp. 1266–1269.

[323] Yamada, M., M. Shimizu, Y. Ohishi, T. Kanamori, and S. Sudo, "Gain vs. Wavelength of 1.3 μm-Band Fluoride Fiber Amplifier," *Institute of Electron., Information and Communication Engineers, General Spring Conf.* (IEICE), Japan, 1994, Paper c-395. (Japanese)

[324] Whitley, T., R. Wyatt, S. Fleming, D. Szebesta, J. R. Williams, and S. T. Davey, "Noise and Cross-Talk Characteristics of a Praseodymium Doped Fluoride Fiber Amplifier," *Proc. OAA'93*, Yokohama, Japan, 1993, Paper TuA2, pp. 244–247.

[325] Sugawa, T., and Y. Miyajima, "Noise Characteristics of Pr^{3+}-Doped Fluoride Fiber Amplifier," *Electron. Lett.*, Vol. 28, 1992, pp. 246–247.

[326] Ohishi, Y., M. Yamada, A. Mori, T. Kanamori, M. Shimizu, and S. Sudo, "Effect of Temperature on Amplification Characteristics of Pr^{3+}-Doped Fluoride Fiber Amplifiers," *Optics Lett.*, Vol. 20, 1995, pp. 383–385.

[327] Layne, C. B., W. H. Lowdermilk, and M. J. Weber, "Multiphonon Relaxation of Rare-Earth Ions in Oxide Glass," *Phys. Rev. B*, 16, 1977, pp. 10–20.

[328] Krummrich, P. M., "Aspects of Pump Wavelength Selection for Pr^{3+}-Doped Fiber Amplifiers," *Proc. OAA'94*, Breckenridge, CO, 1994, Paper WC2.

[329] Yamada, M., M. Shimizu, Y. Ohishi, T. Kanamori, M. Horiguchi, S. Takahashi, J. Temmyo, and M. Wada, "15.1-dB-Gain Pr^{3+}-Doped Fluoride Fiber Amplifier Pumped by High Power Laser Diode Modules," *Proc. ECOC'92*, Berlin, Germany, 1992, Vol. Mo A2. 2, pp. 49–52.

[330] Temmyo, J., M. Sugo, T. Nishiya, M. Yamada, F. Bilodeau, and K. O. Hill, "A 310-mW Fiber-Coupled, Wavelength-Stabilized 1.016-μm InGaAs Quantum Well Laser Using a Fiber Bragg Grating," *Proc. CLEO/QELS'96*, Anaheim, CA, 1996, Paper CTuC6.

[331] Grubb, S. G., G. Nykolak, and P. C. Becker, "Master Oscillator Power Amplifier Diode Laser Pumped Sources for Pr^{3+}-Doped Fiber Amplifier," *Proc. OAA'94*, Breckenridge, CO, 1994, Paper WC3, pp. 25–28.

[332] Yamada, M., M. Shimizu, Y. Ohishi, J. Temmyo, T. Kanamori, and S. Sudo, "Semiconductor Laser-Pumped High-Gain Pr^{3+}-Doped Fiber Amplifier," *Optoelectronics-Devices and Technologies*, Vol. 10, 1995, pp. 109–118.

[333] Fake, M., E. P. Lawrence, and T. J. Simmons: "Practical Praseodymium Power Amplifier with a Saturated Output Power of +18 dBm," *Proc. OFC'94*, San Jose, CA, 1994, Paper ThF2.

[334] Yamada, M., M. Shimizu, Y. Ohishi, J. Temmyo, M. Wada, T. Kanamori, and S. Sudo, "One-LD-Pumped Pr^{3+}-Doped Fluoride Fiber Amplifier Module with a Signal Gain of 23 dB," *Electron. Lett.*, Vol. 29, 1993, pp. 1950–1951.

[335] Shimizu, M., M. Yamada, F. Hanawa, Y. Ohishi, J. Temmyo, M. Wada, Y. Terunuma, T. Kanamori, and S. Sudo, "1.3 μm-Band Pr-Doped Fluoride Fiber Amplifier Module Pumped by Laser Diodes," *Proc. OAA'92*, Santa Fe, NM, 1992, Paper PD3, pp. 9–13.

[336] Kikushima, K., M. Yamada, M. Shimizu, and J. Temmyo, "Distortion and Noise Properties of a Praseodymium-Doped Fluoride Fiber Amplifier in 1.3μm AM-SCM Video Transmission Systems," *IEEE Photon. Tech. Lett.*, Vol. 6, 1994, pp. 440–442.

[337] Suda, H., N. Tomita, S. Furukawa, F. Yamamoto, K. Kimura, M. Shimizu, M. Yamada, and Y. Ohishi, "Application of LD-Pumped Pr-Doped Fluoride Fiber Amplifier to Digital and Analog Signal Transmission," *Proc. OAA'93*, Yokohama, Japan, 1993, Paper PD8.

[338] Yamada, M., M. Shimizu, H. Yoshinaga, K. Kikushima, T. Kanamori, Y. Ohishi, Y. Terunuma, K. Oikawa, and S. Sudo, "Low-Noise Pr^{3+}-Doped Fluoride Amplifier," *Electron. Lett.*, Vol. 31, 1995, pp. 806–807.

[339] Yamada, M., M. Shimizu, T. Kanamori, Y. Ohishi, Y. Terunuma, K. Oikawa, H. Yoshinaga, K. Kikushima, Y. Miyamoto, and S. Sudo, "Low-Noise and High-Power Pr-Doped Fluoride Amplifier," *IEEE Photon. Tech. Lett.*, Vol. 7, 1995, pp. 869–871.

[340] Yamada, M., M. Shimizu, T. Kanamori, Y. Ohishi, Y. Terunuma, and S. Sudo, "Pr^{3+}-Doped Fluoride Fiber Amplifier Module Pumped by a Fiber-Coupled Master Oscillator Power Amplifier Diode Laser," *IEEE Photon. Tech. Lett.*, Vol. 9, 1997, pp. 321–323.

[341] Yoshinaga, H., M. Yamada, and M. Shimizu, "Improved Performance of Pr^{3+}-Doped Fluoride Fiber Amplifier in Subcarrier Multiplexed Multichannel AM-VSB Video Signal Transmission," *Electron. Lett.*, Vol. 30, 1994, pp. 2042–2043.

[342] Ohishi, Y., T. Kanamori, T. Nishi, Y. Nishida, and S. Takahashi, "Quantum Efficiency at 1.3 μm of Pr^{3+} in Infrared Glasses," *Proc. XVI Int. Congress on Glass*, Madrid, Spain, 1992, pp. 73–78.

[343] Becker, P. C., M. M. Broer, V. C. Lambrecht, A. J. Bruce, and C. Nykolak, "Pr^{3+}:La-Ga-S Glass: A Promising Material for 1.3 μm Fiber Amplification," *Proc. OAA'92*, Santa Fe, NM, 1992, Paper PDP5.

[344] Nishida, Y., Y. Ohishi, T. Kanamori, Y. Terunuma, K. Kobayashi, and S. Sudo, "Preparation of Pr^{3+}-Doped InF_3-Based Fluoride Single-Mode Fibers and First Demonstration of 1.3 μm Amplification," *Proc. ECOC'93*, Montreux, Switzerland, 1993, Paper TuC3.1.

[345] Yanagita, H., K. Ito, E. Ishikawa, H. Aoki, and H. Toratani, "26 dB Amplification at 1.31 μm in a Novel Pr3+-Doped InF_3/GaF_3-Based Fiber," *Proc. OFC'95*, San Diego, CA, 1995, Paper PD2.

[346] Kanamori, T., Y. Terunuma, Y. Nishida, K. Hoshino, K. Nakagawa, Y. Ohishi, and S. Sudo, "Fabrication of Fluoride Single-Mode Fibers for Optical Amplifiers," *Tenth Int. Symp. on Non-oxide Glasses*, Corning, NY, 1996, Paper 37, pp. 202–207.

[347] Nishida, Y., T. Kanamori, Y. Ohishi, M. Yamada, K. Kobayashi, and S. Sudo, "Efficient PDFA Module Using PbF_2/InF_3-Based Fluoride Fiber," *Proc. OAA'96*, Monterey, CA, 1996, Paper PD3.

[348] Ohishi, Y., A. Mori, T. Kanamori, K. Fujiura, and S. Sudo, "Fabrication of Praseodymium-Doped Arsenic Sulfide Chalcogenide Fiber for 1.3-μm Fiber Amplifiers," *Appl. Phys. Lett.*, Vol. 65, 1994, pp. 13–15.

[349] Koyamada, Y., N. Ohta, and N. Tomita, "Basic Concepts of Fiber Optic Subscriber Loop Operation Systems," *Proc. Conf. Comm.*, Vol. 4, 1990, pp. 1540–1544.

[350] Takasugi, H., N. Tomita, J. Nakano, and N. Atobe, "Design of a 1.65 μm-Band Optical Time-Domain Reflectometer," *IEEE J. Lightwave Tech.*, Vol. 11, 1993, pp. 1743–1748.

[351] Allain, J. Y., M. Monerie, and H. Poignant, "Tunable CW Lasing Around 0.82, 1.48, 1.88 and 2.35 μm in Thulium-Doped Fluorozirconate Fiber," *Electron. Lett.*, Vol. 25, 1989, pp. 1660–1662.

[352] Sankawa, I., H. Izumita, S. Furukawa, and K. Ishihara, "An Optical Fiber Amplifier for Wideband Wavelength Range Around 1.65 μm," *IEEE Photon. Tech. Lett.*, Vol. 2, 1990, pp. 422–424.

[353] Barnes, W. L., and J. E. Townsend, "Highly Tunable and Efficient Diode Pumped Operation of Tm^{3+} Doped Fiber Laser," *Electron. Lett.*, Vol. 26, 1990, pp. 746–747.

[354] Percival, R. M., D. Szebesta, and S. T. Davey, "Highly Efficient CW Cascade Operation of 1.47 and 1.82 μm Transitions in Tm-Doped Fluoride Fiber Laser," *Electron. Lett.*, Vol. 28, 1992, pp. 1866–1868.

[355] Allen, R., L. Esterowitz, and I. Aggarwal, "An Efficient 1.46 μm Thulium Fiber Laser Via a Cascade Process," *IEEE J. Quantum Electron.*, Vol. 29, 1993, pp. 303–306.

[356] Komukai, T., T. Yamamoto, T. Sugawa, and Y. Miyajima, "Upconversion Pumped Thulium-Doped Fluoride Fiber Amplifier and Laser Operating at 1.47 μm," *IEEE J. Quantum Electron.*, Vol. 31, 1995, pp. 1880–1889.

[357] Sakamoto, T., M. Shimizu, T. Kanamori, Y. Terunuma, Y. Ohishi, M. Yamada, and S. Sudo, "1.4-μm-Band Gain Characteristics of a Tm-Ho-Doped ZBLYAN Fiber Amplifier Pumped in the 0.8-μm Band," *IEEE Photonics Tech. Lett.*, Vol. 7, 1995, pp. 983–985.

[358] Percival, R. M., D. Szebesta, and S. T. Davey, "Thulium Doped Terbium Sensitized CW Fluoride Fiber Laser Operating on the 1.47 μm Transition," *Electron. Lett.*, Vol. 29, 1993, pp. 1054–1056.

[359] Rosenblatt, G. H., R. J. Ginther, R. C. Stoneman, and L. Esterowitz, "Laser Emission at 1.47 μm from Fluorozirconate Glass Doped with Tm^{3+} and Tb^{3+}," *OSA Proc. Tunable Solid State Lasers*, North Falmouth, MA, Vol. 5, 1989, pp. 373–376.

[360] Stoneman, R. C., and L. Esterowitz, "Continuous-Wave 1.50 μm Thulium Cascade Laser," *Opt. Lett.*, Vol. 16, 1991, pp. 232–234.

[361] Pafchek, R., J. Aniano, E. Snitzer, and G. H. Sigel, "Optical Absorption and Emission in Rare Earth Heavy Metal Fluoride Glasses," *Mat. Rea. Soc. Symp. Proc.*, Vol. 172, 1990, pp. 347–352.

[362] Yamamoto, T., Y. Miyajima, T. Komukai, and T. Sugawa, "1.9 μm Tm-Doped Fluoride Fiber Amplifier and Laser Pumped at 1.58 μm," *Electron. Lett.*, Vol. 29, 1993, pp. 986–987.

[363] Percival, R. M., D. Szebesta, C. P. Seltzer, S. D. Perrin, S. T. Davey, and M. Louka, "A 1.6-μm Pumped 1.9-μm Thulium-Doped Fluoride Fiber Laser and Amplifier of Very High Efficiency," *IEEE J. Quantum Electron.*, Vol. 31, 1995, pp. 489–493.

[364] Sakamoto, T., M. Shimizu, M. Yamada, T. Kanamori, Y. Ohishi, Y. Terunuma, and S. Sudo, "35-dB Gain Tm-Doped ZBLYAN Fiber Amplifier Operating at 1.65 μm," *IEEE Photon. Tech. Lett.*, Vol. 8, 1996, pp. 349–351.

[365] Sakamoto, T., M. Shimizu, M. Yamada, T. Kanamori, Y. Ohishi, Y. Terunuma, and S. Sudo, "Tm-Doped ZBLYAN Fiber Amplifier Operating at 1.65 μm," *IEEE J. Quantum Electron.*, to appear.

[366] Sakamoto, T., M. Yamada, M. Shimizu, T. Kanamori, Y. Terunuma Y. Ohishi, and S. Sudo, "Thulium-Doped Fluoride Fiber Amplifiers for 1.4 μm and 1.65 μm Operation," *Proc. OAA'96*, Monterey, CA, 1996, Paper ThC3, pp. 40–43.

CHAPTER 6

Conclusion

As described in the chapters of this book, at the current stage in the development of optical fiber communications, the optical fiber amplifier is having a most exciting effect on optical communication systems and will usher in a worldwide revolution called the "Information Age." However, scanning once more the whole developmental history of glass fiber and fiber amplifier technology, we can clearly recognize some general trends in the advances that have been made.

First, the advances in science and technology have an alternating pattern. One such pattern can be discerned in the alternating development of passive fibers for signal transmission and active fibers for lasing and amplification between the 1960s and the 1990s, as illustrated in Figure 1.17. This pattern can also be recognized over longer historical periods, where there have been many alternating interactions between science and technology. For example, the first half of the present century was one of the most fruitful periods in the development of science, in particular, physics. It saw the theory of relativity, quantum theory, and modern ideas on the structure of atoms, molecules, and solids come into existence. In contrast, the last half of the present century has been a period of rapid technological growth, in particular, in the field of electronics and quantum electronics, which have been developed based on quantum mechanics. The resulting industrial products—televisions, computers, and lasers—have all come into being during this period, whereas quantum mechanics is currently confronting difficulty with regard to its fundamental framework. Moreover, taking a look at the history of science and technology from a much broader perspective, we can recognize that Newtonian mechanics constructed in the 17th century made a marvelous contribution to the growth of technology in the 18th and 19th centuries. At the same time, there was a growing awareness of its limitations and the problems it posed that paralleled this technological growth. We may now be experiencing such a period of alternating interaction between quantum mechanics and the technologies that it has sired. We are making

601

many attractive, technological products based on the scientific results provided by quantum mechanics, whereas quantum mechanics itself is currently facing problems. This means that today we could be nearing the birth of a new science. I think that these inherent ironies have motivated a fascinating, interesting, and alternating growth yet interactive of science and technology.

Second, the advances in science and technology have often been nonuniform and nonlinear. Sometimes a long span of time is needed before a small advance is achieved, whereas at other times great progress can be made very quickly. Some people feel that a period of little progress is like time spent asleep. Interestingly, however, such sleeping time is as important for progress in science and technology as it is for the human brain because, as history has shown, there are many things going on beneath the surface. Take the development of fiber amplifiers as an example; it took such a long time (more than 20 years) to complete the seemingly small step from lasers to amplifiers, but once the breakthrough was made, rapid and dramatic progress has been achieved in a relatively short time period. In fact, during that interval of over 20 years, little progress was made in fiber amplifier technology, but, in contrast, there was remarkable progress in fiber fabrication technology and semiconductor laser technology. Therefore, it can be said that this maturity of the fiber and laser diode technologies has enabled a breakthrough to be made in fiber amplifier technology. The new breakthrough was successful as the result of ambitious efforts to solve pressing problems.

Third, great technological developments and breakthroughs can have great impacts on related areas. In fact, the optical fiber amplifier is having a most exciting effect on optical communication systems, which in turn have had a dramatic impact on our ability to transmit information in today's Information Age. The fiber amplifier will allow information technology to be accessed by a greater portion of the population. No single subject in the field has received more attention than the erbium-doped fiber amplifier. However, the basic mechanism of the optical fiber amplifier is incredibly simple: it amplifies an optical signal in a fiber using the stimulated emission of optically excited rare-Earth ions in the fiber core. Therefore, we can say that the impact of a given technology does not depend on its complexity. Often the simplest is the best, but commonly such simplicity is achieved after a complicated structure has been developed.

In terms of the overall history of mankind, Dr. Arnold Toynbee in *A Study of History* (London, 1946) said that part of the origin of progress lies in the following spirit:

Doloris
Sopitam recreant volnera viva animam.

This means that "a serious wound inspires the sleeping mind." Several of the examples cited here have been brought about by such spirit and mind. Taking a closer look, we find that the key advances in science and technology have commonly been made by a few people. Therefore, it can be said that their ideas and spirit are keys to these advances. In this sense, we will conclude with the words of Leland Stanford to young people. He wrote:

> I attach great importance to general literature for the enlargement of the mind and for giving business capacity. . . . I have noticed that technically educated boys do not make the most successful businessmen. The imagination needs to be cultivated and developed to assure success in life. A man will never construct anything he cannot conceive.

About the Authors

Editor

Shoichi Sudo was born in 1952. He received the B.S. degree in electronic engineering from the University of Kanazawa in 1974.

Since joining NTT Laboratories in 1974, he has been engaged mainly in research on optical fiber fabrication technology. In 1977 he and his colleagues proposed and developed a new fiber preform fabrication method named the Vapor-Phase Axial Deposition (VAD) method. In 1982 he was awarded a Ph.D. from the University of Tokyo for his VAD work. After two years as manager in the research planning department of NTT, Dr. Sudo in 1985 began research on the fabrication of efficient nonlinear optical fibers and their applications. From 1986 to 1987 he was a visiting scholar at Stanford University and was engaged in research on single-crystal fibers and their nonlinear device application in Bob Byer's group. After returning to NTT in 1987, he resumed research on nonlinear optical devices including single-crystal fiber devices for SHG and amplification, short pulse generation by modulation instability in silica fibers and nonlinear couplers. In 1988 he began research that led to the development of a laser diode frequency stabilization scheme using acetylene gas absorption lines. After another two years in a senior management position, Dr. Sudo returned to research as a group leader in the field of optical fiber amplifiers. He placed emphasis on the research and development of 1.3 μm PDFAs and 1.5 μm broadband EDFAs, and also studied various combinations of active ions and host glasses for fiber amplifier use/application. In 1996, he was appointed executive manager in the research planning department of NTT Opto-Electronics Laboratories.

He is a member of the Institute of Electronics, Information, and Communication Engineers of Japan (IEICE), the Japan Society of Applied Physics, OSA, and IEEE (Senior Member). He has been involved in society activities including serving as a program committee member for IPR'91, OEC'92, IOOC'95, CLEO/PR'95, CLEO'94–'96, and serving as a international liaison for OFC'96–'98. He is also serving as editor-in-chief of IEICE technical papers. He is a recipient of the 1979 IEICE Young Engineers Award, the 1980 IEICE Paper Award, the 1982 NTT President Award, the 1982 Kajii Award, and the 1995 IEICE Paper Award.

Dr. Sudo has published three books and over 100 journal articles and has presented more than 150 technical talks, including 10 invited presentations and one tutorial. He has filed more than 350 patents and has been granted more than 100 patents including U.S. patents. He is the author of *Frequency Stabilization of Semiconductor Laser Diodes,* also published by Artech House (1995).

Authors

Yasutake Ohishi was born in 1955. He received the B.S. and M.S. degrees in physics from Tohoku University, Sendai, Japan in 1978 and 1980, respectively.

In 1980, he joined the Electrical Communication Laboratory, Nippon Telegraph and Telephone Public Corporation (now NTT Co.), where he was engaged in research on fluoride fiber technology. In 1988 he was awarded a Ph.D. by Tokyo Institute of Technology, Tokyo. From 1989 to 1990 he was a visiting scholar at Rutgers University where he invented the praseodymium-doped fiber amplifier for 1.3-mm amplification. After returning to NTT in 1990, he started research on fluoride-based fiber amplifiers. Recently, he invented the tellurite-based erbium-doped fiber amplifier for 1.5-mm broadband amplification. His research in this area opened a promising new phase in the field of rare-earth-doped fiber amplifiers. He is now the leader of a group at NTT Opto-Electronics Laboratories working on optical fiber amplifiers.

Dr. Ohishi is a member of the Optical Society of America, the Institute of Electronics, Information, and Communication Engineers of Japan, the Japan Society of Applied Physics, and the Ceramic Society of Japan.

Terutoshi Kanamori was born in 1950. He received the B.E. and M.E. degrees in applied physics from Nagoya University, Nagoya, Japan in 1973 and 1975, respectively.

He joined the Electrical Communication Laboratory, Nippon Telegraph and Telephone Public Corporation (now NTT Co.) in 1975, where he was engaged in research on amorphous materials. Since 1979, he conducted developmental research on infrared optical-fiber fabrication techniques and their characterization, including both fluoride and chalcogenide glass fibers for ultra-low transmission loss. In 1988, he received the D.E. degree in the area of infrared optical-fiber technology from Nagoya University. Since 1990, he has been engaged mainly in research on fluoride single-mode fiber fabrication technology for optical amplification and on chalcogenide single-mode fiber fabrication technology for nonlinear applications.

Dr. Kanamori is a member of the Japan Society of Applied Physics and the Institute of Electronics, Information, and Communication Engineers of Japan.

Makoto Shimizu was born in 1959. He received the B.E. and M.E. degrees in electrical engineering from the University of Electro-Communications, Tokyo, Japan in 1981 and 1983, respectively.

Since joining the Electrical Communication Laboratory, Nippon Telegraph and Telephone Public Corporation (now NTT Co.) in 1983, Mr. Shimizu has been engaged mainly in research on silica-based optical fiber fabrication technology and optical fiber amplifier technology.

Mr. Shimizu is a member of the Japan Society of Applied Physics and the Institute of Electronics, Information, and Communication Engineers of Japan. He received the 1994 Paper Award from the IEICE of Japan.

Makoto Yamada was born in 1960. He received the B.E. and M.E. degrees in electrical engineering from the Technological University of Nagaoka, Niigata, Japan in 1983 and 1985, respectively.

In 1985, he joined NTT Opto-Electronics Laboratories, and engaged in research on guided-wave optical devices. Since 1989, he has been engaged in research on optical fiber amplifiers.

Mr. Yamada is a member of the Institute of Electrical and Electronics Engineers, the Institute of Electronics, Information, and Communication Engineers of Japan, and the Japan Society of Applied Physics. He received the 1994 Paper Award from the IEICE of Japan.

Index

The Artech House Optoelectronics Library

Brian Culshaw and Alan Rogers, *Series Editors*

Amorphous and Microcrystalline Semiconductor Devices, Volume II: Materials and Device Physics, Jerzy Kanicki, editor

Bistabilities and Nonlinearities in Laser Diodes, Hitoshi Kawaguchi

Chemical and Biochemical Sensing With Optical Fibers and Waveguides, Gilbert Boisdé and Alan Harmer

Coherent and Nonlinear Lightwave Communications, Milorad Cvijetic

Coherent Lightwave Communication Systems, Shiro Ryu

Elliptical Fiber Waveguides, R. B. Dyott

Field Theory of Acousto-Optic Signal Processing Devices, Craig Scott

Frequency Stabilization of Semiconductor Laser Diodes, Tetsuhiko Ikegami, Shoichi Sudo, Yoshihisa Sakai

Fundamentals of Multiaccess Optical Fiber Networks, Denis J. G. Mestdagh

Germanate Glasses: Structure, Spectroscopy, and Properties, Alfred Margaryan and Michael A. Piliavin

Handbook of Distributed Feedback Laser Diodes, Geert Morthier and Patrick Vankwikelberge

High-Power Optically Activated Solid-State Switches, Arye Rosen and Fred Zutavern, editors

Highly Coherent Semiconductor Lasers, Motoichi Ohtsu

Iddq Testing for CMOS VLSI, Rochit Rajsuman

Integrated Optics: Design and Modeling, Reinhard März

Introduction to Lightwave Communication Systems, Rajappa Papannareddy

Introduction to Glass Integrated Optics, S. Iraj Najafi

Introduction to Radiometry and Photometry, William Ross McCluney

Introduction to Semiconductor Integrated Optics, Hans P. Zappe

Laser Communications in Space, Stephen G. Lambert and William L. Casey

Optical FDM Network Technologies, Kiyoshi Nosu

Optical Fiber Amplifiers: Design and System Applications, Anders Bjarklev

Optical Fiber Amplifiers: Materials, Devices, and Applications, Shoichi Sudo, editor

Optical Fiber Communication Systems, Leonid Kazovsky, Sergio Benedetto, Alan Willner

Optical Fiber Sensors, Volume Two: Systems and Applicatons, John Dakin and Brian Culshaw, editors

Optical Fiber Sensors, Volume Three: Components and Subsystems, John Dakin and Brian Culshaw, editors

Optical Fiber Sensors, Volume Four: Applications, Analysis, and Future Trends, John Dakin and Brian Culshaw, editors

Optical Interconnection: Foundations and Applications, Christopher Tocci and H. John Caulfield

Optical Measurement Techniques and Applications, Pramod Rastogi

Optical Network Theory, Yitzhak Weissman

Optoelectronic Techniques for Microwave and Millimeter-Wave Engineering, William M. Robertson

Reliability and Degradation of LEDs and Semiconductor Lasers, Mitsuo Fukuda

Reliability and Degradation of III-V Optical Devices, Osamu Ueda

Semiconductor Raman Laser, Ken Suto and Jun-ichi Nishizawa

Semiconductors for Solar Cells, Hans Joachim Möller

Smart Structures and Materials, Brian Culshaw

Ultrafast Diode Lasers: Fundamentals and Applications, Peter Vasil'ev

For further information on these and other Artech House titles, contact:

Artech House
685 Canton Street
Norwood, MA 02062
781-769-9750
Fax: 781-769-6334
Telex: 951-659
email: artech@artech-house.com

Artech House
Portland House, Stag Place
London SW1E 5XA England
+44 (0) 171-973-8077
Fax: +44 (0) 171-630-0166
Telex: 951-659
email: artech-uk@artech-house.com

WWW: http://www.artech-house.com